# Lecture Notes in Applied and Computational Mechanics

## Volume 40

Series Editors

Prof. Dr.-Ing. Friedrich Pfeiffer
Prof. Dr.-Ing. Peter Wriggers

# Lecture Notes in Applied and Computational Mechanics

## Edited by F. Pfeiffer and P. Wriggers

Further volumes of this series found on our homepage: springer.com

**Vol. 40:** Pfeiffer F.
Mechanical System Dynamics
xxxx p. 2008 [978-3-540-79435-6]

**Vol. 20:** Zohdi T.I., Wriggers P.
Introduction to Computational Micromechanics
196 p. 2005 [978-3-540-22820-2]

**Vol. 19:** McCallen R., Browand F., Ross J. (Eds.)
The Aerodynamics of Heavy Vehicles:
Trucks, Buses, and Trains
567 p. 2004 [3-540-22088-7]

**Vol. 18:** Leine, R.I., Nijmeijer, H.
Dynamics and Bifurcations
of Non-Smooth Mechanical Systems
236 p. 2004 [3-540-21987-0]

**Vol. 17:** Hurtado, J.E.
Structural Reliability: Statistical Learning Perspectives
257 p. 2004 [3-540-21963-3]

**Vol. 16:** Kienzler R., Altenbach H., Ott I. (Eds.)
Theories of Plates and Shells:
Critical Review and New Applications
238 p. 2004 [3-540-20997-2]

**Vol. 15:** Dyszlewicz, J.
Micropolar Theory of Elasticity
356 p. 2004 [3-540-41835-0]

**Vol. 14:** Frémond M., Maceri F. (Eds.)
Novel Approaches in Civil Engineering
400 p. 2003 [3-540-41836-9]

**Vol. 13:** Kolymbas D. (Eds.)
Advanced Mathematical and Computational
Geomechanics
315 p. 2003 [3-540-40547-X]

**Vol. 12:** Wendland W., Efendiev M. (Eds.)
Analysis and Simulation of Multifield Problems
381 p. 2003 [3-540-00696-6]

**Vol. 11:** Hutter K., Kirchner N. (Eds.)
Dynamic Response of Granular and Porous Materials
under Large and Catastrophic Deformations
426 p. 2003 [3-540-00849-7]

**Vol. 10:** Hutter K., Baaser H. (Eds.)
Deformation and Failure in Metallic Materials
409 p. 2003 [3-540-00848-9]

**Vol. 9:** Skrzypek J., Ganczarski A.W. (Eds.)
Anisotropic Behaviour of Damaged Materials
366 p. 2003 [3-540-00437-8]

**Vol. 8:** Kowalski, S.J.
Thermomechanics of Drying Processes
365 p. 2003 [3-540-00412-2]

**Vol. 7:** Shlyannikov, V.N.
Elastic-Plastic Mixed-Mode Fracture Criteria and Parameters
246 p. 2002 [3-540-44316-9]

**Vol. 6:** Popp K., Schiehlen W. (Eds.)
System Dynamics and Long-Term Behaviour
of Railway Vehicles, Track and Subgrade
488 p. 2002 [3-540-43892-0]

**Vol. 5:** Duddeck, F.M.E.
Fourier BEM: Generalization
of Boundary Element Method by Fourier Transform
181 p. 2002 [3-540-43138-1]

**Vol. 4:** Yuan, H.
Numerical Assessments of Cracks in Elastic-Plastic Materials
311 p. 2002 [3-540-43336-8]

**Vol. 3:** Sextro, W.
Dynamical Contact Problems with Friction:
Models, Experiments and Applications
159 p. 2002 [3-540-43023-7]

**Vol. 2:** Schanz, M.
Wave Propagation in Viscoelastic
and Poroelastic Continua
170 p. 2001 [3-540-41632-3]

**Vol. 1:** Glocker, C.
Set-Valued Force Laws:
Dynamics of Non-Smooth Systems
222 p. 2001 [3-540-41436-3]

# Mechanical System Dynamics

Friedrich Pfeiffer

With 430 Figures and 19 Tables

Prof. Dr. Friedrich Pfeiffer
TU München
Inst. Mechatronik
LS Angewandte Mechanik (AM)
Boltzmannstr. 15
85747 Garching
Gebäude 1
Germany
pfeiffer@amm.mw.tum.de

ISBN: 978-3-540-79435-6            e-ISBN: 978-3-540-79436-3

Lecture Notes in Applied and Computational Mechanics   ISSN   1613-7736

Library of Congress Control Number: 2008926910

© First Edition 2005. Corrected Second Printing 2008 Springer-Verlag Berlin Heidelberg

This work is subject to copyright. All rights are reserved, whether the whole or part of the material is concerned, specifically the rights of translation, reprinting, reuse of illustrations, recitation, broadcasting, reproduction on microfilm or in any other ways, and storage in data banks. Duplication of this publication or parts thereof is permitted only under the provisions of the German Copyright Law of September 9, 1965, in its current version, and permission for use must always be obtained from Springer. Violations are liable for prosecution under the German Copyright Law.

The use of general descriptive names, registered names, trademarks, etc. in this publication does not imply, even in the absence of a specific statement, that such names are exempt from the relevant protective laws and regulations and therefore free for general use.

*Cover design:* Kirchner, Erich

Printed on acid-free paper

9 8 7 8 6 5 4 3 2 1 0

springer.com

To my wife Ruth

# Preface

Mechanics as a fundamental science in Physics and in Engineering deals with interactions of forces resulting in motion and deformation of material bodies. Similar to other sciences Mechanics serves in the world of Physics and in that of Engineering in a different way, in spite of many and increasing interdependencies. Machines and mechanisms are for physicists tools for cognition and research, for engineers they are the objectives of research, according to a famous statement of the Frankfurt physicist and biologist Friedrich Dessauer. Physicists apply machines to support their questions to Nature with the goal of new insights into our physical world. Engineers apply physical knowledge to support the realization process of their ideas and their intuition. Physics is an analytical Science searching for answers to questions concerning the world around us. Engineering is a synthetic Science, where the physical and mathematical fundamentals play the role of a kind of reinsurance with respect to a really functioning and efficiently operating machine. Engineering is also an iterative Science resulting in typical long-time evolutions of their products, but also in terms of the relatively short-time developments of improving an existing product or in developing a new one. Every physical or mathematical Science has to face these properties by developing on their side new methods, new practice-proved algorithms up to new fundamentals adaptable to new technological developments. This is as a matter of fact also true for the field of Mechanics.

In the 20th century a couple of significant ideas pushed forward the classical field of dynamics, both, with respect to physics and with respect to engineering. In the first half of the 20th century we had, seen from the standpoint of physics, three theories of dynamics, Newtonian dynamics, relativistic dynamics and quantum dynamics [257]. We had four decisive impacts the last hundred years, two with respect to the basis of dynamics and two with respect to applications.

Starting with the at his time revolutionary idea of geometric dynamics by Henri Poincaré [214] some new features of dynamics came up, which should change dynamical arguing considerably, namely the aspects of bifurcations and chaos characterizing the broad field of "Nonlinear Dynamics" [261]. Viewing it from a broader physical basis, Prigogine [218] stated that "we are led from a world of *being* to a world of *becoming*", which indeed fits perfectly well into the concept of engineering. Research in this field comes out with a completely new way of looking at dynamics in form of a topological evolution of mechanical systems with time. Research is going on, there are still open questions related to large systems, to optimization and control. Impact on practical problems is coming up more and more.

The second basic contribution to dynamics consists in creating "non-smooth mechanics", first addressed to by Moreau in Montpellier [160] and by Panagiotopoulos in Thessaloniki [176]. The new theories of non-smooth mechanics added to the idea of bilateral constraints the new idea of unilateral constraints as they appear for example in contacts. The fundamental non-smooth principles are for my opinion comparable to the idea of the principle of d'Alembert-Lagrange and on an equal scientific level. They open large new fields, theoretically and practically. In the meantime progress in that field especially with respect to the classical and non-smooth principles of mechanics clearly indicate, that classical mechanics is embedded in the theories of non-smooth mechanics, it is indeed a subset of it.

Two more application-oriented concepts came up in parallel with the technical development of computers and of space technologies. During the fifties and sixties of the last century the idea of finite element discretization was pushed forward opening new applications especially for new aerospace projects, moving more and more the necessary investigations from experimental fields to computer simulations. Today FEM-codes are commercially available and applicable in all fields of modern technologies. Nevertheless research is going on. The second idea concerns multibody systems, a picture-book model for the application of the constraint ideas well-known since Bernoulli [18]. Therefore quite a number of impressive ideas were published in the 18th and 19th century [43], but a final impact to develop formulations as available today was given by space applications of the fifties and sixties [110]. Also in this field research is going on.

We are living in a world of computers and computing techniques, and we profit from it. Simulations with large and in the meantime very comfortable computer codes allow to establish a virtual world, which, applied in an intelligent way, might give detailed and very helpful insights into the concepts of new products. The development times for new cars or machines have been reduced considerably with the help of computer application. On the other hand we have to be careful. The engineering process requires perfect insight into the physics of a system, thus also into the mechanics of a machine. Engineering thought cannot be replaced by a computer, neither in design, material selection and cost analysis nor in the physical fundamentals like mechanics and others. Every real progress in technology is always accompanied by considerable thought of large depth. Mechanics and mathematics are perfect training areas for such thinking.

In the following we shall consider fundamentals and applications of dynamics, mainly with respect to large dynamical systems typical for modern industry and its products. The bases are models. Models are pictures of thought or constructs of ideas. Using models includes several aspects. Firstly, there are the simple ones, which nevertheless represent the main features of a problem, for example of a vibration problem, in such a good way, that they can be used to give some analytical insight into that problem with regard to dynamics but also with regard to parameter influences. Establishing such models is an art for a very few number of experts. It requires a perfect knowledge of the specific problem under consideration, and it affords intuition and intelligence to reduce such a system to a few parameters. But we often can learn from such models in a couple of days much more than by long-lasting computer simulations.

Secondly, we may establish models by considering as many details as possible. Such models are large and costly regarding computing times. And even in this case we have to investigate very carefully all physical effects for doing the correct neglects without endangering realistic results. Done in a skillful way such models are the basis for physical understanding and for improving design concepts. These two types of models aim at generating some results, which are as realistically as possible related to our real world problem.

Thirdly, if we leave that requirement, we may find models with similar features as our real world case, but only in a more or less qualitative sense. This might help sometimes, but usually it is too far away from practice. Anyway, establishing models represents more an intuitive art than a science. This is mostly underestimated, because only good models in a mechanical sense, at this stage not in a mathematical sense, give access to good solution algorithms and finally to good results. Models should be as simple as possible and so complex as necessary, not more and not less.

As a rule we understand the word model as a theoretical construct. But model and modeling applies in the same way to experimental set-ups. Lack of thought very often identifies experiments with the dogmatic truth of practice, which is only sometimes true. To design and to establish good experiments really related to the practical system under consideration is a difficult task. And it is also a difficult task to find the correct interpretations of measured data. Let us take a measured spectrum of an airborne gas turbine system, and let us find out, if there are hidden self-exited vibrations or not. And if we find them, how does the mechanism of these self-excitations look like? We could continue such examples of open questions. Therefore, comparing theory and measurement requires very much care on both sides, on the side of theory and that of experiments.

The development of computers during the last forty years has created a certain dualism within all Sciences, namely the "Science" itself usually addressed to as being purely theoretical and far away from any computer application, and the "Computational Sciences", which have a clear focus in computer algorithms and applications. In Mechanics we have the same situation, which half a century after the first computers is not only obsolete, but also a bit out-dated and old-fashioned. Computers are used to solve complicated and large mathematical problems from Finite Elements (FEM) over Multibody Systems (MBS) to Computer Aided Design Systems (CAD), to name only a few, or to establish virtual worlds on the basis of topological structures or the like. In all cases we need a lot of fundamentals, and in all cases the colleagues involved in this kind of research are elaborating also some theoretical basis. On the other hand the "pure scientists" apply computers, be it to test their theoretical findings, be it to develop symbolic structures with the help of computers. Therefore the time is overdue, that these dualisms disappear.

The concept of the book is mainly determined by the experiences of the author. Sixteen years aerospace industry with its forerunner role in all technologies, but also in many fundamentals, twenty years academia in the area of mechanics, especially in the areas of dynamics and control, fundamental research on bilaterally and unilaterally constrained systems, on robotics and walking machines and many applications with regard to practical industrial problems have taught me, that for an engineer the combination of physical, mathematical and of practical, empirical knowledge is an indispensable prerequisite for successful professional work. We say, that a good theory is very practical, but we also have to state, that good practice might induce good new theories never known before. Both is important for a good engineer. The book tries to follow these ideas presenting in the first part the theoretical foundation of dynamics and in the second part a collection of industrial examples. Both theory and practice were topics of more than eighty dissertations being carried through during my activities as Professor and head of an Institute at the Technical University of München.

In spite of the fact that many research activities of the last two, three decades include also control design and development, the book will not consider this topic from the theoretical standpoint of view but more within the applications. Doing research in dynamics, especially in dynamical systems, implies also control in the one or other form, because control systems are also dynamical systems with the additional possibility of own decision capabilities. We shall focus on general dynamics, on multibody system dynamics including rigid and elastic components and on the consequences of bilateral and unilateral constraints. Unilateral contacts were and still are a matter of significant research at my former Institute.

Finally I have to thank many people and many Institutions. First of all, I have to thank my doctoral students of the last twenty years, who elaborated in dozens of theses my and their ideas on dynamics providing me with an excellent basis for this book. Many of them can be detected in the literature survey. I have to thank Prof. Christoph Glocker of the ETH Zurich for many fruitful discussions, mainly on non-smooth problems. And I am indebted especially to Dr.-Ing. Martin Foerg, who did some excellent proof-reading and has been for me during his time as doctoral student of my successor a continual contact for many discussions about fundamental problems of dynamics. The same is true for Dr.-Ing. Thomas Geier, who worked on very challenging applications like the pushbelt CVT. I could not have written this book in LaTex without many good advices and ideas concerning LaTex-questions from another doctoral student of my successor, namely Dipl.-Ing. Sebastian Lohmeier, who for his thesis is realizing the new walking machine LOLA, but who is also a real expert in LaTex.

Last but not least I do thank the Technical University of Munich, its Department of Mechanical Engineering and especially my successor, Prof. Heinz Ulbrich, for the opportunity to continue my work at the TU-Munich also after my retirement. A really big help for me were the co-workers of the Springer publishing house in Heidelberg, Dr. Thomas Ditzinger and Dr. Dieter Merkle from the organizational and contract side and Frank Holzwarth for all questions concerning the Springer stylefile. Many thanks to all.

A book cannot be written without any errors. Therefore I would like to motivate the readers of this monograph to give me a message of possible errors, he or she has detected. This concerns also any kind of concept or style aspects. I shall be very grateful for indications of that kind.

Garching, December 2007 *Friedrich Pfeiffer*

# Contents

1 Introduction ................................................. 1
2 Fundamentals ................................................ 5
  2.1 Basic Concepts ......................................... 5
      2.1.1 Mass ............................................. 5
      2.1.2 Cut Principle and Forces ......................... 6
      2.1.3 Constraints and Generalized Coordinates .......... 8
      2.1.4 Virtual Displacements and Velocities ............. 11
  2.2 Kinematics ............................................. 12
      2.2.1 Coordinates ...................................... 12
      2.2.2 Coordinate Transformations ....................... 14
      2.2.3 Velocities and Accelerations ..................... 19
      2.2.4 Transformation Chains and Recurrence Relations ... 25
      2.2.5 Kinematics of Systems ............................ 29
      2.2.6 Parameterized Coordinates ........................ 31
      2.2.7 Relative Contact Kinematics ...................... 36
      2.2.8 Influence of Elasticity .......................... 47
  2.3 Momentum and Moment of Momentum ........................ 53
      2.3.1 Definitions and Axioms ........................... 53
      2.3.2 Momentum ......................................... 54
      2.3.3 Moment of Momentum ............................... 57
      2.3.4 Transformations .................................. 59
  2.4 Energy ................................................. 62
      2.4.1 Introduction ..................................... 62
      2.4.2 Kinetic Energy ................................... 63
      2.4.3 Potential Energy ................................. 66
  2.5 On Contacts and Impacts ................................ 68
      2.5.1 Phenomena ........................................ 68
      2.5.2 Impact Structure ................................. 68
      2.5.3 Basic Laws ....................................... 71
      2.5.4 Impact Models .................................... 74

## XIV Contents

- 2.6 Damping .................................................. 76
  - 2.6.1 Phenomena ......................................... 76
  - 2.6.2 Linear Damping .................................... 77
  - 2.6.3 Nonlinear Damping ................................. 81

# 3 Constraint Systems ......................................... 85
- 3.1 Constraints and Contacts ................................. 85
  - 3.1.1 Bilateral Constraints ............................. 85
  - 3.1.2 Unilateral Constraints ............................ 89
- 3.2 Principles .............................................. 100
  - 3.2.1 Introduction ..................................... 100
  - 3.2.2 Principle of d'Alembert and Lagrange ............. 100
  - 3.2.3 Principle of Jourdain and Gauss .................. 103
  - 3.2.4 Lagrange's Equations ............................. 105
  - 3.2.5 Hamilton's Equations ............................. 110
- 3.3 Multibody Systems with Bilateral Constraints ............ 113
  - 3.3.1 General Comments ................................. 113
  - 3.3.2 Equations of Motion of Rigid Bodies .............. 115
  - 3.3.3 Order(n) Recursive Algorithms .................... 119
  - 3.3.4 Equations of Motion of Flexible Bodies ........... 124
  - 3.3.5 Connections by Force Laws ........................ 128
- 3.4 Multibody Systems with Unilateral Constraints ........... 131
  - 3.4.1 The General Problem .............................. 131
  - 3.4.2 Multibody Systems with Multiple Contacts ......... 134
  - 3.4.3 Friction Cone Linearization ...................... 139
  - 3.4.4 Numerical Aspects ................................ 145
  - 3.4.5 The Continual Benchmark: Woodpecker Toy .......... 150
  - 3.4.6 Some Empirical Conclusions ....................... 155
- 3.5 Impact Systems .......................................... 158
  - 3.5.1 General Features ................................. 158
  - 3.5.2 Classical Approach ............................... 159
  - 3.5.3 Moreau's Measure Differential Equation ........... 170
  - 3.5.4 Energy Considerations ............................ 172
  - 3.5.5 Verification of Impacts with Friction ............ 176
- 3.6 Modeling System Dynamics ................................ 183

# 4 Dynamics of Hydraulic Systems ............................ 187
- 4.1 Introduction ............................................ 187
- 4.2 Modeling Hydraulic Components ........................... 190
  - 4.2.1 Junctions ........................................ 190
  - 4.2.2 Valves ........................................... 193
  - 4.2.3 Hydraulic lines .................................. 198
- 4.3 Hydraulic Networks ...................................... 201
  - 4.3.1 Solutions ........................................ 202
  - 4.3.2 Hydraulic Impacts ................................ 203

|   |   |   |   |
|---|---|---|---|
| | 4.4 | Practical Examples | 204 |
| | | 4.4.1 Hydraulic Safety Brake System | 204 |
| | | 4.4.2 Power Transmission Hydraulics | 207 |
| **5** | **Power Transmission** | | **213** |
| | 5.1 | Automatic Transmissions | 214 |
| | | 5.1.1 Introduction | 214 |
| | | 5.1.2 Drive Train Components | 216 |
| | | 5.1.3 Drive Train System | 227 |
| | | 5.1.4 Measurements and Verification | 229 |
| | | 5.1.5 Optimal Shift Control | 231 |
| | 5.2 | Ravigneaux Gear System | 241 |
| | | 5.2.1 Toothing | 242 |
| | | 5.2.2 Ravigneaux Planetary Gear | 244 |
| | | 5.2.3 Ring Gear | 246 |
| | | 5.2.4 Ring Gear Coupling | 247 |
| | | 5.2.5 Phase Shift of Meshings | 249 |
| | | 5.2.6 Equations of Motion | 250 |
| | | 5.2.7 Implementation | 253 |
| | | 5.2.8 Simulation Results | 254 |
| | 5.3 | Tractor Drive Train System | 257 |
| | | 5.3.1 Introduction | 257 |
| | | 5.3.2 Modeling | 259 |
| | | 5.3.3 Numerical and Experimental Results | 270 |
| | 5.4 | CVT Gear Systems - Generalities | 275 |
| | | 5.4.1 Introduction | 275 |
| | | 5.4.2 The Polygonial Frequency | 278 |
| | 5.5 | CVT - Rocker Pin Chains - Plane Model | 282 |
| | | 5.5.1 Mechanical Models | 282 |
| | | 5.5.2 Mathematical Models | 288 |
| | | 5.5.3 Some Results | 294 |
| | 5.6 | CVT - Rocker Pin Chains - Spatial Model | 301 |
| | | 5.6.1 Introduction | 301 |
| | | 5.6.2 Mechanical Models | 302 |
| | | 5.6.3 Mathematical Models | 307 |
| | | 5.6.4 Some Results | 312 |
| | 5.7 | CVT - Push Belt Configuration | 318 |
| | | 5.7.1 Introduction | 318 |
| | | 5.7.2 Models | 320 |
| | | 5.7.3 Some Results | 327 |

## 6 Timing Equipment ... 329
### 6.1 Timing Gear of a Large Diesel Engine ... 329
#### 6.1.1 Modeling ... 331
#### 6.1.2 Mathematical Models ... 335
#### 6.1.3 Evaluation of the Simulations ... 341
#### 6.1.4 Results ... 342
### 6.2 Timing Gear of a 5-Cylinder Diesel Engine ... 346
#### 6.2.1 Introduction ... 346
#### 6.2.2 Structure and Model of the 5-Cylinder Timing Gear ... 346
#### 6.2.3 Model of the Ancillary Components ... 352
#### 6.2.4 Simulation Results ... 355
### 6.3 Timing Gear of a 10-Cylinder Diesel Engine ... 359
#### 6.3.1 Introduction ... 359
#### 6.3.2 Structure and Model of the 10-Cylinder Timing Gear ... 359
#### 6.3.3 Simulation Results ... 362
### 6.4 Bush and Roller Chains ... 365
#### 6.4.1 Introduction ... 365
#### 6.4.2 Mechanical and Mathematical Modeling ... 366
#### 6.4.3 Results ... 391
### 6.5 Hydraulic Tensioner Dynamics ... 395
#### 6.5.1 Introduction ... 395
#### 6.5.2 Piston/Cylinder Component ... 396
#### 6.5.3 Tube Models ... 397
#### 6.5.4 Leakage Models ... 398
#### 6.5.5 Check Valves ... 402
#### 6.5.6 Tensioner System ... 403
#### 6.5.7 Experiments and Verification ... 407

## 7 Robotics ... 411
### 7.1 Introduction ... 411
### 7.2 Trajectory Planning ... 413
#### 7.2.1 A Few Fundaments ... 413
#### 7.2.2 Parametric Path Planning ... 421
#### 7.2.3 Forces at the Gripper ... 434
#### 7.2.4 Influence of Elasticity ... 437
### 7.3 Dynamics and Control of Assembly Processes with Robots ... 451
#### 7.3.1 Introduction ... 451
#### 7.3.2 Mating with a Manipulator ... 453
#### 7.3.3 Combined Robot and Process Optimization ... 476

## 8 Walking ... 503
### 8.1 Motivation, Technology, Biology ... 503
#### 8.1.1 Motivation ... 503
#### 8.1.2 Technologies ... 505
#### 8.1.3 Biology ... 507

|     |       |                          |     |
| --- | ----- | ------------------------ | --- |
| 8.2 | Walking Dynamics        |                           | 509 |
|     | 8.2.1 | Preliminary Comments     | 509 |
|     | 8.2.2 | Modeling                 | 510 |
|     | 8.2.3 | Equations of Motion      | 523 |
| 8.3 | Walking Trajectories    |                           | 528 |
|     | 8.3.1 | The Problem              | 528 |
|     | 8.3.2 | Trajectory Generation    | 528 |
| 8.4 | The Concept of JOHNNIE  |                           | 536 |
|     | 8.4.1 | Requirements             | 536 |
|     | 8.4.2 | Mechanical Models        | 536 |
|     | 8.4.3 | Sensors                  | 537 |
|     | 8.4.4 | Control Concept          | 539 |
|     | 8.4.5 | Some Results             | 542 |

**References** .......................................................... 547

**Index** ................................................................. 563

# 1

# Introduction

> *Der Verstand vermag nichts anzuschauen, und die Sinne nichts zu denken. Nur daraus, daß sie sich vereinigen, kann Erkenntnis entspringen. (Immanuel Kant, Kritik der reinen Vernunft, Königsberg 1787)*

"Mechanics is the science of motion; we define as its task: to describe *completely* and in the *simplest possible manner* such motions as occur in nature." With respect to Engineering we should complete this statement by "as occur in nature and in technology." Neither Physics nor Engineering are places for garlands, any description must be complete and as simple as possible, in Engineering specifically constrained with respect to the state of knowledge of the technical system under consideration. The above statement of Kirchhoff [130] has not and will not loose its fundamental significance in all areas of Physics and of Engineering. Specifically it describes completely all aspects of Mechanics, motion as such, displacements and deformations and also the limiting case of static systems with a kind of frozen motion, which follows the famous definition of Thomson and Tait [257]: "Keeping in view the proprieties of language, and following the example of the most logical writers, we employ the word *Dynamics* in its true sense as a science which treats of the action of *force*, whether it maintains relative rest, or produces acceleration of relative motion. The two corresponding divisions of Dynamics are thus conveniently entitled *Statics* and *Kinetics*".

The foundation of Mechanics as a Physical Science goes back into the also otherwise very creative and progressive 17th and 18th centuries and is connected with the names of Galilei, Newton, Johann Bernoulli, Euler, d'Alembert and Lagrange (see [43] and [259]). Frequently underestimated and addressed to as old-fashioned the basis created by these great scientists does not only shoulder a large part of classical and modern Physics and Technology, but it is also still a fundamental starting point for many new research topics up to recent times. The 19th and 20th centuries may be characterized by new axioms and principles connected with names like Hamilton, Gauss and Jacobi, where especially in the 20th century a consolidation took place (see books like [93], [181], [180], [1] and [143]).

It is not so long ago that classical mechanics became "real mechanics" [27] by application of the classical principles and foundations directly to large industry problems. This process of the last sixty years was supported by the development of powerful computers, which allowed numerical solutions of complex mathematical structures. But it was accompanied by another necessity, namely a better and clearer formulation of all mechanical foundations as a basis of new methods and algorithms of modern applied and computational mechanics. In spite of very convincing successes in that area we must keep in mind, that mechanics is a physical science and not a mathematical one [67]. To give a citation of Synge [257] with respect to this aspect: "The relationship (*of mathematics and physics*) seems to be based on certain concepts, the names of which provide a common language for all physicists (*for all engineers*), experimental and mathematical. These concepts appear as mathematical concepts in the model and as physical concepts in the direct discussion of nature (*technology*)." (*Italic words added by author*). The way we analyze today complex technical systems follows directly these ideas: we establish first a mechanical model with the goal to achieve some insight into the problem. And only after that we go into the mathematical structures.

By returning to the motion as such we are faced with three fields of interest, which form dynamics: Firstly, the motion of material bodies have to be described geometrically and the change of their position and orientation kinematically. Such a process leads to sets of equations, which usually define the functional performance of our motion system. But motion is caused by forces. Therefore and secondly, we must consider our system not only kinematically but also under the influence of forces. According to Newton [169] forces generate accelerations, and according to Euler [58] torques generate angular accelerations, where the mass or the mass moment of inertia act as a proportional factor. We are accustomed to interpret forces and torques as applied to a point, to a surface or to a volume with the property of unambiguity. With the recently coming up concepts of unilateral contacts in mechanics, mainly due to Moreau ([160] and [161]) and to Panagiotopoulos ([176] and [177]), forces and torques might also possess a set-valued character, which means, they are not unambiguously defined, but only with respect to a certain set as for example the set of sticking contact forces within a friction cone. This interpretation of contacts has influenced efficiently new concepts of mathematics and of unilateral dynamics, which started in France and is spread worldwide the last years.

The third aspect of motion are constraints, which very often are treated as an unpopular appendage to the momentum and moment of momentum equations, but which in reality represent the most important interfaces gluing together masses of bodies and forcing them to a required performance of motion. If momentum and moment of momentum represent the motors of motion, then constraints are the controllers, which tell the dynamical system where to go. This is in perfect correspondence to the geometric and kinematic character of such constraints, which in fact realize the functioning of a machine or any kind of technical or physical system. A complete set of constraints always represents the most abstract description of this functionality, beyond or better before entering kinetic considerations. Adding or deleting constraints to a system is a question of intelligent design, which has a huge amount of experience and examples at its disposal with an impressive outcome of really good solutions. But nevertheless we are still far away from the task to determine on a scientific and systematic basis the best combination of constraints, which realize in an optimal way a given set of requirements. Mechanical technologies without constraints would not exist.

We know bilateral and unilateral constraints, for example perfect joints or guides in the first and closing or opening contacts in the second case. The investigation of bilateral constraints from Bernoulli via d'Alembert [40] to Lagrange [41] represents a pioneering work of the same level as the findings by Newton [169] and Euler [58]. Without the concept of "lost forces", or in modern semantics, of "constraint forces" we would not possess a realistic dynamical foundation, we would not possess an applicable multibody theory, and finally engineers would not be able to evaluate constraint forces for bearings, guides and many other machine elements, which is really essential for all kind of sizing. The projection of motion into the free directions represented by the tangential constraint hyper-surfaces is one of the key essentials of classical and modern dynamics(see [27] and [180]). It includes basic informations of the directions of motion and of the direction of the affiliated constraint forces. Some branches of modern mathematics and physics are based on these ideas.

Unilateral features of mechanics have been treated in a general way throughout the last centuries by Fourier [66], Boltzmann [22] and Signorini [244], to name a few, but it was not before the sixties of the last century that Moreau [160] and Panagiotopoulos [176] developed a new and concise theory on unilateral mechanics, which they called non-smooth mechanics. With respect to contacts the unilateral behavior may be characterized by the property, that for contact dynamics either relative kinematics in the form of relative distances or velocities is zero and the accompanying constraint forces or constraint force combinations are not zero, or vice versa. This establishes a complementarity, which gives access to some mathematical structures valuable for a solution. The consequence from the above contact behavior, which represents by the way a very general property, results in a time-dependent alternation of indicators and of constraints with respect to such a contact. A contact is going to become active, as we say, if relative kinematics as an indicator becomes zero. Being zero, relative kinematics represents a constraint connected with a constraint force. A contact becomes again passive, if the constraint forces become zero and relative kinematics builds up again. Therefore the beginning of an active contact is indicated by magnitudes of relative kinematics, which become a constraint, at the same time establishing the constraint force or a combination of them as indicator for the end of an active contact. For dynamical systems with many contacts, which are typical for technical applications, we would get a combinatorial problem of huge order, which can only be treated by introducing the complementarity idea with its various methods of solution (see [200]).

The triad momentum, moment of momentum and constraints forms an elementary basis for everything going on in dynamics. Even when applying differential or minimal principles [27] we usually come back during the evaluation process to equations of motion, which include all relevant momentum and moment of momentum components. Two basic concepts must be added to the above defined triad, the cut principle first formulated by Euler and the introduction of virtual magnitudes in mechanics, which goes back to Galilei and Johann Bernoulli; it was put into an applicable form by Lagrange. The cut principle, enabling us to "look virtually into the matter, where no eye or experiment can penetrate "[258] and thus establishing a base for continuum mechanics, is one of the most important concepts and tools for all areas of mechanics including of course dynamics. An engineer cannot establish any model of a real plant or machine, if he is not able to delimit the system by proper cuts usually giving him the entrances and exits of the system with the appropriate kinematics and kinetics. When developing a model, the first step will be to choose intelligent cuts deciding on the future model's complexity, capabilities and results. The second concept of virtual kinematic magnitudes and of work or power is indispensable when applying the principles of mechanics ([93] and [180]). It is, similar to the cut principle, a concept of thought opening the possibility to manipulate forces, torques, work or power in a virtual way, in a thought way, necessary for example to derive the equations of motion for a system under consideration. All commercially applied theories like FEM and MBS are based on these concepts.

# 2
# Fundamentals

> *Die Theorie ist das Netz, das wir auswerfen, um "die Welt" einzufangen, — sie zu rationalisieren, zu erklären und zu beherrschen. Wir arbeiten daran, die Maschen dieses Netzes immer enger zu machen.*
>
> *(Karl Popper, "Logik der Forschung", 1935)*
>
> *Theories are nets cast to catch what we call "the world": to rationalize, to explain and to master it. We endeavour to make the mesh ever finer and finer.*
>
> *(Karl Popper, "The Logic of Scientific Discovery", 1959)*

## 2.1 Basic Concepts

### 2.1.1 Mass

We consider dynamics in the sense as discussed in the Introduction. That means we shall not refer to relativistic aspects whatsoever. The only deviation from the classical mass concept consists in the effects generated by rocket systems with their time-dependent masses. Focusing our future considerations mainly to technical artefacts we usually know all relevant mass distributions and can thus define:

- Masses are always positive, also in the time-dependent case, m > 0.
- Masses are
    - either constant with $\dot{m} = 0$,
    - or not constant with $\dot{m} \neq 0$, where ($\dot{m} = \frac{dm}{dt}$).
- Masses can be added and divided into parts.

Another more physically oriented definition of a mass is given by Synge [257]. He states, that a mass is "a quantity of matter in a body, a measure of the reluctance of a body to change its velocity and a measure of the capacity of a body to attract another gravitationally".

Modeling masses depends on the problem under consideration. We might have rigid or elastic masses and in dynamics also interactions with fluid masses. Theoretically we always get as a matter of fact an interdependence of the selected mass model and the results we can achieve with such a model. On the other hand the experience of modeling for a huge amount of practical cases tells us how to choose mass models. Nevertheless it makes sense keeping in mind these features. In the following we shall mainly consider systems with constant masses.

It is interesting to follow the evolution of the mass concept during the centuries [116]. It started as a matter of fact long before Newton, but Newton was the first to give with his *vis inertiae* idea a scientific basis for mass. In the following centuries the development of chemistry influenced the mass concept stating that mass represents a "quantity of matter", which requires a force to be put into motion. Euler developed a new concept defining mass as the fraction of force and acceleration, a conception, which was quickly accepted especially by French representatives of Mathematical Physics. The axiomatization of mechanics in the last two centuries made it necessary to define the mass anew within the framework of deductive and geometric forms. Completely new aspects entered the mass discussion with Einstein's relativistic mechanics.

### 2.1.2 Cut Principle and Forces

Before establishing a model we have to make clear what part of a system we would like to consider. It depends on the results we want to achieve, and this depends on the problem of the system under consideration. In technical artifacts a meaningful set of cuts for a machine or a car, for example, should define the kinematic or kinetic inputs into the system in the form of time-series or of spectra, and it should define the output of the system considering those positions giving typical performance characteristics. For a car's power transmission system, for example, the input might be the oscillations at the motor's crankshaft exciting the transmission system to vibrate, and the output might be the load torques at the tires coming from the environmental conditions like road quality, acceleration and car weight. Between these two cuts we have the complete power transmission, called in Figure 2.1 system, the performance of which we want to know, for example the acoustical performance. To be able to evaluate this performance we must define the limiting areas (points, lines, volumes) in a meaningful manner. In our practical example it is the load on one side (cut 2) and the input as generated by the combustion engine on the entrance side (cut 1). This makes sense from the technical point of view, indicating that positioning cuts results from experience and empirical knowledge more than from scientific arguments.

In order to apply this cut principle we have to generalize it a bit. In mechanics we are interested in the interaction of bodies of any kind with forces or torques. If we therefore separate two bodies by cutting them apart we must at the same time arrange those forces along the cut, which in the original configuration keep the two bodies together. Thus by cutting any system apart we transform internal forces to external ones acting on the cut parts with the same magnitude but opposite sign. This ingenious cut principle, first established by Euler, was characterized by Szabo [258] in a very appropriate way:

## 2.1 Basic Concepts

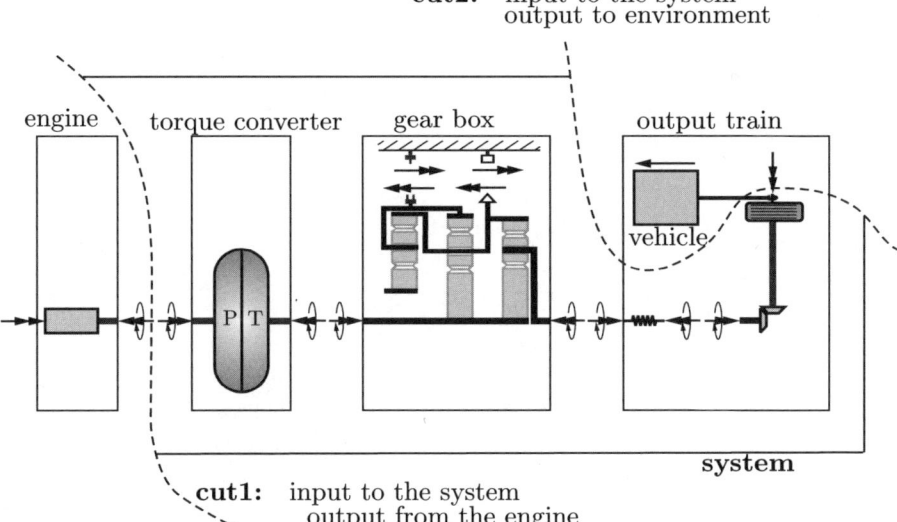

Fig. 2.1: The cut principle for a power transmission

"Euler teaches us with the imagination of an artist to look in thought into the matter, where no eye and no experiment can penetrate. With this he has laid a foundation stone for the only genuine mechanics, namely the continuum mechanics." The cut principle gives us a tool to establish for any part of a system the equations of motion, if we choose the cuts correctly and add to the applied forces and torques also the reaction forces and torques as freed by these cuts. We need in addition a sign definition, which we may choose arbitrarily, but then we must stay with it.

To illustrate the difference of internal and external forces depending on the cut positions we use a simple example [63]. Considering in Figure 2.2 the cut S1 around the three masses we see that all forces within that cut are internal forces possessing no influence on the system S1. Selecting a cut S2 we come out with two external forces $F_{12}$ and $F_{32}$ and with two internal forces $F_{13}$ and $F_{31}$. Finally the cut S3 generates only external forces, namely $F_{21}$ and $F_{31}$.

The mechanical sciences are interested in the interaction of any kind of masses with forces. Dynamics as a part of mechanics is especially interested in those forces, which generate motion. Therefore it makes sense to define as a generic concept that of active and passive forces. Active forces can be moved in their direction of action, and from there they can produce work and power. Passive forces cannot be moved with respect to their point of action. Active forces generate motion, passive forces prevent motion, they are as a matter of fact the consequence of some constraints. All other definitions of forces are subsets of this concept. Internal or external forces, applied or constraint

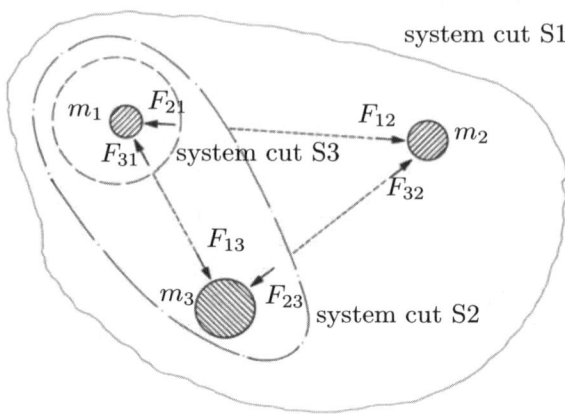

Fig. 2.2: Cut principle: internal and external forces [63]

forces, volume or surface forces, they all might be active or passive, depending on the specific system under consideration.

To give a simple example [208] we consider a block on a plane surface under the influence of an oblique external force. If the block does not move, all forces, the external applied force, the weight force and the contact forces are passive forces. If the external oblique force is big enough to move the block, then the horizontal components of the external applied force and of the contact force are active forces contributing to the motion of the block, whereas the weight force and the vertical components of the external applied force and of the contact force are passive forces adding certain loads to the block and to the ground (see figure 2.3). This simple example demonstrates already, that the property of a force becoming active or passive may depend on the dynamics of the system, which is reasonable especially in the face of unilaterally determined behavior. It sounds complicated, but we shall see later, that this concept is the only workable one with respect to a dynamical theory including all possible types of constraints, bilateral and unilateral ones.

### 2.1.3 Constraints and Generalized Coordinates

Constraints possess a kinematical character. They are the mechanical controllers telling systems where to go and where not to go. In mechanical engineering we do not have any machine or mechanism, which are unconstrained. Constraints realize, at least kinematically, operational requirements and, applied correctly, guarantee the function of a mechanical system. Constraints

## 2.1 Basic Concepts

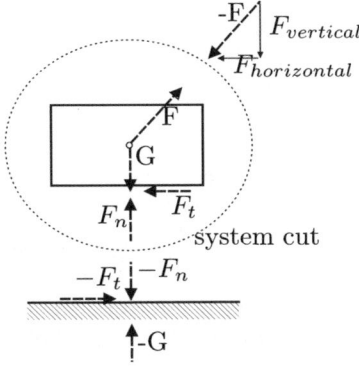

Fig. 2.3: Active and passive forces

might be bilateral or unilateral, representing in the first case an ideal connection between two adjacent bodies or between one body of the system and its environment constraining this connection to a limited number of degrees of freedom, and representing in the second case a connection, which might be open or closed, which might be sticking or sliding, depending on the dynamics of the system under consideration. Mechanical engineering includes as many bilateral as unilateral systems with a clear tendency to unilateral behavior with increasing requirements on modeling details.

Constraints depend on the coordinates of position and orientation, and they depend on velocities. Position- and orientation-dependent constraints are called holonomic, velocity-dependent constraints non-holonomic. They are rheonomic, if they depend on time, and scleronomic, if not. An important non-holonomic property says that such a constraint cannot be integrated to come out with a holonomic constraint. This leads to significant consequences.

The existence of constraints implies two difficulties. The first one concerns the independence of coordinates, which are constrained. Therefore the original coordinate set, for example in some three-dimensional workspace, does not represent the possible number of degrees of freedom. Some of the equations of motion depend on each other. The second difficulty is connected with the forces due to constraints. These constraint forces are not given a priori, they must be evaluated by the solution process. Moreover, the constraint forces do not contribute to the motion of the system, they are internal forces holding the system together, where we should keep in mind that passive forces and motion means passive forces and relative motion. Passive forces may of course move themselves withing the overall system. From the technical standpoint of view we need them as forces in bearings, guides, joints or the like. They determine system design.

Another consequence of constraints is given with the difference of their holonomic and non-holonomic properties. Constraints may be used to generate a set of generalized coordinates representing the degrees of freedom of the system. The elimination of the dependent variables is formally possible, but not necessarily practically. In the case of non-holonomic constraints it is not possible at all to eliminate coordinates of position and orientation, but it is possible ti eliminate velocity coordinates. A well-known example is the rolling disc, which at the same time is an example with the minimal possible number of degrees of freedom for a non-holonomic system, namely three [93]. The rolling condition for rolling without sliding cannot be integrated to come out with a holonomic equality, because a change of the orientation includes also a change in the position. These properties will have significant consequences for the development of the differential principles. A more detailed discussion of constraints can be found in [180], [93], [27] or [63].

Figure 2.4 depicts some typical constraints. The pendulum on the left represents a holonomic constraint depending only on the position of its mass. As long as the mass connection remains under tension we have a bilateral constraint. The sledge example represents also a holonomic, bilateral constraint as long as the sledge does not detach from the ground. In doing so we get a unilateral constraint with contact- and detachment-phases. The wheel example includes a non-holonomic constraint, because in the general case the function $f(\dot{x}, \dot{y}, \dot{\beta}, \alpha) = 0$ cannot be integrated to give then a position-dependent constraint. This is possible only, if the wheel follows exactly a straight line by rolling without sliding. Then we can roll back the wheel coming exactly to the starting point, and only then the constraint can be integrated.

The examples in Figure 2.4 illustrate also the empirical experience, that the constraint forces $F_c$ are in all cases perpendicular to the directions of motion, which later on will give a basis for the principle of d'Alembert-Lagrange.

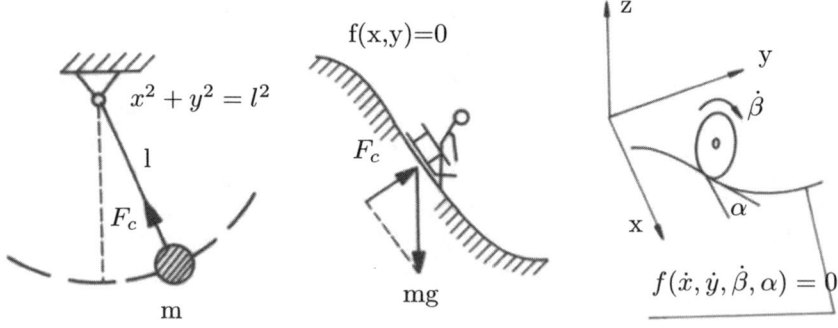

Fig. 2.4: Typical constraint examples

Constraints confine the number of degrees of freedom of a system. In general cases it is not possible to eliminate those degrees of freedom, which are constrained. But for special cases, usually of smaller dimensions, we might succeed in reducing the coordinates to the number of degrees of freedom really existing in the system. We then call these coordinates "minimal coordinates" $\mathbf{q} \in \mathbb{R}^f$, if f are the degrees of freedom. This is the limiting case for generalized coordinates for all constraints being eliminited. We shall not distinguish that in the following. Minimal coordinates include no more constraints in the form of algebraic equations. We then come out with differential equations without any additional constraints in the form of algebraic equations. It is sometimes difficult to interprete the generalized coordinates physically, but in any case they describe the possible motion of the system under consideration. For nonholonomic problems we are usually able to eliminate the generalized velocities, because the relevant constraints are linearly dependent on the velocities, at least in all cases known so far. We shall come back to that.

### 2.1.4 Virtual Displacements and Velocities

The concept of virtual displacements and velocities is not only a very productive one in whole mechanics, it is for certain significant areas of mechanics an indispensable tool, some authors say "axiom", for the development of basic theories, for example for analytical dynamics. Very probable Johann Bernoulli was the first one to use the idea of virtual displacements in the year 1717 and also the word "virtual" [258].

We shall understand as virtual displacements and velocities some thought magnitudes $\delta\mathbf{r}$ or $\delta\dot{\mathbf{r}}$, which necessarily must be compatible with all constraints acting for the time t under consideration. The displacements and velocities are called virtual, firstly because they are thought magnitudes and not real ones, and secondly to distinguish them from real changes $d\mathbf{r}$ and $d\dot{\mathbf{r}}$, which take place during the time interval dt, where forces and constraints might change considerably. Virtual displacements and velocities are considered for a fixed time t, which always means that $\delta t = 0$. As a consequence we take so-to-say a photo of the system at time t and investigate the system's behavior resulting from some virtual changes. The changes of the virtual velocities need not to be necessarily infinitesimal small, they might take on any values, but the above mentioned compatibility with the constraints is a must.

The concept of virtual displacements and velocities forms a basis for the principles of virtual work and power, which play a dominant role in all areas of mechanics.

## 2.2 Kinematics

Kinematics is geometry of motion and its evolution with time. It is the most important foundation of dynamics, as a matter of fact a foundation of any mechanical field. Rigid or elastic bodies must be defined in some suitable coordinate frame, the choice of which is more an art than a science strongly deciding on the complexity, or simplicity, of the mathematical model following from it. The basic movements of a rigid body are translation and rotation, each one described by three coordinates. Considering mechanical systems with many bodies requires the definition of many coordinate frames, body-fixed ones and inertial ones; where again the choice of these coordinates heavily influence the structure of the equations of motion and from there the necessary solution efforts. For most of the applications we apply orthogonal coordinate systems, but sometimes curvilinear coordinates represent the system under consideration in a more elegant way. Contact problems of rigid or elastic bodies are an example.

Dealing with systems, especially with multibody systems, includes vector spaces composed for example by the generalized coordinates. If these generalized coordinates represent the degrees of freedom of our system, then they are linearly independent and form a basis of the $\mathbb{R}$-vector-space with the dimension $\mathbb{R}^f$. The properties of these spaces are indispensable aspects for analyzing dynamical systems [155], [27], [180].

### 2.2.1 Coordinates

We define a coordinate system as a set of orthogonal unit vectors, for example $(\mathbf{e}_x, \mathbf{e}_y, \mathbf{e}_z)$, which form a basis for all vector representations to come. With respect to this base we assign an origin zero (0) or sometimes also an origin zero (O), which we assume to be fixed to some rigid or elastic body under consideration. Only this definition of an origin allows the measurement of distances or dynamic features with the help of such a coordinate system. Depending on the state of motion of the body with the coordinate system $(0, \mathbf{e}_x, \mathbf{e}_y, \mathbf{e}_z)$ we call these coordinates inertial or non-inertial (body-fixed).

The coordinate system $(0, \mathbf{e}_x, \mathbf{e}_y, \mathbf{e}_z)$ possesses the property "inertial", if the base vectors $(\mathbf{e}_x, \mathbf{e}_y, \mathbf{e}_z)$ do not change with time, which means, that such a coordinate system might only move with constant velocity with respect to our not moving postulated space. This is a question of definition. For technical dynamics it is usually sufficient to connect the earth or some building with an inertial coordinate system, for problems of space dynamics the sun might be a more suitable system.

If we connect a coordinate system with a rigid body to define a "body-fixed" coordinate system we have the choice to select for the origin any point, the center of mass or another point convenient for our evaluations, for example the center of mass and one joint in the case of robots. From this we may have several coordinate frames in one body. A basic property of a rigid body consists in the constant distance between two material points. Rigid bodies have six degrees of freedom, three of translation and three of rotation. Therefore the three positions (x,y,z) and the three orientations $(\alpha, \beta, \gamma)$, given for example as Cardan-angles, define unambiguously position and orientation of a rigid body with respect to any coordinate system, inertial or body-fixed (see Figure 2.5).

From this we might define for example the position and the orientation of a body $B_1$ with respect to the inertial system I or with respect to the body-fixed system of $B_2$ by the six magnitudes (x, y, z) and $(\alpha, \beta, \gamma)$, where (x, y, z) are the coordinates of the mass center of the body $B_1$, written in the bases I or $B_2$, and where the Cardan-angles $(\alpha, \beta, \gamma)$ give the orientation between the coordinate system of $B_1$ and those of I or $B_2$, written correspondingly in the I- or $B_2$-bases (see Figure 2.5).

The crucial point in dealing with rotations consists in the fact, that the rotational angles or the orientation angles cannot be evaluated straightforward, but are usually given implicitly by the rotational velocities expressed by the Euler kinematical equations or any other form. Therefore we always have to regard some equations of the form (2.31) to (2.33), which usually are part of the set of differential equations of motion. Only for small rotations, where the rotation angles are allowed to be represented as a vector, the matrices of the mentioned equations degenerate to an identity matrix.

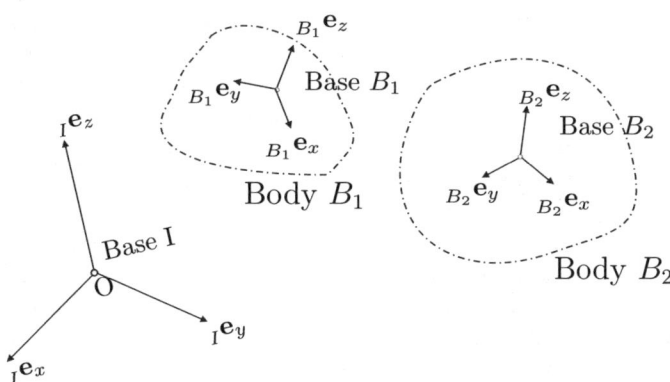

Fig. 2.5: Position and Orientation of Rigid Bodies

From this we introduce for many bodies with large rotations a vector $\mathbf{z}$ for translations and a matrix-vector equation for the rotational velocities

$$\mathbf{z} = (x_1, y_1, z_1, x_2, y_2, z_2, \ldots, x_n, y_n, z_n),$$
$$\boldsymbol{\omega}_i = \mathbf{H}_i(\mathbf{q}_i)\dot{\mathbf{q}}_i, \quad (i = 1, \cdot n), \tag{2.1}$$

where we have chosen a general representation including the rotational velocities $\boldsymbol{\omega}_i = (\omega_1, \omega_2, \omega_3)_i$ and some vector $\mathbf{q}_i$, which may be replaced by Euler or Cardan angles or any other set of orientation angles (see for example [146] and [180]). For many bodies with small rotations we collect all the body coordinates in a vector **z** giving

$$\mathbf{z} = (x_1, y_1, z_1, \alpha_1, \beta_1, \gamma_1, x_2, y_2, z_2, \alpha_2, \beta_2, \gamma_2, \ldots, x_n, y_n, z_n, \alpha_n, \beta_n, \gamma_n), \quad (2.2)$$

which contains n bodies with altogether 6n coordinate elements. Considering only one body we have $\mathbf{z} = \mathbf{r} = (x_1, y_1, z_1, \alpha_1, \beta_1, \gamma_1)$. Instead of **z** we also shall use the vector **r**.

As pointed out already in chapter 2.1.3 all real mechanical systems are constrained, where these constraints might be holonomic or non-holonomic, they might be scleronomic or rheonomic ([180]). In any case they constrain the motion of our system being then described by less coordinates as indicated in equation 2.2, which therefore contains some spare coordinates. Sometimes it is possible to eliminate these spare coordinates and to establish a set of coordinates, which corresponds exactly to the number of degrees of freedom, in many cases though this is not possible. Let us first consider some set of constraints, which might be of any type, but as an example we take into account a number of m holonomic and rheonomic constraint equations in the form

$$\boldsymbol{\Phi}(\mathbf{z}, t) = \mathbf{0}, \qquad \boldsymbol{\Phi} \in \mathbb{R}^m, \qquad m \leq 6n. \quad (2.3)$$

Such a set of m constraint equations reduce the free directions of motion to f = 6n - m, which we shall call in the following the number of degrees of freedom. We assign to these f degrees of freedom the coordinates $\mathbf{q} \in \mathbb{R}^f$, which are the "generalized coordinates". This elimination will not be possible for all applications under consideration leaving us with a certain rest of not fulfilled constraints. We then come out with a set of differential-algebraic equations containing $f_{min}$ differential equations and $m_{min}$ remaining constraints. We then still shall use the name "generalized coordinates" for the vector **q**.

At this point we also should mention the special form of coordinates, which are convenient for the treatment of all kind of trajectory problems and of contact phenomena. It concerns robotics and walking on the one hand and many machine applications on the other one. To give an example: for unilateral contact problems it makes sense to represent the surface coordinates of the bodies in a parametric form (see Figure 2.6 and chapter 2.2.6):

$$\mathbf{r} = \begin{pmatrix} x(u,v) \\ y(u,v) \\ z(u,v) \end{pmatrix} \quad (2.4)$$

### 2.2.2 Coordinate Transformations

The basic elements of kinematics are translation and rotation. In addition, and mostly confined to special cases, we have projections and reflections (see for

## 2.2 Kinematics

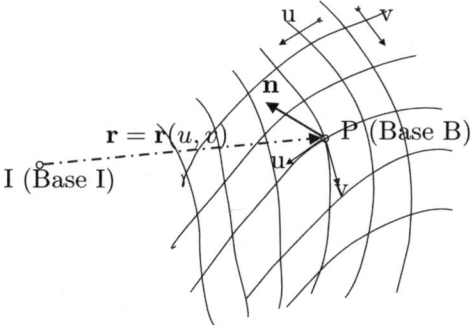

Fig. 2.6: Coordinate systems

example [3] and [155]). We shall focus on the first two movements. Considering mechanical systems requires a precise and unique definition of coordinate frames. In the following we shall use an inertial base I, and several body-fixed bases B or $B_i$ and R or $R_i$ (Figure 2.7). A vector $\mathbf{v}$ is a component of the vector space V, $\mathbf{v} \in V$, and it can be represented in any of the mentioned coordinate systems. From the standpoint of dynamics it is convenient to describe mechanical systems in different frames, and therefore we need a transforma-

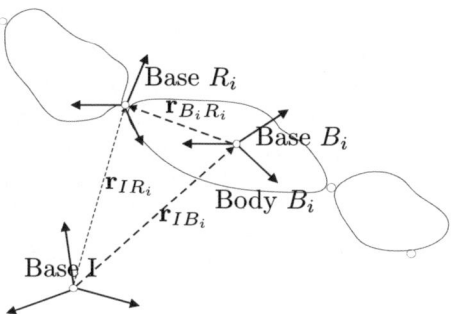

Fig. 2.7: Frame Relations

tion from one frame to any other one. From Figure 2.7 we easily can describe the vector chain in a coordinate-free form

$$\mathbf{r}_{IR_i} = \mathbf{r}_{IB_i} + \mathbf{r}_{B_iR_i}, \tag{2.5}$$

where the indices $(I, R_i, B_i)$ stand for the origins of the bases and for the bases themselves. The property of the necessary coordinate transformation can be nicely illustrated by the transformation triangle, which relates graphically the representation in different coordinate systems (see Figure 2.8). We apply for these representations the convention:

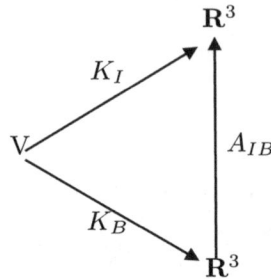

Fig. 2.8: Transformation triangle

$$K_I(\mathbf{v}) = {}_I\mathbf{v} \in \mathbb{R}^3,$$
$$K_B(\mathbf{v}) = {}_B\mathbf{v} \in \mathbb{R}^3,$$
$$K_R(\mathbf{v}) = {}_R\mathbf{v} \in \mathbb{R}^3. \quad (2.6)$$

These definitions indicate that the components of the vector **v** are written in the coordinate frames I,B,R, respectively. Going from one frame to another one we must evaluate the compositions [155]

$$K_I = \mathbf{A}_{IB} \circ K_B,$$
$$K_B = \mathbf{A}_{BI} \circ K_I, \quad (2.7)$$

which according to Figure 2.8 can be performed by a linear transformation with the transformation matrices $\mathbf{A}_{IB}$ or $\mathbf{A}_{BI}$. The index "IB" has to be read from right to left in the sense of transforming the vector **v** from the B-frame to the I-frame, and for "BI" from the I-frame to the B-frame. From equation 2.7 we see immediately, that the matrix-product $\mathbf{A}_{IB}\mathbf{A}_{BI} = \mathbf{E}$ comes out with the unit-matrix, which means, that these matrices **A** are orthogonal:

$$\mathbf{A}_{IB}\mathbf{A}_{BI} = \mathbf{E}. \quad (2.8)$$

A typical situation in multibody system modeling consists in the necessity to

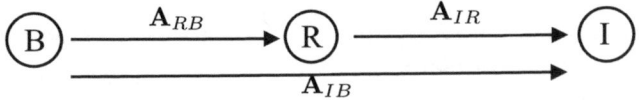

Fig. 2.9: Three successive coordinate systems

choose very many coordinate systems, sometimes several ones for only one of the body-elements. As a consequence we must consider also many successive coordinate frames performing multiple compositions as done in equations 2.7

for one frame only. Let us first consider three coordinate systems B,R,I, and let us go from B to I on the one side and from B to R to I on the other side (see 2.9). According to equation 2.7 we also get successive compositions in the form

$$K_I = \mathbf{A}_{IB} \circ K_B = \mathbf{A}_{IR}\mathbf{A}_{RB} \circ K_B \tag{2.9}$$

From equation 2.9 we get immediately that $\mathbf{A}_{IB} = \mathbf{A}_{IR}\mathbf{A}_{RB}$, which we can easily generalize by introducing a whole chain of intermediate coordinate systems, see Figure 2.10. We come out with the following chain of transforma-

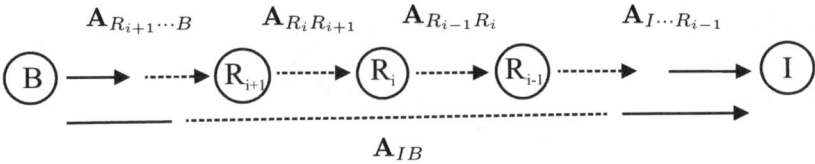

Fig. 2.10: n successive coordinate systems

tions:

$$K_I = \mathbf{A}_{IB} \circ K_B = (\mathbf{A}_{IR_2} \cdots \cdots \mathbf{A}_{R_{i-1}R_i}\mathbf{A}_{R_iR_{i+1}} \cdots \cdots \mathbf{A}_{R_{n-1}B}) \circ K_B$$
$$\mathbf{A_{IB}} = \mathbf{A}_{IR_2} \cdots \cdots \mathbf{A}_{R_{i-1}R_i}\mathbf{A}_{R_iR_{i+1}} \cdots \cdots \mathbf{A}_{R_{n-1}B} \tag{2.10}$$

Numbering the chain of coordinate systems from "1" for "I" and "n" for "B" we also can write the second equation of 2.10 in the form:

$$\mathbf{A}_{1,n} = \prod_{i=1}^{n-1} \mathbf{A}_{i,i+1} \tag{2.11}$$

The basic movements of kinematics are translations and rotations. Translations can be described by vector-chains as given with equation 2.5. Rotations concern a rotation of the complete coordinate frame around its origin. Various angle-triples exist to describe such rotations ([146], [3]). Sometimes it makes sense to apply a four-dimensional representation to avoid singularities, for example in space dynamics. Such quaternions have first been introduced by Hamilton [3]. We shall limit our considerations to Euler- and Cardan-angles, which are used very often in multibody problems. They have their origin in the early works of gyro-dynamics and celestial mechanics. In both cases we must perform three elementary rotations to come from some base B, for example a body-fixed frame, to a base R, which might be an intermediate or an inertial frame.

We start with the Euler angles. In Figure 2.11 we rotate the coordinates from the orientation I to the orientation B by three successive elementary rotations $\psi, \varphi, \vartheta$. In a first step we rotate the I-system around the inertially

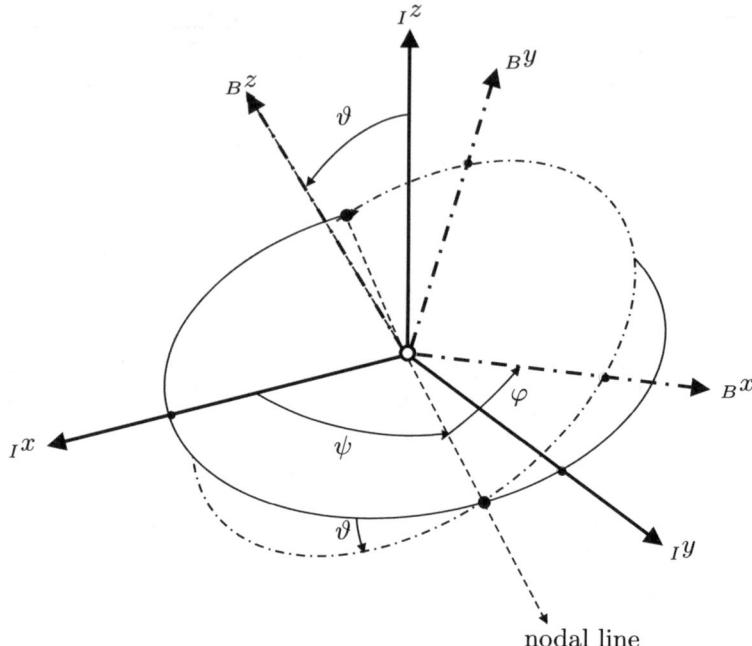

Fig. 2.11: Euler angles

fixed z-axis $_I z$ with the angle $\psi$. This rotation reaches the nodal line and can be represented by the linear elementary transformation

$$\mathbf{A}_{I,\psi} = \begin{pmatrix} \cos\psi & -\sin\psi & 0 \\ \sin\psi & \cos\psi & 0 \\ 0 & 0 & 1 \end{pmatrix}, \qquad (2.12)$$

In a second step we tilt the system around the nodal line applying the angle $\vartheta$. This elementary rotation follows the linear transformation

$$\mathbf{A}_{nodal,\vartheta} = \begin{pmatrix} 1 & 0 & 0 \\ 0 & \cos\vartheta & -\sin\vartheta \\ 0 & \sin\vartheta & \cos\vartheta \end{pmatrix}, \qquad (2.13)$$

Finally, in a third step we rotate around the $_B z$-axis and come out with the final orientation with the index "B", for example as a body-fixed orientation. The corresponding transformation writes

$$\mathbf{A}_{B,\varphi} = \begin{pmatrix} \cos\varphi & -\sin\varphi & 0 \\ \sin\varphi & \cos\varphi & 0 \\ 0 & 0 & 1 \end{pmatrix} \qquad (2.14)$$

The complete transformation then results from the successive products of the three elementary rotations. We get

$$\mathbf{A}_{IB} = \mathbf{A}_{I,\psi} \mathbf{A}_{nodal,\vartheta} \mathbf{A}_{B,\varphi} =$$
$$\begin{pmatrix} \cos\psi\cos\varphi - \sin\psi\cos\vartheta\sin\varphi & -\cos\psi\sin\varphi - \sin\psi\cos\vartheta\cos\varphi & +\sin\psi\sin\vartheta \\ \sin\psi\cos\varphi + \cos\psi\cos\vartheta\sin\varphi & -\sin\psi\sin\varphi + \cos\psi\cos\vartheta\cos\varphi & -\cos\psi\sin\vartheta \\ \sin\vartheta\sin\varphi & \sin\vartheta\cos\varphi & \cos\vartheta \end{pmatrix}.$$
(2.15)

Cardan-angles use a simpler sequence of rotations to go from the coordinate frame I to that of B (see Figure 2.12). We rotate firstly around the $_I x$-axis with the angle $\alpha$ and come to an intermediate axis by the transformation:

$$\mathbf{A}_{I,\alpha} = \begin{pmatrix} 1 & 0 & 0 \\ 0 & \cos\alpha & -\sin\alpha \\ 0 & \sin\alpha & \cos\alpha \end{pmatrix},$$
(2.16)

By a second rotation around the intermediate $_{IB}y$-axis with the angle $\beta$ we come already to the final $_B z$-axis by the transformation:

$$\mathbf{A}_{inter,\beta} = \begin{pmatrix} \cos\beta & 0 & \sin\beta \\ 0 & 1 & 0 \\ -\sin\beta & 0 & \cos\beta \end{pmatrix},$$
(2.17)

A last rotation around the new $_B z$-axis brings us into the final position with $(_B x, {_B y}, {_B z})$:

$$\mathbf{A}_{B,\gamma} = \begin{pmatrix} \cos\gamma & -\sin\gamma & 0 \\ \sin\gamma & \cos\gamma & 0 \\ 0 & 0 & 1 \end{pmatrix}.$$
(2.18)

With Cardan-angles we are able to reach any position of a body-fixed coordinate frame. Their application is especially useful for problems of machine dynamics. The complete transformation matrix is then the result of the above successive rotations. We come out with:

$$\mathbf{A}_{IB} = \mathbf{A}_{I,\alpha} \mathbf{A}_{inter,\beta} \mathbf{A}_{B,\gamma} =$$
$$\begin{pmatrix} \cos\beta\cos\gamma & -\cos\beta\sin\gamma & \sin\beta \\ \cos\alpha\sin\gamma + \sin\alpha\sin\beta\cos\gamma & \cos\alpha\cos\gamma - \sin\alpha\sin\beta\sin\gamma & -\sin\alpha\cos\beta \\ \sin\alpha\sin\gamma - \cos\alpha\sin\beta\cos\gamma & \sin\alpha\cos\gamma + \cos\alpha\sin\beta\sin\gamma & +\cos\alpha\cos\beta \end{pmatrix}$$
(2.19)

### 2.2.3 Velocities and Accelerations

With the knowledge of the coordinates and coordinate transformation we have established a basis for deriving the expressions for velocities and for accelerations in the various possible coordinate systems. Let us first go back to Figure 2.7 and simplify this figure a bit for our purposes (see Figure 2.13). We go from the I-system into the B-system, or vice versa, and we consider

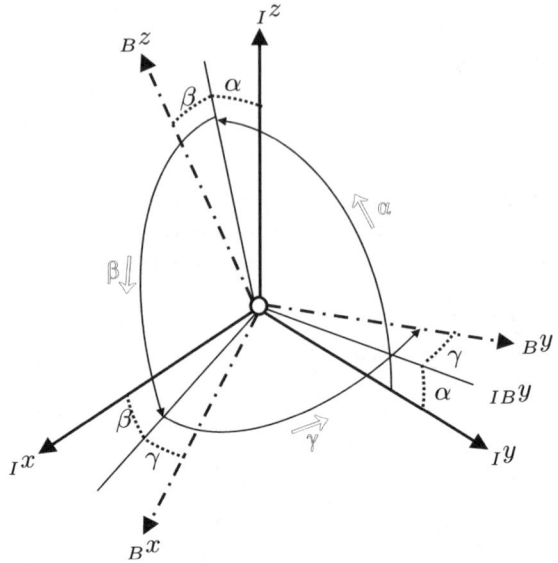

Fig. 2.12: Cardan angles

in both systems the coordinates of the point P. The translation follows from that Figure by

$$\mathbf{r}_{OP} = \mathbf{r}_{OQ} + \mathbf{r}_{QP}. \tag{2.20}$$

According to chapter 2.2.2 the rotation of coordinate system B with respect to I can be described by the matrix $\mathbf{A}_{BI}$, if we go from I to B and by $\mathbf{A}_{IB}$, if we go from B to I. Applying this matrix we can express the coordinates in one frame by those in the other frame, for example

$$\begin{aligned} {}_I\mathbf{r} &= \mathbf{A}_{IB} \, {}_B\mathbf{r} \\ {}_B\mathbf{r} &= \mathbf{A}_{BI} \, {}_I\mathbf{r}, \end{aligned} \tag{2.21}$$

which immediately confirms equation 2.8 and its properties

$$\begin{aligned} \mathbf{A}_{IB}\mathbf{A}_{BI} &= \mathbf{E} \\ \mathbf{A}_{BI} &= \mathbf{A}_{IB}^{-1} = \mathbf{A}_{IB}^{T} \end{aligned} \tag{2.22}$$

Before going to a derivation of the velocities we make a detour by considering the velocity of a point trajectory. It represents for example the case of a robot hand following a prescribed trajectory ([208]). The velocity of a point P along that trajectory is defined as

$$\mathbf{v} = \lim_{\Delta t \to 0} \frac{\mathbf{r}(t+\Delta t) - \mathbf{r}(t)}{\Delta t} = \frac{d\mathbf{r}}{dt} \tag{2.23}$$

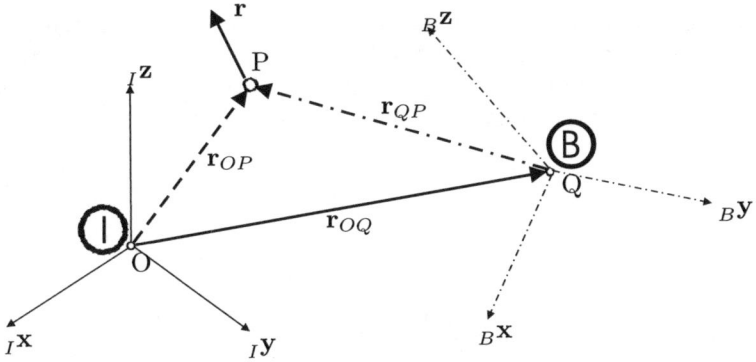

Fig. 2.13: Coordinate Relations

The velocity vector **v** has the direction of the tangent line to the point trajectory of Figure 2.14. If we consider the above defined velocity in an inertial frame as shown in Figure 2.14, we come out with

$$_I\mathbf{v}_{abs} = \frac{\mathrm{d}\,_I\mathbf{r}}{\mathrm{d}\,t} = {_I\dot{\mathbf{r}}}, \tag{2.24}$$

which means the following: The absolute velocity of a moving point P is the derivation with respect to time of a vector $\mathbf{r}(t) = {_I\mathbf{r}(t)}$ represented in a coordinate frame I, which is assumed to be an inertial system. Note that this vector or any other mechanical object can be represented in any coordinate frame without changing its physical properties, for example the magnitude of the vector is always the same. We only look at this vector from another point of view leading to different coordinates, but the vector itself or the object itself remains the same.

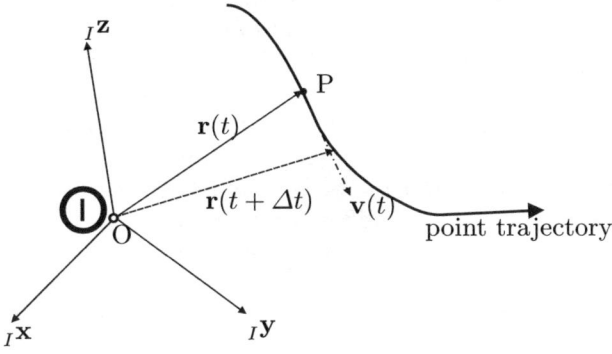

Fig. 2.14: Point Trajectory

Going for example from an inertial system to a body-fixed frame in Figure (2.7) or to a point fixed frame in form of the moving trihedral in Figure 2.14 (see [51]) we must transform the corresponding vectors from one frame to the other one applying the equation 2.21, which yields

$$_I\mathbf{v}_{abs} = {_I\dot{\mathbf{r}}}, \tag{2.25}$$

$$_B\mathbf{v}_{abs} = \mathbf{A}_{BI}\ {_I\dot{\mathbf{r}}}. \tag{2.26}$$

With these transformations we write the equation 2.20

$$_I\mathbf{r}_{OP} = {_I\mathbf{r}_{OQ}} + {_I\mathbf{r}_{QP}} = {_I\mathbf{r}_{OQ}} + \mathbf{A}_{IB} \cdot {_B\mathbf{r}_{QP}}. \tag{2.27}$$

The main reason for doing that lies in two facts, firstly that principally all time derivations in dynamics have to be performed in an inertial frame, which requires the appropriate transformations, and secondly that in many multibody applications it is much more convenient to define some points, some mass elements for example, in body-fixed coordinates than in inertial ones. The additional transformation matrices $\mathbf{A}$ usually can be calculated in a straightforward manner. Deriving 2.27 with respect to time we come out with

$$_I\dot{\mathbf{r}}_{OP} = {_I\dot{\mathbf{r}}_{OQ}} + \frac{\mathrm{d}}{\mathrm{d}t}(\mathbf{A}_{IB} \cdot {_B\mathbf{r}_{QP}}),$$

$$_I\dot{\mathbf{r}}_{OP} = {_I\dot{\mathbf{r}}_{OQ}} + \dot{\mathbf{A}}_{IB} \cdot {_B\mathbf{r}_{QP}} + \mathbf{A}_{IB} \cdot {_B\dot{\mathbf{r}}_{QP}}. \tag{2.28}$$

Equation 2.28 is one of the most important and basic relationships of kinematics. It allows the following interpretation: the left hand term of the second line represents the absolute velocity of the point P defined in the inertial system of Figure 2.13, the first right hand term is the absolute velocity of point Q also defined in the I-system, the second right hand term is the velocity of point P resulting from the rotation of the body-fixed B-system with respect to the inertial I-system, and finally $\dot{\mathbf{r}}_{QP}$ in the third right hand term is the relative velocity of P with respect to the B-system, for example resulting from a deformation, and transformed to the I-system by $\mathbf{A}_{IB}$.

To proceed to the classical formulation we must investigate the properties of these transformation matrices. First we recall the well known fact that the matrices $\mathbf{A}$ are orthogonal (see [155] and [27]). The rows of the matrix $\mathbf{A}_{IB}$ for example, which describes a transformation from the B-system into the I-system, are the unit vectors of the B-system represented in the I-system. We start with equation 2.22 and differentiate it with respect to time resulting in $\dot{\mathbf{A}}_{IB} \cdot \mathbf{A}_{BI} + \mathbf{A}_{IB} \cdot \dot{\mathbf{A}}_{BI} = 0$ and from that we get

$$\dot{\mathbf{A}}_{IB} \cdot \mathbf{A}_{BI} = -\mathbf{A}_{IB} \cdot \dot{\mathbf{A}}_{BI} = -(\dot{\mathbf{A}}_{IB} \cdot \mathbf{A}_{BI})^T. \tag{2.29}$$

Equation 2.29 includes the following facts. Obviously the matrix expression $\dot{\mathbf{A}}_{IB} \cdot \mathbf{A}_{BI}$ is skew-symmetric. Additionally, it must represent angular rotational velocities, which results from the rows of the matrix $\mathbf{A}$ being unit-vectors in the coordinate system under consideration. Any time derivative of unit-vectors can only come out with a rotation, because the magnitudes of the vectors themselves do not change.

If we apply the results of equation 2.29 to the transformation matrices $\mathbf{A}_{IB}$ on the basis of the Euler- and Cardan-angles in the equations 2.15 and 2.19, then after some lengthy calculations and rearranging we come out with the well-known Euler-equations for rotational kinematics

$$_I\boldsymbol{\omega} = \begin{pmatrix} 0 & \cos\psi & \sin\psi\sin\theta \\ 0 & \sin\psi & -\cos\psi\sin\theta \\ 1 & 0 & \cos\theta \end{pmatrix} \begin{pmatrix} \dot\psi \\ \dot\theta \\ \dot\phi \end{pmatrix}, \qquad (2.30)$$

$$_B\boldsymbol{\omega} = \begin{pmatrix} \sin\theta\sin\phi & \cos\phi & 0 \\ \sin\theta\cos\phi & -\sin\phi & 0 \\ \cos\theta & 0 & 1 \end{pmatrix} \begin{pmatrix} \dot\psi \\ \dot\theta \\ \dot\phi \end{pmatrix}, \qquad (2.31)$$

which is the set for the angular velocities using Euler-angles, and

$$_I\boldsymbol{\omega} = \begin{pmatrix} 1 & 0 & \sin\beta \\ 0 & \cos\alpha & -\sin\alpha\cos\beta \\ 0 & \sin\alpha & \cos\alpha\cos\beta \end{pmatrix} \begin{pmatrix} \dot\alpha \\ \dot\beta \\ \dot\gamma \end{pmatrix}, \qquad (2.32)$$

$$_B\boldsymbol{\omega} = \begin{pmatrix} \cos\beta\cos\gamma & \sin\beta & 0 \\ -\cos\beta\sin\gamma & \cos\gamma & 0 \\ \sin\beta & 0 & 1 \end{pmatrix} \begin{pmatrix} \dot\alpha \\ \dot\beta \\ \dot\gamma \end{pmatrix}, \qquad (2.33)$$

which is the set for the angular velocities using Cardan angles. For later considerations we should keep in mind, that the equations (2.31) and (2.33) represent a linear relationship between the rotational velocities $\omega$ and the time derivatives of the orientation angles $\bar{\boldsymbol{\varphi}} \Rightarrow (\psi\theta\phi) \Rightarrow (\alpha\beta\gamma)$ in the form

$$\boldsymbol{\omega} = \mathbf{H}\dot{\bar{\boldsymbol{\varphi}}} \implies \dot{\bar{\boldsymbol{\varphi}}} = \mathbf{H}^*\boldsymbol{\omega} \qquad (2.34)$$

With respect to the time derivatives of the transformation matrices we conclude from the above relations, that the equations 2.29 can only possess the meaning (see [27])

$$\dot{\mathbf{A}}_{IB}\,\mathbf{A}_{BI} = {}_I\tilde{\boldsymbol{\omega}}, \qquad \mathbf{A}_{BI}\,\dot{\mathbf{A}}_{IB} = {}_B\tilde{\boldsymbol{\omega}}, \qquad {}_I\tilde{\boldsymbol{\omega}} = \mathbf{A}_{IB}\,{}_B\tilde{\boldsymbol{\omega}}\,\mathbf{A}_{BI} \qquad (2.35)$$

with the skew-symmetric $\tilde{\boldsymbol{\omega}}$-matrix

$$\tilde{\boldsymbol{\omega}} = \begin{pmatrix} 0 & -\omega_z & +\omega_y \\ +\omega_z & 0 & -\omega_x \\ -\omega_y & +\omega_x & 0 \end{pmatrix} \qquad (2.36)$$

which includes the following helpful properties

$$\boldsymbol{\omega} \times \mathbf{r} = \tilde{\boldsymbol{\omega}}\,\mathbf{r}, \qquad \tilde{\boldsymbol{\omega}}^T = -\tilde{\boldsymbol{\omega}}, \qquad \tilde{\boldsymbol{\omega}}\,\mathbf{r} = -\tilde{\mathbf{r}}\,\boldsymbol{\omega} \qquad (2.37)$$

With these relations in mind we come back to equation 2.28, which gives us the absolute velocity of a point in the inertial frame I. Transforming these equations from I to B by applying the transformation $\mathbf{A}_{BI}$ we come out with the absolute velocity of the point P (figure 2.13) in the body-fixed frame B. We get

$$_B\mathbf{v}_{P,abs} = \mathbf{A}_{BI}\ _I\mathbf{v}_{P,abs} = \mathbf{A}_{BI}\left(_I\dot{\mathbf{r}}_{OQ} + \dot{\mathbf{A}}_{IB}\ _B\mathbf{r}_{QP} + \mathbf{A}_{IB}\ _B\dot{\mathbf{r}}_{QP}\right)$$
$$_B\mathbf{v}_{P,abs} = _B\mathbf{v}_{Q,abs} + _B\tilde{\boldsymbol{\omega}}\ _B\mathbf{r} + _B\dot{\mathbf{r}} \tag{2.38}$$

where the abbreviations $\mathbf{r}_{QP} = \mathbf{r}$, $_I\mathbf{v}_{P,abs} = _I\dot{\mathbf{r}}_{OP}$ and $_B\mathbf{v}_{Q,abs} = \mathbf{A}_{BI}\ _I\dot{\mathbf{r}}_{OQ}$ have been used. Equation 2.38 is of equal importance as equation 2.28, the first one being written in the I-system, the other one in the B-system. Therefore the terms of the last line of equation 2.38 have the following meaning: the left hand term represents the absolute velocity of the point P defined in the body-fixed system of Figure 2.13, the first right hand term is the absolute velocity of point Q also defined in the B-system, the second right hand term is the velocity of point P resulting from the rotation of the body-fixed B-system with respect to the inertial I-system but given in coordinates of the B-system, and finally $_B\dot{\mathbf{r}}$ of the third right hand term is the relative velocity of P with respect to the B-system. Examples might be deformation or the motion of a passenger in a flying airplane. The last line of equation 2.38 is the form usually given in textbooks. It is often addressed to as Coriolis equation.

The accelerations follow in a straightforward manner from the above equations. The acceleration is defined in a similar way as the velocity by

$$\mathbf{a} = \lim_{\Delta t \to 0} \frac{\mathbf{v}(t+\Delta t) - \mathbf{v}(t)}{\Delta t} = \frac{\mathrm{d}\mathbf{v}}{\mathrm{d}t} \tag{2.39}$$

Starting with

$$_I\mathbf{a}_{abs} = \frac{\mathrm{d}\ _I\mathbf{v}_{abs}}{\mathrm{d}t} = _I\ddot{\mathbf{r}}, \qquad _B\mathbf{a}_{P,abs} = \mathbf{A}_{BI} \cdot\ _I\mathbf{a}_{P,abs} \tag{2.40}$$

we get from the above equations the accelerations in the I-system.

$$_I\mathbf{a}_{P,abs} = _I\mathbf{a}_{Q,abs} + \mathbf{A}_{IB} \cdot (_B\dot{\tilde{\boldsymbol{\omega}}}_B\mathbf{r} + _B\tilde{\boldsymbol{\omega}}_B\ _B\tilde{\boldsymbol{\omega}}_B\mathbf{r}) + \mathbf{A}_{IB} \cdot (2_B\tilde{\boldsymbol{\omega}}_B\dot{\mathbf{r}} + _B\ddot{\mathbf{r}}). \tag{2.41}$$

The first term on the right hand side is the absolute acceleration of point Q (Figure 2.13), the second term the angular and the third term the centrifugal acceleration. The first three terms are applied accelerations, the fourth term the Coriolis- and the last term the relative acceleration due to some relative motion within the moving system. Transforming the expression 2.41 into a body-fixed coordinate system we come out with the well-known formula

$$_B\mathbf{a}_{P,abs} = \mathbf{A}_{BI}\ _I\mathbf{a}_{Q,abs} + _B\dot{\tilde{\boldsymbol{\omega}}}_B\mathbf{r} + _B\tilde{\boldsymbol{\omega}}_B\ _B\tilde{\boldsymbol{\omega}}_B\mathbf{r} + 2_B\tilde{\boldsymbol{\omega}}_B\dot{\mathbf{r}} + _B\ddot{\mathbf{r}}. \tag{2.42}$$

The explanations with regard to the individual terms of this expression are the same as above, all these terms are now written in a body-fixed frame.

Coming back again to the point trajectory of figure 2.14 we represent the vector $\mathbf{r}(t)$ in cylinder coordinates and derive from equations 2.23 and 2.24

the corresponding accelerations. We get with $(x,y,z)^T = (r\cos\varphi, r\sin\varphi, z)^T$ the following expressions

$$_I\mathbf{v}_P = {_I\dot{\mathbf{r}}_P} = \begin{pmatrix} \dot{r}\cos\varphi - r\dot\varphi\sin\varphi \\ \dot{r}\sin\varphi + r\dot\varphi\cos\varphi \\ \dot{z} \end{pmatrix},$$

$$_I\mathbf{a}_P = {_I\ddot{\mathbf{r}}_P} = \begin{pmatrix} \ddot{r}\cos\varphi - 2\dot{r}\dot\varphi\sin\varphi - r\ddot\varphi\sin\varphi - r\dot\varphi^2\cos\varphi \\ \ddot{r}\sin\varphi + 2\dot{r}\dot\varphi\cos\varphi + r\ddot\varphi\cos\varphi - r\dot\varphi^2\sin\varphi \\ \ddot{z} \end{pmatrix}. \tag{2.43}$$

For the evaluation of these formulas in a body-fixed frame we have to multiply the last equation with the transformation matrix from I to B, which results in

$$_B\mathbf{a}_P = \mathbf{A}_{BI} \cdot {_I\dot{\mathbf{r}}_P} = \begin{pmatrix} \ddot{r} - r\dot\varphi^2 \\ 2\dot{r}\dot\varphi + r\ddot\varphi \\ \ddot{z} \end{pmatrix}, \quad \text{with } \mathbf{A}_{BI} = \begin{pmatrix} \cos\varphi & \sin\varphi & 0 \\ -\sin\varphi & \cos\varphi & 0 \\ 0 & 0 & 1 \end{pmatrix}. \tag{2.44}$$

### 2.2.4 Transformation Chains and Recurrence Relations

The dynamics of mechanical systems requires as a rule the transformations over a long chain of bodies. In most cases these chains possess a tree-like structure, which is either given by the system structure itself, like in the case of robots, or it can be generated by cutting loops and introducing additional constraints or, if possible, by representing loops by analytical expressions. For developing the kinematics of such systems we establish a recursive algorithm on the basis of the above considerations [208]. In a first step we evaluate the angular velocities of the multibody components. If $\mathbf{\Omega}_{B_{i-1}}$ and $\mathbf{\Omega}_{B_i}$ are coordinate-free representations of the angular velocities of the bodies (i-1) and (i), respectively, then they are connected by the relation (see figure 2.15)

$$\mathbf{\Omega}_{B_i} = \mathbf{\Omega}_{B_{i-1}} + \mathbf{\Omega}_{B_{i-1}B_i}, \tag{2.45}$$

where the last term is the relative velocity between the bodies (i-1) and (i). Following the composition of transformations of figure 2.9 we come from the inertial coordinate system to that of the body $B_{i-1}$ and from there to the body $B_i$ using the matrix $\mathbf{A}_{B_{i-1}B_i}$ for the relative rotation between (i-1) and (i). According to equation 2.9 we get:

$$\mathbf{A}_{IB_i} = \mathbf{A}_{IB_{i-1}} \cdot \mathbf{A}_{B_{i-1}B_i} \quad \text{and} \quad \mathbf{A}_{B_{i-1}B_i} = \mathbf{T}_i. \tag{2.46}$$

The last equation defines the transformation concerning the rotation beteen I and $B_i$, as a convenient abbreviation for the following. With the equations 2.7 and 2.35 we define

$$_{B_i}\tilde{\boldsymbol{\omega}}_{IB_i} = K_{B_i} \circ \mathbf{\Omega}_{B_i} \quad \text{and} \quad _{B_i}\tilde{\boldsymbol{\omega}}_{IB_i} = \mathbf{A}_{B_iI}\,\dot{\mathbf{A}}_{IB_i}, \tag{2.47}$$

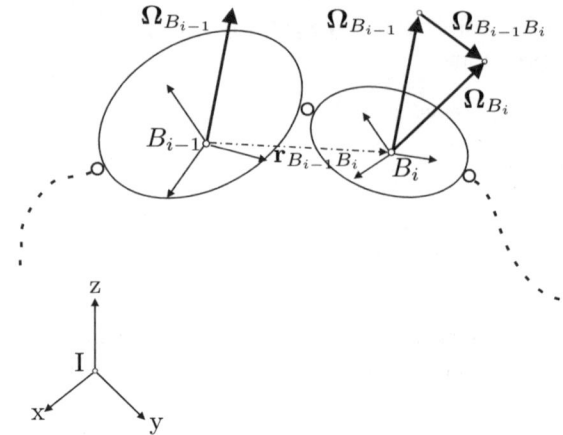

Fig. 2.15: Angular velocities of two successive bodies

which is the angular velocity between $I$ and $B_i$ defined in the $B_i$-coordinate system. With the equations 2.46 and 2.47 we write

$$\begin{aligned}
{}_{B_i}\tilde{\boldsymbol{\omega}}_{IB_i} &= \mathbf{A}_{B_iB_{i-1}}\left(\mathbf{A}_{B_{i-1}I}\,\dot{\mathbf{A}}_{IB_{i-1}}\right)\mathbf{A}_{B_{i-1}B_i} + \left(\mathbf{A}_{B_iB_{i-1}}\,\dot{\mathbf{A}}_{B_{i-1}B_i}\right) \\
&= \mathbf{A}_{B_iB_{i-1}}\left({}_{B_{i-1}}\tilde{\boldsymbol{\omega}}_{IB_{i-1}}\right)\mathbf{A}_{B_{i-1}B_i} + {}_{B_i}\tilde{\boldsymbol{\omega}}_{B_{i-1}B_i} \\
&= \mathbf{T}_i^T\left({}_{B_{i-1}}\tilde{\boldsymbol{\omega}}_{IB_{i-1}}\right)\mathbf{T}_i + \mathbf{T}_i^T\,\dot{\mathbf{T}}_i
\end{aligned} \tag{2.48}$$

The last equation is a recursive relation for going from the body (i-1) to body (i), which can be also written in the more convenient form

$${}_{B_i}\boldsymbol{\omega}_{IB_i} = \mathbf{A}_{B_iB_{i-1}}\left({}_{B_{i-1}}\boldsymbol{\omega}_{IB_{i-1}}\right) + {}_{B_i}\boldsymbol{\omega}_{B_{i-1}B_i} \tag{2.49}$$

In a similar way we can derive such a recurrence relation for the vectors between the bodies. We start with figure 2.16 and consider the coordinate-free relation

$$\mathbf{r}_{IB_i} = \mathbf{r}_{IB_{i-1}} + \mathbf{r}_{B_{i-1}B_i}, \tag{2.50}$$

which we might define in any coordinate frame according to the equations 2.7. For a description in an I-system we get

$$\begin{aligned}
{}_I\mathbf{r}_{IB_i} &= {}_I\mathbf{r}_{IB_{i-1}} + {}_I\mathbf{r}_{B_{i-1}B_i} \\
&= {}_I\mathbf{r}_{IB_{i-1}} + \mathbf{A}_{IB_i}\,{}_{B_i}\mathbf{r}_{B_{i-1}B_i}
\end{aligned} \tag{2.51}$$

Differentiating this expression with respect to time and transforming the result into a body-fixed frame $B_i$ yields after some manipulations the absolute velocities in the form ($I$- and $B_i$-system)

$$\begin{aligned}
{}_I\dot{\mathbf{r}}_{IB_i} &= {}_I\dot{\mathbf{r}}_{IB_{i-1}} + \mathbf{A}_{IB_i}\left({}_{B_i}\tilde{\boldsymbol{\omega}}_{IB_i}\,{}_{B_i}\mathbf{r}_{B_{i-1}B_i} + {}_{B_i}\dot{\mathbf{r}}_{B_{i-1}B_i}\right), \\
{}_{B_i}\dot{\mathbf{r}}_{IB_i} &= \mathbf{A}_{B_iB_{i-1}}[({}_{B_{i-1}}\dot{\mathbf{r}}_{IB_{i-1}} + {}_{B_{i-1}}\dot{\mathbf{r}}_{B_{i-1}B_i}) + \\
&\quad + ({}_{B_{i-1}}\tilde{\boldsymbol{\omega}}_{IB_{i-1}} + {}_{B_{i-1}}\tilde{\boldsymbol{\omega}}_{B_{i-1}B_i})\,{}_{B_{i-1}}\mathbf{r}_{B_{i-1}B_i}],
\end{aligned} \tag{2.52}$$

Equation 2.52 is in combination with equation 2.48 also a recurrence relation, which can be used to build up the vector equations for a chain of bodies.

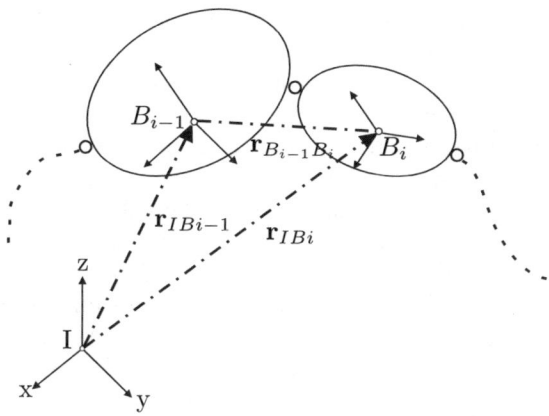

Fig. 2.16: Vectors of two successive bodies

For the equations of motion we need the accelerations. Considering a chain of bodies (Figure 2.16) we define for the body (i)

$p(i) = i - 1$ $\implies$ predecessor body,
$s(i) = i + 1$ $\implies$ successor body,
$i = 0$ $\implies$ inertial system (base body),
$i = n$ $\implies$ end body.

An alternative form of equation (2.51) is given by

$$_{B_i}\mathbf{r}_{IB_i} = \sum_{j=1}^{i} \mathbf{A}_{B_iB_j}({_{B_j}\mathbf{r}_{B_{j-1}B_j}}),  \tag{2.53}$$

where we compose the radius vector $_{B_i}\mathbf{r}_{IB_i}$ by the incremental vectors between the bodies, for example from (j-1) to (j). These inremental vectors $({_{B_j}\mathbf{r}_{B_{j-1}B_j}})$ are defined in the coordinate frame $B_j$ and must consequently transformed to the system $B_i$. In the same way we can determine the rotational velocity by (see equation (2.49) and Figure (2.15))

$$_{B_i}\boldsymbol{\omega}_{IB_i} = \sum_{j=1}^{i} \mathbf{A}_{B_iB_j}({_{B_j}\boldsymbol{\omega}_{B_{j-1}B_j}}). \tag{2.54}$$

The corresponding translational velocity can be derived by a standard process: Transform equation (2.53) into the inertial system by the matrix $\mathbf{A}_{IB_i}$, differentiate quite formally the resulting equation with respect to time and finally transform the result back to the body-fixed coordinates $B_i$. We get

$$_{B_i}\dot{\mathbf{r}}_{IB_i} = \sum_{j=1}^{i} \mathbf{A}_{B_iB_j}[(_{B_j}\dot{\mathbf{r}}_{B_{j-1}B_j}) + {_{B_j}\tilde{\boldsymbol{\omega}}_{IB_j}} \cdot (_{B_j}\mathbf{r}_{B_{j-1}B_j})].\tag{2.55}$$

Before going to the accelerations we recall the indexing. An index on the left side of a magnitude indicates the coordinate system, in which the components of this magnitude are defined, for example the coordinates in $B_i$ for the radius vector $\dot{\mathbf{r}}$ in the equation above. The right side indices indicate positions or orientations (from - to) or (between), for example and again in the above equation $\dot{\mathbf{r}}_{IB_i}$ means the velocity between the inertial system I and the body-fixed system $B_i$, or $\omega_{IB_j}$ means the angular velocity between the inertial system I and the body-fixed frame $B_j$. On the other side we read the transformations from the left side to the right one. For example $\mathbf{A}_{B_iB_j}$ means a transformation from the coordinates $B_j$ into the coordinates $B_i$, which makes sense, because the second right index of these transformation matrices must match with the left index of those magnitudes to be transformed, see again the above equation.

The acceleration will be achieved by the same time-derivation process as discussed above. For the translational and rotational accelerations we come out with

$$_{B_i}\ddot{\mathbf{r}}_{IB_i} = \sum_{j=1}^{i} \mathbf{A}_{B_iB_j}[(_{B_j}\ddot{\mathbf{r}}_{B_{j-1}B_j}) + {_{B_j}\dot{\tilde{\boldsymbol{\omega}}}_{IB_j}}\ (_{B_j}\mathbf{r}_{B_{j-1}B_j})+$$
$$+ {_{B_j}\tilde{\boldsymbol{\omega}}_{IB_j}}\ {_{B_j}\tilde{\boldsymbol{\omega}}_{IB_j}}\ (_{B_j}\mathbf{r}_{B_{j-1}B_j})+$$
$$+ 2{_{B_j}\tilde{\boldsymbol{\omega}}_{IB_j}} \cdot (_{B_j}\dot{\mathbf{r}}_{B_{j-1}B_j})]\tag{2.56}$$

$$_{B_i}\dot{\boldsymbol{\omega}}_{IB_i} = \sum_{j=1}^{i} \mathbf{A}_{B_iB_j}[(_{B_j}\dot{\boldsymbol{\omega}}_{B_{j-1}B_j}) + {_{B_j}\tilde{\boldsymbol{\omega}}_{IB_j}}\ (_{B_j}\boldsymbol{\omega}_{B_{j-1}B_j})].\tag{2.57}$$

Formally we may replace the summation of these formulas by a vector-matrix notation, which writes

$$_{B_i}\ddot{\mathbf{r}}_{IB_i} = \mathbf{A}_i(\Delta\ddot{\mathbf{r}}_i + \text{diag}(\Delta\tilde{\mathbf{r}}_i^T)\dot{\omega}_i) + \mathbf{a}_{Ti}, \qquad _{B_i}\dot{\omega}_{IB_i} = \mathbf{A}_i\Delta\dot{\omega}_i + \mathbf{a}_{Ri}, \qquad (2.58)$$

$$\begin{aligned}
\Delta\mathbf{r}_i &= (_1\mathbf{r}_{01}^T, {}_2\mathbf{r}_{12}^T, {}_3\mathbf{r}_{23}^T, \cdots {}_i\mathbf{r}_{i-1,i}^T)^T = (\ldots [_{B_j}\mathbf{r}_{B_{j-1}B_j}]^T, \ldots)^T, \\
\Delta\omega_i &= (_1\omega_{01}^T, {}_2\omega_{12}^T, {}_3\omega_{23}^T, \cdots {}_i\omega_{i-1,i}^T)^T = (\ldots [_{B_j}\omega_{B_{j-1}B_j}]^T, \ldots)^T, \\
\omega_i &= (_1\omega_{01}^T, {}_2\omega_{02}^T, {}_3\omega_{03}^T, \cdots {}_i\omega_{0i}^T)^T = (\ldots [_{B_j}\omega_{IB_j}]^T, \ldots)^T, \\
\mathbf{a}_{Ti} &= (\mathbf{a}_{T1}^T, \mathbf{a}_{T2}^T, \mathbf{a}_{T3}^T, \ldots \mathbf{a}_{Ti}^T)^T, \\
\mathbf{a}_{Ri} &= (\mathbf{a}_{R1}^T, \mathbf{a}_{R2}^T, \mathbf{a}_{R3}^T, \ldots \mathbf{a}_{Ri}^T)^T, \\
\mathbf{a}_{Tj} &= \mathbf{A}_{B_iB_j}[_{B_j}\tilde{\dot{\omega}}_{IB_j} \cdot {}_{B_j}\tilde{\omega}_{IB_j} \cdot (_{B_j}\mathbf{r}_{B_{j-1}B_j}) + 2{}_{B_j}\tilde{\omega}_{IB_j} \cdot (_{B_j}\dot{\mathbf{r}}_{B_{j-1}B_j})], \\
\mathbf{a}_{Rj} &= \mathbf{A}_{B_iB_j}[_{B_j}\tilde{\omega}_{IB_j} \cdot (_{B_j}\omega_{B_{j-1}B_j})], \qquad (2.59)
\end{aligned}$$

$$\mathbf{A}_i = \begin{pmatrix} \mathbf{E} & \mathbf{0} & \cdots & & & \\ \mathbf{A}_{21} & \mathbf{E} & \mathbf{0} & \cdots & & \\ \mathbf{A}_{31} & \mathbf{A}_{32} & \mathbf{E} & \mathbf{0} & \cdots & \\ \vdots & \vdots & \vdots & & \ddots & \\ \mathbf{A}_{i,1} & \mathbf{A}_{i,2} & \mathbf{A}_{i,3} & \cdots & & \mathbf{E} \end{pmatrix}. \qquad (2.60)$$

We can combine the above relations into one equation of the form

$$\ddot{\mathbf{z}}_i = \bar{\mathbf{A}}_i \Delta\ddot{\mathbf{z}}_i + \mathbf{a}_i, \qquad (2.61)$$

$$\begin{aligned}
\mathbf{z}_i &= (\mathbf{z}_1^T, \mathbf{z}_2^T, \mathbf{z}_3^T, \ldots \mathbf{z}_i^T)^T, & \Delta\mathbf{z}_i &= (\Delta\mathbf{z}_1^T, \Delta\mathbf{z}_2^T, \Delta\mathbf{z}_3^T, \ldots \Delta\mathbf{z}_i^T)^T, \\
\mathbf{a}_i &= (\mathbf{a}_1^T, \mathbf{a}_2^T, \mathbf{a}_3^T, \ldots \mathbf{a}_i^T)^T, & & \\
\bar{\mathbf{A}}_i &= \begin{pmatrix} \mathbf{A}_i & [\mathbf{A}_i(\Delta\tilde{\mathbf{r}}_i^T)\mathbf{A}_i] \\ \mathbf{0} & \mathbf{A}_i \end{pmatrix}, & \ddot{\mathbf{z}}_j &= \begin{pmatrix} {}_{B_j}\ddot{\mathbf{r}}_{IB_j} \\ {}_{B_j}\dot{\omega}_{IB_j} \end{pmatrix}_{j=1,2,\ldots i}, \\
\Delta\ddot{\mathbf{z}}_j &= \begin{pmatrix} {}_{B_j}\Delta\ddot{\mathbf{r}}_{B_{j-1}B_j} \\ {}_{B_j}\Delta\dot{\omega}_{B_{j-1}B_j} \end{pmatrix}_{j=1,2,\ldots i}, & \mathbf{a}_j &= \begin{pmatrix} \mathbf{a}_{Tj} \\ \mathbf{a}_{Rj} \end{pmatrix}_{j=1,2,\ldots i}. \qquad (2.62)
\end{aligned}$$

The matrices $\mathbf{A}_i$ and $\bar{\mathbf{A}}_i$ are both triangular matrices, which offers the possibilty to solve the equations (2.58) or (2.61) in an iterative way starting with the matrix row containing one element only and proceeding step by step until the equation has been solved. This represents a second type of recursion. We shall come back to it.

### 2.2.5 Kinematics of Systems

We go back to chapter 2.2.1 and recall the definitions of coordinates given there. We consider system coordinates $\mathbf{z} \in \mathbb{R}^{6n}$ or $\mathbf{r} \in \mathbb{R}^{6n}$, which in a constrained system do not correspond to the degrees of freedom but are more a representation of the system design and configuration. If we are able to eliminate all m constraints (for example equation 2.3), we come out with a set of generalized coordinates $\mathbf{q} \in \mathbb{R}^f$ with f = 6n - m. If we cannot fulfill all but only a few of the given constraints we remain with a rest of these constraints representing together with the differential equations a set of differential-algebraic equations for the coordinates $\mathbf{q}_{min} \in \mathbb{R}^{f_{min}}$, which is the minimum achievable set of generalized coordinates for the system under consideration.

Anyway, the system coordinates $\mathbf{z}$ are functions of the generalized coordinates $\mathbf{q}$, namely $\mathbf{z} = \mathbf{z}(\mathbf{q}, t)$, and therefore we have the time derivative

$$\dot{\mathbf{z}} = (\frac{\partial \mathbf{z}}{\partial \mathbf{q}}) \cdot \dot{\mathbf{q}} + \frac{\partial \mathbf{z}}{\partial t}, \qquad (\frac{\partial \mathbf{z}}{\partial \mathbf{q}}) \in \mathbb{R}^{6n,f}. \tag{2.63}$$

Depending on the above definitions we call the velocities $\dot{\mathbf{q}}$ the "generalized velocities". From the equation 2.63 we get a very useful and important relationship by differentiating it partially with respect to $\dot{\mathbf{q}}$

$$\frac{\partial \mathbf{z}}{\partial \mathbf{q}} = \frac{\partial \dot{\mathbf{z}}}{\partial \dot{\mathbf{q}}}, \tag{2.64}$$

where the unknown Jacobian $\frac{\partial \dot{\mathbf{z}}}{\partial \dot{\mathbf{q}}}$ can be evaluated from the constraint equations. Going one step further to the acceleration $\ddot{\mathbf{z}}$, we have to differentiate again equation (2.63) with respect to time and receive

$$\ddot{\mathbf{z}} = (\frac{\partial \mathbf{z}}{\partial \mathbf{q}}) \cdot \ddot{\mathbf{q}} + \frac{\partial}{\partial \mathbf{q}}[(\frac{\partial \mathbf{z}}{\partial \mathbf{q}})\dot{\mathbf{q}}]\dot{\mathbf{q}} + 2(\frac{\partial^2 \mathbf{z}}{\partial \mathbf{q} \partial t})\dot{\mathbf{q}} + (\frac{\partial^2 \mathbf{z}}{\partial^2 t}), \tag{2.65}$$

where in most cases of practical relevancy $\mathbf{z} = \mathbf{z}(\mathbf{q})$ and not $\mathbf{z} = \mathbf{z}(\mathbf{q}, t)$. For $\mathbf{z}$ not dependent on the time t the last two terms in equation (2.65) vanish.

Differentiating the constraint equation 2.3 with respect to time we get (see [27])

$$\dot{\mathbf{\Phi}} = (\frac{\partial \mathbf{\Phi}}{\partial \mathbf{z}}) \cdot \dot{\mathbf{z}} + \frac{\partial \mathbf{\Phi}}{\partial t} = \mathbf{0}, \qquad \{= (\frac{\partial \mathbf{\Phi}}{\partial \mathbf{z}})[(\frac{\partial \mathbf{z}}{\partial \mathbf{q}}) \cdot \dot{\mathbf{q}} + \frac{\partial \mathbf{z}}{\partial t}] + \frac{\partial \mathbf{\Phi}}{\partial t}\}, \tag{2.66}$$

which we again differentiate partially with respect to $\dot{\mathbf{q}}$ resulting in

$$(\frac{\partial \mathbf{\Phi}}{\partial \mathbf{z}}) \cdot (\frac{\partial \dot{\mathbf{z}}}{\partial \dot{\mathbf{q}}}) = \mathbf{0} \quad \text{with} \quad (\frac{\partial \mathbf{\Phi}}{\partial \mathbf{z}}) \in \mathbb{R}^{m,6n}, \quad (\frac{\partial \dot{\mathbf{z}}}{\partial \dot{\mathbf{q}}}) \in \mathbb{R}^{6n,f}. \tag{2.67}$$

Equation 2.67 tells us, that the rows of the Jacobian $(\frac{\partial \mathbf{\Phi}}{\partial \mathbf{z}})$ are orthogonal to the columns of the Jacobian $(\frac{\partial \dot{\mathbf{z}}}{\partial \dot{\mathbf{q}}})$. The Jacobian $(\frac{\partial \mathbf{\Phi}}{\partial \mathbf{z}})$ is regular, if we define the constraints in an unambiguous way, which is always possible. In this case we can solve equation 2.67 in a form, which represents a linear relationship of the velocities $\dot{\mathbf{z}}$ and $\dot{\mathbf{q}}$ (see [27])

$$\dot{\mathbf{q}} = (\frac{\partial \dot{\mathbf{q}}}{\partial \dot{\mathbf{z}}}) \cdot \dot{\mathbf{z}} = \{[(\frac{\partial \dot{\mathbf{z}}}{\partial \dot{\mathbf{q}}})^T (\frac{\partial \dot{\mathbf{z}}}{\partial \dot{\mathbf{q}}})]^{-1} (\frac{\partial \dot{\mathbf{z}}}{\partial \dot{\mathbf{q}}})^T\} \cdot \dot{\mathbf{z}} = (\frac{\partial \dot{\mathbf{z}}}{\partial \dot{\mathbf{q}}})^+ \cdot \dot{\mathbf{z}}, \tag{2.68}$$

where the term $(\frac{\partial \dot{\mathbf{z}}}{\partial \dot{\mathbf{q}}})^+$ represents the pseudo-inverse. The equations (2.63), (2.64) and (2.67) are essential concerning multibody theory. We remind of the relation (2.34) and shall come back to them in a later stage.

One of the key points in considering systems consists in the fact, that according to equation 2.63 we always have a linear relationship of the two velocities $\dot{\mathbf{z}}$ and $\dot{\mathbf{q}}$ of the form

$$\dot{\mathbf{z}} = \mathbf{J}(\mathbf{q},t) \cdot \dot{\mathbf{q}} + \mathbf{j}(\mathbf{q},t) \qquad \text{with} \qquad \mathbf{J}(\mathbf{q},t) = \frac{\partial \mathbf{z}}{\partial \mathbf{q}} = \frac{\partial \dot{\mathbf{z}}}{\partial \dot{\mathbf{q}}} \in \mathbb{R}^{6n,f}. \qquad (2.69)$$

The term $\mathbf{j}(\mathbf{q},t)$ comes from external excitation sources. Depending on the cuts in our system it is given by the input or output loads with respect to these cuts (chapter 2.1.2).

The linear relationship of the two velocities $\dot{\mathbf{z}}$ and $\dot{\mathbf{q}}$ also holds for the constraints independent of their type. According to equation 2.66 we have

$$\mathbf{W}^T(\mathbf{z},t)\,\dot{\mathbf{z}} + \mathbf{w}(\mathbf{z},t) = \mathbf{0} \quad \text{with} \quad \mathbf{W}^T(\mathbf{z},t) = \frac{\partial \mathbf{\Phi}}{\partial \mathbf{z}} = \frac{\partial \dot{\mathbf{\Phi}}}{\partial \dot{\mathbf{z}}}, \quad \mathbf{w}(\mathbf{z},t) = \frac{\partial \mathbf{\Phi}}{\partial t}. \qquad (2.70)$$

It is necessary to discuss a bit more in detail the above statement concerning the linear relation of velocities in the constraints. Taking into account first holonomic constraints of the form $\mathbf{\Phi}(\mathbf{z},t) = \mathbf{0} \in \mathbb{R}^m$ (eq. 2.3), which usually represent a set of nonlinear algebraic functions of $(\mathbf{z}(t),t)$, we might differentiate this set as performed in equation (2.70) and receive linear equations in $\dot{\mathbf{z}}$. They hold in an exact way for all holonomic constraints.

But if we have non-holonomic constraints, which are really non-holonomic and not reducible to a position and orientation level, the corresponding constraint equations are usually given in the form

$$\mathbf{\Phi}[\mathbf{z}(t),\dot{\mathbf{z}}(t),t] = \mathbf{0} \quad \in \mathbb{R}^m, \qquad (2.71)$$

which again are nonlinear equations in the arguments $(\mathbf{z}(t),\dot{\mathbf{z}}(t),t)$. It means at least formally that the linear structure of the equations (2.70) does not apply to these non-holonomic constraints. But, on the other hand, practical experience in all areas of technology indicates, that there are no non-holonomic constraints being nonlinear in the velocities $\dot{\mathbf{z}}$. Therefore the linear equations (2.70) possess the quality of an axiom in a physical, in a mechanical sense, at least approximately and as long as there will come no contradiction. The matrix $\mathbf{W}(\mathbf{z},t)$ will be of course completely different for the two constraint types, a Jacobian derived above for the holonomic case, and a Jacobian given by the system configuration for the non-holonomic case.

### 2.2.6 Parameterized Coordinates

The upcoming fields of robotics and of contact dynamics are accompanied by more frequent applications of curvilinear coordinate systems either for robot trajectories or for contact surfaces ([208],[152]). Let us first discuss the idea to project a mechanism with many degrees of freedom like a manipulator or a gear-mechanism onto one suitable degree of freedom connected with a prescribed trajectory or the like. Considering figure 2.17 and borrowing from the theory of spatial curves (see [51]) the most important relationships we introduce the path coordinate s = s(t) as a parameter and write

$$\mathbf{r}(t) = \mathbf{r}[s(t)] \quad \text{with} \quad \mathbf{v}(t) = \dot{\mathbf{r}}(t) = \frac{d\mathbf{r}}{ds} \cdot \frac{ds}{dt} = \mathbf{r}' \cdot \dot{s}. \tag{2.72}$$

The derivation $\dot{\mathbf{r}}$ of the radius vector $\mathbf{r}(t)$ with respect to time divided by the path velocity $\dot{s}$ represents the tangent unit vector $\mathbf{t}$ to the point trajectory:

$$\mathbf{t}(t) = \frac{\dot{\mathbf{r}}}{\dot{s}} = \mathbf{r}' = \frac{d\mathbf{r}}{ds}. \tag{2.73}$$

From the velocity vector of equation 2.72 we derive the acceleration vector by a differentiation with respect to time and get

$$\mathbf{a}(t) = \ddot{\mathbf{r}}(t) = \frac{d(\mathbf{r}' \cdot \dot{s})}{dt} = \frac{d(\mathbf{r}' \cdot \dot{s})}{ds} \cdot \frac{ds}{dt} = \mathbf{r}'' \cdot \dot{s}^2 + \mathbf{r}' \cdot \ddot{s}, \tag{2.74}$$

the first part of this equation being the centripetal and the second part the tangential acceleration. The vector $\mathbf{r}''$ corresponds to the derivation of equation 2.73 and represents thus the derivation of the tangent vector to a certain point of the path, $\mathbf{t}'(t) = \mathbf{r}''$. As $\mathbf{t}(t)$ is a unit vector, its derivative $\mathbf{t}'(t)$ is perpendicular to it. It should be noted that this parameterization of spatial curves must not be carried out by using the time t. Instead we may take any general parameter u, which of course must be selected according to the requirements coming from the dynamical system.

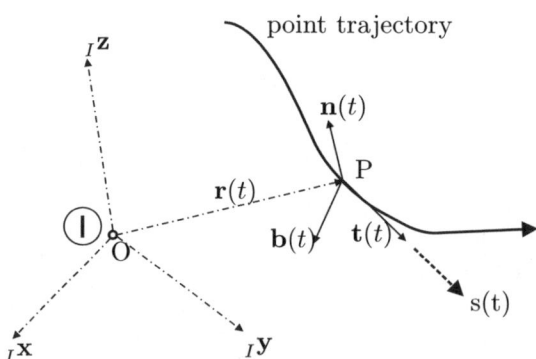

Fig. 2.17: Moving trihedron

For certain applications it makes sense to supplement the path coordinate s(t) and its corresponding unit vector $\mathbf{t}(t)$ by the normal and binormal vectors resulting in an orthogonal trihedron connected with the spatial path curve (see figure 2.17). As pointed out above the principal normal vector $\mathbf{n}(t)$ is perpendicular to the tangent vector $\mathbf{t}(t)$, these two vectors forming the osculating plane for the spatial curve at the point P, and the bi-normal vector is perpendicular to this plane, which means $\mathbf{b}(t) = [\mathbf{t}(t)] \times [\mathbf{n}(t)]$. These properties can be used to evaluate the well-known Frenet-equations, which are a basis for all applications connected with spatial trajectories. They write [51]

$$\frac{d\mathbf{t}}{ds} = +\kappa \cdot \mathbf{n},$$
$$\frac{d\mathbf{n}}{ds} = -\kappa \cdot \mathbf{t} + \tau \cdot \mathbf{b},$$
$$\frac{d\mathbf{b}}{ds} = -\tau \cdot \mathbf{b}. \qquad (2.75)$$

The magnitudes $\kappa$ and $\tau$ are the curvature and the torsion of the spatial trajectory, respectively. It might be helpful to repeat the behavior of the moving trihedron (see [132]). The tangent rotates about the instantaneous binormal direction at the positive angular rate $\kappa$ (curvature at point P). The binormal rotates about the instantaneous tangent direction at the angular rate $\tau$ (torsion at point P). The entire moving trihedron rotates about the instantaneous direction of the Darboux vector $\mathbf{\Omega} = \tau \mathbf{t} + \kappa \mathbf{b}$ at the positive angular rate $\|\mathbf{\Omega}\| = \sqrt{\tau^2 + \kappa^2}$ (total curvature at point P).

For certain robot- or mechanism-problems the advantages are obvious. A system with many degrees of freedom can be projected onto the coordinates of a given path coming out with three degrees of freedom or if we do not consider disturbances around such a path, only with one degree of freedom. We shall discuss an example in one of the applications chapters.

Another and a more demanding type of parameterization is connected with contact problems, where the contacts themselves take place between the surfaces of bodies. Especially if two bodies are approaching it is convenient to express the corresponding kinematics not in minimum but in surface coordinates, which requires some differential geometric transformations for going from the orthogonal world coordinates to the curvilinear surface coordinates (see for example [51], [152], [200] and Figure 2.6). We start with Figure 2.6 and equation 2.4, where a radius vector $\mathbf{r}$ depends on the curvilinear surface coordinates (u,v) by $\mathbf{r}(u,v) = [x(u,v), y(u,v), z(u,v)]^T$. In many cases it is possible to choose the surface coordinates in such a way, that they are mutually orthogonal, which brings some advantages for the evaluation. A partial derivation with respect to these curved coordinates comes out with the tangential vectors in the surface point under consideration.

$$\mathbf{e}_1 = \mathbf{u} = \frac{\partial \mathbf{r}}{\partial u}, \qquad \mathbf{e}_2 = \mathbf{v} = \frac{\partial \mathbf{r}}{\partial v}. \qquad (2.76)$$

The vectors $\mathbf{u}, \mathbf{v}$ span the tangential plane in the point P. The normal unit vector follows from the vector product of $\mathbf{u}$ and $\mathbf{v}$ in the form

$$\mathbf{n} = \frac{\mathbf{u} \times \mathbf{v}}{\|\mathbf{u} \times \mathbf{v}\|}. \tag{2.77}$$

With these vectors, two tangential and one normal, we are able to evaluate the elements of the first fundamental form of a surface, which writes:

$$E = \mathbf{u}^T \cdot \mathbf{u}, \qquad F = \mathbf{u}^T \cdot \mathbf{v}, \qquad G = \mathbf{v}^T \cdot \mathbf{v}. \tag{2.78}$$

If the parameter coordinates u and v are perpendicular to each other, then we get $F = 0$ for $\mathbf{u} \perp \mathbf{v}$. To determine the elements of the second fundamental form of a surface we need the second derivatives of the vector $\mathbf{r}(u,v)$. They characterize the curvature and torsion properties of the surface. We define

$$L = \mathbf{n}^T \cdot \frac{\partial^2 \mathbf{r}}{\partial u^2}, \qquad M = \mathbf{n}^T \cdot \frac{\partial^2 \mathbf{r}}{\partial u \partial v}, \qquad N = \mathbf{n}^T \cdot \frac{\partial^2 \mathbf{r}}{\partial v^2}. \tag{2.79}$$

Applying the equations 2.76 we get an equivalent formulation

$$L = \mathbf{n}^T \cdot \frac{\partial \mathbf{u}}{\partial u}, \qquad M = \mathbf{n}^T \cdot \frac{\partial \mathbf{u}}{\partial v} = \mathbf{n}^T \cdot \frac{\partial \mathbf{v}}{\partial u}, \qquad N = \mathbf{n}^T \cdot \frac{\partial \mathbf{v}}{\partial v}. \tag{2.80}$$

Again, for parameter coordinates (u,v) being mutually orthogonal we have $M = 0$ for $\mathbf{u} \perp \mathbf{v}$. Establishing a general theory of contacts requires additionally some further second derivatives and the derivation of the normal vector as defined in equation 2.77, which will be achieved by the formulas of Weingarten and Gauss in the form (see [51],[132],[152])

$$\frac{\partial \mathbf{n}}{\partial u^\alpha} = -g^{\sigma\gamma} \cdot b_{\gamma\alpha} \cdot \mathbf{e}_\sigma \qquad \text{(Weingarten)},$$

$$\frac{\partial^2 \mathbf{r}}{\partial u^\alpha u^\beta} = \frac{\partial \mathbf{e}_\alpha}{\partial u^\beta} = \Gamma^\sigma_{\alpha\beta} \cdot \mathbf{e}_\sigma + b_{\alpha\beta} \cdot \mathbf{n} \qquad \text{(Gauss)}, \tag{2.81}$$

with the following notations: The magnitudes $\alpha, \beta, \gamma$ and $\sigma$ take on the values 1,2 and follow the summation convention of Einstein. From this $u^\alpha$ means $u = u^1$ and $v = u^2$. The vector $\mathbf{e}_\alpha = \frac{\partial \mathbf{r}}{\partial u^\alpha}$ is defined in equation 2.76. The Christoffel three-index symbol $\Gamma^\sigma_{\alpha\beta} \cdot \mathbf{e}_\sigma$ writes

$$\Gamma^\sigma_{\alpha\beta} \cdot \mathbf{e}_\sigma = \frac{1}{2} g^{\sigma\delta} \left( \frac{\partial g_{\alpha\delta}}{\partial u^\beta} + \frac{\partial g_{\beta\delta}}{\partial u^\alpha} - \frac{\partial g_{\alpha\beta}}{\partial u^\delta} \right). \tag{2.82}$$

Its properties may be seen from [51] and [132]. The elements $g^{\sigma\delta}$ and $g_{\alpha\beta}$ are the fundamental tensor components of a Riemann space. Within the context of surfaces these measure tensors depend on the elements of the first and second fundamental form of a surface (equations 2.78 and 2.80). We get

$$\begin{aligned} &g_{11} = E, & &g_{12} = g_{21} = F, & &g_{22} = G, \\ &g^{11} = \frac{G}{EG - F^2}, & &g^{12} = g^{21} = \frac{-F}{EG - F^2}, & &g^{22} = \frac{E}{EG - F^2}, \\ &b_{11} = L, & &b_{12} = b_{21} = M, & &b_{22} = N. \end{aligned} \tag{2.83}$$

The evaluation of the equations 2.81, 2.82 and 2.83 is straightforward but very lengthy and tedious. It can be found in all details in [152]. As a final result we get

$$\frac{\partial \mathbf{u}}{\partial u} = \Gamma_{11}^1 \mathbf{u} + \Gamma_{11}^2 \mathbf{v} + L\mathbf{n}, \qquad \frac{\partial \mathbf{u}}{\partial v} = \Gamma_{12}^1 \mathbf{u} + \Gamma_{12}^2 \mathbf{v} + M\mathbf{n},$$

$$\frac{\partial \mathbf{v}}{\partial u} = \Gamma_{12}^1 \mathbf{u} + \Gamma_{12}^2 \mathbf{v} + M\mathbf{n}, \qquad \frac{\partial \mathbf{v}}{\partial v} = \Gamma_{22}^1 \mathbf{u} + \Gamma_{22}^2 \mathbf{v} + N\mathbf{n},$$

$$\frac{\partial \mathbf{n}}{\partial u} = \frac{FM - GL}{EG - F^2}\mathbf{u} + \frac{FL - EM}{EG - F^2}\mathbf{v},$$

$$\frac{\partial \mathbf{n}}{\partial v} = \frac{FN - GM}{EG - F^2}\mathbf{u} + \frac{FM - EN}{EG - F^2}\mathbf{v}. \tag{2.84}$$

Obviously $\frac{\partial \mathbf{u}}{\partial v} = \frac{\partial \mathbf{v}}{\partial u}$. The Christoffel symbols can be evaluated in the form [152]

$$\Gamma_{11}^1 = \frac{1}{2(EG - F^2)} \cdot \left( +G\frac{\partial E}{\partial u} - 2F\frac{\partial F}{\partial u} + F\frac{\partial E}{\partial v} \right),$$

$$\Gamma_{11}^2 = \frac{1}{2(EG - F^2)} \cdot \left( -F\frac{\partial E}{\partial u} + 2E\frac{\partial F}{\partial u} - E\frac{\partial E}{\partial v} \right),$$

$$\Gamma_{12}^1 = \frac{1}{2(EG - F^2)} \cdot \left( +G\frac{\partial E}{\partial v} - F\frac{\partial G}{\partial u} \right),$$

$$\Gamma_{12}^2 = \frac{1}{2(EG - F^2)} \cdot \left( -F\frac{\partial E}{\partial v} + E\frac{\partial G}{\partial u} \right),$$

$$\Gamma_{22}^1 = \frac{1}{2(EG - F^2)} \cdot \left( -G\frac{\partial G}{\partial u} - 2G\frac{\partial F}{\partial v} - F\frac{\partial G}{\partial v} \right),$$

$$\Gamma_{22}^2 = \frac{1}{2(EG - F^2)} \cdot \left( +F\frac{\partial G}{\partial u} - 2F\frac{\partial F}{\partial v} + E\frac{\partial G}{\partial v} \right). \tag{2.85}$$

The above equations represent a general set for the description of arbitrary surfaces. They are also a necessary set for the evaluation of the relative kinematics of contacts between arbitrary bodies, where especially the moving trihedron with its rectangular axes of the principal normal, the tangent vector and the binormal is an indispensable requirement for analyzing contact dynamics. The rather costly evaluation for the general case can of course be reduced considerably for special geometries like bodies with rotational symmetry or with at least partly plane surfaces. The combination of bodies with various shapes like a cube on a sphere must be analyzed by considering all possible cases of contact. This is for example a typical problem of assembly processes (see [271], [152]).

As an example we consider bodies with rotational symmetry, which are used quite often in various technologies. Figure 2.18 illustrates the principal situation. The coordinate u is the circumferential coordinate with $u \in (0, 2\Pi)$, and v is the generatrix coordinate. The radius of the body at the position v is f(v). With these definitions we get $((\cdot)' = \frac{d(\cdot)}{dv})$

## 2 Fundamentals

$$\mathbf{r} = \begin{pmatrix} f(v) \cdot \cos u \\ f(v) \cdot \sin u \\ g(v) \end{pmatrix}, \quad \mathbf{u} = \begin{pmatrix} -f(v) \cdot \sin u \\ f(v) \cdot \cos u \\ 0 \end{pmatrix}, \quad \mathbf{v} = \begin{pmatrix} f'(v) \cdot \cos u \\ f'(v) \cdot \sin u \\ g'(v) \end{pmatrix}. \tag{2.86}$$

The principal normal $\mathbf{n}$ and the elements E, F, G of the first order funda-

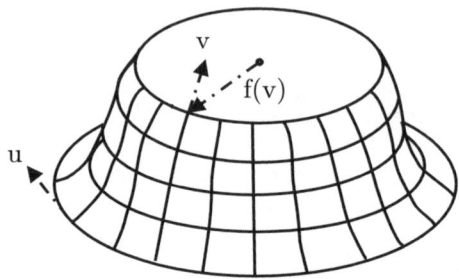

Fig. 2.18: Typical body with rotational symmetry

mental form are then derived by

$$\mathbf{n} = \frac{1}{\sqrt{f'^2(v) + g'^2(v)}} \cdot \begin{pmatrix} g'(v) \cdot \cos u \\ g'(v) \cdot \sin u \\ f'(v) \end{pmatrix}, \quad \begin{matrix} E = & f^2(v), \\ F = & 0, \\ G = & f'^2(v) + g'^2(v). \end{matrix} \tag{2.87}$$

The coordinates $\mathbf{u}$ and $\mathbf{v}$ are perpendicular and therefore F=0 and M=0. The elements of the second fundamental form write

$$L = -\frac{f(v) \cdot g'(v)}{\sqrt{f'^2(v) + g'^2(v)}}, \quad N = \frac{f''(v) \cdot g'(v) - f'(v) \cdot g''(v)}{\sqrt{f'^2(v) + g'^2(v)}}. \tag{2.88}$$

For the specific case of rotational symmetry the Christoffel elements $\Gamma^1_{11}$, $\Gamma^1_{12}$ and $\Gamma^1_{22}$ also become zero. A collection of different contact forms is given in [152] and in [277].

### 2.2.7 Relative Contact Kinematics

#### 2.2.7.1 Plane Case

In order to derive the kinematic contact equations of a whole system we consider first the geometry of a single plane body. Figure 2.19 shows such a body, which may have the rotational velocity $\boldsymbol{\Omega}$ and the rotational acceleration $\dot{\boldsymbol{\Omega}}$. The body-fixed point P moves with a velocity $\mathbf{v}_P$ and has the acceleration $\mathbf{a}_P$. The smooth and planar contour $\Sigma$ is supposed to be strictly convex and can be described in a parametric form by the vector ${}_B\dot{\mathbf{r}}_{P\Sigma}(s)$ in the body-fixed frame B. The parameter s corresponds to the arc length of the body's contour. With these definitions and the formulas of the preceding chapter we can state the moving trihedral $(\mathbf{t}, \mathbf{n}, \mathbf{b})$ in a body-fixed coordinate system B ([86],[200] and [212]).

$$_B\mathbf{t} = {_B}\mathbf{r}'_{P\Sigma}; \quad \kappa \cdot {_B}\mathbf{n} = {_B}\mathbf{r}''_{P\Sigma}; \quad (\cdot)' = \frac{\mathrm{d}}{\mathrm{d}s};$$
$$_B\mathbf{n} = {_B}\mathbf{b} \times {_B}\mathbf{t}; \quad {_B}\mathbf{b} = {_B}\mathbf{t} \times {_B}\mathbf{n}; \quad {_B}\mathbf{t} = {_B}\mathbf{n} \times {_B}\mathbf{b}.$$
$$_B\mathbf{n}' = {_B}\mathbf{b} \times {_B}\mathbf{t}' = {_B}\mathbf{b} \times (\kappa \cdot {_B}\mathbf{n}) = -\kappa \cdot {_B}\mathbf{t}; \quad {_B}\mathbf{t}' = \kappa \cdot {_B}\mathbf{n}. \qquad (2.89)$$

where $\kappa$ denotes the curvature of our convex contour at point $s$, and the normal vector $\mathbf{n}$ is pointing inward. The binormal $_B\mathbf{b}$ is constant for the plane case and thus independent of s. Now we imagine a point $\Sigma$ moving along the contour with the velocity $\dot{s}$. As a consequence the normal and the tangent vectors change their direction with respect to the body-fixed frame B. We describe this effect by differentiating the tangential and the normal vector with respect to time with $\frac{\mathrm{d}}{\mathrm{d}t} = \frac{\mathrm{d}}{\mathrm{d}s} \cdot \frac{\mathrm{d}s}{\mathrm{d}t} = \frac{\mathrm{d}}{\mathrm{d}s} \cdot \dot{s}$ we come out with

$$_B\dot{\mathbf{n}} = {_B}\mathbf{n}' \cdot \dot{s} = -\kappa \dot{s} \cdot {_B}\mathbf{t}, \qquad {_B}\dot{\mathbf{t}} = {_B}\mathbf{t}' \cdot \dot{s} = +\kappa \dot{s} \cdot {_B}\mathbf{n}. \qquad (2.90)$$

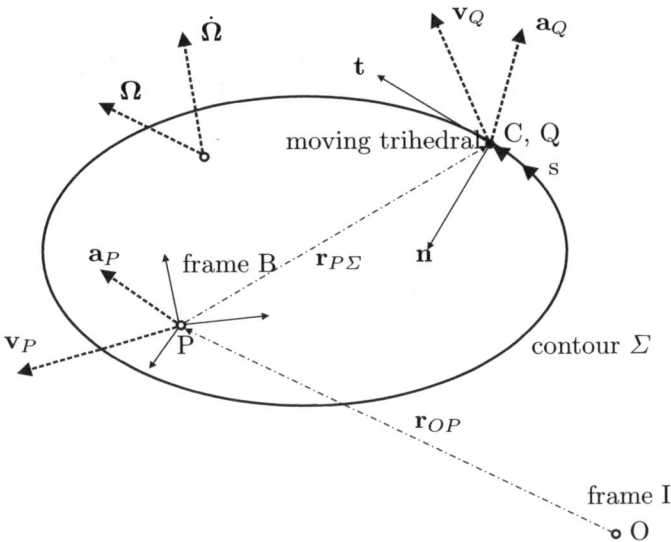

Fig. 2.19: Planar Contour Geometry

On the other hand, the absolute changes of the vectors $_B\mathbf{n}$ and $_B\mathbf{t}$ with respect to time follow from the Coriolis equations 2.38 with $_B\dot{\mathbf{r}} = 0$

$$_B(\dot{\mathbf{n}}) = {_B}\dot{\mathbf{n}} + {_B}\tilde{\Omega} \cdot {_B}\mathbf{n}, \qquad {_B}(\dot{\mathbf{t}}) = {_B}\dot{\mathbf{t}} + {_B}\tilde{\Omega} \cdot {_B}\mathbf{t}, \qquad (2.91)$$

where we have to keep in mind that $_B\boldsymbol{\omega}_{IB} = {}_B\boldsymbol{\Omega}$ for a body-fixed coordinate system expressing the property of rotation of the body-fixed frame with respect to the inertial frame. Combining the equations 2.90 and 2.91 we come out with a coordinate-free representation of the overall changes of the contour normal and tangential vectors

$$\dot{\mathbf{n}} = \tilde{\boldsymbol{\Omega}}\mathbf{n} - \kappa\dot{s} \cdot \mathbf{t}, \qquad \dot{\mathbf{t}} = \tilde{\boldsymbol{\Omega}}\mathbf{t} + \kappa\dot{s} \cdot \mathbf{n}, \tag{2.92}$$

which we might evaluate in any basis. From this the main advantage of equation 2.92 consists in this coordinate-free form and in avoiding the determination of the frame-dependent time-derivatives $_B\dot{\mathbf{n}}$ and $_B\dot{\mathbf{t}}$. In the same way we proceed with the contour vector $\mathbf{r}_{P\Sigma}$. Following the equations 2.90 and 2.91 we write

$$_B\dot{\mathbf{r}}'_{P\Sigma} = {}_B\mathbf{r}'_{P\Sigma} \cdot \dot{s} = \dot{s} \cdot {}_B\mathbf{t}, \qquad _B(\dot{\mathbf{r}}_{P\Sigma}) = {}_B\dot{\mathbf{r}}_{P\Sigma} + {}_B\tilde{\boldsymbol{\Omega}} \cdot {}_B\mathbf{r}_{P\Sigma}, \tag{2.93}$$

and we eliminate $_B\dot{\mathbf{r}}_{P\Sigma}$ resulting in the absolute changes of $\mathbf{r}_{P\Sigma}$ with time

$$\dot{\mathbf{r}}_{P\Sigma} = \tilde{\boldsymbol{\Omega}} \cdot \mathbf{r}_{P\Sigma} + \dot{s}\mathbf{t}. \tag{2.94}$$

Due to Figure 2.19 we have $\mathbf{v}_\Sigma = \mathbf{v}_P + \dot{\mathbf{r}}_{P\Sigma}$, and therefore the absolute velocity of a moving contour point $\Sigma$ is given by

$$\mathbf{v}_\Sigma = \mathbf{v}_P + \tilde{\boldsymbol{\Omega}} \cdot \mathbf{r}_{P\Sigma} + \dot{s}\mathbf{t} = \mathbf{v}_C + \dot{s}\mathbf{t}, \quad \text{with} \quad \mathbf{v}_C = \mathbf{v}_P + \tilde{\boldsymbol{\Omega}} \cdot \mathbf{r}_{P\Sigma}. \tag{2.95}$$

The velocity $\mathbf{v}_C$ results from rigid body kinematics and corresponds to the velocity of a body-fixed point C at the contour, it is the applied velocity in a classical sense.

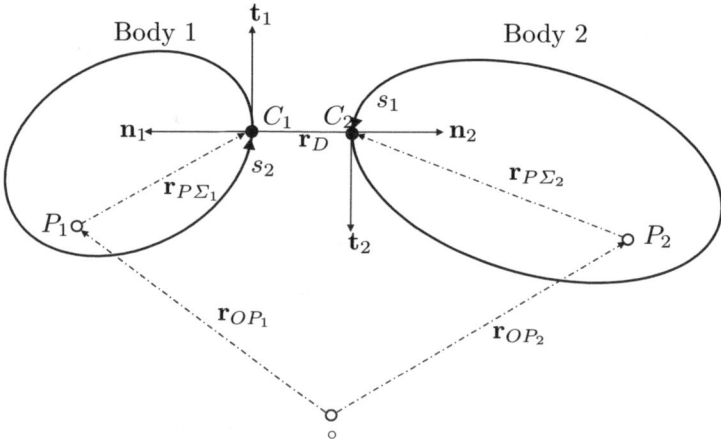

Fig. 2.20: Orientation of two Bodies

In a next step we must evaluate the absolute acceleration of the point C of the body contour. Differentiating equation 2.95 with respect to time we get

$$\dot{\mathbf{v}}_C = \dot{\mathbf{v}}_P + \dot{\tilde{\boldsymbol{\Omega}}} \cdot \mathbf{r}_{P\Sigma} + \tilde{\boldsymbol{\Omega}} \cdot \dot{\mathbf{r}}_{P\Sigma}. \tag{2.96}$$

With the abbreviations $\dot{\mathbf{v}}_C = \mathbf{a}_C$, $\dot{\mathbf{v}}_P = \mathbf{a}_P$ and with $\dot{\mathbf{r}}_{P\Sigma}$ from equation 2.93 we continue with the expressions

$$\begin{aligned}\mathbf{a}_C &= \mathbf{a}_P + \dot{\tilde{\boldsymbol{\Omega}}} \cdot \mathbf{r}_{P\Sigma} + \tilde{\boldsymbol{\Omega}}\tilde{\boldsymbol{\Omega}} \cdot \mathbf{r}_{P\Sigma} + \tilde{\boldsymbol{\Omega}}\mathbf{t}\dot{s} = \mathbf{a}_Q + \tilde{\boldsymbol{\Omega}}\mathbf{t}\dot{s} \qquad \text{with} \\ \mathbf{a}_Q &= \mathbf{a}_P + \dot{\tilde{\boldsymbol{\Omega}}} \cdot \mathbf{r}_{P\Sigma} + \tilde{\boldsymbol{\Omega}}\tilde{\boldsymbol{\Omega}} \cdot \mathbf{r}_{P\Sigma}\end{aligned} \tag{2.97}$$

The acceleration $\mathbf{a}_Q$ results from rigid body kinematics and corresponds to the acceleration of a body-fixed point Q, it is an applied acceleration in analogy to equation 2.95. For later evaluations we need the relative velocities and their time derivatives in normal and tangential directions. For that purpose we consider the corresponding velocities in the form

$$v_n = \mathbf{n}^T \mathbf{v}_C, \qquad v_t = \mathbf{t}^T \mathbf{v}_C, \tag{2.98}$$

with their time derivatives

$$\dot{v}_n = \dot{\mathbf{n}}^T \mathbf{v}_C + \mathbf{n}^T \dot{\mathbf{v}}_C, \qquad \dot{v}_t = \dot{\mathbf{t}}^T \mathbf{v}_C + \mathbf{t}^T \dot{\mathbf{v}}_C. \tag{2.99}$$

With $\dot{\mathbf{n}}, \dot{\mathbf{t}}$ from equation 2.92, $\dot{\mathbf{v}}_C = \mathbf{a}_C$ from equation 2.97 and noting the relations $\mathbf{n}^T \tilde{\boldsymbol{\Omega}} \mathbf{t} = \mathbf{b}^T \boldsymbol{\Omega}$ and $\mathbf{t}^T \tilde{\boldsymbol{\Omega}} \mathbf{t} = 0$ we derive

$$\dot{v}_n = \mathbf{n}^T (\mathbf{a}_Q - \tilde{\boldsymbol{\Omega}} \mathbf{v}_C) - \kappa \dot{s} \mathbf{t}^T \mathbf{v}_C + \dot{s} \mathbf{b}^T \boldsymbol{\Omega}, \tag{2.100}$$

$$\dot{v}_t = \mathbf{t}^T (\mathbf{a}_Q - \tilde{\boldsymbol{\Omega}} \mathbf{v}_C) + \kappa \dot{s} \mathbf{n}^T \mathbf{v}_C. \tag{2.101}$$

With these fundamental equations for one body we are able to consider in a next step two contacting bodies. We are still in a plane. Figure 2.20 gives the nomenclature and the directions. The sense of the contour parameters $s_1$ and $s_2$ are chosen in such a way that the binormals of both moving trihedrals are the same, $\mathbf{b}_1 = \mathbf{b}_2$. The origins of the two trihedrals are connected with the relative distance vector $\mathbf{r}_D$. To determine this distance vector we orient the trihedrals in such a manner, that they are mutually perpendicular, which is always possible and gives the conditions:

$$\mathbf{n}_1^T(s_1) \cdot \mathbf{t}_2(s_2) = 0, \qquad \Leftrightarrow \qquad \mathbf{n}_2^T(s_2) \cdot \mathbf{t}_1(s_1) = 0. \tag{2.102}$$

These conditions require that both, normals and tangents, are parallel (see Figure 2.20). For an evaluation of these equations we need of course only one, because the two are equivalent. The next requirement consists in putting the relative distance $\mathbf{r}_D$ unidirectional with the two normals and thus perpendicular to the two tangent vectors. This gives

$$\mathbf{r}_D^T(s_1, s_2) \cdot \mathbf{t}_1(s_1) = 0, \qquad \Leftrightarrow \qquad \mathbf{r}_D^T(s_1, s_2) \cdot \mathbf{t}_2(s_2) = 0. \tag{2.103}$$

From the four equations 2.102 and 2.103 we need only two, for example the first ones. The solution $(s_1, s_2)$ of these two conditions as nonlinear functions of $(s_1, s_2)$ results in a configuration indicated in Figure 2.20: normal and tangent vectors are antiparallel to each other, and the relative distance vector $\mathbf{r}_D$ is perpendicular to the two surfaces, which at the same time is the shortest possible distance between the two bodies. The values $(s_1, s_2)$ are the "contact parameters" of our problem, and the accompanying points $(C_1, C_2)$ the "contact points". The axes of the two trihedrals are given by

$$\mathbf{n}_1 = -\mathbf{n}_2, \qquad \mathbf{t}_1 = -\mathbf{t}_2, \qquad \mathbf{b}_{12} = \mathbf{b}_1 = \mathbf{b}_2. \tag{2.104}$$

From this we get easily the distance between the two bodies

$$g_N(\mathbf{q}, t) = \mathbf{r}_D^T \mathbf{n}_2 = -\mathbf{r}_D^T \mathbf{n}_1. \tag{2.105}$$

Since the normal vector always points inwards, $g_N$ is positive for separation and negative for overlapping. Thus, a change of sign from positive to negative values indicates a transition from initially separated bodies to contact.

Relative kinematics plays a key role in detecting a change of contact situations, for example such transitions like detachment-contact or stick-slip and vice versa. For impacts we have in addition a non-continuous change of the relative velocities, and if we want to combine these inequality constraints with the equations of motion we need also the relative accelerations. Therefore relative kinematics of contacts include the whole set of position and orientation, of velocities and of accelerations. Some of these magnitudes we get by differentiation. In that case we should not forget the original state of the relative kinematic magnitudes for the special contact event under consideration, for example, in normal direction of a contact the event contact is indicated by the relative distance becoming then a constraint. In tangential direction the relative tangential velocity indicates the stick or slip situation. We shall come back to these properties later.

We consider Figure 2.20 and assume for a while that the equations 2.102 and 2.103 have not yet been fulfilled. The relative distance $\mathbf{r}_D$ is then not perpendicular to the two surfaces $\Sigma_1$ and $\Sigma_2$, but it represents some straight connection between the future contact points $C_1$ and $C_2$. The absolute change of $\mathbf{r}_D$ with time writes

$$\dot{\mathbf{r}}_D = \mathbf{v}_{\Sigma 2} - \mathbf{v}_{\Sigma 1}, \tag{2.106}$$

where the contour velocities $\mathbf{v}_\Sigma$ come from the equations 2.95. For evaluating the relative velocities we need in a further step the contour velocities $\dot{s}$. For this purpose we differentiate the equations 2.102 and 2.103 resulting in

$$\begin{aligned}(\mathbf{r}_D^T \cdot \mathbf{t}_1)^\cdot &= \dot{\mathbf{r}}_D^T \cdot \mathbf{t}_1 + \mathbf{r}_D^T \cdot \dot{\mathbf{t}}_1, \\ (\mathbf{r}_D^T \cdot \mathbf{n}_1)^\cdot &= \dot{\mathbf{r}}_D^T \cdot \mathbf{n}_1 + \mathbf{r}_D^T \cdot \dot{\mathbf{n}}_1, \\ (\mathbf{n}_1^T \cdot \mathbf{t}_2)^\cdot &= \dot{\mathbf{n}}_1^T \cdot \mathbf{t}_2 + \mathbf{n}_1^T \cdot \dot{\mathbf{t}}_2, \end{aligned} \tag{2.107}$$

where $\dot{\mathbf{r}}_D$ is given with the equation 2.106 and the time-derivatives of the unit vectors $\mathbf{n}$ and $\mathbf{t}$ come from the equations 2.92. Together with equation 2.95 we finally get

$$(\mathbf{r}_D^T \cdot \mathbf{t}_1)^\cdot = \mathbf{t}_1^T \cdot (\mathbf{v}_{C2} - \mathbf{v}_{C1} - \mathbf{\Omega}_1 \times \mathbf{r}_D) + \mathbf{t}_1^T \mathbf{t}_2 \dot{s}_2 - \mathbf{t}_1^T \mathbf{t}_1 \dot{s}_1 + \mathbf{r}_D^T \mathbf{n}_1 \kappa_1 \dot{s}_1,$$
$$(\mathbf{r}_D^T \cdot \mathbf{n}_1)^\cdot = \mathbf{n}_1^T \cdot (\mathbf{v}_{C2} - \mathbf{v}_{C1} - \mathbf{\Omega}_1 \times \mathbf{r}_D) + \mathbf{n}_1^T \mathbf{t}_2 \dot{s}_2 - \mathbf{n}_1^T \mathbf{t}_1 \dot{s}_1 + \mathbf{r}_D^T \mathbf{t}_1 \kappa_1 \dot{s}_1,$$
$$(\mathbf{n}_1^T \cdot \mathbf{t}_2)^\cdot = -\mathbf{t}_1^T \mathbf{t}_2 \kappa_1 \dot{s}_1 + \mathbf{n}_1^T \mathbf{n}_2 \kappa_2 \dot{s}_2 + (\mathbf{t}_2 \times \mathbf{n}_1)^T \cdot (\mathbf{\Omega}_2 - \mathbf{\Omega}_1). \tag{2.108}$$

These expressions hold for the general case, where the trihedrals are not yet oriented. They simplify by applying the conditions 2.102 and 2.103, and they can then be taken to evaluate the contour velocities $\dot{s}_1$ and $\dot{s}_2$. After some elementary calculations (see [200]) we come out with

$$\dot{s}_1 = \frac{\kappa_2 \mathbf{t}_1^T(\mathbf{v}_{C2} - \mathbf{v}_{C1}) - \kappa_2 g_N \mathbf{b}_{12}^T \mathbf{\Omega}_1 + \mathbf{b}_{12}^T(\mathbf{\Omega}_2 - \mathbf{\Omega}_1)}{\kappa_1 + \kappa_2 + g_N \kappa_1 \kappa_2},$$
$$\dot{s}_2 = \frac{\kappa_1 \mathbf{t}_1^T(\mathbf{v}_{C2} - \mathbf{v}_{C1}) - \kappa_1 g_N \mathbf{b}_{12}^T \mathbf{\Omega}_2 - \mathbf{b}_{12}^T(\mathbf{\Omega}_2 - \mathbf{\Omega}_1)}{\kappa_1 + \kappa_2 + g_N \kappa_1 \kappa_2}, \tag{2.109}$$

with the "binormal" $\mathbf{b}_{12} = -\mathbf{t}_2 \times \mathbf{n}_1$ to the vectors $\mathbf{n}_1$ and $\mathbf{t}_2$. The relative distance $g_N$ is defined by equation 2.105.

With the above equations we can now evaluate the relative velocities and the relative accelerations of two bodies coming into contact or sliding on each other. Starting with the relative velocities we write (see Figure 2.20 and [200])

$$\dot{g}_N = \mathbf{n}_1^T \mathbf{v}_{C1} + \mathbf{n}_2^T \mathbf{v}_{C2}, \qquad \dot{g}_T = \mathbf{t}_1^T \mathbf{v}_{C1} + \mathbf{t}_2^T \mathbf{v}_{C2}, \tag{2.110}$$

where $\mathbf{v}_{C1}$ and $\mathbf{v}_{C2}$ are the absolute velocities of the potential contact points $C_1$ and $C_2$. These velocities might be expressed by the generalized, or minimal, velocities $\dot{\mathbf{q}}$ using the Jacobians $\mathbf{J}_{C1}$ and $\mathbf{J}_{C2}$ (see [27], [200])

$$\mathbf{v}_{C1} = \mathbf{J}_{C1} \dot{\mathbf{q}} + \tilde{\mathbf{j}}_{C1}, \qquad \mathbf{v}_{C2} = \mathbf{J}_{C2} \dot{\mathbf{q}} + \tilde{\mathbf{j}}_{C2}. \tag{2.111}$$

Combining the last two equations results in

$$\dot{g}_N = \mathbf{w}_N^T \dot{\mathbf{q}} + \tilde{w}_N, \qquad \dot{g}_T = \mathbf{w}_T^T \dot{\mathbf{q}} + \tilde{w}_T, \tag{2.112}$$

with

$$\mathbf{w}_N = \mathbf{J}_{C1}^T \mathbf{n}_1 + \mathbf{J}_{C2}^T \mathbf{n}_2, \qquad \mathbf{w}_T = \mathbf{J}_{C1}^T \mathbf{t}_1 + \mathbf{J}_{C2}^T \mathbf{t}_2,$$
$$\tilde{w}_N = \tilde{\mathbf{j}}_{C1}^T \mathbf{n}_1 + \tilde{\mathbf{j}}_{C2}^T \mathbf{n}_2, \qquad \tilde{w}_T = \tilde{\mathbf{j}}_{C1}^T \mathbf{t}_1 + \tilde{\mathbf{j}}_{C2}^T \mathbf{t}_2, \tag{2.113}$$

which will be used in the sequel as a representation of the relative velocities between neighboring bodies within a whole system of bodies. It should be noticed that a negative value of the relative normal velocity $\dot{g}_N$ corresponds to an approaching process of the neighboring bodies and coincides at vanishing distance $g_N = 0$ with the relative velocity in normal contact direction shortly before a contact, which will be in most cases an impact. In the case of a continual contact with $g_N = 0, \dot{g}_N = 0$ the relative tangential velocity indicates sliding between the two bodies, which we need to determine the tangential transition event from sliding with $\dot{g}_T \neq 0$ to sticking with $\dot{g}_T = 0$.

The relative accelerations are derived by differentiation of the velocity equations 2.112. We get the same structure of the velocity equations

$$\ddot{g}_N = \mathbf{w}_N^T \ddot{\mathbf{q}} + \bar{w}_N, \qquad \ddot{g}_T = \mathbf{w}_T^T \ddot{\mathbf{q}} + \bar{w}_T, \tag{2.114}$$

where the constraint vectors $\mathbf{w}_N, \mathbf{w}_T$ are known from the equations 2.113 and the nonlinear functions $\bar{w}_N, \bar{w}_T$ follow from

$$\begin{aligned}
\bar{w}_N &= \mathbf{n}_1^T (\bar{\mathbf{j}}_{Q1} - \tilde{\boldsymbol{\Omega}}_1 \mathbf{v}_{C1}) - \kappa_1 \dot{s}_1 \mathbf{t}_1^T \mathbf{v}_{C1} + \dot{s}_1 \mathbf{b}_{12}^T \boldsymbol{\Omega}_1 + \\
&\quad \mathbf{n}_2^T (\bar{\mathbf{j}}_{Q2} - \tilde{\boldsymbol{\Omega}}_2 \mathbf{v}_{C2}) - \kappa_2 \dot{s}_2 \mathbf{t}_2^T \mathbf{v}_{C2} + \dot{s}_2 \mathbf{b}_{12}^T \boldsymbol{\Omega}_2, \\
\bar{w}_T &= \mathbf{t}_1^T (\bar{\mathbf{j}}_{Q1} - \tilde{\boldsymbol{\Omega}}_1 \mathbf{v}_{C1}) + \kappa_1 \dot{s}_1 \mathbf{n}_1^T \mathbf{v}_{C1} + \\
&\quad \mathbf{t}_2^T (\bar{\mathbf{j}}_{Q2} - \tilde{\boldsymbol{\Omega}}_2 \mathbf{v}_{C2}) + \kappa_2 \dot{s}_2 \mathbf{n}_2^T \mathbf{v}_{C2}.
\end{aligned} \tag{2.115}$$

All magnitudes of these equations are defined in the preceding equations. For a more detailed description see the literature ([200], [212] and [86]).

### 2.2.7.2 Spatial Case

Spatial contacts require a more complex description. We assume as above that the two approaching bodies are convex at least in the neighborhood of those points, where contacts might occur (Figure 2.21). The two bodies with their coordinate bases $B_1, B_2$ as part of a whole system of bodies move with the velocities $\mathbf{v}_i, \boldsymbol{\Omega}_i$ (i=1,2) and have a certain position and orientation with respect to an inertial frame and with respect to each other. They possess a contour $\Sigma$ in two dimensions, which are parameterized by the curvilinear coordinates s and t. The two bodies may come into contact or they may be already in contact, for example sliding on each other. We assume, that there is only one contact point, but in many technical applications bodies might have several contacts, for example in the fields of power transmission, gears and chains. However, each of these multiple contacts follows the same theory as given below. What we have to do, is some additional indexing.

What we are looking for, is a representation of the relative kinematics in terms of the surface coordinates and their derivatives as a first step, and as a second step we want to express these relative kinematic magnitudes in terms of the body velocities finally arriving at the relative velocities and accelerations depending on the corresponding generalized magnitudes. As a base for all considerations to follow we use chapter 2.2.6. Let us start with a point of the contours $\Sigma_1$ and $\Sigma_2$ with the coordinates $s_i, t_i$ (i=1,2) and the surface unit vectors in these points $\mathbf{n}_i, \mathbf{s}_i, \mathbf{t}_i$ (i=1,2). The unit vectors $\mathbf{s}_i, \mathbf{t}_i$ (i=1,2) span the tangent plane in the potential points of contact, on both sides. These unit vectors are defined by

$$\mathbf{s} = \frac{\partial \mathbf{r}_\Sigma}{\partial s}, \qquad \mathbf{t} = \frac{\partial \mathbf{r}_\Sigma}{\partial t}. \tag{2.116}$$

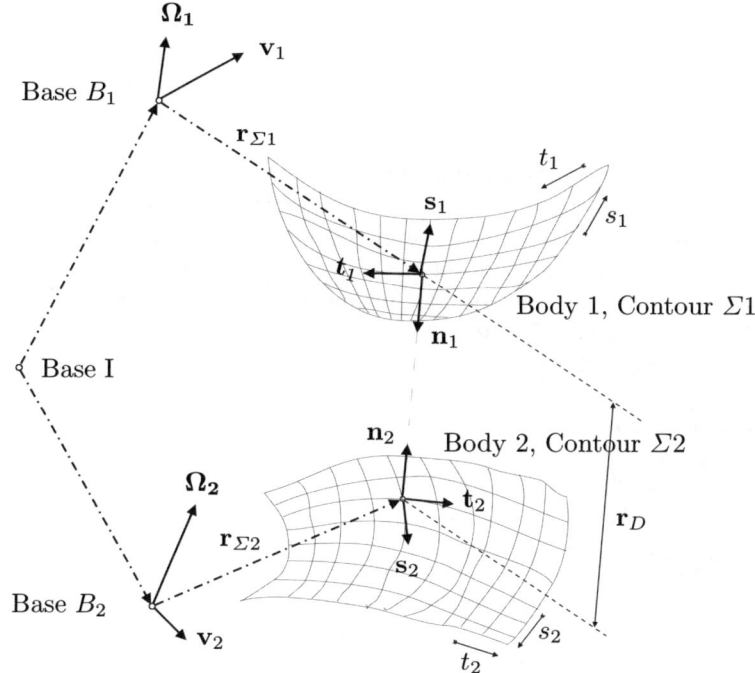

Fig. 2.21: Contact Zones in the Spatial Case

From these basic vectors we are able to determine the fundamental magnitudes of first order

$$E = \mathbf{s}^T \mathbf{s}, \qquad F = \mathbf{s}^T \mathbf{t}, \qquad G = \mathbf{t}^T \mathbf{t}. \tag{2.117}$$

The normal unit vector $\mathbf{n}$ is perpendicular to the surface tangent plane pointing outwardly, therefore

$$\mathbf{n} = \frac{\mathbf{s} \times \mathbf{t}}{\sqrt{EG - F^2}}. \tag{2.118}$$

The fundamental magnitudes of second order follow from the equations 2.79

$$L = \mathbf{n}^T \cdot \frac{\partial^2 \mathbf{r}_\Sigma}{\partial s^2}, \qquad M = \mathbf{n}^T \cdot \frac{\partial^2 \mathbf{r}_\Sigma}{\partial s \partial t}, \qquad N = \mathbf{n}^T \cdot \frac{\partial^2 \mathbf{r}_\Sigma}{\partial t^2} \tag{2.119}$$

For a potential contact point we must achieve some requirements with respect to the directions of the surface unit vectors and the distance vector $\mathbf{r}_D$. From several possibilities we take the following four ones

$$\mathbf{n}_1^T \mathbf{s}_2 = 0, \qquad \mathbf{n}_1^T \mathbf{t}_2 = 0, \qquad \mathbf{r}_D^T \mathbf{s}_2 = 0, \qquad \mathbf{r}_D^T \mathbf{t}_2 = 0. \tag{2.120}$$

This nonlinear problem has to be solved at every time step of the numerical integration by appropriate algorithms, analytical solutions are very unlikely.

Once the solution has been achieved we are able to evaluate the relative distance $g_N$ between the two bodies by

$$g_N = \mathbf{n}_1^T \mathbf{r}_D = -\mathbf{n}_2^T \mathbf{r}_D \qquad (2.121)$$

The relative distance $g_N$ is one of the important contact magnitudes indicating the state "no contact" or the state "contact". It is positive for the first state and negative for the second one in the case of penetration.

For a further definition of the unilateral constraints we need the relative velocities and accelerations. Regarding the spatial case as indicated in Figure 2.21 we have one relative velocity in normal and two relative velocities in tangential directions

$$\begin{aligned}
\dot{g}_N(\mathbf{q},\dot{\mathbf{q}},t) &= \mathbf{n}_1^T \cdot (\mathbf{v}_{\Sigma 2} - \mathbf{v}_{\Sigma 1}), \\
\dot{g}_S(\mathbf{q},\dot{\mathbf{q}},t) &= \mathbf{s}_1^T \cdot (\mathbf{v}_{\Sigma 2} - \mathbf{v}_{\Sigma 1}), \\
\dot{g}_T(\mathbf{q},\dot{\mathbf{q}},t) &= \mathbf{t}_1^T \cdot (\mathbf{v}_{\Sigma 2} - \mathbf{v}_{\Sigma 1}),
\end{aligned} \qquad (2.122)$$

with $\mathbf{v}_{\Sigma 1}$ and $\mathbf{v}_{\Sigma 2}$ defined in the following way: In the plane case we have considered the potential contact points $C_1$ and $C_2$ and expressed their velocities $\mathbf{v}_{C1}$ and $\mathbf{v}_{C2}$ by the generalized or minimal velocities $\dot{\mathbf{q}}$ using some Jacobians for these contact points (see equations (2.95) and (2.111)). The same equations can be used for the points $\Sigma_1$ and $\Sigma_2$. That means the velocities $\mathbf{v}_{\Sigma 1}$ and $\mathbf{v}_{\Sigma 2}$ in the spatial case correspond to the velocities $\mathbf{v}_{C1}$ and $\mathbf{v}_{C2}$ in the plane case. The velocities $\mathbf{v}_{C1}, \mathbf{v}_{C2}$ in the plane and $\mathbf{v}_{\Sigma 1}, \mathbf{v}_{\Sigma 2}$ in the spatial case are those of body-fixed contour points, which momentarily coincide with the potential contact point.

Differentiating the equations 2.122 with respect to time we get the relative accelerations by

$$\begin{aligned}
\ddot{g}_N(\mathbf{q},\dot{\mathbf{q}},t) &= \mathbf{n}_1^T \cdot (\dot{\mathbf{v}}_{\Sigma 2} - \dot{\mathbf{v}}_{\Sigma 1}) + \dot{\mathbf{n}}_1^T \cdot (\mathbf{v}_{\Sigma 2} - \mathbf{v}_{\Sigma 1}), \\
\ddot{g}_S(\mathbf{q},\dot{\mathbf{q}},t) &= \mathbf{s}_1^T \cdot (\dot{\mathbf{v}}_{\Sigma 2} - \dot{\mathbf{v}}_{\Sigma 1}) + \dot{\mathbf{s}}_1^T \cdot (\mathbf{v}_{\Sigma 2} - \mathbf{v}_{\Sigma 1}), \\
\ddot{g}_T(\mathbf{q},\dot{\mathbf{q}},t) &= \mathbf{t}_1^T \cdot (\dot{\mathbf{v}}_{\Sigma 2} - \dot{\mathbf{v}}_{\Sigma 1}) + \dot{\mathbf{t}}_1^T \cdot (\mathbf{v}_{\Sigma 2} - \mathbf{v}_{\Sigma 1}),
\end{aligned} \qquad (2.123)$$

The velocities and accelerations of the contact points follow from the above considerations and by differentiation with respect to time as

$$\begin{aligned}
\mathbf{v}_{\Sigma 1} &= \mathbf{J}_{\Sigma 1}(\mathbf{q},t)\dot{\mathbf{q}} + \tilde{\mathbf{j}}_{\Sigma 1}(\mathbf{q},\dot{\mathbf{q}},t), & \mathbf{v}_{\Sigma 2} &= \mathbf{J}_{\Sigma 2}(\mathbf{q},t)\dot{\mathbf{q}} + \tilde{\mathbf{j}}_{\Sigma 2}(\mathbf{q},\dot{\mathbf{q}},t), \\
\dot{\mathbf{v}}_{\Sigma 1} &= \mathbf{J}_{\Sigma 1}(\mathbf{q},t)\ddot{\mathbf{q}} + \bar{\mathbf{j}}_{\Sigma 1}(\mathbf{q},\dot{\mathbf{q}},t), & \dot{\mathbf{v}}_{\Sigma 2} &= \mathbf{J}_{\Sigma 2}(\mathbf{q},t)\ddot{\mathbf{q}} + \bar{\mathbf{j}}_{\Sigma 2}(\mathbf{q},\dot{\mathbf{q}},t).
\end{aligned} \qquad (2.124)$$

The surface vectors $\dot{\mathbf{n}}_1, \dot{\mathbf{s}}_1$ and $\dot{\mathbf{t}}_1$ can be determined by the formulas of Weingarten and Gauss (equations 2.81), which results in:

## 2.2 Kinematics

$$\dot{\mathbf{n}}_1 = \mathbf{\Omega}_1 \times \mathbf{n}_1 + \frac{\partial \mathbf{n}_1}{\partial s_1}\dot{s}_1 + \frac{\partial \mathbf{n}_1}{\partial t_1}\dot{t}_1,$$

$$\frac{\partial \mathbf{n}_1}{\partial s_1} = \underbrace{\frac{M_1 F_1 - L_1 G_1}{E_1 G_1 - F_1^2}}_{\alpha_1} \mathbf{s}_1 + \underbrace{\frac{L_1 F_1 - M_1 E_1}{E_1 G_1 - F_1^2}}_{\beta_1} \mathbf{t}_1,$$

$$\frac{\partial \mathbf{n}_1}{\partial t_1} = \underbrace{\frac{N_1 F_1 - M_1 G_1}{E_1 G_1 - F_1^2}}_{\alpha'_1} \mathbf{s}_1 + \underbrace{\frac{M_1 F_1 - N_1 E_1}{E_1 G_1 - F_1^2}}_{\beta'_1} \mathbf{t}_1, \qquad (2.125)$$

$$\dot{\mathbf{s}}_1 = \mathbf{\Omega}_1 \times \mathbf{s}_1 + \frac{\partial \mathbf{s}_1}{\partial s_1}\dot{s}_1 + \frac{\partial \mathbf{s}_1}{\partial t_1}\dot{t}_1,$$

$$\frac{\partial \mathbf{s}_1}{\partial s_1} = (\Gamma^1_{11})_1 \mathbf{s}_1 + (\Gamma^2_{11})_1 \mathbf{t}_1 + L_1 \mathbf{n}_1,$$

$$\frac{\partial \mathbf{s}_1}{\partial t_1} = (\Gamma^1_{12})_1 \mathbf{s}_1 + (\Gamma^2_{12})_1 \mathbf{t}_1 + M_1 \mathbf{n}_1, \qquad (2.126)$$

$$\dot{\mathbf{t}}_1 = \mathbf{\Omega}_1 \times \mathbf{t}_1 + \frac{\partial \mathbf{t}_1}{\partial s_1}\dot{s}_1 + \frac{\partial \mathbf{t}_1}{\partial t_1}\dot{t}_1,$$

$$\frac{\partial \mathbf{t}_1}{\partial s_1} = (\Gamma^1_{12})_1 \mathbf{s}_1 + (\Gamma^2_{12})_1 \mathbf{t}_1 + M_1 \mathbf{n}_1,$$

$$\frac{\partial \mathbf{t}_1}{\partial t_1} = (\Gamma^1_{22})_1 \mathbf{s}_1 + (\Gamma^2_{22})_1 \mathbf{t}_1 + N_1 \mathbf{n}_1. \qquad (2.127)$$

The Christoffel symbols $\Gamma^\sigma_{\alpha\beta}$ with $(\alpha\beta=1,2)$ are defined in equation 2.82 (see also [51]). Inserting the above relations for the surface vectors $(\mathbf{n}_1, \mathbf{s}_1, \mathbf{t}_1)$ and $(\dot{\mathbf{n}}_1, \dot{\mathbf{s}}_1, \dot{\mathbf{t}}_1)$ into the equations 2.123 we come out with

$$\begin{aligned}
\ddot{g}_N =& \mathbf{n}_1^T[(\mathbf{J}_{\Sigma 2} - \mathbf{J}_{\Sigma 1})\ddot{\mathbf{q}} + (\bar{\mathbf{j}}_{\Sigma 2} - \bar{\mathbf{j}}_{\Sigma 1})] + \\
& (\mathbf{v}_{\Sigma 2} - \mathbf{v}_{\Sigma 1})^T \cdot [(\mathbf{\Omega}_1 \times \mathbf{n}_1) + ((\alpha_1 \mathbf{s}_1 + \beta_1 \mathbf{t}_1)\dot{s}_1 + (\alpha'_1 \mathbf{s}_1 + \beta'_1 \mathbf{t}_1)\dot{t}_1)],
\end{aligned}$$

$$\begin{aligned}
\ddot{g}_S =& \mathbf{s}_1^T[(\mathbf{J}_{\Sigma 2} - \mathbf{J}_{\Sigma 1})\ddot{\mathbf{q}} + (\bar{\mathbf{j}}_{\Sigma 2} - \bar{\mathbf{j}}_{\Sigma 1})] + \\
& (\mathbf{v}_{\Sigma 2} - \mathbf{v}_{\Sigma 1})^T \cdot [(\mathbf{\Omega}_1 \times \mathbf{s}_1) + ((\Gamma^1_{11})_1 \mathbf{s}_1 + (\Gamma^2_{11})_1 \mathbf{t}_1 + L_1 \mathbf{n}_1)\dot{s}_1 + \\
& \qquad\qquad\qquad\qquad ((\Gamma^1_{12})_1 \mathbf{s}_1 + (\Gamma^2_{12})_1 \mathbf{t}_1 + M_1 \mathbf{n}_1)\dot{t}_1],
\end{aligned}$$

$$\begin{aligned}
\ddot{g}_T =& \mathbf{t}_1^T[(\mathbf{J}_{\Sigma 2} - \mathbf{J}_{\Sigma 1})\ddot{\mathbf{q}} + (\bar{\mathbf{j}}_{\Sigma 2} - \bar{\mathbf{j}}_{\Sigma 1})] + \\
& (\mathbf{v}_{\Sigma 2} - \mathbf{v}_{\Sigma 1})^T \cdot [(\mathbf{\Omega}_1 \times \mathbf{s}_1) + ((\Gamma^1_{12})_1 \mathbf{s}_1 + (\Gamma^2_{12})_1 \mathbf{t}_1 + M_1 \mathbf{n}_1)\dot{s}_1 + \\
& \qquad\qquad\qquad\qquad ((\Gamma^1_{22})_1 \mathbf{s}_1 + (\Gamma^2_{22})_1 \mathbf{t}_1 + N_1 \mathbf{n}_1)\dot{t}_1]. \qquad (2.128)
\end{aligned}$$

The Jacobian matrices $\mathbf{J}_{\Sigma 1}$ and $\mathbf{J}_{\Sigma 2}$ are known from the elastic or rigid body kinematics as discussed in connection with the equations 2.69 and 2.111. The time derivatives $(\dot{s}_1, \dot{t}_1, \dot{s}_2, \dot{t}_2)$ of the surface unit vectors can be evaluated from the surface geometry and its curvilinear coordinates, in a similar way as in the plane case with the equations (2.106) to (2.109). Differentiating the equations (2.120) with respect to time

$$\begin{aligned}
(\mathbf{n}_1^T \mathbf{s}_2)^\cdot &= 0, & (\mathbf{n}_1^T \mathbf{t}_2)^\cdot &= 0, \\
(\mathbf{r}_D^T \mathbf{s}_2)^\cdot &= 0, & (\mathbf{r}_D^T \mathbf{t}_2)^\cdot &= 0,
\end{aligned} \qquad (2.129)$$

which states, that the contact conditions remain the same while the bodies are moving, and which result in a linear system for the unknown contour derivatives $(\dot{s}_1, \dot{t}_1, \dot{s}_2, \dot{t}_2)$

$$\mathbf{A}_C \cdot \mathbf{x}_C = \mathbf{b}_C \qquad \text{with}$$

$$\mathbf{A}_C = \begin{pmatrix} \mathbf{s}_2^T(\alpha_1 \mathbf{s}_1 + \beta_1 \mathbf{t}_1) & \mathbf{s}_2^T(\alpha'_1 \mathbf{s}_1 + \beta'_1 \mathbf{t}_1) & L_2 & M_2 \\ \mathbf{t}_2^T(\alpha_1 \mathbf{s}_1 + \beta_1 \mathbf{t}_1) & \mathbf{t}_2^T(\alpha'_1 \mathbf{s}_1 + \beta'_1 \mathbf{t}_1) & M_2 & N_2 \\ -\mathbf{s}_1^T \mathbf{s}_2 & -\mathbf{s}_1^T \mathbf{s}_2 & \mathbf{s}_2^T \mathbf{s}_2 & \mathbf{s}_2^T \mathbf{t}_2 \\ -\mathbf{s}_1^T \mathbf{t}_2 & -\mathbf{s}_1^T \mathbf{t}_2 & \mathbf{s}_2^T \mathbf{t}_2 & \mathbf{t}_2^T \mathbf{t}_2 \end{pmatrix}$$

$$\mathbf{b}_C = \begin{pmatrix} (\mathbf{s}_2 \times \mathbf{n}_2)^T \cdot (\mathbf{\Omega}_2 - \mathbf{\Omega}_1) \\ (\mathbf{t}_2 \times \mathbf{n}_2)^T \cdot (\mathbf{\Omega}_2 - \mathbf{\Omega}_1) \\ \mathbf{s}_2^T(\mathbf{v}_{\Sigma 2} - \mathbf{v}_{\Sigma 1}) \\ \mathbf{t}_2^T(\mathbf{v}_{\Sigma 2} - \mathbf{v}_{\Sigma 1}) \end{pmatrix} \qquad \mathbf{x}_C = \begin{pmatrix} \dot{s}_1 \\ \dot{t}_1 \\ \dot{s}_2 \\ \dot{t}_2 \end{pmatrix} \qquad (2.130)$$

These linear equations have to be solved at every time step of a numerical simulation. For further developments of the dynamics equations we abbreviate the equations (2.128) in the form

$$\ddot{g}_N = \mathbf{w}_N^T \ddot{\mathbf{q}} + \bar{w}_N, \qquad \ddot{g}_S = \mathbf{w}_S^T \ddot{\mathbf{q}} + \bar{w}_S, \qquad \ddot{g}_T = \mathbf{w}_T^T \ddot{\mathbf{q}} + \bar{w}_T. \qquad (2.131)$$

## 2.2 Kinematics

Combining these equations for a contact (i) of a multibody system results in a form more convenient for the later evaluation of the equations of motion. We get

$$\ddot{g}_{Ni} = \mathbf{w}_{Ni}^T \ddot{\mathbf{q}} + \bar{w}_{Ni}, \qquad \ddot{\mathbf{g}}_{Ti} = \mathbf{W}_{Ti}^T \ddot{\mathbf{q}} + \bar{\mathbf{w}}_{Ti},$$

$$\ddot{\mathbf{g}}_{Ti} = \begin{pmatrix} \ddot{g}_S \\ \ddot{g}_T \end{pmatrix}, \qquad \mathbf{W}_{Ti}^T = \begin{pmatrix} \mathbf{w}_S^T \\ \mathbf{w}_T^T \end{pmatrix}, \qquad \bar{\mathbf{w}}_{Ti} = \begin{pmatrix} \bar{w}_S \\ \bar{w}_T \end{pmatrix}. \qquad (2.132)$$

The vectors $\mathbf{w}_k$ and the scalars $\bar{w}_k$ with (k=N,S,T) follow by comparison with the equations (2.128). The scalar terms usually excitations from external sources.

### 2.2.8 Influence of Elasticity

Machines, mechanisms and structures are always elastic. How "much elastic" depends on the eigenfrequencies of the overall system and of the components. If one of these frequencies is low enough to produce a significant influence on the dynamics within the frequency range we want to consider, then such a component must be modeled elastically, otherwise rigidly. In most of this practical cases we have linear elasticity, which means, that we have approximately a rigid body motion of the multibody system superimposed by small elastic deformations usually in the form of elastic vibrations. Of course we then get a mutual influence of system dynamics and elastic deformations. As many practical systems belong to this class, we shall restrict ourselves to the influence of small elastic deformations of some components in the system. A very good description of linear and nonlinear influences of elasticity in multibody dynamics is given by Bremer [28] and Shabana [242].

We consider small deformations of the body (i) as part of a system of rigid and elastic bodies. These deformations result in a displacement and in a rotation of every mass-element dm of the body (i) (see Figure 2.22) with the base $B_i$. The vector chain from the Base I to the Base $B_{ei}$ includes three vectors, the vector $\mathbf{r}_{IB_i}$ from I to $B_i$, the vector $\mathbf{r}_{B_i B_{ri}}$ from $B_i$ to $B_{ri}$ and the vector $\mathbf{r}_{B_{ri} B_{ei}}$ from the undeformed reference $B_{ri}$ to the deformed reference $B_{ei}$. The three bases with the accompanying coordinate systems may be seen from Figure 2.22. Without touching the principal considerations we could have chosen any other chain of vectors as indicated by the dashed lines in Figure (2.22), where we have depicted specific points for additional coordinates systems. According to the definitions with regard to the coordinates we write the vector chain relation in the coordinates of the body (i):

$$_{B_i}\mathbf{r}_{IB_{ei}} = {}_{B_i}\mathbf{r}_{IB_i} + {}_{B_i}\mathbf{r}_{B_i B_{ri}} + \mathbf{A}_{B_i B_{ei}} \cdot ({}_{B_{ei}}\mathbf{r}_{B_{ri} B_{ei}}), \qquad \text{with}$$

$$_{B_i}\mathbf{r}_{B_i B_{ei}} = {}_{B_i}\mathbf{r}_{B_i B_{ri}} + \mathbf{A}_{B_i B_{ei}} \cdot ({}_{B_{ei}}\mathbf{r}_{B_{ri} B_{ei}})$$
$$\mathbf{A}_{B_i B_{ei}} = \mathbf{A}_{B_{ri} B_{ei}}. \qquad (2.133)$$

The last two terms of the first equation include all influences of elasticity,

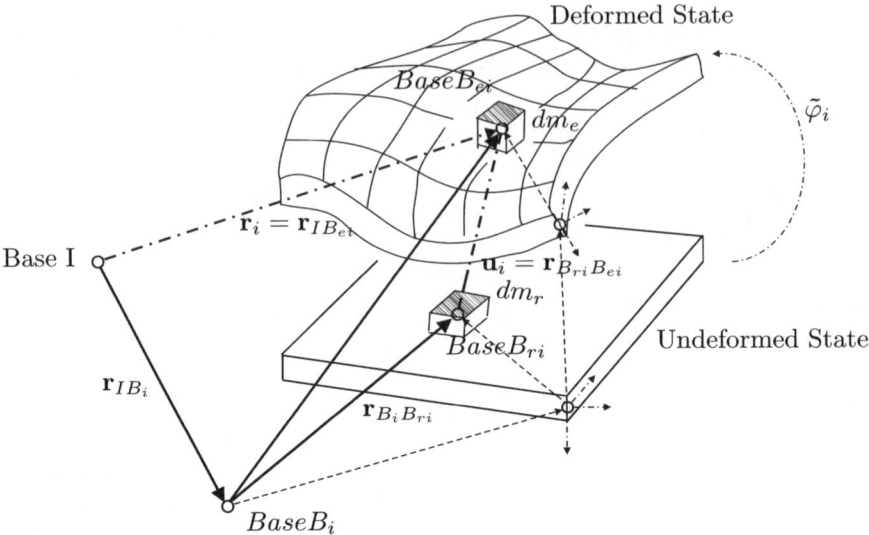

Fig. 2.22: Deformation of a Multibody Component i

the deformation vector $\mathbf{u}_i = \mathbf{r}_{B_{ri}B_{ei}}$ directly, and the rotation matrix $\mathbf{A}_{B_iB_{ei}}$ represents the rotation between the base $B_{ei}$ and the base $B_i$ or $B_{ri}$ due to elasticity. This "elastic rotation" is of the same magnitude for the base $B_{ri}$ and for the base $B_i$. For a rigid body we get $\mathbf{r}_{B_{ri}B_{ei}} = \mathbf{0}$ and $\mathbf{A}_{B_{ri}B_{ei}} = \mathbf{E}$. Assuming small elastic deformations and rotations and using Cardan angles we can evaluate the rotation matrix $\mathbf{A}_{B_iB_{ei}}$ from the formula (2.19) by a Taylor expansion. We get ($hot$ = higher order terms)

$$\mathbf{A}_{B_iB_{ei}} \approx \begin{pmatrix} 1 & -\gamma & +\beta \\ +\gamma & 1 & -\alpha \\ -\beta & +\alpha & 1 \end{pmatrix} + hot = \mathbf{E} + \tilde{\varphi} + hot, \qquad [\varphi = (\alpha, \beta, \gamma)^T]. \quad (2.134)$$

For calculating the absolute velocity in a body-fixed coordinate system we transform the relation (2.133) into inertial coordinates, differentiate with respect to time, and transform the resulting velocity back into body-fixed coordinates. This yields

$$_{B_i}\dot{\mathbf{r}}_{IB_{ei}} =\,_{B_i}\dot{\mathbf{r}}_{IB_i} +\,_{B_i}\dot{\mathbf{r}}_{B_iB_{ri}} +\,_{B_i}\tilde{\omega}_{IB_i} \cdot [_{B_i}\mathbf{r}_{B_iB_{ri}} + \mathbf{A}_{B_iB_{ei}}(_{B_{ei}}\mathbf{u})] + \\ +\, \mathbf{A}_{B_iB_{ei}}[_{B_{ei}}\dot{\mathbf{u}} +\,_{B_{ei}}\tilde{\omega}_{B_{ri}B_{ei}}(_{B_{ei}}\mathbf{u})],$$

$$_{B_i}\dot{\mathbf{r}}_{IB_{ei}} =\,_{B_i}\dot{\mathbf{r}}_{IB_i} +\,_{B_i}\dot{\mathbf{r}}_{B_iB_{ri}} +\,_{B_i}\tilde{\omega}_{IB_i} \cdot [_{B_i}\mathbf{r}_{IB_i} +\,_{B_i}\mathbf{r}_{B_iB_{ri}}] + \\ +\, \mathbf{A}_{B_iB_{ei}}[_{B_{ei}}\dot{\mathbf{u}} +\,_{B_{ei}}\tilde{\omega}_{IB_{ei}}(_{B_{ei}}\mathbf{u})],$$

$$_{B_i}\tilde{\omega}_{IB_{ei}} =\,_{B_i}\tilde{\omega}_{IB_i} +\,_{B_i}\tilde{\omega}_{B_{ri}B_{ei}} =\,_{B_i}\tilde{\omega}_{IB_i} + \mathbf{A}_{B_iB_{ei}}(_{B_{ei}}\tilde{\omega}_{B_{ri}B_{ei}})\mathbf{A}_{B_{ei}B_i} \quad (2.135)$$

In addition to the rigid body motion we get in the same structural way the influence of elasticity. It should be kept in mind, that for the evaluation of the Jacobians as part of the equations of motion we have to develop the kinematic expressions for position and orientation, for translational and rotational velocities and accelerations up to terms of second order with respect to the elastic coordinates. For large systems this is better done either by symbolic or by numerical computer codes.

In a next step we must express the deformation vectors by an elastic model, where we refer mainly to the books of Becker [14], Betten [19], Bremer [28] and Wriggers [278]. We go from some reference configuration of a body $\mathcal{B}$ to a deformed configuration $\varphi(\mathcal{B})$, and we shall use the vectors $\mathbf{X} \in \mathbb{R}^3$ from the origin to the reference state, $\mathbf{x} \in \mathbb{R}^3$ from the origin to the deformed state and the displacement vector $\mathbf{u} \in \mathbb{R}^3$ from the reference to the deformed state (see [278] and Figure 2.23). Any element $d\mathbf{x}$ can then be expressed by a reference element $d\mathbf{X}$ in the form

$$d\mathbf{x} = (\frac{\partial \mathbf{x}}{\partial \mathbf{X}}) \cdot d\mathbf{X} = \mathbf{F} \cdot d\mathbf{X} \quad \text{with} \quad \mathbf{F} = \frac{\partial \mathbf{x}}{\partial \mathbf{X}} = grad_\mathbf{X}(\mathbf{x}) \in \mathbb{R}^{3,3}. \quad (2.136)$$

The deformation gradient $\mathbf{F}$ is one of the fundamental magnitudes of continuum mechanics. It can be expressed as a gradient of the vector $\mathbf{x}$ with respect to $\mathbf{X}$. It is well known that the deformation gradient $\mathbf{F}$ can never be singular and that it can be decomposed by the polar decomposition theorem into a stretching and into a rotational part

$$\mathbf{F} = \mathbf{RU} = \mathbf{VR}, \quad (2.137)$$

where $\mathbf{U}$ is the right stretch or the right Cauchy Green tensor, and $\mathbf{V}$ is the left stretch or the left Cauchy Green tensor. They are defined in the reference or the current configuration, respectively. The rotation tensor $\mathbf{R}$ is orthogonal with $\det(\mathbf{R}) = +1$, the tensors $\mathbf{U}$ and $\mathbf{V}$ are positive definite and possess the same eigenvalues $\lambda_i, (i = 1, 2, 3)$. This is important, because the stretching effects are proportional to these eigenvalues, and they must be independent from the sequences stretching/rotation or rotation/stretching. Applying therefore $d\mathbf{x} = \mathbf{RU} \cdot d\mathbf{X}$ to a mass element results in a stretching of the element with a following rotation, whereas $d\mathbf{x} = \mathbf{VR} \cdot d\mathbf{X}$ comes out with a rotation in a first and a stretching in a second step ([14]).

The usual way to derive the strain tensor consists in considering the difference of the squared elements $(d\mathbf{x}^T d\mathbf{x} - d\mathbf{X}^T d\mathbf{X}) = d\mathbf{X}^T(\mathbf{F}^T\mathbf{F} - \mathbf{E})d\mathbf{X} = 2d\mathbf{X}^T\mathbf{G}d\mathbf{X}$, which represents a suitable deformation measure. The resulting strain tensor $\mathbf{G}$ is then defined as

$$\mathbf{G} = \frac{1}{2}(\mathbf{F}^T\mathbf{F} - \mathbf{E}) \in \mathbb{R}^{3,3}, \quad (2.138)$$

where $\mathbf{E}$ is the unit matrix. The tensor $\mathbf{G}$ is called the Cauchy-Lagrangian strain tensor. It refers to the initial configuration $\mathcal{B}$. From Figure 2.23 we have the property $\mathbf{x} = \mathbf{X} + \mathbf{u}$ and with this the deformation gradient $\mathbf{F} = \mathbf{E} + \frac{\partial \mathbf{u}}{\partial \mathbf{X}}$. With these relations the equation (2.138) writes

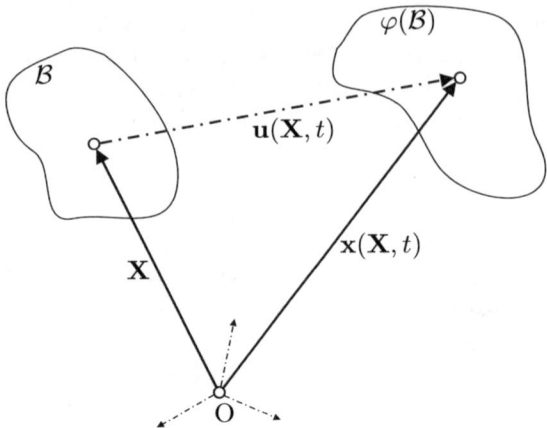

Fig. 2.23: Reference and Deformed States [278]

$$\mathbf{G} = \frac{1}{2}\{(\frac{\partial \mathbf{u}}{\partial \mathbf{X}}) + (\frac{\partial \mathbf{u}}{\partial \mathbf{X}})^T + (\frac{\partial \mathbf{u}}{\partial \mathbf{X}})^T \cdot (\frac{\partial \mathbf{u}}{\partial \mathbf{X}})\}. \tag{2.139}$$

The Green-Lagrangian strain tensor is symmetric and thus a measure for the strain alone, excluding rotations. The nonlinear terms in equation 2.139 are called geometric or kinematic nonlinearities. They can be neglected for very small strains resulting in the so-called kinematic or geometric linearization. Within the frame-work of system dynamics the strain tensor will be needed for defining the potential energy of elastic parts. Some problems do not allow a linearization due to effects of geometrical stiffnesses, which is connected with cases like buckling or tilting of bars or the well-known problem of rotating bars stiffened by centrifugal forces. For these or related cases a linearization comes out with wrong results. Kane and his school called it "premature linearization" ([122], [11], [10]). Anyway, some prudence will be necessary. We shall come back to this matter in later chapters.

Two main ideas characterize the combination of multibody system concepts with the continuum mechanics concepts. The first important point has been considered with the evaluation of the strain tensor $\mathbf{G}$, which usually is defined by the symbol $\epsilon$. The accompanying matrix is given by

$$\epsilon = \mathbf{G} = \begin{pmatrix} \epsilon_{xx} & \epsilon_{xy} & \epsilon_{xz} \\ \epsilon_{yx} & \epsilon_{yy} & \epsilon_{yz} \\ \epsilon_{zx} & \epsilon_{zy} & \epsilon_{zz} \end{pmatrix}. \tag{2.140}$$

From the equations (2.139) and (2.140) as well as from the definitions $\mathbf{X} = (\xi, \eta, \zeta)^T$ and $\mathbf{u} = (u, v, w)^T$ we get the components of the strain tensor in the following form

## 2.2 Kinematics

$$\epsilon_{xx} = \frac{\partial u}{\partial \xi} + \frac{1}{2}[(\frac{\partial u}{\partial \xi})^2 + (\frac{\partial v}{\partial \xi})^2 + (\frac{\partial w}{\partial \xi})^2],$$

$$\epsilon_{yy} = \frac{\partial v}{\partial \eta} + \frac{1}{2}[(\frac{\partial u}{\partial \eta})^2 + (\frac{\partial v}{\partial \eta})^2 + (\frac{\partial w}{\partial \eta})^2],$$

$$\epsilon_{zz} = \frac{\partial w}{\partial \zeta} + \frac{1}{2}[(\frac{\partial u}{\partial \zeta})^2 + (\frac{\partial v}{\partial \zeta})^2 + (\frac{\partial w}{\partial \zeta})^2],$$

$$\epsilon_{xy} = \epsilon_{yx} = \frac{1}{2}\{\frac{\partial u}{\partial \eta} + \frac{\partial v}{\partial \xi} + [\frac{\partial u}{\partial \xi}\frac{\partial u}{\partial \eta} + \frac{\partial v}{\partial \xi}\frac{\partial v}{\partial \eta} + \frac{\partial w}{\partial \xi}\frac{\partial w}{\partial \eta}]\},$$

$$\epsilon_{yz} = \epsilon_{zy} = \frac{1}{2}\{\frac{\partial v}{\partial \eta} + \frac{\partial w}{\partial \xi} + [\frac{\partial u}{\partial \eta}\frac{\partial u}{\partial \zeta} + \frac{\partial v}{\partial \eta}\frac{\partial v}{\partial \zeta} + \frac{\partial w}{\partial \eta}\frac{\partial w}{\partial \zeta}]\},$$

$$\epsilon_{zx} = \epsilon_{xz} = \frac{1}{2}\{\frac{\partial w}{\partial \eta} + \frac{\partial u}{\partial \xi} + [\frac{\partial u}{\partial \zeta}\frac{\partial u}{\partial \xi} + \frac{\partial v}{\partial \zeta}\frac{\partial v}{\partial \xi} + \frac{\partial w}{\partial \zeta}\frac{\partial w}{\partial \xi}]\}, \quad (2.141)$$

which can be easily adapted to the coordinates chosen for the case under consideration. If we are able to really linearize these expressions, for example for problems without large overall motion, then all the nonlinear terms of the equations (2.141) become approximately zero.

The second important connection with the multibody concept concerns the rotation of the mass elements as a consequence of the deformation. We have seen by equation (2.137), that the deformation gradient $\mathbf{F}$ can be split into a stretching and a rotation part, where the rotation tensor $\mathbf{R}$ is orthogonal and its determinant det$(\mathbf{R})=+1$. The effect of this tensor consists in a rigid body rotation. To evaluate this rotation for our multibody purposes we come back to the deformation gradient $\mathbf{F}$ and consider its symmetric and its skew-symmetric part by the decomposition (see equation 2.136)

$$\mathbf{F} = (\frac{\partial \mathbf{x}}{\partial \mathbf{X}}) = \frac{1}{2}\left[(\frac{\partial \mathbf{x}}{\partial \mathbf{X}}) + (\frac{\partial \mathbf{x}}{\partial \mathbf{X}})^T\right] + \frac{1}{2}\left[(\frac{\partial \mathbf{x}}{\partial \mathbf{X}}) - (\frac{\partial \mathbf{x}}{\partial \mathbf{X}})^T\right] = \mathbf{F}_{sym} + \mathbf{F}_{skew}$$

$$\mathbf{F}_{sym} = \frac{1}{2}\left[(\frac{\partial \mathbf{x}}{\partial \mathbf{X}}) + (\frac{\partial \mathbf{x}}{\partial \mathbf{X}})^T\right], \quad \mathbf{F}_{skew} = \frac{1}{2}\left[(\frac{\partial \mathbf{x}}{\partial \mathbf{X}}) - (\frac{\partial \mathbf{x}}{\partial \mathbf{X}})^T\right]. \quad (2.142)$$

Remembering that according to Figure (2.23) the deformation gradient can also be expressed by $\mathbf{F} = \mathbf{E} + \frac{\partial \mathbf{u}}{\partial \mathbf{X}}$, we can write

$$\mathbf{F}_{skew} = \frac{1}{2}\left[(\frac{\partial \mathbf{u}}{\partial \mathbf{X}}) - (\frac{\partial \mathbf{u}}{\partial \mathbf{X}})^T\right] = \frac{1}{2}(\nabla \mathbf{u} - \nabla \mathbf{u}^T). \quad (2.143)$$

This skew-symmetric part of the deformation gradient represents the rotation field (see [264]). The rotation vector itself can be written as

$$\varphi = \frac{1}{2}\text{curl}(\mathbf{u}) \quad \text{with} \quad \tilde{\varphi} = \mathbf{F}_{skew}. \quad (2.144)$$

Returning to our assumption of small deformations superimposed on large rigid body motion we can compare the equations (2.143), (2.144) and (2.134) and come out with the rotation vector

$$\varphi = \begin{pmatrix} \alpha \\ \beta \\ \gamma \end{pmatrix} = \frac{1}{2} \begin{pmatrix} \frac{\partial w}{\partial \eta} - \frac{\partial v}{\partial \zeta} \\ \frac{\partial u}{\partial \zeta} - \frac{\partial w}{\partial \xi} \\ \frac{\partial v}{\partial \xi} - \frac{\partial u}{\partial \eta} \end{pmatrix} \qquad (2.145)$$

With these relations we have established a correlation between the multibody concept of equation (2.134) and the rotational effect of small elastic deformations. With respect to large deformations we recommend the appropriate literature. An explanation for the above formulas is a matter of many undergraduate textbooks. Referring to Figure (2.24) we immediately realize, that the sum of the two angles $\frac{\partial u}{\partial \eta}$ and $\frac{\partial v}{\partial \xi}$ represents the stretch of the element and that the difference of the same two angles represents the rotation of the element, here depicted for the plane $\xi, \eta$.

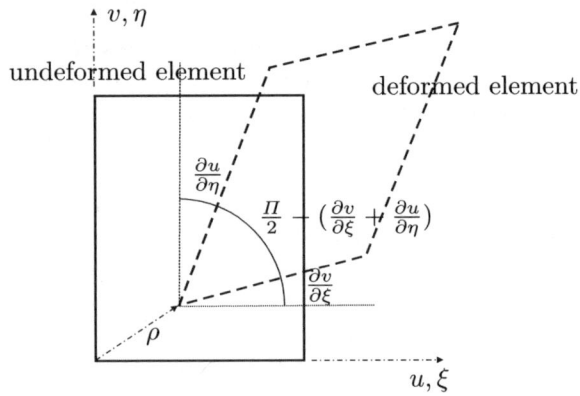

Fig. 2.24: Stretching and Rotating of an Elastic Element

## 2.3 Momentum and Moment of Momentum

### 2.3.1 Definitions and Axioms

Mutual interactions of forces with masses are the basic concept of all mechanical sciences. The result of these interactions may be very different depending on the system considered, it may be motion, deformation or just, as a limiting case, a system at rest, statically in an equilibrium state. The early days of mechanics as a science were therefore characterized by the search for some kind of relationships for force, mass and a kinematic magnitude like acceleration or velocity. The great achievements of Newton [169] must also be seen before the background of his time, where all transportation took place in coaches and carts with wooden wheels and primitive bearings, drawn by horses, giving more the impression, that velocity is much more proportional to forces than acceleration. Newton's laws overcame these popular ideas.

An equally great achievement was the finding of Euler [58], that in addition to Newton's ideas of momentum the laws for the moment of momentum represent independent mechanical statements and cannot be "derived" from the momentum equations, which is sometimes done in older textbooks of mechanics. In the meantime we know, that the moment of momentum equations as usually applied depend on Boltzmann's axiom or the symmetry of Cauchy's stress tensor. For polar materials for example these moment of momentum equations must be supplemented by some expressions including the tensor of moment stresses.

We consider some rigid or elastic body under the influence of active and passive forces (Figure 2.25), where the active forces contribute to the motion and the passive forces not. We know, that the moment equation of Newton and the moment of momentum equation of Euler are independent laws not derivable from each other. We furtheron assume, that there will exist an inertial coordinate system, where these equations become valid. Therefore and following an idea of [63] we define these two equations in the form of two axioms and write

$$\int_{\mathcal{B}} (\ddot{\mathbf{r}} \mathrm{dm} - \mathrm{d}\mathbf{F}_a) = \mathbf{0}, \qquad \int_{\mathcal{B}} \mathbf{r} \times (\ddot{\mathbf{r}} \mathrm{dm} - \mathrm{d}\mathbf{F}_a) = \mathbf{0}, \qquad (2.146)$$

where $\mathbf{r}$ is a vector in an inertial frame I to a mass element dm of the body B, and $(\mathrm{d}\mathbf{F}_a)$ are active forces. With respect to Figure 2.25 we should note, that the passive forces $\mathrm{d}\mathbf{F}_p$ indicated in that Figure are only passive without internal deformations.

We have used the definitions "a" for active and "p" for passive. The idea of active and passive forces being used in continuum mechanics for quite a time is more adequate for our considerations than external and internal forces, though in some cases it means the same. But for unilateral problems the features "active" and "passive" change during the motion, and therefore the definitions "external" and "internal" do not help so much. As a reminder: active forces can be shifted along their lines of action, passive forces cannot. From this we state, that active forces produce work and power, passive forces not.

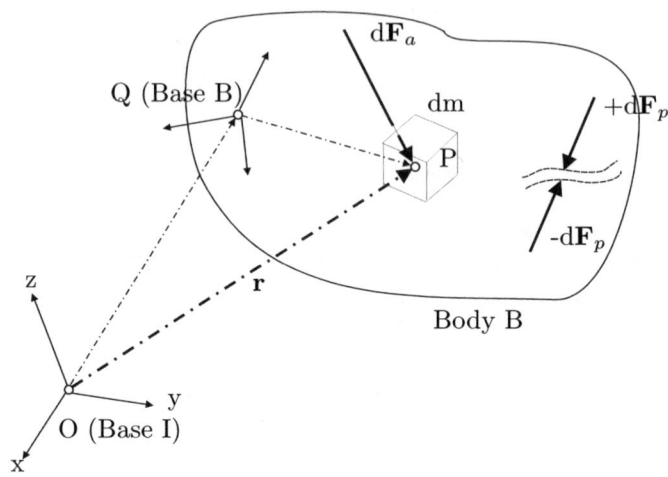

Fig. 2.25: Moment and Moment of Momentum

### 2.3.2 Momentum

According to the equations 2.146 (see also Figure 2.25) we define the momentum by

$$\mathbf{p} = \int_B \dot{\mathbf{r}} \, dm, \tag{2.147}$$

which is a coordinate-free representation. The velocity $\dot{\mathbf{r}}$ is an absolute velocity, and as always, derivations with respect to time have to be performed in an inertial system. On the other hand it is of course possible to transform these equations into any other coordinate system, for example into a body-fixed frame. We come back to this point in chapter 2.3.4.

The fundamental laws considering momentum are the famous three laws of Newton, which possess the quality of axioms. We shall be not so ambitious to give his statements in the original form (for this purpose see [169]). The first axiom writes [175]

**Axiom 1.** *A body at rest remains at rest and a body in motion moves in a straight line with unchanging velocity, unless some external force acts on it.*

## 2.3 Momentum and Moment of Momentum

To illustrate this basic law, which we find already in the statements of Galilei [259], we shall use the notation introduced by Euler for the momentum and moment of momentum laws. Referring to Axiom 1 we have no external, thus no active forces, which means $\int_B d\mathbf{F}_a = \mathbf{0}$ and therefore $\int_B \ddot{\mathbf{r}} dm = \mathbf{0}$ resulting in

$$\mathbf{p} = \int_B \dot{\mathbf{r}} dm = \text{constant}, \tag{2.148}$$

which represents the law of conservation of momentum. Considering the mass center of a body we get

$$\mathbf{p}_C = \mathbf{p}_{C0} = \dot{\mathbf{r}}_C m \quad \text{with} \quad \mathbf{r}_C m = \int_B \mathbf{r} dm \tag{2.149}$$

**Axiom 2.** *The rate of change of the momentum of a body is proportional to the resultant external force that acts on the body.*

For the mass element of Figure 2.25 we get from the first equation 2.146

$$\ddot{\mathbf{r}} dm - d\mathbf{F}_a = \mathbf{0} \tag{2.150}$$

which represents also the momentum budget for a point mass. The time derivative of the momentum is mass times acceleration if we are dealing with a constant mass, as in the above equation. For not constant masses the time derivative of the mass must be considered in addition. In terms of our definitions we may write

$$\frac{d\mathbf{p}}{dt} = \mathbf{F}, \quad \text{with} \quad \mathbf{p} = \int_B \dot{\mathbf{r}} dm, \quad \text{and} \quad \mathbf{F} = \int_B d\mathbf{F}_a. \tag{2.151}$$

Taking again the center of mass of the body we come out with

$$m\left(\frac{d\mathbf{v}_C}{dt}\right) = \mathbf{F}_C. \tag{2.152}$$

The velocity $\mathbf{v}_C$ is defined with respect to an inertial system. It is an absolute velocity. The force vector $\mathbf{F}_C$ is the vector sum of all forces which act on the body. Generally this vector sum does not pass through the center of mass resulting in an additional torque, which has to be regarded in the moment of momentum equation.

Newton's third law writes

**Axiom 3.** *Action and reaction are equal and opposite.*

At the times of Newton this finding was new. But it is very obvious from experience. Wherever any force acts on a body or on the environment we get as a reaction the same force with opposite sign. My feet transfer my weight to the ground, as a reaction the ground is loaded with my weight force in the opposite direction. There is no mechanical interaction without this basic property.

The forces acting on a body might be applied forces, elastic forces or single- and set-valued forces. We shall consider all of them, but concentrate here on forces due to elastic influences. Equation (2.146) then writes

$$\int_B \ddot{\mathbf{r}} dm = \int_{\partial B} \mathbf{p}_\sigma d\mathbf{A} + \int_B \mathbf{f}_a dm, \qquad (2.153)$$

where the first integral on the right hand side is a surface force due to elasticity and the second integral a volume force of some given type. The stress vector $\mathbf{p}_\sigma$ acts on the surface $\partial B$ with the area vector $d\mathbf{A}$. As is well known we can express the stress vector by

$$\mathbf{p}_\sigma = \sigma \mathbf{n}, \qquad (2.154)$$

which indicates, that the stress vector $\mathbf{p}_\sigma$ comes from the surface normal vector $\mathbf{n}$ by a homogenuous and linear transformation with the Cauchy stress tensor $\sigma$ [14]. The Cauchy stress tensor itself is symmetric and describes the stress situation for a homogenuous and isotropic body (Figure (2.26) and ([228], [147])).

$$\sigma = \begin{pmatrix} \sigma_{xx} & \sigma_{xy} & \sigma_{xz} \\ \sigma_{yx} & \sigma_{yy} & \sigma_{yz} \\ \sigma_{zx} & \sigma_{zy} & \sigma_{zz} \end{pmatrix}, \quad \text{with} \quad \sigma_{ij} = \sigma_{ji}, \quad (i,j = x,y,z). \qquad (2.155)$$

Combining these equations and regarding in addition the Gauss theorem for surface and volume integrals results in

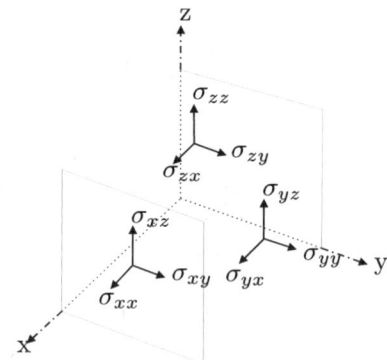

Fig. 2.26: Stresses

$$\int_B \ddot{\mathbf{r}} dm = \int_B \text{div}(\sigma) dV + \int_B \mathbf{f}_a dm, \quad \text{with} \quad \text{div}(\sigma) = \begin{pmatrix} \frac{\partial \sigma_{xx}}{\partial x} + \frac{\partial \sigma_{xy}}{\partial y} + \frac{\partial \sigma_{xz}}{\partial z} \\ \frac{\partial \sigma_{yx}}{\partial x} + \frac{\partial \sigma_{yy}}{\partial y} + \frac{\partial \sigma_{yz}}{\partial z} \\ \frac{\partial \sigma_{zx}}{\partial x} + \frac{\partial \sigma_{zy}}{\partial y} + \frac{\partial \sigma_{zz}}{\partial z} \end{pmatrix}. \tag{2.156}$$

For our purposes and considering only linearly elastic bodies we introduce for a consitutive law the simple material law of Hooke, which writes

$$\sigma = \mathbf{E}\epsilon \tag{2.157}$$

with a constant module of elasticity matrix $\mathbf{E}$ (Young's modulus). Therefore equation (2.156) writes

$$\int_B \ddot{\mathbf{r}} dm = \mathbf{E} \int_B \text{div}(\mathbf{G}) dV + \int_B \mathbf{f}_a dm, \tag{2.158}$$

with the definition of the strain tensor ($\epsilon = \mathbf{G}$) by the equations (2.140) and (2.141), which we shall not evaluate here in combining them with the above equation. For simple structures like bars or plates this can be performed straightforwardly (see [28]).

### 2.3.3 Moment of Momentum

Euler has been the first one to understand the law of moment of momentum as a basic independent law of mechanics. It cannot be "derived" from the second axiom of Newton just by performing the cross-product. We refer to the literature ([27], [180]). From the equations 2.146 we define the moment of momentum by

$$_I\mathbf{L} = \int_B {_I\mathbf{r}} \times {_I\dot{\mathbf{r}}} dm. \tag{2.159}$$

The second axiom of the equations (2.146) may be written in the form

$$\frac{d\mathbf{L}}{dt} = \mathbf{M} \quad \text{with} \quad \mathbf{M} = \int_B \mathbf{r} \times d\mathbf{F}_a \tag{2.160}$$

with all magnitudes represented in an inertial system and the time derivation also performed in an inertial frame. For missing active torques $\mathbf{M}$ the equation (2.160) is $\frac{d\mathbf{L}}{dt} = \mathbf{0}$ and therefore $\mathbf{L} = constant$. This conservation of moment of momentum writes

$$\frac{d\mathbf{L}}{dt} = \mathbf{0}, \quad \mathbf{L} = \mathbf{L}_0 = \int_B \mathbf{r} \times \dot{\mathbf{r}} dm. \tag{2.161}$$

58    2 Fundamentals

We may go two ways. We might assume the equations (2.146) as axioms, which then allows the confirmation of the symmetry of the Cauchy stress tensor from these axioms. Or we might go the way in assuming the symmetry beforehand , that means Boltzmann's axiom, which then comes out with the moment of momentum equation as we know them. We shall pursue the first possibility. Assuming an elastic body as discussed in the chapter (2.3.2) before and evaluating for this body the moment of momentum budget we get the equations (see [14])

$$\frac{d}{dt}\int_B (\mathbf{r}\times\dot{\mathbf{r}})\rho dV = \int_{\partial B}(\mathbf{r}\times\mathbf{p}_\sigma)dA + \int_B (\mathbf{r}\times\mathbf{f}_a)\rho dV, \qquad (2.162)$$

where we have replaced the mass element dm by ($\rho$dV) with the density $\rho$, and where we have used the formula (2.153). The stress vector $\mathbf{p}$ acts on the surface $\partial B$ of a volume element. The applied force $\mathbf{f}_a$ is some given volume force. We assume, that the volume element $B$ has a constant volume dV and a constant surface dA on $\partial B$. The stress vector $\mathbf{p}_\sigma$ was already defined with equation (2.154), namly $\mathbf{p}_\sigma = \sigma\mathbf{n}$. The cross-product can be replaced by a tensor product, $(\mathbf{r}\times\mathbf{p}_\sigma) = (\tilde{\mathbf{r}}\cdot(\mathbf{p}_\sigma))$, allowing us to write

$$\int_{\partial B}(\mathbf{r}\times\sigma\mathbf{n})dA = \int_{\partial B}\tilde{\mathbf{r}}\sigma\mathbf{n}dA, \quad \mathbf{n} = \begin{pmatrix} n_x \\ n_y \\ n_z \end{pmatrix}, \quad \tilde{\mathbf{r}} = \begin{pmatrix} 0 & -z & +y \\ +z & 0 & -x \\ -y & +x & 0 \end{pmatrix}. \qquad (2.163)$$

The first two magnitudes in the above equations can be easily evaluated to give

$$\tilde{\mathbf{r}}\sigma = \frac{1}{2}\begin{pmatrix} (-z\sigma_{yx}+y\sigma_{zx}) & (-2z\sigma_{yy}+y\sigma_{zy}) & (-z\sigma_{yz}+2y\sigma_{zz}) \\ (+2z\sigma_{xx}-x\sigma_{zx}) & (+z\sigma_{xy}-x\sigma_{zy}) & (+z\sigma_{xz}-2x\sigma_{zz}) \\ (-2y\sigma_{xx}+x\sigma_{yx}) & (-y\sigma_{xy}+2x\sigma_{yy}) & (-y\sigma_{xz}+x\sigma_{yz}) \end{pmatrix}. \qquad (2.164)$$

Performing some manipulations of the above relations [14] and applying the Gauss theorem we furtheron can write

$$\int_{\partial B}(\mathbf{r}\times\sigma\mathbf{n})dA = \int_{\partial B}\tilde{\mathbf{r}}\sigma\mathbf{n}dA = \int_B [\text{div}(\tilde{\mathbf{r}}\sigma) + \Delta\mathbf{g}]dV \qquad (2.165)$$

where the term $\Delta\mathbf{g}$ represents a remaining term arising from the evaluation of equation (2.163). Going back to the moment of momentum budget of equation (2.162) and inserting there the momentum balance $\rho\frac{d\dot{\mathbf{r}}}{dt} = \rho\mathbf{f} + \text{div}\sigma$ for the elastic element we get

$$\int_B \tilde{\mathbf{r}}(\text{div}\sigma + \rho\mathbf{f})dV = \int_B \tilde{\mathbf{r}}(\text{div}\sigma + \rho\mathbf{f})dV + \int_B \Delta\mathbf{g}dV. \qquad (2.166)$$

We see [14], that all terms cancel out with one exception, namely the volume integral including $\Delta\mathbf{g}$. The vector $\Delta\mathbf{g}$ writes

$$\Delta\mathbf{g} = \begin{pmatrix} +\sigma_{zy} - \sigma_{yz} \\ -\sigma_{zx} + \sigma_{xz} \\ +\sigma_{yx} - \sigma_{xy} \end{pmatrix} = \mathbf{0}. \tag{2.167}$$

It vanishes for a symmetric Cauchy stress tensor $\sigma$. Only then we are allowed to use the moment of momentum equation for rigid bodies in the classical form. A vanishing $\Delta\mathbf{g}$ confirms that internal forces counterbalance and do not contribute to the motion. For very many materials this is true, but for some classes of materials like polar materials this is not true, and then we have to look for it.

### 2.3.4 Transformations

The above defined expressions for momentum and moment of momentum are to be given in an inertial frame. Also all derivations with respect to time have to be performed in such a coordinate system. But of course it is not forbidden to transform the resulting equations into any other coordinate base, which especially makes sense before carrying out some time derivations. In the following we shall give some formulas transformed from the inertial system into a body-fixed one or vice versa applying the relations developed in the kinematics chapter 2.2.

The momentum equation is defined with (2.147). Considering an additional body-fixed coordinate system on body B and taking into regard the equations (2.28) and (2.38) together with the Figure (2.13) we also can write the momentum definition in the forms

$$_I\mathbf{p} = \int_B (_I\dot{\mathbf{r}}_{OP})dm = \int_B (_I\dot{\mathbf{r}}_{OQ} + \dot{\mathbf{A}}_{IB} \cdot {_B}\mathbf{r}_{QP} + \mathbf{A}_{IB} \cdot {_B}\dot{\mathbf{r}}_{QP})dm,$$

$$_I\mathbf{p} = \int_B (_I\dot{\mathbf{r}}_{OP})dm = \int_B \mathbf{A}_{IB} \cdot (_B\mathbf{v}_{Q,abs} + {_B}\tilde{\boldsymbol{\omega}} \cdot {_B}\mathbf{r} + {_B}\dot{\mathbf{r}})dm. \tag{2.168}$$

The relative velocity $_B\dot{\mathbf{r}}$ is the point, where possible elastic effects might enter the system, or where some particle motion exists. It vanishes for rigid bodies and for bodies with no relatively moved masses.

In a similar way we express the moment of momentum in body-fixed coordinates. Starting with the equations (2.159, 2.28, 2.38) and taking from Figure (2.25) the relation $\mathbf{r}_{OP} = \mathbf{r}_{OQ} + \mathbf{r}_{QP}$ we get

$$_I\mathbf{L} = \int_B (_I\mathbf{r}_{OQ} + {_I}\mathbf{r}_{QP}) \times (_I\dot{\mathbf{r}}_{OQ} + {_I}\dot{\mathbf{r}}_{QP})dm = \mathbf{A}_{IB}\ _B\mathbf{L}. \tag{2.169}$$

The determination of ${}_B\mathbf{L}$ is straightforward, if we define ${}_B\mathbf{v}_{P,abs} = ({}_B\mathbf{v}_{Q,abs} + {}_B\tilde{\boldsymbol{\omega}} \cdot {}_B\mathbf{r} + {}_B\dot{\mathbf{r}})$ with the absolute velocities ${}_I\mathbf{v}_{Q,abs} = {}_I\dot{\mathbf{r}}_{OQ}$, ${}_B\mathbf{v}_{Q,abs} = \mathbf{A}_{BI}{}_I\dot{\mathbf{r}}_{OQ}$ and the abbreviation ${}_B\mathbf{r}_{QP} = {}_B\mathbf{r}$. We come out with

$$
{}_B\mathbf{L} = \int_B ({}_B\tilde{\mathbf{r}}_{OQ} + {}_B\tilde{\mathbf{r}}) \cdot ({}_B\mathbf{v}_{Q,abs} + {}_B\tilde{\boldsymbol{\omega}} \cdot {}_B\mathbf{r} + {}_B\dot{\mathbf{r}})dm,
$$

$$
= \int_B {}_B\tilde{\mathbf{r}}_{OQ} \cdot {}_B\mathbf{v}_{P,abs}dm - {}_B\tilde{\mathbf{v}}_{Q,abs}\int_B {}_B\mathbf{r}dm + (-\int_B {}_B\tilde{\mathbf{r}}_B\tilde{\mathbf{r}}dm)_B\boldsymbol{\omega}
$$

$$
+ \int_B {}_B\tilde{\mathbf{r}}_B\dot{\mathbf{r}}dm,
$$

$$
{}_B\mathbf{L} = \int_B {}_B\tilde{\mathbf{r}}_{OQ} \cdot ({}_B\mathbf{v}_{P,abs}dm) + {}_B\tilde{\mathbf{r}}_{QS}(m_B\mathbf{v}_{Q,abs}) + {}_B\mathbf{I}_B\boldsymbol{\omega} + {}_B\mathbf{L}_Q, \qquad (2.170)
$$

where we have introduced the following abbreviations

$$
\int_B {}_B\mathbf{r}dm = m_B\mathbf{r}_{QS}, \qquad -\int_B {}_B\tilde{\mathbf{r}}_B\tilde{\mathbf{r}}dm = {}_B\mathbf{I}, \qquad \int_B {}_B\tilde{\mathbf{r}}_B\dot{\mathbf{r}}dm = {}_B\mathbf{L}_Q. \quad (2.171)
$$

The meaning is clear. The first term in the last equation (2.170) represents the moment with respect to Q of the absolute momentum $({}_B\mathbf{v}_{P,abs}dm)$ in point P, the second term the moment with respect to the center of mass S of the momentum $(m_B\mathbf{v}_{Q,abs})$ in point Q, the third term is the classical moment of momentum expression for rigid bodies without relative velocities ${}_B\dot{\mathbf{r}}$ and with point Q fixed and finally the last term is a kind of relative moment of momentum for a body with moving parts. If we choose the center of mass S as a reference point instead of Q, we get ${}_B\mathbf{r}_{QS} = \mathbf{0}$, and the second term becomes zero. Though well known from every undergraduate textbook of mechanics we shall repeat here the components of the inertia tensor ${}_B\mathbf{I}$.

$$
\mathbf{I} = -\int_B \tilde{\mathbf{r}}\tilde{\mathbf{r}}dm = \begin{pmatrix} I_{xx} & I_{xy} & I_{xz} \\ I_{yx} & I_{yy} & I_{yz} \\ I_{zx} & I_{zy} & I_{zz} \end{pmatrix}
$$

$$
I_{xx} = A = \int_B (y^2 + z^2)dm, \qquad I_{xy} = I_{yx} = -F = \int_B xydm,
$$

$$
I_{yy} = B = \int_B (z^2 + x^2)dm, \qquad I_{xz} = I_{zx} = -E = \int_B zxdm,
$$

$$
I_{zz} = C = \int_B (x^2 + y^2)dm, \qquad I_{yz} = I_{zy} = -D = \int_B yzdm. \qquad (2.172)
$$

In classical textbooks we find the notation as above, but for the inertia tensor also quite often $\Theta$ instead of $\mathbf{I}$.

Some remarks to the above transformations, which are of course only specific examples for that, what has to be done in building a set of equations of motion and a complete set for all constraints. We establish the kinematics of a system in a first step, performing all relevant transformations with respect to body-fixed and inertial coordinates, we then establish in a second step the kinetic equations also including all necessary coordinate transformations, in a further step we must consider the constraints to finally end with a complete set of the system dynamics. The methods and the style of formulating things turned out to be very helpful also with respect to very large problems. A pars pro toto example might be the introduction of the tilde operation $\tilde{\mathbf{a}} \cdot \mathbf{b} = \mathbf{a} \times \mathbf{b}$, which simplifies all matrix-vector manipulations considerably.

## 2.4 Energy

### 2.4.1 Introduction

If we move a mass element dm under the influence of an active force $d\mathbf{F}_a$ from a point 1 to a point 2 along some arbitrary path (Figure 2.27), then the following work is done

$$dW = \int_{\mathbf{r}_1}^{\mathbf{r}_2} d\mathbf{F}_a^T d\mathbf{r} = dm \int_{\mathbf{r}_1}^{\mathbf{r}_2} \ddot{\mathbf{r}}^T d\mathbf{r} \qquad (2.173)$$

Applying some manipulations to the second term of the above equation we get

$$dm \int_{\mathbf{r}_1}^{\mathbf{r}_2} \ddot{\mathbf{r}}^T d\mathbf{r} = dm \int_{\mathbf{r}_1}^{\mathbf{r}_2} \frac{d\dot{\mathbf{r}}^T}{dt} d\mathbf{r} = \frac{1}{2} dm \int_{\mathbf{r}_1}^{\mathbf{r}_2} d(\dot{\mathbf{r}}^T \dot{\mathbf{r}}) = \frac{1}{2} dm (\dot{\mathbf{r}}_2^2 - \dot{\mathbf{r}}_1^2) = dT_2 - dT_1.$$

$$(2.174)$$

From this the work done by shifting dm from point 1 to point 2 is given by

$$dW = dT_2 - dT_1 = \int_{\mathbf{r}_1}^{\mathbf{r}_2} d\mathbf{F}_a^T \cdot d\mathbf{r} \qquad (2.175)$$

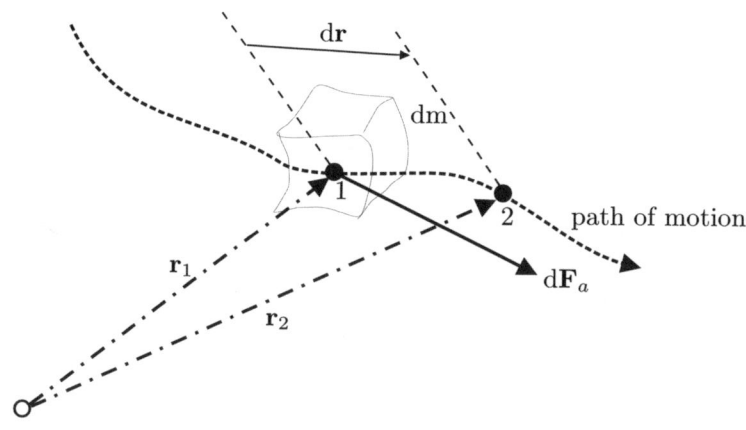

Fig. 2.27: Work and Energy

The work done by the active force is equal to the difference of the kinetic energies. If we move in a force field, where along a closed trajectory s no work is produced, then we call the corresponding system a conservative system with the property

$$\oint_s d\mathbf{F}_a^T \cdot d\mathbf{r} = 0. \tag{2.176}$$

Systems of that kind do not dissipate energy, and they are not supplied with energy. Applying Stokes theorem (see [283]) we can write

$$\oint_s d\mathbf{F}_a^T \cdot d\mathbf{r} = \int_A (\text{curl}(d\mathbf{F}_a)) dA, \tag{2.177}$$

from which we follow that

$$\text{curl}(d\mathbf{F}_a) = 0 \quad \Leftrightarrow \quad d\mathbf{F}_a = -\text{grad}(dV) \tag{2.178}$$

The force field can be derived from a potential V. Therefore we get further on

$$\int_{\mathbf{r}_1}^{\mathbf{r}_2} d\mathbf{F}_a^T \cdot d\mathbf{r} = -\int_{\mathbf{r}_1}^{\mathbf{r}_2} \text{grad}(dV) d\mathbf{r} = -\int_{\mathbf{r}_1}^{\mathbf{r}_2} \frac{\partial V}{\partial \mathbf{r}} d\mathbf{r} = -(dV_2 - dV_1). \tag{2.179}$$

The work done is proportional to the negative difference of the potential function dV between the points 1 and 2 and for a conservative force field as defined by the equations (2.178). Together with equation (2.175) we get

$$dW = -(dV_2 - dV_1) = (dT_2 - dT_1) \quad \Leftrightarrow \quad dT_1 + dV_1 = dT_2 + dV_2 = dT + dV, \tag{2.180}$$

or by integration over a whole body

$$T + V = E_0 = \text{constant}. \tag{2.181}$$

In conservative systems there will be no energy losses, and the energy conservation gives a first integral of motion sometimes useful for applications. In the presence of friction the total energy decreases, and equation (2.176) does not apply. But, on the other hand, a force field including also non-smooth force laws might still be conservative as long as no frictional energy losses occur.

## 2.4.2 Kinetic Energy

Considering a rigid mass-element dm and applying the relevant expressions for the absolute velocity written in a body-fixed frame (equation 2.38) we can express the kinetic energy by

$$T = \frac{1}{2}\int_m \mathbf{v}^T \mathbf{v} \, dm$$

$$= \frac{1}{2}\int_m (_B\mathbf{v}_Q + {_B\tilde{\boldsymbol{\omega}}} \cdot {_B\mathbf{r}})^T \cdot (_B\mathbf{v}_Q + {_B\tilde{\boldsymbol{\omega}}} \cdot {_B\mathbf{r}}) \, dm$$

$$= \frac{1}{2}\int_m [(\mathbf{E}\ \tilde{\mathbf{r}}^T)\begin{pmatrix}\mathbf{v}_Q\\ \boldsymbol{\omega}\end{pmatrix}]^T \cdot [(\mathbf{E}\ \tilde{\mathbf{r}}^T)\begin{pmatrix}\mathbf{v}_Q\\ \boldsymbol{\omega}\end{pmatrix}] \, dm \tag{2.182}$$

The structure of the last equation will be the same in each coordinate system, because we can transform the velocity equation (2.38) into every other base. Continuing in a body-fixed frame the second equation results in

$$T = \frac{1}{2}\int_m (_B\mathbf{v}_Q + {_B\tilde{\boldsymbol{\omega}}} \cdot {_B\mathbf{r}})^T \cdot (_B\mathbf{v}_Q + {_B\tilde{\boldsymbol{\omega}}} \cdot {_B\mathbf{r}}) \, dm$$

$$= \frac{1}{2}\int_m (_B\mathbf{v}_Q^T {_B\mathbf{v}_Q} + 2 {_B\mathbf{r}}^T {_B\tilde{\boldsymbol{\omega}}}^T {_B\mathbf{v}_Q} + {_B\boldsymbol{\omega}}^T(-\tilde{\mathbf{r}}\tilde{\mathbf{r}}) {_B\boldsymbol{\omega}}) \, dm$$

$$= \frac{1}{2}m(_B\mathbf{v}_Q^T {_B\mathbf{v}_Q}) + m {_B\mathbf{r}_S^T} {_B\tilde{\boldsymbol{\omega}}}^T {_B\mathbf{v}_Q} + \frac{1}{2} {_B\boldsymbol{\omega}}^T {_B\mathbf{I}_B}\boldsymbol{\omega}, \tag{2.183}$$

with the already mentioned relations

$$m_B\mathbf{r}_S = \int_m {_B\mathbf{r}} \, dm, \qquad {_B\mathbf{I}} = -\int_m {_B\tilde{\mathbf{r}}} {_B\tilde{\mathbf{r}}} \, dm. \tag{2.184}$$

The last equation of (2.183) contains the well-known terms for a pure translation, a pure rotation and a mixed term, which disappears for $_B\mathbf{r}_S = \mathbf{0}$ in the case of choosing the mass center S as a reference point. In a body-fixed frame the magnitudes $_B\mathbf{r}_S$ and $_B\mathbf{I}$ are constant, which underlines the necessity of using such body-fixed coordinates, but in an inertial frame for example these magnitudes depend on time, because we look at the system so-to-say from outside, from an external point of view. For large multibody systems we must travel through a large number of coordinate systems using the possibilities of chapter (2.2.4), which finally results in structurally similar expressions but multiply augmented by each additional coordinate system and reasonably to be evaluated only by a computer.

To give an idea of the influence of elastic parts on the energy we go back to chapter (2.2.8), consider again the equations (2.133) and Figure (2.22). The absolute velocity of a deformed body point can only be achieved by transforming these equations into an inertial frame and after differentiation by transforming them back into a body-fixed frame, if necessary. The transformation matrix from $B_i$ to I will be called $\mathbf{A}_{IB_i}$. Then we get

$$_I\mathbf{r}_{IB_{ei}} = \mathbf{A}_{IB_i} {_{B_i}\mathbf{r}_{IB_{ei}}}$$

$$= \mathbf{A}_{IB_i}[_{B_i}\mathbf{r}_{IB_i} + {_{B_i}\mathbf{r}_{B_iB_{ri}}} + \mathbf{A}_{B_iB_{ei}} \cdot (_{B_{ei}}\mathbf{r}_{B_{ri}B_{ei}})]. \tag{2.185}$$

## 2.4 Energy

From the equations (2.134) and (2.145) together with Figure (2.22) we have in addition (assuming small elastic deformations)

$$\mathbf{A}_{B_i B_{ei}} \approx \begin{pmatrix} 1 & -\gamma & +\beta \\ +\gamma & 1 & -\alpha \\ -\beta & +\alpha & 1 \end{pmatrix}_i = \mathbf{E}_i + \tilde{\boldsymbol{\varphi}}_i,$$

$$\boldsymbol{\varphi}_i = \begin{pmatrix} \alpha \\ \beta \\ \gamma \end{pmatrix}_i = \frac{1}{2} \begin{pmatrix} \frac{\partial w}{\partial \eta} - \frac{\partial v}{\partial \zeta} \\ \frac{\partial u}{\partial \zeta} - \frac{\partial w}{\partial \xi} \\ \frac{\partial v}{\partial \xi} - \frac{\partial u}{\partial \eta} \end{pmatrix}_i,$$

$$_{B_{ei}}\mathbf{r}_{B_{ri}B_{ei}} =\, _{B_{ei}}\mathbf{u}_i \quad \text{with} \quad _{B_{ei}}\mathbf{u}_i = (u, v, w)_i^T. \tag{2.186}$$

The absolute velocity of the mass element $dm_e$ of Figure (2.22) is then obtained simply by formal differentiation of $_I\mathbf{r}_{IB_{ei}}$. We get

$$_I\dot{\mathbf{r}}_{IB_{ei}} = \dot{\mathbf{A}}_{IB_i}\, _{B_i}\mathbf{r}_{IB_{ei}} + \mathbf{A}_{IB_i}\, _{B_i}\dot{\mathbf{r}}_{IB_{ei}}, \tag{2.187}$$

which we can evaluate a bit further. With $\mathbf{A}_{B_i I}\dot{\mathbf{A}}_{IB_i} = {}_B\tilde{\boldsymbol{\omega}}_{IB}$ and according to the corresponding relations in the kinematics chapter we may write for the absolute velocity in the $B_i$-frame

$$_{B_i}\mathbf{v}_{abs} = \mathbf{A}_{B_i I}\, _I\dot{\mathbf{r}}_{IB_{ei}} = {}_{B_i}\dot{\mathbf{r}}_{IB_{ei}} + {}_{B_i}\tilde{\boldsymbol{\omega}}_{IB_i}\, _{B_i}\mathbf{r}_{IB_{ei}}, \tag{2.188}$$

where $_{B_i}\mathbf{r}_{IB_{ei}}$ is given with equation (2.185) and the velocity $_{B_i}\dot{\mathbf{r}}_{IB_{ei}}$ follows from

$$_{B_i}\dot{\mathbf{r}}_{IB_{ei}} = {}_{B_i}\dot{\mathbf{r}}_{IB_i} + {}_{B_i}\dot{\mathbf{r}}_{B_i B_{ri}} + \mathbf{A}_{B_i B_{ei}}({}_{B_{ei}}\dot{\mathbf{u}}_i + {}_{B_{ei}}\tilde{\boldsymbol{\omega}}_{B_i B_{ei}}\, _{B_{ei}}\mathbf{u}_i) \tag{2.189}$$

with $\dot{\mathbf{A}}_{B_i B_{ei}} = \mathbf{A}_{B_i B_{ei}} \cdot {}_{B_{ei}}\tilde{\boldsymbol{\omega}}_{B_i B_{ei}}$ representing the rotation of the mass element due to the elastic deformation. Setting further

$$_{B_i}\mathbf{u}_i = \mathbf{A}_{B_i B_{ei}}\, _{B_{ei}}\mathbf{u}_i, \quad \mathbf{A}_{B_i B_{ei}}({}_{B_{ei}}\tilde{\boldsymbol{\omega}}_{B_i B_{ei}})_{B_{ei}}\mathbf{u}_i = {}_{B_i}\tilde{\boldsymbol{\omega}}_{B_i B_{ei}}\, _{B_i}\mathbf{u}_i \tag{2.190}$$

we get finally

$$_{B_i}\mathbf{v}_{abs} = {}_{B_i}\dot{\mathbf{r}}_{IB_i} + {}_{B_i}\dot{\mathbf{r}}_{B_i B_{ri}} + {}_{B_i}\tilde{\boldsymbol{\omega}}_{IB_i}({}_{B_i}\mathbf{r}_{IB_i} + {}_{B_i}\mathbf{r}_{B_i B_{ri}})$$
$$+ {}_{B_i}\dot{\mathbf{u}}_i + ({}_{B_i}\tilde{\boldsymbol{\omega}}_{IB_i} + {}_{B_i}\tilde{\boldsymbol{\omega}}_{B_i B_{ei}}) \cdot {}_{B_i}\mathbf{u}_i, \tag{2.191}$$

which confirms the classical approach, that we must add to the velocity the deformation velocity and to the angular velocity the angular velocity due to elasticity [28]. The first line of equation (2.191) represents the translational and rotational motion of the undeformed reference $B_i$ of the body $B_i$. The second line expresses the influence of the elastic deformations, where for example the complete rotation is composed by the rotation between the inertial frame and that of the body $B_i$ depicted by ${}_{B_i}\tilde{\boldsymbol{\omega}}_{IB_i}$ plus the rotation between the undeformed body element $dm_r$ and the deformed element $dm_e$ (see Figure 2.22) given with ${}_{B_i}\tilde{\boldsymbol{\omega}}_{B_iB_{ei}}$, both terms written in the body-fixed frame $B_i$.

All necessary magnitudes are now known from the above relations. The evaluation should of course be done by a computer. The overall kinetic energy of a system writes

$$T = \sum_i \int_{m_i} {}_I\dot{\mathbf{r}}_{IB_{ei}}{}^T {}_I\dot{\mathbf{r}}_{IB_{ei}} dm_{ei} = \sum_i \int_{m_i} {}_B\mathbf{v}_{abs}{}^T {}_B\mathbf{v}_{abs} dm_{ei}, \tag{2.192}$$

where there might be also rigid bodies with ${}_{B_{ei}}\mathbf{u}_i = (u,v,w)_i^T = \mathbf{0}$. It does not change formula (2.192) principally.

Some aspects should be considered in establishing the energy for elastic components: Firstly, one should keep in mind the property, that the transformation $\mathbf{A}_{IB_i}$ from the body $B_i$ to the inertial frame I contains the influences of the elasticities of all bodies between I and $B_i$. As the velocities are generated by multiplication with the transformation matrices it is sufficient to retain only terms up to the second order in the elastic deformations. They are necessary to come out with an exact linearization of the elastic terms in the equations of motion. Secondly, applying a Ritz- or a Galerkin-approach for small elastic deformations we always can separate the integrals of equation (2.192) into a spatial- and a time-dependent part, where the spatial-dependent part can be evaluated beforehand.

### 2.4.3 Potential Energy

Considering in a first step the deformation energy we confine ourselves to the case of linear elastic deformations of isotropic materials including the symmetry of the stress tensor. One definition of this stress tensor is the following (see Figure 2.26)

$$\sigma = \begin{pmatrix} \sigma_{xx} & \sigma_{xy} & \sigma_{xz} \\ \sigma_{yx} & \sigma_{yy} & \sigma_{yz} \\ \sigma_{zx} & \sigma_{zy} & \sigma_{zz} \end{pmatrix}, \quad \text{with} \quad \sigma_{ij} = \sigma_{ji}, \quad (i,j = x, y, z). \tag{2.193}$$

The corresponding strain tensor is given with equation (2.140). Assuming small deformations of the element of Figure (2.26), for example in x-direction an $\epsilon_{xx}dx$ or in y-direction an angular deformation $\gamma_{xy}dx = 2\epsilon_{xy}dx$, we get the work done by these small deformations from the stress forces multiplied by the corresponding strains, in our example $dW_{\epsilon_{xx}} = (\frac{1}{2}\sigma_{xx}dydz)(\epsilon_{xx}dx)$ and $dW_{\gamma_{xy}} = (\frac{1}{2}\sigma_{xy}dydz)(\gamma_{xy}dx)$ with $\gamma_{xy}/2 = \epsilon_{xy}$. For all the other directions we come out with similar expressions.

Collecting all these terms and integrating over the total volume of the elastic body we have the well-known relation (see for example [147] or [28])

$$V = \frac{1}{2}\int_B (\sigma_{xx}\epsilon_{xx}+\sigma_{yy}\epsilon_{yy}+\sigma_{zz}\epsilon_{zz}+\sigma_{xy}\gamma_{xy}+\sigma_{yz}\gamma_{yz}+\sigma_{zx}\gamma_{zx})\mathrm{dx\,dy\,dz}. \quad (2.194)$$

For most engineering problems of dynamics the constitutive relations of Hooke's generalized laws will be sufficient. Confining our considerations to isotropic linearly elastic materials then they write for the strains and the shear strains, respectively,

$$\epsilon_{xx} = \frac{1}{E}[\sigma_{xx} - \mu(\sigma_{yy} + \sigma_{zz})],$$

$$\epsilon_{yy} = \frac{1}{E}[\sigma_{yy} - \mu(\sigma_{zz} + \sigma_{xx})],$$

$$\epsilon_{zz} = \frac{1}{E}[\sigma_{zz} - \mu(\sigma_{xx} + \sigma_{yy})], \quad (2.195)$$

$$\epsilon_{xy} = \frac{\sigma_{xy}}{2G}, \quad \epsilon_{xz} = \frac{\sigma_{xz}}{2G}, \quad \epsilon_{yz} = \frac{\sigma_{yz}}{2G}, \quad (2.196)$$

or in a more general form (see [19] and [228])

$$\epsilon_{ij} = \frac{1}{E}[(1+\mu)\sigma_{ij} - \mu\delta_{ij}\sigma_{nn}], \quad (i,j = x,y,z), \quad (2.197)$$

where we have to summarize over "nn", and $\delta_{ij}$ is the Kronecker-symbol. The material constants have the following meaning: E=Young's modulus, $\mu$=Poisson's ratio and G=shear modulus. The shear modulus can be expressed by $G = \frac{E}{2(1+\mu)}$. Combining the equations (2.194) to (2.197) yields the potential energy in the form

$$V = \int_B [\frac{1}{2E}(\sigma_{xx}^2 + \sigma_{yy}^2 + \sigma_{zz}^2) - \frac{\mu}{E}(\sigma_{xx}\sigma_{yy} + \sigma_{yy}\sigma_{zz} + \sigma_{zz}\sigma_{xx})$$

$$+ \frac{1}{2G}(\sigma_{xy}^2 + \sigma_{yz}^2 + \sigma_{zx}^2)]\mathrm{dxdydz}, \quad (2.198)$$

or again in a more general form (see [19] and [228])

$$V = \int_B \frac{1}{2E}[(1+\mu)\sigma_{ij}\sigma_{ij} - \mu\sigma_{nn}^2]\mathrm{dxdydz} \quad (2.199)$$

In a second step we only want to mention the different features of other potential energies. Any type of springs with linear, nonlinear or non-smooth characteristics possess potential energy if deformed. Gravity and the attraction or the repulsion of masses are connected with potential energies. The same is true for electrostatic and electrodynamic effects. If special problems require these or other forms of potential energy, it is easy to find models in the various literature.

## 2.5 On Contacts and Impacts

### 2.5.1 Phenomena

Impulsive motion takes places for a variety of reasons including the classical contact of two or more bodies, a sudden stop of some fluid flows or a velocity jump due to "dynamic locking" [161]. The last phenomenon was and still is a subject of many discussions and many contributions. A practical example represents the chattering chalk on a blackboard. The contacts and the impacts of two or more bodies will be of the main interest here, and therefore we shall focus on some basic aspects of such collisions.

The interest for understanding impact phenomena was always very large, because impacts possess the possibility to augment considerably the forces for a lot of practical processes like hammering or forcing piles into the ground (in German the "bear"). Therefore all great scientists and engineers worked in the one or other way on impact phenomena. Aristoteles, Galilei, Newton, Marcus Marci, Huygens, Euler, Poisson, Coulomb and many others paved the way to a modern theory of impulsive motion (see [259]).

If two or more bodies collide impulsively and with arbitrary direction, the contact zone will be deformed in normal and in tangential direction thus storing elastic potential energy with respect to these two directions. The deformation in tangential direction depends for a given state before the impact at least to a large extent on the properties of the contacting surfaces, especially on the roughnesses, which are for technical surfaces in the order of magnitude of some micrometers ($\mu$). Under the influence of the relative velocities, friction and the stored energies we get different results.

Firstly and the energy losses being small the bodies might separate again very quickly, where the directions of the separation depend on the velocity before the impact, the frictional features and the impulse storage. We might also get a reversal of the incoming motion depending mainly on the properties in tangential direction [15]. The point of contact, averaged over the deformed contact zone, will be usually different from that point, where the spring forces due to the elastic deformation of the contact zone apply. This has influence on the whole contact process, which becomes significant for pairings of soft materials [15].

### 2.5.2 Impact Structure

Some typical properties of a single impact are indicated in Figure (2.28). Two bodies will impact if their relative distance $\mathbf{r}_D$ becomes zero. This event is then a starting point for a process, which usually is assumed to have an extremely short duration. Nevertheless, deformation of the two bodies occurs, being composed of compression and expansion phases. The forces governing this deformation depend on the initial dynamics and kinematics of the contacting bodies. The impulsive process ends when the normal force of contact vanishes and changes sign, because a contact cannot realize tension forces. We do not consider adhesion phenomena. The condition of zero relative distance cannot be used as an indicator for the end of an impact, because it does not necessarily indicate also a vanishing contact force. In the general case of impact with

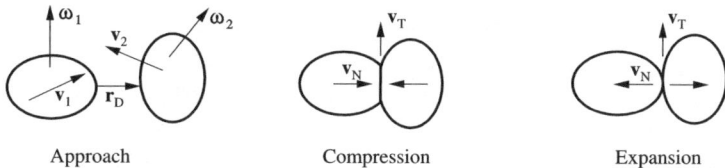

Fig. 2.28: Details of an Impact

friction we must also consider a possible change from sliding to sticking during the impulsive process, or vice versa, which includes frictional aspects as treated later. In the simple case of only normal velocities we sometimes can idealize impacts according to Newton's impact laws, which relate the relative velocity after an impact with that before an impact. Such an idealization can only be performed if the force budget allows it. In the case of impacts by hard loaded bodies we must analyze the deformation in detail. Gear hammering taking place under heavy loads and gear rattling taking place under no load are typical examples [200].

As in all other contact dynamical problems, impacts possess complementarity properties. For ideal classical inelastic impacts either the relative velocity is zero and the accompanying normal constraint impulse is not zero, or vice versa. The scalar product of relative velocity and normal impulse is thus always zero. For the more complicated case of an impact with friction we shall find such a complementarity in each phase of the impact. Friction in one contact only is characterized by a contact condition of vanishing relative distance and by two frictional conditions, either sliding or sticking (see Figure 2.29).

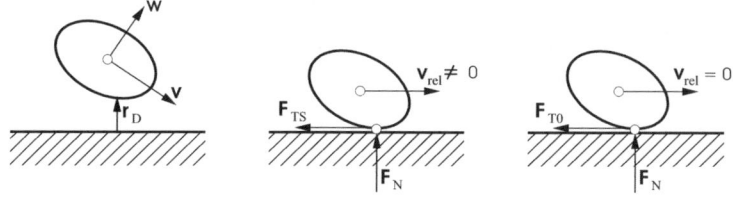

Fig. 2.29: Sliding and Static Friction

A typical property of contacts, whatsoever, is the fact, that kinematic magnitudes indicating the beginning of a contact event become a constraint at that time instant, where a contact becomes "active". For example: a non-zero normal distance between two bodies going to come into contact indicates a "passive" contact state with zero normal constraint force. In the moment it is zero, then the relative distance represents a constraint accompanied by a constraint force, and the contact is "active". In tangential direction it is similar: Non-zero tangential velocity (sliding) means a zero "friction reserve" $\mu_0|\mathbf{F}_N|-|\mathbf{F}_{TC}|=0$ (see Figure 2.30). This tangential relative velocity becomes zero for stiction and represents then a constraint accompanied by a tangential constraint force. The end of a contact event or better of an active contact state will be always indicated by a constraint force or a combination of constraint forces. The normal constraint force becomes zero indicating a separation, and the friction reserve becomes zero indicating a change from sticking to sliding. In more detail this means:

From the contact constraint $\mathbf{r}_D = \mathbf{0}$ we get a normal constraint force $\mathbf{F}_N$ which, according to Coulomb's laws, is proportional to the friction forces, or better vice versa, the friction forces are proportional to the normal force in the contact. For sliding $\mathbf{F}_{TS} = -\mu \mathbf{F}_N \operatorname{sgn}(\mathbf{v}_{rel})$, and for stiction $\mathbf{F}_{T0} = -\mu_0 \mathbf{F}_N$, where $\mu$ and $\mu_0$ are the coefficients of sliding and static friction, respectively. Stiction is indicated by $\mathbf{v}_{rel} = 0$ in tangential direction and by a surplus of the static friction force over the constraint force, $\mu_0|\mathbf{F}_N| - |\mathbf{F}_{TC}| \geq 0$. If this friction reserve becomes zero the stiction situation will end, and sliding will start again with a nonzero relative acceleration $\mathbf{a}_{rel}$ in the tangential direction. Again we find here complementary behavior: Either the relative velocity (acceleration) is zero and the friction reserve (saturation) is not zero, or vice versa. The product of relative acceleration and friction surplus is always zero.

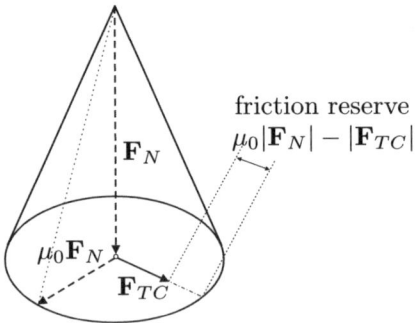

Fig. 2.30: Friction Cone and friction reserve

We may take that in a more classical way. Having stiction we are situated within the friction cone connected with the contact under consideration. The cone boundary represents the static friction as mentioned above, and the friction reserve or the friction surplus is the distance from the tangential force $\mathbf{F}_{TC}$ to the friction cone given with $\mu_0|\mathbf{F}_N|$, see Figure (2.30). If this distance is used up by the changing dynamics of the overall system, then we are on the friction cone with the possibility of tangential sliding. But this sliding in tangential direction can only start, if the vanishing friction reserve will be accompanied by a non-zero acceleration in one of the tangential directions.

### 2.5.3 Basic Laws

All modern efforts to establish contact models on a micro-scale down to the molecular surface structures were not very successful, up to now. Therefore we still rely on a few laws developed by Newton, Poisson and Coulomb. Newton's impact law (1687) relates the relative velocity before an impact to the relative velocity after an impact by a coefficient of restitution $\epsilon$, which must be determined experimentally. A value $\epsilon = 1$ represents a completely elastic and $\epsilon = 0$ a completely plastic impact. In the first case we have no energy losses and in the second case a maximum loss. Newton's law is a kinematic law and refers to the normal contact direction only, whereas the law of Poisson (1835) is a kinetic law relating the impulses after and before an impact. It can easily be extended to general impacts with normal and tangential components of velocities and impulses. The coefficients of restitution must also be measured. For our purposes of multiple impacts in multibody systems with multiple contacts Poisson's law is more general and assures correct results for all cases of technical relevancy, Newton's law not. Poisson's law allows an energy transfer between the normal and tangential directions, and vice versa, Newton's law not. Thus, Poisson's law gives a more realistic approach.

Coulomb's idea of applying a very simple relationship for contacts corresponds exactly to that, what we are doing in engineering sciences in cases where detailed and very sophisticated models are hopeless to realize. We go the simple way. It is really fascinating that these simple laws of Coulomb, which he wrote down about 1780, give such a good approximation also in complicated cases of impulsive motion. He assumes the friction forces in a contact to be proportional to the normal force and introduces friction coefficients for static and for sliding friction, which have to be determined from experiments. Coulomb's law represents the basis for the friction cone and from there, in modern non-smooth mechanics, the starting point for convex analysis.

In chapter (2.5.2) we have discussed some structural aspects by considering one contact only. For the forces we used the expressions $\mathbf{F}_N$ and $\mathbf{F}_T$. With respect to multibody systems we shall have in the chapters to come the expressions $\lambda_N$ and $\lambda_T$ for all kinds of constraints forces. Therefore we shall use these expressions already in the following considerations. To start with we repeat the well-known and above mentioned laws by considering some contact i, which is involved in impulsive motion. For dry friction we shall apply Coulomb's law in the following form:

$$|\lambda_{Ti}| < \mu_{0i}|\lambda_{Ni}| \quad \wedge \quad \dot{g}_{Ti} = 0 \quad \text{sticking,}$$
$$\lambda_{Ti} = +\mu_{0i}\lambda_{Ni} \quad \wedge \quad \dot{g}_{Ti} \leq 0 \quad \text{negative sliding,}$$
$$\lambda_{Ti} = -\mu_{0i}\lambda_{Ni} \quad \wedge \quad \dot{g}_{Ti} \geq 0 \quad \text{positive sliding,} \tag{2.200}$$

where $\dot{g}_{Ti}$ is the relative velocity in contact $i$, and $\lambda_{Ni}, \lambda_{Ti}$ are the relevant constraint forces in normal and tangential direction, respectively. Equation 2.200 can be interpreted as a double corner law as shown in Figure 2.31. We are

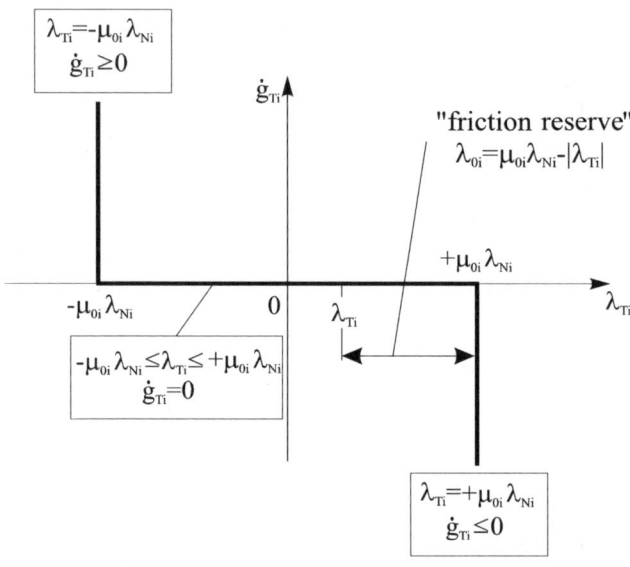

Fig. 2.31: The Friction Law of Coulomb

either within the friction cone with $|\dot{g}_{Ti}| = 0$ and $(-\mu_{0i}\lambda_{Ni} \leq \lambda_{Ti} \leq +\mu_{0i}\lambda_{Ni})$, or we are on the friction cone surface with $|\dot{g}_{Ti}| \neq 0$ and $|\lambda_{Ti}| = \mu_{0i}\lambda_{Ni}$. The friction coefficient $\mu_{0i}$ is defined as (see Figure 2.32)

$$\lim_{\dot{g}_{Ti} \to 0} \mu_i(\dot{g}_{Ti}) = \mu_{0i} \tag{2.201}$$

In connection with this condition it should be noted, that most of the authors working in the field of non-smooth mechanics apply Coulomb's law with the same friction coefficient $\mu$ for sliding and for sticking, which means independent from $\dot{g}_{Ti}$. This is convenient and for many fundamental investigations also sufficient. But with respect to practical contact problems we usually have to consider Stribeck curves (Figure 2.32), which depend on the relative tangential velocity $\dot{g}_{Ti}$. Therefore in applying non-smooth theories the above condition makes sense, because one important kernel of all non-smooth considerations are the various contact transitions.

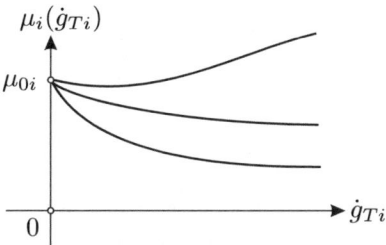

Fig. 2.32: Typical friction characteristics, Stribeck curves

The constraint force $\lambda_{Ni}$ in normal direction results also from a contact law, which might be characterized by a contact-separation-mechanism. If we have a normal relative distance in contact $i$ designated $g_{Ni}$, then the interdependency with the corresponding constraint force $\lambda_{Ni}$ consists in the classical complementarity: Either $g_{Ni} = 0$ and $\lambda_{Ni} \geq 0$ or $g_{Ni} \geq 0$ and $\lambda_{Ni} = 0$, which is depicted in Figure 2.33. From this the product $(g_{Ni} \cdot \lambda_{Ni})$ is always zero. Both contact laws (Figures 2.31, 2.33) include complementary features,

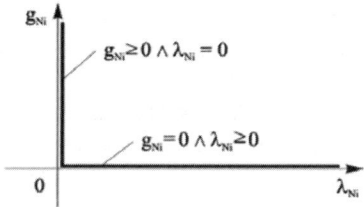

Fig. 2.33: Corner law for normal contacts (Signorini's law)

because also Figure 2.31 can be decomposed in two so-called "unilateral primitives" [87] in form of two simple corners. With the expressions of Figure 2.31 we may say in the case of frictional contacts, that either the relative velocities $|\dot{g}_{Ti}|$ are zero and the expression $\lambda_{0i} = \mu_{0i}\lambda_{Ni} - |\lambda_{Ti}|$ is not zero, or vice versa, the product $(\dot{g}_{Ti}\lambda_{0i})$ is always zero. We shall call that expression either "friction reserve" or "friction surplus", where the first name describes the relevant properties in a better way.

With respect to impact laws we have in classical mechanics two models, Newton's kinematical and Poisson's kinetic relationships. Newton's law connects the relative normal velocity after an impact with that before the impact stating

$$(\dot{g}_{Ni})^+ = -\varepsilon_i(\dot{g}_{Ni})^- ,\qquad(2.202)$$

where (+) means shortly after and (−) shortly before the impact. Losses are approximated by the coefficient of restitution $\varepsilon_i$. Poisson considers impulses $\Lambda = \int \lambda dt$ and relates them by

$$\Lambda_i^+ = -\varepsilon_i^* \Lambda_i^- .\qquad(2.203)$$

The physical idea behind Poisson's law consists in a storage of impulses during compression and a gain connected with losses during expansion of the impact process. Therefore, Poisson's law can be applied in normal and tangential direction of the contact without generating physical inconsistency [15]. In any case the loss coefficients $\varepsilon_i, \varepsilon_i^*$ have to be measured for each specific material pairing.

### 2.5.4 Impact Models

According to the above discussion we have two possibilities to model impulsive motion. Firstly, we may discretize the contact zone, for example by finite elements or by analytical approximative relations [118], which results in a smooth model usually with the problem of stiff differential equations due to the large stiffnesses in the contacts. As a consequence such models come out with very high frequencies either without physical meaning or very often without the need to know them. Secondly, we may use the above discussed complementarities to establish a non-smooth model not including the drawbacks of discretized models, but for the prize of more mathematics and of some numerical difficulties. For large systems, where we do not need the detailed structures of the contacts, non-smooth models possess definitely more advantages than drawbacks.

It should be pointed out, though, that we may also combine the two methods of modeling. For large systems with unilateral contacts it is often more economical to evaluate a non-smooth model and to go that way. In the case where I want to have in addition some detailed informations of some specific contacts with respect to certain criteria, for example maximum contact forces, I always may establish some post-processing including a discretization for the selected contacts thus yielding detailed results of pressure distributions, deformation distributions and the like. The only requirement to be fulfilled consists in adapting the discretization to the results obtained by the "large" non-smooth simulation. For example the averaged impulses must compare with those of the simulation. Altogether this is a simple and straightforward procedure.

Discretized models work with forces and a finite duration of the impulsive process itself. It is not too difficult to include also wave phenomena. Non-smooth rigid-body-models work with an infinitesimal short duration of the

## 2.5 On Contacts and Impacts

impulsive process, and it is difficult to include wave processes. The choice of the model with a best fit to the problem depends of course, as always, on the problem itself. Non-smooth rigid-body-models turn out to be an excellent approximation to a large variety of technical problems. They are governed by the following assumptions:

- The duration of the impact is "very short."
- The impact can be divided into two phases: the compression phase and the expansion phase.
- The compression phase starts at time $t_A$ and ends at time $t_C$. The end of the compression equals the start of the expansion phase. Expansion is finished at time $t_E$, which is also the end of the impact.
- During the short impact duration all magnitudes of the multibody system for position and orientation as well as all nonimpulsive forces and torques remain constant.
- Wave effects are not taken into account.

In multiple-contact problems there might be one impact only in one of the contacts or several impacts in several contacts simultaneously. The existing theories cover both possibilities (see [200], [87], [135]).

## 2.6 Damping

### 2.6.1 Phenomena

From the physical standpoint of view damping is an energy conversion process, in most cases by some kind of friction. From the technological standpoint of view damping is an energy consuming process, in many cases accompanied by useful consequences in form of vibration reduction, for example. All mechanisms of damping require relative motion. Following some systematizing ideas of the German Society of Engineers (VDI - Verein Deutscher Ingenieure, [266]) we have in mechanical systems internal damping like friction in gears or slide ways, external damping by solid-fluid-interactions, material damping by fluid flows or by microplastic deformations of solids and some special forms of damping in connection with electro-dynamical influences, for example. In specific cases of machines and mechanism it is usually not very difficult to localize the damping possibilities, the problem concerns mainly the quantitative evaluation of the damping mechanisms. In spite of very many intelligent theories on damping represented by a huge literature in that area we are mainly concerned with empirical data and thus with experience. It is not by accident, that big companies establish large databases collecting all the experiences available on damping, of course especially with respect to their own products. Nevertheless we shall give some classical and modern results concerning damping and its influence on dynamics of mechanical systems.

Friction is the basis of nearly all types of mechanical damping, friction between solids, between fluids and solids and internal friction of fluids and solids. We can also produce damping by the combination of mechanics and electro-dynamics, like eddy-current brakes, but we shall confine our consideration to mechanical friction effects. Friction between solids might be dry or viscous friction, in the first case governed for example by Coulomb's equations (2.200) and in the second case by some laws usually derived from the boundary layer equations of viscous fluids (see for example [12] and [13]). The viscosity of fluids leads to a large variety of possible representations depending on the environmental parameters like gap size, pressure differences or velocities. In machine dynamics this belongs to the increasing field of fluid film rheology [94]. It should be kept in mind, that fluid films in bearings, between gear teeth or on guide ways have a thickness of some micrometers with usually significantly increasing stiffness with decreasing thickness.

Interactions of solids and fluids generate friction in boundary layers of variable thickness, where the flow might be laminar or turbulent depending on the external conditions. Airplanes, helicopters, buildings and the famous self-excited vibrations of bridges are typical examples. Velocity gradients lead to friction within a fluid, which is of much importance in chemical engineering. For our purposes we need the effects of friction in the form of a linear or nonlinear force law, which we can then implement into the equations of motion. The implementation does of course not solve the data problem, because these force laws contain as a rule some empirical coefficients, which have to be measured. In the following we shall indicate some possibilities of such force laws.

### 2.6.2 Linear Damping

Linear damping is characterized by force laws, which give a linear relation between force and relative velocity. Many engineering applications follow approximately this property. We may include such forces in both, linear or nonlinear equations of motion. In many cases the motion can also be linearized. Therefore we shall consider in the following linear equations of motion including linear damping, and we shall give some simple relations with regard to these damping laws.

Linear system dynamics is a very well established area, which might be taken from a large variety of text books, see for example [174], [157], [148] and [187]. For a mechanical system with f degrees of freedom we get the well-known MDGKN-equations of motion, which write:

$$\mathbf{M}\ddot{\mathbf{y}} + (\mathbf{D}+\mathbf{G})\dot{\mathbf{y}} + (\mathbf{K}+\mathbf{N})\mathbf{y} = \mathbf{g}(t) \qquad \text{with} \quad \mathbf{y} \in \mathbb{R}^f, \mathbf{M} \in \mathbb{R}^{f,f}, \text{etc.} \qquad (2.204)$$

The matrix $\mathbf{M}$ is the symmetric mass matrix, $\mathbf{D}$ the symmetric damping matrix, $\mathbf{G}$ the skew- symmetric gyroscopic matrix, $\mathbf{K}$ the symmetric stiffness matrix and $\mathbf{N}$ the skew-symmetric matrix of non-conservative forces, due to rotational effects, for example. A Laplace transform of these equations comes out with

$$\mathbf{y}(s) = [\mathbf{M}s^2 + (\mathbf{D}+\mathbf{G})s + (\mathbf{K}+\mathbf{N})]^{-1}\mathbf{g}(s), \qquad (2.205)$$

which gives immediately the response chart for the linear system (2.204) under the excitation of $\mathbf{g}(t)$. The existence of a linear damping term ($\mathbf{D}s$) results in a reduction of the amplitudes $\mathbf{y}(s)$. If we put $\mathbf{G} = \mathbf{0}$, $\mathbf{N} = \mathbf{0}$, $\mathbf{g}(t) = \mathbf{0}$ and multiply the remaining equation ($\mathbf{M}\ddot{\mathbf{y}} + \mathbf{D}\dot{\mathbf{y}} + \mathbf{K}\mathbf{y} = \mathbf{0}$) from the left side with $\dot{\mathbf{y}}^T$, we get a kind of power equation in the following form

$$\dot{\mathbf{y}}^T\mathbf{M}\ddot{\mathbf{y}} + \dot{\mathbf{y}}^T\mathbf{K}\mathbf{y} = \frac{d(E_k + E_p)}{dt} = -\dot{\mathbf{y}}^T\mathbf{D}\dot{\mathbf{y}} = -2R. \qquad (2.206)$$

The left hand side of this equation depicts the time derivatives of the kinetic and the potential energies, and the energy term on the right hand side represents the dissipation power produced by the dissipative forces, and R is the Rayleigh dissipation function [148]. For $R > 0$ and thus $\mathbf{D} > \mathbf{0}$ we have *complete damping* reducing the motion energy for any kind of motion. For a positive semidefinite matrix with ($\mathbf{D} \geq \mathbf{0}, det(\mathbf{D}) = 0$) we still have damping, which is called *penetrating damping*. In this case we should have a good parameter adaptation to achieve damping for all types of motion.

Sometimes it makes sense to assume, that the damping matrix $\mathbf{D}$ is proportional to the mass- and the stiffness-matrices, which has some advantages with respect to modal analysis but which on the other hand shifts the data problem to the formula [148]

$$\mathbf{D} = \alpha \mathbf{M} + \beta \mathbf{K}, \tag{2.207}$$

because the coefficients $\alpha$ and $\beta$ are not known and must be estimated. Instead of choosing the form of equation (2.207) with scalar coefficients we also could choose a coefficient matrix, which of course will enlarge the data problem.

The classical results for an oscillator with one degree of freedom only allows some simple interpretations of damping. The equation (2.204) writes for the case f=1

$$\ddot{y} + 2\delta \dot{y} + \omega_0^2 y = g(t), \qquad \delta = \frac{d}{2m}, \qquad \omega_0^2 = \frac{k}{m}. \tag{2.208}$$

For the undamped case ($\delta = 0$) we get from the right hand side of the above equation the sometimes called "undamped natural frequency" $\omega_0 = \sqrt{\frac{k}{m}}$ and for the damped case ($\delta \neq 0$) the "damped natural frequency" $\omega_d = \sqrt{\frac{k}{m} - \frac{d}{2m}^2}$, which indicates the well known reduction of eigenfrequencies with increasing damping. The complete solution possesses two parts, that of the free vibrations, which usually disappear very quickly, and that of the forced vibrations. Assuming a harmonic excitation with ($g(t) = a \cos(\omega t)$) we get as a result for the forced part

$$x_{forced} = R \cos \omega t - \psi,$$

$$R = \frac{a}{\sqrt{(\omega_0^2 - \omega^2)^2 + 4\delta^2 \omega^2}}, \qquad \tan \psi = \frac{2\delta \omega}{\omega_0^2 - \omega^2} \tag{2.209}$$

The amplitude R of the forced vibration and the phase angle $\psi$ depend both on the damping coefficient. Moreover, for vanishing damping we have no phase shift $\psi$. The amplitude R increases with decreasing damping, and its maximum is shifted to lower eigenfrequencies with increasing damping. These effects also can be observed in response curves for systems with many degrees of freedom, mainly due to the fact, that near resonances most large systems with many degrees of freedom behave approximately like a one-degree-of-freedom system. For practical applications it is useful to remember a few qualitative tendencies following from the above equations: Eigenfrequencies decrease with increasing masses and with decreasing stiffness. Resonance amplitudes decrease and are shifted to lower frequencies with increasing damping. These are, for a given configuration, the main parameters to influence vibrations and to avoid resonances.

As already discussed, it is very often straightforward to find a model for damping, but it is a difficult task to get the unknown coefficients included in all force laws of damping. For linear damping this concerns the elements of the matrix $\mathbf{D}$ of equation (2.204). Though we never can do without empirical

## 2.6 Damping

data, at least with respect to the state of the simulation art at the time being, we shall consider some simple laws for an estimation of damping force laws.

Elementary relations for damping are fluid flow boundary layers of various configurations. The plane laminar flow of a Newtonian fluid between a moving and a not moving wall is governed by a parabolic velocity and a linear shear stress distribution. The shear stress at the moving wall and the volume flow through the gap take on the values ([12], [13], see also Figure (2.34)) )

$$\tau_w = \eta \frac{u_w}{h} + \frac{h}{2l}(p_2 - p_1), \qquad \dot{V} = (\frac{u_w h}{2} - \frac{p_2 - p_1}{12\eta l} h^3) b, \qquad (2.210)$$

where $\tau_w$ and $u_w$ is the shear stress and the velocity of the moving wall,

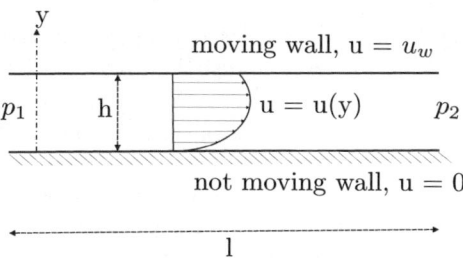

Fig. 2.34: Flow through a Gap

respectively, $\dot{V}$ is the volume flow, h is the gap between the two walls, l the length of the walls and the $(p_i, i = 1, 2)$ are the pressures at the two sides of the gap. The viscosity $\eta$ is the "dynamic" viscosity. It is assumed that the plane flow has a width b. The first part of the first equation (2.210) is proportional to the velocity, the second part is proportional to the pressure difference $(p_2 - p_1)$. For practical cases various combinations are possible.

We first consider an elementary slide way with the length 2l and the width b, see Figure (2.35). The upper sliding structure does not move $u_w = 0$, and the slide way is supplied by an oil volume $\dot{V}$ in the middle of its length 2l. The gap is very small compared with b and l ($h \ll l, h \ll b$), so that we can neglect boundary effects. From the second equation (2.210) we get $(p_2 - p_1 = \frac{12\eta l \dot{V}}{bh^3})$. The pressure distribution in the gap is approximately linear, therefore the averaged pressure difference is $(p_2 - p_1)/2$ acting on the area (2bl). For the normal force $F_n$ in the gap this yields

$$F_n = \frac{12\eta l^2 \dot{V}}{h^3} = (\frac{12\eta A l^2}{h^3}) \cdot v_{rel}, \qquad (2.211)$$

where A is the cross sectional area of the gap and $v_{rel} = \frac{\dot{V}}{A}$ the relative velocity in the gap. Equation (2.211) describes an interesting result for such oil cushions. The force $F_n$ is proportional to $h^{-3}$ resulting in a sharply increasing

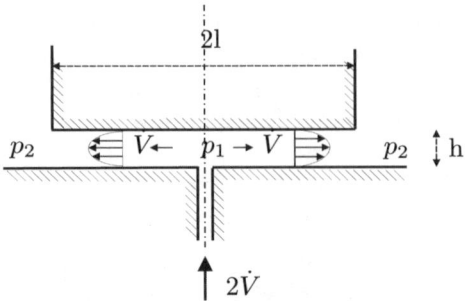

Fig. 2.35: Axial Bearing, Slide Way

pressure difference $(p_2 - p_1)$ without destroying the oil cushion. Thus oil lubrication is maintained even for large loads. We find a similar situation for axial bearings, which is an important component in many machines, for example in power transmissions of ships. Equation (2.211) represents a linear relation of force and velocity.

Another very important application are journal bearings, where the gap between the shaft and the bearing cylinder is very small. Therefore a boundary layer approach gives quite realistic results ([13], [247]). It is well known, that a shaft running exactly symmetrical in a bearing does not develop any force in any direction, because we have around the shaft the same pressure distribution as supplied by some pump, for example (left picture of Figure (2.36)). Putting on a load we get on the upper half of the shaft ($\varphi \in [0, \pi]$) an underpressure with respect to the averaged pressure in the gap, and on the lower half of the shaft ($\varphi \in [\pi, 2\pi]$) an overpressure. The integral over the whole bearing results in a pressure force, exactly vertical upwards, which counterbalances the external load by adapting the "control parameter" $h(\varphi)$ in an appropriate way, see right picture of Figure (2.36). The shaft itself is shifted exactly in a perpendicular direction with respect to the force **F**.

From the simple Reynolds-Sommerfeld-theory we get the following friction torque ([13], [247]):

$$M = \left( \frac{2(1 + 2\varepsilon^2)}{\sqrt{1 - \varepsilon^2}(1 + \frac{\varepsilon^2}{2})} \right) \cdot \left( \frac{\pi r_0^2 b \eta}{a} \right) \cdot u_w. \quad (2.212)$$

The eccentricity $\varepsilon$ describes the offset of the shaft center with respect to the bearing center. For $\varepsilon = 0$ we have the left hand side case of Figure (2.36), that means no eccentricity, and for $\varepsilon = 1$ the other extremum, namely a contact of the shaft at the the left side with $\varphi = 0$. The gap for the load case depends on $\varphi$ and writes: $h(\varphi) = a(1 - \varepsilon \cos \varphi)$. The expression $\pi r_0^2$ is the shaft cross section, and b the bearing length. The gap without load has the value of a, for which we assume $a \ll r_0$. As above, $\eta$ is the dynamic viscosity. It should be noticed, that the damping torque for a disturbance $\Delta u_w$ of the constant

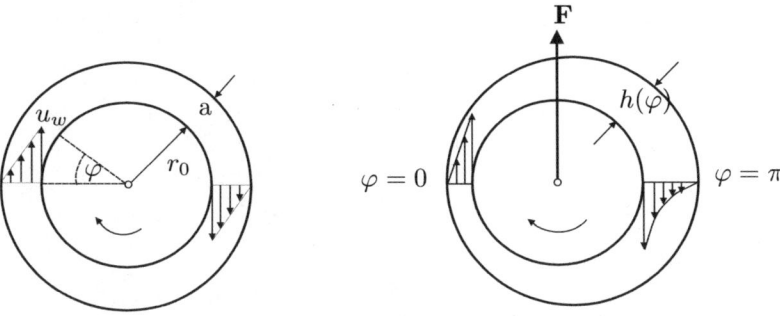

Fig. 2.36: Journal Bearing

shaft speed $u_w$ is given by the same expression as in equation (2.212), we only must replace $u_w$ by $\Delta u_w$.

In practice we are faced with deviations from the above theory, and in the meantime a large body of technical literature exists giving very sophisticated models of journal bearing dynamics and of friction in journal bearings. But nevertheless these simple formulas above give a first estimate of friction and damping in bearings in the form of a linear relation between torque and velocity. For many practical application it fits quite well. We shall come back to these problems in the application part of the book.

### 2.6.3 Nonlinear Damping

Nonlinear damping behavior in machines and mechanisms is very diverse. Material damping, friction in all components having relative motion with respect to other components, damping in all connecting elements like screws or press fits are a few examples. The mounts connecting the motor with the automobile body consist of visco-elastic springs with a progressive characteristic. Surveys of nonlinear damping my be seen from the literature ([47], [266]). A practically reasonable measure of damping is the "relative damping" defined as the friction losses per cycle, which is proportional to the area enclosed by the hysteresis cycle (Figure 2.37).

The mechanical losses due to such a hysteresis behavior is the integral for one cycle

$$W_D = \oint \sigma d\varepsilon, \quad \text{or} \quad W_D = \frac{1}{T} \int\limits_{t}^{t+T} \mathbf{F}_D^T \mathbf{v}_{rel} dt, \qquad (2.213)$$

which represents in the first case the work of damping per volume $[Nm/m^3]$ for arbitrary $\sigma - \varepsilon - curves$, in the second case the damping work in $[Nm]$

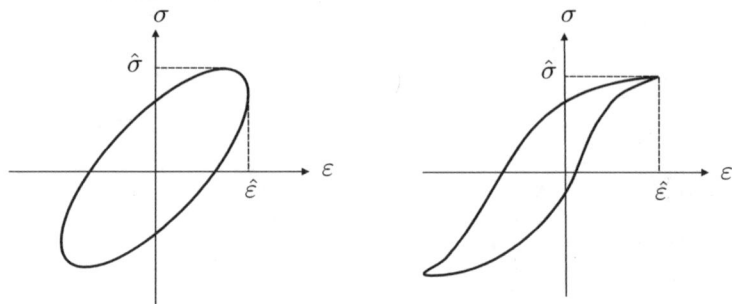

Fig. 2.37: Damping Hysteresis: left-linear material behavior, right-nonlinear material behavior

directly. The first equation is useful only for such cyclic processes with a corresponding friction behavior. The second case is more general by applying the law for the damping force $\mathbf{F}_D$, but we must find a suitable magnitude for the reference time T, usually from simulations.

The equations of motion become nonlinear even in cases, where the dynamics without damping might be linear. Thus we start with the set

$$\mathbf{M}(\mathbf{q},t) - \mathbf{h}(\mathbf{q},\dot{\mathbf{q}},t) = \mathbf{J}_D \cdot \mathbf{F}_D = \mathbf{h}_D(\mathbf{q},\dot{\mathbf{q}},t),$$
$$(\mathbf{q} \in \mathbb{R}^f, \mathbf{M} \in \mathbb{R}^{f,f}, \mathbf{h} \in \mathbb{R}^f, \mathbf{F}_D \in \mathbb{R}^{f_D}, \mathbf{J}_D \in \mathbb{R}^{f,f_D}). \quad (2.214)$$

The vector $\mathbf{q}$ describes the generalized coordinates, f the degrees of freedom, $\mathbf{M}$ the mass matrix, $\mathbf{h}$ all forces with the exception of the damping forces, and the Jacobian $\mathbf{J}_D$ projects the damping forces into the space of the generalized coordinates. The right hand side of the above equation depends on $(\mathbf{q}, \dot{\mathbf{q}}, t)$.

Let us first consider a few examples concerning jointing structures, mainly based on the Coulomb friction law in connection with the Stribeck curve (see chapter 2.5). Coulomb's law for sliding and static friction writes

$$\mathbf{F}_{sliding} = -\mu \mathbf{F}_N \operatorname{sgn}(\mathbf{v}_{rel}), \qquad \mathbf{F}_{stiction} = \mu_0 \mathbf{F}_N, \quad (2.215)$$

where the first equation applies for sliding and the second for stiction. For damping we are interested only in the first case with $\mathbf{F}_{sliding} = \mathbf{F}_S$ in the tangential contact direction, where also $\mathbf{v}_{rel}$ develops between the corresponding surfaces. According to equation (2.201) the validity of Coulomb's law is limited to the area around $\dot{g}_{Ti} = 0$, the further characteristic of the relationship $\mu_i(\dot{g}_{Ti})$ depends on the material pairing and must be measured. Modern material pairings, as applied for example in automotive industry, show partly exotic friction behavior often characterized by increasing and not by falling Stribeck curves. They are indicated in Figure (2.38). From this it follows that for the evaluation of system dynamics we must consider set-valued forces near the origin of Figure (2.38) and apart from the origin some type of Stribeck curve in the form of a smooth force law.

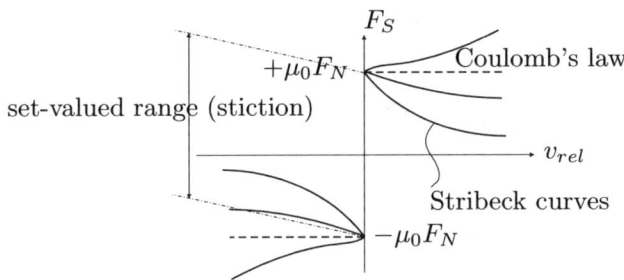

Fig. 2.38: Typical Stribeck Curves

In machine dynamics damping by relative motion in screws, in press fits and in similar contact structures is much larger than material damping, which might be more important in civil engineering. Relative motion in such jointing connections develops as a result of the system dynamics, which can generate loads exceeding the design loads of the connection. It is known from practice, for example, that axial loads on screws oscillating around the mean static value with an amplitude of more than 60% of this mean value will generate sliding within the screws, which must not happen from the standpoint of design, but which results in a damping effect with respect to dynamics. Such damping effects are governed by the nonlinear, non-smooth Coulomb's laws.

Another frequently applied element of machine design are press fits (see [26] and [265]). Such press fits are typically loaded by very large static forces superimposed by vibrations in all directions, spatial bending, torsion and axial vibrations. Most models subdivide the shaft in many elements connected with the hull by a spring-dry-friction- element, which is sometimes called Jenkins element. Figure (2.39) depicts an example from [26], which considers in detail the vibrations of a drive train of a large Diesel-engine with about 4 MW rated power and in this connection especially the influence of press fits on the dynamics of the whole system. These press fits were used to connect several large gear wheels to a shaft. The model of Figure (2.39) was included into the complete system model, and the Jenkins-elements described by the friction laws (2.215).

As a result we get hysteresis curves typical for such machine components. The area circumscribed by these curves increase with increasing load, and the slope of the line connecting the two extremum peaks decreases, both well-known effects from experimental data. The theory was compared with measurements from a laboratory test set-up, correspondence has been very good for all parameter cases. The influence of such press fits on the system dynamics of the complete drive train can be significant. In the case of the 4 MW Diesel engine the reduction of the overall stiffness of the whole train system was nearly 25% with a corresponding modification of the system response and the accompanying resonance frequencies [26].

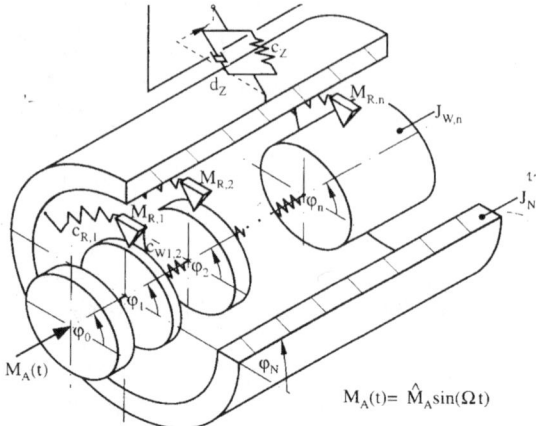

Fig. 2.39: Typical Model of a Press Fit [26]

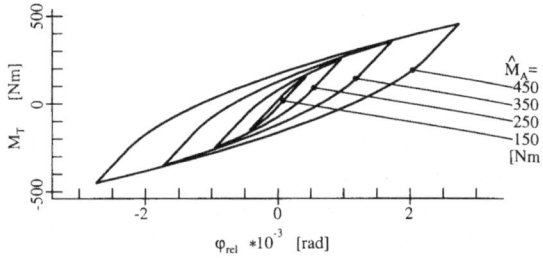

Fig. 2.40: Hyteresis Curves for Figure (2.39), [26]

For some nonlinear systems we might be able to generate a linear approximation by representing the nonlinear hysteresis curves by linear ones. With respect to Figure (2.37) we then must approach the closed hysteresis of the right picture by the ellipse-shaped curve of the left side by watching the condition, that the enclosed area must be the same for both cases, the real nonlinear one and the approximate linear one. Many methods are available to perform such an approximation ([148], [187], [47] or [60]).

# 3
# Constraint Systems

> *Es ist die nächste und in gewissem Sinne wichtigste Aufgabe unserer bewussten Naturerkenntnis, dass sie uns befähige, zukünftige Erfahrungen vorauszusehen, um nach dieser Voraussicht unser gegenwärtiges Handeln einrichten zu können.*
>
> *(Heinrich Hertz, Die Prinzipien der Mechanik, Einleitung, 1894)*
>
> *The most direct, and in a sense the most important task, which our conscious knowledge of nature should enable us to solve is the anticipation of future events, so that we may arrange our present affairs in accordance with such anticipation.*
>
> *(Heinrich Hertz, The Principles of Mechanics Presented in a New Form, authorized English translation by D.E. Jones and J.T. Walley; London, New York, Macmillan, 1899)*

## 3.1 Constraints and Contacts

### 3.1.1 Bilateral Constraints

The concept of constraints together with the accompanying constraint forces is of similar importance as the laws of momentum and of moment of momentum by Newton and Euler. Daniel Bernoulli was possibly the first to recognize this significance by speaking of "lost forces" meaning the constraint forces. A second step was done by d'Alembert requiring, that "all lost forces must be in an equilibrium state", and finally Lagrange put that in a modern form stating, that "lost forces generate no work".

According to classical mechanics (see for example [93] and [180]) we establish the following bilateral constraints as algebraic equations in a kinematic sense:

$$\begin{aligned}
\mathbf{\Phi}(\mathbf{z}) &= \mathbf{0}, & &\text{holonomic and scleronomic} \\
\mathbf{\Phi}(\mathbf{z}, t) &= \mathbf{0}, & &\text{holonomic and rheonomic} \\
\mathbf{\Phi}(\mathbf{z}, \dot{\mathbf{z}}) &= \mathbf{0}, & &\text{non-holonomic and scleronomic} \\
\mathbf{\Phi}(\mathbf{z}, \dot{\mathbf{z}}, t) &= \mathbf{0}, & &\text{non-holonomic and rheonomic}
\end{aligned} \quad (3.1)$$

For n bodies we have 6n system coordinates z, the number of all constraints may be $m = m_h + m_n$ where $m_h$ are the holonomic and $m_n$ the nonholonomic constraints. The real degrees of freedom of the system are given with f=6n-m represented by the generalized coordinates **q**. We have used $\boldsymbol{\Phi} \in \mathbb{R}^m$ for the constraint functions, $\mathbf{z} \in \mathbb{R}^{6n}$ for the system coordinates and $\dot{\mathbf{z}}$ for the system velocities. The constraints form a set of independent equations, not correlated with the equations of motion but restricting them. As indicated already in the section 2.1.3, holonomic constraints are integrable, at least theoretically, non-holonomic constraints are not integrable. This is the basic condition for non-holonomic behavior. Scleronomic constraints are stationary, rheonomic constraints are non stationary.

The constraint equations (3.1) define some constraint hypersurfaces in the space of the system coordinates $\mathbf{z}(\mathbf{q}, t)$, which themselves depend on the generalized coordinates **q** and the time t. The, mostly curvilinear, coordinates on these surfaces are the coordinates $\mathbf{q}(t)$, which represent the possible directions of the free and allowed motion on these surfaces. Requiring $\mathbf{z} = \mathbf{z}(\mathbf{q}, t)$ means nothing else, that the motion with the coordinates **z** must remain on the constraint surfaces and can thus be expressed by the generalized coordinates $\mathbf{q}(t)$. For the further considerations it is important to gain some idea of the derivatives of these magnitudes.

The hypersurface $\boldsymbol{\Phi}(\mathbf{z}, t) = \mathbf{0}$, for example, allows the following derivations:

$$d\boldsymbol{\Phi} = \frac{\partial \boldsymbol{\Phi}}{\partial \mathbf{z}} \cdot d\mathbf{z} + \frac{\partial \boldsymbol{\Phi}}{\partial t} \cdot dt, \qquad \delta\boldsymbol{\Phi} = \frac{\partial \boldsymbol{\Phi}}{\partial \mathbf{z}} \cdot \delta\mathbf{z}, \qquad grad_z(\boldsymbol{\Phi}) = \frac{\partial \boldsymbol{\Phi}}{\partial \mathbf{z}},$$

$$\mathbf{n} = \frac{grad_z(\boldsymbol{\Phi})}{|grad_z(\boldsymbol{\Phi})|}, \qquad \mathbf{z} \in \mathbb{R}^{6n}, \quad \boldsymbol{\Phi} \in \mathbb{R}^{m_h}. \tag{3.2}$$

The gradient $grad_z(\boldsymbol{\Phi})$ of the constraint defines the vector **n** normal to the constraint surface (see Figure 3.1). A real or virtual displacement on this surface takes place only within the tangential plane at the point **z** under consideration [283]. This plane is given with the second and third equation of (3.2). If we succeed in finding a set of generalized coordinates, then $\boldsymbol{\Phi}(\mathbf{z},t) = \boldsymbol{\Phi}(\mathbf{z}(\mathbf{q}),t)$, which formally corresponds to a parametric representation of the constraint. In many cases, frequently also with respect to large problems, we are able to evaluate explicitly the mapping $\mathbf{z} = \mathbf{z}(\mathbf{q}, t)$, which then gives us a parametric form of the constraint $\boldsymbol{\Phi}$. In changing **z** along the minimal coordinates **q** within the constraint surface $\boldsymbol{\Phi}$ we move at least for a very small displacement along the tangent plane of $\boldsymbol{\Phi}$, where the tangent vector is given by $grad_q(\mathbf{z})$ and of course the normal and tangential vectors to this constraint must be orthogonal yielding

$$\left(\frac{\partial \boldsymbol{\Phi}}{\partial \mathbf{z}}\right)\left(\frac{\partial \mathbf{z}}{\partial \mathbf{q}}\right) = \mathbf{0}, \quad \text{with} \quad grad_z(\boldsymbol{\Phi}) = \frac{\partial \boldsymbol{\Phi}}{\partial \mathbf{z}}, \quad grad_q(\mathbf{z}) = \frac{\partial \mathbf{z}}{\partial \mathbf{q}},$$

$$\mathbf{q} \in \mathbb{R}^f, \quad grad_z(\boldsymbol{\Phi}) \in \mathbb{R}^{m_h,6n}, \quad grad_q(\mathbf{z}) \in \mathbb{R}^{6n,f}. \tag{3.3}$$

For the derivation of the equations of motion in the sense of Newton-Euler-Lagrange we must project the relevant momentum and moment of momentum equations of each body into the direction of motion, which are not blocked by the constraint equations. These directions are given with the above derivatives. Thus we find all velocities and accelerations within the tangent spaces given by the constraints and all constraint forces, or better passive forces within the normal spaces perpendicular to the tangential constraint surfaces [283]. For this purpose we consider the form of the equations (3.1), especially

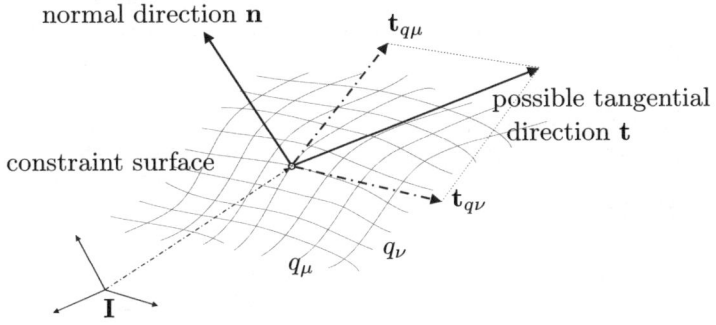

Fig. 3.1: Constraint Surface

the difference between holonomic and non-holonomic properties. For practical purposes it makes sense to distinguish in addition between integrable and non integrable cases.

In the following we shall come back to chapter (2.2.5, Kinematics of Systems) and the relations given there. We especially refer to the two equations(2.69) and (2.70), which confirm the linear relationship of the system velocities $\dot{\mathbf{z}}$ and of the generalized velocities $\dot{\mathbf{q}}$. We start with **implicit holonomic constraints** of the form

$$\mathbf{\Phi}(\mathbf{z}(\mathbf{q}),t) = \mathbf{0}, \qquad \mathbf{z} \in \mathbb{R}^{6n}, \mathbf{q} \in \mathbb{R}^{f}, \mathbf{\Phi} \in \mathbb{R}^{m_h}. \tag{3.4}$$

According to equation (2.68) we are able to express the generalized velocities $\dot{\mathbf{q}}$ by a linear relationship with regard to the $\dot{\mathbf{z}}$. The resolvable parts of the holonomic constraints will be used to eliminate some of the system coordinates and thus to reduce the number of equations of motion. The non resolvable parts form together with the equations of motion a DAE-system of equations, which must be solved numerically ([30], [53]). We may call the resulting coordinates "generalized coordinates", but as long as we are left with some remaining constraints they are not minimal coordinates.

**Implicit non-holonomic constraints** are in reality not completely implicit, because we always assume, that all non holonomic constraints are linear in the velocities. Up to now no case of non holonomic constraints became known which violate this assumption. Therefore we write

$$\mathbf{\Phi}(\mathbf{z},\dot{\mathbf{z}},t) = \mathbf{\Phi}_0(\mathbf{z},t) + \mathbf{\Phi}_1(\mathbf{z},t)\dot{\mathbf{z}} = \mathbf{0}, \qquad \mathbf{\Phi} \in \mathbb{R}^{mn}. \tag{3.5}$$

In many cases we may find a new set of generalized coordinates $\bar{\mathbf{q}}$, which may reduce the system to a certain extend. We get from the above equation

$$\mathbf{\Phi}[\mathbf{z}(\bar{\mathbf{q}}),\dot{\mathbf{z}}(\bar{\mathbf{q}},\dot{\bar{\mathbf{q}}}),t] = \mathbf{\Phi}_0[\mathbf{z}(\bar{\mathbf{q}}),t] + \mathbf{\Phi}_1[\mathbf{z}(\bar{\mathbf{q}}),t](\frac{\partial \mathbf{z}}{\partial \bar{\mathbf{q}}})\dot{\bar{\mathbf{q}}} = \mathbf{0}. \tag{3.6}$$

These equations can be resolved for the velocity $\dot{\bar{\mathbf{q}}}$ provided the matrix $\mathbf{\Phi}_1(\frac{\partial \mathbf{z}}{\partial \bar{\mathbf{q}}})$ possesses full rank indicating that all constraints are independent. Otherwise we must determine the real rank of that matrix resulting in a reduced set of constraints. Usually such a case occurs only for redundant constraints, which has to be avoided by proper modeling.

For most multibody systems we must include rotational relations appearing for example in the equations (2.31) and (2.33) with the general form

$$\dot{\mathbf{z}} = \mathbf{H}(\mathbf{q})\dot{\mathbf{q}}. \tag{3.7}$$

The matrix $\mathbf{H}(\mathbf{q})$ is invertible for many cases of spatial rotations. But in any case we have to make sure, that we have no singularities. Otherwise we better apply other parameters for the finite rotations, for example quaternions ([56], [283]). Within the framework of a large system the combination of the implicit non-holonomic constraints with the momentum and moment of momentum equations leads similar to the case of implicit holonomic constraints finally to a set of DAE-equations, which have to be solved numerically.

Considering **resolvable constraints** of any kind makes the derivation of the equations of motion much simpler. With respect to holonomic constraints positions and orientations can be represented explicitly by the minimal coordinates $\mathbf{q}$ in a form of equation (2.63). With respect to non-holonomic constraints we always have the linear relationship of equation (3.5). Anyway we come out with the important formulas of the equations (2.69) and (2.70), which we repeat here

$$\dot{\mathbf{z}} = \mathbf{J}(\mathbf{q},t) \cdot \dot{\mathbf{q}} + \mathbf{j}(\mathbf{q},t) \qquad \text{with} \qquad \mathbf{J}(\mathbf{q},t) = \frac{\partial \mathbf{z}}{\partial \mathbf{q}} = \frac{\partial \dot{\mathbf{z}}}{\partial \dot{\mathbf{q}}} \in \mathbb{R}^{6n,f},$$

$$\mathbf{W}(\mathbf{q},t) \cdot \dot{\mathbf{q}} + \mathbf{w}(\mathbf{q},t) = \mathbf{0} \quad \text{with} \quad \mathbf{W}(\mathbf{q},t) = \frac{\partial \mathbf{\Phi}}{\partial \mathbf{z}} = \frac{\partial \dot{\mathbf{\Phi}}}{\partial \dot{\mathbf{z}}}, \quad \mathbf{w}(\mathbf{q},t) = \frac{\partial \mathbf{\Phi}}{\partial t}. \tag{3.8}$$

The terms $\mathbf{j}(\mathbf{q},t)$ and $\mathbf{w}(\mathbf{q},t)$ represent given excitations coming from external or internal sources.

### 3.1.2 Unilateral Constraints

#### 3.1.2.1 Unilateral Contact Characteristics

Unilateral constraints of mechanical systems appear with contacts between rigid or elastic bodies. Contacts may be closed, and the contact partners may detach again. Within a closed contact we might have sliding or sticking, both features connected with local friction. If two bodies come into contact, they usually penetrate into each other leading to local deformations. If contacts are accompanied by tangential forces and by tangential relative velocities within the contact plane, we get in addition to normal also tangential deformations. Depending on the dynamical (or statical) environment contacts may change their state, from closure to detachment, from sliding to sticking, and vice versa. We call a contact active, if it is closed or if we have stiction, otherwise we call a contact passive. Constraints are connected with active contacts, not with passive ones.

Active contacts always exhibit contact forces which in the general case of normal and tangential deformations follow from the local material properties of the colliding bodies, from the external (with respect to the contact) dynamics (statics) and from external forces. Considering contacts that way leads to complicated problems of continuum mechanics which as a rule must be solved by numerical algorithms like FEM (finite element method) or BEM (boundary element method). For the treatment of dynamical problems this approach is too costly and in many cases also not adequate. The rigid body approach gives quicker results and applies better to large dynamical systems. Under rigid body approach we understand a contact behavior characterized at least in the local contact zone by no deformations and thus by rigid body properties. The contact process is then governed by certain contact laws like those by Newton and Poisson and appropriate extensions of them. It should be mentioned that the rigidity is assumed only locally not globally for the body under consideration, which might be deformed anyway.

A fundamental law with respect to rigid body models is the complementarity rule, sometimes called corner law or Signorinis's law. It states that in contact dynamics either relative kinematic quantities are zero and the accompanying constraint forces or constraint force combinations are not zero, or vice versa. For a closed contact the relative normal distance and normal velocity of the colliding bodies are zero and the constraint force in normal direction is not zero, or vice versa. For sticking the tangential relative velocity is zero, and the constraint force is located within the friction cone, which means that the difference of the static friction force and constraint force is not zero, or vice versa. The resulting inequalities are indispensable for an evaluation of the transitions between the various contact states.

From these properties follows a well defined indicator behavior giving the transition phases. For normal passive contacts the normal relative distance of the colliding bodies indicates the contact state. If it becomes zero the contact will be active, the indicator "relative distance" becomes a constraint accompanied by the constraint force in normal direction. The end of an active contact is then indicated by the constraint force. If it changes sign, indicating a change from pressure to tension, we get detachment, and the contact again transits to a passive state.

In the case of friction we have in the passive state a non-vanishing tangential relative velocity as an indicator. If it becomes zero there might be a transition from sliding to sticking depending on the force balance within or on the friction cone. The indicator "relative tangential velocity" then becomes a constraint leading to tangential constraint forces. The contact remains active as long as the maximum static friction force is larger than the constraint force, which means that there is a force balance within the friction cone. If this "friction reserve" or "friction surplus" becomes zero, the contact again might go into a passive state with non-zero tangential relative velocity.

If we deal with multibody systems including unilateral constraints the problem of multiple contacts and their interdependences arises. A straightforward solution of these processes would come out with a combinatorial problem of huge dimension [200]. Therefore a formulation applying complementarity rules and the resulting inequalities is a must. To not prosecuting the combinatorial process we need extended contact laws which describe unambiguously the transitions for the possible contact states and which generate only consistent contact configurations.

But even this cannot be carried through in a straightforward manner. To give an example, a relative tangential and vanishing velocity in a contact does not lead necessarily to stiction. In systems with many contacts the nearly impulsively appearing or vanishing constraint forces influence all contacts and of course also the overall system. Therefore a contact going to be active or passive might change its transitional direction due to the events in other contacts, which are coupled with the one under consideration. As a consequence we need contact laws, which describe these transitions in many contacts and their mutual influence on each nother in an unambiguous way.

In a first step we define all contact sets, which can be found in a multibody system:

$$\begin{aligned}
I_A &= \{1, 2, \ldots, n_A\} & &\text{with } n_A \text{ elements} \\
I_C(t) &= \{i \in I_A \,:\, g_{Ni} = 0\} & &\text{with } n_C(t) \text{ elements} \\
I_N(t) &= \{i \in I_C \,:\, \dot{g}_{Ni} = 0\} & &\text{with } n_N(t) \text{ elements} \\
I_T(t) &= \{i \in I_N \,:\, |\dot{\boldsymbol{g}}_{Ti}| = 0\} & &\text{with } n_T(t) \text{ elements}
\end{aligned} \quad (3.9)$$

These sets describe the kinematic state of each contact point. The set $I_A$ consists of the $n_A$ indices of all contact points. The elements of the set $I_C(t)$ are the $n_C(t)$ indices of the unilateral constraints with vanishing normal distance $g_{Ni} = 0$, but arbitrary relative velocity in the normal direction. In the index set $I_N(t)$ are the $n_N(t)$ indices of the potentially active normal constraints which fulfill the necessary conditions for continuous contact (vanishing normal distance $g_{Ni} = 0$ and no relative velocity $\dot{g}_{Ni} = 0$ in the normal direction). The index set $I_N(t)$ includes for example all contact states with slipping. The $n_T(t)$ elements of the set $I_T(t)$ are the indices of the potentially active tangential constraints. The corresponding normal constraints are closed and the relative velocities $\dot{g}_{Ti}$ in the tangential direction are zero. The numbers of elements of the index sets $I_C, I_N$ and $I_T$ are not constant but depend on time due to the variable states of constraints generated by separation-detachment-processes and by stick-slip phenomena.

### 3.1.2.2 Contact Laws and Constraints in Normal Direction

As a next step we must organize all transitions from contact to detachment and from stick to slip and the corresponding reversed transitions. In ***normal direction*** of a contact we find the following situation [200]:

- Passive contact $i$
  $g_{Ni}(\boldsymbol{g},t) \geq 0 \;\; [(\dot{g}_{Ni}, \ddot{g}_{Ni}) \neq 0] \;\; \wedge \;\; \lambda_{Ni} = 0 \;\; \Rightarrow \;\;$ indicator $g_{Ni}$,

- Transition to contact
  $g_{Ni}(\boldsymbol{g},t) = 0 \;\; \wedge \;\; \lambda_{Ni} \geq 0,$

- Active contact $i$
  $g_{Ni}(\boldsymbol{g},t) = 0 \;\; [(\dot{g}_{Ni}, \ddot{g}_{Ni}) = 0] \;\; \wedge \;\; \lambda_{Ni} > 0 \;\; \Rightarrow \;\;$ indicator $\lambda_{Ni}$
  constraint $g_{Ni} = 0$,

- Transition to detachment
  $g_{Ni}(\boldsymbol{q},t) \geq 0 \;\; [(\dot{g}_{Ni}, \ddot{g}_{Ni}) \neq 0] \;\; \wedge \;\; \lambda_{Ni} = 0.$  (3.10)

The kinematical magnitudes $g_{Ni}, \dot{g}_{Ni}, \ddot{g}_{Ni}$ are given with the equations (2.112), (2.114). We shall need the corresponding relations $\dot{g}_{Ni}, \ddot{g}_{Ni}$ when going to a velocity or an acceleration level (chapter 2.2.7.2). The constraint forces $\lambda_{Ni}$ must be compressive forces, and the normal velocity has to meet the impenetrability condition $\dot{g}_{Ni} \geq 0$. If the normal constraint force $\lambda_{Ni}$ changes sign, we get separation. The properties defined above establish a complementarity behavior which might be expressed by $n_N$ (set $I_N$) complementarity conditions (put on an acceleration level)

$$\ddot{\mathbf{g}}_N \geq \mathbf{0} \;;\; \boldsymbol{\lambda}_N \geq \mathbf{0} \;;\; \ddot{\mathbf{g}}_N^T \boldsymbol{\lambda}_N = 0 \;. \tag{3.11}$$

The variational inequality

$$-\ddot{\mathbf{g}}_N^T(\boldsymbol{\lambda}_N^* - \boldsymbol{\lambda}_N) \leq \mathbf{0} \;;\; \boldsymbol{\lambda}_N \in C_N \;;\; \forall \boldsymbol{\lambda}_N^* \in C_N \;, \tag{3.12}$$

is equivalent to the complementary conditions (3.11). The convex set

$$C_N = \{\boldsymbol{\lambda}_N^* : \boldsymbol{\lambda}_N^* \geq \mathbf{0}\} \tag{3.13}$$

contains all admissible contact forces $\lambda_{Ni}^*$ in the normal direction [200], [279].

The complementarity problem defined in (3.11) might be interpreted as a corner law which requires for each contact $\ddot{g}_{Ni} \geq 0$, $\lambda_{Ni} \geq 0$, $\ddot{g}_{Ni}\lambda_{Ni} = 0$. Figure (3.2) illustrates this property.

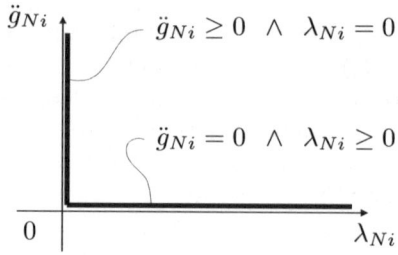

Fig. 3.2: Corner law for normal contacts (Signorini's law)

### 3.1.2.3 Contact Laws and Constraints in Tangential Direction

With respect to the **tangential direction** of a contact we shall apply Coulomb's friction law. The complementary behavior is a characteristic feature of all contact phenomena independent of the specific physical law of contact. Therefore other laws might be used as well. Furthermore we assume that within the infinitesimal small time step for a transition from stick to slip and vice versa the coefficients of static and sliding friction do not change, which may be expressed by

$$\lim_{\dot{g}_{Ti} \to 0} \mu_i(\dot{g}_{Ti}) = \mu_{0i} \tag{3.14}$$

For $\dot{g}_{Ti} \neq 0$ any friction law may be applied (see Figure 3.3 and also 2.38). With this property Coulomb's friction law distinguishes between the two cases

$$\begin{aligned}\text{sticking:} & \quad |\lambda_{Ti}| < \mu_{0i}\lambda_{Ni} \Rightarrow |\dot{\mathbf{g}}_{Ti}| = 0 \quad (\text{Set } I_T) \\ \text{sliding:} & \quad |\lambda_{Ti}| = \mu_{0i}\lambda_{Ni} \Rightarrow |\dot{\mathbf{g}}_{Ti}| > 0 \quad (\text{Set } I_N \setminus I_T)\end{aligned} \tag{3.15}$$

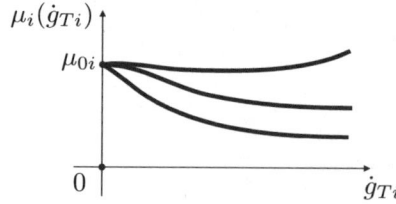

Fig. 3.3: Typical friction characteristic (Stribeck curve)

Equation (3.15) formulates the mechanical property, that we are for a frictional contact within the friction cone if the relative tangential velocity is zero and the tangential constraint force $|\lambda_{Ti}|$ is smaller than the maximum static friction force ($\mu_{0i}\lambda_{Ni}$). Then we have sticking. We are on the friction cone if we slide with $|\dot{\mathbf{g}}_{Ti}| > 0$. At a transition point the friction force is then $(\mu_{0i}\lambda_{Ni})$ (see equation 3.14). In addition we must regard the fact, that in the tangential contact plane we might get one or two directions according to a plane or a spatial contact. From this we summarize in the following way:

- Passive contact $i$ (Sliding, Set $I_N\setminus I_T$)
  $|\dot{\mathbf{g}}_{Ti}| \geq 0 \; [|\ddot{\mathbf{g}}_{Ti}| \neq 0] \quad \wedge \quad |\mu_{0i}\lambda_{Ni}| - |\lambda_{Ti}| = 0 \quad \Rightarrow \quad$ indicator $|\dot{\mathbf{g}}_{Ti}|$,

- Transition Slip to Stick
  $|\dot{\mathbf{g}}_{Ti}| = 0 \; [|\ddot{\mathbf{g}}_{Ti}| = 0] \quad \wedge \quad |\mu_{0i}\lambda_{Ni}| - |\lambda_{Ti}| \geq 0$,

- Active contact $i$ (Sticking, Set $I_T$)
  $|\dot{\mathbf{g}}_{Ti}| = 0 \; [|\ddot{\mathbf{g}}_{Ti}| = 0] \quad \wedge \quad |\mu_{0i}\lambda_{Ni}| - |\lambda_{Ti}| > 0, \quad \Rightarrow$
  $\Rightarrow \quad$ indicator $|\mu_{0i}\lambda_{Ni}| - |\lambda_{Ti}| \quad \wedge \quad$ constraint $|\dot{\mathbf{g}}_{Ti}| = 0$,

- Transition Stick to Slip
  $|\dot{\mathbf{g}}_{Ti}| \geq 0 \; [|\ddot{\mathbf{g}}_{Ti}| \neq 0] \quad \wedge \quad |\mu_{0i}\lambda_{Ni}| - |\lambda_{Ti}| = 0.$ \hfill (3.16)

We should keep in mind, that for the spatial case including two tangential directions according to Figure 3.6 the resulting friction force with respect to the two directions is the quadratic mean value, and then the condition for the friction reserve writes

$$\lambda_{Ti0} = |\mu_{0i}\lambda_{Ni}| - \sqrt{\lambda_{Ti1}^2 + \lambda_{Ti2}^2} = 0 \qquad (3.17)$$

From a numerical standpoint of view we have to check the indicator for a change of sign, which then requires a subsequent interpolation. For a transition from stick to slip one must examine the possible development of a non-zero relative tangential acceleration as a start for sliding. Newer time-stepping algorithms, though, work without such interpolations [198]. We come back to that.

Equation (3.15) put on an acceleration level can then be written in a more detailed form

$$|\lambda_{Ti}| < \mu_{0i}\lambda_{Ni} \wedge \ddot{g}_{Ti} = 0 \; (i \in I_T \; \text{sticking})$$

$$\lambda_{Ti} = +\mu_{0i}\lambda_{Ni} \wedge \ddot{g}_{Ti} \leq 0 \; (i \in I_N\setminus I_T \; \text{negative sliding}) \qquad (3.18)$$

$$\lambda_{Ti} = -\mu_{0i}\lambda_{Ni} \wedge \ddot{g}_{Ti} \geq 0 \; (i \in In\setminus I_T \; \text{positive sliding})$$

This contact law may be represented by a double corner law as indicated in Figure (3.4). To transform the law (3.18) for tangential constraints into a complementarity condition we must decompose the double corner into single ones. A decomposition into four elementary laws is given in [86], a decomposition into two elements in [226] and [87], which we shall discuss shortly.

According to [87] each non-smooth characteristic can be composed by "unilateral primitives", which are nothing else but simple rectangular hooks. Decomposing Figure (3.4) simply results in two such hooks as shown in Figure

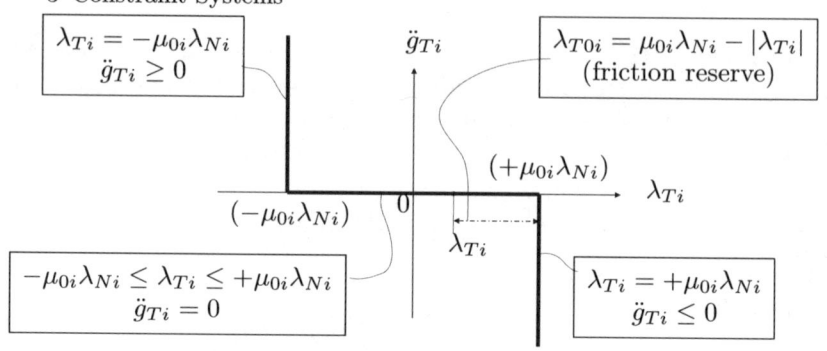

Fig. 3.4: Corner law for tangential constraints

(3.5). Introducing the force law for tangential contacts as a set $\mathcal{T}$, the positive hook as a set $\mathcal{T}^+$ and the negative hook as a set $\mathcal{T}^-$ we can represent the set $\mathcal{T}$ as the intersection of the two sets $\mathcal{T}^+$ and $\mathcal{T}^-$, which is immediately clear from Figure (3.5) namely $\mathcal{T} \in \mathcal{T}^+ \cap \mathcal{T}^-$. The two hooks as the two intersecting sets of the tangential force law can again easily represented by complementarity inequalities, which write

$$(\lambda_{Ti} + \mu_{0i}\lambda_{Ni}) \geq 0, \quad (+\ddot{g}_{Ti}) \geq 0, \quad (\lambda_{Ti} + \mu_{0i}\lambda_{Ni})\ddot{g}_{Ti} = 0, \quad \text{Set } \mathcal{T}^+$$
$$(\mu_{0i}\lambda_{Ni} - \lambda_{Ti}) \geq 0, \quad (-\ddot{g}_{Ti}) \geq 0, \quad (\mu_{0i}\lambda_{Ni} - \lambda_{Ti})\ddot{g}_{Ti} = 0, \quad \text{Set } \mathcal{T}^-$$
$$\mathcal{T} \in \mathcal{T}^+ \cap \mathcal{T}^-. \tag{3.19}$$

Decompositions of that type are always possible, at least for technically relevant force laws, which allows us to reduce such problems to a complementarity formulation as a basis for all further evaluations. For plane contacts we come out with a linear complementarity problem, for spatial contacts with two tangential directions we get a nonlinear complementarity problem, which we shall not treat here (see [87], [279]). For plane contacts we get

$$\mathbf{y} = \mathbf{Ax} + \mathbf{b}, \quad \mathbf{y} \geq \mathbf{0}, \quad \mathbf{x} \geq \mathbf{0}, \quad \mathbf{y}^T\mathbf{x} = \mathbf{0}, \qquad \mathbf{y}, \mathbf{x} \in \mathbb{R}^{n*} \tag{3.20}$$

where $n^* = n_N + 2n_T$ for a decomposition into two unilateral primitives as depicted in Figure (3.5). The quantity $\mathbf{x}$ includes the contact forces and one part of the decomposed accelerations, the quantity $\mathbf{y}$ the relative accelerations and in addition the friction reserve defined as the difference of static friction force and tangential constraint force ($\mu_{0i}\lambda_{Ni} - |\lambda_{Ti}|$). Equation (3.20) describes a linear complementarity problem thus being adequate for plane contacts.

Similar as in the normal case we can represent the contact law eq. (3.20) by a variational inequality of the form

$$\ddot{\mathbf{g}}_{Ti}^T(\lambda_{Ti}^* - \lambda_{Ti}) \geq 0 \; ; \; \lambda_{Ti} \in C_{Ti}, \; \forall \lambda_{Ti}^* \in C_{Ti}. \tag{3.21}$$

The convex set $C_{Ti}$ contains all admissible contact forces $\lambda_{Ti}^*$ in tangential direction

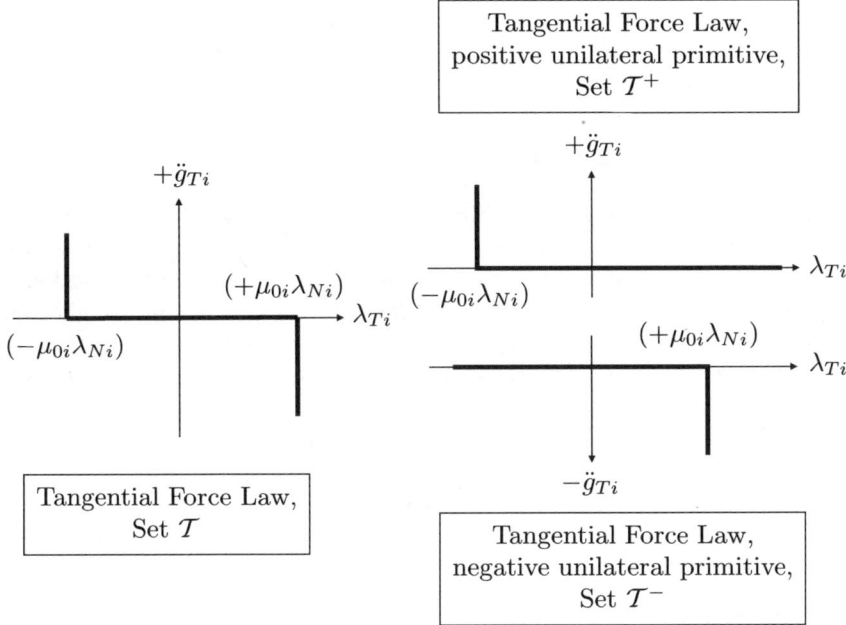

Fig. 3.5: Decomposition of the Tangential Force Law into Unilateral Primitives

$$C_{Ti} = \{\lambda_{Ti}^* | |\lambda_{Ti}| \leq \mu_0 \lambda_{Ni} \; ; \; \forall i \in I_T\} \tag{3.22}$$

#### 3.1.2.4 Convex Mathematical Forms

The theory of rigid body contacts is strongly related to the mathematical theory of convex analysis, see for example ([161], [225], [163], [87], [135], [279]). The above presentation using complementarities follows closely physical arguments and can be interpreted quite easily. Using the mathematical formalism of convex analysis still allows some physical interpretation, but in a much more sophisticated and not in a direct way. Convex analysis has significantly influenced the corresponding theories establishing a rigorous and, at the time being a nearly complete, theoretical fundament in this area. Therefore it makes sense to present the alternative forms to the complementarity problem.

Considering first normal contacts we get (see also Figure (3.6)) five equivalent forms, which we shall explain in more detail together with the corresponding forms of the tangential constraints. For the normal case we come out with

$$\ddot{g}_N \geq \mathbf{0} \ ; \ \lambda_N \geq \mathbf{0} \ ; \ \ddot{g}_N^T \lambda_N = 0,$$
$$-\ddot{g}_N \in \partial \Psi_{C_N}(\lambda_N),$$
$$\lambda_N \in \partial \Psi^*_{C_N}(-\ddot{g}_N),$$
$$\ddot{g}_N^T(\lambda_N^* - \lambda_N) \geq 0, \quad \lambda_N \geq 0, \quad \forall \lambda_N^* \geq 0,$$
$$\lambda_N^T(\ddot{g}_N^* - \ddot{g}_N) \geq 0, \quad \ddot{g}_N \geq 0, \quad \forall \ddot{g}_N^* \geq 0. \tag{3.23}$$

For the tangential case we start again with equation (3.22), which defines all tangential relative velocities in a contact (i) being possible within or on the friction cone (see for example Figure (3.6)). The admissible friction forces $\lambda_{Ti}^*$ are either within the cone for sticking, or they on the cone for sliding. The comparable relation for the normal contact situation is given with equation (3.13) saying that the normal contact force is positive for contact and zero for detachment. The contact forces in normal and in tangential directions form convex sets.

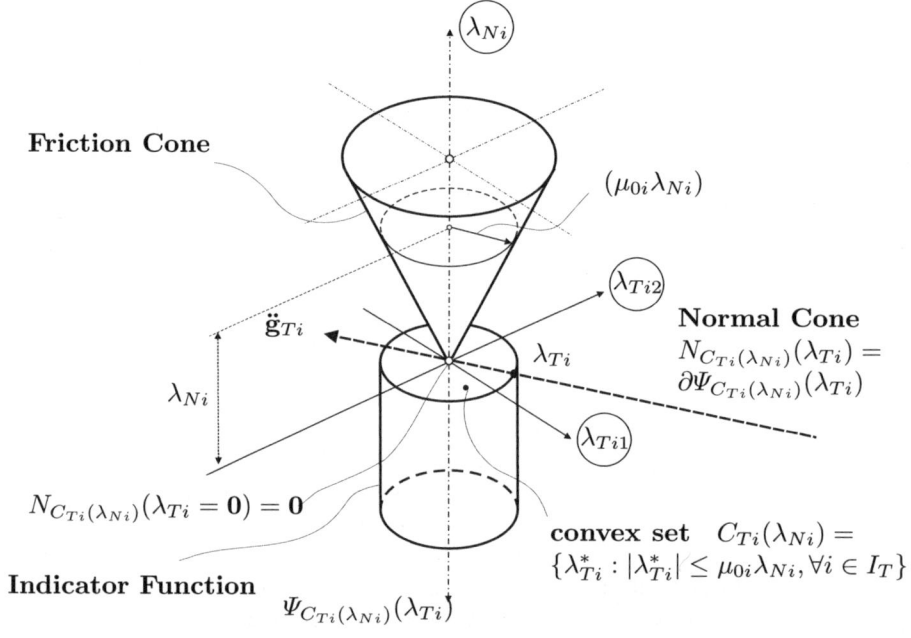

Fig. 3.6: Friction Cone, Normal Cone and Indicator Function for a Contact (i) [279]

The normal cone to a convex set are all vectors which are normal to this set, which makes sense only at the boundary of the set, where these vectors are perpendicular to the boundary. If the boundary possesses a kink, then the normal cone is limited by the two normal directions left and right of the kink thus forming a wedge (Figure 3.7). The normal cone will be zero in the interior

## 3.1 Constraints and Contacts

of the set. Applied to the contact problem we have the normal cone to the convex normal set of equation (3.13) as any straight line perpendicular to the abcsissa with $\lambda_{Ni} \geq 0$, and the normal cone to the convex tangential set of equation (3.22) are all straight lines perpendicular to the circular projection of the friction cone cross section at the height $\lambda = \lambda_{Ni}$. Figure (3.6) gives an impression of these properties, which up to now are nothing else than an alternative formulation of the equations (3.10) and (3.16) in terms of set-oriented mathematics. The normal cone is generally defined by

$$N_C(\mathbf{x}) = \left\{ \mathbf{y} | \mathbf{y}^T(\mathbf{x}^* - \mathbf{x}) \leq 0, \quad \mathbf{x} \in C, \quad \forall \mathbf{x}^* \in C \right\}, \tag{3.24}$$

where we might replace $(\mathbf{x}, \mathbf{y})$ by the magnitudes $[(\lambda_N/\lambda_T), (\dot{\mathbf{g}}_N/\dot{\mathbf{g}}_T)]$.

Figure (3.6) illustrates in addition to the friction cone, the convex tangential set and the normal cone also the indicator function $\Psi_{C_{Ti}(\lambda_{Ni})}(\lambda_{Ti})$, which is defined in [135] and which has nothing to do with the simple indicators as used in the relations (3.10) and (3.16). We also know that the normal cone is identical with the subdifferential of the indicator function. These relations explain the equations (3.23), and they allow us to formulate the tangential force law also in various ways [279]

$$\begin{aligned}
\lambda_T^T \ddot{\mathbf{g}}_T &= -\Psi^*_{C_T}(-\ddot{\mathbf{g}}_T), \quad \lambda_T \in C_T, \\
-\ddot{\mathbf{g}}_T &\in \partial \Psi_{C_T}(\lambda_T), \\
\lambda_T &\in \partial \Psi^*_{C_T}(-\ddot{\mathbf{g}}_T), \\
\ddot{\mathbf{g}}_T^T(\lambda_T^* - \lambda_T) &\geq 0, \quad \lambda_T \in C_T, \quad \forall \lambda_T^* \in C_T, \\
\lambda_T^T(\ddot{\mathbf{g}}_T^* - \ddot{\mathbf{g}}_T) &\geq \Psi^*_{C_T}(-\ddot{\mathbf{g}}_T) - \Psi^*_{C_T}(-\ddot{\mathbf{g}}_T^*), \quad \forall \ddot{\mathbf{g}}_T^*.
\end{aligned} \tag{3.25}$$

In the last 15 years the representations of the equations (3.23) and (3.25) have been replaced by the "proximal point" relation, which includes the above set- and cone-definitions but which turns out to be extremely useful and efficient for the numerical solution of non-smooth problems. We come back to the normal cone defition and use it to consider the proximal point to a convex set. Figure (3.7) illustrates the various definitions. The proximal point of a convex set C to a point $\mathbf{z}$ is the closest point in C to $\mathbf{z}$. A point outside the convex set can be characterized by a distance function which is the shortest distance from the point to the boundary. The following rules are helpful [135]:

$$\text{prox}_C(\mathbf{z}) = \underset{\mathbf{x}^* \in C}{\arg\min} \| \mathbf{z} - \mathbf{x}^* \|, \quad \mathbf{z} \in \mathbb{R}^n,$$

$$\mathbf{x} = \text{prox}_C(\mathbf{z}) \Rightarrow \begin{cases} \mathbf{x} = \mathbf{z} & \text{for } \mathbf{z} \in C, \\ \mathbf{x} \in \text{boundary } C & \text{for } \mathbf{z} \notin C, \end{cases}$$

$$\mathbf{x} = \text{prox}_C(\mathbf{z}) \Leftrightarrow \begin{cases} \mathbf{z} - \mathbf{x} \in \partial \Psi_C(\mathbf{x}), \\ \mathbf{z} - \mathbf{x} \in N_C(\mathbf{x}), \end{cases}$$

$$\text{dist}_C(\mathbf{z}) = \| \mathbf{z} - \text{prox}_C(\mathbf{z}) \|. \tag{3.26}$$

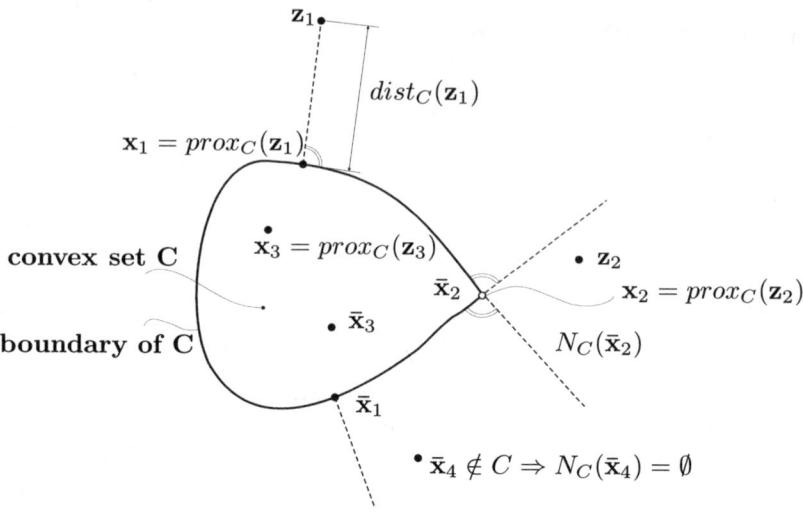

Fig. 3.7: Normal Cone, Proximal Point and Distance Function [135]

Figure (3.7) suggests to us the following explanations. For all points within the convex set C the prox-function defines an identity ($\mathbf{x} = \mathbf{z}$). For all points not in the set C the prox-function puts the point from its external position to the boundary of the set C, in fact following the way indicated in Figure (3.7), where the distance to the boundary is perpendicular to it and given by the distance-function $dist_C(\mathbf{z})$. According to the definitions of the normal cone $N_C(\mathbf{x})$ and the subdifferential $\partial \Psi_C(\mathbf{x})$ we get a one-to-one correspondence of the prox-function with these structures, which again is clear from Figures (3.7) and (3.6). The distance function represents a measure for the closest distance to the boundary of C.

The normal vectors $\mathbf{y}$ perpendicular to the points $\mathbf{x} \in C$ form the normal cone. If we consider equation (3.26) and substitute $\mathbf{z} = \mathbf{x} - r\mathbf{y}$ we get $[(-\mathbf{y} \in N_C(\mathbf{x})) \Leftrightarrow (-r\mathbf{y} \in N_C(\mathbf{x}))]$ for $r > 0$. From this we can establish an important relation for the orthogonal vectors $\mathbf{x}$ and $\mathbf{y}$ in the form

$$[\mathbf{x} = \text{prox}_C(\mathbf{x} - r\mathbf{y}), \quad \text{for } r > 0] \quad \Leftrightarrow \quad [-\mathbf{y} \in N_C(\mathbf{x})]. \tag{3.27}$$

Looking at $\mathbf{x}$ in the sense of constraint forces in normal or tangential directions and at $\mathbf{y}$ in the sense of relative accelerations in these three contact directions we are able to express the laws for the contact forces in terms of prox-functions, which express the same properties as the more "physical" equations (3.10) and (3.16), but which replace the inequalities coming from the complementarity conditions of the equations (3.11), (3.15).

This approach, called the Augmented Lagrangian Method, was first introduced by Hestenes [99] and Powell [216] to solve linear programming problems with equality constraints. Alart [2] applied the method to quasistatic mechanical systems, and Leine [136] used the Augmented Lagrangian approach to simulate the dynamics of the Tippe-Top.

The proximal point to a convex set $C$ as discussed above returns the closest point in $C$ to its argument. With this definition we can express the contact law in normal direction by the equation

$$\lambda_N = \text{prox}_{C_N}(\lambda_N - r\ddot{g}_N), \tag{3.28}$$

where $C_N$ denotes the set of admissible normal contact forces

$$C_N = \{\lambda_N | \lambda_N \geq 0\}. \tag{3.29}$$

In the same manner we can put Coulomb's friction law (3.15) into the form

$$\lambda_T = \text{prox}_{C_T(\lambda_\mathbf{N})}(\lambda_T - r\ddot{g}_T), \tag{3.30}$$

with $C_T$ denoting the set of admissible tangential forces

$$C_T(\lambda_\mathbf{N}) = \{\lambda_T | |\lambda_{Ti}| \leq \mu_i \lambda_{Ni}\}. \tag{3.31}$$

The arbitrary auxiliary parameter $r > 0$ represents the slope of the regularizing function in (3.28) and (3.30).

Considering the classical definitions of (3.1) the above equations and inequalities indicate that with vanishing relative distance in normal direction we get a holonomic constraint. In tangential direction the transition from sticking to sliding is accompanied by a velocity constraint in terms of the relative tangential velocity and thus a non-holonomic constraint in a classical sense. The difference to classical bilateral constraints consists in the fact that the duration of these constraints depend on the overall dynamics of the system. They might appear and disappear during very short time intervals, but nevertheless influencing all other contacts with their active or passive states and their existing or not existing constraints. During these processes the structure of the equations of motion change due to the variable number of degrees of freedom depending on the contact configuration (see equations 3.9).

## 3.2 Principles

### 3.2.1 Introduction

We have seen and we shall see that in more details in the next chapters, that the triad momentum equation, moment of momentum equation and a set of suitable constraints allows to establish the equations of motion. The principles of mechanics offer another possibility to evaluate these equations, though in most cases the methods connected with these principles are better suited for small systems. For large systems the so-called Newton-Euler-equations represent a more adequate basis. We distinguish between differential principles and minimal principles ([180], [27]). The principles of Hamilton and of Gauss may be considered both ways. The idea of minimal principles is closely related to the origin of Euler's variational calculus, which allows in many cases the representation of differential equations by a direct optimization problem. Instead of solving the differential equations one may solve directly the optimization problem, a method, which is often used in continuum mechanics and in all fields of physics.

In the following we shall consider a selection of principles, which are also used in applications of practical relevancy. Beyond such aspects of utilitarianism principles offer a deep insight and understanding of the mechanical fundamentals. We shall not consider these principles in connection with non-smooth dynamical properties, because we prefer more the straightforward way using the moment and moment of momentum equations together with the appropriate constraints for a derivation of the equations of motion [200]. Readers interested in that topic should have a look into the books [87] and [177].

### 3.2.2 Principle of d'Alembert and Lagrange

We apply Newton's second law in the form of Lagrange to the mass element of Figure (3.8), which might be constrained by some surface $\Phi(\mathbf{r}_c) = 0$. We get

$$\ddot{\mathbf{r}} dm = d\mathbf{F} = d\mathbf{F}_a + d\mathbf{F}_p, \tag{3.32}$$

the forces $d\mathbf{F}$ being subdivided into active (applied) forces $d\mathbf{F}_a$ and into passive (constraint) forces $d\mathbf{F}_p$. An integration of this equation over the whole body B yields

$$\int_B \ddot{\mathbf{r}} dm = \int_B d\mathbf{F} = \int_B (d\mathbf{F}_a + d\mathbf{F}_p). \tag{3.33}$$

The subdivision into applied and constraint forces follows an often used classical concept. In many cases, though, it might be more convenient to use the concept of active and passive forces (see chapter 2.1.2). The magnitude $d\mathbf{F}_a$ is an active force, and the magnitude $d\mathbf{F}_p$ a passive one. For contact processes we should keep in mind, however, that passive forces may become active ones or vice versa, for example for transitions stick to slip or slip to stick in the tangential direction of a contact. As already mentioned active forces generate work and power, passive forces not. For simplicity we shall consider the force $d\mathbf{F}_p$ as a passive constraint force, which means with respect to Figure (3.8) the component perpendicular to the constraint surface. The tangential component, if any, we would have to add to the applied force $d\mathbf{F}_a$.

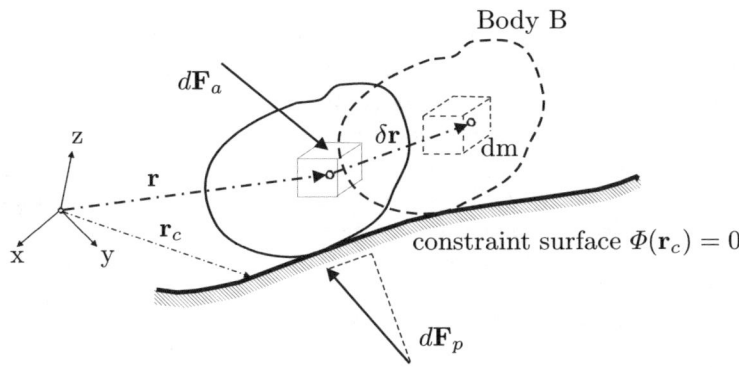

Fig. 3.8: Virtual displacement of a mass element

As indicated in Figure(3.8) we assume that the mass element dm will be shifted by a virtual displacement $\delta\mathbf{r}$. Such a virtual displacement will produce virtual work, and we get together with equation(3.32) for the mass element

$$(d\mathbf{F} - \ddot{\mathbf{r}}dm)^T \delta\mathbf{r} = 0$$
$$(d\mathbf{F}_a - \ddot{\mathbf{r}}dm)^T \delta\mathbf{r} = -(d\mathbf{F}_p)^T \delta\mathbf{r} = -\delta W_p. \tag{3.34}$$

We have assumed in Figure (3.8) a holonomic constraint $\Phi(\mathbf{r}_c) = 0$, which makes the derivations a bit more transparent. We could have chosen of course a more complicated constraint (see for example [180]). Anyway, the constraint does not allow any kind of motion, also not any kind of virtual displacement, but only such displacements which are compatible with

$$\delta\Phi = (\frac{\partial\Phi}{\partial\mathbf{r}})\delta\mathbf{r} = (grad\Phi)\delta\mathbf{r} = \mathbf{0} \quad \text{with} \quad \mathbf{r} \in \mathbb{R}^6, \ \Phi \in \mathbb{R}^m. \tag{3.35}$$

A possible geometric interpretation is obvious. The m constraints $\Phi(\mathbf{r})$ span within the system space of $\mathbf{r}$ altogether m constraint surfaces, the surface normals of which are proportional to $(grad\Phi)$. The vanishing scalar product $[(grad\Phi)\delta\mathbf{r} = \mathbf{0}]$ implies that

$$(grad\Phi) \perp \delta\mathbf{r}. \tag{3.36}$$

The normal vector on this constraint surface is $\mathbf{n} = (grad\Phi)/|(grad\Phi)|$, which together with the two relations (3.34) and (3.36) give the comparison

$$(grad\Phi) \perp \delta\mathbf{r}, \quad \Rightarrow \quad \mathbf{n} \perp \delta\mathbf{r}, \quad \Rightarrow \quad d\mathbf{F}_p \perp \delta\mathbf{r}, \tag{3.37}$$

from which we conclude, that the normal vector $\mathbf{n}$ and the passive force $d\mathbf{F}_p$ are perpendicular to the virtual displacement $\delta\mathbf{r}$ indicating the following: The passive forces have the same direction as the normal vector to the constraint surface, hence, the passive forces are always perpendicular to the constraint surface, and the motion, virtual or real, can only take place within these surfaces. With respect to a certain point, motion takes place within the tangent plane of this point given by equation (3.35) (see also Figure 3.1). This confirms our earlier statement: motion takes place on the constraint surfaces, in the tangent spaces, all passive constraint forces are assembled within the normal spaces.

From these arguments we conclude that the mechanical magnitudes as components of the tangent spaces must be always orthogonal to those of the normal spaces, or more concrete, passive constraint forces are orthogonal to displacements on the tangent plane in a point of the constraint surface. The above arguments form the main basis for the principle of d'Alembert in the setting of Lagrange ([187], [27], [93]):

*passive (constraint) forces do no work:* $\quad \displaystyle\int_B (d\mathbf{F}_p)^T \delta\mathbf{r} = 0.$ (3.38)

Constraint forces are alway passive forces, and they are "lost forces", according to a statement of Daniel Bernoulli, "lost forces" in a sense that they do not contribute to the motion of a mechanical system. But these lost constraint forces keep things together telling the motion where to go. The motion itself will be realized only by the active forces. Therefore we get from the equations (3.33) and (3.34)

$$\int_B (\ddot{\mathbf{r}} dm - d\mathbf{F}_a)^T \delta\mathbf{r} = 0 \tag{3.39}$$

For the static case ($\ddot{\mathbf{r}} = \mathbf{0}$) this equation includes the principle of virtual work ([258], [147])

$$\int_B (d\mathbf{F})^T \delta\mathbf{r} = \delta W = 0 \tag{3.40}$$

Many textbooks of mechanics, especially dynamics, give a nice explanation of the above equations, which possess the character of an axiom. This explanation will be discussed in the following, because it allows a better understanding of constrained dynamics. The terms of the equation

$$\int_B (\ddot{\mathbf{r}} dm - d\mathbf{F}_a) = \int_B d\mathbf{F}_p \tag{3.41}$$

can be split up into components perpendicular ($\perp$) to the constraint surface $\Phi(\mathbf{r})$ and parallel ($\|$) to it, which yields

$$\int_B (\ddot{\mathbf{r}} dm - d\mathbf{F}_a)_\perp + \int_B (\ddot{\mathbf{r}} dm - d\mathbf{F}_a)_\| = \int_B (d\mathbf{F}_p)_\perp + \int_B (d\mathbf{F}_p)_\| \tag{3.42}$$

We have seen that the motion can take place only on the constraint surface $\Phi(\mathbf{r}_c) = 0$, and therefore the components in the two directions must vanish, each one separately. Thus we get

$$\int_B (\ddot{\mathbf{r}} dm - d\mathbf{F}_a)_\| = \int_B (d\mathbf{F}_p)_\| = 0 \quad \text{and} \quad \int_B (\ddot{\mathbf{r}} dm - d\mathbf{F}_a - d\mathbf{F}_p)_\perp = 0. \tag{3.43}$$

This illustrates the fact, that firstly only applied and active forces on the constraint surface contribute to the acceleration and thus to the motion, and that secondly the normal components of all applied forces are passive and in equilibrium with the inertia forces and the constraint forces coming from the external dynamics.

### 3.2.3 Principle of Jourdain and Gauss

We consider again d'Alemberts's principle (3.39) and derive it two times with respect to time, which results in a first step in [180], [27]

$$\int_B \frac{d}{dt}(\ddot{\mathbf{r}} dm - d\mathbf{F}_a)^T \cdot \delta\mathbf{r} + \int_B (\ddot{\mathbf{r}} dm - d\mathbf{F}_a)^T \cdot \frac{d}{dt}(\delta\mathbf{r}) = 0, \tag{3.44}$$

and in a second step we come out with

$$\int_B \frac{d^2}{dt^2}(\ddot{\mathbf{r}} dm - d\mathbf{F}_a)^T \cdot \delta\mathbf{r} + 2 \int_B \frac{d}{dt}(\ddot{\mathbf{r}} dm - d\mathbf{F}_a)^T \cdot \frac{d}{dt}(\delta\mathbf{r})$$
$$+ \int_B (\ddot{\mathbf{r}} dm - d\mathbf{F}_a)^T \cdot \frac{d^2}{dt^2}(\delta\mathbf{r}) = 0. \tag{3.45}$$

According to [180] and [27] we introduce the variations of Jourdain and of Gauss consisting for the Jourdain variation in

$$\delta'\mathbf{r} = \mathbf{0}, \quad \delta'\dot{\mathbf{r}} \neq \mathbf{0}, \quad \delta't = 0, \quad \Rightarrow \quad \delta'\dot{\boldsymbol{\Phi}} = \frac{\partial \dot{\boldsymbol{\Phi}}}{\partial \dot{\mathbf{r}}} \delta'\dot{\mathbf{r}} = \mathbf{0}, \tag{3.46}$$

and for the Gauss variation in

$$\delta''\mathbf{r} = \mathbf{0}, \quad \delta''\dot{\mathbf{r}} = \mathbf{0}, \quad \delta''\ddot{\mathbf{r}} \neq \mathbf{0}, \quad \delta''t = 0, \quad \Rightarrow \quad \delta''\dddot{\mathbf{\Phi}} = \frac{\partial \dddot{\mathbf{\Phi}}}{\partial \ddot{\mathbf{r}}}\delta''\ddot{\mathbf{r}} = \mathbf{0}, \quad (3.47)$$

which could be continued to higher order time derivatives [27], [180]. Applying the Jourdain variation (3.46) to the relation (3.44) we come out with Jourdain's principle in the form [119] [50]

$$\text{passive (constraint) forces generate no power:} \quad \int_B (\ddot{\mathbf{r}}dm - d\mathbf{F}_a)^T \delta'\dot{\mathbf{r}} = 0,$$

(3.48)

which allows an interpretation. We have seen, that the motion has to take place within the local tangent planes of the constraint surfaces, and that the constraint forces as passive forces are perpendicular to them. This property is as a matter of fact not only true for displacements but for any kind of motion, which means also for velocities and for accelerations. Therefore Jourdain's principle tells us also, that passive forces and velocities, again virtual or real, are perpendicular to each other and hence cannot generate any power. Today in most problems of multibody dynamics Jourdain's principle is used, because it is more general and flexible with respect to holonomic as well as to non-holonomic constraints.

Combining in a further step the relations (3.47) and (3.45) we get Gauss principle in the form [50]

$$\text{motion follows trajectories of least constraints:} \quad \int_B (\ddot{\mathbf{r}}dm - d\mathbf{F}_a)^T \delta''\ddot{\mathbf{r}} = 0.$$

(3.49)

The same arguments as above apply of course also to the Gauss principle, but the product of pasive forces and acceleration does not correspond to any physical magnitude, which makes sense. Originally Gauss started with the idea of least constraints resulting directly from the concept of his least square method. The basic idea behind the principle anticipates a motion under the influence of constraints, which will be as near as possible to the free motion without constraints, which as a consequence requires a minimization of the constraints in the sense of Gauss' least square concept. Applying that to the relation (3.41) we may demand, that

$$\int_B (d\mathbf{F}_p)^T (d\mathbf{F}_p) = \int_B (\ddot{\mathbf{r}}dm - d\mathbf{F}_a)^T (\ddot{\mathbf{r}}dm - d\mathbf{F}_a) \quad \Rightarrow \quad \text{min!}, \quad (3.50)$$

which explains also the name of "least constraints". This is an optimization problem, for which the accelerations must be evaluated in such a way that the action of constraints becomes minimal. Applying the Gauss variation equation (3.47) results in

$$\delta'' \int_B (\ddot{\mathbf{r}}dm - d\mathbf{F}_a)^T (\ddot{\mathbf{r}}dm - d\mathbf{F}_a) = 0, \tag{3.51}$$

we recognize that the relation (3.49) represents the variation of equation (3.51) with respect to the acceleration $\ddot{\mathbf{r}}$, the above defined Gauss-variation. Gauss himself explained his principle in the following way [50]: *"The motion of a system of particles connected together in any way, and the motions of which are subject to arbitrary external restrictions, always takes place in the most complete agreement with the free motion or under the weakest possible constraint."*

In manipulating the above equations we have tacitly assumed that the expression $(d\delta \mathbf{r} - \delta d\mathbf{r})$ will be zero, which means $(d\delta \mathbf{r} = \delta d\mathbf{r})$. These relations hold only for generalized coordinates $\mathbf{q}$ and for the system coordinates $\mathbf{r}$ or $\mathbf{z}$. They do not hold for quasi-coordinates related to non-integrable minimal velocities (see for example [27], [180]). As we do not deal with the concept of quasi-coordinates, we can stay with the above equalities.

### 3.2.4 Lagrange's Equations

#### 3.2.4.1 Lagrange's Equations of First Kind

We have seen that constrained mechanical systems can be described by a set of momentum and moment of momentum equations together with a set of constraints. If we are able to find a complete set of generalized coordinates the equations of motion can be reduced to a set with f coordinates $\mathbf{q} \in \mathbb{R}^f$, which correspond to the number of degrees of freedom. If we find only some generalized coordinates we are left with a remaining set of kinematic constraints. We then have to solve some set of differential algebraic equations (DAE). Anyway, motion will take place within the tangent planes of the constraints.

But such a consideration does not provide us with the constraint forces or torques, which we must evaluate from the normal spaces, which means from some magnitudes perpendicular to the constraint surfaces. These forces and torques are necessary for design purposes, because they determine the size if bearings, guideways, connections and the like. We have two possibilities to evaluate such constraint forces. The first one consists in a set of system equations with all constraints, which are not used to reduce the equations of motion, but which are used in connection with a DAE-processor to calculate the constraint forces. Another more analytical method is given by the so-called Lagrange's equations of first kind, which we shall consider in the following. We shall see, that they can be used only under certain conditions. We assume also in the following that the equations are given in a suitable coordinate system.

Starting again with the equations (3.94) and considering a body "i" as a component of a multibody system we get

$$\dot{\mathbf{r}}_i = (\dot{\mathbf{r}}_Q + \tilde{\omega} \cdot \mathbf{r})_i = (\dot{\mathbf{r}}_Q + \tilde{\mathbf{r}}^T \cdot \omega)_i$$

$$\delta \dot{\mathbf{r}}_i = (\mathbf{E}_3 \quad \tilde{\mathbf{r}}^T)_i \begin{pmatrix} \delta \dot{\mathbf{r}}_Q \\ \delta \omega \end{pmatrix}_i \quad (3.52)$$

Applying Jourdain's principle, equation (3.48), for a system of n bodies yields

$$\sum_{i=1}^{n} \int_{B_i} \delta' \dot{\mathbf{r}}_i^T (\ddot{\mathbf{r}}_i dm_i - d\mathbf{F}_{ai} - d\mathbf{F}_{pi}) = 0, \quad (3.53)$$

which can be brought into the form

$$\sum_{i=1}^{n} \int_{B_i} \begin{pmatrix} \delta \dot{\mathbf{r}}_Q \\ \delta \omega \end{pmatrix}_i^T \begin{pmatrix} \mathbf{E}_3 \\ \tilde{\mathbf{r}} \end{pmatrix}_i ([\ddot{\mathbf{r}}_Q + (\dot{\tilde{\omega}}\mathbf{r} + \tilde{\omega}\tilde{\omega})\mathbf{r}]_i dm_i - d\mathbf{F}_{ai} - d\mathbf{F}_{pi}) = \mathbf{0}, \quad (3.54)$$

Using the abbreviations equations (2.171) we get analogous to (2.172)

$$\sum_{i=1}^{n} \begin{pmatrix} \delta \dot{\mathbf{r}}_Q \\ \delta \omega \end{pmatrix}_i^T \begin{pmatrix} m\ddot{\mathbf{r}}_Q + m\tilde{\mathbf{r}}_S^T \dot{\omega} + m\tilde{\omega}\tilde{\omega}\mathbf{r}_S - \mathbf{F}_a - \mathbf{F}_p \\ m\mathbf{r}_S \ddot{\mathbf{r}}_Q + \mathbf{I}\dot{\omega} + \tilde{\omega}\mathbf{I}\omega - \mathbf{M}_a - \mathbf{M}_p \end{pmatrix}_i = \mathbf{0} \quad (3.55)$$

or in a more systematic representation

$$\sum_{i=1}^{n} \delta \begin{pmatrix} \dot{\mathbf{r}}_Q \\ \omega \end{pmatrix}_i^T \left\{ \begin{pmatrix} m\mathbf{E}_3 & m\tilde{\mathbf{r}}_S^T \\ m\tilde{\mathbf{r}}_S & \mathbf{I} \end{pmatrix} \cdot \begin{pmatrix} \ddot{\mathbf{r}}_Q \\ \dot{\omega} \end{pmatrix} + \begin{pmatrix} m\tilde{\omega}\tilde{\omega}\mathbf{r}_S \\ \tilde{\omega}\mathbf{I}\omega \end{pmatrix} \right.$$

$$\left. - \begin{pmatrix} \mathbf{F}_a \\ \mathbf{M}_a \end{pmatrix} - \begin{pmatrix} \mathbf{F}_p \\ \mathbf{M}_p \end{pmatrix} \right\}_i = \mathbf{0}$$

$$\sum_{i=1}^{n} \delta \dot{\mathbf{z}}_i^T \{ \mathbf{M}_i \ddot{\mathbf{z}}_i + \mathbf{h}_i - \mathbf{f}_{ai} - \mathbf{f}_{pi} \} = \mathbf{0}. \quad (3.56)$$

The abbreviations used are obvious. The magnitudes of the second equation are defined in a space $\mathbb{R}^6$. We summarize them to

$$\mathbf{z} := (\mathbf{z}_1^T, \mathbf{z}_2^T, \ldots \mathbf{z}_n^T)^T \in \mathbb{R}^{6n}, \qquad \mathbf{h} := (\mathbf{h}_1^T, \mathbf{h}_2^T, \ldots \mathbf{h}_n^T)^T \in \mathbb{R}^{6n},$$
$$\mathbf{f}_a := (\mathbf{f}_{a1}^T, \mathbf{f}_{a2}^T, \ldots \mathbf{f}_{an}^T)^T \in \mathbb{R}^{6n}, \qquad \mathbf{f}_p := (\mathbf{f}_{p1}^T, \mathbf{f}_{p2}^T, \ldots \mathbf{f}_{pn}^T)^T \in \mathbb{R}^{6n},$$
$$\mathbf{M} := diag(\mathbf{M}_i) \in \mathbb{R}^{6n,6n} \quad (3.57)$$

With these definitions the last equation of (3.56) reduces to

$$\delta \dot{\mathbf{z}}^T (\mathbf{M}\ddot{\mathbf{z}} + \mathbf{h} - \mathbf{f}_a - \mathbf{f}_p) = \mathbf{0}. \quad (3.58)$$

The virtual velocities cannot be chosen freely but only in agreement with the constraints of the equations (3.1), for example

$$\mathbf{\Phi}(\mathbf{z}, \dot{\mathbf{z}}, t) = \mathbf{\Phi}_0(\mathbf{z}, t) + \mathbf{\Phi}_1(\mathbf{z}, t)\dot{\mathbf{z}} = \mathbf{0} \quad \in \mathbb{R}^m, \quad (3.59)$$

where we have assumed a linear relationship with respect to the velocities according to equation (3.5). Setting for the moment $\delta \mathbf{z} = \mathbf{0}$ and $\delta \dot{\mathbf{z}} \neq \mathbf{0}$ the virtual variation of this constraint comes out with

$$\delta \mathbf{\Phi} = \mathbf{\Phi}_1(\mathbf{z},t) \delta \dot{\mathbf{z}} = \mathbf{0} \tag{3.60}$$

with the Jacobian $\mathbf{\Phi}_1(\mathbf{z},t) \in \mathbb{R}^{m,6n}$.

For the derivation of the equations of motion of the multibody system we apply Jourdain's principle (eq. 3.48), which requires that

$$\delta \dot{\mathbf{z}}^T \mathbf{f}_p = \mathbf{f}_p^T \delta \dot{\mathbf{z}} = 0. \tag{3.61}$$

The equations (3.60) and (3.61) indicate, that the constraint forces $\mathbf{f}_p$ and the rows of the constraint Jacobian $\mathbf{\Phi}_1$ are both orthogonal to the virtual velocities $\delta \dot{\mathbf{z}}$. Consequently we can represent the passive constraint forces (and torques) $\mathbf{f}_p$ by a linear relation

$$\mathbf{f}_p = \mathbf{\Phi}_1^T(\mathbf{z},t) \cdot \lambda \quad \text{with} \quad \lambda \in \mathbb{R}^m. \tag{3.62}$$

Including this equation into (3.58) and considering the fact, that the constraint force representation according to (3.62) allows an arbitrary choice of the virtual velocities $\delta \dot{\mathbf{z}}$, we get for the unknowns $\ddot{\mathbf{z}} \in \mathbb{R}^{6n}$ and $\lambda \in \mathbb{R}^m$ altogether (6n+m) equations of the form

$$\mathbf{M}\ddot{\mathbf{z}} + \mathbf{h} - \mathbf{f}_a - \mathbf{W}^T \cdot \lambda = \mathbf{0}, \qquad \mathbf{W}\ddot{\mathbf{z}} + \mathbf{w} = \mathbf{0}, \tag{3.63}$$

where the matrix $\mathbf{W} = \mathbf{\Phi}_1$, and the vector $\mathbf{w}$ follows from a differentiation of the original constraint equation (3.59). We might eliminate $\ddot{\mathbf{z}}$ from (3.63) to get

$$\begin{aligned} \lambda &= (\mathbf{W}\mathbf{M}^{-1}\mathbf{W}^T)^{-1} \cdot [\mathbf{W}\mathbf{M}^{-1}(\mathbf{h} - \mathbf{f}_a) + \mathbf{w}] \\ \ddot{\mathbf{z}} &= -\mathbf{M}^{-1}(\mathbf{h} - \mathbf{f}_a - \mathbf{W}^T \lambda) \end{aligned} \tag{3.64}$$

The constrained mass matrix $(\mathbf{W}\mathbf{M}^{-1}\mathbf{W}^T)^{-1}$ represents the effective mass influence in the directions of the constraint forces $\lambda$. The procedure as used above corresponds to a method introduced by Lagrange for the solution of algebraic equations with side-conditions [155].

### 3.2.4.2 Lagrange's Equations of Second Kind

We assume a given set of generalized coordinates $\mathbf{q}$, and we start our considerations with the principle of d'Alembert-Lagrange (eq. 3.38), which writes for a multibody system with n bodies

$$\sum_{i=1}^{n} \int_{Bi} (\ddot{\mathbf{r}}dm - d\mathbf{F}_a)_i^T \cdot \delta \mathbf{r}_i = 0. \tag{3.65}$$

For further evaluations we need the kinetic energy of the system

$$\sum_{i=1}^{n} \frac{1}{2} \int_{Bi} \dot{\mathbf{r}}_i^T \dot{\mathbf{r}}_i dm_i = \sum_{i=1}^{n} T_i. \tag{3.66}$$

For the following we consider one single rigid body $B_i$ and generalize the results at the end. The acceleration term of equation (3.65) can be manipulated as follows

$$\int_{Bi} \ddot{\mathbf{r}}_i^T dm_i \delta \mathbf{r}_i = \frac{d}{dt} \int_{Bi} \dot{\mathbf{r}}_i^T dm_i \delta \mathbf{r}_i - \int_{Bi} \dot{\mathbf{r}}_i^T dm_i \delta \dot{\mathbf{r}}_i$$

$$= \frac{d}{dt} \int_{Bi} \frac{\partial}{\partial \dot{\mathbf{r}}_i} (\frac{1}{2} \dot{\mathbf{r}}_i^T \dot{\mathbf{r}}_i dm_i) \delta \mathbf{r}_i - \delta \int_{Bi} (\frac{1}{2} \dot{\mathbf{r}}_i^T \dot{\mathbf{r}}_i dm_i) \tag{3.67}$$

Again we used here the relation $d\partial \mathbf{r} - \partial d\mathbf{r} = \mathbf{0}$ [93]. We now introduce the generalized coordinates $\mathbf{q}$ by expressing

$$\delta \mathbf{r}_i = (\frac{\partial \mathbf{r}_i}{\partial \mathbf{q}}) \cdot \delta \mathbf{q} = (\frac{\partial \dot{\mathbf{r}}_i}{\partial \dot{\mathbf{q}}}) \cdot \delta \mathbf{q} \quad \text{with} \quad \mathbf{r}_i \in \mathbb{R}^6, \quad \mathbf{q} \in \mathbb{R}^f, \tag{3.68}$$

which modifies equation (3.67) in the following way

$$\int_{Bi} \ddot{\mathbf{r}}_i^T dm_i \delta \mathbf{r}_i = \frac{d}{dt} \int_{Bi} \frac{\partial}{\partial \dot{\mathbf{r}}_i} (\frac{1}{2} \dot{\mathbf{r}}_i^T \dot{\mathbf{r}}_i dm_i)(\frac{\partial \dot{\mathbf{r}}_i}{\partial \dot{\mathbf{q}}}) \delta \mathbf{q} - \delta T_i$$

$$= \frac{d}{dt} \int_{Bi} \frac{\partial}{\partial \dot{\mathbf{q}}} (\frac{1}{2} \dot{\mathbf{r}}_i^T \dot{\mathbf{r}}_i dm_i) \delta \mathbf{q} - \delta T_i = \frac{d}{dt}(\frac{\partial T_i}{\partial \dot{\mathbf{q}}}) \delta \mathbf{q} - \delta T_i \tag{3.69}$$

Setting $\delta W_i = \int_{Bi} (d\mathbf{F}_a)_i^T \cdot \delta \mathbf{r}_i$ and including the last form of equation (3.69) into equation (3.65) we come out with an important relation, which is sometimes called "central equation" ("Zentralgleichung" in German, see [27])

$$\frac{d}{dt}(\frac{\partial T}{\partial \dot{\mathbf{q}}} \delta \mathbf{q}) - \delta T - \delta W = 0. \tag{3.70}$$

The above equation may be applied for a single body or a system of bodies. In any case it will be very helpful for deriving quite a number of fundamental equations of dynamics. From the above two equations to the second equations of Lagrange is only a small step. We perform the differentiations and the variations, and then we get from the relations (3.69), (3.70) and (3.65) the following results

$$\int_{Bi} \ddot{\mathbf{r}}_i^T dm_i \delta \mathbf{r}_i = \left[\frac{d}{dt}(\frac{\partial T_i}{\partial \dot{\mathbf{q}}}) - (\frac{\partial T_i}{\partial \mathbf{q}})\right] \delta \mathbf{q} + (\frac{\partial T_i}{\partial \dot{\mathbf{q}}})\left[\frac{d}{dt}(\delta \mathbf{q}) - (\delta \dot{\mathbf{q}})\right]$$

$$\delta W_i = \int_{Bi} (d\mathbf{F}_a)_i^T \cdot \delta \mathbf{r}_i = \int_{Bi} (d\mathbf{F}_a)_i^T (\frac{\partial \mathbf{r}_i}{\partial \mathbf{q}}) \cdot \delta \mathbf{q} = \mathbf{Q}_i^T \delta \mathbf{q} \tag{3.71}$$

The last term of the first row will be zero, $[\frac{d}{dt}(\delta \mathbf{q}) - (\delta \dot{\mathbf{q}})] = 0$, due to the fact that for generalized coordinates $(d\delta \mathbf{r} - \delta d\mathbf{r} = 0)$. Combining now the equations (3.71) and (3.65) we get

$$\sum_{i=1}^{n} \left[\frac{d}{dt}(\frac{\partial T_i}{\partial \dot{\mathbf{q}}}) - (\frac{\partial T_i}{\partial \mathbf{q}}) - \mathbf{Q}_i^T\right] \delta \mathbf{q} = 0. \tag{3.72}$$

Finally the second equations of Lagrange write for a single body and for a system of bodies:

- *single body*

$$\frac{d}{dt}(\frac{\partial T}{\partial \dot{\mathbf{q}}}) - (\frac{\partial T}{\partial \mathbf{q}}) - \mathbf{Q}^T = 0, \qquad \mathbf{Q} = \int_B (\frac{\partial \mathbf{r}}{\partial \mathbf{q}})^T (d\mathbf{F}_a)$$

$$T = \int_B \frac{1}{2} \dot{\mathbf{r}}^T \dot{\mathbf{r}} dm \tag{3.73}$$

- *system of bodies*

$$\sum_{i=1}^{n} \left[\frac{d}{dt}(\frac{\partial T_i}{\partial \dot{\mathbf{q}}}) - (\frac{\partial T_i}{\partial \mathbf{q}}) - \mathbf{Q}_i^T\right] = 0, \qquad \mathbf{Q}_i = \int_{Bi} (\frac{\partial \mathbf{r}_i}{\partial \mathbf{q}})^T (d\mathbf{F}_a)$$

$$T_i = \int_{Bi} \frac{1}{2} \dot{\mathbf{r}}_i^T \dot{\mathbf{r}}_i dm_i \tag{3.74}$$

The generalized forces $\mathbf{Q}_i$ might be conservative or non-conservative, in the first case they can be derived from a potential V, which represents the potential energy. With

$$\mathbf{Q}_{conservative} = \mathbf{Q}_{co} = -(\frac{\partial V}{\partial \mathbf{q}})^T = -[grad_q(V)]^T \tag{3.75}$$

and the assumption, that in a real system we always have conservative and non-conservative forces, we get the second equations of Lagrange in the form

- *system of bodies*

$$\sum_{i=1}^{n} \left[\frac{d}{dt}(\frac{\partial T_i}{\partial \dot{\mathbf{q}}}) - (\frac{\partial T_i}{\partial \mathbf{q}}) - (\frac{\partial V_i}{\partial \mathbf{q}})\right] = \sum_{i=1}^{n} \mathbf{Q}_{nc,i}^T, \quad \mathbf{Q}_{nc,i} = \int_{Bi} (\frac{\partial \mathbf{r}_i}{\partial \mathbf{q}})^T (d\mathbf{F}_a),$$

$$T_i = \int_{Bi} \frac{1}{2} \dot{\mathbf{r}}_i^T \dot{\mathbf{r}}_i dm_i, \tag{3.76}$$

where $\mathbf{Q}_{nc,i}$ is a non-conservative force acting on the body i.

## 3.2.5 Hamilton's Equations

### 3.2.5.1 Hamilton's Principle

Principle and equations of Hamilton play a larger role in Physics and Analytical Dynamics than in engineering mechanics. Nevertheless, the increasing significance of Hamilton's concepts are connected with the modern evolution of Nonlinear Dynamics, the methods of which are partly important also in engineering ([261], [159]). We shall consider in a first step the principle of Hamilton by starting with the relations of the last chapter (3.65), (3.70) and (3.76). We assume a conservative, holonomic and scleronomic system and define for the following

$$T = \sum_{i=1}^{n} T_i, \quad V = \sum_{i=1}^{n} V_i, \quad W = \sum_{i=1}^{n} W_i,$$
$$\mathbf{p} = \sum_{i=1}^{n} \mathbf{p}_i, \quad \mathbf{p}_i^T = \frac{\partial T_i}{\partial \dot{\mathbf{q}}}, \tag{3.77}$$

where $\mathbf{p}_i$ is the generalized momentum of body i. The important equation (3.70) then takes on the form

$$\frac{d}{dt}(\mathbf{p}^T \delta \mathbf{q}) - \delta T - \delta W = 0, \tag{3.78}$$

which sometimes is addressed to as Lagrange's Central equation. The work done in a time interval $(t_1, t_2)$ can easily be determined as

$$\int_{t_1}^{t_2} \delta(T + W) dt = (\mathbf{p}^T \delta \mathbf{q})\Big|_{t_1}^{t_2} \tag{3.79}$$

If we assume vanishing virtual displacements $\delta \mathbf{q}(t_1) = \mathbf{0}$ and $\delta \mathbf{q}(t_2) = \mathbf{0}$, we get from the above equation

$$\int_{t_1}^{t_2} \delta(T + W) dt = 0. \tag{3.80}$$

As we presuppose conservative systems, we can apply formula (3.75) and replace the applied force by a gradient of the potential energy V or equivalently the virtual work $\delta W$ by $-\delta V$, which results in Hamilton's principle

$$\int_{t_1}^{t_2} \delta(T - V) dt = \int_{t_1}^{t_2} \delta L^* dt = 0 \quad \text{with} \quad L^* = T - V. \tag{3.81}$$

$L^*$ is called the Lagrangian function. The variations $\delta L^*$ must be possible and compatible with the constraints. For holonomic constraints the Hamilton principle can be reduced to a problem of variational calculus

$$\delta \int_{t1}^{t2} L^* dt = 0 \quad \text{or} \quad \int_{t1}^{t2} L^* dt \Rightarrow stationary \tag{3.82}$$

### 3.2.5.2 Hamilton's Canonical Equations

The basic idea of establishing Hamilton's canonical equations consists in replacing the generalized velocity $\dot{\mathbf{q}}$ by the generalized momenta $\dot{\mathbf{p}}$ (see equation (3.77)). We continue with the assumption of conservative, holonomic and scleronomic systems; for a very detailed and deep discussion also with respect to any constraints see [180]. Going from system space coordinates to generalized coordinates with the help of $\mathbf{z} = \mathbf{z}(\mathbf{q})$, we get the accompanying Jacobian by the transformation $\dot{\mathbf{z}} = \mathbf{J}_z \dot{\mathbf{q}}$, and the kinetic energy writes

$$T = \frac{1}{2}\dot{\mathbf{z}}^T \mathbf{M}\dot{\mathbf{z}} = \frac{1}{2}\dot{\mathbf{q}}^T (\mathbf{J}_z^T \mathbf{M} \mathbf{J}_z)\dot{\mathbf{q}}. \tag{3.83}$$

From this the momenta for the system coordinates and the generalized coordinates can be expressed by

$$\begin{aligned}\frac{\partial T}{\partial \dot{\mathbf{z}}} &= \frac{\partial L^*}{\partial \dot{\mathbf{z}}} = [\mathbf{M}\dot{\mathbf{z}}]^T = \mathbf{p}_z^T \quad &\in \mathbb{R}^{6n} \\ \frac{\partial T}{\partial \dot{\mathbf{q}}} &= \frac{\partial L^*}{\partial \dot{\mathbf{q}}} = [(\mathbf{J}_z^T \mathbf{M} \mathbf{J}_z)\dot{\mathbf{q}}]^T = \mathbf{p}_q^T \quad &\in \mathbb{R}^f \end{aligned} \tag{3.84}$$

Comparing the two equations results in $\mathbf{p}_q^T = \mathbf{J}_z^T \mathbf{p}_z$. Lagrange's equations of motion of second kind write with $L^* = T - V$

$$\frac{d}{dt}(\frac{\partial L^*}{\partial \dot{\mathbf{q}}}) - (\frac{\partial L^*}{\partial \mathbf{q}}) = \mathbf{0} \tag{3.85}$$

and together with equation (3.84)

$$\frac{\partial L^*}{\partial \mathbf{q}} = \dot{\mathbf{p}}_q^T. \tag{3.86}$$

The Lagrangian $L^*$ depends on the generalized coordinates $\mathbf{q}$ and on the generalized velocities $\dot{\mathbf{q}}$, $L^* = L^*(\mathbf{q}, \dot{\mathbf{q}})$. From this the variation writes

$$\delta L^* = (\frac{\partial L^*}{\partial \mathbf{q}})\delta \mathbf{q} + (\frac{\partial L^*}{\partial \dot{\mathbf{q}}})\delta \dot{\mathbf{q}} = \dot{\mathbf{p}}_q^T \delta \mathbf{q} + \mathbf{p}_q^T \delta \dot{\mathbf{q}}. \tag{3.87}$$

On the other side the variation of the momentum-velocity-product yields

$$\delta(\mathbf{p}_q^T \dot{\mathbf{q}}) = \mathbf{p}_q^T \delta \dot{\mathbf{q}} + \dot{\mathbf{q}}^T \delta \mathbf{p}_q. \tag{3.88}$$

Introducing the Hamilton function

$$H(\mathbf{q}, \mathbf{p}_q) = \mathbf{p}_q^T \delta \dot{\mathbf{q}} - L^*, \tag{3.89}$$

performing its virtual variation

$$\delta H = (\frac{\partial H}{\partial \mathbf{q}}) \delta \mathbf{q} + (\frac{\partial H}{\partial \mathbf{p}_q}) \delta \mathbf{p}_q \tag{3.90}$$

and subtracting the equation (3.87) from (3.88) results in

$$\delta(\mathbf{p}_q^T \dot{\mathbf{q}} - L^*) = \dot{\mathbf{q}}^T \delta \mathbf{p}_q - \dot{\mathbf{p}}_q^T \delta \mathbf{q}. \tag{3.91}$$

Comparing the last two equation comes out with

$$\dot{\mathbf{q}}^T = +\frac{\partial H}{\partial \mathbf{p}_q}, \qquad \dot{\mathbf{p}}_q^T = -\frac{\partial H}{\partial \mathbf{q}}. \tag{3.92}$$

These relations are called Hamilton's canonical equations of motion. They form a set of ordinary nonlinear differential equations of first order with the dimension (2f) instead of (f) for the Newton-Euler-equations of motion, for example. For certain mathematical solution methods this might be an advantage.

The Hamiltonian H possesses a nice mechanical property. Combining the equations (3.89) and (3.84) we get

$$H(\mathbf{q}, \mathbf{p}_q) = \mathbf{p}_q^T \delta \dot{\mathbf{q}} - L^*$$
$$= [(\mathbf{J}_z^T \mathbf{M} \mathbf{J}_z) \dot{\mathbf{q}}]^T \dot{\mathbf{q}} - (T - V) = \dot{\mathbf{q}}^T (\mathbf{J}_z^T \mathbf{M} \mathbf{J}_z) \dot{\mathbf{q}} - (T - V)$$
$$H(\mathbf{q}, \mathbf{p}_q) = T + V \tag{3.93}$$

For conservative systems the Hamiltonian H represents the total energy as the sum of kinetic and potential energies.

## 3.3 Multibody Systems with Bilateral Constraints

### 3.3.1 General Comments

All moving systems on earth or in space consist of many material bodies interconnected by some force laws of nearly any type, which means, forces stick the bodies together by single- or multi-valued laws, they might act in a point, along a line or by surface distributions, and forces underly of course modifications with time and space. The rotational motion of planets is balanced by gravitational and gyroscopic forces, the performance of a machine follows from an optimal interaction of forces, torques and constraints, walking of machines or biological systems is governed by a complicated combination of forces, torques, constraints, interactions with the environment and by an intelligent system of very adaptive control structures. But the material side of all these structures are bodies, rigid and elastic bodies, linked by the best possible way to achieve the required performances. Therefore and in considering such systems we have as a first step to decide, how we shall model these forces, how we shall model these constraints and finally how we shall model the masses.

The justifiability of such questions will be immediately clear by considering the picture of a modern power transmission system as depicted in Figure (3.9) consisting of a CVT-belt gear, several gear wheel stages and a power converter forming together a system of considerable complexity. Even when applying modern commercial software it makes very much sense to gain some

Fig. 3.9: Example of a modern multibody system (courtesy Daimler-Chrysler)

physical understanding before doing so. The mechanical model must include all considerations with respect to models of masses, force laws, constraints and neglects. The art of neglects decides on costs and results, it is an art, not a science, and requires a deep understanding of the system under consideration. A good qualitative system comprehension is as important as a quantitative one, otherwise the results from simulation models cannot be interpreted. We always should keep in mind, that Technical Mechanics possesses no deductive structures, whereas Mechanics from the physical standpoint of view may have such properties [63].

We deal with systems of many bodies with bilateral and later on with unilateral constraints. Also the complicated structure of Figure (3.9) can be represented by a structure as indicated in Figure (3.10). We consider rigid or elastic masses, where this decision depends on the operational frequency range of the overall system and the eigenfrequencies of the components, linear elasticity presupposed; and we consider bilateral constraints or single-valued force laws for the interconnections. The concept of elastic multibody systems will be restricted to linear elasticity, which includes the most frequent problems in dynamics. As a consequence we might apply a Ritz-approach for modeling with the help of shape functions coming from anywhere, FEM-analysis, exact solutions or measurements. They only have to fulfill the completeness requirement for approximating some elastic displacement fields ([242], [187], [27]). According to Figure (3.10) every multibody system possesses some inertial reference given by its environment, and its structure may include treelike parts and closed loop parts. As a matter of fact tree-like structures are easier to analyze because in the case of loops we usually must cut these loops and introduce a closing condition, which then modifies the equations of motion to a set of DAE-equations.

The forces can follow any law though most practical problems include springs and dampers of any kind, from linear to nonlinear forms, and sometimes given only by measured characteristics. The evaluation of these laws is very often more cumbersome than the determination of the constraints. Therefore it is also a matter of longer concern to think about the two alternatives: Are detailed force laws really necessary or will it be sufficient to represent the interconnections by constraints? As a résumé we emphasize again, that before going into the evaluation of a mathematical form of the equations of motion it is absolutely necessary to establish a mechanical model and to put enough thoughts into the problem under consideration.

As a result of these efforts we want to have a set of equations, which describe the system dynamics completely and unambiguously. This set will include three parts: ordinary differential equations for rigid multibody components and partial differential equations for elastic multibody components, both of second order, all constraints in a clear and non redundant form and all relations concerning rotational motion, for example in the form of the equations (2.31) and (2.33). In the following we shall give some rules how to establish these equations and how to manipulate them. In spite of these rules the resulting complexity of the equations of motion depend largely on the

## 3.3 Multibody Systems with Bilateral Constraints

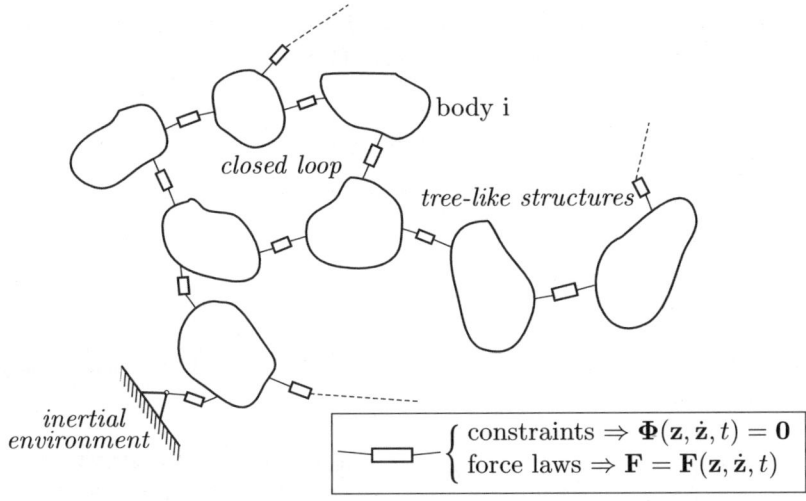

Fig. 3.10: MBS-structure with bilateral connections

skills of the investigator. There is a broad band of possibilities in choosing the minimal coordinates and velocities, the form of the constraints, the rotational coordinates like those from Euler, Kardan, Rodriguous or like quaternions and the solution procedure, for example DAE or not.

Before the background of a worldwide large literature on numerical methods for the treatment of multibody systems we shall not consider this topic here, but refer to two new publications presenting the most modern aspects of numerics in the field, namely [52] and [6].

### 3.3.2 Equations of Motion of Rigid Bodies

We shall focus here on an evaluation of the equations of motion often addressed to as a projective method resulting in a set of Newton-Euler equations of motion [233]. The equations of motion will be bounded by constraints of any form (see equations 3.1) and supplemented by rotational relations of the form $\dot{\mathbf{z}} = \mathbf{H}(\mathbf{q})\dot{\mathbf{q}}$. Combining the principle of Jourdain (3.48) with the absolute velocity of a single rigid body, equation (2.38) with $_B\dot{\mathbf{r}} = \mathbf{0}$, and with the absolute acceleration of a single rigid body, equation (2.42) with the relative dotted magnitudes $_B\dot{\mathbf{r}} = \mathbf{0}$ and $_B\ddot{\mathbf{r}} = \mathbf{0}$, we get in a first step the velocities and accelerations in the form (see also Figure 2.13)

$$_B\dot{\mathbf{r}}_P = {_B\dot{\mathbf{r}}_Q} + {_B\tilde{\boldsymbol{\omega}}} \cdot {_B\mathbf{r}},$$
$$_B\ddot{\mathbf{r}}_P = {_B\ddot{\mathbf{r}}_Q} + {_B\dot{\tilde{\boldsymbol{\omega}}}_B}\mathbf{r} + {_B\tilde{\boldsymbol{\omega}}_B}\tilde{\boldsymbol{\omega}}_B\mathbf{r}. \tag{3.94}$$

As already mentioned all variations ($\delta\mathbf{r}$, $\delta'\dot{\mathbf{r}}$, $\delta''\ddot{\mathbf{r}}$ etc.) must be compatible with the existing constraints by moving within the tangent plane of the constraint surface. Therefore the derivations of the constraints must be included.

# 3 Constraint Systems

For simplicity we assume the constraints in an explicit form $\mathbf{r}_P = \mathbf{r}_P(\mathbf{q}, t)$ and $\dot{\mathbf{r}}_P = \dot{\mathbf{r}}_P(\mathbf{q}, \dot{\mathbf{q}}, t)$ (see chapter (2.2.5)). Then we get

$$\dot{\mathbf{r}}_P = (\frac{\partial \mathbf{r}_P}{\partial \mathbf{q}}) \cdot \dot{\mathbf{q}} + \frac{\partial \mathbf{r}_P}{\partial t}, \qquad \frac{\partial \mathbf{r}_P}{\partial \mathbf{q}} = \frac{\partial \dot{\mathbf{r}}_P}{\partial \dot{\mathbf{q}}}, \qquad (3.95)$$

which can be used to evaluate the variation of the velocity $\dot{\mathbf{r}}$. We get for a body-fixed frame (left index "B" partly omitted)

$$\delta \dot{\mathbf{r}}_P = (\frac{\partial \dot{\mathbf{r}}_P}{\partial \dot{\mathbf{q}}}) \cdot \delta \dot{\mathbf{q}} = {}_B[(\frac{\partial \dot{\mathbf{r}}_Q}{\partial \dot{\mathbf{q}}}) + \tilde{\mathbf{r}}^T(\frac{\partial \boldsymbol{\omega}}{\partial \dot{\mathbf{q}}})]\delta \dot{\mathbf{q}} \qquad (3.96)$$

Combining these equations with Jourdain's principle of equation (3.48) we come out with

$$\int_B \delta' \dot{\mathbf{q}}^T {}_B[(\frac{\partial \dot{\mathbf{r}}_Q}{\partial \dot{\mathbf{q}}}) + \tilde{\mathbf{r}}^T(\frac{\partial \boldsymbol{\omega}}{\partial \dot{\mathbf{q}}})]^T {}_B\{[\ddot{\mathbf{r}}_Q + \dot{\tilde{\boldsymbol{\omega}}}\mathbf{r} + \tilde{\boldsymbol{\omega}}\tilde{\boldsymbol{\omega}}\mathbf{r}]dm - d\mathbf{F}_a\} = \mathbf{0} \qquad (3.97)$$

At this stage a word to the coordinates applied for the equations (3.97). We have chosen a body fixed coordinate system. Instead we could have chosen as well an inertial or any other coordinate system, because the scalar product connected with the principle of virtual power, and of the other principles, does not depend on the choice of the coordinates. This represents a strong advantage of the Newton-Euler approach.

The coordinates $\mathbf{q}$ are generalized coordinates and meet all constraints. Consequently the $\delta' \dot{\mathbf{q}}$ are independent and can be chosen arbitrarily. Introducing the abbreviations

$$\mathbf{F}_a = \int_B d\mathbf{F}_a, \qquad \mathbf{M}_a = \int_B \tilde{\mathbf{r}} d\mathbf{F}_a,$$
$$m\mathbf{r}_S = \int_B \mathbf{r} dm, \qquad \mathbf{I} = -\int_B \tilde{\mathbf{r}}\tilde{\mathbf{r}} dm, \qquad (3.98)$$

we can write the relation (3.97), always with respect to a body-fixed coordinate frame,

$$_B(\frac{\partial \dot{\mathbf{r}}_Q}{\partial \dot{\mathbf{q}}})^T \cdot {}_B[m\ddot{\mathbf{r}}_Q + m(\dot{\tilde{\boldsymbol{\omega}}} + \tilde{\boldsymbol{\omega}}\tilde{\boldsymbol{\omega}})\mathbf{r}_S - \mathbf{F}_a]$$
$$+ {}_B(\frac{\partial \boldsymbol{\omega}}{\partial \dot{\mathbf{q}}})^T \cdot {}_B[m\tilde{\mathbf{r}}_S^T \ddot{\mathbf{r}}_Q + \mathbf{I}\dot{\boldsymbol{\omega}} + \tilde{\boldsymbol{\omega}}\mathbf{I}\boldsymbol{\omega} - \mathbf{M}_a] = \mathbf{0}. \qquad (3.99)$$

The index "S" indicates the center of mass, and "I" is the inertia tensor. The $(\tilde{\cdot})$-tensor has already been defined. To be able to achieve the inertia tensor I, we must consider the quadruple product of equation (3.97) $[\tilde{\mathbf{r}}\tilde{\boldsymbol{\omega}}\tilde{\boldsymbol{\omega}}\mathbf{r} = \mathbf{r} \times (\boldsymbol{\omega} \times (\boldsymbol{\omega} \times \mathbf{r}))]$ and apply some formulas of vector algebra with the result $[\tilde{\mathbf{r}}\tilde{\boldsymbol{\omega}}\tilde{\boldsymbol{\omega}}\mathbf{r} = -\tilde{\boldsymbol{\omega}}\tilde{\mathbf{r}}\tilde{\mathbf{r}}\boldsymbol{\omega}]$.

## 3.3 Multibody Systems with Bilateral Constraints

The equations (3.99) possess some remarkable properties. Firstly we may state that the first term of these equations corresponds to the momentum equation projected into the free directions of motion, due to the constraints and with the help of the Jacobian $(\frac{\partial \dot{\mathbf{r}}_Q}{\partial \dot{\mathbf{q}}})^T$. The second term corresponds to the moment of momentum equations also projected into the free directions with the help of the Jacobian $(\frac{\partial \omega}{\partial \dot{\mathbf{q}}})^T$. These projections force the corresponding equations of momentum and moment of momentum into the tangential planes of the constraint surfaces. For practical problems the above equations are very convenient, because each of the resulting individual equations represents a scalar product, which might be evaluated in different coordinate systems. In many cases it makes sense to use for the momentum equations an inertial and for the moment of momentum equations a body-fixed coordinate system.

Without constraints the Jacobians in equation (3.99) vanish and we get directly the classical equations for a rigid body in the form

$$_B[m\ddot{\mathbf{r}}_Q + m(\dot{\tilde{\omega}} + \tilde{\omega}\tilde{\omega})\mathbf{r}_S - \mathbf{F}_a] = 0,$$
$$_B[m\tilde{\mathbf{r}}_S^T\ddot{\mathbf{r}}_Q + \mathbf{I}\dot{\omega} + \tilde{\omega}\mathbf{I}\omega - \mathbf{M}_a] = 0. \tag{3.100}$$

The equations (3.98) and (3.99) can easily be extended to a system of rigid bodies with altogether n bodies. All magnitudes get an index i and equation (3.99) is summed up over all n bodies. This yields with the abbreviations

$$\mathbf{F}_{ai} = \int_{B_i} d\mathbf{F}_{ai}, \qquad \mathbf{M}_{ai} = \int_{B_i} \tilde{\mathbf{r}}_i d\mathbf{F}_{ai},$$
$$m_i \mathbf{r}_{Si} = \int_{B_i} \mathbf{r}_i dm_i, \qquad \mathbf{I}_i = -\int_{B_i} \tilde{\mathbf{r}}_i \tilde{\mathbf{r}}_i dm_i, \tag{3.101}$$

the equations of motion for a system of rigid bodies

$$\sum_{i=1}^{n} \{_{B_i}(\frac{\partial \dot{\mathbf{r}}_{Q_i}}{\partial \dot{\mathbf{q}}})^T \cdot {}_{B_i}[m_i\ddot{\mathbf{r}}_{Q_i} + m_i(\dot{\tilde{\omega}} + \tilde{\omega}\tilde{\omega})_i \mathbf{r}_{Si} - \mathbf{F}_{ai}]$$
$$+ {}_{B_i}(\frac{\partial \omega_i}{\partial \dot{\mathbf{q}}})^T \cdot {}_{B_i}[m_i\tilde{\mathbf{r}}_{S_i}^T \ddot{\mathbf{r}}_{Q_i} + \mathbf{I}_i \dot{\omega}_i + \tilde{\omega}_i \mathbf{I}_i \omega_i - \mathbf{M}_{ai}]\} = 0. \tag{3.102}$$

The vector $\dot{\mathbf{q}}$ consists of generalized velocities. Introducing

$$_{B_i}\dot{\mathbf{p}}_i = {}_{B_i}[m_i \ddot{\mathbf{r}}_{Q_i} + m_i(\dot{\tilde{\omega}} + \tilde{\omega}\tilde{\omega})_i \mathbf{r}_{S_i}],$$
$$_{B_i}\dot{\mathbf{L}}_i = {}_{B_i}[m_i \tilde{\mathbf{r}}_{S_i}^T \ddot{\mathbf{r}}_{Q_i} + \mathbf{I}_i \dot{\omega}_i + \tilde{\omega}_i \mathbf{I}_i \omega_i],$$
$$_{B_i}\mathbf{J}_{Ti} = {}_{B_i}(\frac{\partial \dot{\mathbf{r}}_{Q_i}}{\partial \dot{\mathbf{q}}}), \qquad _{B_i}\mathbf{J}_{Ri} = {}_{B_i}(\frac{\partial \omega_i}{\partial \dot{\mathbf{q}}}), \tag{3.103}$$

we also can write

$$\sum_{i=1}^{n} [_{B_i}\mathbf{J}_{Ti}{}^T(_{B_i}\dot{\mathbf{p}}_i - {}_{B_i}\mathbf{F}_{ai}) + {}_{B_i}\mathbf{J}_{Ri}{}^T(_{B_i}\dot{\mathbf{L}}_i - {}_{B_i}\mathbf{M}_{ai})] = \mathbf{0}, \qquad (3.104)$$

where it must be kept in mind that the $\dot{\mathbf{p}}_i$ and $\dot{\mathbf{L}}_i$ are abbreviations, not integrable physical magnitudes.

Bremer [28] has introduced a useful method to deal with subsystems. In this case the subsystem generalized velocities $\dot{\mathbf{q}}_k$ depend on the generalized velocities $\dot{\mathbf{q}}$ and therefore we can write

$$\sum_{k=1}^{r} {}_{B_{ki}}\left(\frac{\partial \dot{\mathbf{q}}_k}{\partial \dot{\mathbf{q}}}\right)^T \cdot \sum_{i=1}^{m_k} [_{B_{ki}}\left(\frac{\partial \dot{\mathbf{r}}_{Q_{ki}}}{\partial \dot{\mathbf{q}}_k}\right)^T ({}_{B_{ki}}\dot{\mathbf{p}}_{ki} - {}_{B_{ki}}\mathbf{F}_{a,ki})$$

$$+ {}_{B_{ki}}\left(\frac{\partial \omega_{ki}}{\partial \dot{\mathbf{q}}_k}\right)^T ({}_{B_{ki}}\dot{\mathbf{L}}_{ki} - {}_{B_{ki}}\mathbf{M}_{a,ki})] = \mathbf{0}. \qquad (3.105)$$

According to the above relation we have $r$ subsystems, each with $m_k$ bodies and each of the bodies with an integration domain $B_i(k)$. The generalized velocities $\dot{\mathbf{q}}_k$ can be chosen independently from each other.

For further evaluations we go back to the equations (3.102) with the abbreviations (3.103) and include into (3.102) the velocities and accelerations from the relations (2.63) and (2.65) by setting $\dot{\mathbf{z}} = [(\dot{\mathbf{r}}_{Q_i})^T (\omega_i)^T]^T$. This results in

$$\mathbf{M}\ddot{\mathbf{q}} + \mathbf{h} - \mathbf{f}_a = \mathbf{0} \qquad \text{with}$$

$$\mathbf{M} = \sum_{i=1}^{n} \{\mathbf{J}_{Ti}^T m_i \mathbf{J}_{Ti} + \mathbf{J}_{Ri}^T m_i \tilde{\mathbf{r}}_{S_i}^T \mathbf{J}_{Ti} + \mathbf{J}_{Ti}^T m_i \tilde{\mathbf{r}}_{S_i}^T \mathbf{J}_{Ri} + \mathbf{J}_{Ri}^T \mathbf{I}_i \mathbf{J}_{Ri}\} \in \mathbb{R}^{f,f},$$

$$\begin{pmatrix} \mathbf{J}_{Ti} \\ \mathbf{J}_{Ri} \end{pmatrix} = \left(\frac{\partial \dot{\mathbf{z}}_i}{\partial \dot{\mathbf{q}}}\right) = \frac{\partial}{\partial \dot{\mathbf{q}}}\begin{pmatrix} \dot{\mathbf{r}}_{Qi} \\ \omega_i \end{pmatrix} \in \mathbb{R}^{6n,f},$$

$$\mathbf{h} = \sum_{i=1}^{n} \{\mathbf{J}_{Ti}^T(m_i \tilde{\omega}_i \tilde{\omega}_i \mathbf{r}_{Si}) + \mathbf{J}_{Ri}^T(\tilde{\omega}_i \mathbf{I}_i \omega_i) +$$

$$[(\mathbf{J}_{Ti}^T m_i + \mathbf{J}_{Ri}^T m_i \tilde{\mathbf{r}}_{Si}^T)(\mathbf{J}_{Ti}^T m_i \tilde{\mathbf{r}}_{Si}^T + \mathbf{J}_{Ri}^T \mathbf{I}_i)][\frac{\partial}{\partial \mathbf{q}}(\begin{pmatrix} \mathbf{J}_{Ti} \\ \mathbf{J}_{Ri} \end{pmatrix}\dot{\mathbf{q}})\dot{\mathbf{q}}]\} \in \mathbb{R}^f$$

$$\mathbf{f}_a = \sum_{i=1}^{n} [\mathbf{J}_{Ti}^T \mathbf{F}_{ai} + \mathbf{J}_{Ri}^T \mathbf{M}_{ai}] = \sum_{i=1}^{n} \mathbf{Q}_i \quad \in \mathbb{R}^f,$$

$$\mathbf{z} \in \mathbb{R}^{6n}, \quad \mathbf{q} \in \mathbb{R}^f. \qquad (3.106)$$

For rotations we have to regard in addition the Euler kinematical equtions in some suitable coordinate form, but always representable by a linear relationship

$$\dot{\mathbf{q}} = \mathbf{H}^* \dot{\mathbf{z}} \qquad (3.107)$$

The reduced forces and torques are generalized forces $\mathbf{Q}_i$, and, if we want, $\mathbf{f}_a = \mathbf{Q}$. For the velocities and accelerations of the relations (2.63) and (2.65) the assumption of no dependence on time $\mathbf{z} = \mathbf{z}(\mathbf{q})$ has been used. The above equations are usually applied to multibody systems with resolvable constraints, which concerns quite a lot of such systems. For more general cases, where we cannot include the constraints, we either could reduce the number of the equations of motion as far as possible by the constraints and keep the non-resolvable ones resulting in a DAE-system of equations, or we establish the equations of motion by cutting free all bodies, by deriving the momentum and moment of momentum equations, by coupling them with the help of the reaction forces and torques and by keeping all constraints. The last possibility is mostly the better one for reasons of numerical treatment and program code structuring.

### 3.3.3 Order(n) Recursive Algorithms

One of the problems with respect to large multibody systems is connected with the computing time necessary to simulate such systems. On the other hand and for many cases of practical relevancy the topology of multibody systems allows a division into trees and loops, which can be treated by recurrence processes. They do not require the inversion of the mass matrix $\mathbf{M} \in \mathbb{R}^{6n,6n}$ of the complete system but come out with an inversion of the single mass matrices $\mathbf{M}_i \in \mathbb{R}^{6,6}$ for each recurrence step. For tree-like structures it is possible to develop recurrence relations as already discussed for the kinematics in chapter 2.2.4. Closed loops might be cut into two trees adding some suitable kinematic and kinetic closing conditions. Most of the literature on this topic appeared already in the eighties or earlier, see for example [102], [103], [25], [28], [242]. All recursive algorithms follow principally the same idea including three steps [217], [222]:

In a first step we evaluate the absolute kinematic magnitudes starting from the base link and ending with the end link of a tree and using the results from the relative kinematic magnitudes from the time integration step before. This process is sometimes called the *first forward recursion*, because we go upwards from the base to the end body. We might apply for this step the formulas of chapter 2.2.4.

In a second step we go from the end link to the base link by eliminating body by body the constraint forces of the link connections. Knowing the kinematics we can evaluate all applied forces and torques and are only left with the unknown constraint forces and the accelerations. For the end link we can eliminate the constraint force to the link before, in direction base body, and we can replace the acceleration by the acceleration of the predecessor body by a kinematic recursion. The result will be an equation of motion with exactly the same structure as the end link thus defining a "new end link". We go backwards to the base body and call this process the *backward recursion*. The constraint forces are eliminated completely by this evaluation, and we come out with the absolute acceleration of the base link.

In a third step we calculate the absolute accelerations of all bodies starting again with the base link and finishing with the end link. This process is usually called the *second forward recursion*.

Some authors differentiate between trees and chains. A tree is a topological part of the multibody system under consideration, where every body of that tree might also be the base link of another tree. In this case we must regard the forces coming from this external tree via its last connection. If we have chains things are more simple, because there are no side-trees or chains whatsoever, and the recursions are straightforward. Therefore it is very important to choose a good division of the overall system leading to simple topological substructures.

We again come back to chapter (2.2.4) and start the **first forward-recursion** with the relations (2.52) and (2.49), which write in the coordinate system $B_i$ (see also Figure 3.11)

$$_{B_i}\dot{\mathbf{r}}_{IB_i} = \mathbf{A}_{B_iB_{i-1}}[(_{B_{i-1}}\dot{\mathbf{r}}_{IB_{i-1}} + {}_{B_{i-1}}\dot{\mathbf{r}}_{B_{i-1}B_i}) + $$
$$+ (_{B_{i-1}}\tilde{\omega}_{IB_{i-1}} + {}_{B_{i-1}}\tilde{\omega}_{B_{i-1}B_i}) \cdot {}_{B_{i-1}}\mathbf{r}_{B_{i-1}B_i}],$$
$$_{B_i}\omega_{IB_i} = \mathbf{A}_{B_iB_{i-1}} \cdot (_{B_{i-1}}\omega_{IB_{i-1}}) + {}_{B_i}\omega_{B_{i-1}B_i}. \qquad (3.108)$$

The relative translational and rotational velocities between the bodies $B_{i-1}$ and $B_i$ are ${}_{B_i}\dot{\mathbf{r}}_{B_{i-1}B_i}$ and ${}_{B_i}\omega_{B_{i-1}B_i}$, respectively, written in the $B_i$-system. These two velocities depend directly on the generalized velocities $\dot{\mathbf{q}}_i$ given by the possible degrees of freedom of the connection between the two bodies $B_{i-1}$ and $B_i$. These degrees of freedom depend on the configuration of the interconnection: we may have a rotatory joint with rotational degrees of freedom, very often only with one degree of freedom, or we may have a connection with translational relative motion in the form of linear guideways or the like. In addition some motor system might prescribe the relative motion of the bodies, for example in manipulators. In this case forces in that specific joint are given with the controlled motor torques.

Proceeding to the accelerations we must transform the equations (3.108) into the inertial frame, then differentiate the transformed equations and after that transform them back into the body-fixed coordinate system. This standard process for time derivations results in (see also the equations (2.57))

$$_{B_i}\ddot{\mathbf{r}}_{IB_i} = \mathbf{A}_{B_iB_{i-1}} \cdot {}_{B_{i-1}}[\ddot{\mathbf{r}}_{IB_{i-1}} + (\ddot{\mathbf{r}}_{B_{i-1}B_i} + \dot{\tilde{\omega}}_{IB_i}\mathbf{r}_{B_{i-1}B_i} + $$
$$+ 2\tilde{\omega}_{IB_i}\dot{\mathbf{r}}_{B_{i-1}B_i} + \tilde{\omega}_{IB_i}\tilde{\omega}_{IB_i}\mathbf{r}_{B_{i-1}B_i})]$$

$$_{B_i}\dot{\omega}_{IB_i} = \mathbf{A}_{B_iB_{i-1}} \cdot (_{B_{i-1}}\dot{\omega}_{IB_{i-1}} + {}_{B_{i-1}}\omega_{B_{i-1}B_i} \cdot {}_{B_{i-1}}\omega_{IB_{i-1}}) + $$
$$+ {}_{B_i}\dot{\omega}_{B_{i-1}B_i}. \qquad (3.109)$$

The indices $B_{i-1}$ on the left side of the brackets indicate, that all magnitudes within the brackets are written in the body-fixed coordinate system $B_{i-1}$. The structures of these two equations are obvious, it composes the absolute

## 3.3 Multibody Systems with Bilateral Constraints

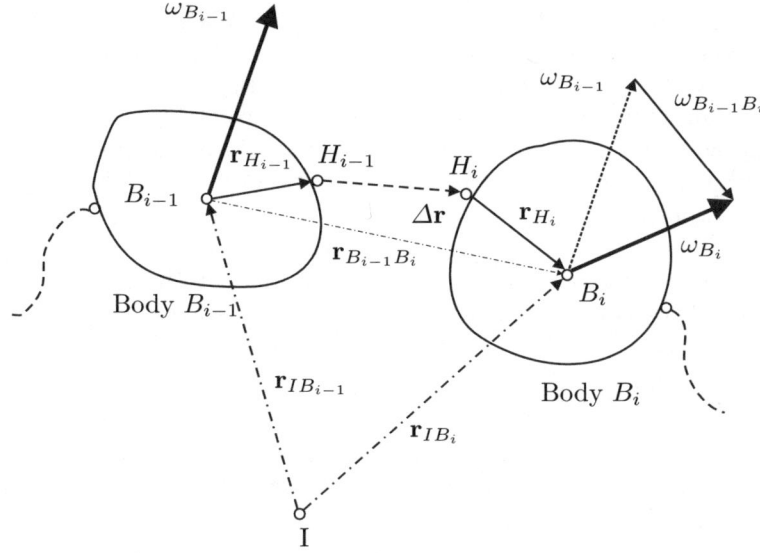

Fig. 3.11: Relative Kinematics for Recursion

accelerations of the predecessor body with the relative accelerations between the two bodies. According to Figure 3.11 and to the equations (2.48) to (2.52) we can establish the following relations

$$_{B_i}\mathbf{r}_{B_{i-1}B_i} =\ _{B_i}[\mathbf{r}_{H_{i-1}} + \Delta\mathbf{r}_{H_{i-1}H_i} + \mathbf{r}_{H_i}] = \mathbf{h_{0i}} + \mathbf{H_{Ti}q_i},$$
$$\mathbf{h_{0i}} = \mathbf{r}_{H_{i-1}} + \mathbf{r}_{H_i}, \qquad \Delta\mathbf{r}_{H_{i-1}H_i} = \Delta\mathbf{r} = \mathbf{H_{Ti}q_i}$$

$$_{B_i}\tilde{\omega}_{IB_i} = \mathbf{A}_{B_iB_{i-1}} \cdot (_{B_{i-1}}\tilde{\omega}_{IB_{i-1}}) \cdot \mathbf{A}_{B_{i-1}B_i} + {}_{B_i}\tilde{\omega}_{B_iB_{i-1}}$$

$$_{B_i}\dot{\mathbf{r}}_{B_{i-1}B_i} = {}_{B_i}\Delta\dot{\mathbf{r}}_{H_{i-1}H_i} = \mathbf{H_{Ti}\dot{q}_i} + \dot{\mathbf{H}}_{Ti}\mathbf{q_i},$$
$$_{B_i}\omega_{B_{i-1}B_i} = {}_{B_i}\Delta\omega = \mathbf{H_{Ri}\dot{q}_i},$$
$$\mathbf{H_{Ti}} \in \mathbb{R}^{3,6}, \quad \mathbf{H_{Ri}} \in \mathbb{R}^{3,6} \quad \mathbf{q_i} \in \mathbb{R}^6, \qquad (3.110)$$

where the vector $\mathbf{q_i}$ represents the joint degrees of freedom and contains accordingly three relative coordinates of translation, for example x,y,z in the case of a local Cartesian frame, and three relative coordinates of rotation, for example $\alpha, \beta, \gamma$ in the case of Cardan angles. The matrices $\mathbf{H_{Ti}}$ and $\mathbf{H_{Ri}}$ take into account the properties of the vectors ${}_{B_i}\mathbf{r}_{B_{i-1}B_i}$ and ${}_{B_i}\omega_{B_{i-1}B_i}$, which in the first case must be composed of the vectors from the bodies' reference points to the hinge points and of the joint vector, and which in the second case result from some finite rotation expressions like the equations (2.31) and (2.33) in chapter (2.2.3).

The **H**-matrices are known, and the $(\mathbf{q}, \dot{\mathbf{q}})$ must be evaluated in the course of the recursion. In many cases of practical relevancy we only have rotatory joints with $\mathbf{H}_{T(i)} = \mathbf{0}$ and $\dot{\mathbf{H}}_{T(i)} = \mathbf{0}$, then the equations for the joint translations vanish, and we have to consider only the rotational part as given with the last equation of (3.110) for ${}_{B_i}\omega_i$. Sometimes it is also possible to choose the body- and joint-fixed coordinates in such a way, that the matrix $\mathbf{H}_{T(i)}$ will be constant with only $\dot{\mathbf{H}}_{T(i)} = \mathbf{0}$, which leads to some simplifications of the above equations.

The above equations are recurrence relations for the absolute velocities in a tree. They can be represented in any coordinate system by appropriate transformations. Together with the equation (2.46)

$$\mathbf{A}_{IB_i} = \mathbf{A}_{IB_{i-1}} \cdot \mathbf{A}_{B_{i-1}B_i} \tag{3.111}$$

we have with the relations (3.108), (3.109) and (3.110) a complete set of recurrence equations for the evaluation of the absolute and relative kinematics in a tree. We start with the base link and go up to the end link. For later applications we combine these equations in the following form

$$\dot{\mathbf{z}}_i = \mathbf{Q}_i(\dot{\mathbf{z}}_{i-1} + \mathbf{H}_i\dot{\mathbf{q}}_i) + \mathbf{v}_i(\mathbf{z}_i, \mathbf{q}_i, t),$$

$$\ddot{\mathbf{z}}_i = \mathbf{Q}_i(\ddot{\mathbf{z}}_{i-1} + \mathbf{H}_i\ddot{\mathbf{q}}_i) + \mathbf{a}_i(\dot{\mathbf{z}}_i, \dot{\mathbf{q}}_i, \mathbf{z}_i, \mathbf{q}_i, t),$$

$$\frac{\partial \mathbf{z}_i}{\partial \mathbf{q}_i} = \frac{\partial \dot{\mathbf{z}}_i}{\partial \dot{\mathbf{q}}_i} = \frac{\partial \ddot{\mathbf{z}}_i}{\partial \ddot{\mathbf{q}}_i} = \mathbf{Q}_i \mathbf{H}_i,$$

$$\dot{\mathbf{z}}_\mathbf{i} = \begin{pmatrix} {}_{B_i}\dot{\mathbf{r}}_{IB_i} \\ {}_{B_i}\omega_{IB_i} \end{pmatrix}, \qquad \overset{\circ}{\Delta}\mathbf{z}_\mathbf{i} = \begin{pmatrix} {}_{B_i}\dot{\mathbf{r}}_{B_{i-1}B_i} \\ {}_{B_i}\omega_{B_{i-1}B_i} \end{pmatrix} = \begin{pmatrix} \mathbf{H}_{Ti} & 0 \\ 0 & \mathbf{H}_{Ri} \end{pmatrix} \cdot \dot{\mathbf{q}}_i,$$

$$\mathbf{Q}_i = \mathbf{A}_{B_i B_{i-1}} \cdot \begin{pmatrix} \mathbf{E} & {}_{B_{i-1}}\tilde{\mathbf{r}}^T_{B_{i-1}B_i} \\ 0 & \mathbf{E} \end{pmatrix}, \qquad \mathbf{H}_\mathbf{i} = \begin{pmatrix} \mathbf{H}_{Ti} & 0 \\ 0 & \mathbf{H}_{Ri} \end{pmatrix}. \tag{3.112}$$

The matrix $\mathbf{Q}_i$ is mainly a transformation from the $B_{i-1}$-system into the $B_i$-system, because the magnitudes in the bracket of the above two first equations are given in the $B_{i-1}$-system. The functions $\mathbf{v}$ and $\mathbf{a}$ depend on kinematical magnitudes, which are at least one level lower than the velocity or the acceleration, respectively. The additional velocity term depends only on position and orientation, and the additional acceleration term depends on some velocities in addition to position and orientation.

With the data from the first forward recursion we are now able to go from the end link to the base link eliminating the constraint forces. This is the second step and called **backward recursion**. We go back to equation (3.100) and write thr equation of motion for the end link $B_i$ in the form

$$_{B_i}[\mathbf{M\ddot{z}} + \mathbf{h} - \mathbf{f}_p] = \mathbf{0}$$

$$_{B_i}\mathbf{M} = \begin{pmatrix} \mathbf{M}_3 & m\tilde{\mathbf{r}}_S^T \\ m\tilde{\mathbf{r}}_S^T & \mathbf{I} \end{pmatrix}, \quad _{B_i}\mathbf{h} = \begin{pmatrix} m\tilde{\omega}\tilde{\omega}\mathbf{r}_S \\ \tilde{\omega}\mathbf{I}\omega \end{pmatrix} - \begin{pmatrix} \mathbf{F}_a \\ \mathbf{M}_a \end{pmatrix}, \quad _{B_i}\mathbf{f}_p = \begin{pmatrix} \mathbf{F}_p \\ \mathbf{M}_p \end{pmatrix}.$$
(3.113)

The mass matrix $\mathbf{M} \in \mathbb{R}^{6,6}$ includes the single body masses and the inertia tensor, the vector $\mathbf{z}$ is defined in equation (3.112), and the vector $\mathbf{h}$ includes all applied forces and $\mathbf{f}_p$ all passive forces and torques. Inserting equation (3.112) into equation (3.113) and applying Jourdain's principle we determine in a first step the unknown generalized accelerations resulting in

$$_{B_i}\mathbf{J}^T{}_{B_i}[\mathbf{M\ddot{z}} + \mathbf{h} - \mathbf{f}_p] = \mathbf{0}, \qquad _{B_i}\mathbf{J}^T{}_{B_i}\mathbf{f}_p = \mathbf{0}, \qquad \mathbf{J} = \frac{\partial \dot{\mathbf{z}}_i}{\partial \dot{\mathbf{q}}_i},$$

$$\ddot{\mathbf{q}}_i = -{}_{B_i}(\mathbf{J}^T\mathbf{MJ})^{-1}{}_{B_i}\mathbf{J}^T{}_{B_i}(\mathbf{MQ\ddot{z}}_{i-1} + \mathbf{Ma} + \mathbf{h}). \tag{3.114}$$

In a second step we apply the original equations of motion (3.113) together with the solution for the generalized acceleration of the relation above to evaluate the constraint force $\mathbf{f}_p$, which comes out with

$$_{B_i}\mathbf{f}_{p_i} = {}_{B_i}\{[\mathbf{E} - \mathbf{MJ}(\mathbf{J}^T\mathbf{MJ})^{-1}\mathbf{J}^T](\mathbf{MQ\ddot{z}}_{i-1} + \mathbf{Ma} + \mathbf{h})\} \tag{3.115}$$

In a third step we consider now the equations of motion of the predecessor body $B_{i-1}$, which write

$$_{B_{i-1}}[\mathbf{M\ddot{z}} + \mathbf{h} - \mathbf{f}_{p_{i-2}} - \mathbf{A}_{B_{i-1}B_i}\mathbf{f}_{p_i} - \sum_k \mathbf{A}_{B_{i-1}B_k}\mathbf{f}_{p_k}] = \mathbf{0}, \tag{3.116}$$

which we have extended by the constraint forces coming from possible branches of the tree, namely $\mathbf{f}_{p_k}$, and by the constraint force $\mathbf{f}_{p_{i-2}}$ going to the successor of $B_{i-1}$. We shall not pursue the branching problem here, because it can be evaluated applying the same process as for one branch. We combine now the equations (3.115) and (3.116), which gives us again the equations of motion in a form, which corresponds exactly to that of the predecessor body. We get

$$_{B_{i-1}}[\mathbf{M\ddot{z}} + \mathbf{h} - \mathbf{f}_{p_{i-2}}] = \mathbf{0},$$

$$_{B_{i-1}}\mathbf{M}_{i-1} := {}_{B_{i-1}}\{\mathbf{M}_{i-1} - \mathbf{A}_{B_{i-1}B_i} \cdot {}_{B_i}[\mathbf{E} - \mathbf{MJ}(\mathbf{J}^T\mathbf{MJ})^{-1}\mathbf{J}^T]_{B_i}(\mathbf{MQ})\}$$

$$_{B_{i-1}}\mathbf{h}_{i-1} := {}_{B_{i-1}}\{\mathbf{h}_{i-1} - \mathbf{A}_{B_{i-1}B_i} \cdot {}_{B_i}[\mathbf{E} - \mathbf{MJ}(\mathbf{J}^T\mathbf{MJ})^{-1}\mathbf{J}^T]_{B_i}(\mathbf{Ma} + \mathbf{h})\}$$
(3.117)

With the above relations we have a recurrence scheme for multibody chains and trees, see also [28] and [217].

The third recursion is the last one, namely the **second forward recursion**, which concerns kinematics again. With the backward recursion we know all accelerations and constraint forces (equations (3.114) and (3.115)) and can thus start again with the base body to evaluate the kinematics of the tree by applying the relations (3.108) to (3.111). If we have branchings we must start with the end body of each branch going back to the tree body where several branches might meet and then go further back to the base body (for example [27], [28], [237]). For the treatment of closed loops see also the references [28], [237], [222] and [103].

As a second possibility we could find also a solution without inverting the whole system matrix by applying equation (2.61) taking advantage of the triangular matrix of this relation. Equation (3.102) together with the abbreviations (3.103) writes

$$\sum_{i=1}^{n} (_{B_i}\mathbf{J}_{Ti}{}^T, {}_{B_i}\mathbf{J}_{Ri}{}^T)_{B_i} \left\{ \begin{pmatrix} m_i & m_i\tilde{\mathbf{r}}_{IB_i}^T \\ m_i\tilde{\mathbf{r}}_{IB_i}^T & \mathbf{I_i} \end{pmatrix} \cdot \begin{pmatrix} {}_{B_i}\ddot{\mathbf{r}}_{IB_i} \\ {}_{B_i}\dot{\omega}_{IB_i} \end{pmatrix} + \begin{pmatrix} m_i\tilde{\omega}_{IB_i}\tilde{\omega}_{IB_i}\mathbf{r}_{IB_i} \\ \tilde{\omega}_{IB_i}\mathbf{I_i}\omega_{IB_i} \end{pmatrix} - \begin{pmatrix} \mathbf{F}_{ai} \\ \mathbf{M}_{ai} \end{pmatrix} \right\} = \mathbf{0}. \quad (3.118)$$

Combining these equations with the recursion kinematics of the equations (2.58) to (2.62) we can generate a set of equations also with triangular matrices, which can then be solved in an iterative way.

### 3.3.4 Equations of Motion of Flexible Bodies

Flexibility in multibody dynamics appears in very different forms. We might have bodies with very large deformations depending on the material and the design; aerospace and comparable light-weight equipment or plants in a storm are examples. And we may have bodies with very small deformations like those in heavy machines or precision tools. Anyway, as long as the eigendynamics of such elastic components lies within the frequency domain of the machine's operational range, we have to consider the influence of these elastic deformations.

In the following we shall restrict ourselves to small elastic deformations. They cover most of the practical applications. For this case we assume, that small elastic movements add to the rigid body motion of a multibody system. Thus elasto-dynamics takes place within the moving system. We shall derive the appropriate equations of motion and give some solution procedures applying for example methods by Ritz and Galerkin with the classical separation of variables. This is one possible and classical way, which in the course of the last decades has been supplemented by algorithms applying finite elements (FEM). They allow the treatment also of large elastic deformations, they offer more flexibility and above all, they solve one of the standard problems, namely how to choose the boundary conditions for elastic bodies within the multibody structure, in an unambiguous way (see for example [242], [237],

[28]). The last mentioned aspect decides on the form of the shape functions necessary for example for a RITZ-approach.

We go back to chapter (2.2.8), to Figure (2.22) and to the equations (2.133) to (2.135) on page 47 and define for the body (i)

$$_{B_i}\mathbf{r}_{ei} = {_{B_i}}(\mathbf{r}_i + \tilde{\varphi}_i \mathbf{u}_i), \qquad _{B_{ei}}\mathbf{r}_{ei} = {_{B_i}}(\mathbf{E} + \tilde{\varphi}_i)_{B_i}\mathbf{r}_{ei},$$

$$\mathbf{r}_{ei} = \mathbf{r}_{IB_{ei}}, \qquad \mathbf{r}_i = \mathbf{r}_{IB_{ri}}, \qquad \mathbf{u}_i = \mathbf{r}_{B_{ri}B_{ei}}$$
$$\tilde{\varphi}_i = \mathbf{F}_{i,skew} = \frac{1}{2}\left(\left(\frac{\partial \mathbf{u}}{\partial \mathbf{X}}\right) - \left(\frac{\partial \mathbf{u}}{\partial \mathbf{X}}\right)^T\right)_i = \frac{1}{2}(\nabla \mathbf{u} - \nabla \mathbf{u}^T)_i \qquad (3.119)$$

where the first equation is written in the $B_i$- and the second equation in the $B_{ei}$-coordinate system. The last equation corresponds to the relation (2.143). The deformation vector $\mathbf{u}_i$ is usually given in the coordinate system of the deformed element. Proceeding to the absolute velocities of a deformed element we get for the two coordinate systems as used above

$$_{B_i}(\dot{\mathbf{r}}_{ei}) = {_{B_i}}\mathbf{v}_{ei} = {_{B_i}}[\mathbf{v}_i + \dot{\mathbf{u}}_i + \tilde{\omega}_i(\mathbf{r}_i + \mathbf{u}_i)],$$
$$_{B_i}\omega_{ei} = {_{B_i}}[\omega_i + \dot{\tilde{\varphi}}_i],$$

$$_{B_{ei}}(\dot{\mathbf{r}}_{ei}) = {_{B_{ei}}}\mathbf{v}_{ei} = {_{B_{ei}}}[\mathbf{E} + \tilde{\varphi}_i]^T \cdot {_{B_i}}\mathbf{v}_{ei},$$
$$_{B_{ei}}\omega_{ei} = {_{B_{ei}}}[\mathbf{E} + \tilde{\varphi}_i]^T \cdot {_{B_i}}\omega_{ei}.$$

$$_{B_i}\mathbf{v}_i = \mathbf{A}_{B_iI} \cdot {_I}(\dot{\mathbf{r}}_{IB_{ri}}), \qquad _{B_i}\mathbf{v}_{ei} = \mathbf{A}_{B_iI} \cdot {_I}(\dot{\mathbf{r}}_{IB_{ri}} + \dot{\mathbf{r}}_{B_{ri}B_{ei}})$$
$$_{B_i}\omega_i = {_{B_i}}\omega_{IB_i} = \mathbf{A}_{B_iI}\dot{\mathbf{A}}_{IB_i} \qquad (3.120)$$

Before deriving the equations of motion for a linearly elastic multibody system we need to define the generalized coordinates, assuming for the sake of simplicity that these coordinates meet all constraints. But we have to define in addition the elastic coordinates, which follow from the Ritz-approach in the form

$$_{B_i}\mathbf{u}_i = \begin{pmatrix} u_i(\mathbf{x}_i, t) \\ v_i(\mathbf{x}_i, t) \\ w_i(\mathbf{x}_i, t) \end{pmatrix} = \begin{pmatrix} \bar{u}_i(\mathbf{x}_i)^T \mathbf{q}_{eui}(t) \\ \bar{v}_i(\mathbf{x}_i)^T \mathbf{q}_{evi}(t) \\ \bar{w}_i(\mathbf{x}_i)^T \mathbf{q}_{ewi}(t) \end{pmatrix} = \bar{\mathbf{u}}_i(\mathbf{x}_i)^T \mathbf{q}_{ei}(t). \qquad (3.121)$$

The vector $\mathbf{x}_i$ is any suitable set of coordinates describing the linear deformations of body (i). The functions $\bar{\mathbf{u}}_i(\mathbf{x}_i)$ must be a complete set of orthonormal functions, which form an orthonormal basis. They have to fulfill certain conditions, for example Parseval's completeness relation (see [132], [283], [187]). For practical purposes cubic splines have proved themselves quite well, and for special cases like bars and plates Fourier expansions represent a good set of such functions [28]. We shall come back to these problems.

Collecting all rigid coordinates in a vector $\mathbf{q_r}$ and all elastic coordinates in a vector $\mathbf{q_e}$ we come out with

$$\mathbf{q} = (\mathbf{q_r}^T, \mathbf{q_e}^T)^T, \qquad \mathbf{v}_i = \mathbf{v}_i(\mathbf{q_r}, \mathbf{q_e}, \dot{\mathbf{q}}_\mathbf{r}, \dot{\mathbf{q}}_\mathbf{e}), \qquad \omega_i = \omega_i(\mathbf{q_r}, \mathbf{q_e}, \dot{\mathbf{q}}_\mathbf{r}, \dot{\mathbf{q}}_\mathbf{e}),$$
$$\mathbf{q_r} \in \mathbb{R}^{f_r}, \qquad \mathbf{q_e} \in \mathbb{R}^{f_e}, \qquad \mathbf{q} \in \mathbb{R}^{f_r + f_e}, \qquad f = f_r + f_e. \tag{3.122}$$

For deriving the equations of motion we first go back to the basic equation of kinematics (2.38), which gives also a rule for the time derivation of any vector with respect to a body-fixed frame. From this velocity and acceleration of a deformed element with respect to the element-coordinates comes out with (see also equation (3.120))

$$_{B_{ei}}\mathbf{v}_{ei} = \mathbf{v}_{ei} = {}_{B_{ei}}(\mathbf{E} + \tilde{\varphi}_i)^T {}_{B_i}[\mathbf{v}_i + \dot{\mathbf{u}}_i + \tilde{\omega}_i(\mathbf{r}_i + \mathbf{u}_i)],$$
$$_{B_{ei}}\omega_{ei} = \omega_{ei} = {}_{B_{ei}}(\mathbf{E} + \tilde{\varphi}_i)^T {}_{B_i}[\omega_i + \dot{\varphi}_i],$$
$$\mathbf{a}_{ei,abs} = \dot{\mathbf{v}}_{ei} + \tilde{\omega}_{ei} \cdot \mathbf{v}_{ei}, \qquad \frac{d}{dt}(\omega_{ei}) = \dot{\omega}_{ei} + \tilde{\omega}_i \cdot \omega_{ei}. \tag{3.123}$$

We start in a second step with Jourdain's principle, equation (3.48) and apply for the absolute velocity the form (eq. 3.52) supplemented by the elastic deformation vector (eqs. 3.120)

$$\sum_{i=1}^{n} \int_{B_{ei}} \delta(\dot{\mathbf{r}}_{ei})^T (\mathbf{a}_{ei} dm - d\mathbf{F}_{ai}) = \mathbf{0}, \qquad \delta \dot{\mathbf{r}}_{ei} = (\mathbf{E}_3 \quad \tilde{\mathbf{r}}^T)_{ei} \begin{pmatrix} \delta \mathbf{v}_{ei} \\ \delta \omega_{ei} \end{pmatrix} \tag{3.124}$$

Jourdain's principle has to be applied in our case to the deformed element, but having assumed only small deformations we shall be able to neglect higher order terms in the considerations to follow. Using the relations (3.120) for the absolute element velocities and regarding the fact, that only the deformation velocities depend on the generalized velocities $\dot{\mathbf{q}}_\mathbf{e}$ we can evaluate the virtual velocity of the deformed element and the accompanying Jacobians to yield

$$\delta \dot{\mathbf{r}}_{ei} = \begin{pmatrix} \delta \mathbf{v}_{ei} \\ \delta \omega_{ei} \end{pmatrix} = \begin{pmatrix} \frac{\partial \mathbf{v}_{ei}}{\partial \dot{\mathbf{q}}_r} & \frac{\partial \mathbf{v}_{ei}}{\partial \dot{\mathbf{q}}_e} \\ \frac{\partial \omega_{ei}}{\partial \dot{\mathbf{q}}_r} & \frac{\partial \omega_{ei}}{\partial \dot{\mathbf{q}}_e} \end{pmatrix} \cdot \delta \begin{pmatrix} \dot{\mathbf{q}}_r \\ \dot{\mathbf{q}}_e \end{pmatrix}$$

$$\frac{\partial \mathbf{v}_{ei}}{\partial \dot{\mathbf{q}}_r} = (\mathbf{E} + \tilde{\varphi}_i)^T \cdot [\frac{\partial \mathbf{v}_i}{\partial \dot{\mathbf{q}}_r} + (\tilde{\mathbf{r}}_i + \tilde{\mathbf{u}}_i)^T \frac{\partial \omega_i}{\partial \dot{\mathbf{q}}_r}]$$

$$\frac{\partial \mathbf{v}_{ei}}{\partial \dot{\mathbf{q}}_e} = (\mathbf{E} + \tilde{\varphi}_i)^T \cdot [\frac{\partial \mathbf{v}_i}{\partial \dot{\mathbf{q}}_e} + \frac{\partial \dot{\mathbf{u}}_i}{\partial \dot{\mathbf{q}}_e} + (\tilde{\mathbf{r}}_i + \tilde{\mathbf{u}}_i)^T \frac{\partial \omega_i}{\partial \dot{\mathbf{q}}_e}]$$

$$\frac{\partial \omega_{ei}}{\partial \dot{\mathbf{q}}_r} = (\mathbf{E} + \tilde{\varphi}_i)^T \cdot [\frac{\partial \omega_i}{\partial \dot{\mathbf{q}}_r}]$$

$$\frac{\partial \omega_{ei}}{\partial \dot{\mathbf{q}}_e} = (\mathbf{E} + \tilde{\varphi}_i)^T \cdot [\frac{\partial \omega_i}{\partial \dot{\mathbf{q}}_e} + \frac{\partial \dot{\varphi}_i}{\partial \dot{\mathbf{q}}_e}] \tag{3.125}$$

At this point we should recall the requirement, that in evaluating the above derivatives the original terms of $\mathbf{v}_{ei}$ and $\omega_{ei}$ have to be developed up to second order terms of the elastic deformations. Performing the derivations we come

## 3.3 Multibody Systems with Bilateral Constraints

back to first order expressions for the above Jacobians, which is absolutely necessary due to forces of zeroth order in the momentum and moment of momentum equations (see the following equations of motion and also the equations (2.133) to (2.135) on the pages 47).

Applying the above result to equation (3.124) and taking the acceleration from equation (3.123) gives for the equations of motion

$$\sum_{i=1}^{n} \int_{B_{ei}} [(\frac{\partial \mathbf{v}_{ei}}{\partial \dot{\mathbf{q}}_r})^T + (\frac{\partial \omega_{ei}}{\partial \dot{\mathbf{q}}_r})^T \cdot \tilde{\mathbf{r}}_{ei}] \cdot [(\dot{\mathbf{v}}_{ei} + \tilde{\omega}_{ei} \cdot \mathbf{v}_{ei}) dm - d\mathbf{F}_{ai}] = \mathbf{0},$$

$$\sum_{i=1}^{n} \int_{B_{ei}} [(\frac{\partial \mathbf{v}_{ei}}{\partial \dot{\mathbf{q}}_e})^T + (\frac{\partial \omega_{ei}}{\partial \dot{\mathbf{q}}_e})^T \cdot \tilde{\mathbf{r}}_{ei}] \cdot [(\dot{\mathbf{v}}_{ei} + \tilde{\omega}_{ei} \cdot \mathbf{v}_{ei}) dm - d\mathbf{F}_{ai}] = \mathbf{0}, \quad (3.126)$$

where the two equations result from the assumption, that the generalized velocity coordinates $\dot{q}_r, \dot{q}_e$ are indepently different from zero, which allows to split Jourdain's equations. As a matter of fact it makes not much sense to evaluate these equations in all details on an analytical level. We shall go on one step further, the rest must be done by computer evaluation.

Combining the equations (3.125) and (3.126) and regarding additionally the definitions of the relations (2.171) and (2.172) we may write the equations (3.126) in the following form (see [29], [196])

$$\sum_{i=1}^{n} \int_{B_{ei}} \{ [\frac{\partial \mathbf{v}_i}{\partial \dot{\mathbf{q}}_r} + (\tilde{\mathbf{r}}_i + \tilde{\mathbf{u}}_i)^T \cdot \frac{\partial \omega_i}{\partial \dot{\mathbf{q}}_r}]^T \cdot (\mathbf{E} + \tilde{\varphi}_i) \cdot [(\dot{\mathbf{v}}_{ei} + \tilde{\omega}_{ei} \cdot \mathbf{v}_{ei}) dm - d\mathbf{F}_{ai}] +$$

$$+ [\frac{\partial \omega_i}{\partial \dot{\mathbf{q}}_r}]^T \cdot (\mathbf{E} + \tilde{\varphi}_i) \cdot [(d\mathbf{I}_i \cdot \omega_{ei} + \tilde{\omega}_{ei} d\mathbf{I}_i \omega_{ei}) +$$

$$+ \tilde{\mathbf{r}}_{ei}^T (\dot{\mathbf{v}}_i + 2\tilde{\omega}_{ei} \dot{\mathbf{r}}_i + \ddot{\mathbf{r}}_i) dm - d\mathbf{M}_{ai}] \} = \mathbf{0},$$

$$\sum_{i=1}^{n} \int_{B_{ei}} \{ [\frac{\partial \mathbf{v}_i}{\partial \dot{\mathbf{q}}_e} + \frac{\partial \mathbf{u}_i}{\partial \dot{\mathbf{q}}_e} + (\tilde{\mathbf{r}}_i + \tilde{\mathbf{u}}_i)^T \cdot \frac{\partial \omega_i}{\partial \dot{\mathbf{q}}_e}]^T \cdot (\mathbf{E} + \tilde{\varphi}_i) \cdot$$

$$\cdot [(\dot{\mathbf{v}}_{ei} + \tilde{\omega}_{ei} \cdot \mathbf{v}_{ei}) dm - d\mathbf{F}_{ai}] +$$

$$+ [\frac{\partial \omega_i}{\partial \dot{\mathbf{q}}_r} + \frac{\partial \dot{\varphi}_i}{\partial \dot{\mathbf{q}}_e}]^T \cdot (\mathbf{E} + \tilde{\varphi}_i) \cdot [(d\mathbf{I}_i \cdot \omega_{ei} + \tilde{\omega}_{ei} d\mathbf{I}_i \omega_{ei}) +$$

$$+ \tilde{\mathbf{r}}_{ei}^T (\dot{\mathbf{v}}_i + 2\tilde{\omega}_{ei} \dot{\mathbf{r}}_i + \ddot{\mathbf{r}}_i) dm - d\mathbf{M}_{ai}] \} = \mathbf{0}. \quad (3.127)$$

The term in the last row of these two equations vanishes if we relate our coordinates to the center of mass, because $\int_{B_{ei}} \mathbf{r}_{ei} dm = m\mathbf{r}_{Si} = \mathbf{0}$ for the mass center. Furtheron, for small deformations we may assume that the definitions for the mass center and the moments of inertia give the same results for the undeformed and the deformed body.

The equations have to be completed by the elastic forces from chapter (2.3.2) given by the equation (2.156). Therefore we supplement in the above relations the forces $\int_{B_{ei}} d\mathbf{F}_{ai}$ by the elastic forces $\int_{\mathcal{B}} div(\sigma)dV$ writing

$$\int_{B_{ei}} d\mathbf{F}_{ai} \Longrightarrow \int_{B_{ei}} d\mathbf{F}_{ai} + \int_{\mathcal{B}_{ei}} div(\sigma)dV,$$

$$\int_{B_{ei}} d\mathbf{M}_{ai} \Longrightarrow \int_{B_{ei}} d\mathbf{M}_{ai} + \int_{\mathcal{B}_{ei}} \tilde{\mathbf{r}} div(\sigma)dV, \qquad (3.128)$$

which can be evaluated by using the relations from the chapters (2.3.2) and (2.2.8), equations (2.153), (2.155) and equations (2.140), (2.141).

The equations of motion (3.127) for a linearly elastic multibody system have been derived in a straightforward way without including any structural considerations. Similar to rigid multibody systems one can establish a recursional topology according to chapter (3.3.3), which in the view of a vast literature on this topic we shall not do here (see for example [27], [28], [242], [237] and others). The problem of the boundary conditions for the shape functions of a Ritz-approach can be solved heuristically for simple cases, but with the danger of bad convergence. A better approach are finite elements or for special body configuration cubic splines, which are local shape functions well adaptable also to complicated cases.

From the authors experience the following considerations are usually successful: The above equations of motion have been derived for a moving system, and the deformation vectors $\mathbf{u}_i$ are also given for the moving undeformed body (i). According to Figure (2.22) we define the deformation vector from a basis within the undeformed but moving body (i). Therefore it makes at least some physical sense to define the boundary conditions within or at the borders of this body, often with respect to the connection to the neighbouring bodies, then to evaluate the shape functions from an Eigenwert-analysis using these boundary conditions. It works in most cases as we shall see in the application part of the book.

### 3.3.5 Connections by Force Laws

Many interconnections of multibody systems may be modelled conveniently using force laws, which act between two bodies. Springs, dampers, fluids and visco-elastic materials are typical examples. The accompanying force laws might be linear or nonlinear depending also on the size of displacements and rotations along the lines of action. In the following we shall assume small displacements and rotations, which makes sense for many machine and automotive applications, but which do not necessarily lead to linear force laws. The visco-elastic elements for connecting the combustion engine with the automobile body are an example.

## 3.3 Multibody Systems with Bilateral Constraints

For the evaluation of such force laws we first must define the line (lines) of action by some unit vector $\psi_L = (\psi_{LT}^T, \psi_{LR}^T)^T \in \mathbb{R}^6$, where the index L stands for the force law and the indices T,R for translation and rotation, respectively. The unit vector is usually given by the configuration of the two bodies and their specific connection. The mesh of teeth represents an example from gears. Secondly we must evaluate the relative displacements and rotations between the two bodies, the positions and orientations of which we know from the integration process of the equations of motion for the multibody system. We also know the velocities and accelerations. Finally in a third step we project the magnitudes of relative kinematics into the force law directions with the help of the above unit vector. Knowing then the relative magnitudes in the line of action we are able to determine the forces and torques from the force law and introduce them as applied forces to the two interconnected bodies with the same absolute value but with opposite signs.

Considering only small displacements and rotations along the lines of action we can describe the above procedure by some simple formulas. Figure (3.12) indicates the most important features. We consider two bodies within a multibody system, and we go from a reference configuration to a displaced and rotated one. The points of connection are $C_{0i}$ and $C_{0p}$, which go to $C_i$

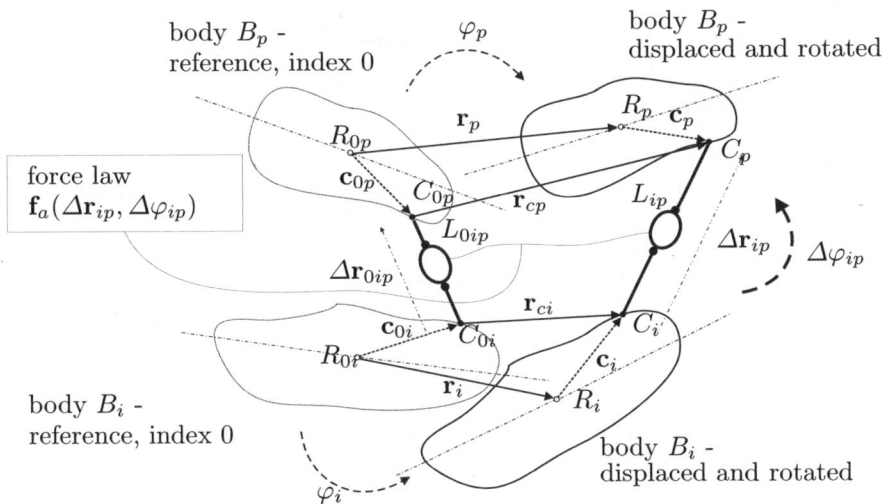

Fig. 3.12: Multibody Interconnection by Force Laws

and $C_p$ by the vectors $\mathbf{r}_{ci}$ and $\mathbf{r}_{cp}$. The vectors $\mathbf{c}$ from the reference points R to the force application points C remain unchanged, they are only rotated by $\mathbf{c} = (\mathbf{E} + \tilde{\varphi})\mathbf{c}_0$, small magnitudes provided. For the small displacements and rotations between $C_i$ and $C_p$ we get

$$\Delta \mathbf{r}_{cip} = \mathbf{r}_{ci} - \mathbf{r}_{cp}, \qquad \Delta \varphi_{ip} = \varphi_i - \varphi_p. \tag{3.129}$$

From Figure (3.12) and the corresponding quadrangles we express the vectors $\mathbf{r}_{ci}$ and $\mathbf{r}_{cp}$ by $\mathbf{r}_{ci} = (-\mathbf{c}_{0i} + \mathbf{r}_i + \mathbf{c}_i)$ and $\mathbf{r}_{cp} = (-\mathbf{c}_{0p} + \mathbf{r}_p + \mathbf{c}_p)$. With the transformation matrices $\mathbf{A}_{Li}$ and $\mathbf{A}_{Lp}$ from the i- and p-systems into the force law system "L" we receive

$$\Delta \mathbf{r}_{ip} = \mathbf{A}_{Li}(\mathbf{r}_i + \tilde{\varphi}_i \mathbf{c}_{0i}) - \mathbf{A}_{Lp}(\mathbf{r}_p + \tilde{\varphi}_p \mathbf{c}_{0p}), \tag{3.130}$$

which can be combined to give (for the indices i and p the same)

$$\Delta \mathbf{z}_{ip} = \mathbf{C}_{Li} \mathbf{z}_i - \mathbf{C}_{Lp} \mathbf{z}_p,$$

$$\Delta \mathbf{z}_{ip} = \begin{pmatrix} \Delta \mathbf{r}_{ip} \\ \Delta \varphi_{ip} \end{pmatrix}, \quad \mathbf{z}_i = \begin{pmatrix} \mathbf{r}_i \\ \varphi_i \end{pmatrix}, \quad \mathbf{C}_{Li} = \begin{pmatrix} \mathbf{A}_{Li} & -\mathbf{A}_{Li} \tilde{\mathbf{c}}_{0i} \\ \mathbf{0} & \mathbf{A}_{Li} \end{pmatrix}. \tag{3.131}$$

We have already introduced the unit vector $\psi_L = (\psi_{LT}^T, \psi_{LR}^T)^T \in \mathbb{R}^6$, which allows us to project the relative displacements $\Delta \mathbf{z}_{ip}$ into the direction of the force law. Therefore they can be expressed by

$$\gamma_L = \psi_L^T \cdot \Delta \mathbf{z}_{ip}, \quad \dot{\gamma}_L = \psi_L^T \cdot \Delta \dot{\mathbf{z}}_{ip}. \tag{3.132}$$

Knowing the relative displacements $\gamma_L$ and displacement velocities $\dot{\gamma}_L$ we can apply directly the given force law in the form

$$\zeta_L = \zeta_L(\gamma_L, \dot{\gamma}_L, t), \tag{3.133}$$

where the time t plays no significant role in most of the practically relevant force laws. Using the above equations we write the forces acting on the bodies i and p

$$\mathbf{f}_{ai} = +\mathbf{C}_{Li}^T \psi_L \zeta_L, \quad \mathbf{f}_{ap} = -\mathbf{C}_{Lp}^T \psi_L \zeta_L, \tag{3.134}$$

## 3.4 Multibody Systems with Unilateral Constraints

### 3.4.1 The General Problem

The most important properties of constraints have been discussed in chapter (3.1) and with respect to unilateral constraints in chapter(3.1.2.1). The basic problem in considering contacts and their unilateral behaviour consists in their kinematical variability, so-to-say in their kinematical richness, which as a consequence leads also on the kinetic side to a very "rich dynamics" characterized by non-smoothness. The fundamental aspect is complementarity, and the fundamental kinetic aspect are set-valued forces in combination with measure equations of motion instead of ordinary differential equations [161]. The concept of set-valued forces as a complement to single-valued forces in classical mechanics has been introduced by Glocker [87].

These ideas are perfectly adapted to the fact that contact forces evolve as a rule according to the kinematics and dynamics of the contact itself, they are not known beforehand. A typical example are Coulomb's laws, where for a sticking contact the contact force vector lies within the friction cone allowing sliding only, if we reach by the dynamics of the system the boundary of this friction cone. Within the cone an arbitraty set of contact forces may develop limited only by the boundary, that is the static friction limit, and by the possible external dynamical forces (see also Figure (3.4) in chapter (3.1.2.1)). Before entering the evaluation of the corresponding equations of motion we shall consider some simple and typical cases, the impact between two bodies, the properties of sliding and static friction and the multiple contact problem.

All classical textbooks on mechanics and most current research concentrate on mechanical systems with a few degrees of freedom and with one impulsive or frictional contact. Books and papers on chaotical properties very often use as mechanical examples impact or stick-slip systems. In the following we shall discuss some ideas [148],[166],[158],[138].

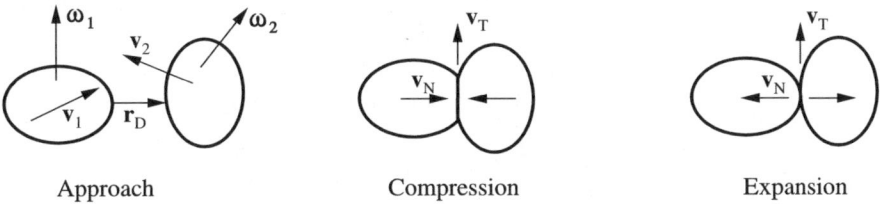

Fig. 3.13: Details of an Impact

Two bodies will collide if their relative distance becomes zero. This event is then a starting point for a process, which usually is assumed to have an extremely short duration. Nevertheless, deformation of the two bodies occurs, being composed of compression and expansion phases (Figure3.13). The forces governing this deformation depend on the initial dynamics and kinematics of the contacting bodies. The impulsive process ends when the normal force of contact vanishes and changes sign. The condition of zero relative distance cannot be used as an indicator for the end of an impact, because the bodies might separate in a deformed state. In the general case of impact with friction we must also consider a possible change from sliding to sticking, or vice versa, which includes frictional aspects as treated later.

In the simple case of only normal velocities we sometimes can idealize impacts according to Newton's impact laws, which relate the relative velocity after an impact with that before an impact. Such an idealization can only be performed if the force budget allows it. In the case of impacts by hard loaded bodies we must analyze the deformation in detail. Gear hammering taking place under heavy loads and gear rattling taking place under no load are typical examples [200].

As in all other contact dynamical problems, impacts possess complementarity properties. For ideal classical inelastic impacts either the relative velocity is zero and the accompanying normal constraint impulse is not zero, or vice versa. The scalar product of relative velocity and normal impulse is thus always zero. For the more complicated case of an impact with friction we shall find such a complementarity in each phase of the impact. A collision with friction in one contact only is characterized by a contact condition of vanishing relative distance and by two frictional conditions, either sliding or sticking (Figure (3.14)).

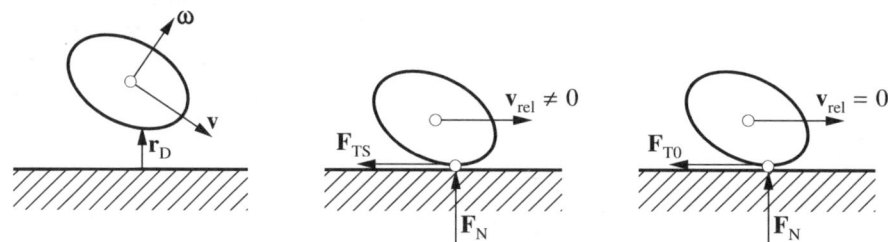

Fig. 3.14: Sliding and Static Friction

From the contact constraint $\mathbf{r}_D = \mathbf{0}$ we get a normal constraint force $\mathbf{F}_N$ which, according to Coulomb's laws, generates friction forces. For sliding $\mathbf{F}_{TS} = -\mu \mathbf{F}_N sgn(\mathbf{v}_{rel})$, and for stiction $\mathbf{F}_{T0} = \mu_0 \mathbf{F}_N$, where $\mu$ and $\mu_0$ are the coefficients of sliding and static friction, respectively. Stiction is indicated by $\mathbf{v}_{rel} = 0$ and by a reserve of the static friction force over the constraint force, which means $\mu_0|\mathbf{F}_N| - |\mathbf{F}_{TC}| \geq 0$. If this friction reserve becomes zero, the stiction situation will end, and sliding will start again with a nonzero relative acceleration $\mathbf{a}_{rel}$. Again we find here a complementary behavior: Either the

## 3.4 Multibody Systems with Unilateral Constraints

relative velocity (acceleration) is zero and the friction reserve is not zero, or vice versa. The product of relative acceleration and friction reserve is always zero.

Things become more complicated if we consider multiple contacts for a multibody system with $n$ bodies and $f$ degrees of freedom. In addition we have $n_G$ unilateral contacts where impacts and friction may occur. Each contact event is indicated by some indicator, for example, the beginning by a relative distance or a relative velocity and the end by a relevant constraint force condition. The constraint equation itself is always a kinematical relationship. If a constraint is active, it generates a constraint force; if it is passive, no constraint force appears.

In multibody systems with multiple contacts these contacts may be decoupled by springs or any other force law, or they may not. In the last case a change of the contact situation in only one contact results in a modified contact situation in all other contacts. If we characterize these situations by the combination of all active and passive constraint equations in all existing contacts, we get a huge combinatorial problem by any change in the unilateral and coupled contacts. Let us consider this problem in more detail [200].

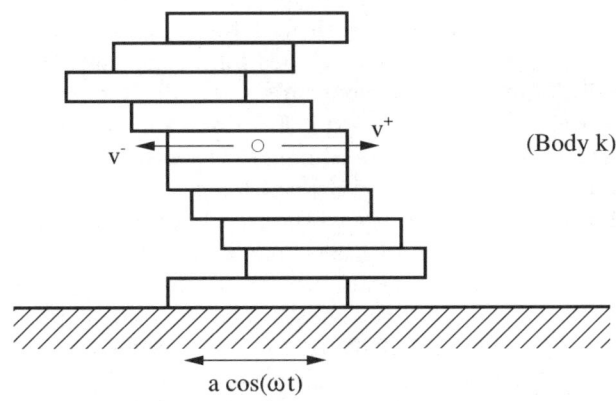

Fig. 3.15: A Combinatorial Problem

Figure (3.15) shows ten masses which may stick or slide on each other. The little mass tower is excited by a periodically vibrating table. Gravity forces and friction forces act on each mass, and each mass can move to the left with $v^-$, to the right with $v^+$, or not move at all. Each type of motion is connected with some passive or active constraint situation. Combining all ten masses, each of which has three possibilities of motion, results in $3^{10} = 59,049$ possible combinations of constraints. But only one is the correct constraint configuration. To find this one configuration is a crucial task by combinatorial search or an elegant way of applying the complementarity idea. We shall focus on this way.

Changes of the contact situation, and thus the constraint configuration, depend on the evolution of the state and, therefore, on the motion itself. They generate a discontinuously varying structure of the equations of motion. Such systems are often called systems with time-variant structure or with time-variant topology. It is also a typical property of all mechanical systems with impacts and friction in unilateral contacts.

### 3.4.2 Multibody Systems with Multiple Contacts

#### 3.4.2.1 Generalities

In chapter (2.2.6) we have considered the relative geometry and the relative kinematics of unilateral contacts, and in chapter (3.1.2) we discussed the significance of unilateral constraints for multibody systems with such contacts. The application of differential geometry is not necessary for all cases, but for many applications it will be a very convenient tool. The complementarity properties of unilateral contacts is inevitable, but can be expressed in various mathematical forms as indicated in chapter (3.1.2). To derive the equations of motion for multibody systems with unilateral contacts we start with the equations (3.106), which we have developed for a rigid multibody system, but which can be evaluated in the same form also for flexible multibody systems. The set of generalized coordinates ($\mathbf{q} \in \mathbb{R}^f$) looses for the unilateral case its character of being "generalized coordinates", because the number of degrees of freedom is not constant but varies with time.

In the following we shall discuss methods of taking into account additional constraints, especially friction-affected contact constraints [200],[279]. For this purpose, we have to include the arising contact forces into the equations of motion, and we have to regard the constrained directions by the kinematical contact equations of chapter (2.2.6). At sliding contacts the directions of the contact forces and the constrained displacements are *not* collinear, which is an important property of friction leading to additional difficulties by using generalized coordinates. Moreover, overconstrained systems, which can be easily handled in the frictionless case, provide uniqueness problems when connected by Coulomb's law. Before discussing these phenomena we first consider the equations of the superimposed constraints.

#### 3.4.2.2 Contact Forces

Two bodies in the state of separation ($g_{ni} > 0$) are depicted in Figure (3.16). A completely inelastic impact or a smooth touchdown bring both bodies into contact with each other. This leads generally to three additional constraints and, hence, to the occurrence of contact forces in the normal and tangential directions, which are shown in Figure (3.17). As a consequence the coordinates

## 3.4 Multibody Systems with Unilateral Constraints

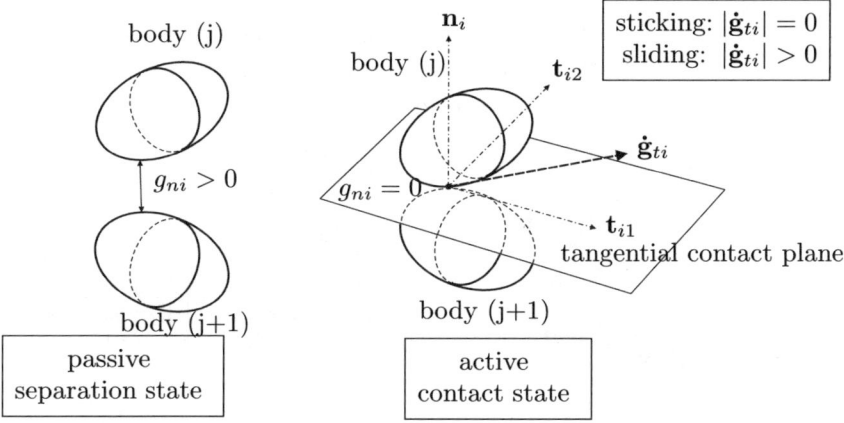

Fig. 3.16: Three-Dimensional Unilateral Constraints in a Multibody Contact (i)

$\mathbf{q}$ are no longer generalized coordinates but only a set of descriptor variables due to the new constraints. This situation can be handled by two different approaches:

Firstly, we could choose a new set of generalized coordinates ($\bar{\mathbf{q}} \in \mathbb{R}^{f-k}$), which is smaller then ($\mathbf{q} \in \mathbb{R}^{f}$), in order to further reduce the system similarly to the method used in Section (3.3.2). The constraint forces of the contact would then not be needed. This is, however, unacceptable for several reasons: Systems with more than one possible contact point would force us to choose for any imaginable contact configuration a certain set of generalized coordinates with varying dimensions. A set of $n$ possible contact points in the frictionless case, for example, would produce $\sum_{k=1}^{n} \frac{n!}{k!(n-k)!} = 2^n - 1$ different sets of generalized coordinates ($\bar{\mathbf{q}} \in \mathbb{R}^{f-k}$). This situation is additionally complicated if the constraints are not independent, or Coulomb friction is considered, where the active tangential friction force depends on the passive normal force in the case of sliding and becomes passive in the case of sticking.

Therefore we suggest a more direct method where the contact forces are included in equations of motion using a Lagrange multiplier approach. After premultiplying the contact forces in Figure (3.17) by the corresponding Jacobians, we get for each of the bodies one additional term which has to be added to the equations of motion (3.106). Before doing so we recall that a dynamical system with additional unilateral constraints possesses a time-varying number of degrees of freedom.

Figure (3.16) depicts a three-dimensional frictional constraint situation with a passive separation state and an active contact state. The constraint forces interact with the bodies only during the active state, in the passive separation state they are zero. To avoid the above mentioned problems with possible generalized coordinates we use a constant set of coordinates ($\mathbf{q} \in \mathbb{R}^{f}$) and regard all active unilateral constraints by additional constraint equations as discussed in chapter (3.1.2), thus leading to a set of differential algebraic equations (DAE's) with additional inequalities resulting from the transition laws with respect to the unilateral contacts.

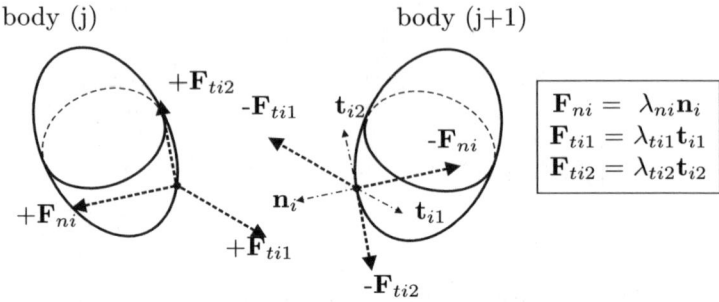

Fig. 3.17: Forces in a Multibody Contact (i)

We arrange the constraint vectors for normal and the constraint matrices for tangential constraints from the equations (2.132) in the following form

$$\mathbf{W}_N = [\cdots \mathbf{w}_{Ni} \cdots] \in \mathbb{R}^{f,n_N}, \quad i \in I_N,$$
$$\mathbf{W}_T = [\cdots \mathbf{W}_{Ti} \cdots] \in \mathbb{R}^{f,2n_T}, \quad i \in I_T. \tag{3.135}$$

The constraint matrices are projection matrices, called Jacobians, which project magnitudes of the constraint space into the space of the generalized coordinates. Vice versa, to go from the space of the generalized coordinates into the constraint space, we need the transpose of the constraint matrices.

According to Figure (3.17) we have for an active contact the following forces in normal and tangential direction

$$\mathbf{F}_{ni} = \lambda_{ni} \mathbf{n}_i, \qquad \mathbf{F}_{ti1} = \lambda_{ti1} \mathbf{t}_{i1}, \qquad \mathbf{F}_{ti2} = \lambda_{ti2} \mathbf{t}_{i2}. \tag{3.136}$$

These forces act in Figure (3.17) on the body (j) in a positive sense, according to the chosen coordinate frame, and this means with $+\mathbf{F}_{ni}$, $+\mathbf{F}_{ti1}$ and with $+\mathbf{F}_{ti2}$, and they act on body (j+1) in a negative sense, according to the usual cutting principle, and this means with $-\mathbf{F}_{ni}$, $-\mathbf{F}_{ti1}$ and with $-\mathbf{F}_{ti2}$. The vectors $\mathbf{n}_i$, $\mathbf{t}_{i1}$ and $\mathbf{t}_{i2}$ are unit vectors, and therefore the absolute values of the constraint forces are given by the three magnitudes $\lambda_{ni}$, $\lambda_{ti1}$ and $\lambda_{ti2}$, one in normal and two in tangential directions. For all active contacts we may collect these $\lambda$-values in the following vectors

$$\lambda_N(t) = \begin{pmatrix} \vdots \\ \lambda_{ni}(t) \\ \vdots \end{pmatrix} \in \mathbb{R}^{n_N}, i \in I_N, \qquad \lambda_T(t) = \begin{pmatrix} \vdots \\ \lambda_{ti}(t) \\ \vdots \end{pmatrix} \in \mathbb{R}^{2n_T}, i \in I_T,$$

$$\lambda_{ti}(t) = [\lambda_{ti1}(t), \lambda_{ti2}(t)]^T \in \mathbb{R}^2. \tag{3.137}$$

### 3.4.2.3 Equations of Motion

With the constraint vectors and matrices $\mathbf{w}_{Ni}$ and $\mathbf{W}_{Ti}$ we are able to project the passive constraint forces into the space of the generalized coordinates, where we include them into the equations of motion eqs. (3.106). It yields

$$\mathbf{M}\ddot{\mathbf{q}} + \mathbf{h} - \sum_{i \in I_N}(\mathbf{w}_{Ni}\lambda_{Ni} + \mathbf{W}_{Ti}\lambda_{Ti}) = \mathbf{0}. \tag{3.138}$$

We may have sliding or sticking frictional contacts. For sliding contacts the sliding friction forces are active forces contributing to the motion, and for sticking contacts the forces are passive forces not contributing to the motion. In the first case the frictional forces are positioned on the friction cone, in the second case they are within the friction cone. The sliding friction forces may follow Coulomb's law, if appropriate. It writes

$$\lambda_{Ti} = -\mu_i(\dot{\mathbf{g}}_i)\frac{\dot{\mathbf{g}}_i}{|\dot{\mathbf{g}}_i|}\lambda_{Ni}, \quad \forall i \in I_N \setminus I_T, \tag{3.139}$$

which defines the $(n_N - n_T)$ sliding friction forces. The coefficients $\mu_i$ usually depend on the relative velocity $\dot{\mathbf{g}}_i$ given by a so-called Stribeck curve (see for example Figure (3.3) on page 92). The negative sign expresses the fact, that the sliding friction force is opposite to the relative sliding velocity.

The friction laws like those of Coulomb or the impact laws of Newton and Poisson are physical relationships for the accompanying forces and are often called constitutive laws. They are purely empirical and must be determined from measurements. Inspite of the fact that especially Coulomb's laws give excellent results for an astonishingly large variety of frictional problems, they are not a dogma. Modern research in contact mechanics developed some extensions, which cover new aspects of friction ([72], [87], [136]), but in any case they must be experimentally verified too. Nevertheless and from the authors experience with all kinds of practical problems, the meaningfulness of Coulomb' laws is very convincing.

Transforming the sliding friction forces into the configuration space (space of $\mathbf{q}$) we have to premultiply $\lambda_{Ti}$ with $\mathbf{W}_{Ti}$ coming out with

$$\mathbf{W}_{Ti}\lambda_{Ti} = -\mu_i(\dot{\mathbf{g}}_i)\mathbf{W}_{Ti}\frac{\dot{\mathbf{g}}_i}{|\dot{\mathbf{g}}_i|}\lambda_{Ni} = \mathbf{W}_{Ri}\lambda_{Ni}, \quad \forall i \in I_N \setminus I_T,$$

$$\mathbf{W}_{Ri} = -\mu_i(\dot{\mathbf{g}}_i)\mathbf{W}_{Ti}\frac{\dot{\mathbf{g}}_i}{|\dot{\mathbf{g}}_i|} \in \mathbb{R}^{f,1} \quad \text{and} \quad \mathbf{W}_R \in \mathbb{R}^{f,n_N} \quad \forall i \in \mathbb{R}^{n_N}.$$

$$\tag{3.140}$$

Combining the equations (3.138) to (3.140) yields

$$\mathbf{M}(\mathbf{q},t)\ddot{\mathbf{q}}(t) + \mathbf{h}(\mathbf{q},\dot{\mathbf{q}}t) - [(\mathbf{W}_N + \mathbf{W}_R) \quad \mathbf{W}_T]\begin{pmatrix}\lambda_N(t)\\\lambda_T(t)\end{pmatrix} = \mathbf{0}, \tag{3.141}$$

with the following dimensions

$$\mathbf{q} \in \mathbb{R}^f, \quad \mathbf{M} \in \mathbb{R}^{f,f}, \quad \mathbf{h} \in \mathbb{R}^f, \quad \lambda_N \in \mathbb{R}^{n_N}, \quad \lambda_T \in \mathbb{R}^{2n_T},$$
$$\mathbf{W}_N \in \mathbb{R}^{f,n_N}, \quad \mathbf{W}_R \in \mathbb{R}^{f,n_N}, \quad \mathbf{W}_T \in \mathbb{R}^{f,2n_T}. \tag{3.142}$$

In equation (3.141) the constraint forces are added with the help of the Lagrange multiplier $\lambda$, where $\lambda_N$ includes all constraint forces in normal and $\lambda_T$ all constraint forces in tangential direction. As indicated above we have for spatial, two-dimensional contacts $\mathbb{R}^{2n_T}$ constraint forces for active contacts with stiction and for plane, one-dimensional contacts $\mathbb{R}^{n_T}$ constraint forces for the active contacts with stiction. The constraint matrix of sliding friction $\mathbf{W}_R$ has the same dimension as the constraint matrix of the active normal contacts $\mathbf{W}_N$, which follows immediately from equation (3.140). The number of the active contacts in the one or two (plane, spatial) tangential directions cannot be larger than those in the normal direction, $n_T \leq n_N$, because a contact must be closed before it becomes active in the tangential direction. Therefore the sliding friction matrix $\mathbf{W}_R$ possesses $(n_N - n_T)$ columns

$$\left[-\mu_i(\dot{\mathbf{g}}_i)\mathbf{W}_{Ti}\frac{\dot{\mathbf{g}}_i}{|\dot{\mathbf{g}}_i|}\right] \qquad \forall i \in I_N \setminus I_T,$$

and the remaining $n_T$ columns contain only zeros (for $I_N, I_T$ see the relations (3.9) on page 90)

The equations of motion (3.141) are of course not complete. We need additionally kinematical side conditions as considered in the chapter on relative contact kinematics (chapter 2.2.7), especially the equations (2.132) on page 47, which represent a relationship for the relative accelerations of the contacts. They concern the potentially active contacts and write

$$\ddot{\mathbf{g}}_N = \mathbf{W}_N^T \ddot{\mathbf{q}} + \bar{\mathbf{w}}_N \quad \in \mathbb{R}^{n_N}, \qquad \ddot{\mathbf{g}}_T = \mathbf{W}_T^T \ddot{\mathbf{q}} + \bar{\mathbf{w}}_T \quad \in \mathbb{R}^{2n_T}. \tag{3.143}$$

Note that these equations are given on an acceleration level, while originally the conditions for normal contacts are formulated on a position level considering relative distances, and the conditions for tangential contacts are given on a velocity level considering relative tangential velocities. This is important for numerical concepts, because any side conditions on an acceleration level generate drift.

Following the above consideration and those of chapter (2.2.7) we collect the dimensions

$$\begin{aligned}
\ddot{\mathbf{g}}_N &= [\cdots \ddot{g}_{Ni} \cdots]^T & \in \mathbb{R}^{n_N}, \quad i \in I_N, \\
\ddot{\mathbf{g}}_T &= [\cdots \ddot{\mathbf{g}}_{Ti}^T \cdots]^T & \in \mathbb{R}^{2n_T}, \quad i \in I_T, \\
\bar{\mathbf{w}}_N &= [\cdots \bar{w}_{Ni} \cdots]^T & \in \mathbb{R}^{n_N}, \quad i \in I_N, \\
\bar{\mathbf{w}}_T &= [\cdots \bar{\mathbf{w}}_{Ti}^T \cdots]^T & \in \mathbb{R}^{2n_T}, \quad i \in I_T.
\end{aligned} \tag{3.144}$$

The relations (3.141) and (3.143) result in a set

## 3.4 Multibody Systems with Unilateral Constraints

$$\mathbf{M}(\mathbf{q},t)\ddot{\mathbf{q}}(t) + \mathbf{h}(\mathbf{q},\dot{\mathbf{q}}t) - [(\mathbf{W}_N + \mathbf{W}_R) \quad \mathbf{W}_T]\begin{pmatrix}\lambda_N(t)\\\lambda_T(t)\end{pmatrix} = \mathbf{0} \quad \in \mathbb{R}^f,$$

$$\ddot{\mathbf{g}}_N = \mathbf{W}_N^T\ddot{\mathbf{q}} + \bar{\mathbf{w}}_N \quad \in \mathbb{R}^{n_N}, \qquad \ddot{\mathbf{g}}_T = \mathbf{W}_T^T\ddot{\mathbf{q}} + \bar{\mathbf{w}}_T \quad \in \mathbb{R}^{2n_T}, \tag{3.145}$$

which altogether comes out with $(f+n_N+2n_T)$ relations for the unknown generalized accelerations $\ddot{\mathbf{q}} \in \mathbb{R}^f$, for the constraint forces ($\lambda_N \in \mathbb{R}^{n_N}$, $\lambda_T \in \mathbb{R}^{2n_T}$) in normal and tangential directions and the accompanying relative accelerations in the contacts ($\ddot{\mathbf{g}}_N \in \mathbb{R}^{n_N}$, $\ddot{\mathbf{g}}_T \in \mathbb{R}^{2n_T}$).

Thus we have at the moment $(f+n_N+2n_T)$ equations for $[f+2(n_N+2n_T)]$ unknowns. The rest of $(n_N+2n_T)$ equations will be provided by the contact laws, which we have discussed in chapter (3.1.2) and formulated in a very compact form on page 95. Adding to the equations (3.145) the contact laws equations (3.28) to (3.31) on page 99 we come out with a complete set of equations of motion for a multibody system with unilateral multiple plane or spatial, dependent or independent contacts in the form

$$\mathbf{M}(\mathbf{q},t)\ddot{\mathbf{q}}(t) + \mathbf{h}(\mathbf{q},\dot{\mathbf{q}}t) - [(\mathbf{W}_N + \mathbf{W}_R) \quad \mathbf{W}_T]\begin{pmatrix}\lambda_N(t)\\\lambda_T(t)\end{pmatrix} = \mathbf{0} \quad \in \mathbb{R}^f,$$

$$\ddot{\mathbf{g}}_N = \mathbf{W}_N^T\ddot{\mathbf{q}} + \bar{\mathbf{w}}_N \quad \in \mathbb{R}^{n_N},$$

$$\ddot{\mathbf{g}}_T = \mathbf{W}_T^T\ddot{\mathbf{q}} + \bar{\mathbf{w}}_T \quad \in \mathbb{R}^{2n_T},$$

$$\lambda_N = \mathrm{prox}_{C_N}(\lambda_N - r\ddot{\mathbf{g}}_N), \qquad C_N = \{\lambda_N | \lambda_{Ni} \geq 0, \forall i \in I_N\},$$

$$\lambda_T = \mathrm{prox}_{C_T(\lambda_\mathbf{N})}(\lambda_T - r\ddot{\mathbf{g}}_T), \qquad C_T(\lambda_\mathbf{N}) = \{\lambda_T | |\lambda_{Ti}| \leq \mu_i\lambda_{Ni}, \forall i \in I_T\},$$

$$\lambda_N \in \mathbb{R}^{n_N}, \qquad \lambda_T \in \mathbb{R}^{2n_T}. \tag{3.146}$$

In addition we have to consider finite rotations by the kinematical equations

$$\dot{\mathbf{q}} = \mathbf{H}^* \cdot \dot{\mathbf{z}} \tag{3.147}$$

in an appropriate way and choosing the appropriate coordinates.

### 3.4.3 Friction Cone Linearization

For the construction of a linear complementarity problem we need the plane case, which corresponds to a vertical cut of the friction cone, or more physically, to a plane contact case with only one tangential contact direction. We may achieve that also for the spatial contact by approximating the corresponding friction cone by a pyramid. Among the first scientists to consider such an approximation of the friction cone by a polygonal pyramid were Pang and Trinkle in the United States [178], [263]. If the friction cone in the case of spatial contacts possesses a circular cross-section - the friction coefficients the same in all directions - the pyramid should be symmetric with each triangular side element the same. If the friction cone possesses for example an elliptic cross-section - the friction coefficients changing with the direction - the pyramid can be also asymmetric with variable triangular side elements.

140     3 Constraint Systems

Inspite of the fact, that we use today other approaches for the numerical solution of non-smooth dynamics problems, we nevertheless shall consider the friction cone approximation as a highly effective method to "linearize" contact problems. We shall consider only the case of independent contacts and refer for dependent contacts to the literature [279].

In the section (3.1) with the Figures (3.4) and (3.6) we have seen, that the friction reserve is defined by (see equation (3.17))

$$\lambda_{T0i}(\lambda_{Ni}, \lambda_{Ti}) = \mu_{0i}\lambda_{Ni} - |\lambda_{Ti}|$$
$$= \mu_{0i}\lambda_{Ni} - \sqrt{\lambda_{Ti1}^2 + \lambda_{Ti2}^2} \geq 0, \quad \forall i \in I_T. \tag{3.148}$$

The meaning is clear. The above formula represents the difference of the static friction force $\mu_{0i}\lambda_{Ni}$ defining the friction cone boundary and the current dynamic friction force $|\lambda_{Ti}|$ as generated by the dynamics of the systems and the relevant stiction constraint. This difference corresponds to the distance from the friction cone boundary thus giving the force reserve before the cone is reached again with the possibility of sliding. Clearly, the friction reserve will be always non-negative. For plane contacts we get only one tangential constraint force $\lambda_{Ti} = \lambda_{Ti1}$, for spatial contacts two constraint forces $\lambda_{Ti} = (\lambda_{Ti1}, \lambda_{Ti2})^T$. Consequently we have for plane contacts a Linear Complementarity Problem (LCP) and for spatial contacts a nonlinear one (NLCP), as we shall see immediately.

To avoid the nonlinear complementarity we approximate the friction cone by a l-sided pyramid, where for each side we can derive again a linear complementarity relationship. Figure (3.18) depicts the method for a contact (i). The linearized friction cone is a convex set, boundless in direction of the normal constraint force $\lambda_{Ni}$. Such a set is called a generalized polytope according to convex analysis. The friction reserve for each of the side elements (j) can be written as

$$\lambda_{T0ij}(\lambda_{Ni}, \lambda_{Ti}) = \mu_{0ij}\lambda_{Ni} - \lambda_{Ti1}\cos\alpha_j - \lambda_{Ti2}\sin\alpha_j \geq 0,$$
$$\forall j \in (1, 2, \cdots l), \tag{3.149}$$

which can be summarized in one single vector equation

$$\lambda_{\mathbf{T0i}} = \begin{pmatrix} \lambda_{T0i1} \\ \vdots \\ \lambda_{T0il} \end{pmatrix} = \begin{pmatrix} \mu_{0i1} \\ \vdots \\ \mu_{0il} \end{pmatrix} \lambda_{Ni} - \begin{pmatrix} \cos\alpha_1 & \sin\alpha_1 \\ \vdots & \vdots \\ \cos\alpha_l & \sin\alpha_l \end{pmatrix} \lambda_{\mathbf{Ti}}$$

$$= \bar{\mathbf{G}}_{Ni}^T \lambda_{Ni} - \bar{\mathbf{G}}_{Ti}^T \lambda_{Ti} \in \mathbb{R}^l \quad \forall i \in I_T. \tag{3.150}$$

The definitions of $\bar{\mathbf{G}}_{Ni}$ and $\bar{\mathbf{G}}_{Ti}$ are obvious (see also Figures 3.19 and 3.20). The l angles for the polytopes are simply

$$\alpha_j = (j-1)\frac{2\pi}{l}, \quad \forall j \in \{1, 2, \cdots l\}, \tag{3.151}$$

3.4 Multibody Systems with Unilateral Constraints    141

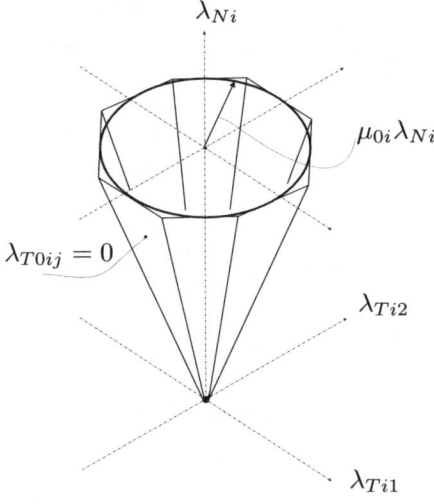

Fig. 3.18: Approximation of the Friction Cone by a Polygonal Pyramid

which is illustrated by Figure (3.19). From the two Figures 3.19 and 3.20 we can give an approximate definition, approximate with respect to the friction cone but not to the polytopes, of the permissible tangential friction forces with

$$C_{Tij}(\lambda_{Ni}) = \{\lambda_{Ti}^* \in \mathbb{R}^2 | -\lambda_{T0ij}(\lambda_{Ni}, \lambda_{Ti}^*) \leq 0\} \quad \forall j \in \{1, 2, \cdots l\}, \quad (3.152)$$

which is a convex set forming a half-space at the boundary (j) of the polytope pyramid. The intersection $(C_{Ti}(\lambda_{Ni}) = C_{Ti1} \cap C_{Ti2} \cap \cdots \cap C_{Til})$ includes all permissible tangential constraint forces for the friction pyramid. It writes

$$C_{Ti}(\lambda_{Ni}) = \{\lambda_{Ti}^* \in \mathbb{R}^2 | -\lambda_{T0i}(\lambda_{Ni}, \lambda_{Ti}^*) \leq 0\} \quad \forall i \in I_T. \quad (3.153)$$

We go now back to the equations (3.23) and (3.25) on the pages 96 and 97 using the equivalence of the subdifferential definition with the complementarity inequality (see [86], [279] and [135]) and watching that the subdifferential for the set (3.153) containing a linear relationship for the friction reserve $\lambda_{T0i}$ with respect to the tangential force $\lambda_{Ti}$ degenerates to a classical gradient. For the overall system we come to the following complementarity inequalities:

$$\lambda_{T0} \geq \mathbf{0}, \quad \ddot{\mathbf{g}}_T \geq \mathbf{0}, \quad \lambda_{T0}^T \ddot{\mathbf{g}}_T = 0. \quad (3.154)$$

These are altogether $(l \cdot n_T)$ complementarity inequalities for all friction polytopes and thus for all active tangential constraints $(i \in I_T)$. For simplicity we have assumed here l the same for all friction cones and also isotropic friction behaviour in all contacts, which is of course not necessary. The vectors $\ddot{\mathbf{g}}_T$ are the accelerations within the polytopes to the boundary of the polytopes. According to the above considerations we get the following definitions:

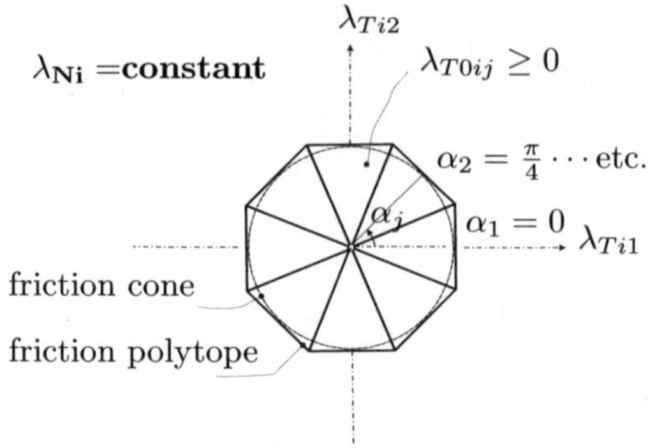

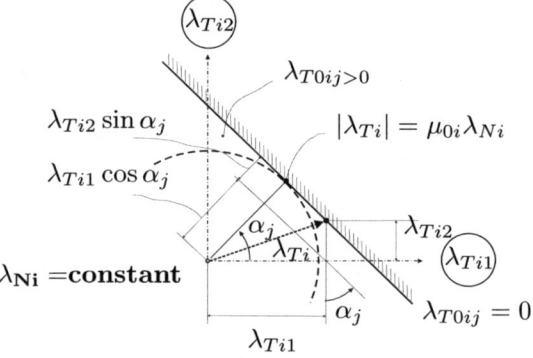

Fig. 3.19: Cut of the friction cone/pyramid at $\lambda_{Ni} = $ constant

Fig. 3.20: Linearization of the Friction reserve [279]

$$\ddot{\bar{\mathbf{g}}} = [\cdots \ddot{\bar{\mathbf{g}}}_{Ti}^T \cdots ]^T \in \mathbb{R}^{ln_T}, \quad i \in I_T,$$
$$\lambda_{T0} = [\cdots \lambda_{T0i}^T \cdots ]^T \in \mathbb{R}^{ln_T}, \quad i \in I_T. \tag{3.155}$$

The friction reserves $\lambda_{T0}$ can be put together according to the equation (3.148) in the form

$$\lambda_{T0} = \bar{\mathbf{G}}_N^T \lambda_N - \bar{\mathbf{G}}_T^T \lambda_T \in \mathbb{R}^{ln_T} \quad \forall i \in I_T. \tag{3.156}$$

with the following matrices and vectors

## 3.4 Multibody Systems with Unilateral Constraints

$$\lambda_N = [\cdots \lambda_{Ni} \cdots]^T \in \mathbb{R}^{n_N}, \quad i \in I_N,$$

$$\lambda_T = [\cdots \lambda_{Ti}^T \cdots]^T \in \mathbb{R}^{2n_T}, \quad i \in I_T,$$

$$\bar{\mathbf{G}}_N = \{\bar{\mathbf{G}}_{N\alpha\beta}\} \in \mathbb{R}^{n_N, l n_T}, \quad \alpha = (1, 2, \cdots n_N), \quad \beta = (1, 2, \cdots n_T),$$

$$\{\bar{\mathbf{G}}_{N\alpha\beta}\} = \begin{cases} \bar{\mathbf{G}}_{Ni} \in \mathbb{R}^{1,l} & \text{for } i \in I_N \cap I_T \\ 0 \in \mathbb{R}^{1,l} & \text{for } i \in I_N \setminus I_T \end{cases}$$

$$\bar{\mathbf{G}}_T = \text{diag}[\cdots \bar{\mathbf{G}}_{Ti} \cdots] \in \mathbb{R}^{2n_T, l n_T}, \quad i \in I_T. \tag{3.157}$$

We have defined above the relative tangential accelerations $\ddot{\bar{\mathbf{g}}}$ as the accelerations within the friction polytopes and with the direction to the polytope boundary (see Figure 3.20). These accelerations are given by

$$\ddot{\mathbf{g}} = -\bar{\mathbf{G}}_T \ddot{\bar{\mathbf{g}}} \in \mathbb{R}^{2n_T}, \quad \text{with} \quad \ddot{\mathbf{g}} = [\cdots \ddot{\mathbf{g}}_{Ti} \cdots]^T \in \mathbb{R}^{2n_T}, \quad i \in I_T \tag{3.158}$$

The friction laws as given with the relations (3.154) to (3.25) provide us with a sufficient number of equations for the case with independent contact constraints, they cannot be applied for the cases with dependent constraints. As a next step we shall carry together all relations for evaluating the complete equations of motion for the independent constraint case. Replacing in the equations (3.146) the prox-formulation by the complementarity equations (3.23) (first equation) on page 96 and (3.154) we finally get a complete set of equations of motion including complementarity inequalities.

$$\mathbf{M}(\mathbf{q},t)\ddot{\mathbf{q}}(t) + \mathbf{h}(\mathbf{q},\dot{\mathbf{q}}t) - [(\mathbf{W}_N + \mathbf{W}_R) \quad \mathbf{W}_T]\begin{pmatrix} \lambda_N(t) \\ \lambda_T(t) \end{pmatrix} = \mathbf{0} \quad \in \mathbb{R}^f,$$

$$\ddot{\mathbf{g}}_N = \mathbf{W}_N^T \ddot{\mathbf{q}} + \bar{\mathbf{w}}_N \quad \in \mathbb{R}^{n_N},$$

$$\ddot{\mathbf{g}}_T = \mathbf{W}_T^T \ddot{\mathbf{q}} + \bar{\mathbf{w}}_T \quad \in \mathbb{R}^{2n_T},$$

$$\ddot{\mathbf{g}}_N \geq \mathbf{0}, \quad \lambda_N \geq \mathbf{0}, \quad \ddot{\mathbf{g}}_N^T \lambda_N = 0,$$

$$\lambda_{T0} \geq \mathbf{0}, \quad \ddot{\bar{\mathbf{g}}}_T \geq \mathbf{0}, \quad \lambda_{T0}^T \ddot{\bar{\mathbf{g}}}_T = 0. \tag{3.159}$$

With the equations (3.159), (3.158) and (3.156) we have a complete set for all $(f + 2n_N + 2(l+2)n_T)$ unknowns, which are the following: The generalized accelerations $\ddot{\mathbf{q}} \in \mathbb{R}^f$, the constraint forces $\lambda_N \in \mathbb{R}^{n_N}$ and $\lambda_T \in \mathbb{R}^{2n_T}$, the relative contact accelerations $\ddot{\mathbf{g}}_N \in \mathbb{R}^{n_N}$ and $\ddot{\mathbf{g}}_T \in \mathbb{R}^{2n_T}$, the friction reserves $\lambda_{T0} \in \mathbb{R}^{l n_T}$ and finally the relative tangential acceleration $\ddot{\bar{\mathbf{g}}} \in \mathbb{R}^{l n_T}$. The magnitude pairs ($\lambda_N$ and $\ddot{\mathbf{g}}_N$), ($\lambda_{T0}$ and $\ddot{\bar{\mathbf{g}}}$) are complementary pairs and must be evaluated for constructing the complementarity inequalities. For reducing the above mentioned equations to the complementary pairs only we have to eliminate the magnitudes $\ddot{\mathbf{q}}$, $\lambda_T$ and $\ddot{\mathbf{g}}_T$ from these equations.

Taking in a first step the generalized accelerations $\ddot{\mathbf{q}}$ from the first equation (3.159) and inserting them into the second and the third equations we get

$$\begin{pmatrix} \ddot{\mathbf{g}}_N \\ \ddot{\mathbf{g}}_T \end{pmatrix} = \begin{pmatrix} \mathbf{W}_N^T \\ \mathbf{W}_T^T \end{pmatrix} \cdot \mathbf{M}^{-1}[(\mathbf{W}_N + \mathbf{W}_R) \quad \mathbf{W}_T] \begin{pmatrix} \lambda_N \\ \lambda_T \end{pmatrix}$$
$$- \begin{pmatrix} \mathbf{W}_N^T \\ \mathbf{W}_T^T \end{pmatrix} \cdot \mathbf{M}^{-1}\mathbf{h} + \begin{pmatrix} \bar{\mathbf{w}}_N \\ \bar{\mathbf{w}}_T \end{pmatrix},$$
$$\begin{pmatrix} \ddot{\mathbf{g}}_N \\ \ddot{\mathbf{g}}_T \end{pmatrix} = [\mathbf{W} + \mathbf{W}_R^*] \begin{pmatrix} \lambda_N \\ \lambda_T \end{pmatrix} + \bar{\mathbf{w}} \tag{3.160}$$

with the abbreviations

$$\mathbf{W} = \begin{pmatrix} \mathbf{W}_N^T \\ \mathbf{W}_T^T \end{pmatrix} \cdot \mathbf{M}^{-1}[\mathbf{W}_N \quad \mathbf{W}_T], \quad \in \mathbb{R}^{n_N+2n_T, n_N+2n_T}$$
$$\mathbf{W}_R^* = \begin{pmatrix} \mathbf{W}_N^T \\ \mathbf{W}_T^T \end{pmatrix} \cdot \mathbf{M}^{-1}[\mathbf{W}_R \quad \mathbf{0}], \quad \in \mathbb{R}^{n_N+2n_T, n_N+2n_T}$$
$$\bar{\mathbf{w}} = - \begin{pmatrix} \mathbf{W}_N^T \\ \mathbf{W}_T^T \end{pmatrix} \cdot \mathbf{M}^{-1}\mathbf{h} + \begin{pmatrix} \bar{\mathbf{w}}_N \\ \bar{\mathbf{w}}_T \end{pmatrix} \quad \in \mathbb{R}^{n_N+2n_T} \tag{3.161}$$

The equation (3.160) can be solved for the constraint forces, which yields

$$\begin{pmatrix} \lambda_N \\ \lambda_T \end{pmatrix} = [\mathbf{W} + \mathbf{W}_R^*]^{-1} \left\{ \begin{pmatrix} \ddot{\mathbf{g}}_N \\ \ddot{\mathbf{g}}_T \end{pmatrix} - \bar{\mathbf{w}} \right\} \tag{3.162}$$

Such a solution will be only possible, if the matrix $[\mathbf{W} + \mathbf{W}_R^*]$ is regular and thus its determinant not zero, $\det[\mathbf{W} + \mathbf{W}_R^*] \neq 0$. For independent contacts this property is always guaranteed, because as a consequence the rows and columns of the constraint matrix are also independent. It has full rank. Dependent constraints do not possess such features.

In a second step we combine the two relations (3.162) and (3.156) to achieve a set of equations for the complementary pair $(\lambda_N/\lambda_{T0})$ in the form

$$\begin{pmatrix} \lambda_N \\ \lambda_{T0} \end{pmatrix} = \begin{pmatrix} \mathbf{E} & \mathbf{0} \\ \bar{\mathbf{G}}_N^T & -\bar{\mathbf{G}}_T^T \end{pmatrix} \begin{pmatrix} \lambda_N \\ \lambda_T \end{pmatrix} = (\mathbf{I} + \bar{\mathbf{G}}) \begin{pmatrix} \lambda_N \\ \lambda_T \end{pmatrix}$$
$$\begin{pmatrix} \lambda_N \\ \lambda_{T0} \end{pmatrix} = (\mathbf{I} + \bar{\mathbf{G}})[\mathbf{W} + \mathbf{W}_R^*]^{-1} \left\{ \begin{pmatrix} \ddot{\mathbf{g}}_N \\ \ddot{\mathbf{g}}_T \end{pmatrix} - \bar{\mathbf{w}} \right\}$$

$$\mathbf{E} = \mathrm{diag}\{1,1,1,\cdots,1,1\} \quad \in \mathbb{R}^{n_N, n_N}$$
$$\mathbf{I} = \begin{pmatrix} \mathbf{E} & \mathbf{0} \\ \mathbf{0} & \mathbf{0} \end{pmatrix} \quad \in \mathbb{R}^{n_N+ln_T, n_N+2n_T}$$
$$\bar{\mathbf{G}} = \begin{pmatrix} \mathbf{0} & \mathbf{0} \\ \bar{\mathbf{G}}_N^T & -\bar{\mathbf{G}}_T^T \end{pmatrix} \quad \in \mathbb{R}^{n_N+ln_T, n_N+2n_T}$$
$$\bar{\mathbf{G}}_0 = \begin{pmatrix} \mathbf{0} & \mathbf{0} \\ \mathbf{0} & -\bar{\mathbf{G}}_T^T \end{pmatrix} \quad \in \mathbb{R}^{n_N+ln_T, n_N+2n_T} \tag{3.163}$$

where we need $\bar{\mathbf{G}}_0$ for the next eleimination step. Finally we can replace the relative acceleration by the equation (3.158) and come to the standard complementarity condition in the form

$$\begin{pmatrix} \lambda_N \\ \lambda_{T0} \end{pmatrix} = (\mathbf{I} + \bar{\mathbf{G}})[\mathbf{W} + \mathbf{W}_R^*]^{-1} \left\{ (\mathbf{I} + \bar{\mathbf{G}}_0)^T \begin{pmatrix} \ddot{\mathbf{g}}_N \\ \ddot{\mathbf{g}} \end{pmatrix} - \bar{\mathbf{w}} \right\}$$

$$\begin{pmatrix} \lambda_N \\ \lambda_{T0} \end{pmatrix} \geq \mathbf{0}, \qquad \begin{pmatrix} \ddot{\mathbf{g}}_N \\ \ddot{\mathbf{g}} \end{pmatrix} \geq \mathbf{0}, \qquad \begin{pmatrix} \lambda_N \\ \lambda_{T0} \end{pmatrix}^T \begin{pmatrix} \ddot{\mathbf{g}}_N \\ \ddot{\mathbf{g}} \end{pmatrix} = \mathbf{0}. \qquad (3.164)$$

This represents a linear standard complementarity problem for independent contacts and thus for regular constraint matrices.

### 3.4.4 Numerical Aspects

Refering to the discussion in the chapters (3.1) and (3.4) with respect to the evolution of the mathematical description of unilateral systems we state, that this evolution was very much oriented towards the computational needs for certain applications. The theories concerning unilateral multibody systems as well as those for bilateral ones are strongly related to the relevant solution procedures thus confirming the experience that saving computing time requires a "algorithm-friendly" theory.

Theory goes from linear and nonlinear complementarity problems with their corresponding inequalities and indicator interpolations via the augmented Lagrangian method to the prox-description including the augmented Lagrangian idea. All these theories and their accompanying numerical processes may be characterized by a decreasing effort in computing, in spite of the fact that computing times are still very large with respect to system problems. However, the introduction of the prox-description by Alart and Curnier [2] has been a real revolution with respect to computing, though it took some time before the scientific community recognized the really significant advantages connected with this idea.

In the following we shall follow the evolution as indicated above and start with event-driven solutions connected with the complementarity inequalities.

#### 3.4.4.1 Event-Driven Algorithms

Event-driven schemes detect changes of the constraints (events), for example stick-slip transitions, and resolve the exact transition times by an interpolation, for example using the regula falsi. Between the events the motion of the system is smooth and can be computed by a standard ODE/DAE-integrator with root-finding. If an event occurs, the integration stops, and the computation of the contact forces is performed by solving a LCP or a NLCP. While this approach is very accurate, the event-detection can be time consuming, especially in case of frequent transitions. Therefore the approach is only recommended for systems with few contacts. Another drawback consists in the fact that the constraints are only fulfilled at the acceleration level, which results in numerical drift effects. Woesle [279] gives a detailed overview of event-driven integration schemes in combination with different contact formulations like LCPs, NLCPs and the Augmented Lagrangian method.

### 3.4.4.2 Time-stepping schemes

Time-stepping schemes in connection with unilateral dynamics have been first proposed by Moreau [161] as a direct consequence of his measure differential equations. In a dissertation Stiegelmeyr applied these ideas to multibody systems and gives solutions for the dynamics of roller coasters and chimney dampers. Parts of the following representation are based on his numerical methods [252], [253]. Time stepping is based on a time-discretization of the system dynamics including the contact conditions in normal and tangential direction. The whole set of discretized equations and constraints is used to compute the next state of the motion. In contrast to event-driven schemes these methods need no event-detection. Moreover, time stepping allows to satisfy the unilateral constraints at the position and velocity level without any correction step [198].

We start with equation (3.159) on page 143 and use for a discretization the simple explicit Euler formula. The discretized acceleration then writes

$$\ddot{\mathbf{q}} = \frac{\mathbf{u}^{l+1} - \mathbf{u}^l}{\Delta t} \tag{3.165}$$

with the velocities $\mathbf{u}^l = \dot{\mathbf{q}}(t)$, $\mathbf{u}^{l+1} = \dot{\mathbf{q}}(t + \Delta t)$ and the time step $\Delta t$. The indices (l) and (l+1) stand for (t) and ($t + \Delta t$), respectively. The discretization on a position/orientation and velocity level is performed by an implicit Euler formula yielding

$$\mathbf{q}^{l+1} = \mathbf{q}^l + \mathbf{u}^{l+1}\Delta t \tag{3.166}$$

with the same abbreviations as above. The combination of an explicit and implicit Euler scheme is consistent with respect to the discretization of the complementarity problems [253], which would be not the case for two explicit formulas. As we want to give only an impression of a numerical scheme including the complementary inequalities, we restrict our considerations to the plane case, which results in linear complementarity problems. An extension to the spatial case is straightforward [253], [279].

The equations (3.165) and (3.166) represent Taylor expansions up to the first order only and are of course numerical approximations to the real solution, they underly discretization errors during the numerical evaluation, which must be controlled by standard methods. Time stepping possesses the advantage that we are able to consider the constraints in their original form on a position/orientation-level in normal and on a velocity-level in tangential directions, whereas the formulation in a classical way (equations 3.159) works on an acceleration-level requiring drift corrections. Therefore we start with the kinematical relations of chapter 2.2.7.1 on page 36 regarding especially the equations (2.112) on page 41 for the normal and equations (2.114) on page 42 for the tangential directions. They write

$$\dot{g}_N = \mathbf{w}_N^T \dot{\mathbf{q}} + \tilde{w}_N, \qquad \ddot{g}_T = \mathbf{w}_T^T \ddot{\mathbf{q}} + \hat{\mathbf{w}}_T^T \dot{\mathbf{q}} + \hat{w}_T. \tag{3.167}$$

## 3.4 Multibody Systems with Unilateral Constraints

The second equation has been modified by the term $\hat{\mathbf{w}}_T^T \dot{\mathbf{q}}$, which is included in the original magnitude $\bar{w}_T$ of equation (2.114) and can be evaluated by using the relations (2.115) with (2.107), (2.108) and (2.109). Including all this into the relations (3.159) results in:

$$\mathbf{u}^{l+1} = \mathbf{u}^l - \mathbf{M}^{-1}\{\mathbf{h} - [(\mathbf{W}_N + \mathbf{W}_R) \quad \mathbf{W}_T]\begin{pmatrix}\lambda_N \\ \lambda_T\end{pmatrix}\}\Delta t$$

$$\mathbf{q}^{l+1} = \mathbf{q}^l + \mathbf{u}^{l+1}\Delta t$$

$$\mathbf{g}_N^{l+1} = \mathbf{g}_N^l + \mathbf{W}_N^T(\mathbf{q}^{l+1} - \mathbf{q}^l) + \hat{\mathbf{w}}_N \Delta t + \mathbf{R}_N$$

$$\dot{\mathbf{g}}_T^{l+1} = \dot{\mathbf{g}}_T^l + \mathbf{W}_T^T(\mathbf{u}^{l+1} - \mathbf{u}^l) + \hat{\mathbf{W}}_T^T(\mathbf{q}^{l+1} - \mathbf{q}^l) + \hat{\mathbf{w}}_T \Delta t + \mathbf{R}_T$$

$$\mathbf{g}_N^{l+1} \geq \mathbf{0}, \qquad \lambda_N \geq \mathbf{0}, \qquad \mathbf{g}_N^{l+1}\lambda_N = 0,$$

$$\begin{cases} \dot{\mathbf{g}}_T^{(+)} \geq \mathbf{0}, & \lambda_{T0}^{(+)} \geq \mathbf{0}, & (\dot{\mathbf{g}}_T^{(+)})^T \lambda_{T0}^{(+)} = 0, \\ \dot{\mathbf{g}}_T^{(-)} \geq \mathbf{0}, & \lambda_{T0}^{(-)} \geq \mathbf{0}, & (\dot{\mathbf{g}}_T^{(-)})^T \lambda_{T0}^{(-)} = 0. \end{cases}_{l+1} \qquad (3.168)$$

The first equation corresponds to the discretized equations of motion, where the implicit form for the generalized coordinates is given by the second equation. The third relation is the normal relative distance in a contact, and the fourth equation represents the tangential relative velocity, both relations put on a system level by the **W**-matrices. The remaining three equations are complementarities in normal and tangential directions, to be evaluated at the time (l+1). This makes sense, because we start at time (l) with a physically consistent contact configuration and passing a contact event we want to have at the end of this event again a physically consistent contact configuration, which is assured by the last three complementarities in the equations (3.168). The tangential complementarities are split up into two unilateral primitives according to chapter 3.1.2.3 with the Figure 3.5 on page 95.

As next steps we introduce the abbreviations

$$\mathbf{\Lambda}_{N,T} = \lambda_{N,T}\Delta t, \qquad \mathbf{\Lambda}_{T0}^{(\pm)} = \lambda_{T0}^{(\pm)}\Delta t, \qquad \mathbf{H} = \mathbf{h}\Delta t, \qquad (3.169)$$

where the magnitudes $\mathbf{\Lambda}$ are shocks, "standard inelastic shocks" according to Moreau [161], and we establish a LCP (Linear Complementarity Problem) formulation by eliminating all unknowns not possessing a complementary partner. Looking at the equations (3.168) we see, that $\mathbf{q}^{l+1}$, $\mathbf{u}^{l+1}$ and $\mathbf{\Lambda}_T$ do not have such partners. Without going into the details of the elimination process by combining the relevant equations of the set (3.168) we come out with the results

$$\mathbf{g}_N^{l+1} = \mathbf{g}_N^l + [(\mathbf{G}_{NN} - \mathbf{G}_{NT}\mu) \quad \mathbf{G}_{NT}] \begin{pmatrix} \mathbf{\Lambda}_N \\ \mathbf{\Lambda}_{T0}^{(+)} \end{pmatrix} \} + \mathbf{G}_N \mathbf{H} + \mathbf{r}_N$$

$$\dot{\mathbf{g}}_T^{l+1} = \dot{\mathbf{g}}_T^l + [\mathbf{G}_{TN} \quad \mathbf{G}_{TT}] \begin{pmatrix} \mathbf{\Lambda}_N \\ \mathbf{\Lambda}_T \end{pmatrix} + \mathbf{G}_T \mathbf{H} + \mathbf{r}_T$$

$$\mathbf{\Lambda}_T = \mathbf{\Lambda}_{T0}^{(+)} - \mu \mathbf{\Lambda}_N \tag{3.170}$$

$$\mathbf{G}_N = \mathbf{W}_N^T \mathbf{M}^{-1}, \qquad \mathbf{G}_T = (\mathbf{W}_T + \hat{\mathbf{W}}_T \Delta t)^T \mathbf{M}^{-1},$$
$$\mathbf{r}_N = (\mathbf{W}_N^T \mathbf{u}^l + \hat{\mathbf{w}}_N)\Delta t + \mathbf{R}_N, \qquad \mathbf{r}_T = (\hat{\mathbf{W}}_T^T \mathbf{u}^l + \hat{\mathbf{w}}_T)\Delta t + \mathbf{R}_T. \tag{3.171}$$

Summarizing these results we get finally the following LCP (Linear Complementarity Problem):

$$\begin{pmatrix} \mathbf{g}_N^{l+1} \\ (\dot{\mathbf{g}}_T^{(+)})^{l+1}\Delta t \\ \mathbf{\Lambda}_{T0}^{(-)}\Delta t \end{pmatrix} = \begin{pmatrix} (\mathbf{G}_{NN} - \mathbf{G}_{NT}\mu) & \mathbf{G}_{NT} & \mathbf{0} \\ (\mathbf{G}_{TN} - \mathbf{G}_{TT}\mu) & \mathbf{G}_{TT} & \mathbf{E} \\ 2\mu & -\mathbf{E} & \mathbf{0} \end{pmatrix} \begin{pmatrix} \mathbf{\Lambda}_N \Delta t \\ \mathbf{\Lambda}_{T0}^{(+)}\Delta t \\ (\dot{\mathbf{g}}_T^{(-)})^{l+1}\Delta t \end{pmatrix} + $$
$$+ \begin{pmatrix} \mathbf{g}_N^l + \mathbf{G}_N \mathbf{H}\Delta t + \mathbf{r}_N \\ (\dot{\mathbf{g}}_T^l + \mathbf{G}_T \mathbf{H} + \mathbf{r}_T)\Delta t \\ \mathbf{0} \end{pmatrix},$$

$$\begin{pmatrix} \mathbf{g}_N^{l+1} \\ (\dot{\mathbf{g}}_T^{(+)})^{l+1}\Delta t \\ \mathbf{\Lambda}_{T0}^{(-)}\Delta t \end{pmatrix} \geq \mathbf{0}, \qquad \begin{pmatrix} \mathbf{\Lambda}_N \Delta t \\ \mathbf{\Lambda}_{T0}^{(+)}\Delta t \\ (\dot{\mathbf{g}}_T^{(-)})^{l+1}\Delta t \end{pmatrix} \geq \mathbf{0},$$

$$\begin{pmatrix} \mathbf{g}_N^{l+1} \\ (\dot{\mathbf{g}}_T^{(+)})^{l+1}\Delta t \\ \mathbf{\Lambda}_{T0}^{(-)}\Delta t \end{pmatrix}^T \begin{pmatrix} \mathbf{\Lambda}_N \Delta t \\ \mathbf{\Lambda}_{T0}^{(+)}\Delta t \\ (\dot{\mathbf{g}}_T^{(-)})^{l+1}\Delta t \end{pmatrix} = \mathbf{0}. \tag{3.172}$$

The third line in (3.172) represents the definition of the friction reserve. More precisely it is the sum of the two parts of the friction reserve multiplied by $\Delta t^2$. Expression (3.172) is a LCP in standard form and can be solved directly by a pivoting algorithm like Lemke's method or iteratively, with a Block-Gauss-Seidel relaxation scheme. Stiegelmeyr [253] considers in addition the numerical dissipation behaviour of first order time stepping, as above, and second order time stepping. It turns out that the "numerical energy of damping $E_{ND}$" is of second order small in the first order case, which means $E_{ND} \sim (\Delta t)^2$, but in the second order discretization of time stepping the "numerical energy" might lead as well to numerical damping as to numerical excitation. Therefore a second order approach does not apply reasonably well to impulsive problems.

#### 3.4.4.3 Augmented Lagrangian formulation

The "Augmented Lagrangian" formulation represents a completely different approach to obtain a solution of non-smooth problems. Originally presented

## 3.4 Multibody Systems with Unilateral Constraints

by Hestenes [99] and Powell [216] within the framework of nonlinear optimisation and then also considered by Rockafeller [225] it was applied by Alart and Curnier [2] to quasistatic mechanical problems. Leine [135] gives a nice description of the theory behind it and applies it to the Tippe-Top problem [136]. The name "Augmented Lagrangian" originates from the necessary extensions of smooth potentials by additional non-smooth terms.

To describe the numerical procedure in principle we start with the equations of motion given already in the form including prox-functions (equations (3.146) on page 139, see also [279] and [198]). In doing so we have to take care for not using the constraints in their original form, that means the normal constraints on a position/orientation level and the tangential constraints on a velocity level. Foerg [64] has shown that such a discretization includes a singular time step problem $\Lambda_N \sim \frac{1}{\Delta t}$ known from DAE systems of index 3 [5], which can be only avoided by formulating the discretized constraints in terms of velocities, both in normal and tangential directions. Taking care with respect to that problem we discretize (3.146) in the following form:

$$\mathbf{M}\Delta\mathbf{u} + \mathbf{h}\Delta t - [(\mathbf{W}_N + \mathbf{W}_R) \quad \mathbf{W}_T]\begin{pmatrix}\mathbf{\Lambda}_N \\ \mathbf{\Lambda}_T\end{pmatrix} = \mathbf{0},$$

$$\Delta \mathbf{q} = \mathbf{q}^{l+1} - \mathbf{q}^l = (\mathbf{u}^l + \Delta\mathbf{u})\Delta t$$

$$\left.\begin{array}{l}\dot{\mathbf{g}}_N = \mathbf{W}_N^T(\mathbf{u}^l + \Delta\mathbf{u}) + \bar{\mathbf{w}}_N \Delta t \\ \dot{\mathbf{g}}_T = \mathbf{W}_T^T(\mathbf{u}^l + \Delta\mathbf{u}) + \bar{\mathbf{w}}_T \Delta t\end{array}\right\}_{l+1}$$

$$\left.\begin{array}{l}\mathbf{\Lambda}_N - \text{prox}_{C_N}(\mathbf{\Lambda}_N - r\dot{\mathbf{g}}_N) = \mathbf{0} \\ \mathbf{\Lambda}_T - \text{prox}_{C_T(\mathbf{\Lambda}_N)}(\mathbf{\Lambda}_T - r\dot{\mathbf{g}}_T) = \mathbf{0}\end{array}\right\}_{l+1} \quad (3.173)$$

As before the constraints and the relative velocities are evaluated at the time (l+1). The abbreviations for the $\mathbf{\Lambda}$'s and the $\mathbf{u}$'s follow from the equations (3.166) and (3.169).

The numerical solution of these equations have to start with a choice of the factors (r) in such a way that the iteration process is stable and rapidly converging. Iteration procedures evaluating such a factor for better convergence are well known. They choose in most cases the "relaxation-factor" by an optimization problem looking for the minimum spectral radius of the iteration matrix [283]. Applied to our problem we have to reduce the numerical set of equations (3.173) to a fixed-point form which represents a nonlinear point mapping set of equations [65].

To indicate the procedure and to achieve a better overview we modify the equations (3.173) a bit by putting $\mathbf{W}_N \Rightarrow (\mathbf{W}_N + \mathbf{W}_R)$ and by assigning $\mathbf{u} \Rightarrow \mathbf{u}^{l+1} \Rightarrow (\mathbf{u}^l + \Delta\mathbf{u})$. Inserting the equations of motion (first equation of (3.173)) into the contact velocity equations (second and third equations of (3.173)) and these into the constraints (the last two equations of (3.173)) we finally receive

$$\begin{aligned}\mathbf{\Lambda}_N &= \text{prox}_{C_N}[(\mathbf{I}-r\mathbf{G}_{NN})\mathbf{\Lambda}_N - r\mathbf{G}_{NT}\mathbf{\Lambda}_T - r\mathbf{w}_N]\\ \mathbf{\Lambda}_T &= \text{prox}_{C_T}[(\mathbf{I}-r\mathbf{G}_{TT})\mathbf{\Lambda}_T - r\mathbf{G}_{TN}\mathbf{\Lambda}_N - r\mathbf{w}_T]\end{aligned} \qquad (3.174)$$

with the mass action matrix $[\mathbf{G}_{ij} = \mathbf{W}_i^T \mathbf{M}^{-1} \mathbf{W}_j, \quad (i,j) = (N,T)]$ and the abbreviations $[\mathbf{w}_i = \mathbf{W}_i^T \mathbf{M}^{-1}(\mathbf{u}+\mathbf{h}\Delta t), \quad (i) = (N,T)]$. The above equation can be interpreted as a fixed point equation of the form

$$\mathbf{\Lambda} = \begin{pmatrix}\mathbf{\Lambda}_N\\ \mathbf{\Lambda}_T\end{pmatrix} = \begin{pmatrix}\text{prox}_{C_N}[(\mathbf{I}-r\mathbf{G}_{NN})\mathbf{\Lambda}_N - r\mathbf{G}_{NT}\mathbf{\Lambda}_T - r\mathbf{w}_N]\\ \text{prox}_{C_T}[(\mathbf{I}-r\mathbf{G}_{TT})\mathbf{\Lambda}_T - r\mathbf{G}_{TN}\mathbf{\Lambda}_N - r\mathbf{w}_T]\end{pmatrix} = \mathbf{F}(\mathbf{\Lambda}) \qquad (3.175)$$

allowing according to [283] and [65] the following fixed point iteration

$$\mathbf{\Lambda}^{l+1} = \mathbf{F}(\mathbf{\Lambda}^l). \qquad (3.176)$$

This iteration converges locally, if the spectral radius of $\frac{\partial \mathbf{F}}{\partial \mathbf{\Lambda}}$ remains limited and smaller one, which can be realized by a corresponding optimization with respect to the factor (r)( for details see [65] and [64]).

The numerical procedure has been applied to many large technical systems with excellent results. The amazing fact consists in a very rapid convergence for nearly all problems, which can be substantiated by the structure of these equations and especially by the structure of the mass action matrix $\mathbf{G}$ with its projected masses into the not constrained directions (see also [64]).

### 3.4.5 The Continual Benchmark: Woodpecker Toy

The woodpecker toy has been from the very beginning the main motivation to look into non-smooth theories available at that time and to try to transfer these findings into multibody theory. The first investigation was published in 1984 [182] including a woodpecker analysis, where the impacts with friction were approximated by a semi-empirical method mainly based on measurements. The next step was then done by Glocker [86], who introduced a concise theoretical model of impacts with friction allowing a description without empirical approximations. He improved his results in [89]. The last step, at least at the time being, is a contribution of Zander [281], who analyzed the woodpecker going down an elastic bar based on a theory of elastic continua with non-smooth contacts.

A woodpecker toy hammering down a pole is a typical system combining impacts, friction and jamming. It consists of a sleeve, a spring and the woodpecker. The hole of the sleeve is slightly larger than the diameter of the pole, thus allowing a kind of pitching motion interrupted by impacts with friction. The motion of the woodpecker can be described by a limit cycle behavior as illustrated in Figure 3.21. The gravitation represents an energy source, the energy of which is transmitted to the woodpecker mass by the $y$-motion. The woodpecker itself oscillates and possesses a switching function by the beak for

## 3.4 Multibody Systems with Unilateral Constraints

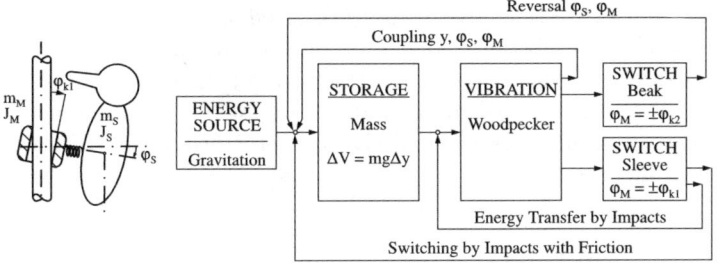

Fig. 3.21: Self-Sustained Vibration Mechanism of a Woodpecker Toy [200]

quick $\varphi_S$ reversal and by the jammed sleeve, which transmits energy to the spring by jamming impacts.

A typical sequence of events is portrayed in Figure 3.21. We start with jamming in a downward position, moving back again due to the deformation of the spring, and including a transition from one to three degrees of freedom between phases 1 and 2. Step 3 is jamming in an upward position (1 DOF) followed by a beak impact which supports a quick reversal of the $\varphi$-motion. Steps 5 to 7 are then equivalent to steps 3 to 1.

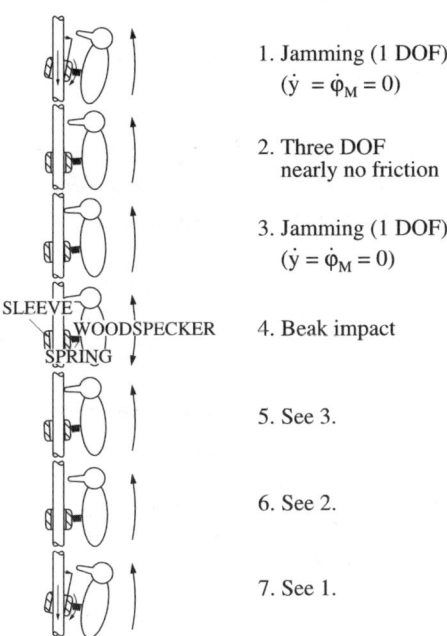

Fig. 3.22: Sequence of Events for a Woodpecker Toy [200]

The system possesses three degrees of freedom $q = (y, \varphi_M, \varphi_S)^T$, where $\varphi_S$ and $\varphi_M$ are the absolute angles of rotation of the woodpecker and the sleeve, respectively, and $y$ describes the vertical displacement of the sleeve (Figure 3.23): Horizontal deviations are negligible. The diameter of the hole in the sleeve is slightly larger than the diameter of the pole. Due to the resulting clearance, the lower or upper edge of the sleeve may come into contact

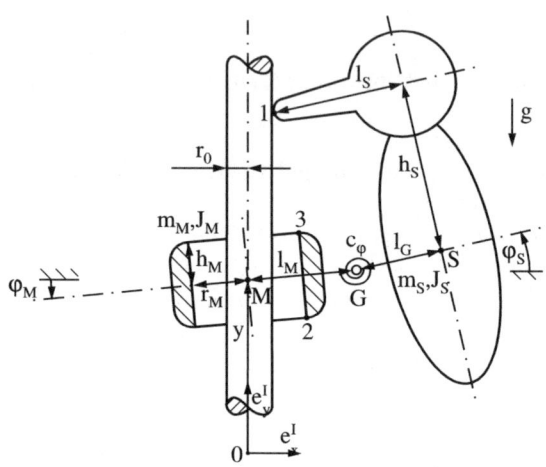

Fig. 3.23: Woodpecker Model

with the pole. This is modeled by constraints 2 and 3. Further contact may occur when the beak of the woodpecker hits the pole, which is expressed by constraint 1. The special geometrical design of the toy enables us to assume only small deviations of the displacements. Thus a linearized evaluation of the system's kinematics is sufficient and leads to the dynamical terms and constraint magnitudes listed below. For the dynamics of the woodpecker we apply the theory section 3.5 on page 158 for impacts with friction, but we assume that no tangential impulses are stored during the impulsive processes. The mass matrix $M$, the force vector $h$ and the constraint vectors $w$ follow from Figure 3.23 in a straightforward manner. They are

$$M = \begin{pmatrix} (m_S + m_M) & m_S l_M & m_S l_G \\ m_S l_M & (J_M + m_S l_M^2) & m_S l_M l_G \\ m_S l_G & m_S l_M l_G & (J_S + m_S l_G^2) \end{pmatrix}$$

3.4 Multibody Systems with Unilateral Constraints    153

$$\boldsymbol{h} = \begin{pmatrix} -(m_S + m_M)g \\ -c_\varphi(\varphi_M - \varphi_S) - m_S g l_M \\ -c_\varphi(\varphi_S - \varphi_M) - m_S g l_G \end{pmatrix} ; \quad \boldsymbol{q} = \begin{pmatrix} y \\ \varphi_M \\ \varphi_S \end{pmatrix}$$

$$\boldsymbol{w}_{N1} = \begin{pmatrix} 0 \\ 0 \\ -h_S \end{pmatrix} ; \quad \boldsymbol{w}_{N2} = \begin{pmatrix} 0 \\ h_M \\ 0 \end{pmatrix} ; \quad \boldsymbol{w}_{N3} = \begin{pmatrix} 0 \\ -h_M \\ 0 \end{pmatrix} \quad (3.177)$$

$$\boldsymbol{w}_{T1} = \begin{pmatrix} 1 \\ l_M \\ l_G - l_S \end{pmatrix} ; \quad \boldsymbol{w}_{T2} = \begin{pmatrix} 1 \\ r_M \\ 0 \end{pmatrix} ; \quad \boldsymbol{w}_{T3} = \begin{pmatrix} 1 \\ r_M \\ 0 \end{pmatrix} .$$

For a simulation we consider theoretically and experimentally a woodpecker toy with the following data set:

Dynamics: $m_M = 0.0003$; $J_M = 5.0 \cdot 10^{-9}$; $m_S = 0.0045$; $J_S = 7.0 \cdot 10^{-7}$; $c_\varphi = 0.0056$; $g = 9.81$.
Geometry: $r_0 = 0.0025$; $r_M = 0.0031$; $h_M = 0.0058$; $l_M = 0.010$; $l_G = 0.015$; $h_S = 0.02$; $l_S = 0.0201$.
Contact: $\varepsilon_{N1} = 0.5$; $\varepsilon_{N2} = \varepsilon_{N3} = 0.0$; $\mu_1 = \mu_2 = \mu_3 = 0.3$; $\varepsilon_{T1} = \varepsilon_{T2} = \varepsilon_{T3} = \nu_1 = \nu_2 = \nu_3 = 0.0$.

Using these parameters, the contact angles of the sleeve and the woodpecker result in $|\varphi_M| = 0.1$ rad and $\varphi_S = 0.12$ rad, respectively. Before discussing the dynamical behavior obtained by a numerical simulation, some results from an analytical investigation of the system may be presented.

Firstly, we assume that constraint 2 is sticking. The coordinates $\varphi_M$ and $y$ are then given by certain constant values ($\varphi_M = -0.1$ rad), and the system has only one degree of freedom ($\varphi_S$) with an equilibrium position at $\varphi_{S0} = -0.218$ rad. Sticking at that position is only possible if $\mu_2 \geq 0.285$. Such values of $\mu_2$ correspond at the same time to a jamming effect of the system in the sense, that no vertical force, acting on the woodpecker's center of mass, could lead to a transition to sliding, however large it would be. Undamped oscillations around this equilibrium with a frequency of 9.10 Hz influence the contact forces and lead the system to change into another state if the amplitudes are large enough.

The second analytically investigated system state is the unconstrained motion with three degrees of freedom. Besides the fourfold zero eigenvalue which describes the rotational and translational free-body motion, a complex pair of eigenvalues with a frequency of $f = 72.91$ Hz exists. The corresponding part of the eigenvector related to the coordinates $\boldsymbol{q} = (y, \varphi_M, \varphi_S)^T$ is given by $\boldsymbol{u} = (-0.086, 10.7, -1.0)^T$ and shows the ratio of the amplitudes.

The limit cycle of the system, computed by a numerical simulation, is depicted in Figure 3.24. We start our discussion at point (6) where the lower edge of the sleeve hits the pole. This completely inelastic frictional impact leads to continual contact of the sleeve with the pole. After a short episode of sliding (6)–(7) we observe a transition of the sleeve to sticking (7). The angle of the woodpecker is now large enough to ensure continual sticking of the sleeve by the self-locking mechanism. In that state the system has only one degree of freedom, and the 9.10-Hz oscillation can be observed where the woodpecker swings down and up until it reaches point (1).

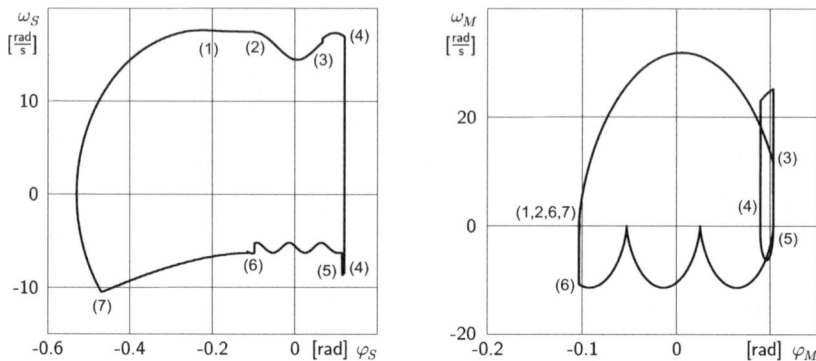

Fig. 3.24: Phase Space Portraits [89]

At (1) the tangential constraint becomes passive and the sleeve slides up to point (2) where contact is lost. Note that the spring is not free of stresses in this situation; thus during the free-flight phase (2)–(3) the high-frequency oscillation ($f = 72.91$ Hz) of the unbound system occurs in the phase space plots. In this state the sleeve moves downward ($y$ decreases), and the first part of the falling height $\Delta y$ at one cycle is achieved.

At (3) the upper edge of the sleeve hits the pole with a frictional, completely inelastic impact. Contact, however, is not maintained due to the loaded

| Phase Plot Point | State Transitions | |
|---|---|---|
| (1) Constraint 2 | Sticking → Sliding | |
| (2) Constraint 2 | Sliding → Separation | |
| (3) Constraint 3 | Separation → Separation | First upper sleeve impact |
| (4) Constraint 1 | Separation → Separation | Beak impact |
| (5) Constraint 3 | Separation → Separation | Second upper sleeve impact |
| (6) Constraint 2 | Separation → Sliding | Lower sleeve impact |
| (7) Constraint 2 | Sliding → Sticking | |

Fig. 3.25: Table of the Possible Transitions

| Change in Potential Energy | $\Delta V = 2.716 \cdot 10^{-4}$ | 100.00% |
|---|---|---|
| First upper sleeve impact | $\Delta T_{(3)} = -0.223 \cdot 10^{-4}$ | 8.21% |
| Second upper sleeve impact | $\Delta T_{(5)} = -0.046 \cdot 10^{-4}$ | 1.69% |
| Beak impact | $\Delta T_{(4)} = -1.370 \cdot 10^{-4}$ | 50.44% |
| Lower sleeve impact | $\Delta T_{(3)} = -1.032 \cdot 10^{-4}$ | 38.00% |
| Phases of sliding | $\Delta T_{(G)} = -0.045 \cdot 10^{-4}$ | 1.66% |

Fig. 3.26: Amounts of Dissipated Energy

spring. Point (4) corresponds to a partly elastic impact of the beak against the pole. After that collision the velocity $\dot{\varphi}_S$ is negative and the woodpecker starts to swing downward. At (5) the upper edge of the sleeve hits the pole a second time with immediate separation. Then the system is unbound and moving downward (5)–(6), where the second part of the falling height is achieved and the 72.91-Hz frequency can be observed once more.

Table 3.25 summarizes all of the state transitions during one cycle, and Table 3.26 compares the amounts of dissipated energy. The main dissipation results from the beak impact and lower sleeve impact, which contribute 88% of the dissipation. The remaining 12% are shared by the upper sleeve impacts and phases of sliding which are nearly negligible. The frequency of the computed limit cycle in Figure 3.24 amounts to $f = 8.98$ Hz and is slightly different from the measured value of $f = 9.2$ Hz. The total falling height during one cycle can be seen in the left diagram of Figure 3.24. The computed and measured values are $\Delta y = 5.7$ mm and $\Delta y = 5.3$ mm, respectively.

### 3.4.6 Some Empirical Conclusions

The application of all multibody theories, bilateral and unilateral, may be characterized by a permanent fight against large computing times. In spite of the overwhelming evolution of computer systems the requirements with regard to large system dynamics were and still are always a bit ahead of the computing possibilities. On the other hand computer performance is only one aspect, the generation of better and with respect to numerical procedures faster theories and algorithms is another aspect. Therefore the contents of this book reflect also a bit these trends.

We have considered for the unilateral case complementarity representations and the ideas of the Augmented Lagrangian solutions in connection with the prox-functions. We did not discuss the application of the so-called complementarity functions as introduced for example by Mangasarian [151], which replace the complementarity problem by a set of nonlinear functions. Such functions have been an alternative way of considering unilaterality, because they avoid the inequalities. We shall try to find an assessment with respect to some of the existing methods, the more popular ones like complementarity and prox-approach, and the not so popular complementarity functions. A comparison is carried out for example in [279].

As it is always difficult to establish for one problem various numerical solutions we shall estimate the numerical expenditures by considering the dimensions of a problem. Let us define a hypothetical example with the following data:

$$\left.\begin{array}{rl} f = & 6 \quad \text{degrees of freedom} \\ n_N = & 3 \quad \text{active normal constraints} \\ n_T = & 3 \quad \text{active tangential constraints} \\ l = & 8 \quad \text{faces of the polygonal pyramid} \end{array}\right\} \tag{3.178}$$

Our hypothetical problem has 6 degrees of freedom (DOF), 3 normal and 3 tangential active constraints, and the friction cones will be approximated by a pyramid with 8 lateral faces.

According to chapter 3.4.3 we get for the above spatial case with a friction cone linearization of an 8-sided pyramid a dimension of the linear complementarity problem of $2(n_N + 2ln_T) = 102$. It increases linearly with the number of lateral faces for the polygonal pyramid, for a 12-faced pyramid we would have a dimension of 150. On the other side the computing time to solve a linear complementarity problem by a pivot-algorithm increases with the third power of the dimension, in our example this means for going from 8 faces to 12 faces of the pyramid an increase by a factor of 3.18 or 318%. These properties should be seen before the background that the approximation of the friction cone will be improved only by more lateral pyramid faces. We conclude that the complementarity solution with the Lemke algorithm is not well suited for large systems and applicable only for special and smaller problems.

Considering the approach with complementarity functions ([151], [279]) we receive for the example (3.178) altogether $f + 2n_N + 5n_T = 27$ nonlinear algebraic equations, which are solved either by a Newton-Raphson- or by a homotopy-method. The recently introduced approach using prox-functions (equations 3.146 on page 139) comprises $f + n_N + 2n_T = 15$ nonlinear equations. Computing time for a numerical solution of nonlinear equations is approximately proportional to the square of the number of equations, which represents of course an additional advantage with respect to complementarity solutions.

In summary we state, that for the example (3.178) we get a linear complementarity problem regarding an eight-faced pyramid approach with a LCP-dimension 102 including a third-power dependency of this dimension regarding computing time, for the complementarity functional approach (Mangasarian [151]) we receive 27 and for the prox-approach 15 nonlinear algebraic equations with only a second-power dependency on these numbers of equations with respect to computing time. This second-power dependency on the problem size can additionally and significantly reduced by skilful use of the equations' structure, especially in the case of the prox-function approach ([65],[64]), where the computing time is proportional to $(f + n_N + 2n_T)^p$ with $2 \geq p \geq 1$.

All this results in very large advantages for the Augmented-Lagrangian-prox formulation, not alone with respect to computing time, but also with respect to numerical stability and convergence in connection with an ease of application to large practical problems.

## 3.5 Impact Systems

### 3.5.1 General Features

Impacts are contact processes of very short duration time, which generate in a small zone of the contact some typical deformations. Assuming the general case of two arbitrarily approaching bodies with normal and tangential relative velocities with respect to the future impact area we get at the instant of the impact also normal and tangential impact forces, which deform the two bodies in normal and tangential directions. These deformations are accompanied by a partial conversion of the incoming kinetic energy into potential energy of the deformations, which act similarly as a kind of spring on the bodies when moving apart again. The process is accompanied by energy losses, which in special cases might be large enough, that the bodies do not separate again.

Modelling such processes requires the knowledge of the local stiffnesses near the point of contact, an inclusion of the corresponding forces into the equations of motion usually leading to stiff differential equations and finally a numerical integration with very small time steps. Such models are very often used for problems, where we need to know the details of the impact deformations [118], [255]. For mechanical systems with many contacts the method results in very large computing times due to the elastic resolution of the individual impacts.

An alternative approach consists in rigid body models. We go from the above force/acceleration level of the equations of motion to an impulse/velocity level by integrating the equations of motion during the very small assumed duration time. Or better, we use the concept of measure differential equations instead the classical equations of motion as suggested by Moreau [161]. Working on an impulse-velocity-level instead on a force-acceleration-level might be a problem for some practical applications. But from the engineering standpoint of view it is always possible to develop for the worst cases results from some local models to estimate the impact duration time and then to evaluate the forces. In many cases, for example for noise or wear problems, forces are not directly needed, but more the influence of contact processes on the multibody system as a whole.

For "rigid body contacts" the following assumptions will be made [276]:

- The duration of the impact is so short, that the mathematical description may assume a zero impact time.
- As a consequence we neglect wave processes, which would take place in a finite time interval.
- Following these assumptions the mass distribution of the body does not change during the impact, the bodies remain rigid or elastic as a whole.
- All positions and orientations of the impacting bodies remain constant. The translational and rotational velocities of the bodies are finite and may change jerkyly during the impact.

- Accordingly the position of the impact point and that of the normal and tangential vectors remain constant.
- All forces and torques, which are not impulsive forces and torques, remain also constant during the impact.
- All during the impact evolving impulses act during the impact in a constant direction. Their lines of action do not change and correspond to the normal and tangential vectors in the impact point.
- The impact can be divided into two phases: the compression phase and the expansion phase.
- The compression phase starts at time $t_A$ and ends at time $t_C$. The end of the compression equals the start of the expansion phase. Expansion is finished at time $t_E$, which is also the end of the impact.

### 3.5.2 Classical Approach

#### 3.5.2.1 Newton's Impacts

Newton's impact law is a kinematical law, which connects the relative velocities in normal direction before an impact with those after an impact (see chapter 2.5 on page 68). In multiple-contact problems there might be one impact only in one of the contacts or several impacts in several contacts simultaneously. The theory presented will cover both possibilities. The locations of impacts are given by the $n_A$ contact points of $I_A$ (see equation (3.9) on page 90, chapter 3.1.2). For each of them we can write the distance $g_{Ni}(\mathbf{q}, t)$ in the normal direction. If one or more of these indicators becomes zero at one time instant $t_A$ and the corresponding relative velocities $\dot{g}_{Ni}$ are less than zero, an impact occurs. The impact contacts are then closed and the unilateral constraints are active. The set of constraints which participate in the impact is then given by

$$I_C^* = \{i \in I_A \mid g_{Ni} = 0; \ \dot{g}_{Ni} \leq 0\} \quad \text{with } n_C^* \text{ elements.}$$

From equation (3.138) we get the equations of motion for a constrained system without friction:

$$\mathbf{M}\ddot{\mathbf{q}} + \mathbf{h} - \sum_{i \in I_S^*} (\mathbf{w}_N \lambda_N)_i = \mathbf{0} \quad \in \mathbb{R}^f, \tag{3.179}$$

where the terms $(\mathbf{w}_N \lambda_N)_i$ result from a projection of the normal contact forces into the space of the generalized coordinates. As a second equation we use the relative velocity in the normal direction (eq. 3.143):

$$\dot{g}_{Ni} = \mathbf{w}_{Ni}^T \dot{\mathbf{q}} + \tilde{w}_{Ni}; \quad i \in I_C^*. \tag{3.180}$$

Equations (3.179) and (3.180) can be stated in matrix notation:

$$\mathbf{M}\ddot{\mathbf{q}} + \mathbf{h} - \mathbf{W}_N \lambda_N = \mathbf{0}; \qquad \dot{\mathbf{g}}_N = \mathbf{W}_N^T \dot{\mathbf{q}} + \tilde{\mathbf{w}}_N, \tag{3.181}$$

where we will assume independent constraints, rank $\mathbf{W}_N = n_C^* \leq f$. Furthermore $\mathbf{M} = \mathbf{M}(\mathbf{q}, t)$, $\mathbf{h} = \mathbf{h}(\dot{\mathbf{q}}, \mathbf{q}, t)$ and $\mathbf{W}_N = \mathbf{W}_N(\mathbf{q}, t)$, $\tilde{\mathbf{w}}_N = \tilde{\mathbf{w}}_N(\mathbf{q}, t)$.

Next, we integrate over the time interval of the impact to achieve a representation of the equations of motion on the impulse level. Let $t_A$ and $t_E$ denote the time instances at the beginning and end of the impact, respectively, and let

$$\dot{\mathbf{q}}_A = \dot{\mathbf{q}}(t_A); \qquad \dot{\mathbf{q}}_E = \dot{\mathbf{q}}(t_E)$$

be the generalized velocities at these instances. The relative velocities in the normal direction are then given by

$$\dot{\mathbf{g}}_{NA} = \dot{\mathbf{g}}_N(t_A); \qquad \dot{\mathbf{g}}_{NE} = \dot{\mathbf{g}}_N(t_E),$$

and the integration of the dynamics equation (3.181) over the impact yields

$$\lim_{t_E \to t_A} \int_{t_A}^{t_E} (\mathbf{M}\ddot{\mathbf{q}} + \mathbf{h} - \mathbf{W}_N \lambda_N) \, dt = \mathbf{0}. \tag{3.182}$$

During this integration, only terms that can rise to infinity have to be taken into account. The vector $\mathbf{h}$ consists of finite nonimpulsive terms and therefore vanishes. Under the assumption of constant displacements we get

$$\mathbf{M}(\dot{\mathbf{q}}_E - \dot{\mathbf{q}}_A) - \mathbf{W}_N \mathbf{\Lambda}_N = \mathbf{0} \quad \text{with} \quad \mathbf{\Lambda}_N = \lim_{t_E \to t_A} \int_{t_A}^{t_E} \lambda_N \, dt \tag{3.183}$$

with $\mathbf{\Lambda}_N$ being the impulses transferred by the contacts during the impact. Finally, we state the relative velocities in eqs. (3.181) at the instances $t_A$ and $t_E$ to be

$$\dot{\mathbf{g}}_{NA} = \mathbf{W}_N^T \dot{\mathbf{q}}_A + \tilde{\mathbf{w}}_N; \qquad \dot{\mathbf{g}}_{NE} = \mathbf{W}_N^T \dot{\mathbf{q}}_E + \tilde{\mathbf{w}}_N, \tag{3.184}$$

and express (3.184) for convenience as a sum and a difference:

$$\begin{aligned}\dot{\mathbf{g}}_{NE} + \dot{\mathbf{g}}_{NA} &= \mathbf{W}_N^T (\dot{\mathbf{q}}_E + \dot{\mathbf{q}}_A) + 2\tilde{\mathbf{w}}_N \\ \dot{\mathbf{g}}_{NE} - \dot{\mathbf{g}}_{NA} &= \mathbf{W}_N^T (\dot{\mathbf{q}}_E - \dot{\mathbf{q}}_A).\end{aligned} \tag{3.185}$$

After the elimination of the $f$-vector $(\dot{\mathbf{q}}_E - \dot{\mathbf{q}}_A)$ with the help of eq. (3.183), the second equation of (3.185) consists of $n_C^*$ relations for the $2n_C^*$ unknowns $(\dot{\mathbf{g}}_{NE}, \mathbf{\Lambda}_N)$:

$$\dot{\mathbf{g}}_{NE} - \dot{\mathbf{g}}_{NA} = \mathbf{G}_N \mathbf{\Lambda}_N; \qquad \mathbf{G}_N = \mathbf{W}_N^T \mathbf{M}^{-1} \mathbf{W}_N, \tag{3.186}$$

thus $n_S^*$ conditions are missing to determine the transferred impulses $\mathbf{\Lambda}_N$ and the relative velocities $\dot{\mathbf{g}}_{NE}$ at the end of the impact. These missing conditions

## 3.5 Impact Systems

are the impact laws of the problem. Here we will use Newton's law, which connects the relative velocities before and after the impact by the relation

$$\dot{\mathbf{g}}_{NE} = -\bar{\bar{\boldsymbol{\varepsilon}}}_N \dot{\mathbf{g}}_{NA} \tag{3.187}$$

where $\bar{\bar{\boldsymbol{\varepsilon}}}_N$ is a diagonal matrix, $\bar{\bar{\boldsymbol{\varepsilon}}}_N = \text{diag}\{\varepsilon_{Ni}\}$, which contains the $n_C^*$ coefficients of restitution $0 \leq \varepsilon_{Ni} \leq 1$. The value $\varepsilon_{Ni} = 0$ means a completely inelastic shock where both collision partners remain in contact, and $\varepsilon_{Ni} = 1$ describes fully reversible behavior. Inserting (3.187) into (3.186) yields

$$-\left(\mathbf{E} + \bar{\bar{\boldsymbol{\varepsilon}}}_N\right)\dot{\mathbf{g}}_{NA} = \mathbf{G}_N \boldsymbol{\Lambda}_N \tag{3.188}$$

which determines the transferred impulses $\boldsymbol{\Lambda}_N$:

$$\boldsymbol{\Lambda}_N = -\mathbf{G}_N^{-1}\left(\mathbf{E} + \bar{\bar{\boldsymbol{\varepsilon}}}_N\right)\dot{\mathbf{g}}_{NA}. \tag{3.189}$$

If only one contact participates in the impact, then the matrix $\mathbf{G}_N^{-1}$ reduces to the scalar $\mathbf{G}_N^{-1} = 1/(\mathbf{w}_N^T \mathbf{M}^{-1} \mathbf{w}_N)$. Under these circumstances eq. (3.189) determines the impulses of an equivalent system where the condensed mass $\mathbf{G}_N^{-1}$ bounces against a rigid wall.

The term $\mathbf{G}_N = \mathbf{W}_N^T \mathbf{M}^{-1} \mathbf{W}_N$ corresponds to that reduced mass of our multibody system which is effective in the impact direction. We shall call it "mass action matrix", because in all multibody systems with unilateral characteristics this mass action matrix represents exactly those mass effects as allowed by the constraints of the system. It has first been introduced by [182] in connection with frictionless impacts of multibody systems, for example for problems of gear-rattling and the like ([183], [184] and [206]). A further application is the contact of robots with the environment [271], [137].

After resubstituting the impulses (3.189) into eq. (3.183) we get the generalized velocities $\dot{\mathbf{q}}_E$ at the end of the impact:

$$\dot{\mathbf{q}}_E = \dot{\mathbf{q}}_A - \mathbf{M}^{-1}\mathbf{W}_N \mathbf{G}_N^{-1}\left(\mathbf{E} + \bar{\bar{\boldsymbol{\varepsilon}}}_N\right)\dot{\mathbf{g}}_{NA}, \tag{3.190}$$

where $\dot{\mathbf{g}}_{NA}$ is given with eq. (3.184). Note that the impulses $\boldsymbol{\Lambda}_N$ must act with a compressive magnitude ($\boldsymbol{\Lambda}_N \geq 0$) in the physical sense.

The use of Newton's impact law ensures that no penetration of the bodies after the impact can occur. This is easy to see, since by the construction of $I_C^*$ only contacts with $\dot{g}_{NAi} \leq 0$ are considered. With eq. (3.187) and the property of the coefficients of restitution, $0 \leq \varepsilon_{Ni} \leq 1$, it is obvious that the bodies separate after the impact, $\dot{g}_{NEi} \geq 0$. Contacts, however, which have been removed from $I_C^*$ due to the reasons described above, are excluded from the evaluation of Newton's law and therefore are not expected to have final relative velocities which avoid penetration. Thus, impulses of contacts in $I_C^*$ and relative velocities of contacts which have been removed from $I_C^*$ must be checked with respect to physical correctness. All these difficulties can be handled by a unilateral formulation of the impact laws including Poisson's impact hypothesis.

## 3.5.2.2 Poisson's Impacts

Poisson's friction law is a kinetic law, which connects the impulses after an impact with those before the impact. In contrast to Newton's law the friction law of Poisson allows an energy transfer between the normal and tangential directions representing thus a more realistic situation of an impact. For the formulation of the impacts the same assumptions are made as before, but we now allow friction at the contact points and take into account all active unilateral constraints, which means constraints that are elements of $I_C$ in the set equation (3.9) on page 90, chapter 3.1.2.

$$I_C = \{i \in I_A \mid g_{Ni} = 0\} \quad \text{with } n_C \text{ elements.} \tag{3.191}$$

It is noteworthy that eq. (3.191) contains all the sliding and sticking continuous-contact constraints ($\dot{g}_{Ni} = 0$) as well as the impact contacts ($\dot{g}_{Ni} < 0$). This enables us to examine whether a contact separates under the influence of an impact at another location in the multibody system. We start with the derivation of the impact equations using the first three equations of (3.146) on page 139. They write in a more condensed form

$$\mathbf{M}(\mathbf{q},t)\ddot{\mathbf{q}}(t) + \mathbf{h}(\mathbf{q},\dot{\mathbf{q}}t) - [(\mathbf{W}_N + \mathbf{W}_R) \quad \mathbf{W}_T]\begin{pmatrix}\lambda_N(t)\\ \lambda_T(t)\end{pmatrix} = \mathbf{0} \quad \in \mathbb{R}^f,$$

$$\begin{pmatrix}\ddot{\mathbf{g}}_N\\ \ddot{\mathbf{g}}_T\end{pmatrix} = \begin{pmatrix}\mathbf{W}_N^T\\ \mathbf{W}_T^T\end{pmatrix}\ddot{\mathbf{q}} + \begin{pmatrix}\bar{\mathbf{w}}_N\\ \bar{\mathbf{w}}_T\end{pmatrix} \quad \in \mathbb{R}^{n_N + 2n_T} \tag{3.192}$$

We assume as noted above that an impact takes place in an infinitesimal short time and without any change of position, orientation and all non-impulsive forces. Nevertheless and virtually we zoom the impact time, establish the equations for a compression and for an expansion phase and then apply these equations again for an infinitesimal short time interval. The evaluation has to be performed on a velocity level which we realize by formal integration of the equations of motion, the constraints and the contact laws. Denoting the beginning of an impact, the end of compression and the end of expansion by the indices $A, C, E$, respectively, we get for $\Delta t = t_E - t_A$

$$\mathbf{M}(\dot{\mathbf{q}}_C - \dot{\mathbf{q}}_A) - (\mathbf{W}_N \quad \mathbf{W}_T)\begin{pmatrix}\mathbf{\Lambda}_{NC}\\ \mathbf{\Lambda}_{TC}\end{pmatrix} = \mathbf{0},$$

$$\mathbf{M}(\dot{\mathbf{q}}_E - \dot{\mathbf{q}}_C) - (\mathbf{W}_N \quad \mathbf{W}_T)\begin{pmatrix}\mathbf{\Lambda}_{NE}\\ \mathbf{\Lambda}_{TE}\end{pmatrix} = \mathbf{0},$$

$$\text{with} \quad \mathbf{\Lambda}_i = \lim_{t_E \to t_A} \int_{t_A}^{t_E} \lambda_{Ni}\, dt \quad i = \{NC, TC, NE, TE\}. \tag{3.193}$$

Here $\mathbf{\Lambda}_{NC}, \mathbf{\Lambda}_{TC}$ are the impulses in the normal and tangential direction which are transferred during compression, and $\mathbf{\Lambda}_{NE}, \mathbf{\Lambda}_{TE}$ those of expansion. In addition the abbreviations $\dot{\mathbf{q}}_A = \dot{\mathbf{q}}(t_A)$, $\dot{\mathbf{q}}_C = \dot{\mathbf{q}}(t_C)$ $\dot{\mathbf{q}}_E = \dot{\mathbf{q}}(t_E)$ have been introduced. With these notations we write the relative velocities in the various phases of the impact in the form

$$\begin{pmatrix} \dot{\mathbf{g}}_{NA} \\ \dot{\mathbf{g}}_{TA} \end{pmatrix} = \begin{pmatrix} \mathbf{W}_N^T \\ \mathbf{W}_T^T \end{pmatrix} \dot{\mathbf{q}}_A + \begin{pmatrix} \tilde{\mathbf{w}}_{NA} \\ \tilde{\mathbf{w}}_{TA} \end{pmatrix}$$

$$\begin{pmatrix} \dot{\mathbf{g}}_{NE} \\ \dot{\mathbf{g}}_{TE} \end{pmatrix} = \begin{pmatrix} \mathbf{W}_N^T \\ \mathbf{W}_T^T \end{pmatrix} \dot{\mathbf{q}}_E + \begin{pmatrix} \tilde{\mathbf{w}}_{NE} \\ \tilde{\mathbf{w}}_{TE} \end{pmatrix}$$

$$\begin{pmatrix} \dot{\mathbf{g}}_{NC} \\ \dot{\mathbf{g}}_{TC} \end{pmatrix} = \begin{pmatrix} \mathbf{W}_N^T \\ \mathbf{W}_T^T \end{pmatrix} \dot{\mathbf{q}}_C + \begin{pmatrix} \tilde{\mathbf{w}}_{NC} \\ \tilde{\mathbf{w}}_{TC} \end{pmatrix} \qquad (3.194)$$

Considering in a first step the compression phase and combining the equations (3.193) and (3.194) we come out with

$$\begin{pmatrix} \dot{\mathbf{g}}_{NC} \\ \dot{\mathbf{g}}_{TC} \end{pmatrix} = \underbrace{\begin{pmatrix} \mathbf{W}_N^T \\ \mathbf{W}_T^T \end{pmatrix} \mathbf{M}^{-1} \begin{pmatrix} \mathbf{W}_N \\ \mathbf{W}_T \end{pmatrix}^T}_{\mathbf{G}} \cdot \begin{pmatrix} \boldsymbol{\Lambda}_{NC} \\ \boldsymbol{\Lambda}_{TC} \end{pmatrix} + \begin{pmatrix} \dot{\mathbf{g}}_{NA} \\ \dot{\mathbf{g}}_{TA} \end{pmatrix}$$

$$\mathbf{G} = \begin{pmatrix} \mathbf{W}_N^T \\ \mathbf{W}_T^T \end{pmatrix} \mathbf{M}^{-1} \begin{pmatrix} \mathbf{W}_N \\ \mathbf{W}_T \end{pmatrix}^T = \begin{pmatrix} \mathbf{G}_{NN} & \mathbf{G}_{NT} \\ \mathbf{G}_{TN} & \mathbf{G}_{TT} \end{pmatrix} \quad \text{with}$$

$$\mathbf{G}_{ij} = \mathbf{W}_i^T \mathbf{M}^{-1} \mathbf{W}_j, \quad \text{and} \quad \mathbf{G}_{ij} = \mathbf{G}_{ji}^T, \qquad i,j = \{N,T\}. \qquad (3.195)$$

where $\mathbf{G}$ is the mass action matrix as defined above. It consists of four blocks $\mathbf{G}_{NN} \ldots \mathbf{G}_{TT}$ with the above defined properties.

We start with the **compression phase and the normal impact direction** [15]. At the end of compression the relative normal velocity is either zero or non-negative, $\dot{g}_{Ni} \geq 0$. The contact process for impacts follows the same complementarity ideas as for contacts on a force/acceleration level. In normal direction either the relative velocity is zero ($\dot{g}_{Ni} = 0$) and the normal impulse is not zero ($\Lambda_{Ni} \neq 0$), or vice versa. Hence we may write (see also equation 3.159)

$$\dot{\mathbf{g}}_N \geq \mathbf{0}, \qquad \boldsymbol{\Lambda}_N \geq \mathbf{0}, \qquad \dot{\mathbf{g}}_N^T \boldsymbol{\Lambda}_N = 0, \qquad (3.196)$$

which has to be evaluated by components. This means, that we start with the left characteristic of Figure (3.27), which is of course well known from chapter 3.1.2.2 on page 91.

The next step must consider the **compression phase and the tangential impact direction**. The tangential compression phase is characterized mainly by friction. At the end of compression we may have three states:

- Sliding in a positive tangential direction ($\dot{g}_{NC} > 0$): The tangential impulse acts during this phase in opposite direction with $\Lambda_{TC} = -\mu \Lambda_{NC}$.

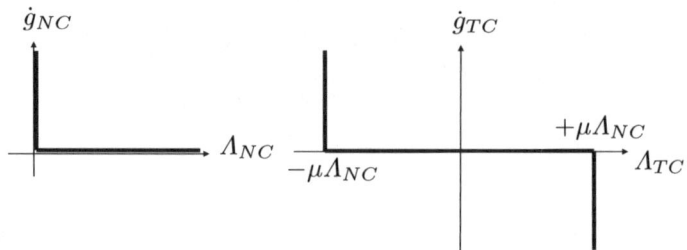

Fig. 3.27: Contact Laws for Impacts

- Sticking at the end of compression ($\dot{g}_{NC} = 0$): The tangential impulse is small enough to generate sticking during the whole compression phase. It is located within the friction cone with the possible values ($-\mu\Lambda_{NC} < \Lambda_{TC} < +\mu\Lambda_{NC}$) defining a set-valued impact law.
- Sliding in a negative tangential direction ($\dot{g}_{NC} < 0$): The tangential impulse acts during this phase in opposite direction with $\Lambda_{TC} = +\mu\Lambda_{NC}$.

Analogously to the force-acceleration-level of the equations (3.18) on page 93 we can summarize this impact behaviour in tangential direction to give

$$|\mathbf{\Lambda}_{Ti}| < \mu_{0i}\Lambda_{Ni} \wedge \dot{g}_{Ti} = 0 \quad (i \in I_T \text{ sticking})$$

$$\Lambda_{Ti} = +\mu_{0i}\Lambda_{Ni} \wedge \dot{g}_{Ti} \leq 0 \quad (i \in I_N \backslash I_T \text{ negative sliding}) \qquad (3.197)$$

$$\Lambda_{Ti} = -\mu_{0i}\Lambda_{Ni} \wedge \dot{g}_{Ti} \geq 0 \quad (i \in I_N \backslash I_T \text{ positive sliding})$$

These definitions correspond to the right chart of Figure (3.27). At this point some remarks on the validity of Coulomb's law for impacts are necessary. From the standpoint of mechanics there exists no hard argument why the classical friction laws should not apply to impact processes, see [86], [200], [15]. Moreover, the theories based on that law have been verified very carefully be systematic experiments, which confirms the approach nearly perfectly [15]. And finally all simulations of large industrial projects result in a very good correspondence of theory and measurement. Therefore as long as no better law appears, these are the best possible friction laws.

The set-valued impulse law on the velocity-impulse-level possesses the same properties as the set-valued force law on the acceleration-force-level of chapter (3.4.2) on page 134. For active contacts it represents an impulse law with kinematical constraints, for passive contacts an indicator for contact events to come. The equations (3.195) allow to calculate the relative velocities $\dot{g}_{NC}$ and $\dot{g}_{TC}$ at the end of the compression phase, depending from the velocities at the beginning of the impact $\dot{\boldsymbol{g}}_{NA}$ and $\dot{\boldsymbol{g}}_{TA}$ under the influence of the contact impulses $\boldsymbol{\Lambda}_{NC}$ and $\boldsymbol{\Lambda}_{TC}$. To calculate these impulses two impact laws in normal and tangential direction are necessary. As already indicated magnitudes of relative kinematics and constraint forces (here impulses) are

## 3.5 Impact Systems

complementary quantities. In normal direction these are $\dot{g}_{NC}$ and $\Lambda_{NC}$. In tangential direction we have the relative tangential velocity vector $\dot{g}_{TC}$ and the friction reserve $(\Lambda_{TC} - (\text{diag}\mu_i)\Lambda_{NC})$. Decomposing the tangential behavior we obtain [15]:

$$\begin{aligned}\Lambda_{TCV,i} &= \Lambda_{TC,i} + \mu_i \Lambda_{TN,i} \\ \dot{g}_{TC,i} &= \dot{g}^+_{TC,i} - \dot{g}^-_{TC,i} \\ \Lambda^{(+)}_{TCV,i} &= \Lambda_{TCV,i} \\ \Lambda^{(-)}_{TCV,i} &= -\Lambda_{TCV,i} + 2\mu_i \Lambda_{NC,i}\end{aligned} \quad (3.198)$$

Together with equation (3.195) this results in a Linear Complementary Problem (LCP) in standard form $y = Ax + b$ with $x \geq 0, y \geq 0$ and $x^T y = 0$:

$$\underbrace{\begin{pmatrix} \dot{g}_{NC} \\ \dot{g}^+_{TC} \\ \Lambda^{(-)}_{TCV} \end{pmatrix}}_{y} = \underbrace{\begin{pmatrix} G_{NN} - G_{NT}\mu & G_{NT} & 0 \\ G_{TN} - G_{TT}\mu & G_{TT} & E \\ 2\mu & -E & 0 \end{pmatrix}}_{A} \underbrace{\begin{pmatrix} \Lambda_{NC} \\ \Lambda^{(+)}_{TCV} \\ \dot{g}^-_{TC} \end{pmatrix}}_{x} + \underbrace{\begin{pmatrix} \dot{g}_{NA} \\ \dot{g}_{TA} \\ 0 \end{pmatrix}}_{b} \quad (3.199)$$

$\mu$ is a diagonal matrix, containing the friction coefficients of the contacts. The problem can be solved numerically. The velocities $\dot{g}_{NC}, \dot{g}_{TC}$ and the impulsions $\Lambda_{NC}, \Lambda_{TC}$ are either part of the result or can be obtained by transformation (3.198) and by $\Lambda_{TC} = \Lambda^{(+)}_{TCV} - \mu\Lambda_{NC}$. The generalized velocities at the end of the compression phase can be evaluated with the help of equation (3.193) and refering to Figure (3.28), which depicts the decomposition structure. We get

$$\dot{q}_C = \dot{q}_A + \mathbf{M}^{-1}(\mathbf{W}_N \Lambda_{NC} + \mathbf{W}_T \Lambda_{TC}) \quad (3.200)$$

Knowing the state at the end of the compression phase we are able to evaluate the expansion phase, in a first step the **expansion phase in the normal impact direction**. Generally, impulse has been stored during compression, or in terms of energy, kinetic energy has been partly converted into potential energy by the deformation of the impacting bodies of the local contact zones. These stored impulses are released during expansion, or again in terms of energy, the potential deformation energy will be partly converted again into kinetic energy during expansion more or less by spring effects of the deformed local zone. We assume that this proces is governed, similar as in compression, by Poisson's law relating the impulses after with those before the impact, and we assume furtheron, that Poisson's law can be applied in normal and in tangential directions.

Concentrating first on the normal direction Poisson's law has the form

$$\Lambda_{NE} = \varepsilon_N \Lambda_{NC}, \quad (3.201)$$

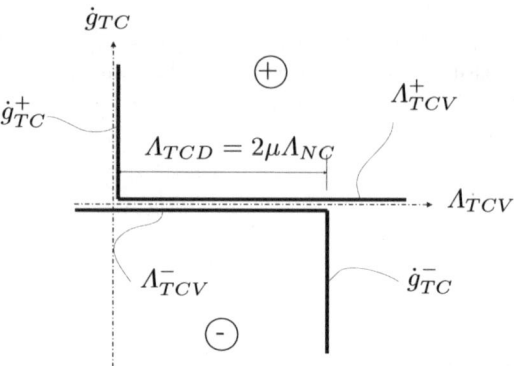

Fig. 3.28: Decomposition of the Tangential Impact Law

which relates the normal impulse at the end of the impact with the impulse at the end of compression. The impact coefficient $\varepsilon_N$ must be measured. Its value is between zero and one, $0 \leq \varepsilon_N \leq 1$. For practical problems we always have $\varepsilon_N < 1$, which means loss of energy during the impact. Poisson's impact law possesses one drawback due to its formulation on a kinetic level. It cannot exclude a penetration of the bodies, which is given on a kinematic level. But we can require instead, that the normal impulse during expansion must always be equal or larger than the impulse, which we can get back from the stored impulse during compression. Releasing impulses is accompanied with losses, therefore the minimum we might get back is $\varepsilon_N \Lambda_{NC}$, but the normal impulse during expansion might be larger ($\Lambda_{NE} \geq \varepsilon_N \Lambda_{NC}$). As a consequence we have to deal with a shifted characteristic during expansion as shown in Figure (3.29).

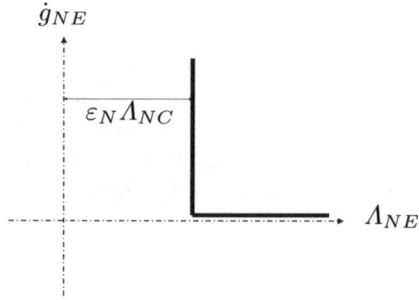

Fig. 3.29: Shifted Normal Characteristic for Impact Expansion

Figure (3.29) indicates that for a detachment at the end of compression we have an impulse according to equation (3.201), otherwise it can become arbitrarily large to avoid penetration. To evaluate a complementarity condition we therefore transform the individual impulses, for example for body (i)

$$\Lambda_{NPi} = \Lambda_{NEi} - \varepsilon_{Ni}\Lambda_{NCi} \tag{3.202}$$

For the transformed impulses we then get the complementarity condition in normal direction

$$\dot{\mathbf{g}}_{NE} \geq \mathbf{0}, \quad \boldsymbol{\Lambda}_{NP} \geq \mathbf{0}, \quad \dot{\mathbf{g}}_{NE}^T \boldsymbol{\Lambda}_{NP} = \mathbf{0}, \tag{3.203}$$

which must be evaluated by components.

The *expansion phase for the tangential impact direction* is governed by friction and the corresponding restoring mechanism. Consequently we must find some realistic models for the friction acting in the tangential direction of the contact and for the release mechanism of the impulses stored during compression. In a way, we have two actions in series: the contact partners touch each other in a contact point, where Coulomb's law applies; and in addition we have the stored contact energy within the elastically deformed contact zone, which is partly released during expansion and acting in series with the Coulomb friction as a second force element [15]. These physical/mechanical properties must be modeled as realistically as possible coming out with a basis for further model developments.

Restoring the tangential impulse affords some additional considerations. According to Poisson's law we get back the stored tangential impulse $\Lambda_{TCi}$ of the (i)th contact with a certain loss, that is $(\varepsilon_{Ti}\Lambda_{TCi})$, where Poisson's losses are defined in the range $(0 \leq \varepsilon_{Ti} \leq 1)$. The tangential friction coefficient $\varepsilon_{Ti}$ must be measured. But this contains not all losses during expansion. The restoration of the tangential impulse possesses another quality compared with the restoration of the normal impulse, because it cannot take place independently from the normal impulse, which as a matter of fact represents the driving constraint impulse for the generation of tangential friction forces. Therefore we shall assume, that the restoration of the tangential impulse is additionally accompanied by losses in "normal direction" expressed by $\varepsilon_{Ni}$. The complete restoration then follows the law

$$\Lambda_{TEi} = \varepsilon_{Ni}\varepsilon_{Ti}\Lambda_{TCi}. \tag{3.204}$$

The above formulation avoids also that the reversible impulse will not be larger than the maximum impulse, which can be transmitted by friction and is illustrated by the following inequalities

$$\varepsilon_{Ni}\varepsilon_{Ti}|\Lambda_{TCi}| \leq \varepsilon_{Ni}|\Lambda_{TCi}| \leq \varepsilon_{Ni}\mu\Lambda_{NC} \leq \mu\Lambda_{NE}. \tag{3.205}$$

Figure (3.30) depicts the most important features for the case $(\Lambda_{TC} > 0)$, other combinations are considered in [86] and [200]. The left limit of the characteristic is given by equation (3.204) and the restored impulse $\varepsilon_{Ni}\varepsilon_{Ti}\Lambda_{TCi}$. At this point we might get sticking or positive sliding with $+\dot{g}_{TE}$. The right limit $\Lambda_{TER} = \mu\Lambda_{NE}$ is again sticking or sliding with a negative velocity

$-\dot{g}_{TE}$. Between $\varLambda_{TEL}$ and $\varLambda_{TER}$ we are within the friction cone with sticking if $\dot{g}_{TE0} = 0$ or with sliding if $\dot{g}_{TE0} \neq 0$. This effect requires some explanation.

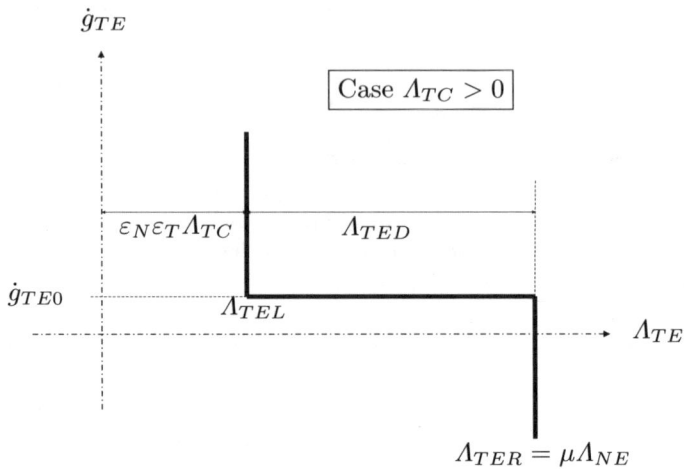

Fig. 3.30: Shifted Tangential Characteristic for Impact Expansion

As already pointed out above we have a serial combination of friction impulse and of the impulsive effects of the local deformation. The corresponding impulses do not apply at the same points with the consequence, that we may have stiction at the contact point but small sliding at the point of the spring force application. For materials like steel this will be no problem, but for soft materials like rubber there exists a difference between these two points which can be measured [15]. In such cases it makes sense to introduce a correction with $\dot{g}_{TE0}$. For this purpose we assume at the end of compression a zero tangential velocity $\dot{g}_{TC} = 0$ and evaluate the equations (3.194) and (3.195) together with the above transformations with the result

$$\dot{\mathbf{g}}_{TE0} = \mathbf{G}_{TN}\varepsilon_N \mathbf{\Lambda}_{NC} + \mathbf{G}_{TT}\varepsilon_N \varepsilon_T \mathbf{\Lambda}_{TC},$$
$$\varepsilon_N = \mathrm{diag}\{\cdots \varepsilon_{Ni} \cdots\}, \qquad \varepsilon_T = \mathrm{diag}\{\cdots \varepsilon_{Ti} \cdots\}. \tag{3.206}$$

It should be noted that independent of the combination of sliding or sticking in the compression and expansion phases tangential impulse will be stored during compression and restored during expansion, also for sliding.

For the compression phase we have developed a complementarity representation with the equations (3.199). We shall do the same for the expansion phase. We know the relative velocities and the impulses $(\dot{\mathbf{g}}_{NC}, \dot{\mathbf{g}}_{TC}, \mathbf{\Lambda}_{NC}, \mathbf{\Lambda}_{TC})$ at the end of compression. We know also the necessary parameters like $(\varepsilon_N, \varepsilon_T, \mu, \mathbf{G})$. We are now looooking for the magnitudes $(\dot{\mathbf{g}}_{NE}, \dot{\mathbf{g}}_{TE}, \mathbf{\Lambda}_{NE}, \mathbf{\Lambda}_{TE})$ at the end of expansion and thus at the end of the impact. For this purpose we follow a presentation given in [15], see also [86], [200].

For the normal direction we have shifted the characteristics into the origin of the coordinates to be able to generate a Linear Complementarity Problem (LCP). Considering Figure (3.30) we get

$$\Lambda_{TEVi} = \Lambda_{TEi} - \Lambda_{TELi}, \qquad \dot{g}_{TEVi} = \dot{g}_{TEi} - \dot{g}_{TE0i} \qquad (3.207)$$

with $\Lambda_{TELi} = \varepsilon_{Ni}\varepsilon_{Ti}\Lambda_{TCi}$. These transformation shifts the left corner of the characteristic in Figure (3.30) into the origin. Collecting now all these transformations of the equations (3.202), (3.207) and combining them with the momentum equations (3.194), (3.195) results in two equations for the normal and the tangential directions

$$\dot{\mathbf{g}}_{NE} = \mathbf{G}_{NN}(\mathbf{\Lambda}_{NP} + \varepsilon_N\mathbf{\Lambda}_{NC}) + \mathbf{G}_{NT}(\mathbf{\Lambda}_{TEV} + \mathbf{\Lambda}_{TEL}) + \dot{\mathbf{g}}_{NC}$$
$$\dot{\mathbf{g}}_{TEV} = \mathbf{G}_{TN}(\mathbf{\Lambda}_{NP} + \varepsilon_N\mathbf{\Lambda}_{NC}) + \mathbf{G}_{TT}(\mathbf{\Lambda}_{TEV} + \mathbf{\Lambda}_{TEL}) + \dot{\mathbf{g}}_{TC} - \dot{\mathbf{g}}_{TE0}$$
$$(3.208)$$

Replacing $\dot{\mathbf{g}}_{TE0}$ in the last equation by (3.206) yields for the tangential direction

$$\dot{\mathbf{g}}_{TEV} = \mathbf{G}_{TN}\mathbf{\Lambda}_{NP} + \mathbf{G}_{TT}(\mathbf{\Lambda}_{TEV} + \mathbf{\Lambda}_{TEL} - \varepsilon_N\varepsilon_T\mathbf{\Lambda}_{TC}) + \dot{\mathbf{g}}_{TC} \qquad (3.209)$$

To construct a linear complementarity problem we have to decompose the tangential double hook into two unilateral primitives by the additional definitions

$$\dot{g}_{TEVi} = \dot{g}^+_{TEVi} - \dot{g}^-_{TEVi}$$
$$\Lambda^+_{TEVi} = \Lambda_{TEVi}, \qquad \Lambda^-_{TEVi} = -\Lambda_{TEVi} + \Lambda_{TEDi}. \qquad (3.210)$$

The complementary pairs of the shifted and decomposed system are $(\dot{\mathbf{g}}_{NE}, \mathbf{\Lambda}_{NP})$, $(\dot{\mathbf{g}}^+_{TEV}, \mathbf{\Lambda}^+_{TEV})$ and $(\dot{\mathbf{g}}^-_{TEV}, \mathbf{\Lambda}^-_{TEV})$. For the abbreviations $\mathbf{\Lambda}_{TEL}$ and $\mathbf{\Lambda}_{TED}$ (Figure 3.30) we write

$$\mathbf{\Lambda}_{TEL} = \mathbf{S}^+\varepsilon_N\varepsilon_T\mathbf{\Lambda}_{TC} - \mathbf{S}^-\mu(\mathbf{\Lambda}_{NP} + \varepsilon_N\mathbf{\Lambda}_{NC})$$
$$\mathbf{\Lambda}_{TED} = \mu(\mathbf{\Lambda}_{NP} + \varepsilon_N\mathbf{\Lambda}_{NC}) - \varepsilon_N\varepsilon_T|\mathbf{\Lambda}_{TC}|$$

$$\mathbf{S}^+ = \text{diag}[\frac{1}{2}(1 + \text{sign}(\Lambda_{TCi}))]$$
$$\mathbf{S}^- = \text{diag}[\frac{1}{2}(1 - \text{sign}(\Lambda_{TCi}))] \qquad (3.211)$$

The matrices $\mathbf{S}^+$ and $\mathbf{S}^-$ regard the different cases of the direction of the tangential momentum during the compression phase. They are modified unit matrices, which possess as diagonal elements either "0" or "1". Combining now the equations (3.208) to (3.211) we finally come out with the following complementarity inequalities

$$\begin{pmatrix} \dot{\mathbf{g}}_{NE} \\ \dot{\mathbf{g}}_{TEV}^+ \\ \mathbf{\Lambda}_{TEV}^- \end{pmatrix} = \mathbf{A} \begin{pmatrix} \mathbf{\Lambda}_{NP} \\ \mathbf{\Lambda}_{TEV}^+ \\ \dot{\mathbf{g}}_{TEV}^- \end{pmatrix} + \mathbf{b}$$

$$\mathbf{A} = \begin{pmatrix} \mathbf{G}_{NN} - \mathbf{G}_{NT}\mathbf{S}^-\mu & \mathbf{G}_{NT} & 0 \\ \mathbf{G}_{TN} - \mathbf{G}_{TT}\mathbf{S}^-\mu & \mathbf{G}_{TT} & \mathbf{E} \\ \mu & -\mathbf{E} & 0 \end{pmatrix}$$

$$\mathbf{b} = \begin{pmatrix} \mathbf{G}_{NN}\varepsilon_N\mathbf{\Lambda}_{NC} + \mathbf{G}_{NT}\mathbf{S}^+\varepsilon_N\varepsilon_T\mathbf{\Lambda}_{TC} - \mathbf{G}_{NT}\mathbf{S}^-\mu\varepsilon_N\mathbf{\Lambda}_{NC} + \dot{\mathbf{g}}_{NC} \\ \mathbf{G}_{TT}(\mathbf{S}^- - \mathbf{E})\varepsilon_N\varepsilon_T\mathbf{\Lambda}_{TC} - \mathbf{G}_{TT}\mathbf{S}^-\mu\varepsilon_N\mathbf{\Lambda}_{NC} + \dot{\mathbf{g}}_{TC} \\ \mu\varepsilon_N\mathbf{\Lambda}_{NC} - \varepsilon_N\varepsilon_T|\mathbf{\Lambda}_{TC}| \end{pmatrix}$$

(3.212)

For recalculating the original magnitudes after the solution of the above complementarity problem we apply the formulas in the other way around.

$$\dot{\mathbf{g}}_{TE} = \dot{\mathbf{g}}_{TEV}^+ - \dot{\mathbf{g}}_{TEV}^- + \dot{\mathbf{g}}_{TE0}$$
$$\mathbf{\Lambda}_{NE} = \mathbf{\Lambda}_{NP} + \varepsilon_N\mathbf{\Lambda}_{NC}$$
$$\mathbf{\Lambda}_{NP} = \mathbf{\Lambda}_{NE} - \varepsilon_N\mathbf{\Lambda}_{NC}$$
$$\mathbf{\Lambda}_{TE} = \mathbf{\Lambda}_{TEV}^+ + \mathbf{\Lambda}_{TEL} = \mathbf{\Lambda}_{TEV}^+ + \mathbf{S}^+\varepsilon_N\varepsilon_T\mathbf{\Lambda}_{TC} - \mathbf{S}^-\mu\mathbf{\Lambda}_{NE}$$

$$\dot{\mathbf{q}}_E = \dot{\mathbf{q}}_A + \mathbf{M}^{-1}[\mathbf{W}_N(\mathbf{\Lambda}_{NC} + \mathbf{\Lambda}_{NE}) + \mathbf{W}_T(\mathbf{\Lambda}_{TC} + \mathbf{\Lambda}_{TE})] \quad (3.213)$$

The last relation gives the generalized velocities at the end of an impact in dependence of those at the beginning. With the two sets of equations (3.199) and (3.213) we have the complete set for the evaluation of impacts in a multibody system with plane not dependent contacts. It is at the same time an illustrative example of the complexity of such problems solving them on the basis of complementarities.

### 3.5.3 Moreau's Measure Differential Equation

One of the great merits of Moreau [161] in establishing a theory of non-smooth mechanics consists in categorizing the physical features with respect to existing modern mathematics and in creating necessary extensions especially concerning convex analysis. The mathematical fundaments of measure and integral theory correspond for example to the properties of friction cones and of impacts ([283] [132]). Any impact between two bodies is accompanied by a finite jump of the velocities, and, as a rule, system dynamics shortly before and shortly after an impact is going to be smooth. The velocities $\mathbf{u}^-$ shortly before an impact represents a left-limit and the velocity $\mathbf{u}^+$ shortly after an impact represents a right-limit. These limits always exist and their difference $\Delta\mathbf{u} = \mathbf{u}^+ - \mathbf{u}^-$ is always finite and of bounded variation.

Following Moreau [161] Glocker [87] introduces a decomposition of the velocity $\mathbf{u}$ into three parts, an absolute continuous part $d\mathbf{u}_L = \dot{\mathbf{u}}dt$ with the

## 3.5 Impact Systems

Lebesgue measure dt, a discontinuous part $d\mathbf{u}_A = (\mathbf{u}^+ - \mathbf{u}^-)d\eta$ in the form of a step function with the atomic measure $d\eta = \sum_i d\delta_i$ and a singular part $d\mathbf{u}_C$, which we shall not consider here. Collecting these ideas we get

$$d\mathbf{u} = d\mathbf{u}_L + d\mathbf{u}_A \quad \text{with} \quad d\mathbf{u} = \ddot{\mathbf{q}}dt \tag{3.214}$$
$$d\mathbf{u}_L = \dot{\mathbf{u}}dt,$$
$$d\mathbf{u}_A = (\mathbf{u}^+ - \mathbf{u}^-)d\eta,$$

$$d\eta = \sum_i d\delta_i \quad \text{with} \quad \int_{I_{kl}} d\delta_i = \begin{cases} 1 & \text{for } i \in I_{kl} \\ 0 & \text{for } i \notin I_{kl} \end{cases} \tag{3.215}$$

where $I_{kl}$ portrays the time interval of an impact, for example $I_{kl} \in [t^-, t^+]$. In a similar way we may split up the forces in a continuous and thus Lebesgue-measurable part $\lambda dt$ and in an atomic part $\boldsymbol{\Lambda} d\eta$

$$d\boldsymbol{\Lambda} = \boldsymbol{\lambda} dt + \boldsymbol{\Lambda} d\eta. \tag{3.216}$$

The part $\boldsymbol{\lambda} dt$ contains all contact reactions due to non-impulsive contacts and the part $\boldsymbol{\Lambda} d\eta$ all impulsive contact reactions. Correspondingly and considering the equations of motion (3.146) we may decompose these equations of motion into two classical parts:

$$\mathbf{M}d\mathbf{u} + \mathbf{h}dt - \mathbf{W}d\Lambda = \mathbf{0} \iff \begin{cases} \mathbf{M}\dot{\mathbf{u}} + \mathbf{h} - \mathbf{W}\boldsymbol{\lambda} = \mathbf{0} & (t \neq t_i) \\ \mathbf{M}(\mathbf{u}^+ - \mathbf{u}^-) - \mathbf{W}\Lambda = \mathbf{0} & (t = t_i) \end{cases} \tag{3.217}$$

The time $t_i \in I_{kl}$ represents one of the instants (i), where an impact takes place. The vector $\mathbf{h}$ includes all non-impulsive and applied forces, whatsoever, and for multibody systems without closed loops we also include in the generalized coordinates $(\mathbf{q}, \dot{\mathbf{q}})$ all bilateral constraints. If we have closed loops, it makes sense to include into the prox-functions also the bilateral constraints, which simplifies the solution procedure (see [65] and [64]).

Applying these decompositions to the set of equations of motion, equations (3.146) on page 139, we get the following set of measure differential equations representing a more general and convenient form of the equations of motion including impacts

$$\mathbf{M}d\mathbf{u} + \mathbf{h}dt - [(\mathbf{W}_N + \mathbf{W}_R) \quad \mathbf{W}_T] \begin{pmatrix} d\boldsymbol{\Lambda}_N(t) \\ d\boldsymbol{\Lambda}_T(t) \end{pmatrix} = \mathbf{0},$$

$$d\dot{\mathbf{g}}_N = \mathbf{W}_N^T d\mathbf{u} + \bar{\mathbf{w}}_N dt,$$
$$d\dot{\mathbf{g}}_T = \mathbf{W}_T^T d\mathbf{u} + \bar{\mathbf{w}}_T dt,$$

$$\lambda_N = \text{prox}_{C_N}(\lambda_N - r g_N), \quad \boldsymbol{\Lambda}_N = \text{prox}_{C_N}(\boldsymbol{\Lambda}_N - r g_N),$$
$$\lambda_T = \text{prox}_{C_T}(\lambda_T - r\dot{\mathbf{g}}_T), \quad \boldsymbol{\Lambda}_T = \text{prox}_{C_T}(\boldsymbol{\Lambda}_T - r\dot{\mathbf{g}}_T). \tag{3.218}$$

The prox-functions as the laws for forces and shocks have been split up directly for the Lebesgue-part and the atomic part and additionally been put on the original kinematical level, which means position and orientation for the normal and velocities for the tangential contact directions [64]. The corresponding convex sets are given as above by the relations

$$C_N = \{\lambda_N | \lambda_{Ni} \geq 0, \forall i \in I_N\},$$
$$C_T = \{\lambda_T | |\lambda_{Ti}| \leq \mu_i \lambda_{Ni}, \forall i \in I_T\}. \tag{3.219}$$

The equations (3.218) can be discretized directly, which gives this type of presentation a very practical property, though it might look at a first glance rather complicated. We refer to section 3.4.4 on page 145.

### 3.5.4 Energy Considerations

All contact processes are accompanied by losses due to the energy conversion mechanisms taking place within the contact zone. The loss of energy is the difference of the total system energy after an impact and before an impact. In terms of the generalized velocities $\dot{\mathbf{q}}$ we write (see also [88])

$$\Delta T = T_E - T_A \leq 0$$
$$\Delta T = \frac{1}{2}\dot{\mathbf{q}}_E^T \mathbf{M} \dot{\mathbf{q}}_E - \frac{1}{2}\dot{\mathbf{q}}_A^T \mathbf{M} \dot{\mathbf{q}}_A = \frac{1}{2}(\dot{\mathbf{q}}_E + \dot{\mathbf{q}}_A)^T \mathbf{M} (\dot{\mathbf{q}}_E - \dot{\mathbf{q}}_A). \tag{3.220}$$

These are expressions considering scleronomic systems without an excitation by external kinematical sources and consequently do not take into account the **w**-terms of the equations (3.194), for example. Using the equations (3.193) to (3.195) we can express the generalized velocities by the relative velocities $\dot{\mathbf{g}}$ in the contacts or by the contact impulses $\mathbf{\Lambda}$. We get

$$2\Delta T = +2 \begin{pmatrix} \dot{\mathbf{g}}_{NE} \\ \dot{\mathbf{g}}_{TE} \end{pmatrix}^T \mathbf{G}^{-1} \left[ \begin{pmatrix} \dot{\mathbf{g}}_{NE} \\ \dot{\mathbf{g}}_{TE} \end{pmatrix} - \begin{pmatrix} \dot{\mathbf{g}}_{NA} \\ \dot{\mathbf{g}}_{TA} \end{pmatrix} \right] -$$
$$- \left[ \begin{pmatrix} \dot{\mathbf{g}}_{NE} \\ \dot{\mathbf{g}}_{TE} \end{pmatrix} - \begin{pmatrix} \dot{\mathbf{g}}_{NA} \\ \dot{\mathbf{g}}_{TA} \end{pmatrix} \right]^T \mathbf{G}^{-1} \left[ \begin{pmatrix} \dot{\mathbf{g}}_{NE} \\ \dot{\mathbf{g}}_{TE} \end{pmatrix} - \begin{pmatrix} \dot{\mathbf{g}}_{NA} \\ \dot{\mathbf{g}}_{TA} \end{pmatrix} \right]$$

$$2\Delta T = +2 \begin{pmatrix} \dot{\mathbf{g}}_{NE} \\ \dot{\mathbf{g}}_{TE} \end{pmatrix}^T \left[ \begin{pmatrix} \mathbf{\Lambda}_{NC} \\ \mathbf{\Lambda}_{TC} \end{pmatrix} + \begin{pmatrix} \mathbf{\Lambda}_{NE} \\ \mathbf{\Lambda}_{TE} \end{pmatrix} \right] -$$
$$- \left[ \begin{pmatrix} \mathbf{\Lambda}_{NC} \\ \mathbf{\Lambda}_{TC} \end{pmatrix} + \begin{pmatrix} \mathbf{\Lambda}_{NE} \\ \mathbf{\Lambda}_{TE} \end{pmatrix} \right]^T \mathbf{G} \left[ \begin{pmatrix} \mathbf{\Lambda}_{NC} \\ \mathbf{\Lambda}_{TC} \end{pmatrix} + \begin{pmatrix} \mathbf{\Lambda}_{NE} \\ \mathbf{\Lambda}_{TE} \end{pmatrix} \right] \tag{3.221}$$

The second term of these two energy equations is a quadratic form and for itself always positive or zero. The matrix $\mathbf{G}$ is at least positive semi-definite, which is also true for its inverse $\mathbf{G}^{-1}$. The energy loss has to be negative, which will be decided by the first term of the above relations. If this term is negative or at least zero, the condition $\Delta T \leq 0$ will hold. Therefore we shall concentrate on these first terms which writes in more detail (see equation (3.195) for the matrix $\mathbf{G}$ and the abbreviations $\mathbf{G}^{-1} = \bar{\mathbf{G}}$, $\Delta T = \Delta T_1 + \Delta T_2$ for the two energy terms in eq. (3.221))

$$\Delta T_1 = + \begin{pmatrix} \dot{\mathbf{g}}_{NE} \\ \dot{\mathbf{g}}_{TE} \end{pmatrix}^T \bar{\mathbf{G}} \left[ \begin{pmatrix} \dot{\mathbf{g}}_{NE} \\ \dot{\mathbf{g}}_{TE} \end{pmatrix} - \begin{pmatrix} \dot{\mathbf{g}}_{NA} \\ \dot{\mathbf{g}}_{TA} \end{pmatrix} \right] =$$
$$= \{\dot{\mathbf{g}}_{NE}^T [\bar{\mathbf{G}}_{NN}(\dot{\mathbf{g}}_{NE} - \dot{\mathbf{g}}_{NA}) + \bar{\mathbf{G}}_{NT}(\dot{\mathbf{g}}_{TE} - \dot{\mathbf{g}}_{TA})] +$$
$$+ \dot{\mathbf{g}}_{TE}^T [\bar{\mathbf{G}}_{TN}(\dot{\mathbf{g}}_{NE} - \dot{\mathbf{g}}_{NA}) + \bar{\mathbf{G}}_{TT}(\dot{\mathbf{g}}_{TE} - \dot{\mathbf{g}}_{TA})]\}$$

$$\Delta T_1 = + \begin{pmatrix} \dot{\mathbf{g}}_{NE} \\ \dot{\mathbf{g}}_{TE} \end{pmatrix}^T \left[ \begin{pmatrix} \mathbf{\Lambda}_{NC} \\ \mathbf{\Lambda}_{TC} \end{pmatrix} + \begin{pmatrix} \mathbf{\Lambda}_{NE} \\ \mathbf{\Lambda}_{TE} \end{pmatrix} \right] =$$
$$= \{\dot{\mathbf{g}}_{NE}^T (\mathbf{\Lambda}_{NC} + \mathbf{\Lambda}_{NE}) + \dot{\mathbf{g}}_{TE}^T (\mathbf{\Lambda}_{TC} + \mathbf{\Lambda}_{TE})\} \tag{3.222}$$

To estimate the sign of these terms we need to look at the contact laws, for example the complementarities of the relations (3.196) and (3.197). For this consideration the second form of the energy losses is more convenient than the first form, we only have to find out the signs of the expression $\{\dot{\mathbf{g}}_{NE}^T(\mathbf{\Lambda}_{NC} + \mathbf{\Lambda}_{NE}) + \dot{\mathbf{g}}_{TE}^T(\mathbf{\Lambda}_{TC} + \mathbf{\Lambda}_{TE})\}$. For this purpose we investigate the possible impact cases, namely compression and expansion either with sticking or sliding, which makes altogether four cases under the assumption, that we have always contact and no detachment *during* the impact.

At this point we must discuss a bit our model concept. We consider the beginning of an impact with index A, a compression phase with index C and an expansion phase with index E, all indices expressing the end of the corresponding phase. Therefore we get from our model and the evaluations of the preceding chapters the C-magnitudes at the end of compression and the E-magnitudes at the end of expansion, the last ones being the magnitudes after the impact. This is all clear, theoretically and experimentally often verified, and gives correct results. For the consideration of an impact we do not need the internal details of compression and expansion. But we need them for an energy consideration.

We need to know, for example, how and where a transition sticking/sliding or vice versa occurs within the structure of the impact. As we do not have some means to determine that, we say, transitions occur always at the end of the phases compression and expansion in an infinitesimal short instant of time not influencing the impact dynamics but only going from one branch of the corner laws of the Figures (3.27), (3.29) and (3.30) to another branch, which means, transitions take place in the corners of the contact laws. This model concept has significant influence on the energy evaluation.

So it can be shown, that the first term $\dot{\mathbf{g}}_{NE}^T(\mathbf{\Lambda}_{NC} + \mathbf{\Lambda}_{NE})$ of the energy equation (3.222), last line, is not zero due to positive normal impulses ($\mathbf{\Lambda}_{NC} + \mathbf{\Lambda}_{NE}$) and due to a non-zero end velocity $\dot{\mathbf{g}}_{NE}$ after the impact, which is physically reasonable for a separation of the two contacting bodies. But on the other hand sliding during expansion requires a zero normal relative velocity

174    3 Constraint Systems

$\dot{g}_{NE} = 0$ in the contact, which makes the above mentioned term to zero. The solution can only consist in a model concept, where the change from contact to detachment takes place at the very last end of the expansion phase. The ($\mathbf{\Lambda}_{NE}$)-value slips into the corner of Figure (3.29) allowing the system to build up the necessary separation velocity.

As a result of the last condition of continual contact during the impact we get for compression and expansion $\mathbf{\Lambda}_N > 0$ and $\dot{\mathbf{g}}_N = 0$, which is also part of the complementarity eq. (3.196), and therefore simply

$$2\Delta T_1 = 2\dot{\mathbf{g}}_{TE}^T(\mathbf{\Lambda}_{TC} + \mathbf{\Lambda}_{TE}), \tag{3.223}$$

the sign of which we have to investigate. Before doing so we consider the sliding cases. All sticking cases are governed by set-valued impulse laws, all sliding cases by a single-valued impulse law, the one by Coulomb. Accordingly, the sliding impulse is proportional to the normal constraint impulse and opposite to ($\dot{\mathbf{g}}_{Tk}$). Therefore

$$\mathbf{\Lambda}_{Tk} = -\text{diag}(\mu)\text{sign}(\dot{\mathbf{g}}_{Tk})\mathbf{\Lambda}_{Nk}, \qquad (k = C, E) \tag{3.224}$$

For sliding ($\dot{\mathbf{g}}_{Tk}$) $\neq \mathbf{0}$, always, and therefore we get for the second expression of the energy term $\Delta T_1$ in equation (3.223) together with equation (3.224) the following result

$$\dot{\mathbf{g}}_{TE}^T \mathbf{\Lambda}_{TE} = -\text{diag}(\mu)|\dot{\mathbf{g}}_{TE}|\mathbf{\Lambda}_{NE} \leq \mathbf{0} \tag{3.225}$$

due to the fact of continual contact and thus $\mathbf{\Lambda}_{NE} > \mathbf{0}$. With these results in mind we come to the four impact cases:

- **sticking during compression, sticking during expansion**

  The tangential impulses have to be within the appropriate friction cones. The tangential velocities are zero, therefore we need not to consider the magnitudes of the impulses. For the definitions see also the Figures 3.27 and 3.30.
  $$-\text{diag}(\mu_0)\mathbf{\Lambda}_{NC} \leq \mathbf{\Lambda}_{TC} \leq +\text{diag}(\mu_0)\mathbf{\Lambda}_{NC}, \quad \mathbf{\Lambda}_{TEL} \leq \mathbf{\Lambda}_{TE} \leq \mathbf{\Lambda}_{TER}$$
  $$\implies \dot{\mathbf{g}}_{TE}^T(\mathbf{\Lambda}_{TC} + \mathbf{\Lambda}_{TE}) = 0$$

- **sliding during compression, sliding during expansion**

  Sliding means single-valued impulse laws according to equation (3.224). Some difficulties will appear for the cases with reversed sliding, that means, with a tangential relative velocity the sign of which is different during compression and during expansion. Therefore we have to consider the two cases without and with tangential reversibility. For the first case we do not have a change of sign of the relative tangential velocity, which gives $\text{sign}(\dot{\mathbf{g}}_{TC}) = \text{sign}(\dot{\mathbf{g}}_{TE})$. This comes out with the relations:
  $$\dot{\mathbf{g}}_{TE}^T \mathbf{\Lambda}_{TC} = -\dot{\mathbf{g}}_{TE}^T[\text{diag}(\mu)\text{sign}(\dot{\mathbf{g}}_{TE})\mathbf{\Lambda}_{NC}] = -\text{diag}(\mu)|\dot{\mathbf{g}}_{TE}|\mathbf{\Lambda}_{NC} \leq \mathbf{0},$$

$$\implies \quad \dot{\mathbf{g}}_{TE}^T(\mathbf{\Lambda}_{TC} + \mathbf{\Lambda}_{TE}) < 0$$

The case with tangential reversibility is more complicated, because it includes a change of sign of the tangential relative velocity and thus at least an extremely short stiction phase, which we put exactly at the point (end of compression)/(beginning of expansion). The sliding velocity during compression decreases until it arrives at one of the corners of Figure 3.27, then we get an extremely short shift from this corner to the other one, which allows the contact to build up a tangential velocity with an opposite sign, then valid for the expansion phase. Only by such a short stiction phase a reversal of tangential velocity is possible. On the other hand such a transition from stick to slip, as short as it might be, follows the same process as for the next case sticking/sliding. Therefore it is dissipative:

$$\implies \quad \dot{\mathbf{g}}_{TE}^T(\mathbf{\Lambda}_{TC} + \mathbf{\Lambda}_{TE}) < 0$$

- **sticking during compression, sliding during expansion**

  The transition from sticking in compression and sliding in expansion follows the mechanism (Figure 3.27): If $\mathbf{\Lambda}_{TC} \gtrless \mathbf{0}$, then sliding is only possible for being at the very end of compression on the friction cone boundary with $\mathbf{\Lambda}_{TC} = \pm\text{diag}(\mu)\mathbf{\Lambda}_{NC}$ and $\dot{\mathbf{g}}_{TC-at} \lessgtr \mathbf{0}$ (at = after transition stick-slip). This results always in a negative sign of the expression $(\dot{\mathbf{g}}_{TE}^T\mathbf{\Lambda}_{TC})$. For the rest we assume a continuation of the signs after going from stick to slip $[\text{sign}(\dot{\mathbf{g}}_{TE}) = \text{sign}(\dot{\mathbf{g}}_{TC-at})]$. Then we arrive at:

  $$\implies \quad \dot{\mathbf{g}}_{TE}^T(\mathbf{\Lambda}_{TC} + \mathbf{\Lambda}_{TE}) < 0$$

- **sliding during compression, sticking during expansion**

  This case is again simpler, because we get sticking at the end with a zero relative tangential velocity. Therefore we need not to consider the impulses.

  $$\implies \quad \dot{\mathbf{g}}_{TE}^T(\mathbf{\Lambda}_{TC} + \mathbf{\Lambda}_{TE}) = 0$$

- **summarized result for all cases**

  $$\implies \quad \dot{\mathbf{g}}_{TE}^T(\mathbf{\Lambda}_{TC} + \mathbf{\Lambda}_{TE}) \leq 0 \implies \Delta T_1 \leq 0 \implies \Delta T \leq 0$$

One may object that the above considerations assume in the case of multiple impacts the same impact structure for all simultaneously appearing impacts, which is usually not true. But even any combination of the above four cases for simultaneous impacts gives a loss of energy. Practical experience indicates in addition that the simultaneous appearance of impacts is extremely scarce, it is an event, which nearly does not happen. A very comprehensive elaboration of impact structures are presented in [72].

As a final result we may state that the above evaluation confirms the physical argument, that any impact processes are accompanied by energy losses. This confirms also the well-known statement of Carnot, that *"in the absence of impressed impulses, the sudden introduction of stationary and persistent constraints that change some velocity reduces the kinetic energy. Hence, by the collision of inelastic bodies, some kinetic energy is always lost"*[180].

### 3.5.5 Verification of Impacts with Friction

With respect to large technical applications impacts with friction play such an important role, that a really applicable theory came amazingly late. The first ideas regarding large dynamical systems with frictional impacts are contained in the famous contribution of Moreau in the year 1988 [161], but then more or less applied to static or quasi-static (smaller) problems. During the nineties these ideas were included into multibody theory by ([86], [15], [200]) and applied to large industrial problems, which without exception confirmed the relevant theories. In addition a thourough and systematic experimental proof of the theory has been performed by Beitelschmidt [15].

In the following we shall focus on these experiments. In designing a test set-up for measuring impacts with friction a first principal decision had to be made with respect to the experiments and to the geometrical type of impact, plane or spatial. Colliding bodies moving in a plane include linear complementarity problems, spatial contacts generate nonlinear complementarities. Therefore motion in a plane was considered where one body is a disc and the other one the ground. On this basis some further requirements had to be defined:

- maximum translational velocity     10 m/s
- maximum rotational velocity     40 rps
- throw direction     $0° - 90°$
- release time     $< 12$ ms
- encoder main axis     1600 points
- encoder momentum axis     400 points
- throwing disc     diameter 50 mm
  - thickness 20 mm
  - weight 300 g
- continuous variable velocity control
- translation and rotation decoupled
- disturbance-free support and release of disc
- mass balance, statically and dynamically
- electric drives (pulse width modulation with 250 steps)
- automatic control for the throwing process, the release of stroboscope and camera

As a result, the machine of Figure (3.31) was designed and realized, which meets all requirements. A release unit containing the disc is mounted at the end of a rotating arm with mass balance. The unit itself drives the disc giving it a prescribed rotational speed. The arm drive and momentum drive are decoupled allowing to control the two speeds independently. The rotation of the arm can be used to generate a translation, the rotation of the release

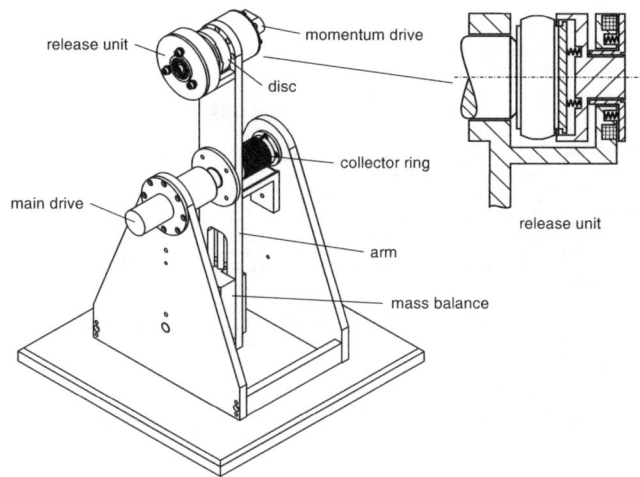

Fig. 3.31: Principle of the Throwing Machine

unit realizes a rotation of the disc. Both mechanisms require an extremely precise time management of the release process. The flight of the body is photographed under stroboscopic exposure in a dark room before and after hitting his target. From the evaluation of the photographs one can calculate the velocities and the position of the body immediately before and after the impact.

Figure (3.32) depicts the structure of the test set-up. A computer performs all control calculations, processes sensor data, evaluates control torques, releases stroboscope and camera and records all measured data. Within this overall structure we find for each drive an individual control concept, which has thoroughly been optimized with regard to the above requirements [15]. Also, a typical sequence of events for the test procedure can be seen from Figure (3.33). All computer codes have been realized in C++, which was feasible due to the fact that the PC-Mode activities are not critical with respect to time.

The evaluation of the measurements as recorded by the camera and the processor was straightforward. Figure (3.34) illustrates the method and depicts additionally two photographs of experiments. Especially the rubber disc experiment shows nicely a reversal of the trajectory due to the disc's rotation. The experimental process provided thus a very precise and well reproducible basis for determining the properties of impacts with friction.

In the following we shall give only a few examples out of more than 600 experiments performed with axisymmetric and with eccentric discs. In all cases the comparisons with theory come out with a good to excellent correspondence

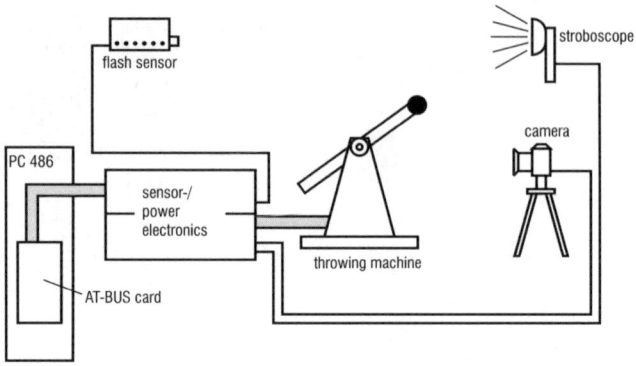

Fig. 3.32: Structure of the Complete Test Set-Up

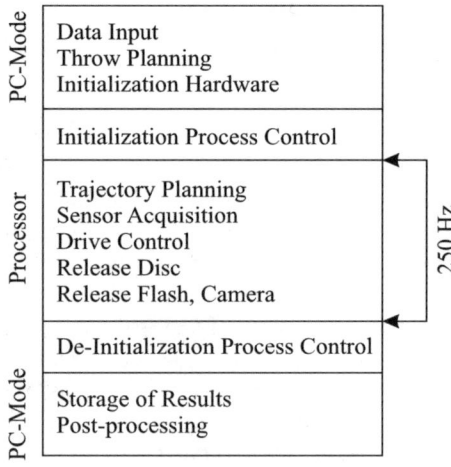

Fig. 3.33: Sequence of Events of the Throwing Machine Control

[15]. In the subsequent diagrams we shall use dimensionless velocities and impulses defined by

$$\gamma = \frac{\dot{g}_{TA}}{-\dot{g}_{NA}}, \quad \gamma_{NC} = \frac{\dot{g}_{NC}}{-\dot{g}_{NA}}, \quad \gamma_{TC} = \frac{\dot{g}_{TC}}{-\dot{g}_{NA}},$$
$$\gamma_{NE} = \frac{\dot{g}_{NE}}{-\dot{g}_{NA}}, \quad \gamma_{TE} = \frac{\dot{g}_{TE}}{-\dot{g}_{NA}}, \quad \gamma_{TE0} = \frac{\dot{g}_{TE0}}{-\dot{g}_{NA}}, \quad (3.226)$$

where the indices $N, T$ refer to normal and tangential directions. The indices $A, C, E$ are the beginning and the end of the compression phase, and the end of the expansion phase, respectively. The kinematical magnitude $\dot{g}$ is a relative velocity in the contact zone. Experiments usually generate a negative normal velocity $(-\dot{g}_{NA})$ at the beginning.

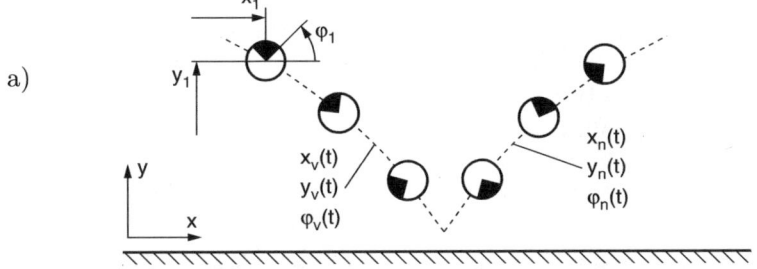
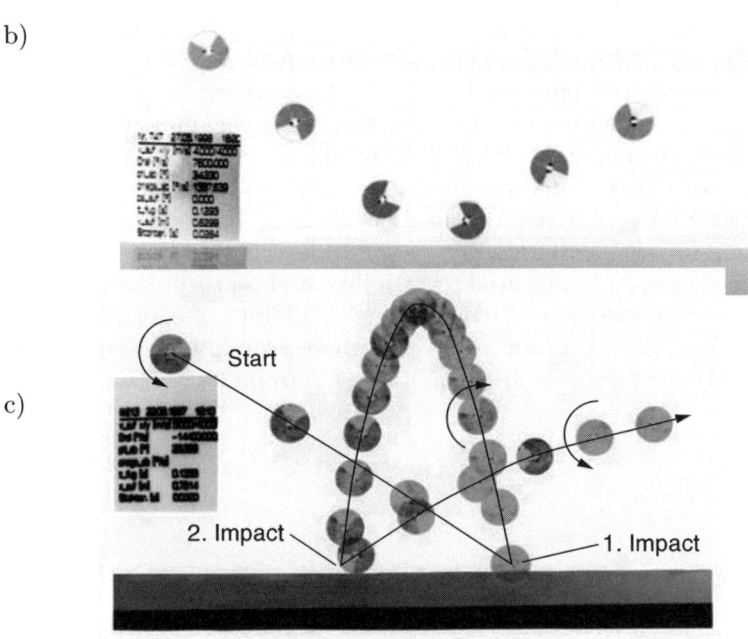

Fig. 3.34: Disc Trajectory During an Experiment, a) method of evaluation, b) photograph steel, c) photograph rubber

Figure (3.34) also indicates the evaluation process for all experimental results. For every small part of the trajectory we perform three stroboscope flashes thus achieving a certain redundancy for the measurements. The trajectory is approximately a parabola, and the velocity possesses a positive component in x- and a negative component in y-direction. The stroboscopic measurements in connetion with the marked sectors of the discs allow a safe evaluation of the translational and rotational velocities of the discs. To find the time and the point of impact, the measurements before and after such an

180    3 Constraint Systems

impact are represented by a statistical interpolation scheme, which allows to determine the impact together with the dispersion of the results.

Figure (3.34) gives two examples. The material pairing steel on steel behaves conventional and in such a way, which one could expect. The picture (b) shows the steel disc approaching the ground with a translational and rotational velocity and leaving the ground with a mirrored but more or less a similar trajectory.

The part (c) of Figure (3.34) represents a spectacular case. The rubber disc appears from the left side with a horizontal velocity of 5 m/s and a vertical velocity of 4 m/s in negative y-direction. The rotational velocity in a counterclockwise direction amounts to 40 rps (2400 rpm). This results in a tangential relative velocity of 12.5 m/s at the point of impact. The impact process is depicted by the Figure part (c). After the first contact the velocities reverse by the impact jump, and the disc flies backwards with a clockwise rotation. At the second impact the velocities change again, and the disc flies forward with the original direction of rotation.

As a result we may state, that firstly for the rubber case the impact coefficient of restitution in normal direction depends much more on the velocities at collision than for stiff materials, that secondly we get a typical characteristic behavior in the sense of tangential reversibility, and that thirdly for soft materials like rubber we may have friction coefficients larger than one ($\mu > 1$). The theory describes this behavior very well, where especially for soft materials a correction is advantageous (see equation (3.206) on page 168). The Figure

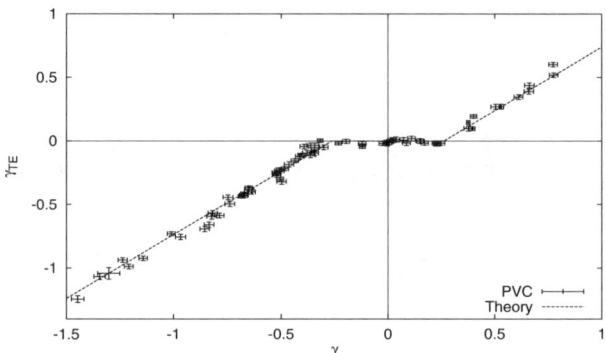

Fig. 3.35: Dimensionless Tangential Relative Velocity, after vs. before the impact, PVC-body

(3.35) shows results of experiments with a PVC test body. The experiments are marked by crosses, the dotted line shows the theoretical result. For small tangential relative velocities before the impact, sticking occurs, and the rolling constraint between disc and ground is fulfilled after the impact. If the relative velocity is big enough, the body slides throughout the impact and has a redcuced tangential relative velocity at the end of the impact. No tangential reversion occurs. In the area around zero tangential speed we get sticking.

A similar diagram for a rubber-body is shown in Figure (3.36). For most of the impacts the tangential relative velocity has changed during the impact: the bodies collide with a negative relative velocity and separate with a positive velocity. The inclination of the line through the origin is $-\varepsilon_N \varepsilon_T$. If $\varepsilon_N$ is known from another simple experiment one can evaluate the coefficient of tangential reversibility from this plot. For this series of experiments the parameters $\varepsilon_N = 0.75$ and $\varepsilon_T = 0.9$ were identified. If the tangential relative velocity increases

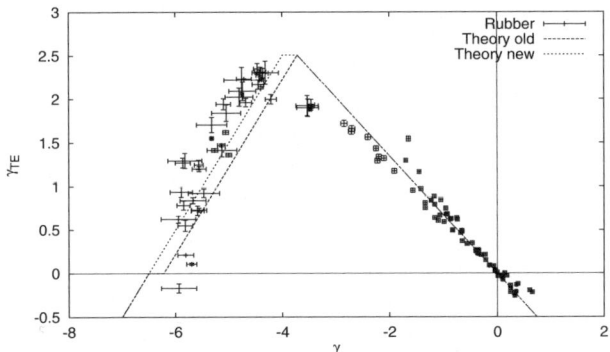

Fig. 3.36: Dimensionless Tangential Relative Velocity, after vs. before the impact, rubber-body

further, sliding occurs in the contact point during the impact. Then it is not possible to restore the elastic potential energy during the phase of expansion. For very high velocities the rubber body slides during the whole impact and the effect of tangential reversibility is not further visible.

In Figure (3.36) two lines are plotted for comparing theory with experiment. What is called "Theory old" corresponds to the original theory of impacts with friction as presented in Glocker's dissertation [86]. What is called "Theory new" includes the extension as given by Beitelschmidt [15], which applies mainly for very soft material pairings. If we consider the contact point of two bodies, where Coulomb's friction applies, and that point of the contact zone, where the spring force resulting from the storage of impulse applies, we come out with two force laws in series. This gives a modification of the complementarities with respect to the friction cone, and thus a modification of the final results (equation (3.206) on page 168).

Impacts with friction play an essential role in machines, mechanisms and also in biology. Therefore we need good models being verified by sound experiments. Such experiments have been performed by an especially designed throwing machine. The experimental results compare excellently with calculated values from existing theories. A slight improvement of the theory could be deduced from the experimental findings. It concerns the frictional complementarities during expansion, which have been slightly modified and are mainly applicable for soft material pairings.

An additional verification of the theory is steadily been performed by applying it to a large variety of industrial problems, where measurements were carried through within an industrial environment. Some typical examples are given by [200], [188], [199], [210], [91], [80], [23] or [195].

## 3.6 Modeling System Dynamics

Models comprise two important elements, the mechanical model and the mathematical model following from it. Heinrich Hertz [98] postulated, that a physical model, he called it a picture or a fictitous picture, should be logically permissible, correct and simple. The first point concerns the requirement, that a mechanical picture representing for example a large machine must be consistent, and it must be a clear picture of the real world. These pictures must be correct in an unambiguous sense, not least because of their property of being the starting point for the mathematical models. Finally mechanical models should be simple, which relates to the efforts needed to solve such problems. My own experience tells me, that models should be as complicated as necessary to achieve a good relation to reality, but as simple as possible to keep expenditure small.

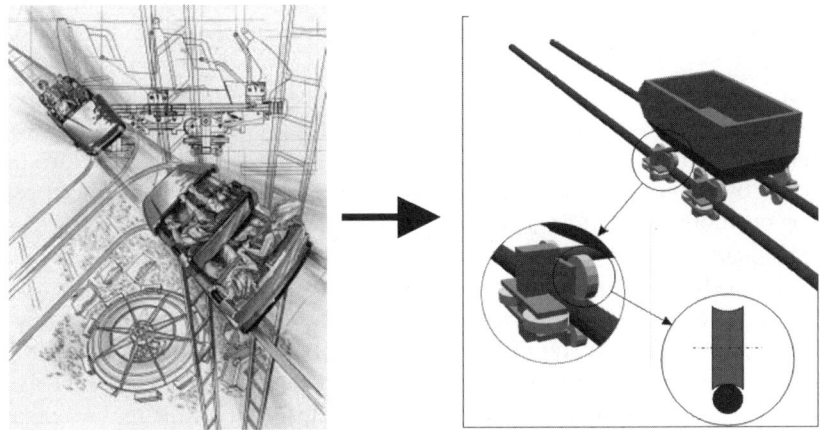

Fig. 3.37: An Example of Reality and Model

Figure 3.37 illustrates a typical case. On the left side we see an artist's impression of a roller coaster system, called the "wild mouse", which is in operation on many German fairs. The loads on the wheel packages and the strength of the wheels caused some problems. Therefore we had to model the contacts of the wheels with the track to evaluate the contact forces. Each car possesses four wheel packages with six wheels each. Driving along the track, a rather exciting track as partly indicated in the Figure, puts on the wheels a quickly varying set of load cycles, which led to wheel damages. The model should focus on the wheel loads but also consider the overall dynamics of the car going along the track. Therefore we modeled all four wheel packages with all details, especially with all contact details according to the right side of Figure 3.37. The model went along the track finally providing us with the necessary information to suggest a better wheel design [253]. The example illustrates that model complexity can be reduced sometimes by focussing on the most problematic components, in this case the wheels, and modeling the other system parts around that in a more or less rough and simplified way.

All methods presented in this book generate mathematically a set of ordinary second order differential equations together with certain sets of linear or non-linear, smooth or non-smooth side conditions. The equations of motion are linear in the accelerations, but non-linear with respect to positions /orientations and velocities. A model of that kind is the mathematical counterpart of that, what we have put into the mechanical model, not less and not more. Mathematics cannot give more information than what the mechanical model comprises. Therefore, constructing a mechanical model of a real world machine or mechanism requires special care, instinctive feeling and experience and definitely a complete understanding of the technical and physical features behind the problem. At this point of all problem solving actions we can produce for the future a lot of additional expenditure, but we also can try to reduce the problem to its hard kernel with the chance to achieve a fast solution.

The minimal possible size of the mathematical model is determined by the set of generalized, or sometimes called minimal, coordinates. Additional reductions might then be realized by linearizations, possible only for special cases, or by bringing in partial solutions, for example by regarding possible energy or momentum integrals. For large systems it often makes sense to linearize around some given operating point to achieve the eigenbehavior or the local stability. Any way we go, it will be absolutely necessary to find out as many as possible generalized coordinates, which is not always easy to perform, and to reduce the original set of equations of motion. We shall not always succeed in finding the complete minimal coordinate set, which leaves us with some equations in the original form together with additional constraints. Depending on the specific problem under consideration it might be even useful to maintain all constraints, for example in some modern gear problems. But normally this should be neither our goal nor the general rule.

Another point of considerable significance is sometimes forgotten in times of computational sciences. Deriving and establishing the set of equations of motion usually gives by the process itself some important insights into the problem we have to solve, so-to-say without computational efforts and by mere thinking. So this meaningful step should not be left out, because besides mechanical and mathematical insight and understanding we get an idea of the best solution procedures, which are the next step. This remains true also before the background of better possibilities of presenting simulation results by moving pictures, by combinations of CAD with FEM, MBS or the like, which as a matter of fact makes the assessment of difficult R&D-problems much easier, but does not solve them really. Such aids are very useful, but do not replace thought.

It is of course a well-known matter of experience that computing time can be reduced by two measures, firstly by a more computer-friendly formulation of the equations of motion and secondly by better numerical algorithms. The evolution from complementarity inequalities to prox-functions for non-smooth constraints and the evolution from second order differential equations to measure differential equations are very good examples for the first aspect, and the evolution from contact event interpolation with piecewise Runge-Kutta integration to time stepping is a good example for the second aspect. From this we may conclude that the choice of the methods, theoretical as well as numerical, possesses a decisive influence on the solution expenditures of the problem and needs to be handled carefully.

Whatever methodology we choose, for considering mechanical system dynamics we have to start with kinematics. We need to select coordinate frames, to investigate them with respect to their best placement, inertial fixed, body-fixed, interconnection-fixed, and we have to formulate the kinematical magnitudes, orientations and positions, velocities and accelerations, in terms of these coordinate systems, requiring at this stage already an iteration with respect to coordinate selection and the resulting kinematical relations. Again, they should be as simple as possible but as complicated as necessary. Next we have to explore possible sets of generalized coordinates. Also these sets reflect the coordinate selection and can be modified by a modified choice of coordinates. Only the basis of a very carefully built up kinematics opens the chance to establish the equations of motion and the appropriate constraint equations in a consistent way.

Large systems are interconnected. Seen from the standpoint of kinetics they can be dealt with by several approaches. To derive the equations of motion we may apply the momentum and moment of momentum equations together with the cutting principle or together with constraints, we may apply the relations of Lagrange I or II together with the system's energies, or we may use some of the basic principles directly. Again, our choice depends on our goals. It will be significantly different for aiming at basic research or at the solution of practical problems. We shall focus a bit on the second task.

A putative straightforward approach would be the usage of the momentum and moment of momentum equations together with the cutting principle. We cut all our system's bodies apart thus laying bare all unknown cut forces and torques, which then appear with opposite signs for neighbouring bodies, and solve this set for the unknown accelerations and forces. For large systems, though, it will be very hard to eliminate the unknown cut forces and torques and to construct some reasonable systematic evaluation. But for small systems this approach might be easy to apply and possesses the advantage of straightforwardness.

Applying the Lagrange I or II equations requires the evaluation of energies. In many cases this is simpler than describing a lot of constraints and forming the corresponding Jacobians, but on the other side we have to derive the energies with respect to the generalized coordinates for establishing the equations of motion. The kinetic energy for example needs the absolute velocities dependent on the generalized velocities and coordinates, which we have to know presupposing the elimination of all constraints for arriving at the generalized and minimal kinematical magnitudes. The argument against Lagrange I or II for large systems is connected with the derivation of energies, which is more costly than the evaluation of the Jacobians. Nevertheless for small systems Lagrange I or II might be a very fast and simple approach. It should be noted however, that in the last years some theory based on differential geometry and using energies has been elaborated, which partly see the above arguments in relative terms [149] [150].

Apart from many indications concerning a systematic formulation of multibody system dynamics by famous scientists in the past a first concrete step into that direction has been done about half a century ago with the upcoming space research and technologies. After decades of very vivacious discussions and quarrelling between many scientists we have in the meantime a consolidated approach based on Jourdain's principle. It requires the set of momentum and moment of momentum equations together with the appropriate set of constraints, and it allows a very clear and efficient form of the equations of motion, a form, which is also computer-friendly. Again, the basic hypothesis, that constraint forces are "lost forces" (Daniel Bernoulli), that they do not produce work (d'Alembert), and that they do not generate power (Jourdain), turns out to be one of the most important ideas of all mechanical sciences. In the meantime this Newton-Euler approach, as it is called today, forms the basis of nearly all commercial codes in the field.

# 4
# Dynamics of Hydraulic Systems

> *Die Methode der Wissenschaft ist die Methode der kühnen Vermutungen und der erfinderischen und ernsthaften Versuche, sie zu widerlegen.*
>
> *(Karl Popper, "Objektive Erkenntnis", 1984)*
>
> *The method of science is the method of bold conjectures and ingenious and severe attempts to refute them.*
>
> *(Karl Popper, "Objective Knowledge", 1972)*

## 4.1 Introduction

Hydraulic systems represent a steadily growing area of mechanical engineering with large applications in the automotive industry, in aerospace industry, in building and agricultural machinery and in machine tools, to name the most important fields. Large pressures, large forces generated by very compact equipment and for many cases acceptable control frequencies characterize hydraulics, which fundamentally comes from fluid mechanics. As a consequence we have various simulation tools on the market, which simulate hydraulics, though usually accompanied by large computing times due to a detailed consideration of all components, especially with respect to fluid compressibility. Compressibility generates stiff differential equations and steep characteristics. The basic idea of our treatment replaces steep characteristics by complementarities, where non-smooth set-valued force laws take the place of smooth, but steep single-valued forces [23].

In hydraulic networks we find such a complementarity behavior in connection with check valves, with servo valves and with cavitation in fluid-air-mixtures. A check valve for example might be open, then we have approximately no pressure drop, but a certain amount of flow rate. Or a check valve might be closed, then we have a pressure drop, but no flow rate. A small amount of air in the fluid will be compressed by a large pressure to a neglectable small air volume, but for a very small pressure the air will expand in a nearly explosive way, a behavior, which can be approximated by a complementarity. The replacement of steep characteristics by complementarities in connection with the neglection of fluid compressibility for small volumes reduces computing time by 3-4 orders of magnitudes, which has been proven several times by simulation of very large hydraulic systems. We shall give examples ([195], [192]).

With respect to compressibility we need to be cautious, nevertheless. Fluid nets can be approximately represented by two-terminal elements or by quadrupoles depending on the compressibility influence. The characteristic compressibility measure in form of the fluid capacitance is proportional to the fluid volume and inversely proportional to the fluid pressure of the fluid line under consideration, or the other way around, proportional to the fluid line volume and inversely proportional to the fluid density multiplied by the square of the fluid velocity of sound. The fluid volumes of hydraulic nets are usually very small leading also to very small capacitances, several orders of magnitude smaller than fluid resistance or fluid inductance. Therefore it makes not much sense to model such capacitances as representatives of compressibility, only to be on the safe side. We must check it and consider only really large fluid volumes with respect to compressibility (see also [174]).

Borchsenius presents a nice example for illustrating the influence of compressibility [23]. According to Figure 4.1 we have a simple system with an oil storage, a fluid line and a hydraulic cylinder. We want to investigate the influence of compressibility. The fluid storage has 15 bar, compressibility in the fluid line will be neglected, in the cylinder not, and the position of the piston is given with the coordinate x, its pressure is p. The volume of the cylinder couples the fluid line with the piston thus exhibiting features of a hydraulic joint. The piston is loaded by two forces, the pressure force on the left and the spring force on the right side. The simulation starts at x=0, the left end of the cylinder.

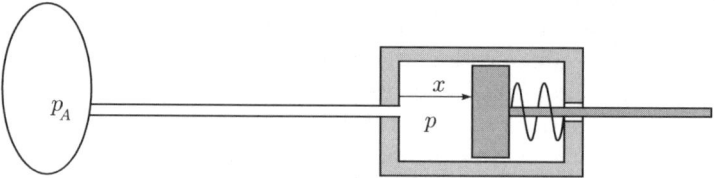

Fig. 4.1: A Simple Example for the Compressibility Effects

We model the system with and without compressibility. Due to the pressure in the storage we get a flow Q into the cylinder, which has the volume V. The piston with the cross-secton $A_P$ moves with velocity $v_P$. Compressibility is taken into account by the module E. For the pressure change $\dot{p}$ we get

$$\dot{p} = \frac{E}{V}(Q - A_P v_P) = \frac{1}{\varepsilon}(Q - A_P v_P) \qquad \text{with} \qquad \varepsilon = \frac{V}{E} \qquad (4.1)$$

The compressibility module of oil is about $E = 10^4 \text{bar} = 10^9 \text{Pa}$. Therefore the value of $\varepsilon = \frac{V}{E}$ becomes very small, at least for most of the fluid volumes in a hydraulic system, finally leading to very high frequencies. On the other hand we get for the incompressible case no pressure change with time and a simple quasistatic equation

$$Q - A_P v_P = 0, \tag{4.2}$$

which corresponds to Kirchhoff's nodal equation as known from electrical nets.

The simulation results reflect the above arguments (Figure 4.2). Compressibility does not influence the piston position, and it produces only some small oscillations around the mean value of the pressure. These high frequency oscillations are usually far away from the frequency range interesting for the system's operation. From this point of view they can be neglected for most practical cases. But it should be examined by some simple estimates as those of the two equations (4.1) and (4.2).

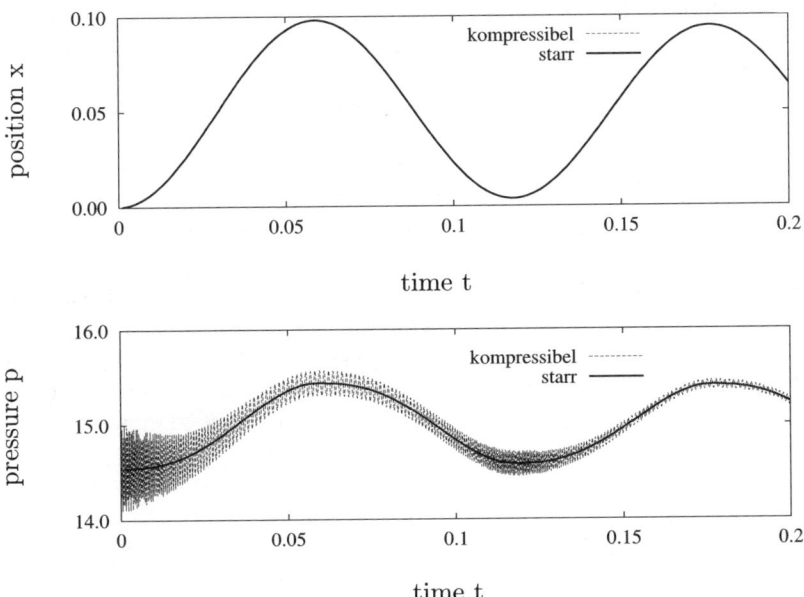

Fig. 4.2: Results for the Example of Figure 4.1, dotted line-compressible, solid line-incompressible

In summarizing we find a potential for reducing computing time with respect to hydraulic systems by establishing another type of theoretical models and by introducing other approximate neglections compared to classical theories.

The first aspect is compressibility, which does not possess influence for small volumes generating there extremely large frequency oscillations, which in most cases can be neglected. Criteria are the volume size of a component in comparison with adjoining volumes, for example fluid lines in comparison

# 4 Dynamics of Hydraulic Systems

with cylinder volumes, and the generation of high frequency oscillations far away from the operating frequencies of the system.

The second aspect concerns smooth non-linearities appearing in components like valves, orifice plates, leading or guiding edges and the like. Very often it is possible to replace such nonlinear characteristics by a linear approach with respect to the operating point. Smooth nonlinearities are also connected with a change of pressure and temperature of the fluid itself. Density, viscosity and compressibility depend on that, especially in domains of small pressure.

The third aspect concerns non-smooth, or better approximate non-smooth features of hydraulic systems. Examples are cylinders with a stop for the piston, valves and guiding edges, friction in cylinders leading to stick-slip phenomena and cavitation. They all can be approximated by complementarities, which we shall do in the following sections.

## 4.2 Modeling Hydraulic Components

In order to set up a mathematical model we assume, that the hydraulic system can be considered as a network of basic components. These components are connected by nodes. In conventional simulation programmes these nodes are very often assumed to be elastic. In the case of relatively large volumes this assumption is reasonable whereas for very small volumes incompressible junctions, with unilateral or bilateral behavior, are a better approach. Complex components like control valves can be composed of elementary components like lines, check valves and so forth. In the following a selection of elementary components is considered. It is shown how the equations of motion are derived and how they are put together to form a network.

### 4.2.1 Junctions

Junctions are hydraulic volumes filled with oil. The volumes may be considered as constant volumes or variable volumes as show in Figure 4.3. Junctions with variable volume are commonly used for hydraulic cylinders.

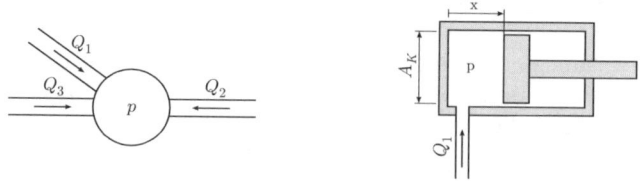

Fig. 4.3: Hydraulic junctions with constant and variable volume

### 4.2.1.1 Compressible Junctions

Assuming compressible fluid in such a volume leads to a nonlinear differential equation for the pressure $p$. Introducing the pressure-dependent bulk modulus

$$E(p) = -V \frac{dp}{dV} \tag{4.3}$$

yields a differential equation for the pressure in a constant volume

$$\dot{p} = \frac{E}{V} \sum Q_i \tag{4.4}$$

and

$$\dot{p} = \frac{E}{V} (Q_1 - A_K \dot{x}) \tag{4.5}$$

in a variable volume, respectively. A common assumption with respect to the fluid properties considers a mixture of linear elastic fluid with a low fraction of air. Figure 4.4 shows the calculated specific volume of a mixture of oil and 1 % air (at a reference value of 1 bar). For high pressure values the air is compressed to a neglectable small volume whereas the air expands abruptly for low pressure values, see Figure 4.4 with the pressure p versus the specific volume v. This figure illustrates also that the curve for the pressure in dependency of the specific volume can be very well approximated by a unilateral characteristic. If we would choose a smooth model we would get stiff differential equations 4.4, 4.5 for very small volumes $V \to 0$.

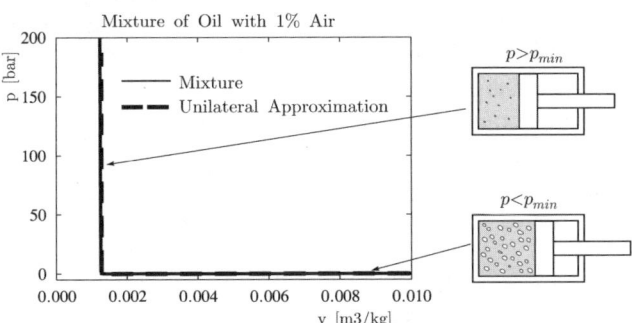

Fig. 4.4: Fluid expansion for low pressures

### 4.2.1.2 Incompressible Junctions

To avoid stiff differential equations for small volumes it is obviously possible to substitute the differential equations by algebraic equations. Assuming a

## 4 Dynamics of Hydraulic Systems

constant specific volume of the incompressible fluid yields for a constant and a variable volume V, respectively, the following algebraic equations:

$$\sum Q_i = 0, \quad \text{(constant V)}, \qquad \sum Q_i - A_K \dot{x}_K = 0, \quad \text{(variable V)}. \qquad (4.6)$$

These equations consider neither the elasticity nor the unilaterality of the fluid properties. A fluid model covering both elasticity and unilaterality is described in section 4.2.1.1. In the case of neglectable small volumes the fluid properties can be approximated by a unilateral characteristic. As illustrated in Figure 4.4 a unilateral law can be established by introducing a state variable

$$\bar{V} = -\int_0^t \sum Q \, d\tau \qquad (4.7)$$

which represents the total void volume in a fluid volume. Obviously this void volume is restricted to be positive, $\bar{V} \geq 0$. As long as the pressure value is higher than a certain minimum value $p_{min}$, the void volume is zero. A void formation starts when the pressure $p$ approaches the minimum value $p_{min}$. This idealized fluid behavior can be described by a complementarity.

$$\bar{V} \geq 0, \quad \bar{p} \geq 0, \quad \bar{V}\bar{p} = 0. \qquad (4.8)$$

The pressure reserve $\bar{p}$ is defined by $\bar{p} = p - p_{min}$. The complementarity can be put by differentiation on a velocity level.

$$\dot{\bar{V}} = \sum Q_i \geq 0, \qquad (4.9)$$

The equality sign represents the Kirchhoff equation stating that the sum of all flow rates into a volume is equal to the sum of all flow rates out of the volume. If the outflow is higher than the inflow the void volume increases, $\dot{\bar{V}} > 0$. Substituting the flow rates $Q$ into a fluid volume by the vector of the velocities within the connecting lines and their corresponding areas,

$$\boldsymbol{v} = \begin{pmatrix} v_1 \\ v_2 \\ \vdots \\ v_i \end{pmatrix}, \quad \boldsymbol{W} = \begin{pmatrix} A_1 \\ A_2 \\ \vdots \\ A_i \end{pmatrix} \qquad (4.10)$$

results in a unilateral form of the junction equation

$$-\boldsymbol{W}^T \boldsymbol{v} \geq 0. \qquad (4.11)$$

It is evident that fluid volumes with non-constant volume can be put also into this form by extending the velocity and area vectors by the velocity and the area of the piston, respectively. As long as the pressure is higher than the minimum value, $\bar{p} > 0$, the unilateral equation 4.11 can be substituted by a bilateral equation $\boldsymbol{W}^T \boldsymbol{v} = 0$. In this case it is necessary to verify the validity of the assumption $\bar{p} > 0$ because the bilateral constraint alone does not prevent negative values of $\bar{p}$.

### 4.2.2 Valves

In the following we shall give some examples of modelling elementary valves and more complex valves as a network of basic components. Physically, any valve is a kind of controllable constraint, wether the working element be a flapper, ball, needle or the like.

#### 4.2.2.1 Orifices

Orifices with variable areas are used to control the flow in hydraulic systems by changing the orifice area. As illustrated in Figure 4.5 the pressure drop in an orifice shows a nonlinear behavior. The classical model to calculate the pressure drop $\Delta p$ in dependency of the area $A_V$ and the flow rate $Q$ is the Bernoulli equation.

$$\Delta p = \frac{\rho}{2}\left(\frac{1}{\alpha A_V}\right)^2 Q|Q| \tag{4.12}$$

The factor $\alpha$ is an empirical magnitude with regard to geometry- and

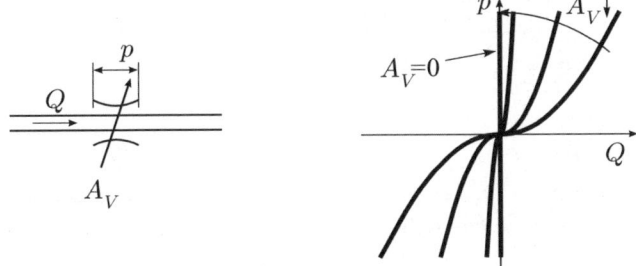

Fig. 4.5: Pressure drop in an orifice

Reynoldsnumber-depending pressure losses. It must be determined experimentally.

As long as the valve is open the pressure drop can be calculated as a function of the flow rate and the valve area, eq. 4.12. As shown in Figure 4.5 the characteristic becomes infinitely steep when the valve closes. In most commercial simulation programmes this leads to numerical ill-posedness and stiff differential equations for very small areas. In order to avoid such numerical problems the characteristic for the pressure drop of closed valves can be replaced by a simple constraint equation:

$$Q = Av = 0, \quad \text{or} \quad A\dot{v} = 0. \tag{4.13}$$

This constraint has to be added to the system equations when the valve closes. In the case of valve opening it has to be removed again. It leads to a time-varying set of constraint equations. In order to solve the system equations one has to distinguish between active constraints (closed valves) and passive constraints (opened valves). The last ones can be removed. The constraint equations avoid stiff differential equations. On the other hand they require to define active and passive sets.

### 4.2.2.2 Check Valves

Check valves are directional valves that allow flow in one direction only. It makes no sense trying to describe all existing types, so only the basic principle and the mathematical formulation is presented. Figure 4.6 shows the principle

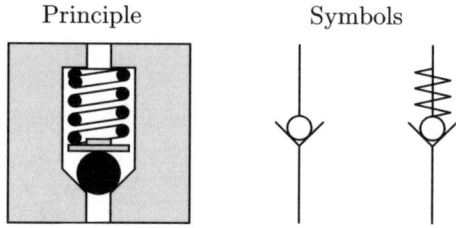

Fig. 4.6: Check valve

of a check valve with a ball as working element. Assuming lossless flow in one direction and no flow in the other direction results in two possible states:

- Valve open: pressure drop $\Delta p = 0$ for all flow rates $Q \geq 0$
- Valve closed: flow rate $Q = 0$ for all pressure drops $\Delta p \geq 0$

The two states define also a complementarity in the form

$$Q \geq 0, \quad \Delta p \geq 0, \quad Q\Delta p = 0. \tag{4.14}$$

Prestressed check valves with springs show a modified unilateral behavior, see Figure 4.7. The pressure drop curve of a prestressed check valve can be split into an ideal unilateral part $\Delta p_1$ and a smooth curve $\Delta p_2$ considering the spring tension and pressure losses, see Figure 4.8

### 4.2.2.3 Combined Components

Many hydraulic standard components are combinations of basic elements. Since the combination of unilateral and smooth characteristics yields either non-smooth or smooth behavior it is worth to consider such components with

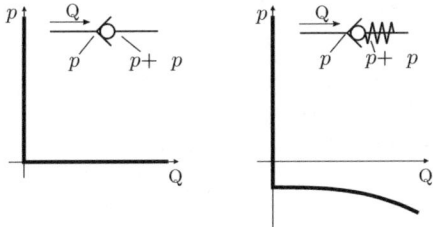

Fig. 4.7: Check valve characteristics

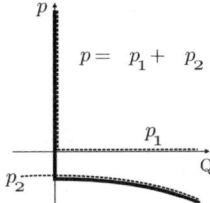

Fig. 4.8: Superposition of unilateral and smooth curves

a smooth characteristic separately. As an example we consider a typical combination of a throttle and a check valve. Figure 4.9 shows the symbol and the characteristics of both components. Since the flow rate of the combined component is the sum of the flows in the check valve and the throttle, the sum of the flow rates is a smooth curve. In such cases it is convenient to model the combined component as a smooth component (in the mechanical sense as a smooth force law).

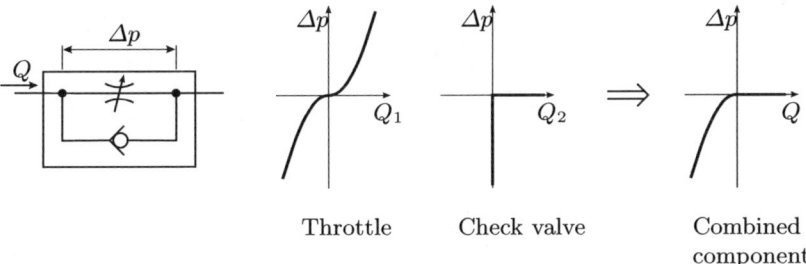

Fig. 4.9: Combination of smooth and non-smooth components

### 4.2.2.4 Servovalves

As an example for a servovalve we consider a one-stage 4-way-valve. It is a good example for the complexity of the networks representing such components like valves, pressure control valves, flow control valves and related valve systems. Multistage valves can be modelled in a similar way as a network consisting of servovalves and pistons, which themselves are working elements of a higher stage valve. Figure 4.10 shows the working principle of a 4-way valve. Moving the control piston to the right connects the pressure inlet P with the output B and simultaneously the return T with the output A. If one connects the outputs A and B with a hydraulic cylinder, high forces can be produced with small forces acting on the control piston. The valve works like a hydraulic amplifier.

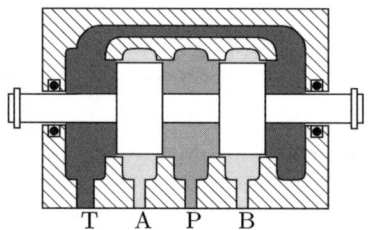

Fig. 4.10: 4-way valve

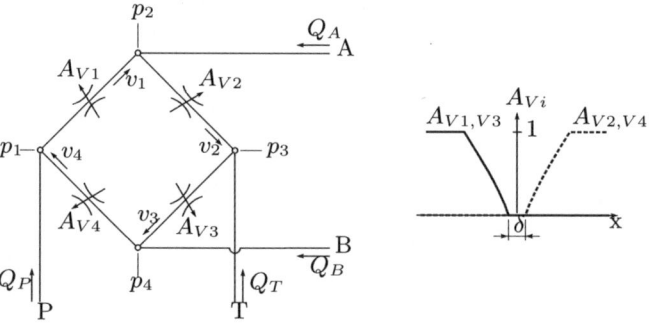

Fig. 4.11: Network model of a 4-way valve

Figure 4.11 shows a network model of the 4-way valve. The areas of the orifices $A_{V1} \ldots A_{V4}$ are controlled by the position $x$ of the piston. The orifice areas are assumed to be known functions of the position $x$. The parameter $\delta$

covers a potential deadband. To derive the equations of motion the lines in the network are assumed to be flow channels with cross sectional areas $A_1 \ldots A_4$. The fluid is incompressible since the volumes are usually very small, and the bulk modulus of the oil is very high. The oil masses in the lines are $m_1 \ldots m_4$. Denoting the junction pressures with $p_i$ and the pressure drops in the orifices with $\Delta p_i$, we get the equations of momentum as

$$\begin{aligned}
m_1 \dot{v}_1 - A_1 p_1 + A_1 p_2 + A_1 \Delta p_1 &= 0 \\
m_2 \dot{v}_2 - A_2 p_2 + A_2 p_3 + A_2 \Delta p_2 &= 0 \\
m_3 \dot{v}_3 - A_3 p_3 + A_3 p_4 + A_3 \Delta p_3 &= 0 \\
m_4 \dot{v}_4 + A_4 p_1 - A_4 p_4 + A_4 \Delta p_4 &= 0
\end{aligned} \quad (4.15)$$

which can be expressed as

$$\boldsymbol{M\dot{v} + Wp + W_V \Delta p = W_a \Delta p_a}, \quad (4.16)$$

where $\boldsymbol{v}$ is the vector of flow velocities, $\boldsymbol{p}$ the vector of junction pressures, $\Delta \boldsymbol{p}$ the vector of pressure drops in the closed orifices and $\Delta \boldsymbol{p}_a$ the vector of pressure drops in the open orifices. The mass matrix $M = diag(m_i)$ is the diagonal matrix of the oil masses. The matrix

$$\boldsymbol{W} = \begin{bmatrix} -A_1 & A_1 & 0 & 0 \\ 0 & -A_2 & A_2 & 0 \\ 0 & 0 & -A_3 & A_3 \\ A_4 & 0 & 0 & -A_4 \end{bmatrix} \quad (4.17)$$

is used to calculate the forces acting on the oil masses in the channels resulting from the junction pressures $\boldsymbol{p}$. The junction equations are given by

$$\begin{pmatrix} Q_P \\ Q_A \\ Q_T \\ Q_B \end{pmatrix} + \begin{bmatrix} -A_1 & 0 & 0 & A_4 \\ A_1 & -A_2 & 0 & 0 \\ 0 & A_2 & -A_3 & 0 \\ 0 & 0 & A_3 & -A_4 \end{bmatrix} \begin{pmatrix} v_1 \\ v_2 \\ v_3 \\ v_4 \end{pmatrix} = \boldsymbol{0} \quad (4.18)$$

which can be written in the form

$$\boldsymbol{Q}_{in} + \boldsymbol{W}^T \boldsymbol{v} = \boldsymbol{0} . \quad (4.19)$$

In order to determine the pressure drops $\Delta p_i$ one has to distinguish between open and closed orifices to avoid stiff equations, see section 4.2.2.1. In case of open orifices the pressure drop can be calculated directly subject to the given flow rates and the orifice area, whereas closed orifices are characterized by a constraint equation.

$$\begin{aligned} \Delta p_{ai} &= f(v_i, A_{Vi}(x)) & \text{open orifices i} \\ A_j v_j &= 0 & \text{closed orifices j} \end{aligned} \quad (4.20)$$

The constraint equations for the closed orifices are collected to give

$$\boldsymbol{W}_V^T \boldsymbol{v} = \boldsymbol{0}, \quad (4.21)$$

where the number of columns of $\boldsymbol{W}_V$ is the number of closed orifices. Note that this matrix has to be updated every time an orifice opens or closes.

### 4.2.3 Hydraulic lines

Hydraulic lines or hoses are used to connect components. For long lines the dynamics of the compressible fluid has to be taken into account. In order to get a precise system model, it is necessary to investigate pressure wave phenomena as well as the pipe friction. The pipe friction is rather complicated since the velocity profile is not known a priory. In the case of laminar flow it is possible to derive analytical formulas for a uniform fluid transmission line in the Laplace domain. The so-called 4-pole-transfer-functions relate the pressure and the flow at the input and at the output of the line in dependency of Bessel functions. Many attempts have been made to approximate the transfer functions with rational polynomial functions which can be re-transformed into the time domain. Unfortunately the form of the equations of these models is not compatible with the equations in the framework of this paper, because the coupling with constraint equations might lead to numerical instability due to violation of the principle of virtual work.

In the following a time domain modal approximation is presented. This model can be extended to cover frequency dependent friction as well. The starting point are the linearized partial differential equations for one-dimensional flow. The coordinates are shown in Figure 4.12. Partial derivatives of a arbitrary coordinate $q$ are denoted by $\frac{\partial q}{\partial t} = \dot{q}$ and $\frac{\partial q}{\partial x} = q'$, respectively.

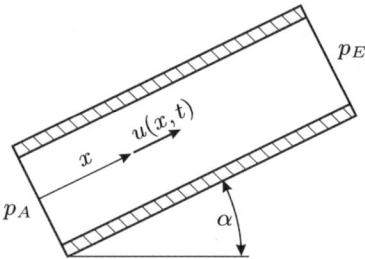

Fig. 4.12: Coordinates for one-dimensional flow

The mass balance

$$\dot{p} + \frac{E}{A}Q' = 0 \qquad (4.22)$$

with the flow rate $Q = Au$ and the introduced state variable

$$\tilde{x} = \frac{1}{A}\int_0^t Q d\tau \; ; \quad \dot{\tilde{x}} = \frac{Q}{A} \; ; \quad \ddot{\tilde{x}} = \frac{\dot{Q}}{A} \qquad (4.23)$$

can be solved analytically with respect to the pressure $p$:

## 4.2 Modeling Hydraulic Components

$$p(x,t) = -\frac{E}{A}\int_0^t Q' d\tau + p_0(x) = -E\int_0^t \dot{\tilde{x}}' d\tau + p_0(x) = -E\tilde{x}' + p_0(x). \quad (4.24)$$

The term $p_0(x)$ represents the initial pressure distribution in the line. The equation of momentum

$$\rho \dot{u} + p' + f_R + f_g = 0 \quad (4.25)$$

with a friction force $f_R$ and a gravity force $f_g$ can be transformed with eq. 4.23 to

$$\rho \ddot{\tilde{x}} - E\tilde{x}'' + p_0' + f_R + f_g = 0 \,. \quad (4.26)$$

Multiplying eq. 4.26 with an arbitrary test functions $w(x)$ and integrating over the length $L$ of the line yields the weak formulation

$$\begin{aligned}\rho \int_0^L \ddot{\tilde{x}} w(x) dx - E \int_0^L \tilde{x}'' w(x) dx + \int_0^L p_0'(x) w(x) dx + \\ + \int_0^L f_R w(x) dx + \int_0^L f_g w(x) dx = 0 \quad \forall\; w(x)\end{aligned} \quad (4.27)$$

where the term $w\bar{x}''$ can be integrated by parts to fit the boundary conditions.

$$\begin{aligned}\int_0^L w\bar{x}'' dx &= w\,\bar{x}'\big|_0^L - \int_0^L w'\bar{x}' dx = \\ &= w_L \tfrac{1}{E}(p_0(L) - p_E) - w_0 \tfrac{1}{E}(p_0(0) - p_A) - \int_0^L w'\bar{x}' dx\end{aligned} \quad (4.28)$$

$p_A$, $p_E$ are boundary pressures and $w_0$, $w_L$ are the values of the test function $w(x)$ at the boundaries ($x = 0, x = L$). For the sake of simplicity the initial pressure distribution is assumed to be uniform, $p_0(x) = p_0 = const$. In order to approximate the partial differential equations by a set of ordinary differential equations we introduce spatial shape functions $w(x)$ and a separation of the variables $x$ and $t$,

$$\bar{x} \approx \boldsymbol{q}(t)^T \boldsymbol{w}(x) \quad (4.29)$$

According to Galerkin's method the shape functions $\boldsymbol{w}(x)$ are the same functions used in the weak formulation, eq. 4.27. The discretized equations of motion are then

$$\begin{aligned}\rho \int_0^L \boldsymbol{ww}^T dx \ddot{\boldsymbol{q}} + E \int_0^L \boldsymbol{w'w'}^T dx \boldsymbol{q} + \int_0^L f_R \boldsymbol{w} dx + \int_0^L f_g \boldsymbol{w} dx = \\ = \boldsymbol{w}_0 p_A - \boldsymbol{w}_L p_E - (\boldsymbol{w}_0 - \boldsymbol{w}_L) p_0\end{aligned} \quad (4.30)$$

which can be transformed with $\boldsymbol{w}_A = A\boldsymbol{w}_0$, $\boldsymbol{w}_E = A\boldsymbol{w}_L$ to yield

$$\boldsymbol{M}\ddot{\boldsymbol{q}} + \boldsymbol{K}\boldsymbol{q} + \boldsymbol{W}\boldsymbol{p} = -A\int_0^L f_R \boldsymbol{w} dx - A\int_0^L f_g \boldsymbol{w} dx - (\boldsymbol{w}_A - \boldsymbol{w}_E) p_0 \quad (4.31)$$

with the abbreviations

$$M = \rho A \int_0^L ww^T x \quad \text{(mass matrix)}$$
$$K = EA \int_0^L w'w'^T \quad \text{(stiffness matrix)} \tag{4.32}$$
$$W = (-w_A \; w_E) \; .$$

The structure of the relation (4.31) corresponds to the equations of motion of a mechanical system. Suitable functions $w(x)$ are harmonic functions and B-spline functions, as numerical experiments confirmed.

### 4.2.3.1 Frequency dependant friction

In case of laminar flow the cross-sectional velocity profile of oscillatory flow can be calculated analytically as functions of Bessel functions. It turns out that the gradient of the velocity becomes higher with increasing frequencies $\omega$. Figure 4.13 shows calculated profiles for dimensionless frequencies $\frac{\omega}{\nu}R$ where $R$ is the radius and $\nu$ the kinematic viscosity.

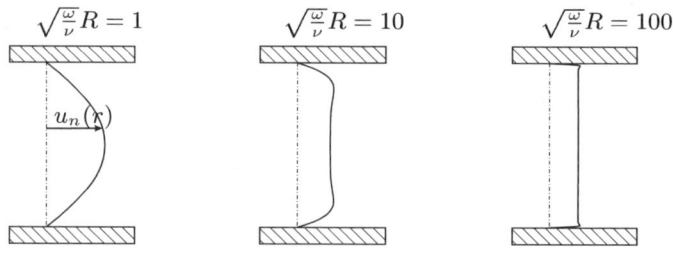

Fig. 4.13: Normalized velocity profiles for oscillatory flow

For low frequencies the profile is the well-known parabolic Hagen-Poiseuille profile for stationary flow. Since the friction force depends on the gradient at the pipe wall the friction force becomes higher with increasing frequency. Therefore the increasing friction has to be taken into account by a correction of the steady-state friction factor. It can be shown that the pipe friction for a parabolic velocity profile can be covered by a damping matrix $D_0$

$$M\ddot{q} + D_0\dot{q} + Kq = h \tag{4.33}$$

with

$$D_0 = 8\mu\pi \int_0^L ww^T dx. \tag{4.34}$$

Since the parabolic velocity profile is valid only for low frequencies this damping matrix has to be corrected. In the Laplace domain a friction correction factor can be derived as

$$N = \frac{\frac{U}{r}\big|_R}{-\frac{4}{R}} = -\frac{R}{8}\gamma\left(\frac{J_{-1}(\gamma R) - J_1(\gamma R)}{J_2(\gamma R)}\right) \tag{4.35}$$

where

$$\gamma = \sqrt{\frac{-\rho s}{\mu}} \ . \tag{4.36}$$

With the dimensionless frequency $k_H = \sqrt{\frac{\omega}{\nu}}R$ the following approximation of the real part of eq. 4.35 can be given (see [23]):

$$N(k_H) = \begin{cases} 1 + 0.0024756 \cdot k_H{}^{3.0253322} & k_H \leq 5 \\ (N(5) - 0.175 \cdot 5) + 0.175 \cdot k_H & k_H > 5 \end{cases} \tag{4.37}$$

In order to correct the damping matrix the eigenvalues $\lambda_i$ and the eigenfrequencies

$$\omega_i = \sqrt{\lambda_i} \tag{4.38}$$

of the matrix $\boldsymbol{M}^{-1}\boldsymbol{K}$ are calculated. The eigenvectors are collected in the orthogonal modal matrix

$$\boldsymbol{\Phi} = (\boldsymbol{\Phi}_1, \boldsymbol{\Phi}_2, \ldots, \boldsymbol{\Phi}_n) \ . \tag{4.39}$$

With a diagonal correction matrix $\bar{\boldsymbol{N}} = diag(N(\omega_i))$ a corrected modal damping matrix is introduced in the form

$$\boldsymbol{D}_m = \boldsymbol{\Phi}^T \boldsymbol{D}_0 \boldsymbol{\Phi} \cdot \bar{\boldsymbol{N}} \tag{4.40}$$

which can be re-transformed to yield

$$\boldsymbol{D} = \left(\boldsymbol{\Phi}^T\right)^{-1} \boldsymbol{D}_m \boldsymbol{\Phi}^{-1} \tag{4.41}$$

With this matrix the friction term in equation (4.33) can be substituted:

$$A \int_0^L f_R \boldsymbol{w} dx = \boldsymbol{D}\dot{\boldsymbol{q}} \tag{4.42}$$

## 4.3 Hydraulic Networks

The above collection of hydraulic components represents a selection of some important elements only. This selection may always be extended. For our purposes it is sufficient to present a principal method for establishing hydraulic nets. Hydraulic components are described by the state variables velocity, piston position, pressure and the like. We collect these variables in two state vectors.

- All state variables which are given by ordinary differential equations without any constraints are collected in a vector $\boldsymbol{x} \in \mathbb{R}^{n_x}$
- The vector $\boldsymbol{v} \in \mathbb{R}^{n_v}$ contains velocities which are described by momentum equations with unilateral and bilateral constraints.

Summarizing the equations of all components results in the following set of equations.

*Momentum equations for $\boldsymbol{v}$*

$$\boldsymbol{M}\dot{\boldsymbol{v}} - \boldsymbol{W}\boldsymbol{p} - \boldsymbol{W}_V \boldsymbol{p}_V - \bar{\boldsymbol{W}}^* \bar{\boldsymbol{p}} = \boldsymbol{f}(t, \boldsymbol{v}, \boldsymbol{x}) \in \mathbb{R}^{n_v} \qquad (4.43)$$

*Ordinary differential equations for $\boldsymbol{x}$*

$$\dot{\boldsymbol{x}} = \boldsymbol{g}(t, \boldsymbol{x}, \boldsymbol{v}) \in \mathbb{R}^{n_x} \qquad (4.44)$$

*Bilateral junction equations*

$$\boldsymbol{W}^T \boldsymbol{v} + \boldsymbol{w}(t) = \boldsymbol{0} \in \mathbb{R}^{n_I} \qquad (4.45)$$

*Constraints for closed valves/orifices*

$$\boldsymbol{W}_V^T \boldsymbol{v} = \boldsymbol{0} \in \mathbb{R}^{n_{Va}} \qquad (4.46)$$

*Unilateral constraints*

$$\bar{\boldsymbol{Q}} \geq \boldsymbol{0}, \quad \bar{\boldsymbol{p}} \geq \boldsymbol{0}, \quad \bar{\boldsymbol{Q}}\bar{\boldsymbol{p}} = 0, \quad \text{with} \quad \bar{\boldsymbol{Q}} = \bar{\boldsymbol{W}}^T \boldsymbol{v} \in \mathbb{R}^{\bar{n}_a} \qquad (4.47)$$

### 4.3.1 Solutions

In a first step we combine the active bilateral constraint equations 4.45, 4.46 to give

$$\boldsymbol{W}_G^T \boldsymbol{v} + \boldsymbol{w}_G = \boldsymbol{0} . \qquad (4.48)$$

The number of independent constraints is the rank of the matrix $\boldsymbol{W}_G$,

$$r = rank(\boldsymbol{W}_G) \qquad (4.49)$$

The constraint equations can be fulfilled by introducing minimal coordinates

$$\boldsymbol{v}_m \in \mathbb{R}^{n_{min}} \text{ where } n_{min} = n_v - r \qquad (4.50)$$

in the form

$$\boldsymbol{v} = \boldsymbol{v}(\boldsymbol{v}_m, t) = \boldsymbol{J}\boldsymbol{v}_m + \boldsymbol{b}(t) . \qquad (4.51)$$

The Jacobian $\boldsymbol{J}$ and the vector $\boldsymbol{b}$ can be calculated numerically by a singular value decomposition of the matrix $\boldsymbol{W}_G$ with the benefit that also dependent constraints can be handled. Since the column vectors of the Jacobian are orthogonal to the column vectors of $\boldsymbol{W}_G$ we can write the momentum equations as

$$\boldsymbol{J}^T \boldsymbol{M} \left( \boldsymbol{J} \dot{\boldsymbol{v}}_m + \dot{\boldsymbol{b}} \right) - \boldsymbol{J}^T \bar{\boldsymbol{W}}^* \bar{\boldsymbol{p}} = \boldsymbol{J}^T \boldsymbol{f}(t, \boldsymbol{x}, \boldsymbol{v}) \tag{4.52}$$

The square matrix $\boldsymbol{J}^T \boldsymbol{M} \boldsymbol{J} \in \mathbb{R}^{n_{min}, n_{min}}$ is the invertible mass action matrix.

As a next step we derive a linear complementary problem (LCP) for the unilateral constraint equations. For this purpose we put all active unilateral constraints on a velocity level and transform eq. 4.47 with eq. 4.51 to

$$\dot{\bar{\boldsymbol{Q}}} \geq \boldsymbol{0}, \quad \bar{\boldsymbol{p}} \geq \boldsymbol{0}, \quad \dot{\bar{\boldsymbol{Q}}}^T \bar{\boldsymbol{p}} = 0, \quad \text{with} \quad \dot{\bar{\boldsymbol{Q}}} = \bar{\boldsymbol{W}}^T \dot{\boldsymbol{v}} = \bar{\boldsymbol{W}}^T \boldsymbol{J} \dot{\boldsymbol{v}}_m + \bar{\boldsymbol{W}}^T \dot{\boldsymbol{b}} \tag{4.53}$$

Together with eq. 4.52 we obtain

$$\dot{\bar{\boldsymbol{Q}}} = \underbrace{\bar{\boldsymbol{W}}^T \boldsymbol{J} (\boldsymbol{J}^T \boldsymbol{M} \boldsymbol{J})^{-1} \boldsymbol{J}^T \bar{\boldsymbol{W}}^*}_{\boldsymbol{A}_{LCP}} \bar{\boldsymbol{p}} + \tag{4.54}$$

$$+ \underbrace{\bar{\boldsymbol{W}}^T \boldsymbol{W}^T \boldsymbol{J} (\boldsymbol{J}^T \boldsymbol{M} \boldsymbol{J})^{-1} \boldsymbol{J}^T \left( \boldsymbol{f} - \boldsymbol{M} \dot{\boldsymbol{b}} \right) + \bar{\boldsymbol{W}}^T \dot{\boldsymbol{b}}}_{\boldsymbol{b}_{LCP}} \tag{4.55}$$

This is a standard LCP in the form

$$\dot{\bar{\boldsymbol{Q}}} = \boldsymbol{A}_{LCP} \bar{\boldsymbol{p}} + \boldsymbol{b}_{LCP} \quad \text{and} \quad \dot{\bar{\boldsymbol{Q}}} \geq \boldsymbol{0}, \quad \bar{\boldsymbol{p}} \geq \boldsymbol{0}, \quad \dot{\bar{\boldsymbol{Q}}}^T \bar{\boldsymbol{p}} = 0, \tag{4.56}$$

which can be solved using a standard Lemke algorithm. Experience shows that the Lemke algorithms works reliable in many cases, also for large systems, but the computing times are long. In the meantime better algorithms are available, which are based on the ideas of time-stepping and of the Augmented Lagrange method (see chapter 3.4.4 on page 145). Numerical experiences in other areas indicate, that on this new basis computing time can be shortened significantly.

With this solution of the complementarity problem the time derivative of $\boldsymbol{v}$ can be calculated. The evolution of $\boldsymbol{x}$ and $\boldsymbol{v}$ with respect to time is obtained by a numerical integration scheme, for example a Runge-Kutta-scheme.

### 4.3.2 Hydraulic Impacts

Some examples for possible impacts in hydraulic systems are:

- Valve closure
- Impacts of mechanical components, for example piston/housing contact
- Condensation of vapor (waterhammer), cavitation

In connection with the use of constraint equations instead of stiff elasticities we must consider multiple impact situations. Due to the algebraic relationship between some components sudden velocity changes in one component may cause also velocity jumps in other components. In order to calculate the velocities after an impact it is necessary to solve the impact equations for the complete system. The starting point are the momentum equations 4.52. We assume as usual an infinitesimal short impact time $\Delta t \to 0$ without any position changes. The integration of the momentum equation 4.52 over the duration of the impact yields:

$$\int_t^{t+\Delta t} \left(\boldsymbol{J}^T \boldsymbol{M} \boldsymbol{J} \dot{\boldsymbol{v}}_m\right) dt = \int_t^{t+\Delta t} \left(\boldsymbol{J}^T \bar{\boldsymbol{W}}^* \bar{\boldsymbol{p}} + \boldsymbol{J}^T \boldsymbol{f} - \boldsymbol{J}^T \boldsymbol{M} \dot{\boldsymbol{b}}\right) dt \qquad (4.57)$$

Denoting the beginning of the impact with $^-$ and the end with $^+$, we can rewrite eq. 4.57 as

$$\boldsymbol{J}^T \boldsymbol{M} \boldsymbol{J} \left(\boldsymbol{v}_m^+ - \boldsymbol{v}_m^-\right) = \boldsymbol{J}^T \bar{\boldsymbol{W}}^* \bar{\boldsymbol{p}} \Delta t + \boldsymbol{J}^T \boldsymbol{f} \Delta t - \boldsymbol{J}^T \boldsymbol{M} \dot{\boldsymbol{b}} \Delta t \ . \qquad (4.58)$$

Introducing unilateral pressure impulses

$$\bar{\boldsymbol{P}} = \int_t^{t+\Delta t} \bar{\boldsymbol{p}} \ dt = \bar{\boldsymbol{p}} \Delta t \qquad (4.59)$$

and letting $\Delta t \to 0$ we come out with

$$\boldsymbol{J}^T \boldsymbol{M} \boldsymbol{J} \left(\boldsymbol{v}_m^+ - \boldsymbol{v}_m^-\right) = \boldsymbol{J}^T \bar{\boldsymbol{W}}^* \bar{\boldsymbol{P}} \ . \qquad (4.60)$$

The unilateral constraint equations can be set up for the end of the impact in the form of a linear complementarity problem

$$\bar{\boldsymbol{Q}}^+ = \underbrace{\bar{\boldsymbol{W}}^T \boldsymbol{J} (\boldsymbol{J}^T \boldsymbol{M} \boldsymbol{J})^{-1} \boldsymbol{J}^T \bar{\boldsymbol{W}}^*}_{\boldsymbol{A}_{LCP}} \bar{\boldsymbol{P}} + \underbrace{\bar{\boldsymbol{W}}^T \left(\boldsymbol{J} \boldsymbol{v}_m^- + \boldsymbol{b}\right)}_{\boldsymbol{b}_{LCP}}$$

$$\bar{\boldsymbol{Q}}^+ \geq 0, \quad \bar{\boldsymbol{P}} \geq 0, \quad (\bar{\boldsymbol{Q}}^+)^T \bar{\boldsymbol{P}} = 0,$$

$$\bar{\boldsymbol{Q}}^+ = \bar{\boldsymbol{W}}^T \boldsymbol{v}^+ = \bar{\boldsymbol{W}}^T \boldsymbol{J} \boldsymbol{v}_m^+ + \bar{\boldsymbol{W}}^T \boldsymbol{b} \ , \qquad (4.61)$$

With the solution vector $\bar{\boldsymbol{P}}$ of this LCP we can calculate $\boldsymbol{v}_m^+$ from eq. 4.60. Finally we get the velocity vector at the end of the impact as $\boldsymbol{v}^+ = \boldsymbol{J} \boldsymbol{v}_m^+ + \boldsymbol{b}$.

## 4.4 Practical Examples

### 4.4.1 Hydraulic Safety Brake System

As one example from industry we consider the hydraulic safety brake system of a fun ride, the free fall tower [23], [192]. It is manufactured by the company Maurer Söhne GmbH, Munich, Germany. Figure 4.14 shows the tower. Under normal operation conditions the cabin with the passengers is lifted by a cable winch to a height of about 60 m. Subsequently the cabin is released and falls

## 4.4 Practical Examples

Fig. 4.14: Free fall tower

down nearly undamped. Before reaching the ground the normal brake system stops the cabin softly via the cable winch. For safety reasons a redundant brake system is necessary. In the case of a failure of the normal operating brake or a cable rupture the safety brake system has to catch the cabin even under disadvantageous conditions.

The safety brake system is a hydraulic system which moves brake-blocks via hydraulic cylinders. For this purpose steel blades are fixed at the cabin. The steel blades fall into a guide rail, where the brake-blocks are fixed. The safety brake system consists of up to 7 identical modules with 4 hydraulic cylinders each. The modules are arranged upon each other. Figure 4.15 shows the model of the system. The simulations were carried out with the computer programme *HYSIM*. As the Figure indicates, the programme *HYSIM* allows multi-hierarchical modelling of systems. It means that some components can be put together to form a group, which itself can be used and duplicated like any elementary component. Under normal operating conditions the brake-blocks are moved outwards the guide rail quickly so that the system is decelerated softly by the winch. Only under bad conditions the safety brakes stay closed and the cabin will be stopped by the friction forces. Due to the very short opening time of the brake the hydraulic cylinders move quickly and reach the stop position with high velocities. The resulting impact forces caused in some cases damage of the cylinders. In order to reduce the impact forces simulations were carried out. The aim was to find new values of adjustable parameters such that the impact velocities are reduced and the opening time does not exceed a certain maximum value.

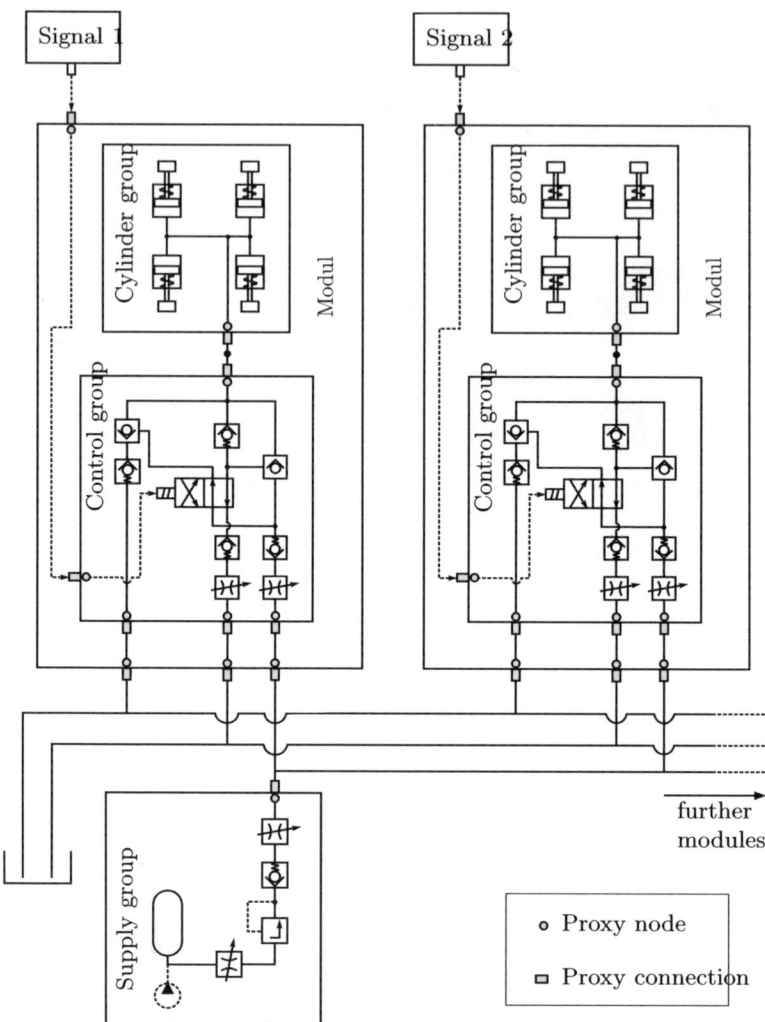

Fig. 4.15: HYSIM-model of the hydraulic system

Figure 4.16 shows some simulation results. The impact velocity and thus the resulting force is significantly high for the Cylinder 1. The reason for this behavior lies in the different length of the supply line and the different movement direction (up/down) of the cylinders, i.e. hydraulic and mechanical asymmetries. The impact velocities can be reduced significantly by increasing restriction parameters at some valves, as fig. 4.16 shows. This simple measure has been adopted to the real system and no problems occurred since then.

A comparison of measurement and simulation is given in Figure 4.17. As the cylinders start moving the pressure increases due to the increasing spring

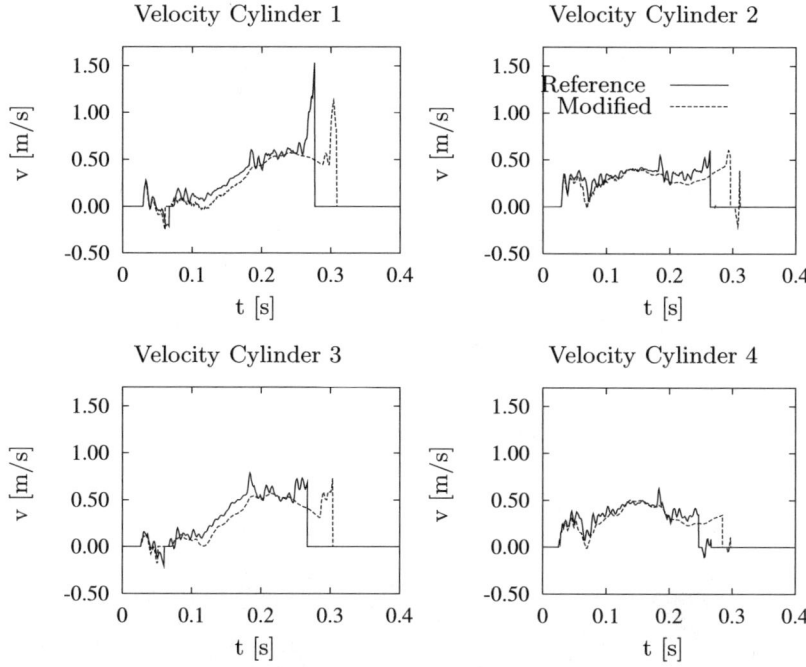

Fig. 4.16: Cylinder movement

force of the cylinders. At $t \approx 0,37s$ the last cylinders reaches the stop position. The impact causes a pressure jump.

The model for the drop tower consists of nearly 1000 hydraulic degrees of freedom, computing time was in the range of less than one hour. A comparison of the computing time with that of convential and commercial codes results in a factor of about 6000 to 10000. This large difference for the new model has been also confirmed by many other problems from industry.

### 4.4.2 Power Transmission Hydraulics

With growing requirements with respect to fuel economy of automobiles the significance of automatic transmissions will also increase. As a classical solution of such transmissions we have the automated gears, today with up to eight stages, and more recently the applications of CVTs, **C**ontinuous **V**ariable **T**ransmissions, which compete with the classical configurations. Modeling such components always means modeling the complete system, because the neglection of components might result in an only partly realistic output, though we may model the complete drive for example in a more rough way so-to-say around the automatic transmission. Figure 4.18 gives an impression of a five-stage transmission, and Figure 4.19 depicts the principal configuration of the

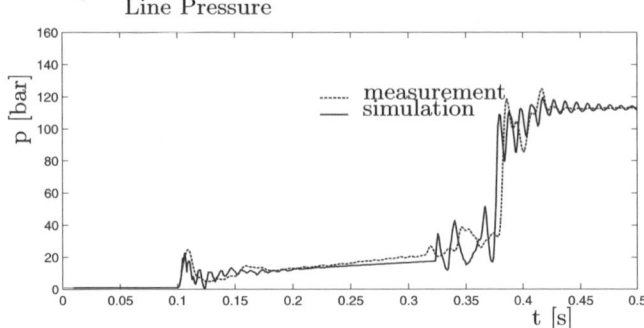

Fig. 4.17: Measured and Calculated Pressure in Supply Line of Cylinders

complete powertrain with the accompanying and necessary gear and motor management systems on an electronic basis.

Fig. 4.18: Five-Stage Automatic Transmission 5HP24 (Zahnradfabrik Friedrichshafen ZF [284])

Automatic transmissions, as many other components in cars, are controlled by complicated hydraulic systems, which supply the switching elements with the necessary oil pressure [91]. Such hydraulic control units include hydraulic lines, pressure reducers, control valves, pilot valves, gate valves and dampers, all arranged within a die cast metal box with an extremely complicated topology, see Figure 4.20. Additionally we have some valve types, which convert electric signals into hydraulic control signals. These electro-hydraulical converters are used as pressure control valves and controlled by pulse-width-modulated electric signals. Magnetically operated valves close and open fluid lines by two switching positions, closed and open, and thus by a simple electrical control.

## 4.4 Practical Examples

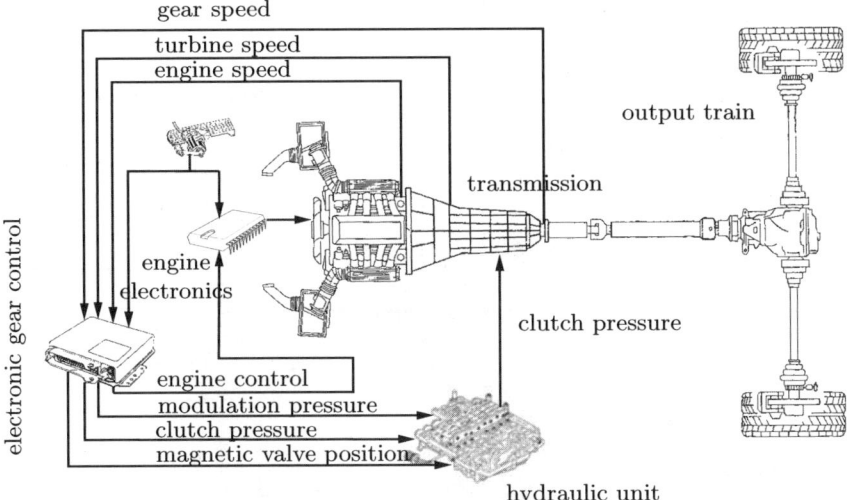

Fig. 4.19: Complete Power train System

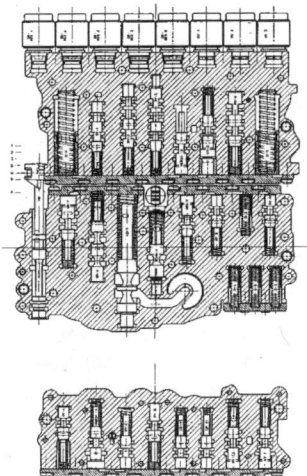

Fig. 4.20: Hydraulic Control Unit A6S 440Z/5HP24 (ZF) [91]

For the illustration of the working principles of a hydraulic control unit we consider switching from the first to the second gear stage. The power flows are depicted in Figure 4.21. Going from the first to the second stage we close hydraulically the multiple disc clutch (E) and open automatically the one-way clutch (FL). The clutch (A) remains closed during this process. For this purpose we have left in Figure 4.22 only those elements of the hydraulic system, which are needed for this specific switching process under consideration.

Figure 4.22 comprises three units, a power supply unit, a general control unit and a clutch control unit. The power supply unit includes a pump, for bringing the oil from the oil sump to the hydraulic system, and four pressure reducers. The pressure valve (HDV) controls the oil pressure on a level necessary for the system's operation. It is additionally supported by the valve (MOD-V), which will be controlled on its own side and in a load-dependent way by the electro-hydraulic element (EDS1). Finally the pressure reducers (DRV1) and (DRV2) take care for the correct pressures for the adjoining elements.

210    4 Dynamics of Hydraulic Systems

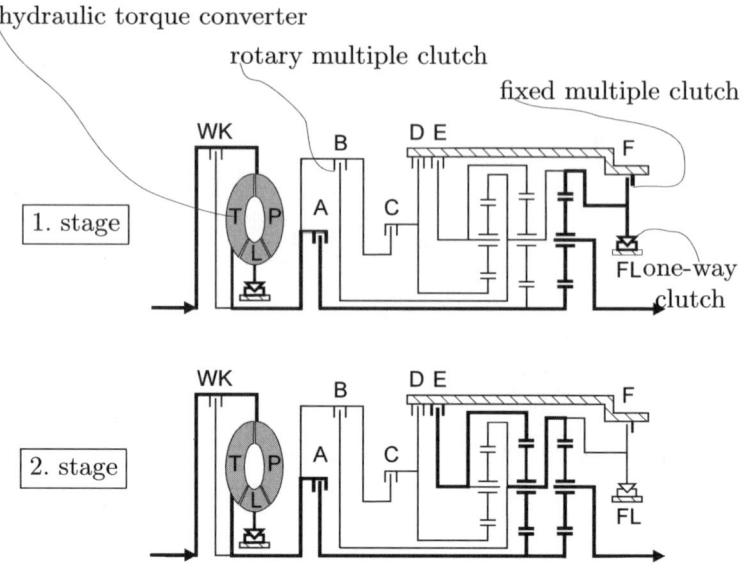

Fig. 4.21: Force Flow in the First and Second Stage

The second unit is the general control unit. It comprises the electrohydraulic converters (EDS1) and (EDS5), the two magnetically operated valves (MV1) and (MV2) and the switching valve (SV1). The components (EDS1), (EDS5) and (MV2) are controlled directly by the electronic transmission control system (EGS), they then activate the valves (KVE) and (HVE) for the clutch (E), Figure 4.22.

These two valves (KVE) and (HVE) are the elements of the third unit, which includes also the damper (D5). The clutch valve (KVE) represents a variable pressure reducer tuning the clutch (E) pressure, dependent on the pressure coming from (EDS5). The holding valve (HVE), also controlled by (EDS5), puts the clutch pressure after the switching process to the system's pressure (holding pressure). The damper (D5) takes care of non-smooth pressure effects and damps them out.

The complete mechanical and hydraulic system together with the power transmission electronic control system is presented and modeled in all details in [91]. The models apply the theory as presented above, which due to its small computing time allows comprehensive parameter investigations. Figure 4.23 depicts some typical results for switching the first to the second gear. The measurements have been performed by the company ZF, Zahnradfabrik Friedrichshafen, Germany. The first graph of Figure 4.23 shows the pressure signals coming from (EDS1) and (EDS5), the second graph the piston positions of the valves (KVE) and (HVE), the third and the fourth graphs com-

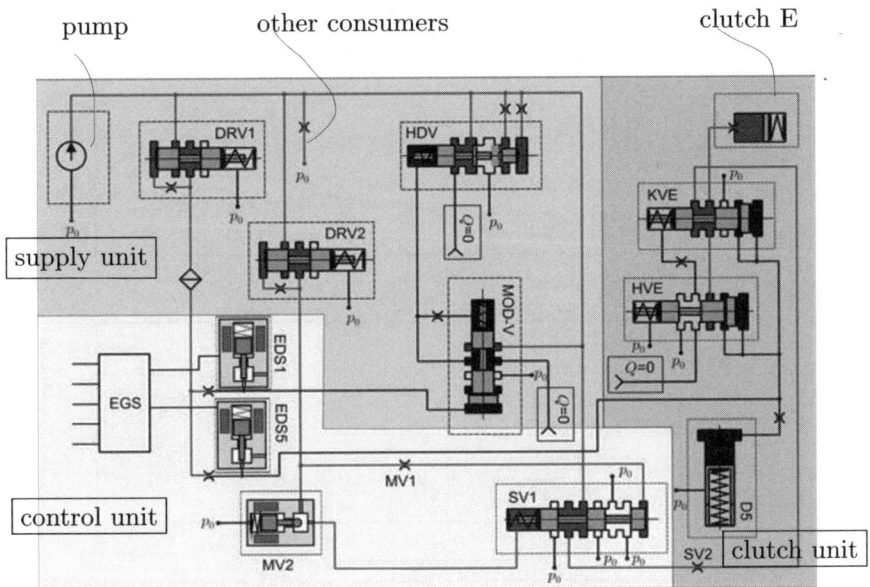

Fig. 4.22: Hydraulic Scheme for a First/Second Gear Switch

parisons of theory and measurements of the clutch (E) pressure and its piston position.

Starting with the first graph we recognize that at the beginning of switching the system pressure will be enlarged to the modulation pressure by the EDS1-valve (about 2.8 sec, 8 to 13.5 bar). This pressure does not apply to the clutch at that time, because the line is still closed by (KVE). The EDS5 pressure is enlarged to a so-called quick-acting pressure, which acts on one side of the clutch valve piston thus opening a guiding edge and allowing oil to flow into the valve cylinder. Due to its spring force the piston of the clutch valve does not move before a sufficiently large pressure difference has developed to overcome the spring force and the static friction effects. After that the clutch valve cylinder is filled until the multiple clutch system is closed, the rest is mainly controlled by the two valves (EDS5) and (EDS1).

The last two graphs of Figure 4.23 give also comparisons between theory and measurements. The theory comprises for that case a multibody system approach for the complete power transmission system combined with the transmission hydraulics as shown above and furtheron combined with the electronic control of the transmission. The curves for clutch pressure and piston position compare very well, as approved also by many other examples [91], see also the sections under "power transmission" starting on page 213.

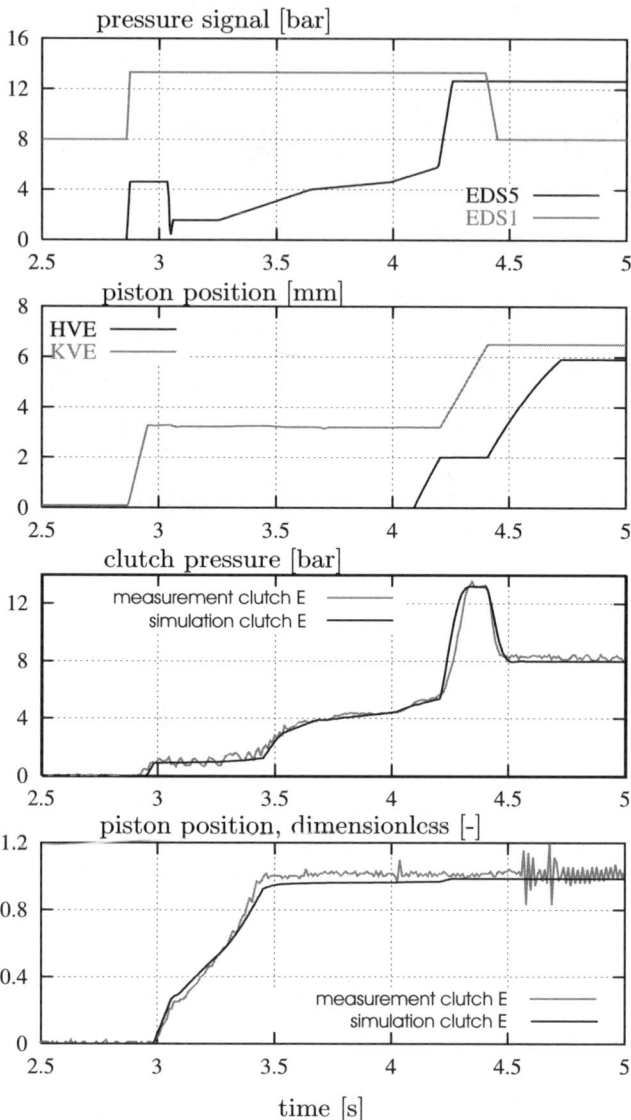

Fig. 4.23: Simulation and Measurements for Switching First/Second Gear with Clutch (E)

# 5

# Power Transmission

> *Die Theorie ist ein Werkzeug, das wir durch Anwendungen erproben und über dessen Zweckmäßigkeit wir im Zusammenhang mit seiner Anwendung entscheiden.*
>
> *(Karl Popper, Logik der Forschung, 1935)*
>
> *A theory is a tool which we test by applying it, and which we judge as to its fitness by the results of its application.*
>
> *(Karl Popper, The Logic of Scientific Discovery, 1959)*

In the following we shall consider some practical systems originating from a large variety of industrial applications. It is a selection of problems from industry, which requires more or less all the elements presented in this book. All these problems are in the one or other form a matter of ongoing applied research, which due to their practical aspects take place in both, academia and industry. Problems coming from that environment are an indispensable measure for theories and their solutions, at least from the engineering standpoint of view. Our examples will come from mechanical engineering industry and with a certain focus from automotive industry.

At the time being we see a competition with respect to several solutions appearing in the field of power transmission. The classical automatic transmissions with four and up to eight stages compete with the CVTs (**C**ontinuous **V**ariable **T**ransmission) in various forms, like rocker pin chains, push belt systems and toroidal gears. Practitioners estimate, that we shall get a share of about 16% for CVTs, the rest mainly for automatic gears either in standard form or as Dual Clutch Transmissions (DCT). In spite of their larger complexity, including more than 1000 structural parts, automatic gear systems possess a better efficieny and a large flexibility with respect to possible stage combinations, especially for the dual clutch configurations. In all probability the automatic gear boxes will retain their dominating position [269]. In the following we shall consider a standard automatic gear box, its Ravigneaux-component and two types of CVTs, the rocker pin and the push belt configurations.

## 5.1 Automatic Transmissions

### 5.1.1 Introduction

As already indicated, establishing a model of an important machine component means establishing a model of the complete machine. We can limit to a certain extent these models by considering the component we want to have in all details and by modeling the surrounding machine with a few degrees of freedom, which nevertheless have to cover all influential frequencies. Modeling the automatic transmission [91] of Figure 4.18 on page 208 therefore comprises the complete drive train as shown in Figure 4.19 on page 209, the representation of which is self-explaining. The hydraulic unit has already been considered in chapter 4.4.2.

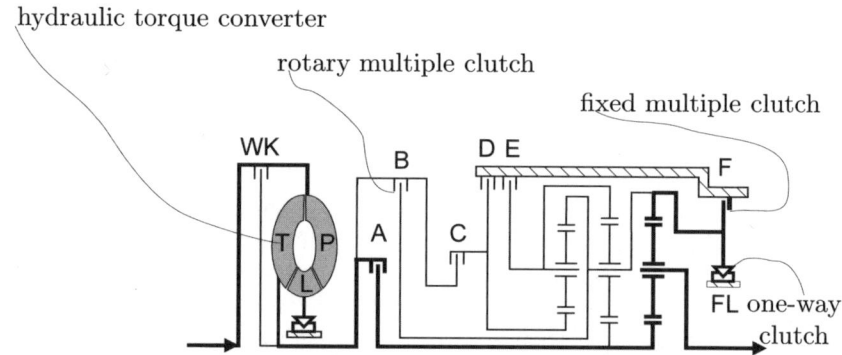

Fig. 5.1: Five-Speed Automatic Transmission, Example 1. Stage,[91]

| Clutch | A | B | C | D | E | F |
|---|---|---|---|---|---|---|
| 1st Gear | x | - | - | - | - | - |
| 2nd Gear | x | - | - | - | x | - |
| 3rd Gear | x | - | - | x | - | - |
| 4th Gear | x | x | - | - | - | - |
| 5th Gear | - | x | - | x | - | - |
| R-Gear | - | - | x | - | - | x |

Table 5.1: Gear-Clutch Table

Gear shift operations are carried out by engaging and disengaging different shift elements, which are normally wet clutches and one-way clutches. The gear box, schematically shown in Figure 5.1, consists of three planetary sets, which are connected to each other and to the gear housing by shafts, clutches and one-way clutches. Each ratio change is performed by engaging one clutch while disengaging another as shown in Table 5.1. We distinguish between two different kinds of gear shift operations. In the gear box presented above, the one-way clutch which locks in the first gear, unlocks automatically during the engagement of the clutch (E). This kind of gear shift operation is called one-way clutch gear shifting. The case that the ratio change is performed only by engaging and disengaging wet clutches, is termed overlapping gear shifting. In both cases engagement and disengagement of shift elements are performed simultaneously. Hence, the driving torque in the drive train is not interrupted while changing the gear ratio.

Some explanations with respect to Figure 5.1 have already been given in connection with the hydraulic unit of the automatic transmission (Figure 4.21 on page 210). Table 5.1 presents a scheme of the clutches, which have to be closed for establishing the various stages (character "x" for closed). Together with Figure 5.1 the possible combinations are obvious.

To model the complete drive train we work on the following assumptions, which have been derived from the real system as shown in Figure 4.18 on page 208. A drive train with automatic transmission generally consists of five main components: engine, torque converter, gear box, output train and vehicle. Each component of the drive train can be considered as a rigid multibody system. The partitioning into single bodies is often given by their technical function. In case of an elastic shaft, a discretization of the body is performed using the stiffness and mass distribution as criteria. Thereby, only torsional degrees of freedom are considered. The rigid bodies are connected to the inertial environment and to each other by ideal rigid joints, clutches and force elements.

The engine for example is modelled as a rotating rigid body. The drive train excitation caused by the engine is described by its torque, which can be interpolated from a measured two-dimensional characteristic map as a function of the throttle opening and the angular velocity of the engine. The hydrodynamic torque converter consists of a pump connected to the input shaft, a turbine connected to the output shaft, and an impeller born in the housing by a one-way clutch. Investigations of the vibrational behavior of the converter can be performed using a detailed model with four degrees of freedom based on a model given in literature, which is still one of the really realistic models of limited size [97].

The gear box consists of three planetary sets, which are connected to each other and to the gear housing by shafts, clutches or one-way clutches. Assuming that the gear wheels are rigid, the kinematics of the planetary gear can be described using the conditions of pure rolling. The output train consists of the cardan shaft, the differential gear, the output shafts and the wheels. The differential gear is used to divert the rotating motion from the drive shaft longitudinal axis to the output shaft longitudinal axis. The vehicle model takes into account the vehicle mass, the rolling friction, the driving resistance, the tire elasticity and damping.

In the following we shall derive in a first step the component models and then in a second step the system model. The section closes with some typical results and with comparisons with measurements.

## 5.1.2 Drive Train Components

With respect to the *engine* model we rely on performance maps and approximate motor dynamics by two degrees of freedom. Figure 5.1.2 depicts the basic model. With the help of a measured performance map, for example Figure 5.3, we determine a drive torque $M_{KF}$ depending on the accelerator position $\alpha_{DK}$ of the driver and on the engine speed $\dot{\varphi}_M$. This torque $M_{KF}$ may be reduced during switching by the motor engagement angle $\beta$ by using the digital engine electronics (DME), which works in connection with the gear control electronics (EGS). The measure of reduction depends on the engine speed and the torque $M_{KF}$ itself.

Due to delays in connection with changes of the throttle valve position, of the motor control and of the motor speed the engine generates the engine torque $M_M$ also in a delayed manner, an effect, which can be approximated by a first order delay element with delay time $T_M$, see Figure 5.1.2. This effect

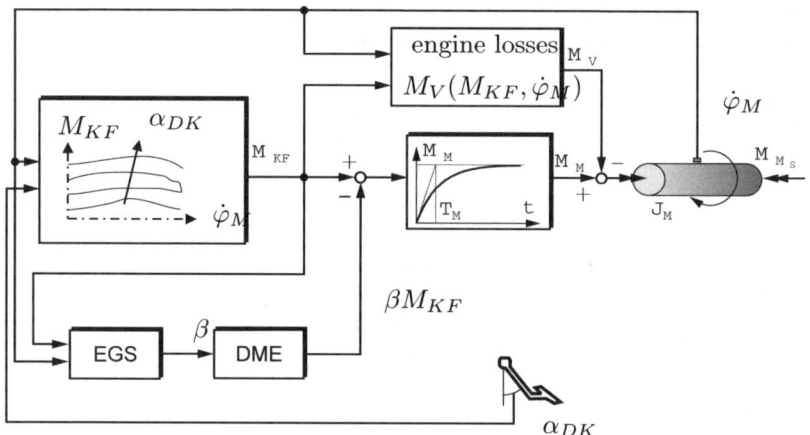

Fig. 5.2: Model of the Combustion Engine [91]

is important especially with respect to starting and shifting processes. The engine losses comprise friction in bearings and the drive losses of all auxiliary equipment. They are expressed by the loss torque $M_V$ and of course indirectly included in the performance characteristics of Figure 5.3. As a final result we have the torque $(M_M - M_V)$ which represents one part of the engine shaft load, the other part being the torque of cut $M_{M_S}$ to the adjacent components.

The moment of inertia of the engine shaft includes the inertia of the shaft itself, but also in a summarized way the projected inertias of the crankshaft, the pistons and the piston rods and of all relevant auxiliary equipment. With this in mind we get for the equations of motion of the combustion engine the simplified set

## 5.1 Automatic Transmissions

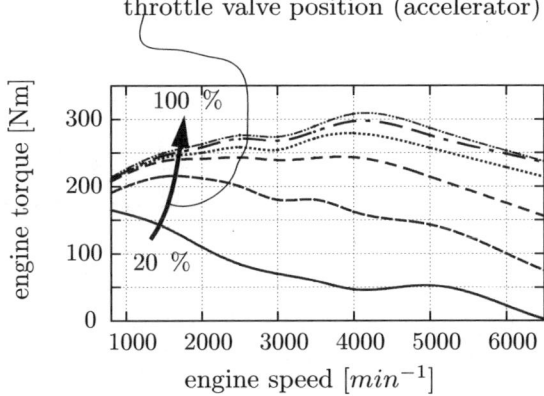

Fig. 5.3: Engine Performance Characteristic [91]

$$\underbrace{\begin{bmatrix} J_M & 0 \\ 0 & T_M \end{bmatrix}}_{\boldsymbol{M}_M} \underbrace{\begin{pmatrix} \ddot{\varphi}_M \\ \dot{M}_M \end{pmatrix}}_{\ddot{\boldsymbol{q}}_M} = \underbrace{\begin{pmatrix} M_M - M_V \\ (1-\beta)M_{KF} - M_M \end{pmatrix}}_{\boldsymbol{h}_M} + \underbrace{\begin{pmatrix} -M_{M_S} \\ 0 \end{pmatrix}}_{\boldsymbol{h}_{M_S}}. \quad (5.1)$$

**Hydrodynamic converters** are basic components of automatic gear boxes. They have three elements, a pump, a turbine and an impeller. Their specific design allows a balancing of even large speed differences, for example in the case of car starting. The impeller is then fixed to the housing by a one-way clutch, which opens for small speed differences. Since nearly three decades hydrodynamic converters include also a special clutch bridging the converter and operating in a closed, open and in a slip state. With an appropriate control such a torque converter lock-up augment comfort and reduce fuel consumption.

We may go now two ways for modeling the torque converter, either applying a known theory based on a stream tube approach [97], which is quite frequently used in industry, or applying performance maps based on measurements. We shall go both ways.

The first model according to [97] considers a stream tube for all three wheels and averages the partial differential equations for the fluid motion over the relevant cross-sections. The result of such a one-dimensional stream tube theory consists in a model with four degrees of freedom, which are indicated in Figure 5.4. The dynamics approximated by the pump speed $\dot{\varphi}_P$, the turbine speed $\dot{\varphi}_T$, the impeller speed $\dot{\varphi}_L$ and the oil volume flow $\dot{V}$ can be described by the following set of equations

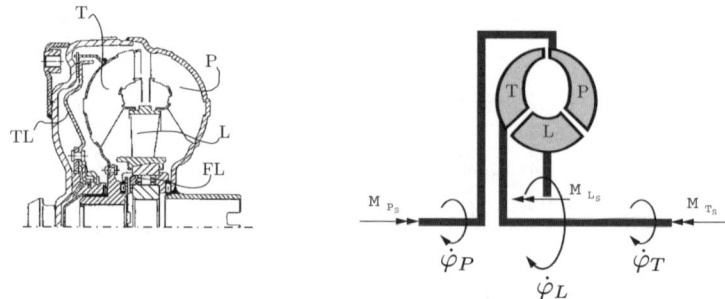

Fig. 5.4: Torque Converter and Model, P=pump wheel, T=turbine wheel, L=impeller, FL=one-way clutch (free wheel), TL=torque converter lock-up, [91]

$$\underbrace{\begin{bmatrix} a_{11} & 0 & 0 & a_{13} \\ 0 & a_{22} & 0 & a_{23} \\ 0 & 0 & a_{38} & a_{33} \\ a_{41} & a_{42} & a_{50} & a_{43} \end{bmatrix}}_{\boldsymbol{M}_W} \underbrace{\begin{pmatrix} \ddot{\varphi}_P \\ \ddot{\varphi}_T \\ \ddot{\varphi}_L \\ \ddot{V} \end{pmatrix}}_{\ddot{\boldsymbol{q}}_W} =$$

$$= \underbrace{\begin{pmatrix} -a_{14}\dot{V}\dot{\varphi}_P - a_{17}\dot{V}\dot{\varphi}_L - a_{16}\dot{V}^2 - M_{PV} \\ -a_{24}\dot{V}\dot{\varphi}_P - a_{25}\dot{V}\dot{\varphi}_T - a_{26}\dot{V}^2 \\ -a_{35}\dot{V}\dot{\varphi}_T - a_{37}\dot{V}\dot{\varphi}_L^2 - a_{36}\dot{V}^2 \\ h_V \end{pmatrix}}_{\boldsymbol{h}_W} + \underbrace{\begin{pmatrix} M_{P_S} \\ -M_{T_S} \\ -M_{L_S} \\ 0 \end{pmatrix}}_{\boldsymbol{h}_{W_S}} \quad (5.2)$$

with the abbreviation

$$h_V = -a_{44}\dot{V}\dot{\varphi}_P - a_{45}\dot{V}\dot{\varphi}_T - a_{46}\dot{V}^2 - a_{47}\dot{\varphi}_P^2 - a_{48}\dot{\varphi}_T^2 \\ -a_{48}\dot{\varphi}_P\dot{\varphi}_T - a_{51}\dot{V}\dot{\varphi}_L - a_{52}\dot{\varphi}_P\dot{\varphi}_L - a_{53}\dot{\varphi}_T\dot{\varphi}_L - a_{54}\dot{\varphi}_L^2 \ .$$

The torques $M_{P_S}$, $M_{T_S}$ and $M_{L_S}$ arise from the cuts to the automatic gear box. All constant values of the $a_{ij}$ follow from the converter theory [97], they depend on geometry, material data, impact-, diffusion- and power losses within the converter. Their analytical expressions are given in [91]. The torque $M_{PV}$ is also a loss due to the oil pump of the hydraulic control unit and depending on the pump speed.

The second model uses measured performance charts. This results in a simplified model with two degrees of freedom, the pump speed $\dot{\varphi}_P$ and the turbine speed $\dot{\varphi}_T$. The equations of motion write

$$\underbrace{\begin{bmatrix} J_P & 0 \\ 0 & J_T \end{bmatrix}}_{\mathbf{M}_W} \underbrace{\begin{pmatrix} \ddot{\varphi}_P \\ \ddot{\varphi}_T \end{pmatrix}}_{\ddot{\mathbf{q}}_W} = \underbrace{\begin{pmatrix} -M_P - M_{PV} \\ M_T \end{pmatrix}}_{\mathbf{h}_W} + \underbrace{\begin{pmatrix} M_{P_S} \\ -M_{T_S} \end{pmatrix}}_{\mathbf{h}_{W_S}}. \qquad (5.3)$$

The torques $M_P$ and $M_T$ depend on the characteristics of $M_{PC}(\nu)$ and $\mu(\nu)$. The value of $(\nu)$ is the ratio of the turbine and the pump speed $(\nu) = \frac{\dot{\phi}_T}{\dot{\phi}_P}$, and $\dot{\phi}_{PC}$ is the turbine speed as generated by an electromotor. All data can be seen from Figure 5.5.

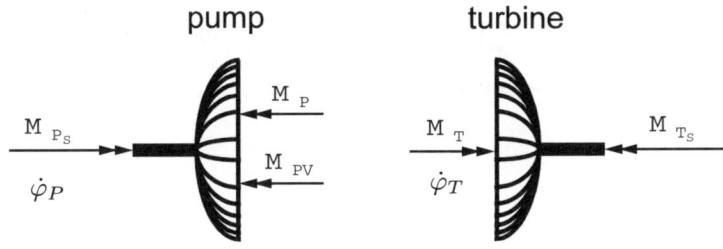

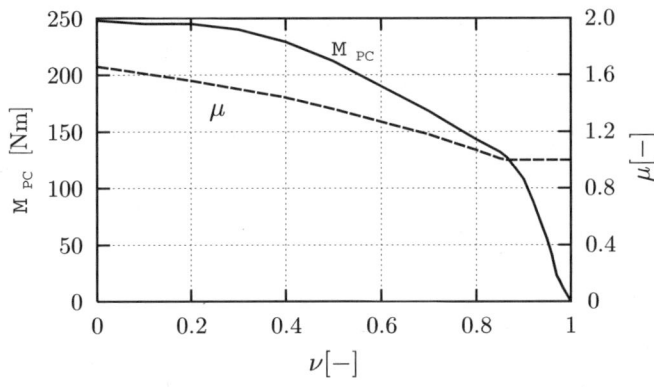

Fig. 5.5: Simplified Converter Model [91]

Automatic transmissions include several **planetary gears systems**, which offer a large variety of stage combinations. Furtheron, their rotational symmetry, their large power density and their compact, coaxial type of construction represent additional advantages. In the following we shall consider simple models of classical planetary gears and of Ravigneaux gears, two types which are frequently used in automatic gear boxes. In the next chapter we shall discuss Ravigneaux gears in more detail.

A *planetary set* consists of the sun-wheel S), the internal gear or annulus (H) and the planet gears (PL), see Figure 5.6. Sun wheel and internal gear are usually connected with a central shaft. The planet carrier represents a third central shaft with an appropriate arrangement of the planet gears. One of the shafts is used as a support element, the other two shafts as input and output

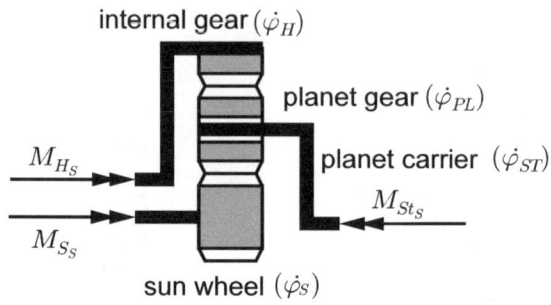

Fig. 5.6: Classical Planetary Gear [91]

shafts. The three teeth numbers $z_H, z_S$ and $z_P$ for the internal, the sun and the planet gears, respectively, determine the possible gear ratios. From this we can evaluate one unknown gear speed from two given ones, which we choose to be the sun wheel and the internal gear wheel speeds $\varphi_S$ and $\varphi_H$. We then can express all other speeds by these two ones yielding

$$\underbrace{\begin{pmatrix} \dot\varphi_S \\ \dot\varphi_{ST} \\ \dot\varphi_H \\ \dot\varphi_{PL} \end{pmatrix}}_{\dot q_{PS_{max}}} = \underbrace{\begin{pmatrix} 1 & 0 \\ a_1 & a_2 \\ 0 & 1 \\ a_3 & a_4 \end{pmatrix}}_{J_{PS}} \underbrace{\begin{pmatrix} \dot\varphi_S \\ \dot\varphi_H \end{pmatrix}}_{\dot q_{PS}} \qquad (5.4)$$

with $a_1 = \dfrac{z_S}{z_H + z_S}$, $a_2 = \dfrac{z_H}{z_H + z_S}$, $a_3 = \dfrac{-z_S}{z_H - z_S}$, $a_4 = \dfrac{z_H}{z_H - z_S}$.
The equations of motion for this simplified model result from a projection of the momentum and moment of momentum equations into the space of the minimal coordinates with the help of the Jacobian. We come out with

$$\underbrace{\begin{bmatrix} J_S + a_1^2 \bar J_{ST} + a_3^2 n J_{PL} & a_1 a_2 \bar J_{ST} + a_3 a_4 n J_{PL} \\ a_1 a_2 \bar J_{ST} + a_3 a_4 n J_{PL} & J_H + a_2^2 \bar J_{ST} + a_4^2 n J_{PL} \end{bmatrix}}_{M_{PS}} \underbrace{\begin{pmatrix} \ddot\varphi_S \\ \ddot\varphi_H \end{pmatrix}}_{\ddot q_{PS}} =$$

$$= \underbrace{\begin{pmatrix} M_{S_S} - a_1 M_{ST_S} \\ M_{H_S} - a_2 M_{ST_S} \end{pmatrix}}_{h_{PS_S}} \qquad (5.5)$$

with $\bar{J}_{ST} = J_{ST} + n\, m_{PL}\, r_{ST}^2$. The magnitudes $J_i, M_k$ are the moment of inertia terms and the torques for the elements of the automatic gear box, $m_{PL}$ is the mass of a planet wheel and $r_{ST}$ the radius of the planet carrier.

For establishing the equations of motion of an overall automatic gear it makes sense to derive for the individual components mechanical and mathematical models which are "as simple as possible but as comprehensive as necessary" to achieve a good representation of the reality. Therefore we shall use in a first step for the *Ravigneaux gear set* an approach with two degrees of freedom resulting from the assumption of kinematically coupled gear wheels, but in a further step a more detailed description (see chapter 5.2). Figure

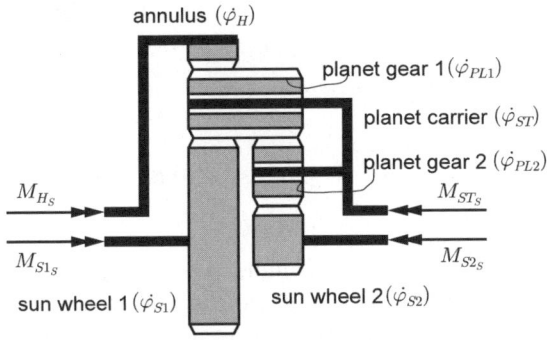

Fig. 5.7: Structure of a Ravigneaux Planetary Gear Set [91]

5.7 gives an impression of the simplified kinematical Ravigneaux model. The principle of such gears is so-to-say a 1.5-planetary gear set. The planet gear wheels on the left side of Figure 5.7 have been designed so wide that they mesh as well with the annulus as with the second planet wheels. Instead of meshing with a second annulus the second planet gears mesh with the first planet gears. The combination of two sun wheels, one annulus and two sets of planet gears results in a large variety of transmission ratio possibilities, especially with the additional constructive advantage of a compact, rotationally centered and large torque transmitting unit.

Choosing the minimal velocities $\mathbf{q}_{RS} = (\dot\varphi_{S1}, \dot\varphi_{ST})^T$ we get for all remaining speeds

$$\underbrace{\begin{pmatrix} \dot\varphi_{S1} \\ \dot\varphi_H \\ \dot\varphi_{S2} \\ \dot\varphi_{ST} \\ \dot\varphi_{PL1} \\ \dot\varphi_{PL2} \end{pmatrix}}_{\dot{\mathbf{q}}_{max_{RS}}} = \underbrace{\begin{pmatrix} 1 & 0 \\ -b_1 & 1+b_1 \\ -b_2 & 1+b_2 \\ 0 & 1 \\ -b_3 & 1+b_3 \\ b_4 & 1-b_4 \end{pmatrix}}_{\mathbf{J}_{RS}} \underbrace{\begin{pmatrix} \dot\varphi_{S1} \\ \dot\varphi_{ST} \end{pmatrix}}_{\dot{\mathbf{q}}_{RS}} \qquad (5.6)$$

with $b_1 = \frac{z_{S1}}{z_H}$, $b_2 = \frac{z_{S1}}{z_{S2}}$, $b_3 = \frac{z_{S1}}{z_{P1}}$ and $b_4 = \frac{z_{S1}}{z_{P2}}$.

Analoguously to the simple planetary gear set we may reduce these equations to

$$\underbrace{\begin{bmatrix} J_{11} & J_{12} \\ J_{21} & J_{22} \end{bmatrix}}_{\boldsymbol{M}_{RS}} \underbrace{\begin{pmatrix} \ddot{\varphi}_{S1} \\ \ddot{\varphi}_{ST} \end{pmatrix}}_{\ddot{\boldsymbol{q}}_{RS}} = \underbrace{\begin{pmatrix} M_{S1_S} - b_1 M_{H_S} + b_2 M_{S2_S} \\ (1+b_1) M_{H_S} - (1+b_2) M_{S2_S} - M_{ST_S} \end{pmatrix}}_{\boldsymbol{h}_{RS_S}} \tag{5.7}$$

with the abbreviations

$$J_{11} = J_{S1} + b_1^2 J_H + b_2^2 J_{S2} + b_3^2 n_1 J_{PL1} + b_4^2 n_2 J_{PL2}$$
$$J_{12} = J_{21} = -(b_1 + b_1^2) J_H - (b_2^2 + b_2) J_{S2} - (b_3^2 + b_3) n_1 J_{PL1} - (b_4^2 - b_4) n_2 J_{PL2}$$
$$J_{22} = (1+b_1)^2 J_H + (1+b_2)^2 J_{S2} + \bar{J}_{ST} + (1+b_3)^2 n_1 J_{PL1} + (1-b_4)^2 n_2 J_{PL2}$$
$$\bar{J}_{ST} = J_{ST} + n_1 m_{PL1} r_{ST1}^2 + n_2 m_{PL2} r_{ST2}^2$$

$J_i$ are the mass moments of inertia, $n_1$ and $n_2$ the number of the two planet gear wheels, $m_{PL_1}$ and $m_{PL_2}$ their masses and $r_{ST_1}, r_{ST_2}$ the radii of the planet carriers.

The component **output train** includes the output shaft, the tires and the car mass. We shall consider three resistances, the inclination resistance, the resistance to rolling and the air drag resistance. They write

$$F_{St} = m_{car} g \sin\alpha \tag{5.8}$$
$$F_{Roll} = \mu_R(\dot{x}_{car}) \operatorname{sign}(\dot{x}_{car})(m_{car} g \cos\alpha - F_{cV} - F_{cH}) \tag{5.9}$$
$$F_L = \frac{1}{2} c_w A_w \rho_L \dot{x}_{car}^2 \operatorname{sign}(\dot{x}_{car}) \tag{5.10}$$

with the front- and rear wheel lift forces $F_{cV}$ und $F_{cH}$

$$F_{cV} = \frac{1}{2} c_V A_w \rho_L \dot{x}_{car}^2, \quad F_{cH} = \frac{1}{2} c_H A_w \rho_L \dot{x}_{car}^2. \tag{5.11}$$

The magnitudes $c_V$ und $c_H$ are the lift coefficients of front and rear wheel, $\rho_L$ is the atmospheric density, $\alpha$ the road inclination, $\mu_R$ the roll resistance coefficient, $c_w$ the air drag coefficient, $A_w$ the front area of the car and $\dot{x}_{car}$ the speed of the car. The tire will be modeled by a linear force element with spring and damper (spring stiffness $c_R$, damping coefficient $d_R$). The tire torque writes correspondingly

$$M_R = (\varphi_R - \frac{x_{car}}{r_R}) c_R + (\dot{\varphi}_R - \frac{\dot{x}_{car}}{r_R}) d_R. \tag{5.12}$$

Braking is governed by the maximal possible brake torque $M_{B_{max}}$ and a brake pedal coefficient $k_B$, which represents a suitable measure for the brake pedal actuation and has to be measured. The brake torque then writes

## 5.1 Automatic Transmissions

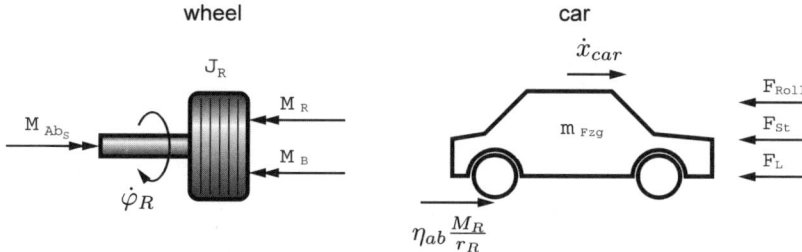

Fig. 5.8: Model of the Output Train including the Car Mass [91]

$$M_B = k_B \, M_{B_{max}} \, \text{sign}(\dot{x}_{car}) \,. \tag{5.13}$$

Combining these equations results in the equations of motion for the output train model according to Figure 5.8

$$\underbrace{\begin{bmatrix} J_R & 0 \\ 0 & m_{car} \end{bmatrix}}_{M_{Ab}} \underbrace{\begin{pmatrix} \ddot{\varphi}_R \\ \ddot{x}_{car} \end{pmatrix}}_{\ddot{q}_{Ab}} = \underbrace{\begin{pmatrix} -M_R - M_B \\ \eta_{Ab} \dfrac{M_R}{r_R} - F_{Roll} - F_{St} - F_L \end{pmatrix}}_{h_{Ab}} + \underbrace{\begin{pmatrix} M_{Ab_S} \\ 0 \end{pmatrix}}_{h_{Ab_S}} .$$

(5.14)

The efficiency $\eta_{Ab}$ takes into account all losses of the output train.

**Shafts** are very fundamental elements of all machinery. They are something like a blood circulation system distributing and passing on torques within a mechanical system. According to the various design possibilities shafts possess also various influence on the dynamics of the overall system, especially with respect to the eigenbehaviour expressed by eigenfrequencies and eigenfunctions. The longitudinal dynamics of a car is mainly concerned by shifting gears on the one and influenced by the rotational behaviour of the shafts on the other side. Therefore we shall focus on that rotational behaviour, where we have to distinguish rigid and elastic shafts.

*Rigid shafts* are the simplest possible elements of a machine including an input, an out put torque and a rotational inertia (Figure 5.9). Therefore the equation of motion is simply

$$\underbrace{J}_{M_{SW}} \underbrace{\ddot{\varphi}}_{\ddot{q}_{SW}} = \underbrace{M_{an_S} - M_{ab_S}}_{h_{SW_S}} \,. \tag{5.15}$$

*Elastic shafts* represent the simplest possible case of an elastic multibody element, as far as rotational linear elasticity is concerned. Assuming only linear elastic deformations gives us two modeling alternatives, namely applying some

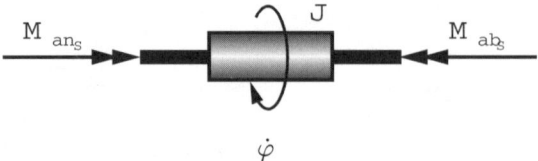

Fig. 5.9: Rigid Shaft Model

Ritz approach according to chapter 3.3.4 on page 124 or just discretizing the shaft into a limited number of shaft elements. Anyway, the number of shape functions of a Ritz approach as well as the number of the shaft elements depend on the frequency range of the system under consideration. In our case we discretize into n elements interconnected by springs and dampers, see Figure 5.10.

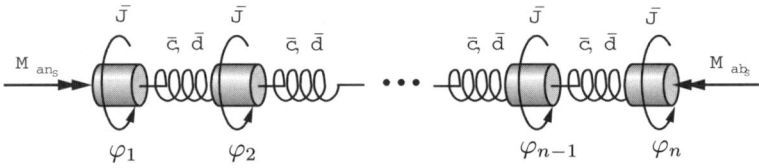

Fig. 5.10: Elastic Shaft Model

The equations of motion for such a chain of torsional elements with (n) equal bodies and (n-1) equal springs and dampers can be written in the form

$$\underbrace{\begin{pmatrix} \bar{J} & 0 & \cdots & 0 \\ 0 & \bar{J} & \cdots & 0 \\ \vdots & \vdots & \ddots & \vdots \\ 0 & 0 & \cdots & \bar{J} \end{pmatrix}}_{\boldsymbol{M}_{EW}} \underbrace{\begin{pmatrix} \ddot{\varphi}_1 \\ \ddot{\varphi}_2 \\ \vdots \\ \ddot{\varphi}_{n-1} \\ \ddot{\varphi}_n \end{pmatrix}}_{\ddot{\boldsymbol{q}}_{EW}} =$$

$$= \underbrace{\begin{pmatrix} (\varphi_2 - \varphi_1)\,\bar{c} + (\dot{\varphi}_2 - \dot{\varphi}_1)\,\bar{d} \\ (\varphi_3 - 2\varphi_2 + \varphi_1)\,\bar{c} + (\dot{\varphi}_3 - 2\dot{\varphi}_2 + \dot{\varphi}_1)\,\bar{d} \\ \vdots \\ (\varphi_{n+1} - 2\varphi_n + \varphi_{n-1})\,\bar{c} + (\dot{\varphi}_{n+1} - 2\dot{\varphi}_n + \dot{\varphi}_{n-1})\,\bar{d} \\ (\varphi_{n-1} - \varphi_n)\,\bar{c} + (\dot{\varphi}_{n-1} - \dot{\varphi}_n)\,\bar{d} \end{pmatrix}}_{\boldsymbol{h}_{EW}} + \underbrace{\begin{pmatrix} M_{an_S} \\ 0 \\ \vdots \\ 0 \\ -M_{ab_S} \end{pmatrix}}_{\boldsymbol{h}_{EW_S}}.$$

(5.16)

**One-way clutches** allow a relative rotational motion in one direction only, whereas the other direction is blocked. They are passive components, which connect shafts with shafts or shafts with the housing. Within an automatic power transmission one-way clutches allow shifting without an interruption of the traction forces and without an influence on the shifting quality itself, and that with a really simple technology. Figure 5.11 depicts a model of

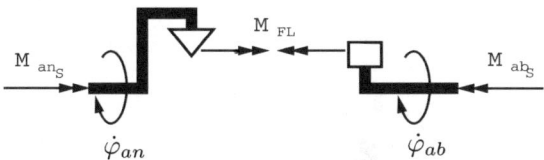

Fig. 5.11: Model of a One-Way Clutch [91]

such a one-way clutch, which includes some inertias of the input and output parts and in addition unilateral constraints. The state of the clutch is given by its relative rotational speed

$$\dot{g}_{FL} = \dot{\varphi}_{an} - \dot{\varphi}_{ab}, \tag{5.17}$$

with non-negative values in the free direction. The one-way clutch possesses two states, which exclude each other. If we have the same speed on both sides, input and output, the clutch is blocked and transmits a positive constraint torque $M_{FL} > 0$ from input to output. For this case $\dot{g}_{FL} = 0$. If we have a positive relative speed $\dot{g}_{FL} > 0$, then the one-way clutch moves without load and therefore $M_{FL} = 0$. This exclusion can be formulated as a complementarity in the form

$$\dot{g}_{FL} \geq 0; \quad M_{FL} \geq 0; \quad M_{FL}\dot{g}_{FL} = 0 \ . \tag{5.18}$$

Together with this inequality constraint we have the equations of motion

$$\underbrace{\begin{pmatrix} J_{an} & 0 \\ 0 & J_{ab} \end{pmatrix}}_{\boldsymbol{M}_{FL}} \underbrace{\begin{pmatrix} \ddot{\varphi}_{an} \\ \ddot{\varphi}_{ab} \end{pmatrix}}_{\ddot{\boldsymbol{q}}_{FL}} = \underbrace{\begin{pmatrix} M_{an_S} \\ -M_{ab_S} \end{pmatrix}}_{\boldsymbol{h}_{FL_S}} + \underbrace{\begin{pmatrix} 1 \\ -1 \end{pmatrix}}_{\bar{\boldsymbol{w}}_{FL}} M_{FL} \ , \tag{5.19}$$

where the constraint vector $\bar{\boldsymbol{w}}_{FL}$ follows from a differentiation of equation (5.17)

$$\bar{\boldsymbol{w}}_{FL} = \left(\frac{\partial \dot{g}_{FL}}{\partial \dot{\boldsymbol{q}}_{FL}}\right)^T . \tag{5.20}$$

The quantities $J_{an}$ and $J_{ab}$ are the mass moments of inertia of the input and output sides of the one-way clutch, respectively.

Every automatic transmission possesses a variety of multiple ***clutches***. We shall consider such clutches in detail in another chapter (see [91]). For the system analysis performed here it will be sufficient to descibe multiple clutches by a simple three DOF model, including oil effects. Multiple clutches are controled hydraulically, they transmit torques between gear components. We shall approximate this behaviour by the model of Figure 5.12, which includes inertias $J_{an}$ and $J_{ab}$ on both sides and a friction constraint. The relative

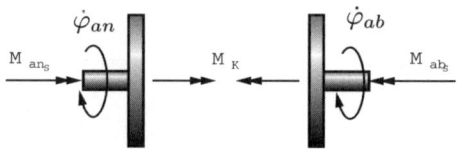

Fig. 5.12: Simple Model of a Multiple Clutch

rotational speed

$$\dot{g}_K = \dot{\varphi}_{an} - \dot{\varphi}_{ab}, \tag{5.21}$$

is zero for no-slip ($\dot{g}_K = 0$) and not zero for sliding ($\dot{g}_K \neq 0$). For the no-slip stuation the clutch torque can be evaluated from the oil pressure and a suitable friction law, for example by Stribeck graphs. We get

$$M_K = -\mu_G(\dot{g}_K)(pA_K - F_{TF})r_m n_R \, \text{sign}(\dot{g}_K). \tag{5.22}$$

The quantity $A_K$ is the effective piston area, $F_{TF}$ the restoring force of the cup spring, $r_m$ the averaged radius of the friction plane, $n_r$ the number of the friction planes and $\mu_G(\dot{g}_K)$ the friction coefficient of sliding. The constraint torque for no-slip follows from a set-valued force law, which in addition has its limitation by the oil pressure in the cylinder defining some maximum possible torque:

$$-\mu_H(pA_K - F_{TF})r_m n_R \leq M_K \leq \mu_H(pA_K - F_{TF})r_m n_R \tag{5.23}$$

$\mu_H$ is the no-slip coefficient.

According to chapter 4.2 and equation (4.4) on page 191 we can approximate the relation of oil pressure and oil flow by

$$\dot{p} = \frac{E}{V_K} Q \, , \tag{5.24}$$

where $V_K$ is the oil volume, E the oil compressibility depending on the pressure and the air percentage solved in the oil. Altogether the simplified clutch model is described by the following equations of motion

$$\underbrace{\begin{pmatrix} J_{an} & 0 & 0 \\ 0 & J_{ab} & 0 \\ 0 & 0 & 1 \end{pmatrix}}_{\boldsymbol{M}_K} \underbrace{\begin{pmatrix} \ddot{\varphi}_{an} \\ \ddot{\varphi}_{ab} \\ \dot{p} \end{pmatrix}}_{\ddot{\boldsymbol{q}}_K} = \underbrace{\begin{pmatrix} M_{an_S} \\ -M_{ab_S} \\ 0 \end{pmatrix}}_{\boldsymbol{h}_{K_S}} + \underbrace{\begin{pmatrix} 0 \\ 0 \\ \frac{E}{V_K} Q \end{pmatrix}}_{\boldsymbol{h}_K} + \underbrace{\begin{pmatrix} 1 \\ -1 \\ 0 \end{pmatrix}}_{\bar{\boldsymbol{w}}_K} M_K \qquad (5.25)$$

with the unilateral friction law

$$\begin{aligned} \dot{g}_K = 0 &\implies |M_K| \leq \mu_H (pA_K - F_F) r_m n_R, \\ \dot{g}_K < 0 &\implies M_K = \mu_G(\dot{g}_K)(pA_K - F_F) r_m n_R, \\ \dot{g}_K > 0 &\implies M_K = -\mu_G(\dot{g}_K)(pA_K - F_F) r_m n_R, \end{aligned} \qquad (5.26)$$

which is as a matter of fact a set-valued torque law. The constraint vector $\bar{\boldsymbol{w}}_K$ is defined by

$$\bar{\boldsymbol{w}}_K = \left( \frac{\partial \dot{g}_K}{\partial \dot{\boldsymbol{q}}_K} \right)^T. \qquad (5.27)$$

### 5.1.3 Drive Train System

The ***system equations of motion*** and the appropriate equality and inequality constraints follow from a suitable combination of the equations of motion of all components [91]. We consider a drive train with n components, which are interconnected by m constraints. Each component has been modeled by a set of equations

$$\boldsymbol{M}_i \ddot{\boldsymbol{q}}_i = \boldsymbol{h}_i + \boldsymbol{h}_{i_S} + \boldsymbol{W}_i \boldsymbol{\lambda}_i, \qquad (5.28)$$

see the relations evaluated above. Then the overall equations of motion write quite formally

$$\underbrace{\begin{pmatrix} \boldsymbol{M}_1 & \cdots & 0 & \cdots & 0 \\ \vdots & \ddots & \vdots & \ddots & \vdots \\ 0 & \cdots & \boldsymbol{M}_i & \cdots & 0 \\ \vdots & \ddots & \vdots & \ddots & \vdots \\ 0 & \cdots & 0 & \cdots & \boldsymbol{M}_n \end{pmatrix}}_{\boldsymbol{M}_m} \underbrace{\begin{pmatrix} \ddot{\boldsymbol{q}}_1 \\ \vdots \\ \ddot{\boldsymbol{q}}_i \\ \vdots \\ \ddot{\boldsymbol{q}}_n \end{pmatrix}}_{\boldsymbol{q}_m} = \underbrace{\begin{pmatrix} \boldsymbol{h}_1 \\ \vdots \\ \boldsymbol{h}_i \\ \vdots \\ \boldsymbol{h}_n \end{pmatrix}}_{\boldsymbol{h}_m} + \underbrace{\begin{pmatrix} \boldsymbol{h}_{S_1} \\ \vdots \\ \boldsymbol{h}_{S_i} \\ \vdots \\ \boldsymbol{h}_{S_n} \end{pmatrix}}_{\boldsymbol{h}_{S_m}} +$$

$$+ \underbrace{\begin{pmatrix} \boldsymbol{W}_1 & \cdots & 0 & \cdots & 0 \\ \vdots & \ddots & \vdots & \ddots & \vdots \\ 0 & \cdots & \boldsymbol{W}_i & \cdots & 0 \\ \vdots & \ddots & \vdots & \ddots & \vdots \\ 0 & \cdots & 0 & \cdots & \boldsymbol{W}_n \end{pmatrix}}_{\boldsymbol{W}_m} \underbrace{\begin{pmatrix} \boldsymbol{\lambda}_1 \\ \vdots \\ \boldsymbol{\lambda}_i \\ \vdots \\ \boldsymbol{\lambda}_n \end{pmatrix}}_{\boldsymbol{\lambda}}. \qquad (5.29)$$

We have two possibilities to interconnect the components, either by cut forces or directly by kinematic constraints. We choose the second possibility. In the case of drive trains these connecting constraints are not only, as always, of kinematic nature, but also linear in the generalized coordinates describing the system. As a rule we are not able to find a complete set of generalized coordinates describing a large system. We usually can define a set $\mathbf{q}$ of independent coordinates and a second set $\mathbf{q}_d$ of dependent coordinates. Furtheron we define $\mathbf{q}_m = \mathbf{q} + \mathbf{q}_d$.

With the assumption, or better the experience, of a linear relationship of the constraints $\boldsymbol{\Phi}(\mathbf{q}_m)$ with respect to the coordinates $\mathbf{q}_m$ we can simply write

$$\boldsymbol{\Phi}(\mathbf{q}_m) = \frac{\partial \boldsymbol{\Phi}(\mathbf{q}_m)}{\partial \mathbf{q}_m} \mathbf{q}_m = \mathbf{K} \mathbf{q}_m = \mathbf{0} \quad \text{and} \quad \mathbf{K}_1 \mathbf{q} + \mathbf{K}_2 \mathbf{q}_d = \mathbf{0} , \quad (5.30)$$

which enables us to express $\mathbf{q}_d$ as a function of $\mathbf{q}$, namely $\mathbf{q}_d = -\mathbf{K}_2^{-1} \mathbf{K}_1 \mathbf{q}$. On the other hand we may express the coordinate vector $\mathbf{q}_m$ by the relation

$$\mathbf{q}_m = \mathbf{P} \begin{pmatrix} \mathbf{q} \\ \mathbf{q}_d \end{pmatrix} = \underbrace{\mathbf{P} \begin{pmatrix} \mathbf{E} \\ -\mathbf{K}_2^{-1} \mathbf{K}_1 \end{pmatrix}}_{\mathbf{J}} \mathbf{q} , \quad (5.31)$$

which gives an expression for the overall Jacobian. The matrix $\mathbf{P}$ represents a permutation matrix arranging the coordinates $\mathbf{q}$ and $\mathbf{q}_d$ in the right way. After that we project the equations of motion (5.29) into the space of the minimal coordinates $\mathbf{q}$, which yields

$$\underbrace{\mathbf{J}^T \mathbf{M}_m \mathbf{J}}_{\mathbf{M}} \ddot{\mathbf{q}} = \underbrace{\mathbf{J}^T \mathbf{h}_m}_{\mathbf{h}} + \mathbf{J}^T \mathbf{W}_m \boldsymbol{\lambda} . \quad (5.32)$$

These equations include the constraints and exclude the constraint forces at the various cut points [91].

The equations of motion (5.32) have to be supplemented by the constraints. With respect to equality constraints this is no problem, with respect to inequality constraints we need to take into consideration the rules as discussed in the chapters 3.1.2 on page 89, 3.4 on page 131, 3.5 on page 158 and 3.4.4 on page 145. With regard to a more detailed investigation of the hydraulic control system we also must consider the equations of chapter 4 on page 187. The combination of all constraints is usually straightforward, but needs special attention concerning the choice of the numerical algorithms involved. Haj Fraj [91] still worked with Lemke's algorithm, and had success with it. Today we would apply the Augmented Lagrange method in combination with the prox-approach [64].

It should be noted that the above presentation gives only an impression how to model a drive train system. Haj Fraj [91] goes one step further and establishes a program code control system, which allows to investigate any combination of mechanics, hydraulics and electronics of a drive train with an automatic transmission, as well in an extremely detailed form as in an extremely simplified way. In the following we shall give some verification results using a medium sized model.

### 5.1.4 Measurements and Verification

We consider the system of Figure 5.13, which describes the components necessary to achieve some realistic results for the longitudinal dynamics of an automobile. We shall evaluate a shifting process from standstill to the third gear stage under full load. The overall system includes four modules: driver, control, hydraulics and drive train mechanics. All components have been explained in the preceding chapters, also the hydraulics in chapter 4 on page 187.

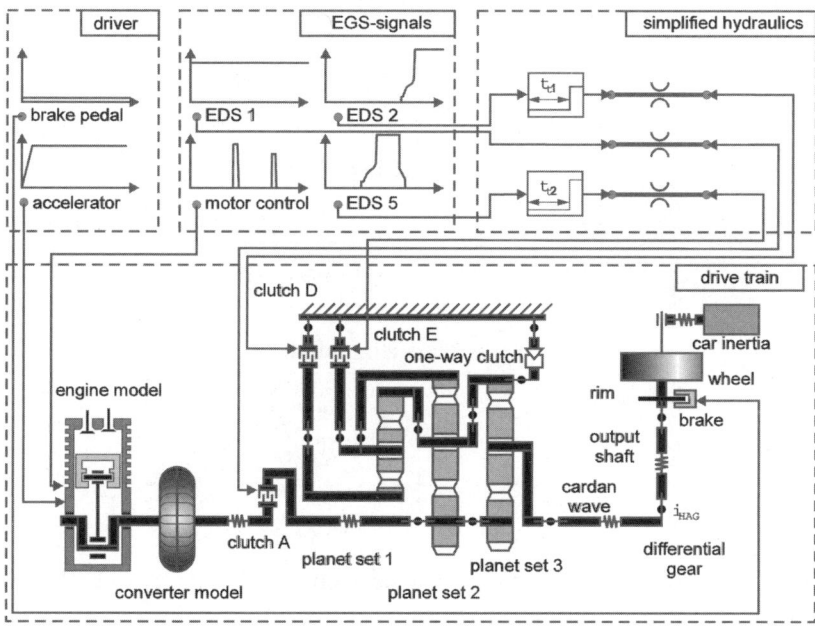

Fig. 5.13: Medium Sized Model of the Longitudinal Dynamics [91]

These components are a bit simplified to a degree sufficient for modeling the shifting process. The clutch A connects the turbine shaft with the gear input shaft. The one-way clutch is blocked for the first gear stage and opens for the second one, while the clutch E will be closed. Shifting from the second to the third stage requires closing the clutch D and opening the clutch E. Therefore we have for this example two types of shifting, one with the one-way clutch to clutch E and one with two multiple clutches $E \Rightarrow D$.

The driver model is also rather simple including the braking and acceleration behaviour. The control module (EGS-signals) comprises all types of information processing with respect to the automatic transmission. The hydraulic model is reduced to two delay time elements connected to the oil lines, which has turned out to be sufficient for the shifting problem on hand. A more sophisticated approach is presented in chapter 4.4.2 on page 207. A simplified clutch model is used, and all existing compliances are taken into consideration.

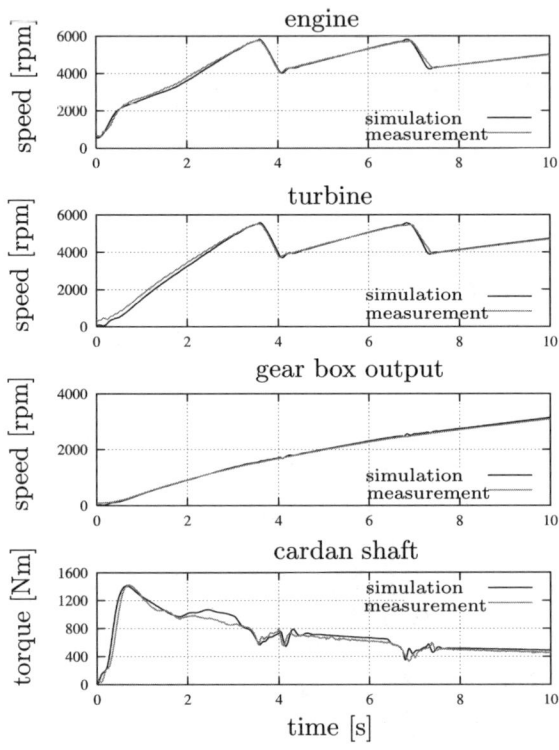

Fig. 5.14: Comparison Simulation/Measurement for shifting Standstill⇒Third Gear [91]

Figure 5.14 depicts some simulation results and also a comparison with measurements, performed in industry. The agreement is excellent and confirms this way of modeling. The first three diagrams show the rotational speeds of the engine, the converter turbine and the gear output, the fourth diagram presents the output torque, which is proportional to the longitudinal car dynamics. At the beginning we have a steep ascent of the torque due to the large speed difference of engine and converter turbine, which decreases with increasing car speed. After that we recognize the two shifting events accompanied by a speed ascent and some torque oscillations around the shifting event. They need to be small, otherwise they would influence comfort considerably.

### 5.1.5 Optimal Shift Control

Shift control can only be performed during a short phase, which allows relative speed between the participating clutches. The control possibilities end with the synchronous point. But as a consequence of the shifting process we get acceleration changes, defined by the time derivative of the acceleration called jerk, and these changes have to be kept small for comfort reasons. We achieve that by control, but with the problem that beyond the synchronous point it is not possible to control jerk.

The problem can be reduced considerably by the idea, to implement exactly at the end of the controllable phase, namely at the synchronous point, a kind of a moment of momentum kick, which influences the dynamics after the synchronous point in such a way, that jerk is suppressed as far as possible [91]. In the following we shall present an approach based on Bellman's dynamic programming theory [16], [92], [190].

Designing such a model-based control requires a simplified model of the drive train including all components necessary to describe the shifting process, but neglecting all parts and components not contributing very much to this process. We choose the model of Figure 5.15, which consists of five main components: engine, torque converter, gear box, output train and vehicle. Each component of the drive train can be considered as a rigid multibody subsystem. The rigid bodies are connected to each other by ideal rigid joints, clutches and force elements.

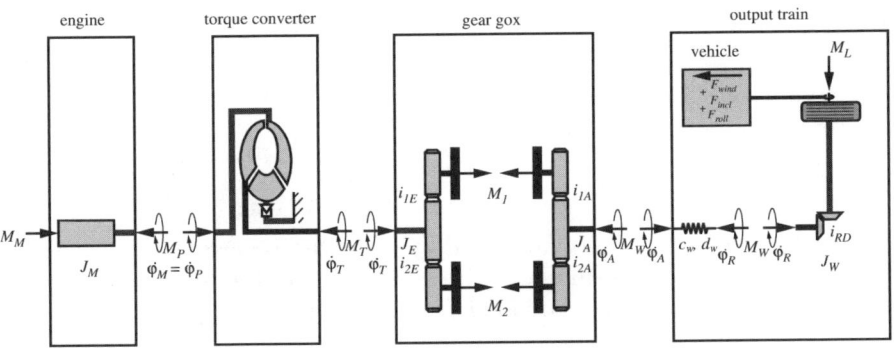

Fig. 5.15: Power train model

With the component-models described above the simplified power train can be described by the five states $\dot{\varphi}_M, \dot{\varphi}_T, \dot{\varphi}_A, \dot{\varphi}_R$, and $\varphi_A - \varphi_R$. For the comfort evalutation the jerk will be necessary. Therefore, we add the acceleration as an additional component to the state vector which can later be used to calculate the jerk. The state vector is defined as

$$\boldsymbol{x}^T = (x_1, x_2, x_3, x_4, x_5, x_6) = (\dot{\varphi}_M, \dot{\varphi}_T, \dot{\varphi}_A, \dot{\varphi}_R, (\varphi_A - \varphi_R), a) \tag{5.33}$$

with the vehicle acceleration

$$a = \frac{r_R}{i_{RD}} \ddot{\varphi}_R. \tag{5.34}$$

The equations of motion of the mechanical model (Figure 5.15) can be formulated as

$$\begin{pmatrix} \dot{x}_1 \\ \dot{x}_2 \\ \dot{x}_3 \\ \dot{x}_4 \\ \dot{x}_5 \\ \dot{x}_6 \end{pmatrix} = \begin{pmatrix} \frac{1}{J_M}((1-\beta)M_M - M_P) \\ \frac{1}{J_E}\left(M_T + \frac{1}{i_{1_E}}M_1 + \frac{1}{i_{2_E}}M_2\right) \\ \frac{1}{J_A}(-M_W - i_{1_A}M_1 - i_{2_A}M_2) \\ x_6 \frac{i_{RD}}{r_R} \\ x_3 - x_4 \\ f_6 \end{pmatrix} \tag{5.35}$$

with

$$f_6 = \frac{r_R}{J_W i_{RD}} \left[ c_w(x_3 - x_4) + d_w\left(\dot{x}_3 - x_6\frac{i_{RD}}{r_R}\right) - \rho c_w A \frac{r_R^2}{i_{RD}^2} x_4 x_6 \right] \tag{5.36}$$

Introducing the control vector consisting of the reduction factor for the engine torque and the clutch pressure

$$\boldsymbol{u} = \begin{pmatrix} \beta \\ p \end{pmatrix} \tag{5.37}$$

and eliminating $\dot{x}_3$ in the sixth equation by the third equation in eq. (5.35) we obtain the compact form

$$\dot{\boldsymbol{x}} = \boldsymbol{f}(\boldsymbol{x}, \boldsymbol{u}) \ . \tag{5.38}$$

While the torques $M_M, M_P, M_T, M_W$ and $M_L$ can be calculated at each instant of time during the simulation, the evaluation of the torques $M_1$ and $M_2$ depends on the operating state of the shift elements

- one-way clutch
  $\Delta\ddot{\varphi}_1 = 0 \Rightarrow M_1 \geq 0$
  $\Delta\ddot{\varphi}_1 > 0 \Rightarrow M_1 = 0$
- wet clutch
  $\Delta\dot{\varphi}_2 \neq 0 \Rightarrow M_2 = -\text{sign}(\Delta\dot{\varphi}_2)\mu_C A_p z r_m p$
  $\Delta\dot{\varphi}_2 = 0 \Rightarrow |M_2| \leq \mu_{C_0} A_p z r_m p$

The comparison between simulations performed with the presented model and measurements carried out during a gear shift process from the first to the second gear shows again a very good agreement (Fig. 5.16). In the simulation, the control strategy used in the car, was implemented and used as reference in the following sections. Therefore, the mechanical model of the power train can be used to develop a model-based optimal control for the gear shift process in an automatic transmission.

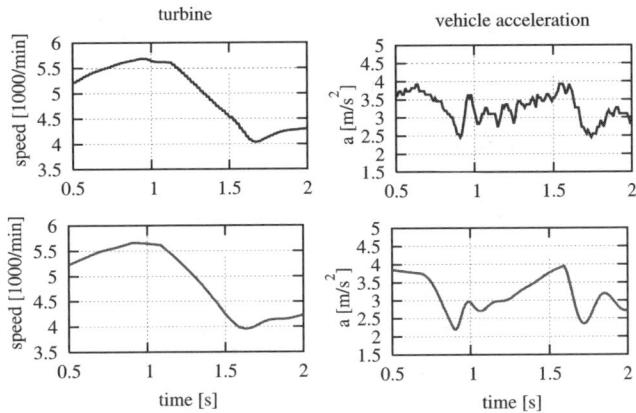

Fig. 5.16: Comparison of simulation (bottom) and measurements (top)

To solve the dynamic programming problem of the gear shift process analytically, the equation of motion (5.35) must be available in a discrete linear form. This can be achieved in two steps: the nonlinear system equations have to be first linearized and then discretized.

The gear shift operation results in a nonsmooth dynamical behaviour which is described by a set of equations with time-varying structure. Because of large changes of some state variables like the engine and turbine speed during the process, the linearization must be performed with respect to a reference trajectory $(\boldsymbol{x}_0(t), \boldsymbol{u}_0(t))$ rather than with respect to a constant reference operating point $(\boldsymbol{x}_0, \boldsymbol{u}_0)$. Therefore we consider for a given load case $\alpha_{TH}$ a reference control $\boldsymbol{u}_0(t)$ which yields the state trajectory $\boldsymbol{x}_0(t)$ according to the equations of motion (5.35). This reference control is the same one used to verify the model.

The **linear equations of motion** with respect to the reference state vector $\boldsymbol{x}_0(t)$ and reference control $\boldsymbol{u}_0(t)$ can then be obtained using the series expansion

$$\dot{\boldsymbol{x}}(t) = \boldsymbol{A}[\boldsymbol{x}_0(t), \boldsymbol{u}_0(t)]\boldsymbol{x}(t) + +\boldsymbol{B}[\boldsymbol{x}_0(t), \boldsymbol{u}_0(t)]\boldsymbol{u}(t) + \boldsymbol{e}(t) \tag{5.39}$$

where

$$A[\boldsymbol{x}_0(t), \boldsymbol{u}_0(t)] = \left(\frac{\partial \boldsymbol{f}}{\partial \boldsymbol{x}}\right)_0, \qquad B[\boldsymbol{x}_0(t), \boldsymbol{u}_0(t)] = \left(\frac{\partial \boldsymbol{f}}{\partial \boldsymbol{u}}\right)_0,$$

$$\boldsymbol{e}(t) = \dot{\boldsymbol{x}}_o(t) - \{A[\boldsymbol{x}_0(t), \boldsymbol{u}_0(t)]\boldsymbol{x}_0(t) + B[\boldsymbol{x}_0(t), \boldsymbol{u}_0(t)]\boldsymbol{u}_0(t)\} \quad (5.40)$$

We may derive the **time-varying discrete state equations** by considering the gear shift process as a sequence of equal time increments $T$. We get with the linear equations of motion (5.39) the discrete form

$$\boldsymbol{x}_{k+1} = \boldsymbol{A}_k \boldsymbol{x}_k + \boldsymbol{B}_k \boldsymbol{u}_k + \boldsymbol{e}_k \qquad (5.41)$$

with the definitions

$$\boldsymbol{A}_k = \left(\boldsymbol{E} + \boldsymbol{A}(kT)T + \frac{1}{2}\boldsymbol{A}^2(kT)T^2 + \ldots + \frac{1}{m!}\boldsymbol{A}^m(kT)T^m\right)$$

$$\boldsymbol{B}_k = \left(T\boldsymbol{E} + \frac{1}{2}\boldsymbol{A}(kT)T^2 + \ldots + \frac{1}{(m+1)!}\boldsymbol{A}^m(kT)T^{m+1}\right)\boldsymbol{B}(kT)$$

$$\boldsymbol{e}_k = \boldsymbol{x}_{0k+1} - (\boldsymbol{A}_k \boldsymbol{x}_{0k} + \boldsymbol{B}_k \boldsymbol{u}_{0k}) \qquad (5.42)$$

For the formulation of the performance function we need the jerk $\dot{a}$ of the vehicle during the gear shift process, which can be included by adding a new component to the discrete state space vector (5.33)

$$(x_7)_{k+1} = (x_6)_k = a_k \qquad (5.43)$$

The jerk can be calculated as output variable

$$y_k = \dot{a}_k = \frac{(x_7)_k - (x_6)_k}{T} \qquad (5.44)$$

With the extended state space $\boldsymbol{x}_k \in \mathbb{R}^7$ the discrete description of the system writes

$$\boldsymbol{x}_{k+1} = \boldsymbol{A}_k \boldsymbol{x}_k + \boldsymbol{B}_k \boldsymbol{u}_k + \boldsymbol{e}_k, \qquad y_k = \boldsymbol{c}^T \boldsymbol{x}_k,$$

$$\boldsymbol{c}^T = \begin{bmatrix} 0 & 0 & 0 & 0 & 0 & \frac{1}{T} & -\frac{1}{T} \end{bmatrix} \qquad (5.45)$$

The dimension of the matrices $\boldsymbol{A}_k, \boldsymbol{B}_k$ and $\boldsymbol{e}_k$ must be extended according to equation (5.43).

We shall consider the gear upshifting from the first to the second gear. In order to develop an **optimal control strategy** we split up the gear shift process into three phases as shown in Figure 5.17. These phases can be found in every gear upshifting and downshifting process.

- The first phase starts with the shifting signal which indicates the beginning of the gear change into the next gear. After a delay due to the dead time of the electric and hydraulic actuators of the system the pressure of the upcoming clutch is raised continuously. During this phase the off-going clutch, in this case the one-way clutch, remains blocked. Although the gear shift process has already begun, the transmission ratio corresponds to the previous gear, in this case the first gear.

- The second phase begins when the torque of the one-way clutch becomes zero. The one-way clutch releases and the gear box has no fixed transmission ratio because there is no determined kinematic relationship between the gear box input and output speed. The beginning of this phase is accompanied by a turnaround in the turbine speed.
- The third phase starts when the upcoming clutch sticks. This point is called the synchronous point and is reached when the relative speed of the wet clutch becomes zero, and the actual clutch torque is less then the maximal transmittable torque.

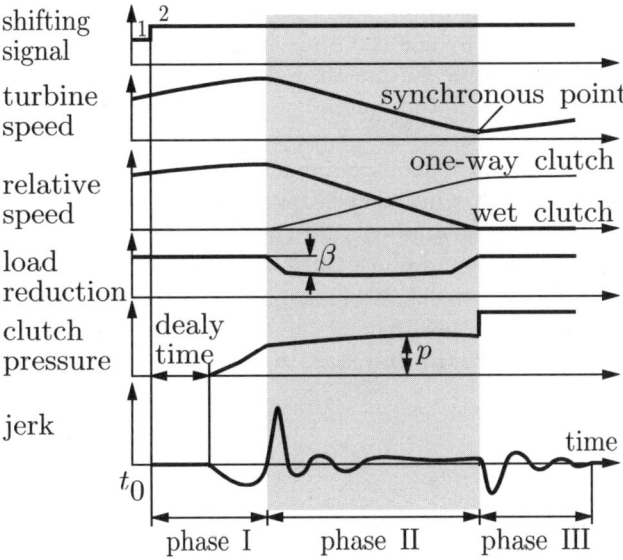

Fig. 5.17: Control strategy of the gear shifting

A reasonable optimal control can only be applied when the gear box has two degrees of freedom, one for the input shaft and one for the output shaft, which are kinematically independent. This state is only given for the second phase, when the relative speeds of both shift elements are unequal zero and the gear box has no fixed transmission ratio. The first phase must be used as a pre-control phase to put the gear box in a controllable state. This can be achieved by applying a feedforward control

$$\boldsymbol{u}(t) = \begin{pmatrix} 0 \\ p(t) \end{pmatrix} \tag{5.46}$$

The pressure of the upcoming clutch is raised smoothly after a time delay. Since a part of the driving input torque can now be transmitted by the wet clutch, the torque of the one-way clutch decreases. When its torque disappears, the one-way clutch releases. Now an optimal control law in the form

$$\boldsymbol{u}(\boldsymbol{x}(t),t) = \begin{pmatrix} \beta(\boldsymbol{x}(t),t) \\ p(\boldsymbol{x}(t),t) \end{pmatrix} \tag{5.47}$$

can be applied to the power train in order to achieve the gear shift in a given shift time. The control law is optimal in terms of minimizing some performance criteria, which will be specified in the following. It should be noted that once the target gear is engaged, the wet clutch can not be controlled anymore because the clutch torque can not be influenced by the clutch pressure in the sticking phase. Furthermore the load reduction must be finished at the synchronous point. This avoids the undesirable excitation of the power train by the change of the engine torque after the new gear is engaged.

The dynamic torque of the clutch, just before the synchronous point, is generally greater than the static torque just after it. An abrupt drop in the clutch torque at the synchronous point leads to an excitation of the output train and causes an undesired vehicle jerk during the third phase. Since no control intervention can be applied after the synchronization of the clutch, the control law during the second phase must be designed such that the jerk after the gear change is suppressed. This requires a predictive control approach. We determine the control law driving the power train system (5.45) from the initial state $\boldsymbol{x}_0 = \boldsymbol{x}(t_0)$ to the end state $\boldsymbol{x}_K = \boldsymbol{x}(t_e)$ with the constraint

$$g(\boldsymbol{x}(t_e)) = \Delta\dot{\varphi}_2 = \dot{\varphi}_A i_{2_A} - \frac{\dot{\varphi}_T}{i_{2_E}} = \boldsymbol{w}^T \boldsymbol{x}_K = 0 \; ,$$

$$\boldsymbol{w}^T = (0 \quad -\frac{1}{i_{2_E}} \quad i_{2_A} \quad 0 \; 0 \; 0 \; 0) \tag{5.48}$$

by minimizing the performance measure

$$J = \lambda\theta(\boldsymbol{x}_K) + \sum_{k=0}^{K-1} \Phi(\boldsymbol{x}_k, \boldsymbol{u}_k, k) \; . \tag{5.49}$$

The acceleration change is the most critical issue which affects the passengers comfort during the gear shift process. Therefore we define a cost function

$$\Phi(\boldsymbol{x}_k, \boldsymbol{u}_k, k) = \dot{a}_k^2(\boldsymbol{x}_k) + \boldsymbol{u}_k^T \boldsymbol{R}_k \boldsymbol{u}_k = \boldsymbol{x}_k^T \boldsymbol{c}\,\boldsymbol{c}^T \boldsymbol{x}_k + \boldsymbol{u}_k^T \boldsymbol{R}_k \boldsymbol{u}_k. \tag{5.50}$$

This cost function keeps the jerk close to zero without excessive expenditure of the control effort.

Although the first term in the performance measure (5.49) depends only on the end state, it can describe a behaviour which extends over a time interval after the synchronous point. We choose a criterion which evaluates the jerk after the end of the gear shift over a certain period of time. For this purpose we need a prediction function which estimates the jerk at each discrete step considered after the synchronous point as a function of the end state $x_K$. This can be accomplished by making use of the discrete linear form of the equation of motion of the system. We come out with

$$x_{K+1+i}(x_{K-1}, u_{K-1}) = \left(\prod_{l=-1}^{i} A_{K+l}\right) x_{K-1} + \left(\prod_{l=0}^{i} A_{K+l}\right) B_{K-1} u_{K-1} +$$

$$+ \sum_{j=0}^{i} \left(\prod_{l=i-j}^{i} A_{K+l}\right) e_{K-1+i-j} + e_{K+i} \,. \qquad (5.51)$$

Assuming that the interval considered for the evaluation of the jerk in the third phase has $N$ equal increments, $T$, the cost function can be written as

$$\theta(x_K) = \theta(x_{K-1}, u_{K-1}) = \sum_{i=K}^{K+N} x_i^T c\, c^T x_i \,. \qquad (5.52)$$

Note that the performance measure $\theta$ permits a prediction of the jerk over the next $N+1$ discrete time increments, which results from applying a given control $u_{K-1}$ to a given state $x_{K-1}$ at the discrete $(K-1)$-th stage.

The **optimization of a gear shift operation** can be formulated as a classical problem of optimal control with the following structure. Minimize the performance function eq. (5.49) by additionally considering the process dynamics eq. (5.45), the initial state $x_0 = x(t_0)$, the final state $x_K = x(t_e)$ and the constraint at $(t = t_e)$ eq. (5.48). Moreover, the optimization process must watch the constraint state and control spaces

$$x(k) \in X(k) = \{x(k)|\, (|x_i(k)| \le x_{i,\max})\}$$
$$x(u) \in U(k) = \{u(k)|\, (|u_i(k)| \le u_{i,\max})\} \qquad (5.53)$$

Applying Bellman's principle of optimality and the dynamic programming algorithm, we have to optimize at the $k$-th stage [16]

$$J_k = \lambda \vartheta(x_K) + \sum_{\kappa=k}^{K-1} \Phi(x_\kappa, u_\kappa, \kappa) \,. \qquad (5.54)$$

where $\Phi$ is defined by eq.(5.50) and $\vartheta$ by eq.(5.52). The magnitude $\lambda$ represents an externally selectable weighting measure. Regarding the stage oriented decision process according to Bellman we come out with his well-known recursion equation for the $k$-th stage:

$$J_K^*(\boldsymbol{x}_k, k) = \min \{\Phi(\boldsymbol{x}_k, \boldsymbol{u}_k, k) + J_{k+1}^*(\boldsymbol{x}_{k+1}, k+1)\},$$
$$\boldsymbol{u}_k \in \boldsymbol{U}$$
$$\boldsymbol{x}_k \in \boldsymbol{X} \tag{5.55}$$

where $\boldsymbol{x}_{k+1}$ can be replaced by eq.(5.45). In the case of gear shift operations the minimization is performed analytically applying [91]

$$\frac{\partial J_k^*}{\partial \boldsymbol{u}_k} = \boldsymbol{0}, \quad \frac{\partial^2 J_k^*}{\partial \boldsymbol{u}_k^2} > \boldsymbol{0}. \tag{5.56}$$

The first equations result in the optimal $(\boldsymbol{u}_k^*)$, and the second equation can be proved analytically [91].

Altogether we could achieve the following **results**. Figure (5.18a) and (5.18b) show the acceleration and jerk for the load case (100 % throttle opening) resulting from considering only the function $\Phi$ in performance measure (5.49) by setting $\lambda$ to zero. As expected the jerk is kept close to zero during the second phase. The acceleration change becomes consequently very smooth. Since the cost function $\theta$ is not considered the jerk after the synchronization remains unimproved. This changes as soon as the function $\theta$ is considered in the performance measure (5.49) by using an adequate weighting factor $\lambda$ (Figure (5.18c) and (5.18d)). The applied control drives the system to the synchronous point in the given shift time by minimizing the jerk during the second and third phase. During the first phase the applied pre-control is the same as the reference. Therefore the acceleration remain unchanged.

The proposed approach can be applied for any load case of the car. Figure 5.19 shows further results for the 80 % and 40 % throttle openings. In both cases the jerk and acceleration smoothness are apparently improved by keeping the desired shift time. Moreover the control approach is found to be robust with respect to varying the desired shift time. In Figure 5.20 some results are presented for the full load case where the desired shift time was reduces by 20 % with respect to the reference. The plot of the relative speed of the wet clutch (Figure (5.20c)) shows that the synchronization is achieved at the desired time. The acceleration and jerk are still smooth (Figure (5.20a) and (5.20b)). Furthermore the frictional losses in the wet clutch (Fig. (5.20d)) are reduced, which improves the life expectancy of the friction discs.

The above example of an optimal control approach for gear shift operations in vehicle automatic transmissions illustrates the following procedure and the following results. Firstly, a mechanical model is developed for the whole power train and verified by measurements. After the discretization of the equations of motion and making use of the dynamic programming method the gear shift operation is considered as a multistage process with constraints. Furthermore a suitable performance measure for evaluating the gear shift comfort during the process is formulated. The analytical solution of the dynamic programming problem leads to an explicit discrete optimal control law for the gear shift process. The application of the derived optimal control to the verified

## 5.1 Automatic Transmissions

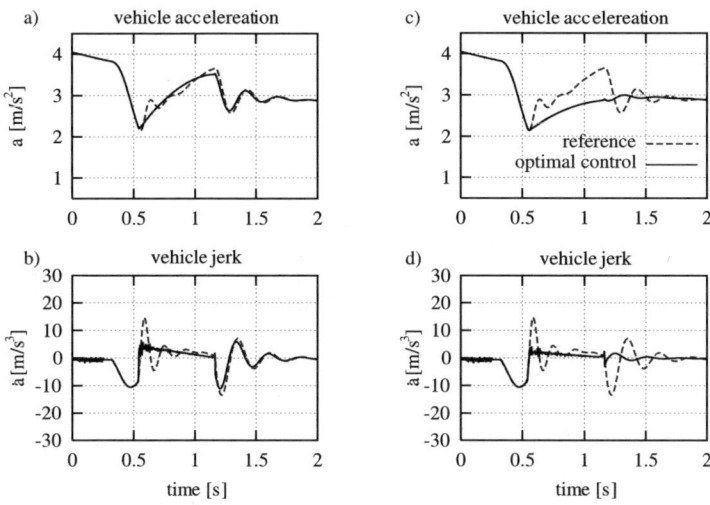

Fig. 5.18: Results for 100 % throttle opening: a) and b) without the end cost function $\theta$ ($\lambda = 0$), c) and d) with the end cost function $\theta$ ($\lambda \neq 0$) [91].

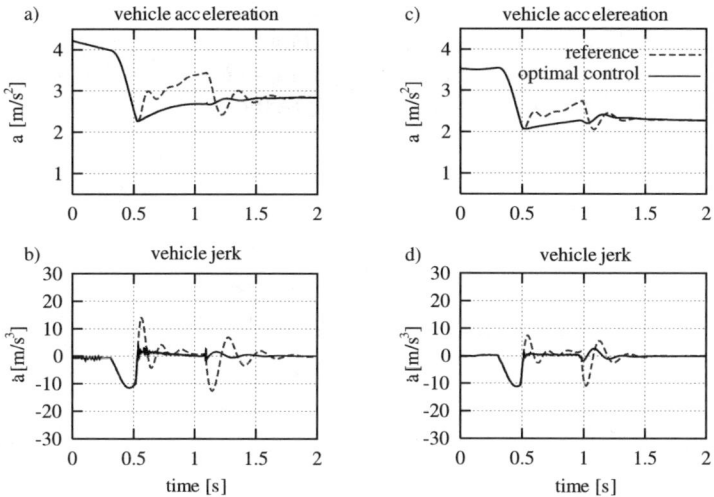

Fig. 5.19: Results for 80 % throttle opening: a) and b), and 40 % throttle opening: c) and d) [91]

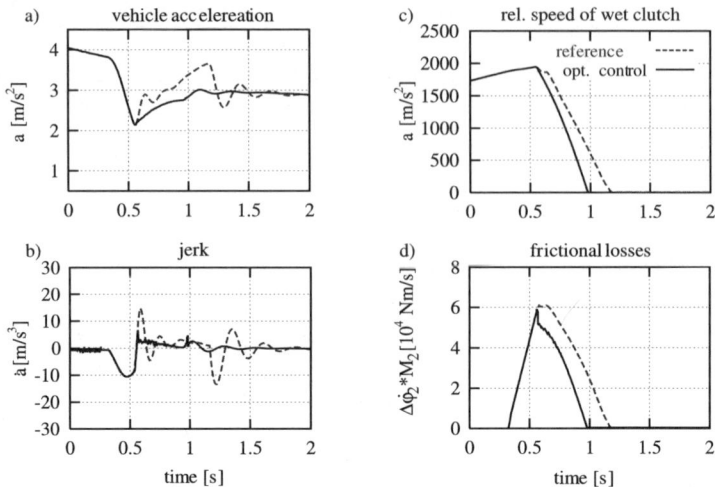

Fig. 5.20: Results for 100 % throttle opening with a reduction of the shift time of 20 % with respect to the reference [91]

nonlinear model in computer simulations shows major improvements in terms of the passengers comfort for throttle openings. Moreover the shift time and the frictional losses during the process are reduced.

## 5.2 Ravigneaux Gear System

Ravigneaux gears are planetary gears used preferably in automatic power transmission systems. They represent an integration of two planetary gears, where the two planet carriers are combined, and one ring gear is omitted. They possess a complicated mesh structure and are a source of parameter-excited vibrations in the interior of an automatic transmission. For modeling the dynamics of such a gear in a correct way, the dynamics of the complete transmission and in addition at least the approximate dynamics of the complete driveline system must be considered. This might be performed by many commercial computer codes on the market. But for complex standard problems it is sometimes more useful to develop special system adapted programs which are designed for the analysis of particular tasks. This has been done in the case of the Ravigneaux gears and will be presented here as another practical example of the application of multibody system theory [215], [194].

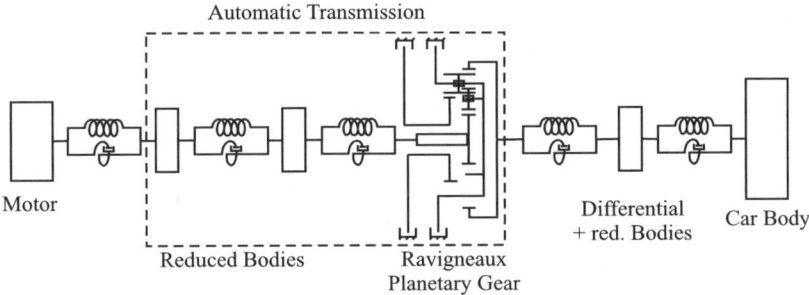

Fig. 5.21: Model of a Drive Train with Automatic Transmission

Figure 5.21 shows the mechanical model of a driveline of a passenger car. It consists of the main components motor, gearbox, cardan shaft and differential gear. Since in this case the effects of the parameter excitation caused by the meshing of the gear wheels of the Ravigneaux planetary gear are studied, a more detailed model of the gearbox is necessary. For that reason most of the parts of the Ravigneaux gear are represented as individual bodies in the mechanical model of that component. The connection between the single bodies can be provided either by force elements or by unilateral and bilateral constraints. Due to the structure of the total system and in order to ensure the adaptability and a variable depth of modeling, multibody system theory is applied. This technique allows the derivation of a structured and closed description of the systems consisting of rigid and elastic bodies [28], [87], [200]. Within the following sections the mechanical model and the mathematical description of the main parts, components and connections are presented in more detail.

## 5.2.1 Toothing

Gear trains are the most common mechanical elements in drive technology for the transformation of rotational speed and torque. Depending on the requirements of the simulation model a gear train can be described using different mechanical substitute models. Basically there can be differentiated between the three versions:

- Rigid transmission
- Force element with clearance
- Engagement with impacts

The last one will be used for gear trains with backlashes, which can generate rattling or hammering phenomena [200]. The examination of the effects of

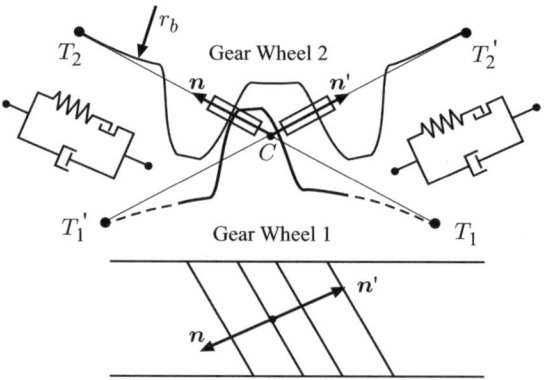

Fig. 5.22: Mechanical Model of a Toothing

parameter excitations induced by the meshing gear wheels requires a meticulous consideration of the elasticity of the toothing. Furthermore some swaying of the rotational speed of the shafts may cause a change in the direction of force transmission. For that reason the force element with clearance has to be chosen for the toothing inside the planetary gear. Figure 5.22 schematically illustrates the design of this widely used force element [126], [217]. Parallel to the stiffness, which is effective only for contacting teeth, a damper generates a force along the direction $n$ of the line of action. If the load is changing, and the back flanks of the tooth profiles are coming into contact, the force changes its direction to $n'$. The forces acting on the points $T_1$ and $T_2$ can be computed by

$$\boldsymbol{F}_{T1} = -\boldsymbol{n}\lambda_N, \quad \boldsymbol{F}_{T2} = \boldsymbol{n}\lambda_N, \quad \lambda_N = \lambda_N(g_N, \dot{g}_N) = \lambda_{Nc} + \lambda_{Nd} \qquad (5.57)$$

where $\lambda_N$ is the scalar force depending on the relative distance $g_N$ between the tooth profiles and the approaching velocity $\dot{g}_N$. The fraction $\lambda_{Nc}$ results from the stiffness $c$ of the engaging teeth and is a function of $g$ and the clearance $s$.

The relative distance in normal direction follows from classical kinematical theory of toothing [170]

$$g_N = \bar{\mathbf{w}}_N^T(\mathbf{q},t)\mathbf{q}, \qquad \dot{g}_N = \mathbf{w}_N^T(\mathbf{q},t)\dot{\mathbf{q}}, \qquad (5.58)$$

where $\bar{\mathbf{w}}_N$ and $\mathbf{w}_N$ are different constraint vectors. The relations pictured in

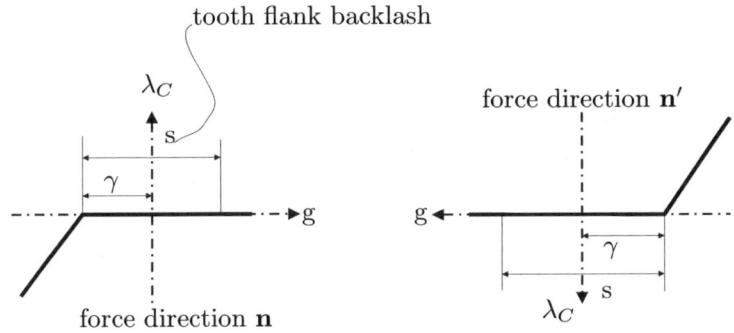

Fig. 5.23: Tooth Flank Backlash

Figure 5.23 may conveniently expressed by applying the well-known Heaviside function yielding

$$\lambda_{Nc} = H(-[g_N + \tfrac{s}{2}])\, c(x)[g_N + \tfrac{s}{2}]$$

$$H(-[g_N + \tfrac{s}{2}]) = \begin{cases} 0, & g_N > -\tfrac{s}{2} \\ 1, & g_N \le -\tfrac{s}{2} \end{cases}. \qquad (5.59)$$

that makes sure that there is no force acting within the clearance $s$. As apparent the stiffness $c$ of the force element is a function of the dimensionless rotational distance

$$x = \frac{r_b \varphi}{p_t}, \qquad (5.60)$$

which is equivalent to the angle of twist $\varphi$ of the gear train. The magnitude $p_t$ denotes the base pitch of the gear with the base radius $r_b$. The periodical variation of the total and the single stiffnesses of a gear train is shown in Figure 5.24 and illustrates the main source of the parameter excitation. The main influences on the run of the curve are the geometry of a single tooth, the material properties and the contact ratio factor.

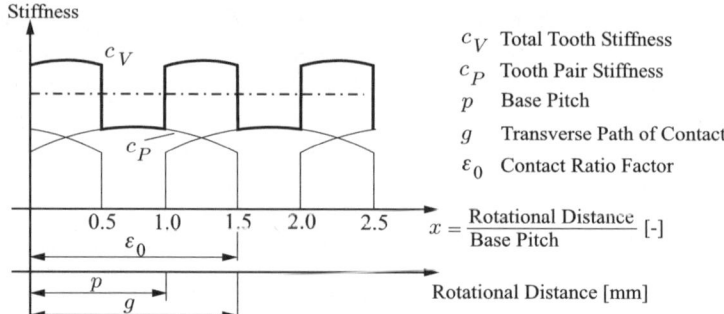

Fig. 5.24: Stiffness of a Gear Train

The velocity dependent fraction $\lambda_{Nd}$ of the force element is computed according to the damping model of [217] via

$$\lambda_{Nd} = d(g_N)\dot{g}_N. \qquad (5.61)$$

It takes into account that due to the clearance between the tooth profiles and the resulting variation of the thickness of the oil film the damping properties change subject to the distance $g_N$ between the contacting surfaces. As pictured in figure 5.25, detail 3 or 6, the damping coefficient $d$ attains the constant value $d_0$ for two gear wheels in contact. Within the limits of the clearance $s$ the damping coefficient $d$ is approximated by an exponential curve which leads to a steady transition into the contact area.

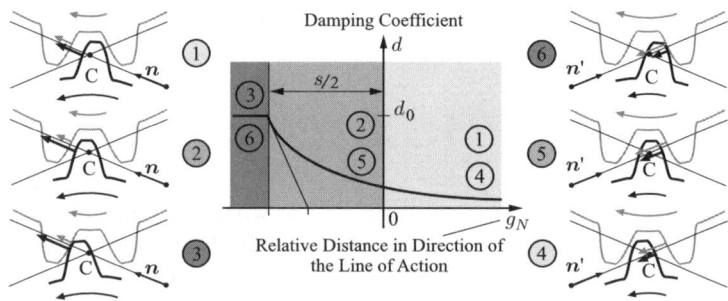

Fig. 5.25: Curve of the Damping Coefficient

### 5.2.2 Ravigneaux Planetary Gear

Ravigneaux type planetary gears are often used in passenger cars since they offer a multitude of gear-stage combinations and are at the same time very

## 5.2 Ravigneaux Gear System

compact requiring only small space. Figure 5.26 illustrates the basic design of such a planetary gear, the main components and their function for the first gear respectively:

- Small and large sun gear S1 and S2
- Ring gear R
- Planet carrier PC
- Inner (short) planets P1 and outer (long) planets P2
- Input shaft A
- Free coupling shaft F

A Ravigneaux planetary gear is designed using two conventional planetary gears whereby the two planet carriers are combined, and one ring gear is omitted. In the first gear, which is treated in this paper, the shaft A drives the sun wheel S1 which is meshing the short planets P1. The planets P1 are in gear with the long planets P2 and both are pivoted in the planet carrier PC that is inertially fixed by a rotational force element which approximately substitutes the effect of a multiple disk clutch. The ring gear is engaged with the planets P2 and transmits the load by the coupling described later. Other transmissions can be selected by changing the input shaft and the fixation of shafts and the planet carrier respectively. The connection between the components above-mentioned is performed by the force elements pictured on the right side of figure 5.26. The toothing is modeled according to the equations given in the section "Toothing", where the geometrically caused phase shifts between the

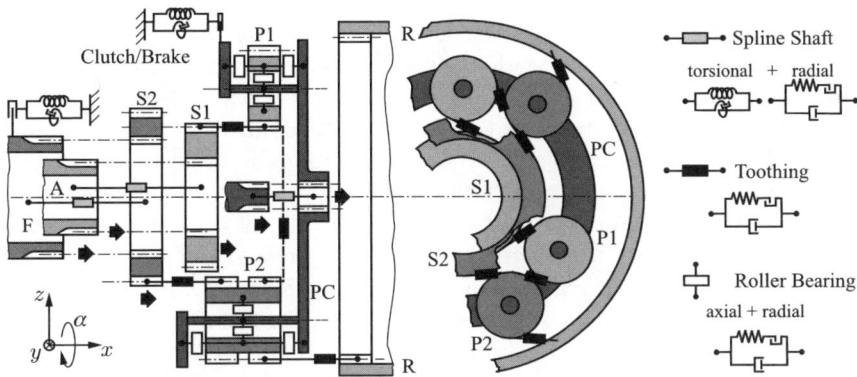

Fig. 5.26: Components of a Ravigneaux Planetary Gear

curves of the tooth stiffnesses of the single gear trains have to be taken into account. The vector of the generalized coordinates may contains either 4 or 6

246    5 Power Transmission

degrees of freedom depending on the pitch motion under consideration. Due to the stiffness of the shafts and the small clearance in the bearings the 4 degrees of freedom $\boldsymbol{q}^T = (x, y, z, \alpha)$ were selected in this case.

### 5.2.3 Ring Gear

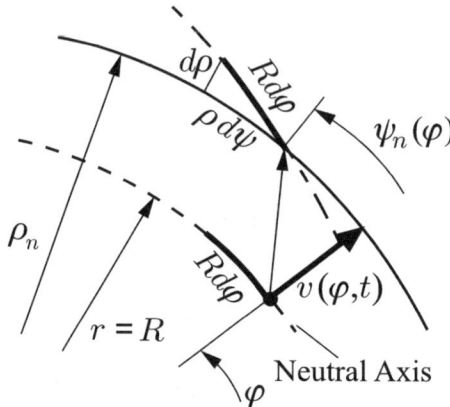

$r, \varphi$    Undeformed Position
$\rho_n, \psi_n$ Deformed Position

Fig. 5.27: Deformation of the Neutral Axis

The way of modeling the ring gear R is influenced by the connection between the ring gear and the output shaft O (see figure 5.29), due to various design configurations which lead to different dynamical behavior of the ring gear. If ring gear and output flange are welded together the ring gear can be regarded as stiff having 4 or 6 degrees of freedom. Otherwise if the ring gear like in this case is coupled to the output flange by the carrying toothing pictured in figure 5.29, the small thickness of the ring gear and the clearance between R and O require an elastic model that allows the correct reproduction of the movements within the tooth-work. Lachenmayr presents in [134] an elastic model of a ring gear which provides a function for the planar deformation of the neutral axis in radial and circumferential direction starting from the rigid body position like illustrated in figure 5.27. Complying with the model of the Rayleigh beam the neutral axis is not stretched and the circumferential deformations $\Delta \psi$ cancel out:

$$\int_0^{2\pi} \Delta \psi d\varphi = \int_0^{2\pi} [\psi_n(0) - \psi_n(\varphi)] d\varphi = 0 \tag{5.62}$$

Therefore the scalar radial deformation is uniquely described by the coordinate $v$. Using a modified Ritz approximation with an additional quadratic term, $v$ can be separated into the vector $\bar{\mathbf{v}}$ of the shape functions depending on space only and the vector $\mathbf{q}_{el}$ of the time depending elastic degrees of freedom, see also section 3.3.4 on page 124:

$$v(\varphi,t) = \bar{\mathbf{v}}^T \mathbf{q}_{el} + \frac{1}{R}\mathbf{q}_{el}^T \mathbf{V}\mathbf{q}_{el} \quad , \quad \mathbf{V} = \begin{bmatrix} \frac{1-i^2}{4} & 0 \\ 0 & \frac{1-i^2}{4} \end{bmatrix}_D \tag{5.63}$$

The extension of the Ritz approximation is necessary since the chosen $i\pi$-

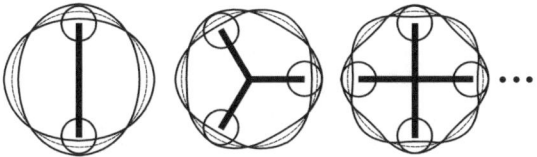

Fig. 5.28: Shape Functions of the Ring Gear

periodic harmonic shape functions with $i \geq 2$ illustrated in figure 5.28

$$\bar{\mathbf{v}} = (\cos(2\varphi), \cdots, \cos(n\varphi), \sin(2\varphi), \cdots, \sin(n\varphi))^T = [\cos(i\varphi)\ \sin(i\varphi)]^T \tag{5.64}$$

do not fulfil the closing condition $\psi(0) = \psi(2\pi)$. In practice a small number of shape functions is sufficient because the oscillations with high frequencies are not of interest for the dynamic behaviour of the total system. Based on the deformation coordinates the kinematic correlations and the equations of motions of the elastic ring gear are derived.

### 5.2.4 Ring Gear Coupling

The alternative to the welding of ring gear and output is the connection by the toothing shown in Figure 5.29 [73] which transmits the torque by the elastic contact of the teeth. The clearance between the teeth in radial and circumferential direction enables an additional relative movement between ring gear and output shaft. Figure 5.29 makes clear, that the induced friction forces possess a remarkable influence on the deformation of the ring gear, contingent on the number of teeth and the size of the clearance. Therefore the couplings between ring gear and output shaft are modelled in detail for every single tooth as force elements with clearance and radial friction forces according to figure 5.30. Again the relative normal and tangential distances and velocities within the contacts of Figure 5.30 can be evaluated by elementary considerations from the kinematics of gear meshing [215],[170]. We get

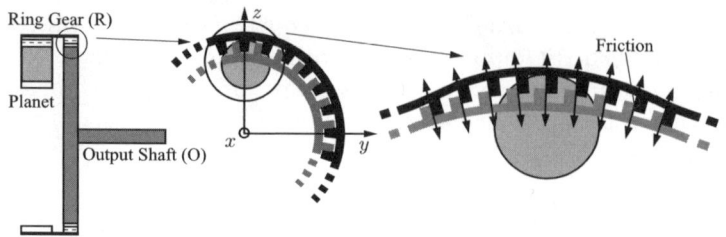

Fig. 5.29: Coupling of Ring Gear and Output Drive

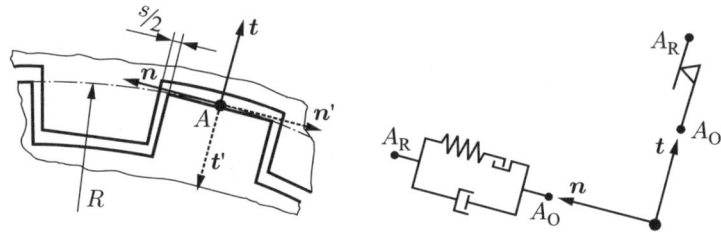

Fig. 5.30: Force Directions of the Contact

$$g_N = \bar{\mathbf{w}}_N^T(\mathbf{q}, t)\mathbf{q}, \qquad \dot{g}_N = \mathbf{w}_N^T(\mathbf{q}, t)\dot{\mathbf{q}} + \hat{\mathbf{w}}_N$$
$$g_T = \bar{\mathbf{w}}_T^T(\mathbf{q}, t)\mathbf{q}, \qquad \dot{g}_T = \mathbf{w}_T^T(\mathbf{q}, t)\dot{\mathbf{q}} + \hat{\mathbf{w}}_T \tag{5.65}$$

Accordingly the resulting forces acting on output flange and ring gear can also be calculated from normal and tangential fractions in the form

$$\mathbf{F}_{A,O} = -\mathbf{n}\lambda_N - \mathbf{t}\lambda_T, \qquad \mathbf{F}_{A,R} = \mathbf{n}\lambda_N + \mathbf{t}\lambda_T . \tag{5.66}$$

where $\lambda_N$ is divided into the fraction $\lambda_{Nc}$ caused by the contact stiffness

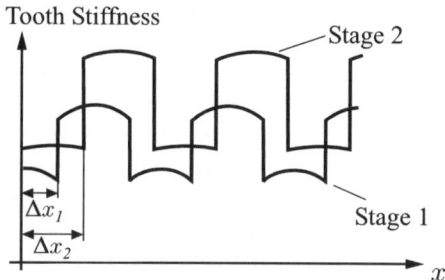

Fig. 5.31: Phase Shift

and the fraction $\lambda_{Nd}$ caused by the damping properties, equivalent to the toothings. The force $\lambda_T$ in radial direction is computed employing the viscous friction coefficient $\mu_v$ and the relative velocity $\dot{g}_T$:

$$\lambda_T(\dot{g}_T) = \mu_v \dot{g}_T \ . \tag{5.67}$$

### 5.2.5 Phase Shift of Meshings

Phase shift is the offset $\Delta x$ of the varying tooth stiffness along the transverse path of contact of a gear train as pictured in figure 5.31. It is of high importance for the parameter excitation in multistage or planetary gears. The curve of the tooth stiffness is always specified for a special point, in this case for the (A) marked position of the first contact point on the line of action as shown in Figure 5.32. Depending on the number of teeth and the geometrical config-

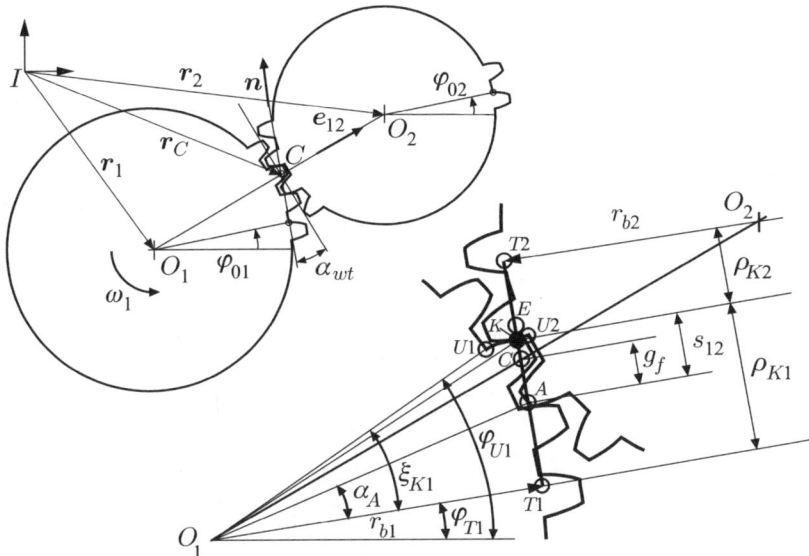

Fig. 5.32: Geometrical Derivation of the Phase Shift

uration, which means the tilt of a gear train respective to the antecedent one, the contact point of two engaging tooth profiles will normally not be placed in point A for the starting position. The quotient of the phase shift $s_{12}$ of Figure 5.32 and the base pitch $p_t$ leads to the phase shift $\Delta x$ used for the calculation of the current tooth stiffness. The following equations show the derivation of the phase shift for a single gear train and can be transferred to single stage and Ravigneaux planetary gears.

The location of the pitch point $C$ of a mesh can be determined by the pitch circle diameter $d_w$ which is depending on axle base and the number of teeth of both gear wheels:

$$r_C = r_1 + \frac{d_{w1}}{2} e_{12} \qquad (5.68)$$

Several geometrical characteristics of the toothing define the length $g_f$ of the line from the pitch point $C$ to the point of action $A$

$$r_A = r_C - g_f\, n \quad . \qquad (5.69)$$

The initial rotation angle $\varphi_{01}$ in figure 5.32 provides the position of the first tooth of the driving gear 1 versus the y-axis and follows from the prior meshing and the resulting status of the gear wheel. In combination with the position $\varphi_A$ of point $A$, the following equation calculates the angle $\varphi_{U1}$ of the root point of the first tooth within the transverse path of action from $A$ to $E$

$$\varphi_{U1} = \frac{2\pi}{z_1}\,\text{floor}(z_1 \frac{\Delta\varphi}{2\pi} + 1) + \varphi_{01},$$

$$\Delta\varphi = \begin{cases} \varphi_A - \varphi_{01}, & \varphi_A \geq \varphi_{01} \\ 2\pi + \varphi_A - \varphi_{01}, & \varphi_A < \varphi_{01} \end{cases} . \qquad (5.70)$$

The function floor($\bullet$) rounds the argument down toward the next smaller integer. The difference $\xi_{K1} = \varphi_{U1} - \varphi_{T1}$ supplies the radius of curvature of the involute in the contact point $K$ of two tooth profiles by the equation

$$\rho_{K1} = \xi_{K1}\frac{d_{b1}}{2} \quad . \qquad (5.71)$$

Eventually the length of the distance $A - K$ is

$$s_{12} = g_f + \rho_{K1} - \sqrt{\frac{d_{w1}^2 - d_{b1}^2}{4}} \qquad (5.72)$$

and the dimensionless phase shift $\Delta x$ of the tooth stiffness function according to Figure 5.24 can be computed using the transverse pitch $p_t$ by $\Delta x = 1 - s_{12}/p_t$ . Knowing the contact point $K$ it is possible to determine the bending radius and the root point of the opposite tooth flank. Consequently the position of the gear wheel is defined and the resulting initial rotation $\varphi_{02}$ can be used for the calculation of the phase shift of the successive gear trains. The same correlations as presented for this external gear pair are also valid for internal gear pairs. Thus the phase shifts of a complete planetary gear can be computed starting from a gear train including the sun gear.

### 5.2.6 Equations of Motion

Starting with the principles of linear and angular momentum or with Jourdain's principle of virtual power, the equations of motion for rigid and elastic bodies are derived, see chapters 3.3 on page 113 and 3.4 on page 131. Combining the equations of the single components the differential equations of motion of the whole multibody system can be set up.

## 5.2 Ravigneaux Gear System

The equations of motion of a *rigid body* represent a base for the description of a multibody system. Figure 5.33 shows a rigid body with mass $m$, inertial tensor $I_S$ regarding the center of gravity $S$ and external force $F_A \in \mathbb{R}^3$ and torque $M_A \in \mathbb{R}^3$ acting in point $A$ separated from $S$ by the vector $r_{SA}$. The translational velocity of the center of gravity is called $v_S \in \mathbb{R}^3$ and the associated acceleration $a_S = \dot{v}_S$. The rotational speed is $\omega \in \mathbb{R}^3$. Introducing the vectors of the generalized coordinates $q$, velocities $\dot{q}$ and accelerations $\ddot{q}$ the principles of linear and angular momentum yield the classical multibody equations of motion

$$\begin{pmatrix} J_S \\ J_R \end{pmatrix}^T \begin{pmatrix} mE & 0 \\ 0 & I_S \end{pmatrix} \begin{pmatrix} J_S \\ J_R \end{pmatrix} \ddot{q} +$$
$$+ \begin{pmatrix} J_S \\ J_R \end{pmatrix}^T \left[ \begin{pmatrix} 0 \\ \tilde{\omega} I_S \omega \end{pmatrix} \begin{pmatrix} mE & 0 \\ 0 & I_S \end{pmatrix} \begin{pmatrix} \bar{\iota}_S \\ \bar{\iota}_R \end{pmatrix} \right] - \begin{pmatrix} J_A \\ J_R \end{pmatrix} \begin{pmatrix} F_A \\ M_A \end{pmatrix} = 0. \tag{5.73}$$

The dimension of these equations is f, according to the number f of degrees of

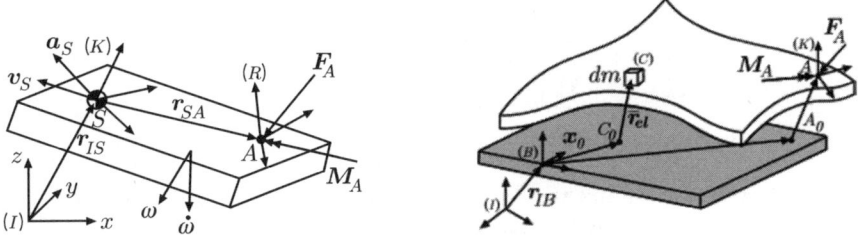

Fig. 5.33: Rigid and Elastic Bodies

freedom, $q \in \mathbb{R}^f$. The Jacobians are defined by $J_S = \frac{\partial v_S}{\partial \dot{q}}$ and $J_R = \frac{\partial \omega}{\partial \dot{q}}$. The magnitudes $\bar{\iota}_S$ and $\bar{\iota}_R$ are generated by the partial time derivation $\bar{\iota}_S = \dot{J}_S \dot{q}$ and $\bar{\iota}_R = \dot{J}_R \dot{q}$.

*Elastic bodies* can be integrated into the structure of multibody systems [28]. Figure 5.33 shows an elastic body in undeformed and deformed state as well as several coordinate systems associated with the body: the inertial system $I$, a reference frame $B$ coinciding with the undeformed body and a body fixed system $K$ for the deformed body. The external force $F_A \in \mathbb{R}^3$ and the torque $M_A \in \mathbb{R}^3$ are acting in point $A$. Jourdain's principle of virtual power leads to the equations of motion of an elastic body in the form

$$\int_{m_i} \left( \frac{\partial_B v}{\partial \dot{q}} \right) (_B a - {}_B f_A) dm = 0 \tag{5.74}$$

where $a$ is the absolute acceleration of the mass element $dm$ and $\boldsymbol{f}_A \in \mathbb{R}^6$ the vector of the external forces and moments on the mass element. Choosing a notation in the system $B$ makes the subsequent search for shape functions much easier. Making use of the position vector from "I" to "C"

$$_B\boldsymbol{r}_{IC} = {}_B\boldsymbol{r}_{IB} + {}_B\boldsymbol{x}_0 + {}_B\overline{\boldsymbol{r}}_{el} \tag{5.75}$$

the velocity ${}_B\boldsymbol{v}_C$ and the acceleration ${}_B\boldsymbol{a}_C$ of the mass element $dm$ can be calculated by two time derivation. The vector of the elastic deformation $\overline{\boldsymbol{r}}_{el}(\boldsymbol{x}_0,t)$ possesses dependencies on both time and space and can be separated using a Ritz approach in order to get ordinary differential equations instead of the partial ones of equation (5.74)

$$\overline{\boldsymbol{r}}_{el}(\boldsymbol{x}_0,t) = \boldsymbol{W}(\boldsymbol{x}_0)\boldsymbol{q}_{el}(t) \,, \quad \boldsymbol{W} \in \mathbb{R}^{3,n_{el}} \,, \quad n_{el} \in \mathbb{N} \quad . \tag{5.76}$$

Thus the dependency on space is represented by $n_{el}$ shape functions which are collocated within the matrix $\boldsymbol{W}$ and weighted by the time dependent elastic coordinates in the vector $\boldsymbol{q}_{el}$. Combining the Ritz approach with equation (5.74) results in the equations of motion

$$\sum_{i=1}^{n} \int_{m_i} \left[ \boldsymbol{J}_T^T + \boldsymbol{J}_R^T [\widetilde{\boldsymbol{x}}_0 + (\widetilde{\boldsymbol{W}\boldsymbol{q}_{el,i}})] + \boldsymbol{J}_E^T \boldsymbol{W}^T \right] \cdot$$

$$\cdot [\boldsymbol{a} + \widetilde{\boldsymbol{\omega}}\widetilde{\boldsymbol{\omega}}(\boldsymbol{x}_0 + \boldsymbol{W}\boldsymbol{q}_{el,i}) + \dot{\widetilde{\boldsymbol{\omega}}}(\boldsymbol{x}_0 + \boldsymbol{W}\boldsymbol{q}_{el,i}) +$$

$$+ 2\widetilde{\boldsymbol{\omega}}\boldsymbol{W}\dot{\boldsymbol{q}}_{el,i} + \boldsymbol{W}\ddot{\boldsymbol{q}}_{el,i} - \boldsymbol{f}_A] dm = \boldsymbol{0}. \tag{5.77}$$

The rigid body Jacobians are $\boldsymbol{J}_T$ and $\boldsymbol{J}_R$. The elastic Jacobians $\boldsymbol{J}_E$ come out by a similar derivation with respect to the elastic coordinates

$$\dot{\boldsymbol{q}}_{el} = \boldsymbol{J}_E \dot{\boldsymbol{q}} \,, \quad \ddot{\boldsymbol{q}}_{el} = \boldsymbol{J}_E \ddot{\boldsymbol{q}} \quad \text{with} \quad \boldsymbol{J}_E = \frac{\partial \dot{\boldsymbol{q}}_{el}}{\partial \dot{\boldsymbol{q}}} \quad . \tag{5.78}$$

After the calculation of the integral matrices in equation (5.77) and their linearization with respect to the small elastic coordinates, the well known formulation of discrete mechanical systems

$$\boldsymbol{M}(\boldsymbol{q},t)\ddot{\boldsymbol{q}} - \boldsymbol{h}_g(\boldsymbol{q},\dot{\boldsymbol{q}},t) - \boldsymbol{h}_f(\boldsymbol{q},\dot{\boldsymbol{q}},t) = 0 \tag{5.79}$$

is obtained, where $\boldsymbol{M}$, $\boldsymbol{h}_g$ and $\boldsymbol{h}_f$ denote mass matrix, vector of gyroscopic and external forces of the elastic body, respectively.

We write the external forces $\boldsymbol{F}_A$ or torques $\boldsymbol{M}_A$ in the form

$$\boldsymbol{f}_A = \begin{pmatrix} \boldsymbol{F}_A \\ \boldsymbol{M}_A \end{pmatrix} = \boldsymbol{w}\lambda \in \mathbb{R}^6 \tag{5.80}$$

where $\boldsymbol{w}$ is the generalized force direction represented by the Jacobian matrix at the point of the application and $\lambda$ a scalar value of the force. Three different cases for the expression $\boldsymbol{w}\lambda$ have to be considered:

- The scalar force value $\lambda$ depends on the relative displacement $g$ and the relative velocity $\dot{g}$ of the force element: $\lambda = \lambda(g, \dot{g})$
  The force calculation of the toothing in chapter 5.22 is performed that way.

- The expression may also represent a bilateral constraint. On displacement level it can be described by: $g = 0$, $\lambda$ any.

- Substructures are connected by a unilateral constraint. Contact phenomena of rigid bodies are typical examples for this type of linking. A unilateral contact in normal direction can be described by a complementary problem: $g_N \geq 0$, $\lambda_N \geq 0$, $g_N \lambda_N = 0$.

### 5.2.7 Implementation

The application of the methods of rigid and elastic multibody systems for the mathematical description of the total system matches the object oriented method of product development. According to Figure 5.34 the analogies between a multibody system and a correlating class diagram of an object model can be identified at first sight. Furthermore the technique of object oriented modelling facilitates the developing process by the utilisation of synergies, the encapsulation of components and the heredity of equivalent properties of different elements. Mechanical modelling, mathematical description and the implementation of the software have to be treated in common.

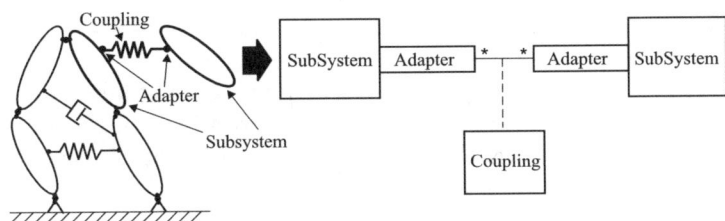

Fig. 5.34: Multibody System – Analogy of Structure and Object Model

The implementation of the derived mathematical models into the simulation program **DynAs** was achieved using the "Unified Modeling Language (UML)" [227]. The structure of the system is split into two independent units: the kernel which contains the implementation of the basic equations and controls the solution of the equations of motion, and the model library which treats the specific equations and can be easily upgraded for the integration of new or refined models [215].

254    5 Power Transmission

### 5.2.8 Simulation Results

Based on the modelling presented above a data set for simulations of the driveline pictured in figure 5.21 containing the Ravigneaux planetary gear was built up.

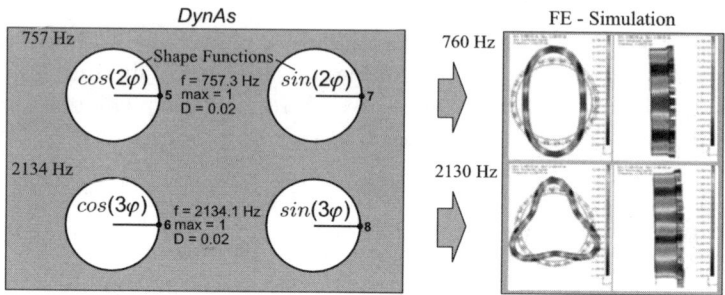

Fig. 5.35: Comparison of the Planar Eigenforms

The *verification of the mechanical model of the elastic ring gear* was achieved by comparing the planar eigenmodes of the linearized equations of motion of the uncoupled ring with the results of a Finite Element analysis. Figure 5.35 illustrates the well-fitting match of the important first two eigenforms and eigenfrequencies using the modified harmonic shape functions presented with equation (5.64) and Figure 5.28.

The detailed modelling of the Ravigneaux planetary gear permits the examination of the parameter excited oscillations caused by the varying total tooth stiffnesses of the gear pairs. The curve of the tooth stiffness according to Figure 5.24 can be represented by the Fourier series (which is a very popular method in industry in spite of the fact, that details of the non-smooth curves of Figure 5.24 are more or less lost)

$$f_A = \begin{pmatrix} F_A \\ M_A \end{pmatrix} = w\lambda \in \mathbb{R}^6, \qquad c_V = c_0 + \sum_{i=1}^{n} c_i \sin(i(x - \Delta x) + \varphi_i) \quad . \quad (5.81)$$

Furthermore the number of teeth $z_R$ of the ring gear and the number of planets $n_{Pl}$ of one gear stage have a remarkable influence on the effect of the parameter excitation. If the result of the quotient $z_R/n_{Pl}$ specifying the phase shift of the gear pairs is an integer, all meshings between planets and ring gear are phased (figure 5.36, case 1). This leads to the time depending total torque acting on the ring gear pictured in the lower left part of figure 5.36 whereas the resulting planar force on the centre of gravity S is constantly zero. If, on the other hand, the term $modulo(z_R/n_{Pl})$ is unequal to zero the phase shifted inequalities e.g. in the dominating 1st order (figure 5.36, case 2a) effect a constant total torque but a varying force moving S out of the central position

## 5.2 Ravigneaux Gear System

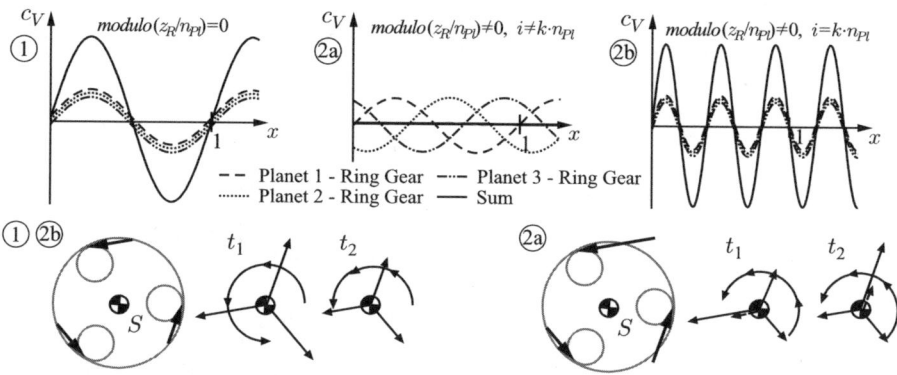

Fig. 5.36: Effects of the Phasing Ring Gear–Planets

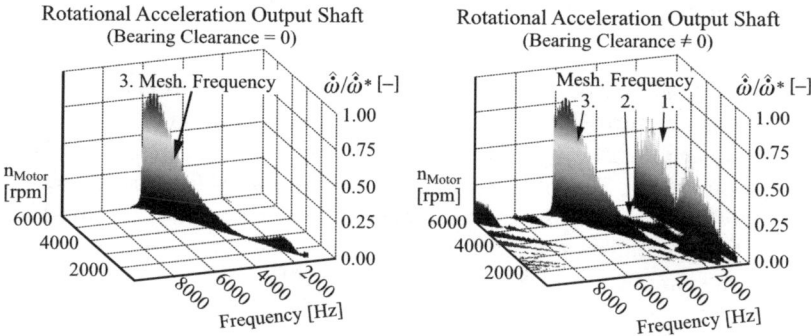

Fig. 5.37: Effect of the Bearing Clearance, $modulo\left(\frac{z_R}{n_{Pl}}\right) \neq 0$

as depicted down right in figure 5.36. Case 2b shows an exception to the correlations mentioned above. If the order $i$ of interest is a whole-numbered multiple of the number of planets $n_{Pl}$ the runs of the single tooth stiffnesses are phasing again and the effects are equivalent to case 1.

These considerations are confirmed by the simulation results. Figure 5.37 pictures in the left diagram the rotational accelerations for a run-up of a Ravigneaux gear with 2x3 planets and $n_R = 85$. As expected, accelerations can be recognised in the 3rd order of the meshing frequency since $i = 3$ is the first multiple of $n_{Pl} = 3$. Accelerations of higher orders (e.g. 6, 9) could not occur since the Fourier series of equation (5.81) was limited to order $i = 5$. In contrast, the right graph of figure 5.37 presents simulation results if the clearances of all bearings within the planetary gear are considered. Several amplitudes of orders $i \neq 3$ can be identified which were also detected in measurements of a test gear. Due to the clearances the rotational axis of planet carrier, ring gear and output shaft may be relocated under the influence of gravity which causes differing tooth forces and the acceleration amplitudes of orders $i \neq k \cdot n_{Pl}$.

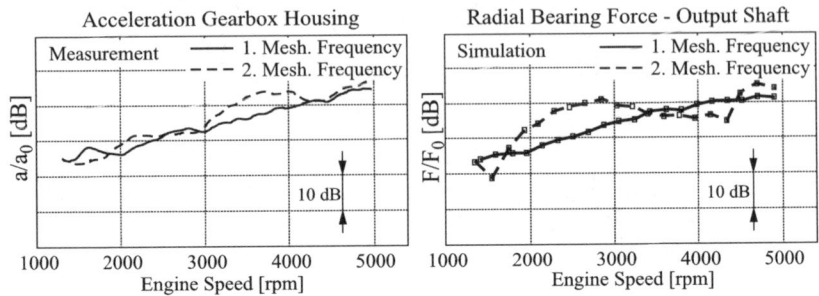

Fig. 5.38: Comparison Surface Acceleration – Bearing Force

The effects of the parameter excitation can also be observed outside the gearbox housing since the movements of the single components are transmitted by the bearings. Figure 5.38 shows a comparison between the accelerations measured on the surface of the gearbox housing and the bearing force of the output shaft calculated using **DynAs**. The amplitudes are pictured for the first and the second meshing frequency. The basic tendencies can be reproduced by the simulation program, especially for the dominating first meshing frequency.

## 5.3 Tractor Drive Train System

### 5.3.1 Introduction

Power tansmissions of tractors represent the most complicated drive train units of nearly all transportation systems. They must meet a large variety of very hard requirements, and in the last one, two decades the demands for more versatility, flexibility and comfort for the driver increase continuously. Tractors are supposed to do all kind of agricultural work, but also to fulfill many requirements of forestry, and additionally tractors should be able to move along normal roads with reasonable speed and safety. Figure 5.39 pictures a typical modern tractor concept with a very sophisticated drive train system.

Fig. 5.39: Typical Modern Tractor Concept (courtesy AGCO/Fendt)

This altogether leads to very complex drive trains, for which modern CVT-concepts offer significant advantages. They may operate with some given speed independently from the engine speed thus running the motor at the point of best fuel efficiency. It offers advantages for all implements by generating an optimal power distribution between driving and working.

The German company Agco/Fendt as one of the leading enterprises of modern tractor technologies developed for that purpose a very efficient power transmission including for low speeds a hydrostatic drive and for larger speeds a gear system with the possibility of mixing the power transmission according to external requirements. The hydrostat system possesses the great advantage to develop very large torques especially at low speeds and at standstill.

Figure 5.3.1 ilustrates the kernel of the power transmission, and Figure 5.41 presents a sketch of the overall system. Following this sketch we recognize that the torque of the Diesel engine is transmitted via a torsional vibration damper (1) to the planet carrier (5) of the planetary gear set (2). The planet gear distributes the power to the sunwheel (4) and the ring gear (3). This ring gear drives via a cylindrical gear pair the hydraulic pump (6), which itself powers the two hydraulic motors (7), where the oil flow from pump to motor depends on the pump displacement angle $\alpha$. The hydraulic motors (7) generate a torque according to the oil flow and the motor displacement angle $\beta$. Furtheron, the sun wheel transmits its own torque via a gear pair to a collecting shaft (8), which adds the torques coming directly from the Diesel engine via the sun wheel on the one and coming indirectly from the chain ring gear-hydraulic pump-hydraulic motors on the other side. Thus the collecting shaft (8) combines the mechanical and the hydraulic parts of the

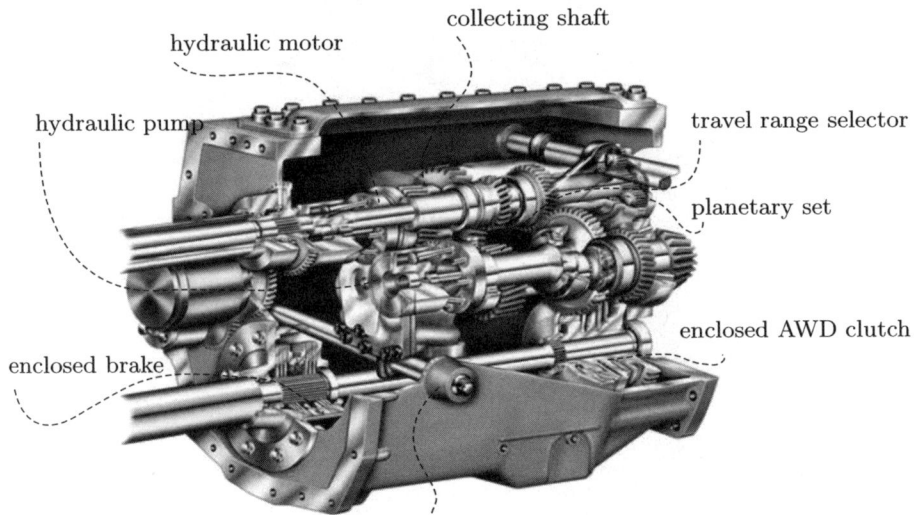

Fig. 5.40: ML Power Transmission of the VARIO series (courtesy AGCO/Fendt)

torque. This splitting of mechanical and hydraulic power in an optimal way is the basic principle of the stepless VARIO system. As an additional option we may also connect the planet carrier (5) with the rear power take off (PTO) system, which drives all possible rear implements for performing processing of agriculture and forestry. The travel range selector (9) allows switching between slow and fast operation. The outgoing torque will be transmitted to the rear axle by a pinion and to the front axle by a gear pair.

The system allows an adaptation of the tractor speed only by an appropriate combination of the displacement angles $\alpha$ and $\beta$ without changing the engine's speed, which means, that the engine's speed may be kept constant at an optimal fuel efficiency point in spite of varying tractor speeds, of course within a limited speed range. This stepless VARIO-concept represents a type of CVT-system (**C**ontinuous **V**ariable **T**ransmission), which results in significant improvements with respect to handling and working performance.

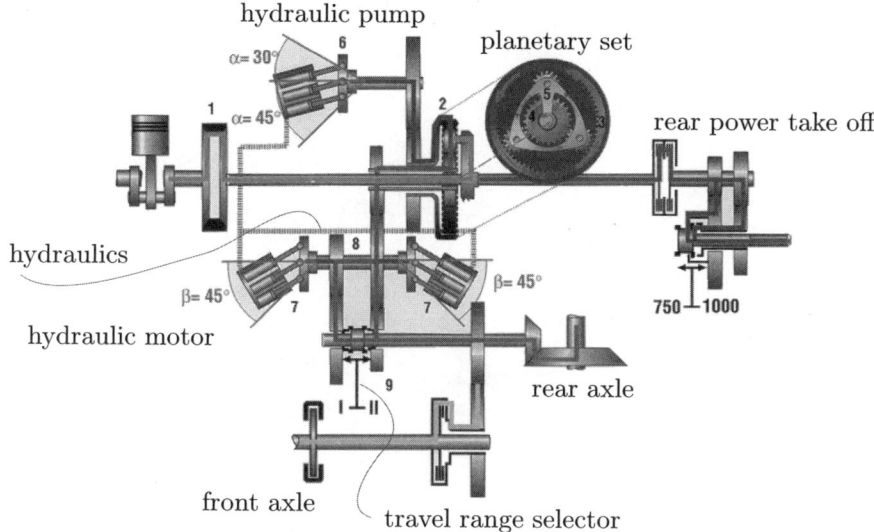

Fig. 5.41: Sketch of VARIO-ML System (courtesy AGCO/Fendt)

### 5.3.2 Modeling

For the investigation of dynamic loads of certain critical components and additionally for the assessment of the vibrational behaviour an appropriate model has to be established, which includes as most important elements the mechanical parts, the Cardan shafts and the hydrostat components, and also the hydraulics. Many mechanical parts like shafts, elastic connection and the like can be assumed to behave linearly, but the Cardan shafts, the hydraulic pumps and motors as well as the hydraulics itself will generate non-linearities. Therefore we shall focus on those components. We start with a consideration of the complete system representing afterwards the equations of the components, as far as necessary [24].

Figure 5.3.2 depicts the model of the **complete power transmission** of the tractor under consideration. Partly and as far as possible we model the mesh of two gear wheels by two rigid bodies with a fixed transmission ratio resulting in a reduction to one degree of freedom (DOF). The crankshaft of the Diesel engine is described by a rigid body (2) with one DOF, which is loaded by the torque from the combustion pressure. They must be determined from measurements of the combustion process and correctly projected to the crankshaft model. The front PTO (power take off, 20, 21) is connected to the engine by a vibration damper. Also the planetary set (7) and all auxiliary equipment (5) are driven more or less directly by the engine. The model follows the diagrammatic sketch of Figure 5.41 with some extensions. They concern mainly the two power take off systems, front PTO and rear PTO, and also the front and rear axles with the tires. We shall come back to all components.

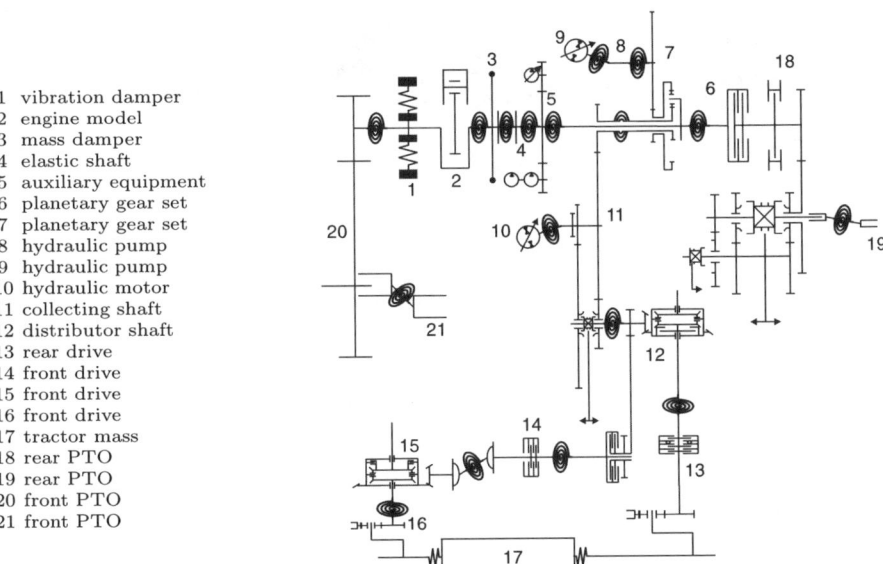

1 vibration damper
2 engine model
3 mass damper
4 elastic shaft
5 auxiliary equipment
6 planetary gear set
7 planetary gear set
8 hydraulic pump
9 hydraulic pump
10 hydraulic motor
11 collecting shaft
12 distributor shaft
13 rear drive
14 front drive
15 front drive
16 front drive
17 tractor mass
18 rear PTO
19 rear PTO
20 front PTO
21 front PTO

Fig. 5.42: Mechanical Model of the VARIO Power Transmission

Figure 5.3.2 pictures a very classical mechanic-hydraulic system, where the physical relations are obvious. Therefore it represents a good example how to establish a mechanical (or physical) model, which is equivalent to the real world problem. As already mentioned, engineering mechanics nor engineering physics are not deductive sciences thus requiring as a rule bundles of assumptions and neglections without destroying the principal information base of a system. From this we shall try also in this case to come to simple equations.

## 5.3 Tractor Drive Train System

Dealing with **components** we start with rigid and elastic shafts. For the problem of loads and vibrations we assume for the rigid shaft a single rotating mass with one degree od freedom, and for an elastic shaft simply two rotating masses connected by a spring with two degrees of freedom. We get

- for the rigid shaft

$$J\ddot{\varphi} = \sum_i M_i \quad \text{with} \quad M_i = M_i(t, \varphi_k, \dot{\varphi}_k). \tag{5.82}$$

$J$ is the mass moment of inertia, $\varphi$ the absolute rotational angle and $M_i$ an external torque depending on time, the angles and the angular velocities of some neighboring bodies or the environment.

- for the elastic shaft

$$\begin{pmatrix} J_1 & J_{12} \\ J_{12} & J_2 \end{pmatrix} \begin{pmatrix} \ddot{\varphi}_1 \\ \ddot{\varphi}_2 \end{pmatrix} + \begin{pmatrix} d_T & -d_T \\ -d_T & d_T \end{pmatrix} \begin{pmatrix} \dot{\varphi}_1 \\ \dot{\varphi}_2 \end{pmatrix} + \\ + \begin{pmatrix} c_T & -c_T \\ -c_T & c_T \end{pmatrix} \begin{pmatrix} \varphi_1 \\ \varphi_2 \end{pmatrix} = \begin{pmatrix} \sum_k M_{1,k} \\ \sum_k M_{2,k} \end{pmatrix}. \tag{5.83}$$

The quantities in this equation follow from

$$c_T = \left(\frac{1}{c_1} + \frac{1}{c_2}\right)^{-1}, \quad \text{with} \quad c_i = \frac{GJ_p}{l_i}, \quad l_2 = l - l_1,$$

$$\begin{pmatrix} J_1 & J_{12} \\ J_{12} & J_2 \end{pmatrix} = \frac{1}{3}\rho J_p l \begin{pmatrix} 1 & \frac{1}{2} \\ \frac{1}{2} & 1 \end{pmatrix},$$

$$d_T = 2D\sqrt{c_T J}, \tag{5.84}$$

where c and d are spring and damper coefficients, respectively, J are mass moments of inertia, D is the Lehr attenuation constant (practically D ≈ 0.02-0.05 for our case), G the modulus of shear and $\rho$ the material density. The torqes $M_{1,k}, M_{2,k}$ come again from couplings to the neighboring bodies or environment.

All interconnections with elasticity must be examined with respect to the magnitude of these elasticities. In our case the stiffness of all tooth meshings is so large, that the coressponding frequencies exceed the frequency range of interest by far. Therefore we are able to model such gear meshes as a purely kinematical connection. Figure 5.43 illustrates such a connection. The two parts are under the load of the contraint force $F_{12}$ and the torques $M_{1,k}, M_{2,k}$. If we have a kinematical connection with a ratio ($i_{1,2} = \frac{\varphi_1}{\varphi_2} = \frac{r_2}{r_1}$), for example, we are able to define a Jacobian by

$$\dot{q} = \begin{pmatrix} \dot{\varphi}_1 \\ \dot{\varphi}_2 \end{pmatrix} = \begin{pmatrix} 1 \\ \frac{1}{i_{1,2}} \end{pmatrix} \dot{\varphi}_1 = Q \dot{\varphi}_1 \tag{5.85}$$

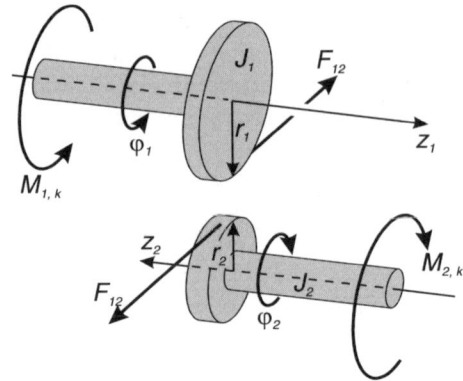

Fig. 5.43: Kinematic Connection, for example Gear Mesh

The equations of motion of the configuration of Figure 5.43 write

$$\left. \begin{array}{l} J_1 \ddot{\varphi}_1 = M_{1,k} + F_{12} r_1 \\ J_2 \ddot{\varphi}_2 = M_{2,k} - F_{12} r_2 \end{array} \right\} \; \hat{=} \; \boldsymbol{M}\ddot{\boldsymbol{q}} = \boldsymbol{h}, \qquad (5.86)$$

which can be transformed with the help of the Jacobian (eq. (5.85)) and with the reduced mass moment of inertia $J = J_1 + \frac{J_2}{i_{1,2}^2}$ to yield

$$\boldsymbol{Q}^T \boldsymbol{M} \boldsymbol{Q}\, \ddot{\boldsymbol{q}} = \boldsymbol{Q}^T \boldsymbol{h} \;\; \Longrightarrow \;\; J\ddot{\varphi}_1 = M_{1,k} + \frac{M_{2,k}}{i_{1,2}}, \qquad (5.87)$$

The constraint force $F_{12}$ eliminates by the multiplication with the Jacobian $\boldsymbol{Q}$.

The tractor power transmission system includes several **Cardan shafts**, especially for all PTO systems. Cardan shafts are sources of parameter excited vibrations with their sub- and super-harmonic resonances, which can become dangerous. Therefore good models are obligatory. An excellent survey is given in [240]. We choose a model with four bodies interconnected by springs and dampers (see Figure 5.44). The four equations of motion write

$$\left. \begin{array}{l} J_1 \ddot{\varphi}_1 = +M_{G12,1} + M_1 \\ J_2 \ddot{\varphi}_2 = -M_{G12,2} + c(\varphi_3 - \varphi_2) \\ J_3 \ddot{\varphi}_3 = +M_{G34,3} + c(\varphi_2 - \varphi_3) \\ J_4 \ddot{\varphi}_4 = -M_{G34,4} + M_4 \end{array} \right\} \; \hat{=} \; \boldsymbol{M}\ddot{\boldsymbol{q}} = \boldsymbol{h} \qquad (5.88)$$

with $M_{G12,2} = M_{G12,1} \frac{\sin^2 \varphi_1 + \cos^2 \varphi_1 \cos^2 \alpha_{12}}{\cos \alpha_{12}}$

and $M_{G34,3} = M_{G34,4} \frac{\sin^2 \varphi_4 + \cos^2 \varphi_4 \cos^2 \alpha_{34}}{\cos \alpha_{34}}$

To not loose a good overview we have left out the damping terms. The complete force element of the shaft between "1" and "2" comprises a linear damper d and linear spring c. Between the shaft parts (1 - 2) and (3 - 4) we have a kinematical relation of the well-known form ([240], [49])

$$\varphi_{i+1} = \varphi_{i+1}(\varphi_i) = \arctan(\frac{\tan \varphi_i}{\cos \alpha}), \tag{5.89}$$

where $\alpha$ is the Cardan angle ($\alpha = \varphi_{12} = \varphi_{34}$). The above relation reduces the four degrees of freedom to two, for example $\mathbf{q} = (\varphi_1, \varphi_2, \varphi_3, \varphi_4) \Longrightarrow (\varphi_1, \varphi_4) = \mathbf{q}_{red}$. The corresponding Jacobian follows from equation (5.89):

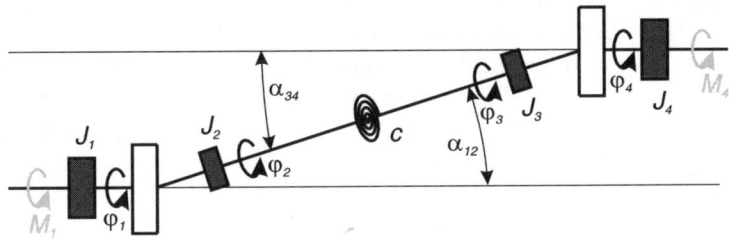

Fig. 5.44: Model of a Cardan Shaft

$$\mathbf{Q}^T = \begin{pmatrix} 1 & \frac{\cos \alpha_{12}}{\sin^2 \varphi_1 + \cos^2 \varphi_1 \cos^2 \alpha_{12}} & 0 & 0 \\ 0 & 0 & \frac{\cos \alpha_{34}}{\sin^2 \varphi_4 + \cos^2 \varphi_4 \cos^2 \alpha_{34}} & 1 \end{pmatrix}$$

$$\dot{\mathbf{q}} = \begin{pmatrix} \dot{\varphi}_1 \\ \dot{\varphi}_2 \\ \dot{\varphi}_3 \\ \dot{\varphi}_4 \end{pmatrix} = \mathbf{Q}\dot{\mathbf{q}}_{red} = \mathbf{Q} \begin{pmatrix} \dot{\varphi}_1 \\ \dot{\varphi}_4 \end{pmatrix} \tag{5.90}$$

With these relations we receive the final equations of motion of the Cardan shaft

$$\mathbf{M}_{red}\ddot{\mathbf{q}}_{red} + \mathbf{b}_{red} = \mathbf{h}_{red}$$

$$\mathbf{M}_{red} = \begin{pmatrix} J_1 + J_2(\frac{\cos \alpha_{12}}{\sin^2 \varphi_1 + \cos^2 \varphi_1 \cos^2 \alpha_{12}})^2 & 0 \\ 0 & J_3(\frac{\cos \alpha_{34}}{\sin^2 \varphi_4 + \cos^2 \varphi_4 \cos^2 \alpha_{34}})^2 + J_4 \end{pmatrix}$$

$$\mathbf{b}_{red} = \begin{pmatrix} J_2 \frac{\cos^2 \alpha_{12} \sin 2\varphi_1 (\cos^2 \alpha_{12} - 1)}{(\sin^2 \varphi_1 + \cos^2 \varphi_1 \cos^2 \alpha_{12})^3} \dot{\varphi}_1^2 \\ J_3 \frac{\cos^2 \alpha_{34} \sin 2\varphi_4 (\cos^2 \alpha_{34} - 1)}{(\sin^2 \varphi_4 + \cos^2 \varphi_4 \cos^2 \alpha_{34})^3} \dot{\varphi}_4^2 \end{pmatrix}$$

$$\mathbf{h}_{red} = \begin{pmatrix} \frac{\cos \alpha_{12}}{\sin^2 \varphi_1 + \cos^2 \varphi_1 \cos^2 \alpha_{12}} \, c \, (\arctan(\frac{\tan \varphi_4}{\cos \alpha_{34}}) - \arctan(\frac{\tan \varphi_1}{\cos \alpha_{12}})) + M_1 \\ \frac{\cos \alpha_{34}}{\sin^2 \varphi_4 + \cos^2 \varphi_4 \cos^2 \alpha_{34}} \, c \, (\arctan(\frac{\tan \varphi_1}{\cos \alpha_{12}}) - \arctan(\frac{\tan \varphi_4}{\cos \alpha_{34}})) + M_4 \end{pmatrix}$$
$$\tag{5.91}$$

The dependency on the rotational angles enters these equations by the Jacobian $\mathbf{Q}$ (equation (5.90)). All quantities of the equations of motion change with

## 5 Power Transmission

the rotational angles and speeds. Furtheron they depend strongly on the kink angles $\alpha_{12}$ and $\alpha_{34}$. For these angles being zero, the dependency disappears, and with increasing angles their influence increases nonlinearly.

The homokinematical configuration [49] gives a balancing effect for $\alpha_{12} = \alpha_{34}$. The parameter influences disappear for equal rotational angles $\varphi_2 = \varphi_3$, which is only possible for a completely rigid shaft ($c \Rightarrow \infty$). But Cardan shafts are usually very elastic leading to oscillations between the two shaft masses and thus to parameter excitation. The largest influence comes from the nonuniformity of the stiffness terms in $\mathbf{h}_{red}$ of equation (5.91). Nevertheless we can expect small angular displacements, which allows us to develop a linearized shaft stiffness. From equation (5.91, last term) and a linearization with respect to the difference of the angles $\varphi_1$ and $\varphi_4$ and finally with the assumption $\alpha_{12} = \alpha_{34} = \alpha$ we come out with:

$$c_{lin} = \left( \frac{\cos \alpha}{\sin^2 \varphi_1 + \cos^2 \varphi_1 \cos^2 \alpha} \right)^2 c, \qquad M_c = c_{lin}(\varphi_4 - \varphi_1). \tag{5.92}$$

Figure 5.3.2 confirms the model and allows to use for the simulation of the PTO Cardan shafts the linearized approach. The graph on the right side depicts the dependency of the Cardan shaft stiffness on the angular orientation $\varphi$ and on the kink angle $\alpha$. For an angle $\alpha = 10°$ the difference of minmal and maximum stiffness is 6%, for $\alpha = 20°$ about 10% and for $\alpha = 30°$ already 30%. The parameter excitation influences as a matter of fact the whole power transmission system of the tractor.

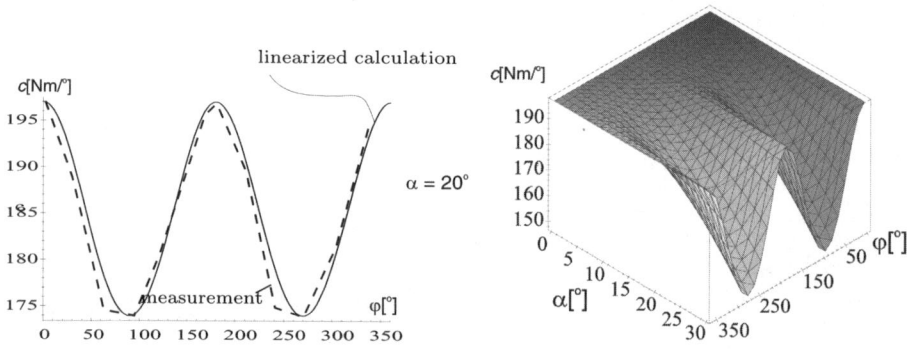

Fig. 5.45: Linearized Stiffness of the Cardan Shaft

The heart of the VARIO power drive is the **hydrostatic system** with its hydraulic pumps and motors operating on the basis of a piston type machine. They are heavily loaded, especially for processes like ploughing and mulching. Therefore it makes sense to know these loads already during the design phase to find the correct lay-out. Figure 5.46 shows a drawing of the pump/motor configuration as used in the VARIO system. The piston drum is rotationally displaced with respect to the shaft axis by an angle $\alpha$, the magnitude of which determines the oil fluid flow. The translational motion of the nine pistons within the corresponding cylinder liners increase with increasing pivot angle.

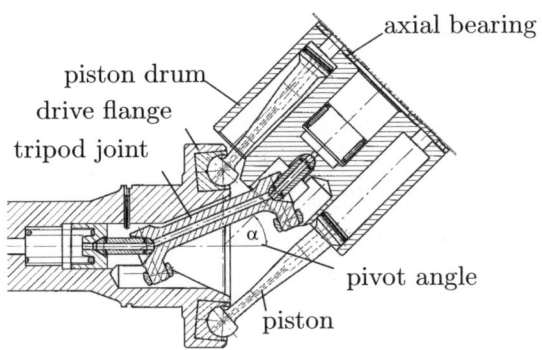

Fig. 5.46: Pump/Motor Configuration of the Hydrostat

Drive flange, pistons and piston drum can be modeled as rigid bodies. The tripod joint is an elastic part, it transmits the rotational motion of the drive flange to the piston drum. It is coupled to the drive flange and to the piston drum by a kind of three "feet", which allow relative motion in axial and radial but not in circumferential tangential direction. These three feet consist of three pins perpendicular to the tripod axis and three rings with a spherical outer surface moving in corresponding bushings in the flange or the drum, which allow axial and radial motion, but which can transmit the full torsional torques. Therefore the tripod joint is a really critical part, which has to be designed properly.

Figure 5.47 depicts a tripod model with the relative kinematics of the tripod axis, the position and orientation of which we need to know. As a first step we state, that the contact points of the six tripod feet in the drum and in the flange can be described twice, using the body coordinates of the flange and the drum on the one and using the body coordinates of the tripod itself on the other side. Putting $\mathbf{r}_{drum,flange} = \mathbf{r}_{tripod}$ results altogether in (6x3 = 18) algebraic equations, because everyone of the six $\mathbf{r}$ possesses of course three components.

These 18 equations correspond to the unknown displacements $\xi_{ij}$ and $\zeta_{ij}$ of the contact points to the tripod axis and to the front sides of drum and flange, respectively (i=1 for drum, i=2 for driving flange, j=1,2,3 for the three contact points of the three feet on each side of the tripod). They correspond furtheron to the three unknown positions (x, y, z) of the tripod joint, where the axial displacement x can be neglected, it corresponds to its tilt angles $\gamma$ and $\beta$, its rotational orientation $\varphi_T$ and finally to the rotational orientation

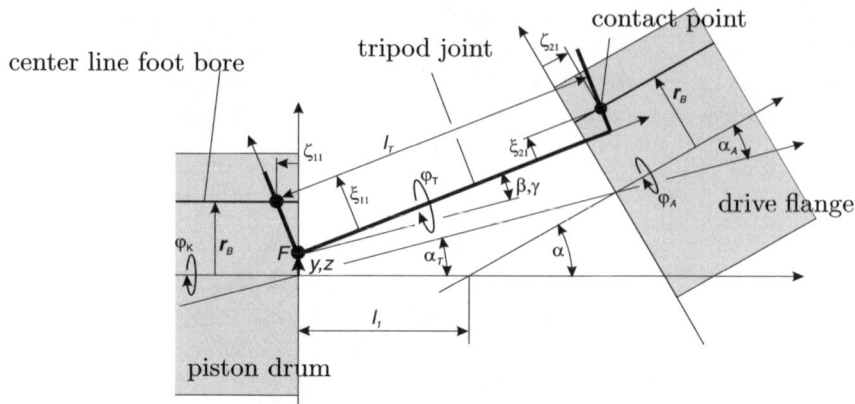

Fig. 5.47: Pump/Motor Model

$\phi_K$ of the drum. These are 18 unknowns covered by 18 algebraic equations of the form

$$y \sin\alpha_T + l_T(i-1) - \xi_{ij}\gamma \cos[\varphi_T + \frac{2\pi}{3}(j-1)] + \xi_{ij}\beta \sin[\varphi_T + \frac{2\pi}{3}(j-1)] =$$
$$= l_1(i-1)\cos\alpha_T + \zeta_{ij}\cos(\alpha_{i-1} - \alpha_T) - r_B \sin(\alpha_{i-1} - \alpha_T)\cos[\varphi_i +$$
$$+ \frac{2\pi}{3}(j-1)]$$

$$y \cos\alpha_T + l_T(i-1) - \xi_{ij}\gamma \cos[\varphi_T + \frac{2\pi}{3}(j-1)] =$$
$$= -l_1(i-1)\sin\alpha_T + \zeta_{ij}\sin(\alpha_{i-1} - \alpha_T) - r_B \cos(\alpha_{i-1} - \alpha_T)\cos[\varphi_i +$$
$$+ \frac{2\pi}{3}(j-1)]$$

$$z - \beta l_T(i-1) - \xi_{ij}\gamma \sin[\varphi_T + \frac{2\pi}{3}(j-1)] = r_B \cos[\varphi_i + \frac{2\pi}{3}(j-1)] \qquad (5.93)$$

Again we have i=1 for drum, i=2 for driving flange, j=1,2,3 for the three contact points of the three feet on each side of the tripod.

The unknown quantities $\xi_{ij}$ and $\zeta_{ij}$ are contained in the equations (5.93) in a linear form. Therefore we are able to reduce the set to one with six degrees of freedom only, namely to the six quantities y, z, $\beta$, $\gamma$, $\varphi_T$ and $\varphi_K = \varphi_1$. The evaluation is performed numerically. The considerations and relations are equally valid for the hydromotors and the hydropump, where in both cases the kinematics of the tripod joint is of special interest.

The tripod kinematics depends mainly on the displacement angle $\alpha$ of the piston drum and on the rotational orientation $\varphi_T$ of the driving flange. The three tripod feet bring in a periodicity of 120° for all tripod results. The translational quantities (y, z) and the tilt angles ($\beta, \gamma$) are 90° dephased. The Figures 5.48 and 5.49 illustrate the kinematical properties. The influence of these characteristics on the dynamics of the tripod itself is not very large, which corresponds to experience. Therefore a detailed analysis will be only necessary for special evaluations required for the tripod joint.

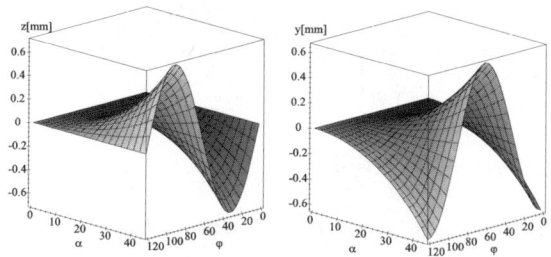

Fig. 5.48: Translational Displacements of a Tripod Foot

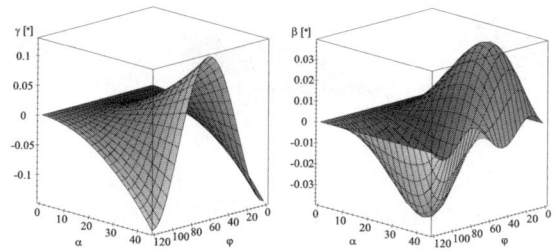

Fig. 5.49: Rotational Displacements of a Tripod Foot

The **hydrostat piston kinematics** produces oil flow and oil pressure. The pistons perform a stroke motion within the drum cylinders, while rotating around the tripod joint. This motion depends on the displacement angle $\alpha$, which at the same time represents a control quantity for flow and pressure. According to Figure 5.50 we get the following vector chain

$$\boldsymbol{r}_{TF} = \boldsymbol{h} + \boldsymbol{r}_K + \boldsymbol{l}_K, \tag{5.94}$$

which can be decomposed to the three equations for the unknowns (h, $\beta, \gamma$)

$$\gamma = \arcsin\left(\frac{r_{TF}}{l_K}(\sin\varphi_{TF}\cos\varphi_K - \cos\varphi_{TF}\sin\varphi_K\cos\alpha)\right)$$

$$\beta = \arcsin\left(\frac{r_K - r_{TF}(\sin\varphi_{TF}\sin\varphi_K + \cos\varphi_{TF}\cos\varphi_K\cos\alpha)}{l_K\cos\gamma}\right)$$

$$h = r_{TF}\cos\varphi_{TF}\sin\alpha + l_K\cos\beta\cos\gamma \tag{5.95}$$

The results of these equations are illustrated in Figure 5.3.2. To get an idea of

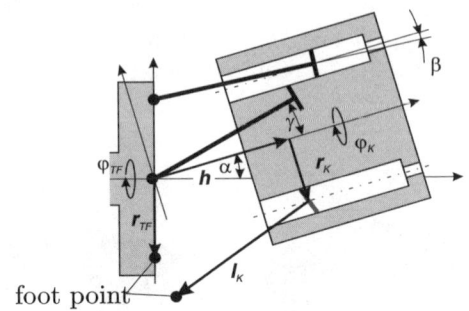

Fig. 5.50: Piston Kinematics

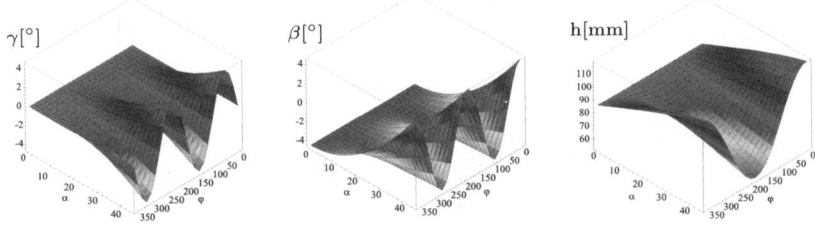

Fig. 5.51: Results of Piston Kinematics

the forces resulting from the above kinematical consideration we investigate a simplified case with no rotational speed difference, $\varphi_T - \varphi_K = 0$, and a piston pressure according to Figure 5.52. The pistons possess at one end spherical bearings, which allow a force tranfer only in the direction of the piston axis to the driving flange. This direction for one piston is defined by

$$_D\mathbf{e}_{K,k} = \begin{pmatrix} -\cos\beta\cos\gamma \\ -\sin\beta\cos\gamma \\ \sin\gamma \end{pmatrix}_k \implies {}_D\mathbf{F}_{K,k} = F_{p,k} \begin{pmatrix} -\cos\beta\cos\gamma \\ -\sin\beta\cos\gamma \\ \sin\gamma \end{pmatrix}_k,$$

$$(k = 1, 2, \cdots 9) \tag{5.96}$$

The reference coordinates are piston drum fixed. The forces can be decomposed into three components and then summarized for all nine pistons, regarding their individual position and orientation. Before showing the results we transform the forces from the piston drum "D" to the driving flange "F" by the transformation (Figure 5.52)

$$_F \boldsymbol{F}_{K,k} = F_{p,k} \begin{pmatrix} -\cos\alpha\cos\beta\cos\gamma \\ -\sin\alpha\cos\beta\cos\gamma\cos\varphi \\ +\sin\alpha\cos\beta\cos\gamma\sin\varphi \end{pmatrix}_k, \qquad (k = \quad 1,2,\cdots 9) \qquad (5.97)$$

A numerical evaluation indicates, that axial forces of the pistons are large on

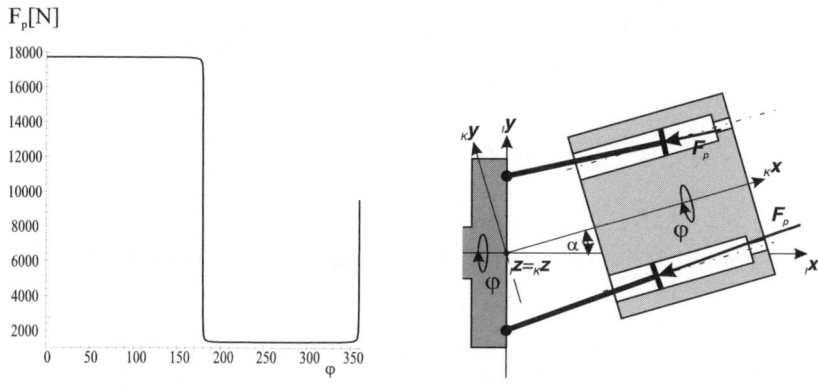

Fig. 5.52: Load Example

both, drum and flange; whereas tangential and radial forces are only large on the driving flange, see Figures 5.3.2 and 5.3.2. To establish the equations of

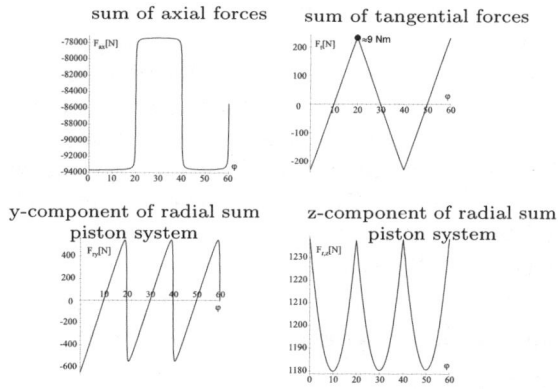

Fig. 5.53: Summarized Piston Forces on the Drum

motion of **the complete system** we have to combine the above models and to supplement them by the appropriate models for the oil hydraulics and the axial drum bearing. Such models follow the methods as discussed in chapter 4 on 187 and in section 2.6 on page 76. It will not be presented here. In

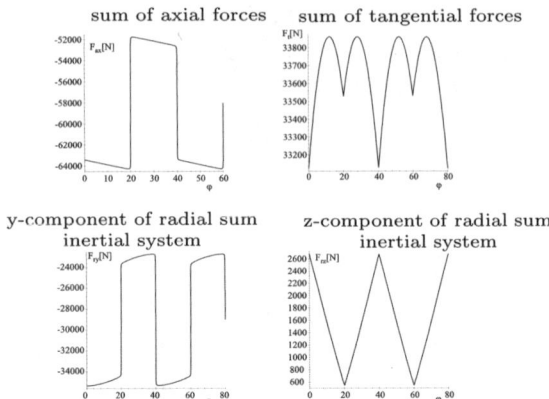

Fig. 5.54: Summarized Piston Forces on the Driving Flange

spite of the fact that the system is complicated we can avoid both, bilateral and unilateral constraints, expressing instead all interconnections by linear or nonlinear force laws. In a first step this comes out with

$$\mathbf{M\ddot{q}} - \mathbf{h}(\mathbf{q}, \mathbf{\dot{q}}, t) = \mathbf{0}, \tag{5.98}$$

which can be approximated by linear and nonlinear parts yielding

$$\mathbf{M\ddot{q}} + \mathbf{D\dot{q}} + \mathbf{Kq} = \mathbf{h}(\mathbf{q}, \mathbf{\dot{q}}, t), \qquad \mathbf{\dot{q}}_{hydraulic} = \mathbf{h}_{hydraulic}. \tag{5.99}$$

Nonlinear terms come mainly from the hydrostat system and from the hydraulics. Many elastic parts are linear. The above separation makes sense, because for stationary operation the nonlinear terms do not have much influence, and the linear part of the equations of motion can be used simply for evaluating the eigen-behaviour, which gives important informations of what components oscillate with respect to other ones and where potential resonances could be expected. This presupposes a linearization around the operation point under consideration. The eigenvalues and the eigenforms result from $\mathbf{M\ddot{q}} + \mathbf{Kq} = \mathbf{0}$. We shall give examples.

### 5.3.3 Numerical and Experimental Results

Experiments were performed by the company AGCO/Fendt at all PTO systems. Figure 5.3.3 depicts a comparison of simulation and measurements for the rear PTO [24]. These measurements were performed with a very stiff Cardan shaft and Cardan angles of $\alpha \leq 14°$. The process considered is mulching. The stiff Cardan shaft leads to large shaft eigenfrequencies of about 40 Hz, which corresponds to an engine speed of about 2400 rpm. This value exceeds the operational speed range and is therefore not dangerous. Simulations and experiments agree very well.

## 5.3 Tractor Drive Train System

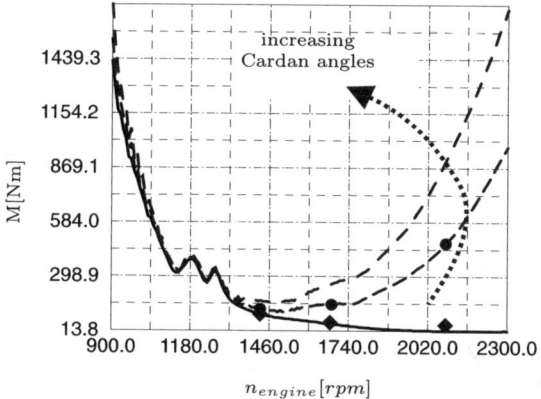

Fig. 5.55: Torque Amplitudes at the Rear PTO, Comparison Simulation/Measurements (●,♦), (courtesy AGCO/Fendt)

The equations of motion (5.99) were linearized around a stationary operation point, which gives for our case a reasonable approximation of the real dynamic behaviour including nonlinearities. With $\mathbf{M\ddot{q} + Kq = 0}$ we are able to evaluate eigenforms and eigenfrequencies, which has been done for 20 eigenfrequencies up to $\approx 2000$ Hz. We shall consider only a few of them.

The matrix $\mathbf{K}$ contains all linearized stiffnesses of the system, which are of particular interest for the hydrostat, because its stiffness depends on the inclination angles $\alpha$ and $\beta$ (Figure 5.50) and thus on the speed ratio $i_s = n_{collectingshaft}/n_{engine} = 1/i$. As a consequence all eigenforms and eigenfrequencies resulting from an analysis of $\mathbf{M\ddot{q} + Kq = 0}$ depend also on the ratio $i_s$. Figure 5.3.3 depicts the depencies of some typical eigenfrequencies on the speed ratio. With the exception of hydropump and hydromotor themselves the influence on all other components is very small.

For system design one need to know the topological vibration behaviour, that means what components vibrate with respect to other ones. We shall focus our consideration on front and rear PTO loads. They possess significant influence on the second, the third, the 17th and the 18th eigenforms and -frequencies. While these loads influence both, the second and the third eigenmodes, they have an effect on the 17th and 18th eigenmode only locally, front load on front PTO system and rear load on rear PTO system. Therefore we show only one case for the second eigenmode (19.7 Hz without load, ≤10 Hz with load) in Figure 5.57, the numbering of which follows Figure 5.3.2 on page 260. The masses of Figure 5.3.2 are represented by small rhombi, and the difference between solid and dotted lines indicate the vibration amplitudes. The loads reduce of course the eigenfrequencies, and the complete transmission system vibrates down to the collecting shaft (11) (Figure 5.3.2).

With respect to the higher eigenmodes with about 1600 Hz without load we have similar tendencies, drastic reduction of frequencies to values of 30-

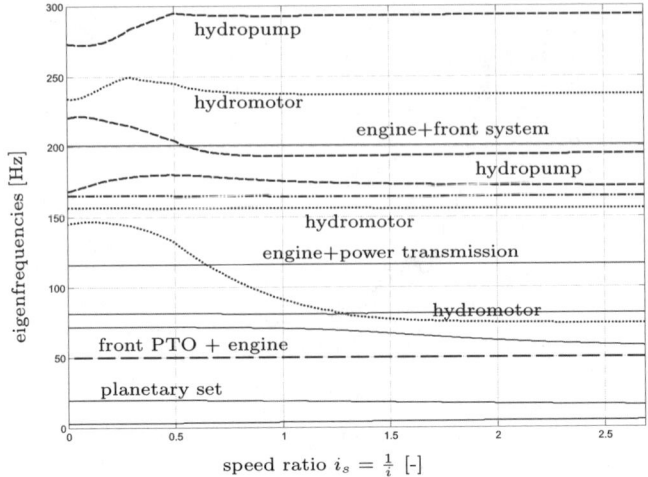

Fig. 5.56: Influence of the Speed Ratio on some Eigenfrequencies

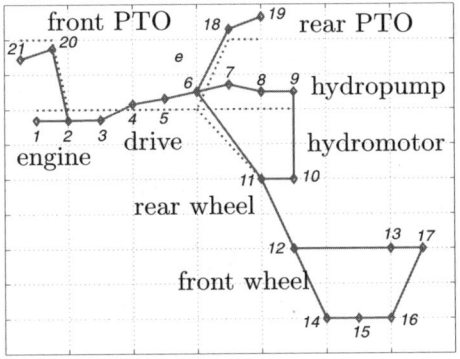

Fig. 5.57: Second Eigenform (both PTO moment of inertia $J \geq 2\text{kgm}^2$)

50 Hz and local vibrations at front and rear PTO's. Figure 5.58 illustrates the situation of the eigenforms. Loads at the front or rear PTO's excite only vibrations of the front or rear regime, which do not influence each other.

At this point a general remark for practical applications: The Figures 5.57 and 5.58 represent eigenvectors of a linearized system, which describe parts of the approximated eigenbehaviour of the tractor drive. These curves are characterized by the fact, that the eigenforms are not those of a continuum but of a discrete system with discrete masses. It is sometimes very convenient to use such eigenvectors, also in the case of a discrete mass system, as shape functions for a Ritz-approach, because we get then a very simple first-step approach to very complicated system structures. Usually eigenbehaviour analysis is performed only for continuum systems.

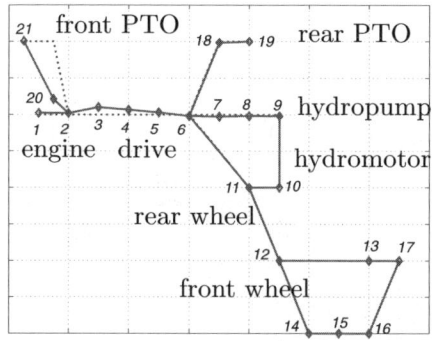 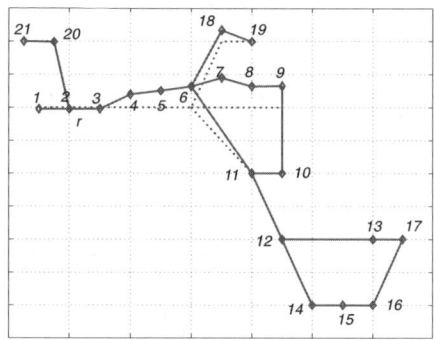

Fig. 5.58: PTO Eigenforms at 30-50 Hz (both PTO moment of inertia $J \geq 2\text{kgm}^2$), left figure for front load, right figure for rear load

As a last result we present some Campbell diagrams illustrating the excitation structure in certain areas of the system. We consider again the case of a rear load at a Cardan-angle of 22° and 25°. The engine is a four-stroke configuration. Figure 5.3.3 illustrates the torque at the rear PTO shaft. We recognize the second order influence of the rear PTO and the second and fourth engine order. Higher engine orders do not posses an influence. Due to the relatively small stiffness of the Cardan shaft for the case considered the second PTO shaft order is dominant. The corresponding load on the tripod is shown by Figure 5.3.3. Also here and to a larger extent the second PTO system order dominates the vibrations. From these results we conclude that the internal excitation by the PTO systems at the front and rear side of a tractor lead to significant vibrations. This is confirmed also by additional simulations and by practical experience.

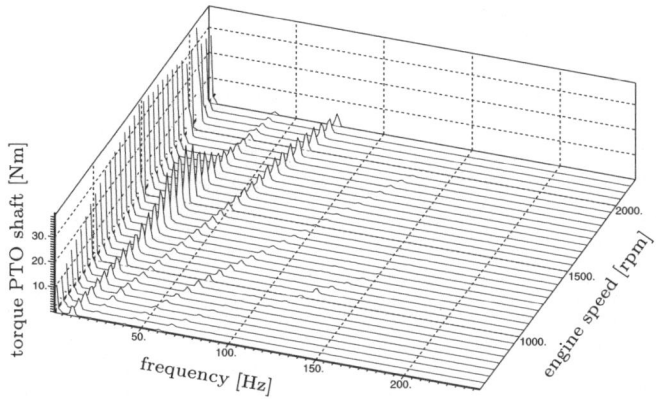

Fig. 5.59: Campbell Diagram of the Rear PTO Torque (with rear implement and a PTO shaft with 22° and 25°)

274    5 Power Transmission

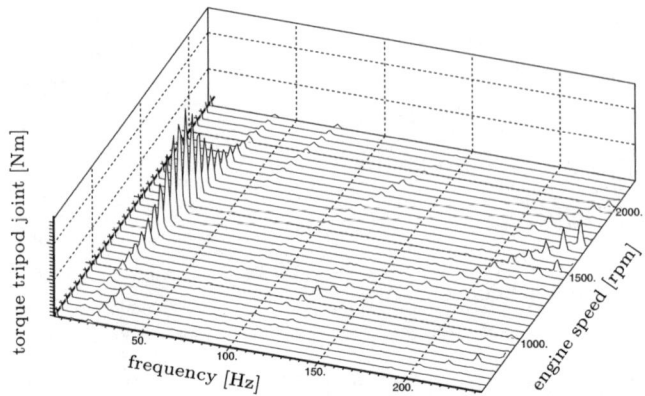

Fig. 5.60: Campbell Diagram of the Tripod Joint Torque of the Hydraulic Motor (with rear implement and a PTO shaft with 22° and 25°)

## 5.4 CVT Gear Systems - Generalities

### 5.4.1 Introduction

Power transmission in automotive systems is classically carried out by gear trains, which transmit power by form-closure. In recent times an increasing number of continuous variable transmissions (CVT) are applied. They transmit power by friction, and at the time being they compete more and more with automated gears or with hand-shifted gears. Experts state, that there might be a share of CVT-solutions of 15% to 20% with respect to the power transmission market considered in the longer term. The advantage of gear trains with gear wheels consists in a better component efficiency due to power transmission by form closure, the disadvantage in an only stepwise approximation of the drag-velocity hyperbola. This disadvantage is significantly reduced step by step by introducing automatic gear boxes with up to eight gear stages. The advantage of a CVT configuration consists in a perfect adaptation to the drag-velocity hyperbola, the disadvantage in a lower efficiency due to power transmission by friction and in a somehow limited torque transmission. An additional advantage of the CVT's is the possibility of very smoothly changing the transmission ratio without any danger of generating jerk. Figure 3.9 on page 113 depicts an example from industry.

The following pictures give an impression of the systems and the components. They are of general importance though the configurations shown correspond to the LUK/PIV chain system. The operation of a chain- or belt-driven CVT is for the various configurations always the same. Figure 5.61 depicts the main features. The chain or the belt moves between two pulleys with conically shaped sheaves. One side of these pulleys possesses a movable sheave controlled by a hydraulic system. The other side id fixed. Reducing or increasing the distance between the pulley sheaves forces the chain or belt to

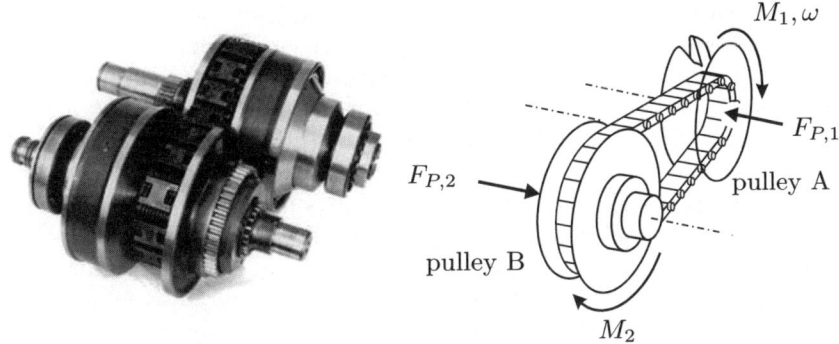

Fig. 5.61: CVT-Drive, Example LUK/PIV System

move radially upwards or downwards thus changing the transmission ratio of the CVT. We have a driving pulley A (Figure 5.61) with an incoming torqe $M_1$, with a pressure force $F_{P,1}$ and a resulting rotational speed $\omega$, and we have a driven pulley B with the outgoing torque $M_2$ and a pressure force $F_{P,2}$. The control of these hydraulic pressure forces represents one of the crucial points for CVT operation.

Some important components of CVT drives are pictured in Figure 5.62. On the left one pulley is depicted with its axially fixed sheave with a mechanical torque sensor on the same side, with the movable sheave and the hydraulic chamber, and as indicated with a rocker pin or a push belt element between the sheaves. As can been seen the structural differences between the

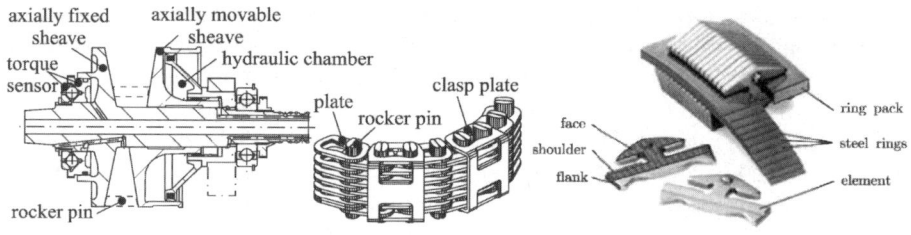

Fig. 5.62: CVT Chain Drive Components

LUK/PIV chain with its rocker pins and the VDT push belt (VDT = van Doorne Transmission) with its very small elements are significant, and from that the differences in modeling these two configurations are also very large. In the case of the LUK/PIV rocker pin chain we have to consider contacts between the rocker pins and the pulley sheaves, between the two rocker pin parts and between the rocker pins and the plates. The VDT push belt situation is much more complicated. Contacts are between the elements and the pulley sheaves, the elements and the elements, between the elements and the rings and within the steel ring package. As most simulations of such systems refer to vibrations, noise and wear, very many of the components need to be modeled as elastic bodies, though linear elastic bodies, but nevertheless imbedded in a very complicated model structure.

At the time being we have three types of chains or belts, see Figure 5.63. The LUK/PIV rocker pin chain, originating from the system Reimers, possesses the plate-rocker-pin structure somehow related to the classical roller chain. The Borg Warner chain, mainly used in the US, includes similar structural elements but with add-on pins generating also additional internal torques. The VDT push belt comprises very many small elements hold together by two steel ring packages and designed in such a way, that one of the strands is able to act as a kind of a strut supporting the transmission of forces. In the following we shall give some typical data for the LUK chain and the VDT belt, because we shall focus on these two configurations.

LUK/PIV  VDT/Bosch  BWD

Fig. 5.63: Types of Chains and Belts

|  | LUK chain | VDT belt |
|---|---|---|
| number of elements | 63 | 382 |
| element length, thickness | 9.85 mm | 1.80 mm |
| element width | 36 mm | 29.6 mm |
| polygonial frequency | 550 Hz | 3000 Hz |
| elements/sec at 1000 rpm, i=1 | 550 | 3000 |
| lowest eigenfrequency | 90 Hz | 120 Hz |
| rigid DOF | 300-400 | 1000-2000 |
| elastic DOF | 200-300 | 300-400 |
| number of contacts | 100 | 1500-2000 |

The number of the degrees of freedom (DOF) and of the possible contacts indicate already the larger complexity of the push belt configuration. The elasticity of the rings in connection with the unilateral contacts of the elements on the ring require a one-dimensional continuum theory including unilateral contact events. From the table we conclude in addition that the frequency range to be considered has to be extended to the kilo-Hertz range, which influences the size of potential Ritz approaches.

The literature in the area of CVT-chains and belts is very numerous starting already at the early days of the mechanical sciences. Euler was the first to establish the well-known theory for a rope wrapped around a cylinder, where the rope force follows an exponential law along the angle of wrap. Eytelwein [59] applied Euler's theory to V-belts and flat belts, which has been an important contribution in those days with machines driven by a lot of very large and dangerous belts. Grashof [90] investigated 1883 flat belts defining those arc areas, where the belt tension and thus the belt force are constant. He called that an equilibrium arc, which lost its significance at least for V-belts with the finding of Dittrich [46] that any elasticity at right angles to a V-belt generates radial displacements within the pulley. With the assumption of a logarithmic spiral he could achieve good comparisons with measurements.

In the years to come many efforts have been undertaken to improve the existing approaches. Lutz and his group ([144], [145]) introduced the "orthogonal point" with radial sliding only, but neglected longitudinal elasticity. Hartmann [96] developed a theory with longitudinal and with crosswise elasticities, but his method did not really bring an improvement with respect to existing theories. A satisfying method was then established by Gerbert and his school in Sweden ([83], [84]), who took care in a very efficient way of the numerical problems involved.

Japan is the country with the most frequent applications of CVT-gears. Therefore activities in the field are widepread. We cite Ide and his colleagues, who established very valuable approaches for CVT's based mainly on experiments ([114], [113], [112]). In Germany a nice and astonishingly simple theory has been published by Sattler [230], which nevertheless gives a good representation of many important features of CVT-belts and -chains. We used it in a modified form as a starting algorithm for estimating the state of our very detailed models and the correspondingly complex numerics. Sattler used, by the way, quite a lot of ideas of Gerbert's theory on CVT's.

Design and functionality aspects have been considered within the larger area of automotive power transmission by Tenberge [260] and Höhn ([101], [100]), both scientists connected more to the practical problems of mechanical engineering. Tenberge developed new concepts of power transmission including CVT's, and Höhn is certainly one of the fathers of the hybrid power transmissions including rocker pin CVT's. Applications of CVT systems can be found in cars of AUDI, Daimler and of course in many Japanese automotive systems.

At the author's former Institute we started with plane models of rocker pin chains ([250], [249]) including all details like elastic deformations of the pulleys, the pins and the plates. A second dissertation ([239], [238]) extended the plane theory to the spatial case including small deviations from the plane pulley-chain-configuration, thus taking into account manufacturing tolerances as well as the geometric-kinematical effects for transient states. All components are modelled with all spatial degress of freedom and again most parts in a linear elastic form. We shall present these results. Parallel to the rocker pin chains the problem of the push belts have been considered, in a first step by certain approximations, see ([32], [31], [77]), and quite recently in a more concise way using set-valued laws and contacts with the one-dimensional continua of the rings ([80], [81]). Research is going on for the three-dimensional push-belt case. All theories and algorithms have been compared with experiments from industry, with convincing results.

### 5.4.2 The Polygonial Frequency

One important feature distinguishing classical flat belts and V-belts from rocker pin chains and push belts consists in the polygon effect generated by the pins or the push belt elements when entering or leaving the pulley [250]. This process is a source of internal parameter excitation, which usually does not become dangerous in the form of resonances and the like, but which represents one of the major sources of noise and wear. Neighboring pins or elements possess a certain distance to each other depending on the specific design of the chain or belt. Therefore the chain or belt does not enter or leave the pulley continuously but in a discrete manner as characterized by Figure 5.64.

We give an approximate description assuming a transmission ratio of one and stiction in the contacts. The arc of wrap is a polygon line, where the points of contact are given by the pins or the elements, approximately with equal distances $l_G$ and with a constant rotational speed $\omega_R$. All points of

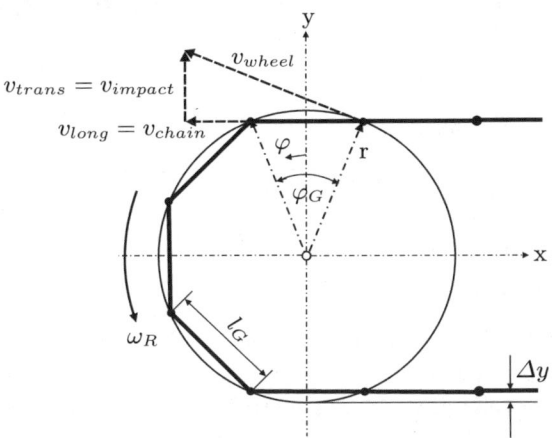

Fig. 5.64: Polygon Effect

contact are assumed to be on a circle with radius r. The element length $l_G$ corresponds to an angle $\varphi_G$ by

$$\varphi_G = 2\arcsin\left(\frac{l_G}{2r}\right). \tag{5.100}$$

For real chains and belts the value of $\varphi_G$ results from practical experience including aspects like noise, wear and the possibilities of manufacturing. In addition the polygon length $l_G$ is not constant for all elements, but slightly different to reduce noise. Nevertheless, for a first estimate but not for the theories to follow, the simplified formulas are useful. We get for the polygon frequency

$$f_{Polygon} = \frac{\omega_R}{2\pi\varphi_G}, \tag{5.101}$$

which is comparable with the mesh of tooth frequency for gears.

All components of the CVT are excited by this frequency and by integral multiples or integral parts of this frequency. It concerns also the free strands. To derive again a first estimate, we assume a transmission ratio i=1 implying the free strand to be parallel to the x-axis. The first contact of the pin (element) with the pulley takes place at $\varphi = -\varphi_G/2$ defining the initial point of the polygon course within the pulley. The last contact takes place at $\varphi = +\varphi_G/2$ defining the final point of the polygon course within the pulley.

According to Figure 5.64 we get for the transverse and for the longitudinal velocities of the pin or element contact point the expressions

$$v_{trans} = r\omega_R \sin\varphi, \qquad v_{long} = r\omega_R \cos\varphi. \qquad (5.102)$$

Performing the average of these velocities over the partial arc $(r\varphi_G)$ associated with the pin or element distance we get

$$\bar{v}_{trans} = \left(\frac{\omega_R r}{\varphi_G}\right) \int_{-\varphi_G/2}^{+\varphi_G/2} \sin\varphi \, d\varphi = 0,$$

$$\bar{v}_{long} = \left(\frac{\omega_R r}{\varphi_G}\right) \int_{-\varphi_G/2}^{+\varphi_G/2} \cos\varphi \, d\varphi = (\omega_R r)\frac{\sin\varphi_G/2}{\varphi_G/2}. \qquad (5.103)$$

The fluctuations of the incoming velocity in longitudinal and transverse directions (Figure 5.64) can then be calculated by the differences of the real and the averaged velocities. They generate the following velocity components

$$\Delta v_{trans} = v_{trans} - \bar{v}_{trans} = r\omega_R \sin\varphi,$$

$$\Delta v_{long} = v_{long} - \bar{v}_{long} = r\omega_R \left(\cos\varphi - \frac{\sin(\varphi_G/2)}{\varphi_G/2}\right). \qquad (5.104)$$

As a rough estimate we may say, that $\varphi_G$ is usually very small, which results in $\frac{\sin(\varphi_G/2)}{\varphi_G/2} \approx 1$. The angle $\varphi$ is also very small, so that $\sin\varphi \sim \varphi$ and $\cos\varphi - 1 \sim (\varphi)^2$ indicating, that the tranverse fluctuations grow with $\varphi$ and the longitudinal fluctuations with the square of $\varphi$. These angles are again roughly proportional to the pitch.

The longitudinal velocity remains continuous during an impulsive entrance of a new pin or element, but not the transverse velocity. It undergoes a velocity jump leading to the impact at the entrance or the exit of the pulley. Assuming that the entering (or leaving) pin possesses the same velocity as the initial (end) point of the polygon course, we can write the components of the velocity changes as

$$\Delta v_{impact,t,r} = 2r\omega_R \sin\frac{\varphi_G}{2} \cos\frac{\varphi_G}{2} \cos\vartheta,$$

$$\Delta v_{impact,t,a} = 2r\omega_R \sin^2\frac{\varphi_G}{2},$$

$$\Delta v_{impact,n} = 2r\omega_R \sin\frac{\varphi_G}{2} \cos\frac{\varphi_G}{2} \sin\vartheta. \qquad (5.105)$$

The indices n, t, r, a stand for normal, tangential, radial and azimuthal directions, respectively, see Figure 5.65. The impact velocity $\Delta v_{impact}$ comprises normal and tangential components with respect to the pulley surface, of which the tangential radial part includes also an axial component due to the cone angle of the pulley. Both, the azimuthal and the radial velocity components, generate a frictional impact, where the azimuthal component $\Delta v_{impact,t,a}$ accelerates the chain link, and the radial component $\Delta v_{impact,t,r}$ decelerates it.

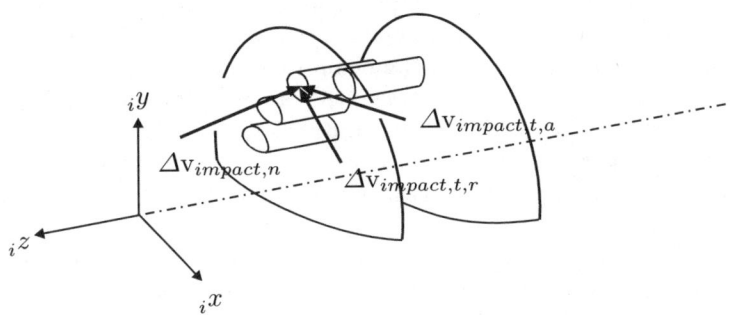

Fig. 5.65: Simplified Impact Velocities

Additionally we have a velocity component $\Delta v_{impact,n}$ perpendicular to the pulley surface, which represents a normal impact. These impacts are especially in connection with large rotational speeds a major source of the noise, they have in addition a large influence on the wear of the chain or the belt. For disadvantageous transmission and speed configurations the chain or belt might even detach and come into contact again by a second and smaller impact [250].

The polygon effect is always accompanied by a large rotational impact, because the rotational link velocities in the two free strands are approximately zero, and a link within the polygon course of the pulley rotates with $\omega_R$. The rotational velocity jump is therefore

$$\Delta \omega_{impact} = \omega_R, \tag{5.106}$$

which takes place with the polygon frequency $f_{Polygon}$, equation (5.101), and thus in a time interval of $\Delta t = 1/f_{Polygon}$. According to the data given above this results for a chain approximately in $\Delta t \approx 2 \cdot 10^{-3}$s and for the push belt in $\Delta t \approx 3 \cdot 10^{-4}$s and from that for a speed of 1000 rpm and a ratio of i=1 in a rotational impact acceleration of $\dot{\omega} \approx 55.000 s^{-2}$ in the first and $\dot{\omega} \approx 300.000 s^{-2}$ in the second case. Multiplied by a pulley radius of 50 mm we would get 275 g in the first and 1500 g in the second case on each element. This results in additional noise and wear.

## 5.5 CVT - Rocker Pin Chains - Plane Model

In a first step we shall consider a CVT system with a rocker pin chain according to the system Reimers (PIV Antriebe, Werner Reimers, Bad Homburg, Germany), which is in the meantime taken over and produced by the company LUK (Bühl bei Baden-Baden, Germany). The components of such a system are pictured in the Figures 5.61, 5.62 and 5.63, they comprise the two pulley sets with conical sheaves and the rocker pin chain, see [250] and [249]. The chain consists of inner plates, clasp plates and rocker pins. All plates transmit the tractive power, while the clasp plates orientate in addition the rocker pins perpendicular to the direction of the chain motion. The rocker pins transmit the frictional and normal loads between the pulley sheaves on the one and the chain on the other side, utilizing the contacts between the pin faces and the conical sheave surfaces. They connect within the chain structure the plates thus acting as a kind of shaft with a rotational degree of freedom for the bending chain.

The rocker pins and thus the chain are clamped between the sheaves of the two pulley sets. Each pulley set consists of one axially fixed sheave and one axially movable sheave. The axially movable sheaves are control elements, which adjust the transmission ratio and change it by producing a transient state using oil pressure.

### 5.5.1 Mechanical Models

Models require in a first step the definition of the model boundaries. The boundaries of the system are the two pulley sets, for which the boundary conditions are a constant speed of the driving pulley and a constant torque acting on the driven pulley. Additionally, the oil pressure force of the output pulley is prescribed, whereas the force acting on the axially movable sheave of the driving pulley is specified by the transmission ratio.

Due to its very large loads, the *pulley set* needs to be modeled elastically. As the elastic deformations are very small, nevertheless influencing heavily the contact processes, we shall consider a RITZ-approach following the theory discussed in chapter 3.3.4 on pages 124. The shape functions we evaluate from a FEM-calculation for the pulleys. It turnes out however, that the dynamic influence of the deformations themselves are very small. This allows us to replace the Ritz-approach by Maxwell's method of influence numbers representing practically the sheave elasticity by an areal spring, the areal stiffness distribution of which is calculated by a static FEM-analysis.

We start with the rigid pulley set (Figure 5.66). The whole set is supported by two elastic bearings with the same stiffness and damping in the two bearing directions. The corresponding forces write

$$F_{bearing,x} = c(x_R - x_{R,0}) + d\dot{x}_R,$$
$$F_{bearing,y} = c(y_R - y_{R,0}) + d\dot{y}_R. \tag{5.107}$$

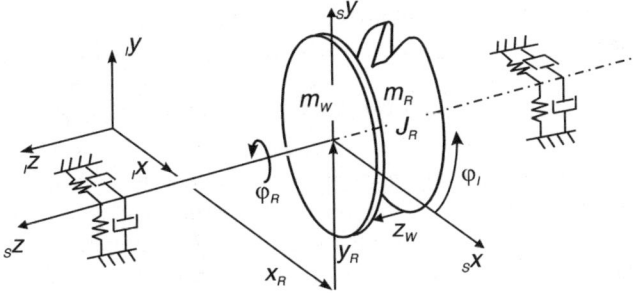

Fig. 5.66: Rigid Body Model of the Pulley [249]

The pulley set and the shaft possess the same rotational speed, if we assume nearly no backlash in circumferential direction between the shaft and the movable sheave. But the backlash between movable sheave and pulley shaft must be considered, because it allows a translational motion and some tilt (Figure 5.67). The corresponding force law is a linear spring with backlash.

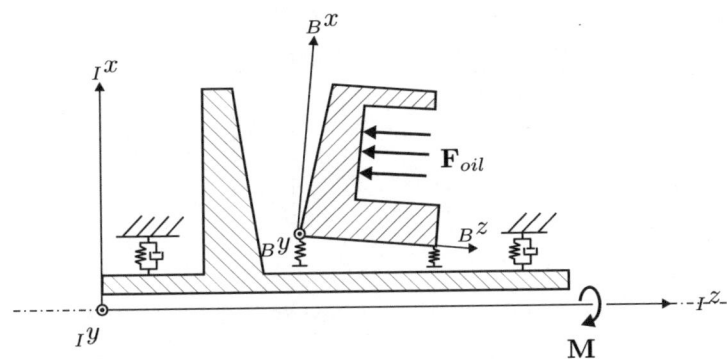

Fig. 5.67: Movable Sheave with Backlash [249]

The oil pressure $p_{oil}$ itself has two parts, the pressure $p_0$ coming from the oil supply and the pressure generated by centrifugal forces. We get

$$p_{oil} = p_0 + \frac{1}{2}\rho\omega_R^2\left(r^2 - r_0^2\right) \tag{5.108}$$

with the appropriate values from the design charts. The force due to the oil pressure follows from an integration over the surface A and gives

$$F_{oil} = \int_{r_0}^{r_a}\int_0^{2\pi} pr\,d\varphi\,dr = Ap_0 + k_{centri}\omega_R^2, \tag{5.109}$$

which means, that the oil pressure force is proportional to the pressure in the oil chamber and to the square of the rotational speed of the shaft-pulley-system.

It makes sense to include into these considerations of the rigid pulley model the areal spring model of the sheaves. Originally it followed from a rather extensive Ritz-approach-analysis, which indicated the possibilities for simplification [250]. The main deformation of the pulley sheaves is a tilt motion, which superimposes the tilt originating from the backlash. Furtheron and by regarding the deformation structure, we find a linear increase of the sheave deformation with the contact radii and in circumferential direction a cosine-form. This is plausible for physical reasons, because the heaviest load on the sheave surface comes along the arc of wrap of the chain. Here we get the maximum deformations. Therefore the following assumptions are reasonable.

The deformations of the sheave are proportional to the contact radius in radial direction and to a cos-function in circumferential direction. They are linearly dependent on the oil pressure, and they can be expressed by a tilt angle $\Delta\vartheta$ (Figure 5.68). The maximum depends on the position of the contact forces, and we assume a symmetric stiffness distribution with respect to the

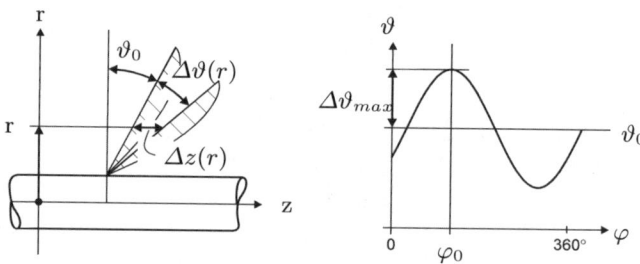

Fig. 5.68: Approximation of Sheave Deformation [249]

two sheaves of the pulley. We then can establish the following formulas

$$\Delta\vartheta = \Delta\vartheta_{max}(F_{oil}) \cos(\varphi - \varphi_0(T_{tilt})), \qquad \Delta\vartheta_{max}(F_{oil}) = c_E(r,\varphi)_{oil} F_{oil}$$

$$\varphi_0 = \frac{\pi}{2} + \arccos \frac{T_{tilt,x}}{\sqrt{T_{tilt,x}^2 + T_{tilt,y}^2}}, \qquad \begin{pmatrix} T_{tilt,x} \\ T_{tilt,y} \\ T_{rot} \end{pmatrix} = \sum_i \tilde{\mathbf{r}}_{SC_i} \mathbf{F}_{C_i}$$

$$\vartheta(\varphi, F_{oil}, T_{tilt}) = \vartheta_0 + \Delta\vartheta(\varphi, F_{oil}, T_{tilt}) \tag{5.110}$$

The maximum tilt $\Delta\vartheta_{max}$ depends linearly on the oil pressure force $F_{oil}$, which is plausible. The vectors $\mathbf{r}_{SC_i}$ and $\mathbf{F}_{C_i}$ are radial positions and forces at the contact points of the pins within the arc of wrap, respectively. The rest is clear from Figure 5.68.

## 5.5 CVT - Rocker Pin Chains - Plane Model

Proceeding with the elastic pulleys via Ritz-approach modelling we have to establish suitable nets of the meshes and to evaluate the eigenforms, which we shall use as shape functions. For each pulley set we consider two elastic bodies, the fixed sheave together with the pulley shaft and the movable sheave alone, which has to be coupled with the shaft within the framework of elastic multibody dynamics. Figure 5.69 gives an impression of the undeformed and deformed configuration of the pulley-shaft-system.

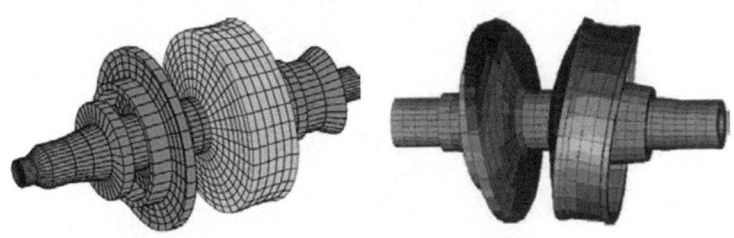

Fig. 5.69: FEM Model of the Pulley with Shaft [249]

The Ritz-shape-functions for both bodies are the eigenvectors of the FEM-based modal analysis. From this we have no continuous functions describing the bodies' deformations, but only nodal informations. The masses of the bodies are concentrated in the nodes and, hence, integrals over the body's volume have to be transformed to summations over the nodes. The nodal structure makes also a certain interpolation scheme for the contact points mandatory, because we cannot expect the contact points to be positioned at the nodes. In a first run eleven eigenmodes were used to achieve some feeling of the eigenbehaviour. The methods applied are those of chapter 3.3.4 on pages 124.

With respect to the operational frequencies one or two eigenmodes would have been sufficient, but finally the above presented approximation comes out with results not much different from those for a Ritz-approximation with eleven shape functions. The main reason for this kind of investigation comes from the intention to get some insight into the elastic behaviour and to explore the possibilities for simplification.

For the evaluation of the dynamic behaviour of the *rocker pin chain* itself we need a model of the chain elements, which includes all the mass effects and also the elastic deformations necessary to calculate stresses and strains. As already mentioned, the chain and its links are responsible for the internal parameter excitations of the system, and considering in addition the slight geometric differences of the chain elements for the purpose of noise reduction, we need to model every individual chain element which is one reason of large computing times. Therefore the chain link models have to be simple, compact, but nevertheless realistic.

To start with the model we consider Figure 5.70. Every chain link represents a rigid body, consisting of several plates and a pair of rocker pins. It has three degrees of freedom $\mathbf{q}_L^T = (x_L, y_L, \alpha_L)$, which is the maximum number of degrees of freedom in a plane. Therefore, the chain links are kinematically

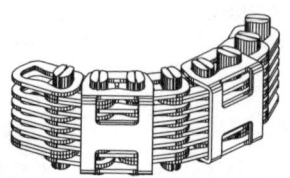

Fig. 5.70: Rocker Pin Chain

decoupled. The chain links are connected by force elements, acting between the point $D$ of one chain link and the point $B$ of its successor (lower part of Figure 5.71). These force elements take into account the elasticity and damping of the link and the joint. When a link comes into contact with a pulley, the frictional and normal contact forces act on the bolts of the chain link and, therefore, on the link at the point $B$. Hence, all contact forces acting on one pair of rocker pins are assigned to one link. In Figure 5.71 a left positioned

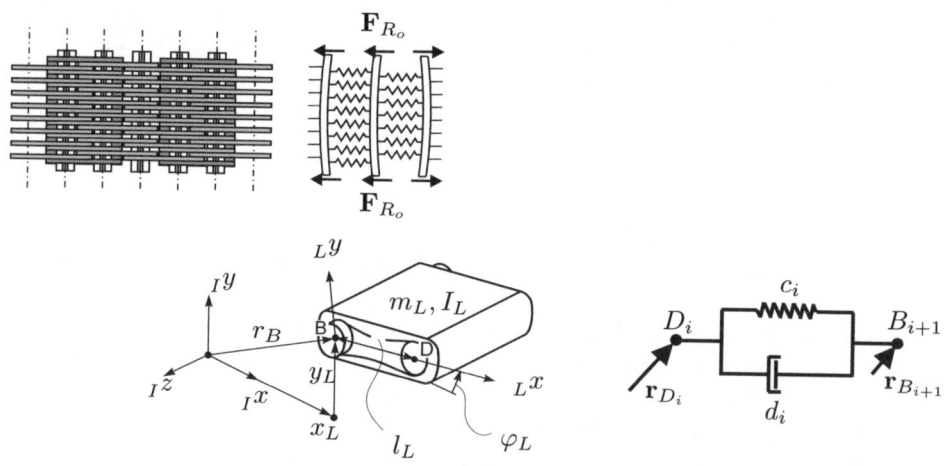

Fig. 5.71: Model of a Rocker Pin Chain Link [249]

index refers to the coordinate systems, for example "I" for an inertial and "L" for a link-fixed coordiante system. A right index are the coordinates, or angles, or length, themselves. The magnitudes $(m_L, I_L)$ are the link mass and mass moment of inertia, respectively.

## 5.5 CVT - Rocker Pin Chains - Plane Model

CVT's transmit power by friction in the **contact** of the rocker pins with the pulley surfaces. Therefore these contacts need to be considered very carefully. Phenomenologically rocker pins between the pulley sheaves build up in normal direction certain normal forces, which are generated by the elastic deformations of the pins and the sheaves on the one and by the contact constraints on the other side. In reality we have a unilateral contact in an elastic multibody environment. Principally such contacts might detach again but it is very unprobable, that under the tensile forces of the chain such an event occurs.

After the normal contact direction will be closed, we get relative velocities between pins and sheave surfaces in two directions, in radial and in circumferential directions. The radial relative velocities are closely related to the sheave deformations. In the sense of chapter 3.4 we have a nonlinear complementarity problem, which has to be treated by a friction cone linearization, for example. In the meantime the prox-algorithm makes such an approach unnecessary. For the tangential situation we might choose a model on the basis of unilateral multibody systems on the basis of chapter 3.4 on the pages 131, or we might choose a regularized force law approximating Coulomb's law. In the first case we are able to describe stick-slip effects, in the second case not. For the planar case Srnik [250] developed both models, unilateral and regularized ones and compared the two with much success. The contact situation as a whole is pictured by Figure 5.72.

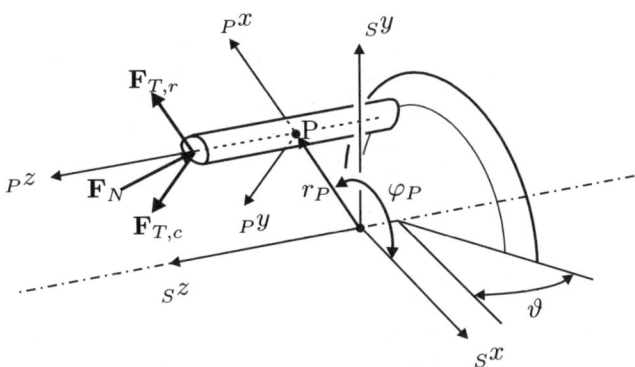

Fig. 5.72: Contact Rocker Pin - Sheave [249]

Neglecting its eigendynamics the pair of rocker pins can be modeled as one single massless spring acting only perpendicular to the model plane, which is identical with the pulley-chain plane. Figure 5.73 shows the model and the forces acting in the contact plane. The oblique contact planes are symmetrical with respect to the two pulley sheaves. The pin danamics can be neglected thus allowing a quasi-static investigation. The normal contact force $\mathbf{F}_N$ acts perpendicular to the contact plane, and the frictional forces $\mathbf{F}_{T,r}$ and $\mathbf{F}_{T,c}$ are parallel to this plane.

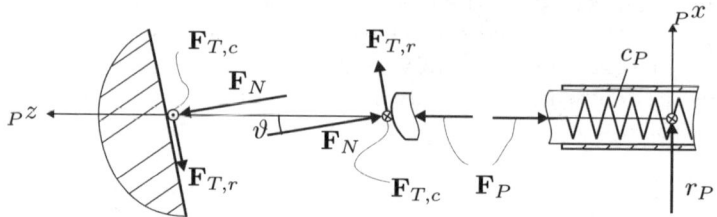

Fig. 5.73: Rocker Pin Model [249]

The pins change their length due to elastic effects, which have two parts. The pin as an elastic strut will be shortened by the pressure force $\mathbf{F}_P$ on the one, and it will be compressed in its contact together with the sheave surface by the same force on the other side (Figure 5.73). From this we get

$$\Delta l_P = \Delta l_{strut} + 2\Delta l_{Hertz}, \quad \text{with} \quad \Delta l_{strut} = \frac{f_P}{c_{strut}},$$

$$\Delta l_{Hertz} = \frac{f_P^{\frac{2}{3}}}{c_{Hertz}}. \tag{5.111}$$

The coefficients $c_{strut}$ and $c_{Hertz}$ may be taken from some standard text books, for example [147] and [118]. To solve the above equation for the force $\mathbf{F}_P$ is a bit cumbersome. Therefore we use a least square approximation in the from [250]

$$F_P = c_1 \Delta l_P + c_2 \Delta l_P^{\frac{3}{2}} + c_3 \Delta l_P^2. \tag{5.112}$$

### 5.5.2 Mathematical Models

With respect to mathematical modeling we refer to the chapters 3.3 on the pages 113 and 3.4 on the pages 131, where the fundamental equations and the necessary algorithms are discussed in some detail. We apply these relations to our case of a planar CVT-system with rocker pin chains ([250], [249]), though presented here in a reduced form. Details are in the cited literature.

For the derivation of the chain drive's equations of motion the equations of momentum and the equations of moment of momentum of each single body have to be considered. They are transformed into the corresponding space of the minimal coordinates by the Jacobian matrices. We always come out with a form corresponding to the equations (3.106) on page 118 for the bilateral and to the equations (3.159) on page 143 for the unilateral case. In the following we shall discuss the elements of these equations.

## 5.5 CVT - Rocker Pin Chains - Plane Model

The model of the chain drive includes the chain links as ***rigid bodies***. Their sets of minimal coordinates $\mathbf{q}_{B_i}^T = (x_{B_i}, y_{B_i}, \alpha_{B_i})$, index $(B_i)$ for body $(B_i)$, define the planar motion of the links. The simplified relations (3.106) for this case and only for body $(B_i)$ with index (i) write:

$$\mathbf{J}_i^T \begin{pmatrix} m\mathbf{E} & m\tilde{\mathbf{r}}_{BS} \\ m\tilde{\mathbf{r}}_{BS} & \mathbf{I}_B \end{pmatrix}_i \mathbf{J}_i \ddot{\mathbf{q}} + \mathbf{J}_i^T \begin{pmatrix} m\tilde{\omega}\tilde{\omega}\tilde{\mathbf{r}}_{BS} \\ \tilde{\omega}\mathbf{I}_B\omega \end{pmatrix}_i = \mathbf{J}_i^T \begin{pmatrix} \mathbf{F} \\ \mathbf{M} + \tilde{\mathbf{r}}_{BF}\mathbf{F} \end{pmatrix}_i, \quad (5.113)$$

where we use the following definition for the single link and the overall system:

$$\mathbf{J}_i = \begin{pmatrix} \frac{\partial \mathbf{v}}{\partial \dot{\mathbf{q}}} \\ \frac{\partial \omega}{\partial \dot{\mathbf{q}}} \end{pmatrix}_i, \quad \mathbf{v}_i = \begin{pmatrix} \dot{x} \\ \dot{y} \\ 0 \end{pmatrix}_i, \quad \omega_i = \begin{pmatrix} 0 \\ 0 \\ \dot{\varphi} \end{pmatrix}_i, \quad (5.114)$$

We define the above equations for example in a body-fixed coordinate frame and transform it to inertial coordinates or vice versa. The most convenient way consists in writing these equations with respect to the reference point $(B_i)$, which is the hinge point of a chain link. The point $(S_i)$ marks the center of mass. Hence, the vector $\mathbf{r}_{BS_i}$ points from $(B_i)$ to $(S_i)$. $\mathbf{I}_{B_i}$ is the matrix of the mass moments of inertia, $m_{B_i}$ is the mass, $\omega_i$ the rotational speed of the body, and $\mathbf{F}_i$ and $\mathbf{M}_i$ are all active forces and torques acting on the body, including the contact forces between the link and the pulley set and the joint forces between the links. The last one is modeled as a linear spring-damper law with the coefficient of elasticity $c$ and the damping coefficient $d$ (see Figure 5.71):

$$\mathbf{F} = c\left(\mathbf{r}_{B_{i+1}} - \mathbf{r}_{D_i}\right) + d\frac{\mathbf{r}_{B_{i+1}} - \mathbf{r}_{D_i}}{|\mathbf{r}_{B_{i+1}} - \mathbf{r}_{D_i}|}\left(\mathbf{v}_{B_{i+1}} - \mathbf{v}_{D_i}\right) \quad (5.115)$$

The point $D_i$ belongs to the link $i$ under consideration and the point $B_{i+1}$ to it's successor $i+1$. Building all individual equations for each link and combining them in an appropriate way we get finally for the rigid body $B_i$:

$$\mathbf{M}_i \ddot{\mathbf{q}} = \mathbf{h}_i(\mathbf{q}, \dot{\mathbf{q}}, t) \quad (5.116)$$

These equations of motion for each body are decoupled kinematically as long as no stiction forces occur. Therefore, the mass matrix of the entire system has a block-diagonal structure enabling a symbolic inversion.

For ***elastic bodies*** like the pulley-shaft system of Figure 5.69 we go back to the equations of motion formulated in chapter 3.3.4 on page 124 mainly represented by the relations (3.120), (3.121) and (3.127) on the pages starting with page 125. The shape functions $\bar{\mathbf{u}}_i$ are determined from a modal FEM-analysis. Performing the evaluations given by the relations (3.127) on page 127 comes out with the same formal set of equations as given with (5.116), but with an extended meaning (see [250]).

According to equation (3.120) the elastic deformations enter the equations of motion(3.127) using the Ritz approach of the equations (3.121). We have

290   5 Power Transmission

to take care that for the derivations of the Jacobians second order terms are regarded in the original kinematical relations. For linear elastic multibody systems the equations of motion in the form (5.116) include only linear elastic term of first order. The mass matrix consists only of 0-th order terms, while the minimal accelerations $\ddot{\mathbf{q}}_i$ as well as the right hand side vector $\mathbf{h}_i$ comprise also 1-th order terms. The corresponding integrals depend only on space and not on time and can be evaluated once only at the beginning of every simulation.

One of the problems connected with the consideration of an elastic pulley set consists in the then necessary interpolation of the contact points. Figure 5.74 illustrates the situation. The discretization of the pulley sets in order to

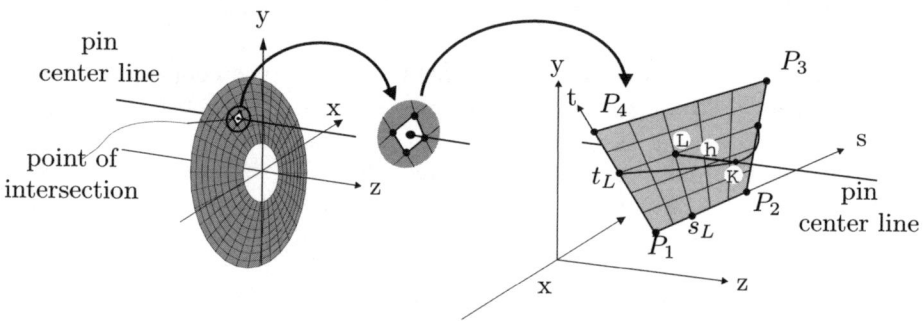

Fig. 5.74: Contact Point Interpolation [249]

evaluate the eigenvectors leads also to a truncated description of the surface of the cone sheaves. These surfaces are the contact zones with the pins of the chain link. Hence, an exact description is inevitable to generate the correct contact forces. For this purpose we apply the following procedure.

We know approximately the position and orientation of the chain within the pulley set, which allows us to select a data set of potential nodes situated near some possible contact point. We store these data together with the elastic node deformations. As we know also the position of the pin centerline we can determine the four nodes $P_1$ to $P_4$, which span a bilinearly defined plane including the point of intersection (Figure 5.74). The nodes $P_1$ to $P_4$ belong to the sheave surface. In a first step we then calculate the exact point of intersection of the pin center line and the bilinearly assumed area spanned by the four nodes $P_1$ to $P_4$. This is point L. It is not positioned on the sheave surface, but has a distance h to it. To evaluate this distance h we assume in a second step, that the value of h in the deformed is not so much different from its value in the undeformed state, which allows an analytical calculation of h on the basis of no deformation. With the intersection point L and the approximate distance h we know also the contact point K for the deformed state, for which we evaluate all kinematical and kinetic magnitudes. For the details see [250].

## 5.5 CVT - Rocker Pin Chains - Plane Model

For the derivation of the **contact forces** it is necessary to quantify the pin's spring force. It depends on the pin's length $l_P$ and stiffness $c_P$ as well as on the local distance $s$ of the two sheave surfaces of a pulley set (Figure 5.75 left). We get

$$F_P = \begin{cases} c_P(l_P - s) = c_P \Delta l_P & \forall \ s \leq l_P \\ 0 & \forall \ s > l_P \end{cases} \tag{5.117}$$

Neclecting the dynamics of the pins along their center line we can evaluate the normal forces approximately from the static equilibrium of forces perpendicular to the model plane. Taking into account equation (5.117) it depends on the minimal coordinates of the corresponding chain link and the contact force $F_{R_r}$ in radial direction:

$$F_N = F_{R_r} \tan \vartheta + \frac{F_P}{\cos \vartheta} = F_{R_r} \tan \vartheta + \frac{c_P(l_P - s)}{\cos \vartheta} \tag{5.118}$$

Coulomb's friction law is used in order to determine the remaining frictional forces as function of the normal force and the relative velocity $\dot{g}$:

sliding: $\quad \dot{g} \neq 0 \quad \Rightarrow \quad \mathbf{F}_R = -\mu_0 F_N \dfrac{\dot{\mathbf{g}}}{|\dot{\mathbf{g}}|}$,

stiction: $\quad \dot{g} = 0 \quad \Rightarrow \quad F_R = \sqrt{F_{R_r}^2 + F_{R_t}^2} \leq \mu_0 F_N.$ (5.119)

In the case of sliding, a single-valued dependency for the frictional forces exists, and all contact forces can be calculated. By inserting equation (5.119) into equation (5.118) all contact forces are a function of the minimal coordinates of the bodies under consideration.

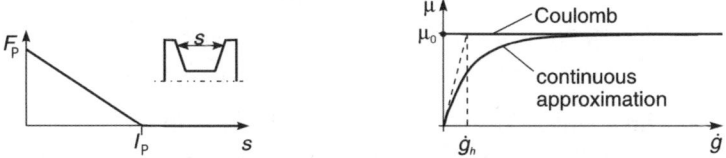

Fig. 5.75: Characteristics of the Pin's Force (left) and of the Friction forces (right)

If the relative velocity $\dot{g}$ vanishes, a transition from sliding to stiction is going to take place. In this case, the friction law (5.119) gives only an upper limit for the magnitude of the stiction force. After it is reached, a transition from stiction to sliding is possible. By eliminating the normal force with the help of equation (5.118), one obtains an elliptical cone in dependency of the pin's force $F_P$, representing the convex set of the valid stiction area (Figure 5.76):

$$\frac{(F_{R_r} - F_{R_r,M})^2}{R_r^2} + \frac{F_{R_t}^2}{R_t^2} \leq 1 \tag{5.120}$$

with

$$R_r = \frac{\mu_0 F_P}{\cos\vartheta(1 - \mu_0^2 \tan^2\vartheta)}, \qquad \text{radial semi axis,}$$

$$R_t = \frac{\mu_0 F_P}{\cos\vartheta\sqrt{1 - \mu_0^2 \tan^2\vartheta}}, \qquad \text{azimuthal semi axis,}$$

$$F_{R_r,M} = \frac{\mu_0^2 \sin\vartheta\, F_P}{\cos^2\vartheta(1 - \mu_0^2 \tan^2\vartheta)}, \qquad \text{radial displacement of the center.} \tag{5.121}$$

In the case of a sticking contact, the vector of the frictional force points to an inner point of the stiction cone, whereas in the sliding case its tip is positioned on the surface.

The semi axis as well as the radial displacement of the center (equations (5.121)) indicate, that the cone angle enlarges the maximal frictional forces in radial and circumferential directions. The radial displacement of the center

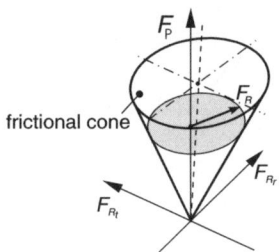

Fig. 5.76: Convex Set for the Pin/Sheave Contact

results from the coupling of radial and normal forces in equation (5.118) and leads to higher frictional forces in radial direction for a movement towards the axis of the sheaves than away from this axis.

The stiction force is a constraint force. It can be calculated by taking into account the appropriate kinematic condition of vanishing relative velocity $\dot{g} = 0$ at the contact points. For the implementation into the equations of motion it is necessary to formulate this condition on an acceleration level ($\ddot{g} = 0$). In this case one can distinguish between sticking and transition to sliding. For a transition to sliding the relative acceleration and the frictional force point in opposite directions:

$$\text{stiction:} \quad F_R \leq \mu_0 F_N \quad \wedge \quad \ddot{\mathbf{g}} = 0,$$
$$\text{transition to sliding:} \quad F_R = \mu_0 F_N \quad \wedge \quad \ddot{\mathbf{g}} = -\rho \mathbf{F}_R, \qquad \rho > 0 \tag{5.122}$$

## 5.5 CVT - Rocker Pin Chains - Plane Model

A stiction force is a constraint to the system. Therefore, each independent sticking contact reduces the number of degrees of freedom by two and the equations of motion have to be modified. With the methods discussed in chapter 3.4 on the pages 131 the system dynamics can be described by a set of nonlinear differential-algebraic equations for the unknowns $\ddot{\mathbf{q}}$ and $\lambda_i$:

$$\mathbf{M}\ddot{\mathbf{q}} = \mathbf{h} + \sum \mathbf{W}_i \lambda_i$$

$$\ddot{g}_i = 0 \quad \wedge \quad \lambda_{0_i} \geq 0, \quad \ddot{g}_i \geq 0 \quad \wedge \quad \lambda_{0_i} = 0, \quad \ddot{g}_i \lambda_{0_i} = 0,$$
$$\lambda_{0_i} = \mu_0 F_{N_i} - |\lambda_i| \geq 0 \tag{5.123}$$

This is a complementarity problem being solved at that time by Lemke's algorithm. $\lambda_i$ are the frictional forces of the $i$-th sticking contact. It is transformed by the constraint matrix of the contact point $\mathbf{W}_i$ into the corresponding space of the minimal coordinates. Chains and belts transmit power by sliding friction in the contact pins/pulley or elements/pulley. Sticking contacts do not occur very frequently. Therefore the possible maximal number of such sticking events are of minor interest.

For the calculation of the stiction forces $\lambda_i$ of the $i$-th potentially active tangential constraint, additional nonlinear complementary conditions have to be taken into account. They correspond to the contact law on an acceleration level. We call $\lambda_{0_i}$ the frictional saturation or the friction reserve. For the determination of the normal forces no additional constraints have to be taken into account, because they are known by equation (5.118).

Due to the transitions from stiction to sliding, and vice versa, and the corresponding changes of the number of degrees of freedom, the equations of motion (5.123) possess a time-variant structure. The transition from sliding to stiction in a contact between a chain link and a pulley can produce a discontinuity of the contact forces leading to a jump in the accelerations of this two bodies. If the pulley is modeled as one rigid body the acceleration jump can lead to changes in the contact configuration of the other contacts of this pulley. However, because of the pulley's large moment of inertia, these couplings are week and may be neglected. Nevertheless, the corresponding transition points have to be determined, which leads in combination with the large number of contacts to large simulation times.

We have two possibilities for reducing the contact simulation efforts. The first one uses a continuous chain model, something like a belt-shaped structure, for example [230]. For such a model the contacts are reduced to one for each pulley. This neglects the polygonal effect and the corresponding excitation mechanisms, which are a major subject of the current studies. So this possibility will not be considered further. The second possibility for reducing the system's order is the application of a continuous approximation of Coulomb's friction law by a regularized characteristic (right part of Figure 5.75), the equations of which write

$$\mathbf{F}_R = -\mu F_N \frac{\dot{\mathbf{g}}}{|\dot{\mathbf{g}}|}, \qquad \text{with} \qquad \mu = \mu_0(1 - e^{-\frac{\dot{g}}{\dot{g}_h}}), \qquad \dot{g} = |\dot{\mathbf{g}}|. \qquad (5.124)$$

The factor $\dot{g}_h$ defines the gradient of the curve for $\dot{g} = 0$ and is a measure for the deviation of the approximation from the exact solution. By this friction law the frictional forces are uniquely determined in dependency of the normal force for any relative velocity. It is a single-valued force law, not a set-valued one. This property implies, that stick-slip phenomena will not take place, which is very near to the practical operation of CVT-systems. They transmit power by sliding friction. A further outcome of this regularization consists in a set of equations of motion without complementarities, which can be solved in a conventional way.

It might be of interest to present a comparison between the two methods, the one with and the one without complementarities. The differences are presented in Figure 5.77. It depicts the radius of a chain link as well as the tensile force in this chain link while it is in the arc of wrap of the driving pulley. Generally both contact models provide results, which are next to each

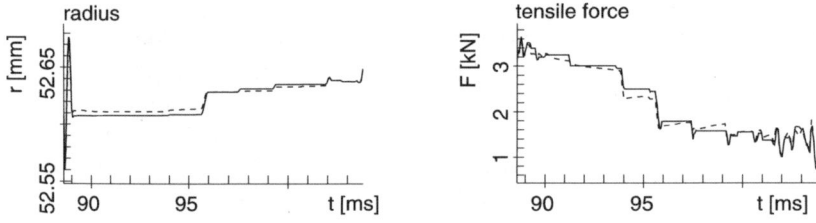

Fig. 5.77: Radius (left) and Tensile Force (right) of a Chain Link in the Arc of Wrap of the Driving Pulley [249]; Coulombs Friction Law (Set-Valued): ——, regularized friction characteristic (Single-Valued): - - - -

other. However, for the continuous model, the phases of stiction, which are recognized in the discontinuous model by exact constant values, are only approximately constant. On the other side the calculation time decreases by at least one order of magnitude. Thereby, larger values of the coefficient $\dot{g}_h$ in equation (5.124) lead to larger simulation times and to a more exact approximation of the unsteady case. This is a point, where the regularization requires some trade-off feeling. As the correspondence of both models is proven the faster simulations are used for further investigations.

### 5.5.3 Some Results

The following results are computed for a uniform motion with constant driving speed $\omega_{driving}$, an output torque $T_{driven} = 150Nm$, a geared level $i = 1$ and a pressure force of $F_{pressure,driving} = 20kN$ on the driving side, which

## 5.5 CVT - Rocker Pin Chains - Plane Model

corresponds to a pressure force $F_{pressure,driven} = 27.4kN$ on the driven side (Figure 5.78). The chain drive consist of two pulleys and 63 chain links. To

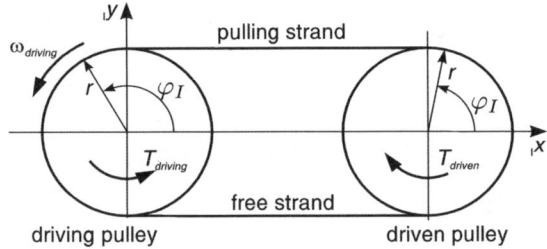

Fig. 5.78: Configuration of the Simulated System

analyse the influence of the **pulley's deformation** two kinds of simulations have been performed. One with the real elastic pulley sets and one, which neglects any elastic deformations of the pulley sets. The elastic deformations of the fixed pulley sheave must be considerably smaller than that for the movable sheaves, because in that case we get a superposition of the movable sheave tilt and the elastic deformations. The displacements of a surface point depend on its position on the sheave surface and on time. For stationary operation however and neglecting for this consideration the polygonal effects we get more or less constant functional properties of the load by the contact and pressure forces, seen from an inertial position. As a consequence the functional characteristics of the pulley deformations are also always the same for an inertial standpoint of view, they can be interpreted as a kind of attachment modes with respect to a possible Ritz-approach.

The position of a (static) deformation point can be described by a circumferential (azimuthal) angle $\varphi$ and by a radial position r. The radius is varied from the innermost radius $r = 22.5$mm to the maximum radius $r = 72.9$mm, while the chain runs on a radius of $r \approx 52.5$mm. Figure 5.79 shows the deformations of the four sheaves as a function of the azimuthal angle $\varphi_I$ (Figure 5.78) on several radius groups. The plots show, that the sheaves are bent outward in the contact zones, represented by a positive sign for the axially movable sheave and by a negative sign for the axially fixed sheave. In the zones with no chain contact the sheaves are bent inward, while the transition occurs for all radius groups of one sheave at the same azimuthal position. The amplitudes of the axially movable sheaves are larger than those of the axially fixed sheaves, because of an additional tilting of the axially movable sheaves, which again is due to the clearance between shaft and sheave. The maximum of the deformation is located adjacent to the exit area of the chain, for both pulley sets. This causes high normal contact forces right after the maximum deformation and, hence, right before the exit phase. In addition the chain link,

296    5 Power Transmission

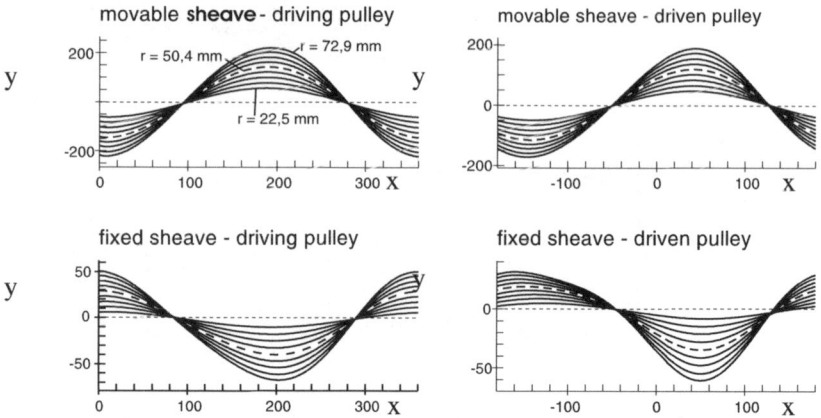

Fig. 5.79: Deformation of the Axially Movable Sheaves (top) and the Axially Fixed Sheave (bottom) of the Driving Pulley (left) and of the Driven Pulley (right) in Dependency of the Azimuthal Angle $\varphi_I$ on Different Radii

which is on a minimum radius at the maximum deformation point, can not move outward fast enough to avoid jamming between the sheaves.

The *pulley's deformation* decisively influences the path of motion withing the angle of wrap and the efficiency of the complete chain drive. The corresponding effects can be seen in an especially illustrative way by considering the radius of a chain link on its path around the pulley, one time for rigid and one time for elastic sheaves (Figure 5.80). The graphs for the rigid sheaves

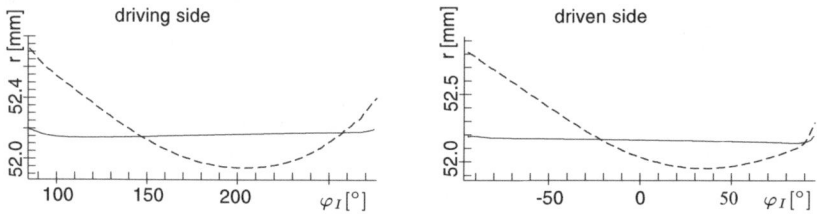

Fig. 5.80: Radius of a Chain Link in the Driving Pulley (left) and in the Driven Pulley (right); rigid sheaves: ———, flexible sheaves: - - - - -

show minimal radial movements. In the contact phase of a chain link with the driving pulley, the tensile force in the link declines from the value of the pulling strand to the one of the free strand. The contact forces in the corresponding pin have correlated values. After the entry phase, which is marked by a decline of the radius, the radius increases again. The exit phase finishes the circulation of the link by an abrupt increase of the radius. A chain link contacting the driven pulley shows the same behavior in the entry and exit

## 5.5 CVT - Rocker Pin Chains - Plane Model

phase. Nevertheless, the plots differ, because the tensile force increases from the value of the free strand to that of the pulling strand and according to this, the radius decreases.

The lengths of the radial paths of a chain link in contact with flexible sheaves are about 15 times larger than the radial paths of a chain link in contact with rigid sheaves. The larger radial motion, preferably going ahead with energy dissipation, leads to a deterioration of the calculated efficiency by about 2%. The radial paths are determined mainly by the deformation function, which has its maximum at a wrap angle of 200° for the driving pulley and of 30° for the driven pulley. In this region the chain link has its minimal radius and the radial movement changes its sign.

As long as the chain link is part of a strand no contact forces act on its rocker pins. When the link comes into contact with one of the pulleys the pins are pressed between the two sheaves and, hence, the normal force increases. Its amplitude depends on the geometry of the sheaves and the transmitted power. Figure 5.81 illustrates an example. The frictional force is a function

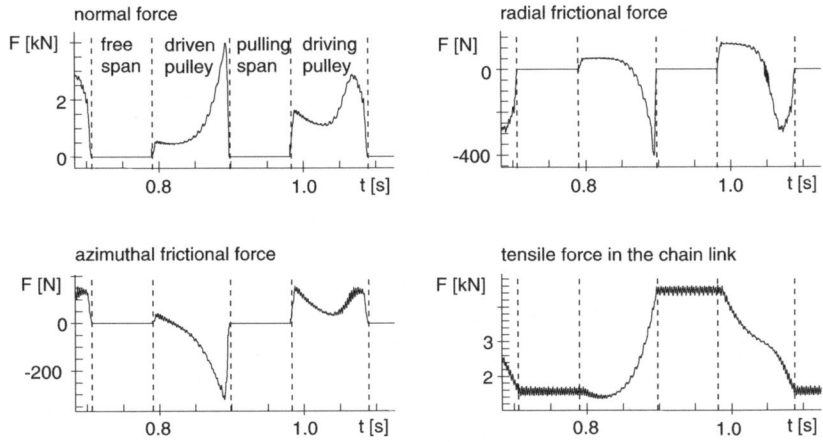

Fig. 5.81: Forces acting on a Chain Link during one Circulation

of the normal force and the relative velocity between the pulley and the pins. It is split into one radial and one azimuthal component. The radial contact force coincides with a radial movement of the chain link, which results in a dissipation of energy. In contrast to this the azimuthal contact force causes the changes of the tensile force in the corresponding chain link, leading to different tensile force levels in the two strands which agree with the transmitted torque.

**Measurements** have been performed at the Technical University of Munich at the Lehrstuhl für Landmaschinen [231]. In order to compare these measurements with simulation results, it is necessary to determine the distribution of the tensile force on the plates of the chain links. Therefore, an elastostatic model of the chain, including all its components, was established, see Figure 5.71 on page 286. For this purpose the pairs of rocker pins are modeled as bending beams and the plates as linear springs. Neglecting second order effects, only the azimuthal frictional forces and the stretched chain length are adopted from the dynamic simulation as boundary conditions. With this model we obtain the right graph of Figure 5.82 for the tensile force in the clasp plate.

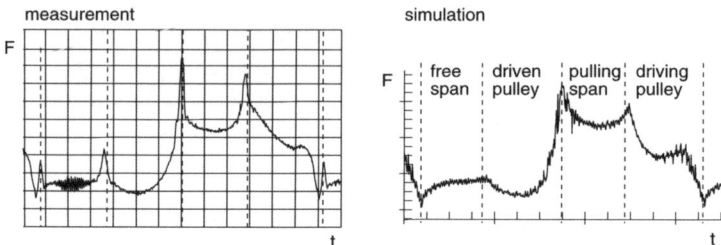

Fig. 5.82: Tensile Force in a Clasp Plate: Comparison of Measurements (left)[231] and Simulation (right)[250]

The results differ a bit from the results of the model, due to the deformation of the link components. The plate forces are usually significantly different. Especially the force peaks of the elastostatic model do not exist in the model. The reason for this is, that in the model the tensile force is an integral value of all plates, summing up the deformat effects of the link's components. The force peaks result from pin bending, which depends on the azimuthal frictional forces. These forces are subject to large changes in this area (Figure 5.81). Therefore, the plates have to compensate the bending differences between successive bolts, leading to high tensional forces in the outer plates.

The force peaks appear also in the measurement, even though they are more distinctive. During the contact phases measurement and simulation correspond to each other. This indicates that the assumptions for the sheave's deformations are correct, which is one of main influences on the tensile forces in the contact zones. The measured vibrations of the free span are due to the measurement system. Hence, they do not appear in the simulation. The two upward peaks of the measurement at the beginning and the end of the free span are missing in the simulation. Nevertheless the principal course in the free span, including the downward peak in the beginning, as well as the force difference between the two spans, coincides. Further comparisons confirm the mechanical and mathematical models ([250], [239]).

The **pulley clamping forces** and the efficiency offer additional possibilities for a comparison of simulations and measurements. These magnitudes are easy to measure, and from the standpoint of the theory they cover a wide operational range. The ratio of the contact pressures $\zeta_p = p_{driving}/p_{driven}$ is a measure for the pressure system efficiency [232]. Hence, various measurements exist, describing the global chain drive's behaviour. Figure 5.83 shows the measured and calculated contact pressures for a typical data set.

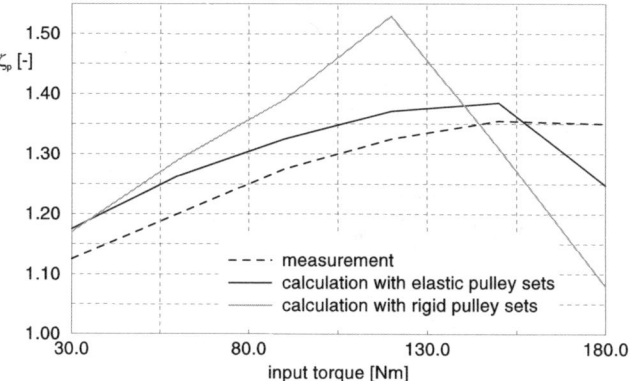

Fig. 5.83: Ratio of Contact Pressures for a Chain Drive on a Geared Level $i = 1$ and a Driving Speed $\omega_{driving} = 2000$ rpm; Measurements from [231]

The calculated curve for elastic pulley sets shows a good correspondence with the measured curve, qualitatively and quantitatively. The maximum relative error is less than 10%. In contrast to this, the calculations with the rigid pulley sets have no accordance with the measurements, their maximum value is much to high and the following decline occurs to early, indicating sliding of the chain at low torques. Obviously a model of a CVT-chain drive with rigid pulley sets cannot describe the dynamic system behavior, whereas a model with elastic pulley sets is able to reflect the reality.

An important quantity for the evaluation of transmission system components is the coefficient of efficiency. Therefore, it should be possible to meet predictions for this indicator by simulation. Figure 5.84 shows measured and calculated values. A simulation, which neglects losses of the bearings, gaskets and the hydraulic unit, shows the efficiency of the chain drive itself. In this case the efficiency is relatively high. It is above 95% in a wide operating range. It decreases only for small torques, when the losses of the frictional forces have a decisive influence, and for high torques, when the slippage increases. Taking into account the losses of bearings and gaskets, but neglecting again the influence of the hydraulic unit, we reach coefficients of efficiency $\eta \approx 95\%$ in the main operating range. The correspondence between this calculation and the measurement is very good and the relative error is less than 5%.

Considering the global as well as the detailed accordances of measurements and simulations, we can again infer that the comparisons confirm the mechanical model.

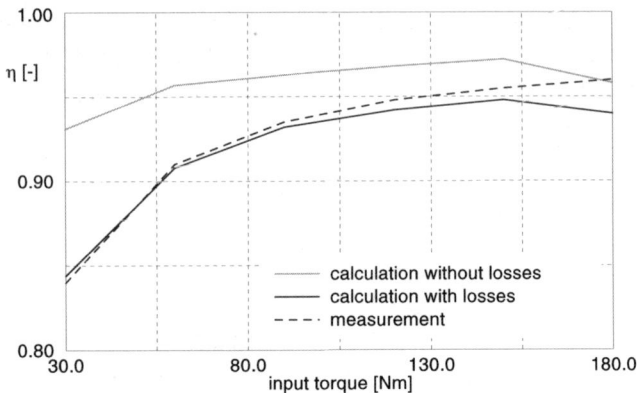

Fig. 5.84: Coefficient of Efficiency for a Chain Drive on Geared Level $i = 1$ and a Driving Speed $\omega = 2000$ rpm; Measurements from [231]

## 5.6 CVT - Rocker Pin Chains - Spatial Model

### 5.6.1 Introduction

Mainly due to geometric incompatibilities all CVT-chains do not perfectly move in a plane but show also out-of-plane effects, which for some cases of predesign considerations cannot be neglected. In addition the elastic behaviour of all components play a crucial role concerning three-dimensional motion. With respect to modeling this increases the number of degrees of freedom considerably leading to growing simulation times. On the other hand design or design improvements of chains, especially with respect to wear and noise, are only possible applying a 3D-theory.

The most important effects for misalignment are of geometrical nature. The sheaves must be controled, and this is achieved by moving axially the movable sheaves. To avoid already misalignments by the arrangement of these movable sheaves, they are arranged in an asymmetric way, for the two pulleys the axially movable parts on opposite sides. This assures that the chain moves during a transient change of the transmission ratio for both pulleys in the same direction, where the axial position is given by the not movable sheaves. But due to different chain radii within the two pulleys we get a misalignment $\Delta z = z_2 - z_1$, which becomes zero only for a transmission ratio i=1 [173]. For a ratio i=1 we have in both pulleys the same radius of wrap and thus also the same inclination angles of the sheave surfaces. But for large transmission ratios the situation changes dramatically, as can be seen from the Figure 5.85.

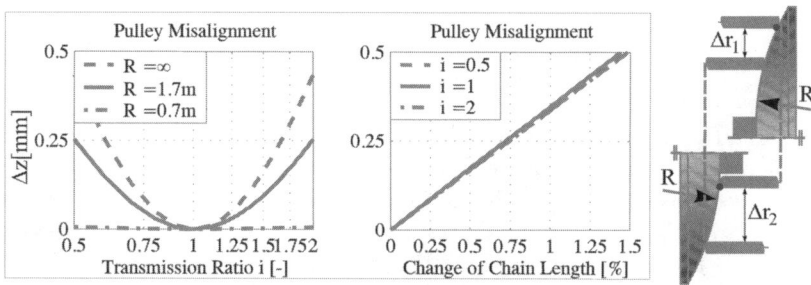

Fig. 5.85: Pulley Misalignment [239]

It is possible to influence that kind of misalignment by the curvature of the pulley sheaves. But this is very much limited, because on the other hand too large curvatures would lead to edge-carrying effects of the pins, which have to be avoided [239]. In order to limit the contact pressure between pins and sheaves the disc curvature radius $R$ must have a lower bound. These arguments indicate, that chain misalignment represents definitely a typical system's pecularity being not only a result of erroneous manufacturing.

302   5 Power Transmission

In the following we shall continue the discussions of the preceding chapters concerning a plane model of a rocker pin chain. We extend the coresponding theory to the spatial case, where "spatial" means only small deviations from the planar configuration. On the other hand we shall recognize, that even very small deviations result in large changes of the chain forces as a consequence of an extreme sensitivity of all contact processes to slight changes of the contact conditions. The chapter is mainly based on the research findings of Sedlmayr [239].

### 5.6.2 Mechanical Models

The contact forces between the chain and the pulleys cause a considerable deformation of the *pulley's sheaves*. The eigenfrequency of the sheaves is much higher than the operating frequencies. Therefore and according to [230] and [249] the mass forces of the elastic deformations can be neglected in this

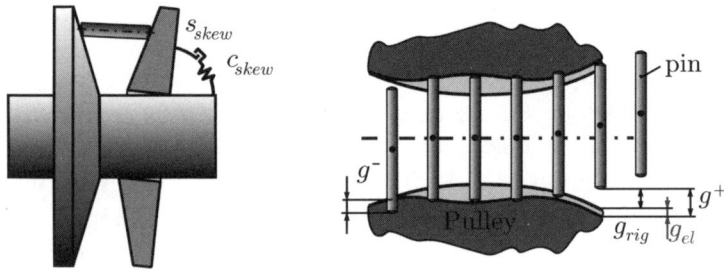

Fig. 5.86: Lateral Buckling and Pulley Deformation [239]

case. Using the degrees of freedom $(\Delta\vartheta_{F/M}, \gamma_{0,F/M})$ the lateral buckling of the movable sheaves may be approximated by a sine function $\Delta\vartheta_{F/M}\sin(\gamma_{c,F/M} - \gamma_{0,F/M})$ (see chapter 5.5.1). The magnitude $\Delta\vartheta$ is the angle of inclination of the sheaves including rigid and elastic parts. The gliding angle $\gamma$ describes the difference of the pin's motion along the disc and the exact circular direction. The indices $F$ and $M$ denote the fixed and the movable sheave, respectively, and $\gamma_{c,F/M}$ the contact location. The amplitude $\Delta\vartheta_{F/M}$ consists of a backlash $s_{skew}$ and an elastic deformation between the movable sheave and the shaft, Figure 5.86.

With a FEM-analysis the sheave deformation can be calculated with the Reciprocal Theorem of Elasticity (Betti/Maxwell). Together with Hooke's law applied at the pin this results in a Linear Complementary Problem (LCP) in a standard form [200]. The meaning is clear: If the pins get into contact, we have constraint forces, otherwise not (Figure 5.86). Both pulley sets have one rotational ($\varphi$), two translational in-plane ($x, y$) and one axial out-of-plane

($z_M$) degrees of freedom. All degrees of freedom can be collocated in $\boldsymbol{q}_P = (x, y, z_M, \varphi, \Delta\vartheta_F, \gamma_{0,F}, \Delta\vartheta_M, \gamma_{0,M})^T$.

The relative distances and the complementary behavior, as indicated in Figure 5.86, can then be formulated as follows:

$$g_i = g_{rig,i} + g_{el,i} = g_i^+ - g_i^-, \qquad g_i^- = c_{rp}^{-1} F_{rp,i}, \qquad g_{el,i} = \sum_j w_{ij} F_{rp,j}$$

$$g_i^+ = \sum_j w_{ij} F_{rp,j} + c_{rp}^{-1} F_{rp,i} + g_{rig,i}$$

$$\mathbf{g}^+ = \mathbf{W}\mathbf{F}_{rp} + \mathbf{g}_{rig}, \qquad \text{with} \qquad \mathbf{F}_{rp} \geq \mathbf{0}, \quad \mathbf{g}^+ \geq \mathbf{0}, \quad \mathbf{F}_{rp}^T \mathbf{g}^+ = \mathbf{0} \qquad (5.125)$$

Each **link** represents an elastic body with three translational rigid body degrees of freedom and in addition the elastic degrees of freedom. The angles $\beta_L$ and $\gamma_L$, shown in Figure 5.87, kinematically depend on the translational position of the successor link. In order to describe the orientation and elastic

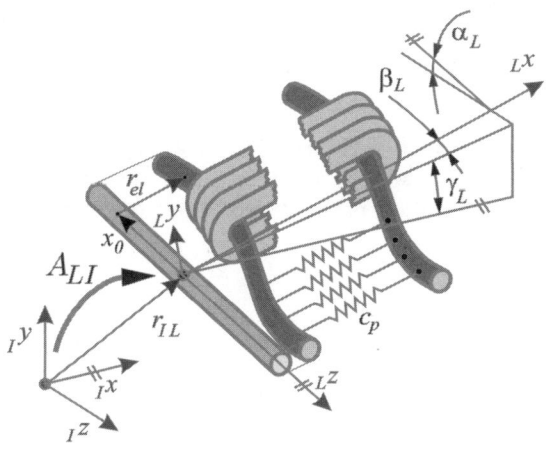

Fig. 5.87: Chain Link Model

deformation of a pin some more degrees of freedom $\mathbf{q}_{el} = (\mathbf{q}_{el,x}^T \; \mathbf{q}_{el,y}^T)^T$ have to be introduced. We distinguish between the radial ($y$) and azimuthal ($x$) directions. Thus the set of generalized coordinates can be written as $\mathbf{q}_L = (x_L \; y_L \; z_L \; \mathbf{q}_{el}^T)^T$. The links are kinematically interconnected by pairs of rocker pins. The elasticity and the translational damping of the joint is taken into account by the link force element whereas the rotational damping and the axial friction between the pair of pins is considered by the joint force element. The link force element takes into account every plate as a spring. The effect of moving contact points relative to the plate spring reference points between a rocker pin and an adjoining plate are modeled as a contact torque. This effect has turned out to be very important for an optimization of the link elements, therefore we shall consider it in more detail (see [168], [167]).

The rocker pins are composed of two elements of the same kind, which roll on each other without sliding. In order to be able to describe the **kinematics of a rolling rocker pin joint** we need in a first step a specification of the inner contour of a rocker pin, because the two pins of a joint are rolling one on the other along their inner contour. It consists of two symmetrical halves, each is an involute to a circle, and is defined by the two parameters $r_0$ and $r$ (Fig. cvts-contour). The following equations hold true for the upper half

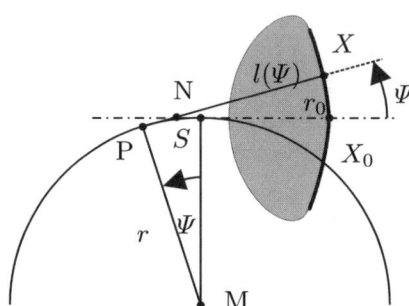

Fig. 5.88: Inner Contour of a Rocker Pin: Two symmetric halves, each an involute to a circle. The pictured circle is only used for the construction of the upper half. The lower half is the reflection of the upper half with respect to the symmetry line $SX_0$.

$(0 \leq \Psi \leq \Psi_{max})$:

$$r_0 := \overline{SX_0},$$
$$l(\Psi) = \overline{PX} = r_0 + r\Psi,$$
$$\overline{PN} = \overline{NS} = r \tan \frac{\Psi}{2}. \tag{5.126}$$

The lower half is the reflection of the upper half with respect to the symmetry line $SX_0$.

As illustrated in Fig. 5.89 the rocker pins are fixed to the links with their outer contours. It should be noticed, that the symmetry axis of the rocker pin is not parallel to the link plate axis, but it is rotated by the constant angle $\delta$. This fact has been neglected in former models [239], but should be considered when studying the joint kinematics more in detail. Figure 5.89 shows an angled rocker pin joint. The most important kinematical effect consists in an offset $\Delta a_i$ between the intersection point $X_0$ of the symmetry axis with the contour of the pin and the intersection point $B_i$ of the two link axes. For common rotational joints ($r_0 = r = 0$) this offset vanishes. The connection between

## 5.6 CVT - Rocker Pin Chains - Spatial Model

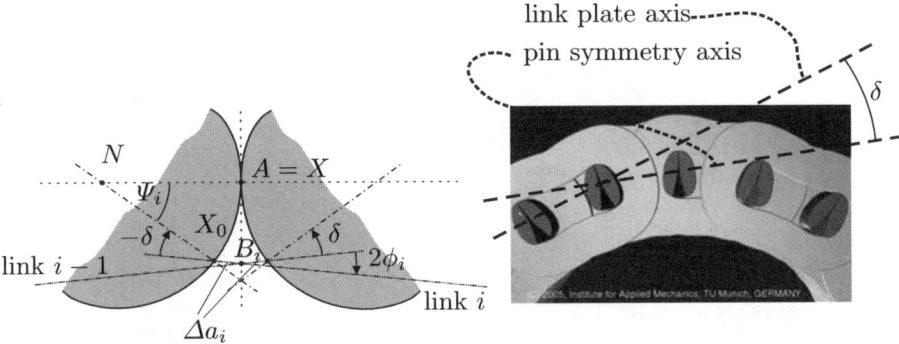

Fig. 5.89: Kinematics of an Angled Rocker Pin Joint

the joint angle $2\phi_i$ and the offset $\Delta a_i$ is calculated by using Eqs. 5.126 and applying some trigonometrical operations:

$$\psi_i = \phi_i - \delta, \qquad \Psi_i = |\psi_i|$$

$$\overline{AN} = l(\Psi_i) - r \tan \frac{\Psi_i}{2}, \qquad \overline{NX_0} = r_0 + r \tan \frac{\Psi_i}{2}$$

$$\Delta a_i = \overline{X_0 B_i} = \frac{1}{\cos \phi_i} \left[ r_0 \left(1 - \cos \Psi_i\right) + r \left(\Psi_i - \sin \Psi_i\right) \right] \tag{5.127}$$

Rolling without sliding or detachment occurs in the contact between each rocker pin pair. So the coordinates of the intersection points $B_i$ of neighbouring links can be chosen as generalized coordinates, as they describe the complete configuration of the chain ($n = 63$):

$$\mathbf{q} = (\mathbf{q}_1^T, \mathbf{q}_2^T, \ldots, \mathbf{q}_n^T)^T \in \mathbb{R}^{2n}, \qquad \text{with} \qquad \mathbf{q}_i = (y_i, z_i)^T. \tag{5.128}$$

For developing the chain equations of motion we consider Figure 5.90. We collect the proportionate masses $m_{ji}$ of the chain (link plates and pins) in the point $P_{ji}$. The point $K_{ji}$ denotes the position of the potential contact points $K_{ji}$ on the rocker pins to the pulleys. The position vectors of $P_{ji}$ and $K_{ji}$ can be stated by referring to the angles $\alpha_i$ of the link axes (Figure 5.90). Furtheron we introduce the offset $\Delta a_i$ resulting from the rocker pin joint kinematics (Eq. 5.127, Fig. 5.89) and the constant $\Delta b_i := \overline{P_{ji} K_{ji}}$, which describes the position of the body fixed contact point on the pin. The vectors to the above points then write

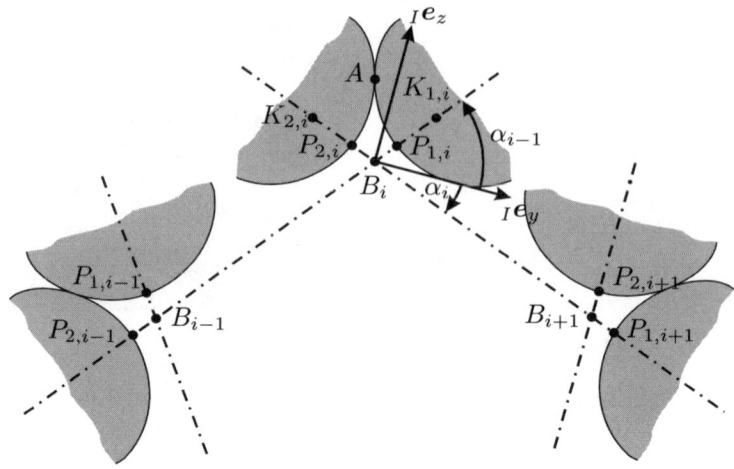

Fig. 5.90: Model of the Rocker Pin Chain

$$\mathbf{r}_{P_{1i}} = \mathbf{q}_i + \Delta a_i \begin{pmatrix} \cos\alpha_{i-1} \\ \sin\alpha_{i-1} \end{pmatrix}$$
$$\mathbf{r}_{P_{2i}} = \mathbf{q}_i - \Delta a_i \begin{pmatrix} \cos\alpha_i \\ \sin\alpha_i \end{pmatrix} \tag{5.129}$$

$$\mathbf{r}_{K_{1i}} = \mathbf{r}_{P_{1i}} + \Delta b_i \begin{pmatrix} \cos(\alpha_{i-1}+\delta) \\ \sin(\alpha_{i-1}+\delta) \end{pmatrix}$$
$$\mathbf{r}_{K_{2i}} = \mathbf{r}_{P_{2i}} - \Delta b_i \begin{pmatrix} \cos(\alpha_i-\delta) \\ \sin(\alpha_i-\delta) \end{pmatrix} \tag{5.130}$$

The equations (5.129) refer to the mass points and the equations (5.130) to the contact points. Each of the two rocker pin components possesses its own contact point, which turns out to be quite realistic. The link plates are regarded as massless spring damper elements (stiffness $c_i$, damping ratio $d_i$) and connect the points $P_{2,i}$ and $P_{1,i+1}$. The resulting link forces $\boldsymbol{F}_{L_i}$ write

$$\mathbf{F}_{L_i} = \left[c_i\left(l_i - l_{i,0}\right) + d_i \dot{l}_i\right] \begin{pmatrix} \cos\alpha_i \\ \sin\alpha_i \end{pmatrix}, \tag{5.131}$$

where $l_i$ denotes the length of link $i$, and $\dot{l}_i$ is its time derivative. Together with the gravitational parts (gravitational acceleration $g$) the following forces act on pin $ji$ at $P_{ji}$:

$$\mathbf{F}_{P_{1i}} = -\mathbf{F}_{L_{i-1}} + m_{1i}\begin{pmatrix} 0 \\ g \end{pmatrix}, \quad \mathbf{F}_{P_{2i}} = +\mathbf{F}_{L_i} + m_{2i}\begin{pmatrix} 0 \\ g \end{pmatrix}, \tag{5.132}$$

Additional frictional forces $\boldsymbol{F}_{K_{ji}}$ due to the contact between the discs and the rocker pin have to be taken into account. On part of the chain they depend on the position and the velocity of the potential contact point $K_{ji}$ (details can be found in [249]). In each joint there arise damping moments $\boldsymbol{M}_{\phi_i}$, proportional to the angular velocity in the joint

$$\mathbf{M}_{\phi_i} = -d_{\phi_i}(\dot{\alpha}_i - \dot{\alpha}_{i-1}) \tag{5.133}$$

The **chain pulley contacts** are rocker pin pulley contacts, either modeled with one contact point [239] or with two contact points [167]. As was pointed out already, we may get edge carrying situations for extreme transmission ratios. This is also confirmed experimentally by the dissertation [105]. The three-dimensional dynamics of the chain generates additional shear and torsional deformations superimposed to the bending of the rocker pins, which appears anyway. Therefore the elasto-hydrodynamic oil film is squeezed out of the contact zone and something like mixed friction will be built up. Altogether it turned out that Coulomb's friction law represents a good approximation. In order to take into account Hertzian deformation a nonlinear spring law in the chain pulley contact will be introduced, already presented for plane CVT-dynamics with equation (5.112) on page 288.

### 5.6.3 Mathematical Models

Applying the principle of virtual power (Jourdain) the equations of motion for a rigid body as well as for an elastic body can be derived in a form, which has already been discussed in chapter 3.4 in connection with the equations (3.159) on page 143. They write

$$\mathbf{M}(\mathbf{q},t)\ddot{\mathbf{q}}(t) + \mathbf{h}(\mathbf{q},\dot{\mathbf{q}}t) - [(\mathbf{W}_N + \mathbf{W}_R) \quad \mathbf{W}_T] \begin{pmatrix} \lambda_N(t) \\ \lambda_T(t) \end{pmatrix} = \mathbf{0} \tag{5.134}$$

The mechanical quantities appearing in these equations have been already explained on page 143 and the following. The special derivation of the equations of motion for the rocker pin chain considers therefore mainly the pecularities for this case.

For **rigid bodies** we start with the equations of momentum and the equations of moment of momentum, which yields

$$\mathbf{J}^T \begin{pmatrix} m\mathbf{E} & m\tilde{\mathbf{r}}_{BS} \\ m\tilde{\mathbf{r}}_{BS} & \mathbf{I}_B \end{pmatrix} \mathbf{J}\ddot{\mathbf{q}} + \mathbf{J}^T \begin{pmatrix} m\tilde{\boldsymbol{\Omega}}\tilde{\boldsymbol{\Omega}}\tilde{\mathbf{r}}_{BS} \\ \tilde{\boldsymbol{\Omega}}\mathbf{I}_B\boldsymbol{\Omega} \end{pmatrix} = \mathbf{J}^T \begin{pmatrix} \mathbf{F} \\ \mathbf{T} + \tilde{\mathbf{r}}_{BF}\mathbf{F} \end{pmatrix}$$

$$\mathbf{J}^T = (\mathbf{J}_T^T \quad \mathbf{J}_R^T) = \left[ (\frac{\partial \mathbf{v}_B}{\partial \dot{\mathbf{q}}})^T \quad (\frac{\partial \boldsymbol{\Omega}}{\partial \dot{\mathbf{q}}})^T \right] \tag{5.135}$$

These equations of motion refer to a point $B$ of the rigid body. The point $S$ marks the center of mass. The vector $\mathbf{r}_{BS}$ points from $B$ to $S$. $\mathbf{I}_B$ denotes the matrix of moments of inertia, $m$ the mass, $\mathbf{\Omega}$ the rotational speed, $\mathbf{v}_B$ the velocity, $\mathbf{F}$ the active forces and $\mathbf{T}$ the active torques. The Jacobian $\mathbf{J}$ transforms the equations of motion from the space of rigid body motion to the hyper-spaces orientated tangential to the constraint surface called configuration space. The rigid body degrees of freedom of the pulleys are given with the equations (5.135). The rigid body motion is superposed by elastic deformations.

Starting from an inertial reference the vector to a mass point of a *deformed body* is composed by rigid parts $(\mathbf{r}_{IL} + \mathbf{x}_0)$ and by the elastic deformation parts $\mathbf{r}_{el}$. The vector $\mathbf{r}_{IL}$ denotes the distance from the inertially fixed point $I$ to the origin of the reference system of coordinates $L$, $\mathbf{x}_0$ denotes the vector from this origin to a mass element $dm$ in the undeformed configuration and $\mathbf{r}_{el}$ the elastic displacement vector (see for example Figure 2.22 on page 48). Assuming small displacements one can introduce a Ritz approach for the elastic deformation in a time and space separated formulation.

$$\mathbf{r}_i = (\mathbf{r}_{IL} + \mathbf{x}_0 + \mathbf{r}_{el})_i,$$

$$(\mathbf{r}_{IL})_i = (x_L \ \ y_L \ \ z_L)_i^T, \qquad \mathbf{A}_{IR_i} = \begin{pmatrix} \cos\gamma_L & -\sin\gamma_L \\ \sin\gamma_L & \cos\gamma_L \end{pmatrix}_i,$$

$$(\mathbf{r}_{el})_i = \left\{ \left[ \mathbf{A}_{IR} \left( \mathbf{w}_x^T(\mathbf{x}_0)\mathbf{q}_{el,x}(t) \ \ \mathbf{w}_y^T(\mathbf{x}_0)\mathbf{q}_{el,y}(t) \right)^T \right]^T \ \ \mathbf{0} \right\}_i^T. \tag{5.136}$$

The equations of motion for an elastic body may be derived from Jourdain's principle with $\mathbf{a}$ the absolute acceleration and $\mathbf{f}$ the applied forces:

$$\int \left(\frac{\partial \mathbf{v}}{\partial \dot{\mathbf{q}}}\right)^T (\mathbf{a}\,dm - d\mathbf{f}) = \mathbf{0} \tag{5.137}$$

For the evaluations of these equations we have to go back to the relations (3.127) on page 127 in connection with the elasto-kinematical considerations of chapter (2.2.8) on page 47. These evaluations are very lengthy and tedious, and they have to be performed for every component of the spatial CVT system as presented before (see [239]). Fortunately, the results can be partly simplified and to a certain extend also decoupled by the fact, that the out-of-plane motion is very small compared to the nominal motion. Nevertheless it should be noted, that this refers to kinematics not to forces. Very small geometric changes produce very large changes in the forces, especially contact forces.

The determination of the **contact forces between pulley and rocker pins** we need some definitions according to Figure 5.91. We shall use (L) for the movable, F) for the fixed pulley side and (B) for the pin. Furtheron the indices (I, C, K) stand for inertial, contact point and middle point of the rocker pins. The angles between the corresponding coordinates are also given in Figure 5.91. We additionally consider in the following only one contact point per rocker pin end surface [239] and not two contact points as presented in [167].

## 5.6 CVT - Rocker Pin Chains - Spatial Model

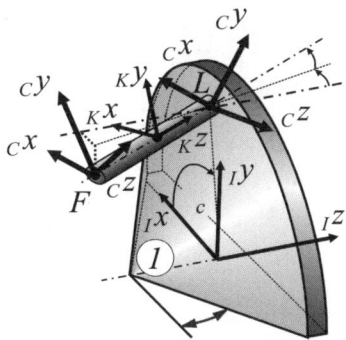

Fig. 5.91: Contact Pin/Sheave for the Movable Side

With these definitions the contact forces between the sheaves and the front faces of the rocker pins are determined by the geometrical gap function $g(\mathbf{q})$ being defined in the axial direction of the undeformed rocker pin. A system of coordinates $C$ is introduced on the cone surface in the middle of the rocker pin end, defined by the triple of circumferential, radial and normal direction $(t, r, n)$ and their forces $(F_t, F_r, F_n)$. From this we get (see also [250])

$$F_{B,i} = -c_B(g_i^-)\, g_i^- \approx c_{B_1}|g_i^-|^{p_{B_1}} + c_{B_2}|g_i^-|^{p_{B_2}}, \qquad (5.138)$$

which is a nonlinear empirical force law for the pins. In dependency of the sign $s_z = \text{sign}(_I z_C)$ of the inertial position of the contact we define the quasistatic equilibrium in the form

$$_K\mathbf{F}_{L/F,i} = \begin{pmatrix} * \\ * \\ s_z F_B \end{pmatrix}_{L/F,i} = \mathbf{A}_{K_i C_{L/F,i}}\, F_{n,L/F,i} \begin{pmatrix} \mu \frac{\dot{g}_{T,a}}{\dot{g}_T} \\ \mu \frac{\dot{g}_{T,r}}{\dot{g}_T} \\ s_z 1 \end{pmatrix}_{L/F,i} \qquad (5.139)$$

From these equations we are able to derive the normal forces and the friction forces acting on the pulley sheaves. We come out with [239]

$$F_{N,i} = \left( \frac{F_{B,i}}{\cos\vartheta_c + \sin\vartheta_c\, \mu \frac{\dot{g}_{T,r}}{\dot{g}_T} + s_z\, \psi_c\, \mu \frac{\dot{g}_{T,a}}{\dot{g}_T}} \right)_{L/F,i},$$

$$F_{R,a,i} = F_{N,i}\, \mu_i \frac{\dot{g}_{T,a,i}}{\dot{g}_{T,i}}, \qquad F_{R,r,i} = F_{N,i}\, \mu_i \frac{\dot{g}_{T,r,i}}{\dot{g}_{T,i}}, \qquad (5.140)$$

where (a) and (r) define the azimuthal and radial directions, repectively. The various magnitudes necessary to evaluate these relations are the transformation matrix $\mathbf{A}_{K_i C_{L/F,i}}$, the gap velocities $\dot{g}_T$ and the angles $\psi_{c,i}, \vartheta_{c,i}$ of the contact points under consideration (see Figure 5.137). We get

$$\boldsymbol{A}_{K_iC_{L/F,i}} = \begin{pmatrix} 1 & -s_z\psi_c\sin\vartheta_c & -\psi_c\cos\vartheta_c \\ 0 & \cos\vartheta_c & -s_z\sin\vartheta_c \\ \psi_c & s_z\sin\vartheta_c & \cos\vartheta_c \end{pmatrix}_{L/F,i}, \qquad s_z = \text{sign}(_Iz_c) ,$$

$$\begin{pmatrix} \dot{g}_{T,a} \\ \dot{g}_{T,r}\cos\vartheta \end{pmatrix} = \begin{pmatrix} \sin(\varphi_c) & -\cos(\varphi_c) \\ \cos(\varphi_c) & \sin(\varphi_c) \end{pmatrix} \left[ \begin{pmatrix} \dot{x}_G \\ \dot{y}_G \end{pmatrix} - \begin{pmatrix} \dot{x} \\ \dot{y} \end{pmatrix}_F \right] +$$
$$+ \frac{l_B}{2}\begin{pmatrix} s_z\dot{\psi} \\ -s_z\dot{\alpha} \end{pmatrix} + \begin{pmatrix} r_L\omega_L \\ 0 \end{pmatrix} ,$$

$$\psi_{c,i} = \psi_{G,i} + s_z\left(\frac{\partial[\vartheta(r,\varphi_c)(r - r_{Kipp})]}{\partial[\varphi_c\, r]}\right)_i + \Delta\psi_W(\varphi_c),$$

$$\vartheta_{c,i} = [\vartheta(r,\varphi_c)]_i + s_z\,\alpha_i + \Delta\vartheta_W(\varphi_c) , \qquad (5.141)$$

The angles $\Delta\psi_W$ and $\Delta\vartheta_W$ describe the pulley shaft obliqueness due to some given tolerances. With respect to the friction we may introduce as a nearly perfect approximation a smooth law, because for all types of chains and belts stick events are extremely seldom and can thus be neglected. Therefore, the relation

$$\mu_i(\dot{g}_i) = \mu_0\left[1 - \exp\left(-|\frac{\dot{g}_i}{v_c}|\right)\right] \qquad (5.142)$$

will be sufficient for modeling friction between pins and sheaves. For further evaluations we need the contact forces in an inertial system, which we get from

$$_I\boldsymbol{F}_{c,L/F,i} = \begin{pmatrix} \sin(\varphi_c) & \cos(\varphi_c) & 0 \\ -\cos(\varphi_c) & \sin(\varphi_c) & 0 \\ 0 & 0 & 1 \end{pmatrix}_i \begin{pmatrix} 1 & 0 & \psi_G \\ 0 & 1 & -\alpha_G \\ -\psi_G & \alpha_G & 1 \end{pmatrix}_i \boldsymbol{A}^{-1}_{K_iC_{L/F,i}} \begin{pmatrix} F_{R,a} \\ F_{R,r} \\ s_zF_N \end{pmatrix}_i .$$
$$(5.143)$$

The normal and tangential contact forces are represented in the C-coordinate frame, the $_Cz$-axis of which is perpendicular to the contact surface. The pin force possesses the direction of the $_Kz$-axis, see Figure 5.91.

Considering the motion of the contact points, we introduce **contact torques** in an exponential approximation as a function "con" of the normal force $F_c$, the distance $l_e$ of the edge of a contact body to the reference point, the radius $r_c$ of the contact surface, the rotational contact stiffness $c_c$ and the relative rotational displacement $\varphi_c$.

$$\text{con}(F_c, l_e, c_c, \varphi_c) = F_c l_e \text{sign}\left[1 - \exp\left(-\frac{|\varphi_c|c_c}{F_c l_e}\right)\right],$$
$$c_c = \begin{cases} c_{c,max} & \forall \quad F_c : c_{c,max} < F_c r_c \\ F_c r_c & \forall \quad F_c : c_{c,max} \geq F_c r_c \end{cases} \qquad (5.144)$$

## 5.6 CVT - Rocker Pin Chains - Spatial Model

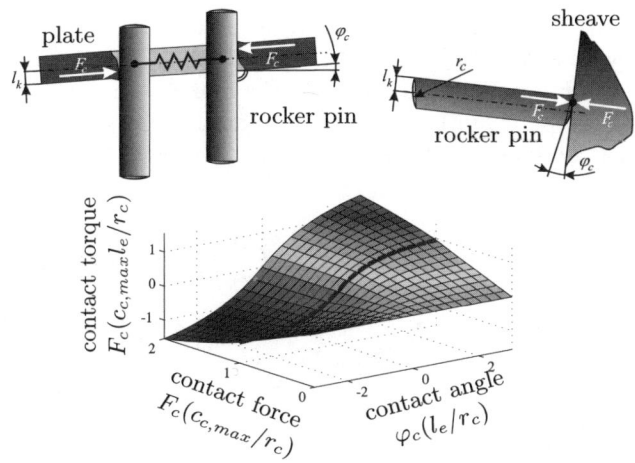

Fig. 5.92: Contact Torque by Relative Angle Displacement

By increasing the contact forces $F_c$ over the limit $c_{c,max}/r_c$ the contact area reaches the cross section area, see Figure 5.92. Then the stiffness is independent from the contact radius $r_c$ and becomes the constant value of $c_{c,max}$. Thus the projected torques $\boldsymbol{T}_{F/M,i}$ and $\boldsymbol{T}_{pl,i}$ between a rocker pin pair $i$ and a sheave ($F/M$) and between the plates ($pl$) and the pin $i$ are

$$\boldsymbol{T}_{F/M,i} = -2 \begin{pmatrix} \boldsymbol{w}'_x \boldsymbol{A}_{IC_{F/M}} \cdot \mathrm{con}(F_n/2, l_h/2, c_h, \vartheta_c) \\ \boldsymbol{w}'_y \boldsymbol{A}_{IC_{F/M}} \cdot \mathrm{con}(F_n/2, l_w/2, c_w, \vartheta_c) \\ 0 \end{pmatrix}_i ,$$

$$\vartheta_c \approx \vartheta_0 + (\Delta\vartheta \sin(\gamma_c - \gamma_0))_{F/M} \mp \alpha_L ,$$

$$\psi_c \approx \pm (\Delta\vartheta \cos(\gamma_c - \gamma_0))_{F/M} \pm \psi_L ,$$

$$\boldsymbol{T}_{pl,i} = \int_{-1}^{+1} \boldsymbol{w}'_x(\xi)(\mathrm{con}(f_i, \frac{b_{pl}}{2}, \frac{c_{pl}}{l_{rp}}, \beta_{L,i} - \psi_i) -$$

$$- \mathrm{con}(f_{i-1}, \frac{b_{pl}}{2}, \frac{c_{pl}}{l_{rp}}, \beta_{L,i-1} - \psi_i))d\xi, \quad (5.145)$$

respectively, where $(l_h, l_w)$ denote the height and width of a single pin and $(c_h, c_w)$ the contact stiffness in their rotational direction. In equation (5.145) the elastic pin deformations are neglected. To take into account the friction forces between the pair of rocker pins $i$ we calculate the torque

$$\boldsymbol{T}_{rp,i} = -\mu l_w \int_{-1}^{+1} \boldsymbol{w}'_x(\xi) f_i(\xi) \mathrm{sign}(\boldsymbol{w}'^T_x \dot{\boldsymbol{q}}_{el,x,i}) d\xi \quad \text{with} \quad f_i(\xi) = \frac{c_L}{l_{rp}}(a_{el,i}(\xi) - a_i)$$

(5.146)

as the line load of the pin. The parameter $a_i$ is the kinematically unstressed and $a_{el,i}$ is the kinematically stressed length of a link. The forces and torques are visualized in Figure 5.93. In order to take into account the position of the contact line between a pin pair, we have to add an in-plane bending torque $T_\gamma$. Regarding the clasp plate stiffness $c_{clasp}$, we have to project the shear

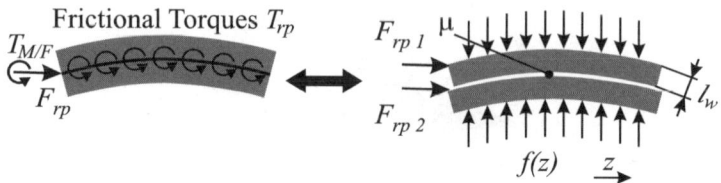

Fig. 5.93: Contact Reactions Acting on the Rocker Pin Pair

torque $T_{clasp}$ into the configuration space of a link containing a clasp plate. Figure 5.94 shows the joint kinematics and their parameters. Distinguishing

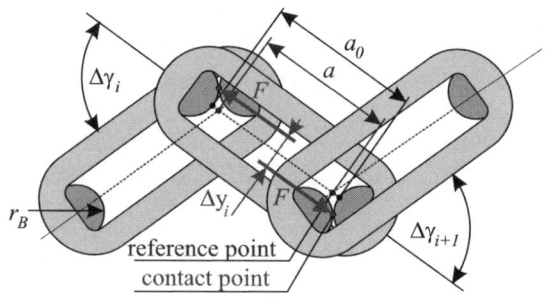

Fig. 5.94: Joint Kinematics

between the reference point and the contact point the link length and contact torque are determined. The rocker pin radius $r_{rp}$ can change according to the following equation with regard to the angle $\Delta\gamma$.

$$T_\gamma = c_L(a_{el} - a)\Delta y = F_{Chain}\Delta y, \qquad T_{clasp,\beta} = -T_{clasp,\psi} = c_{clasp}(\psi_L - \beta_L),$$

$$r_{rp} = r_{rp,0} + \frac{\partial r_{rp}}{\partial \Delta\gamma}\Delta\gamma \qquad (5.147)$$

### 5.6.4 Some Results

In close cooperation with a few colleagues and supported significantly by the German Research Foundation (DFG) a large variety of practical cases have been considered, evaluated and verified by measurements. All these efforts confirm the theory and, what seems to me the most important issue, they confirm the ideas of the models behind the theory. We shall give a few examples, see also [239], [167], [250], [249], [238], [199], [207]. In a first step it

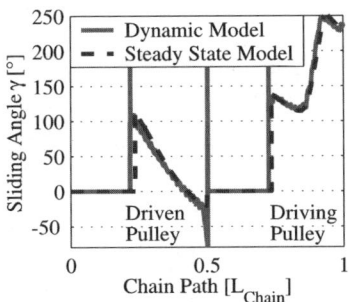

Fig. 5.95: Model Comparison

makes sense to compare the quite popular models representing the chains by a quasistatic belt models and the dynamic models as presented above. Due to the lateral buckling of the sheaves and the elastic deformation of sheaves and pins we get in addition to belt creep and azimuthal belt slip a radial movement between the chain and the sheaves. Introducing the sliding angle $\gamma$ we can define the direction of the relative velocity between pins and sheaves. The sliding angle determines the equilibrium between the two pulley clamping forces and also the chain forces.

Thus a comparison of the sliding angle is suitable for the verification of chain models. Figure 5.95 depicts the sliding angle of the discrete dynamic model and the continuous steady state model of Sattler [230]. In the dynamic model the contact arcs are longer than by the model of Sattler where they are 180 Grad. In the continuous model the polygonal excitation is missing. Only for low rotational speeds the comparison is good. For high speeds the continuous model does not describe all effects.

The results of the dynamic model presented in the following are computed for an uniform motion with constant driving speed $n_1$ and an external output torque $T_2$. The Figures (5.96) show the tensile force of an outer plate of a chain with clasp plates for two different pulley misalignments ($T = 150 Nm, n = 600 rpm, i = 1$). The comparison of simulation and measurement [231] confirms the mechanical model. Due to the bending forces the misalignment induces a large gradient of the tensile forces in the spans. Entering a pulley the shape of a pin changes abruptly because of the sudden growing contact forces. Thus especially at the beginnings and the ends of the pulling spans great force peaks appear.

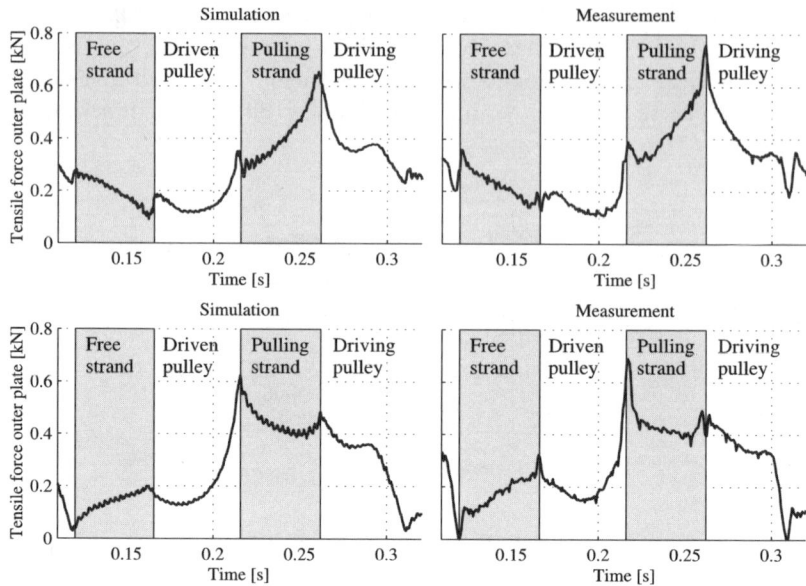

Fig. 5.96: Comparisons of Simulation and Measurements for the Tensile Forces of an Outer Plate with Clasp Plate: Figure top with $\Delta z = +1.5mm$, Figure bottom with $\Delta z = -0.5mm$, [231], [239]

In the following we consider some parameter studies with the goal to show the possibility of influencing the chain performance significantly by even small changes of chain component geometry. For this purpose we concentrate only on the forces of the outer plates, which are the most stressed plates on the one side, but represent also a good measure of the chain performance on the other side. With longer rocker pins and more plates the tensile forces of a plate can be reduced because of a load splitting on more plates. But with the same cross section of the pins due to the pin bending the load splitting on the plates becomes worse. Thus the gain connected with more plates is not large (Figure 5.97).

By changing the design of a plate without altering the tensile strength we influence the stiffness $c_L(\pm 20\%)$ and the mass $m_L(\pm 10\%)$ of a link. The resistance to the pin bending with low stiffness $c_L$ is dominated by the bending stiffness of the pin. As a result the static components of the tensile forces of both outer plates become smaller with softer plates, see Figure 5.98. Additionally the dynamic and centrifugal force components get smaller because of the lower link mass $m_L$. The previous calculation in the figures above are executed with no specification of the assemblage of the plates and thus with a continuously approximation.

In Figure 5.99 we have three chains with different assemblages of the plates. But in each chain the link configuration is repeated after the third following

## 5.6 CVT - Rocker Pin Chains - Spatial Model

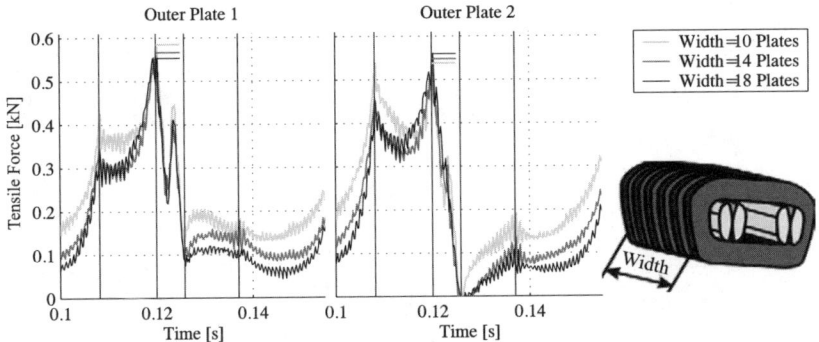

Fig. 5.97: Variation of the Number of Plates of Each Link
($i_{red} = 1, i_{CVT} = 2.3, n_1 = 4000 rpm$)

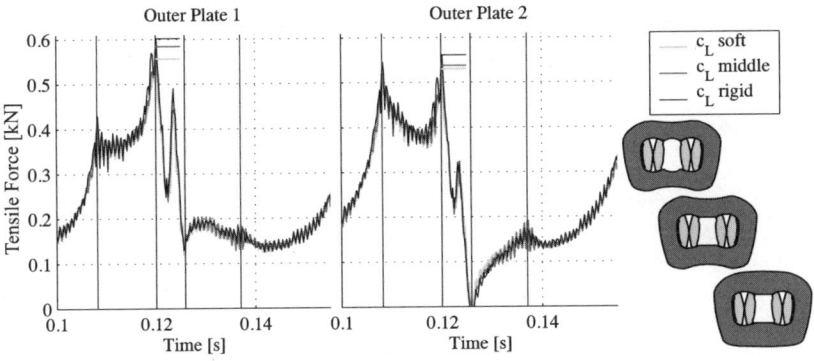

Fig. 5.98: Variation of the Chain Link Stiffness $c_L$
($i_{red} = 1, i_{CVT} = 2.3, n_1 = 4000 rpm$)

link. In Figure 5.99 the brightest, middle and darkest link is link number 1, 2 and 3, respectively. In both cases examined with a symmetric link configuration pattern the load on all outer plates are nearly equal, whereas the maximum tensile force of the right outer plate of link 2 is larger in the case of the asymmetric traditional arrangement than the highest plate forces of the links with symmetric configuration patterns. There are even more design parameters than discussed above. For example for a greater axle-base the tensile forces of the chain are smaller [270]. By enlarging the curvature of the pin ends and lowering the contact stiffness between pins and plates the outer plate tensile forces can be reduced [238]. The polygonal excitation causes some resonance of the chain drive. Altering the kinematics of the excitation we influence the resonance. On the right hand side of Figure 5.100 we see the excitation

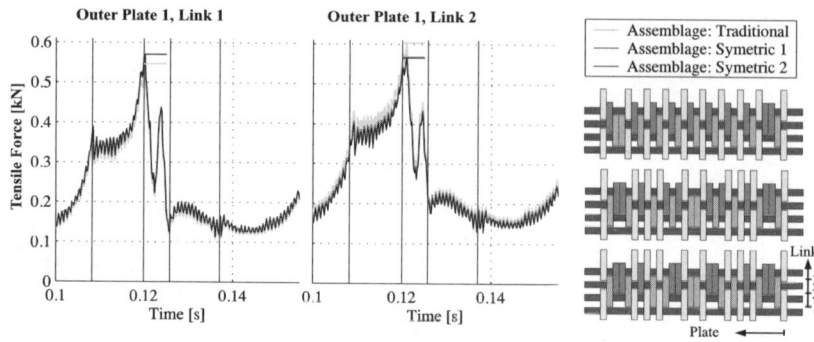

Fig. 5.99: Variation of the Assemblage of the Plates
($i_{red} = 1, i_{CVT} = 2.3, n_1 = 4000 rpm$)

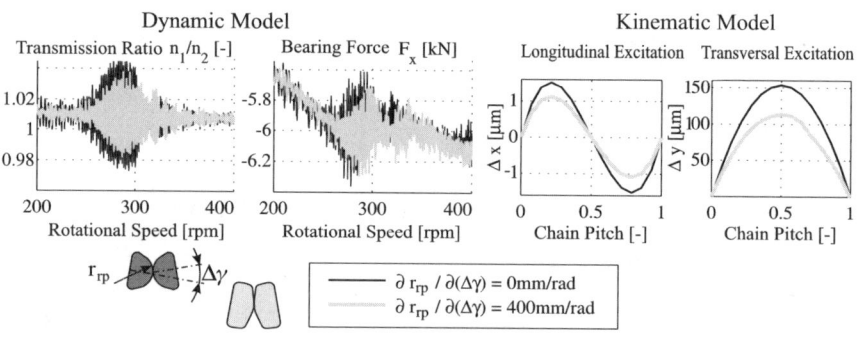

Fig. 5.100: Influence of the Rocker Pin Kinematics

at the entrance and the exit areas of the sheaves assuming the contact arc as an ideal circle. On the left hand side the dynamic response of the excitation with all details of the model is illustrated. The oscillation of the transmission induces rotational vibration in the drive train whereas the vibrating bearing forces impair the gear acoustic.

With low torques the ratio of the axial clamping forces is near one, because there is nearly no difference between the driving and driven pulley. At adequate torques the ratio is high due to different direction of the lateral buckling of the sheaves. Near the slipping border the radial friction forces vanish and therefore the ratio decreases. At high torques before the slippage of the chain the losses of the gear including the bearing and excluding the hydraulics are low in comparison with the gear power. The pulley thrust ratio in Figure 5.101 is changing with the pulley misalignment because of axial components of the chain tensile forces, whereas the efficiency shows no influence. At transient operating state the global behavior of the gear, the forces, thrust ratio and the efficiency change. In Figure 5.102 the effects are illustrated for a transient

## 5.6 CVT - Rocker Pin Chains - Spatial Model

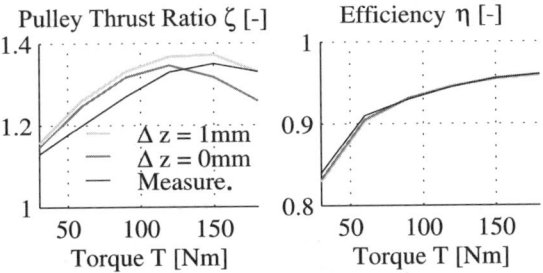

Fig. 5.101: Global Parameters Influenced by Pulley Misalignment

transmission ratio $i$ of the CVT. It is obvious that the pulley forces have to differ to induce a large gradient of the transmission ratio. At a positive gradient the great losses due to the clamping effect at the chain exit of the driving pulley becomes significantly smaller. As a result the CVT-efficiency increases. Due to this clamping effect the slip at the driving pulley is smaller than at the driven pulley.

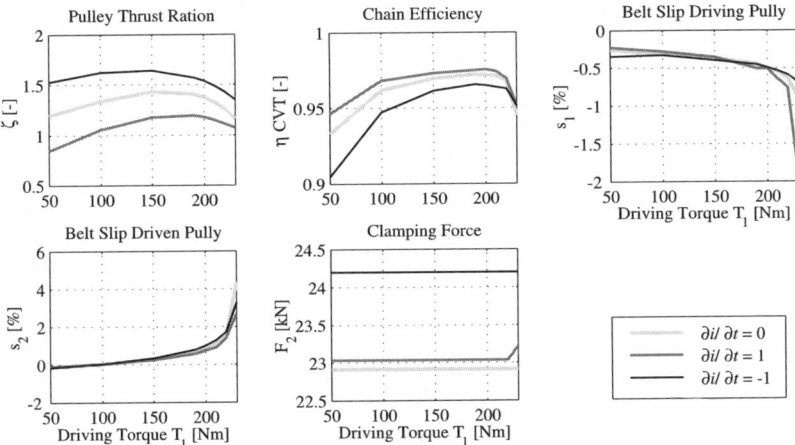

Fig. 5.102: Global Parameters Influenced by a Transient Transmission Ratio ($i_{CVT} = 1, n_1 = 2000 rpm$)

318    5 Power Transmission

## 5.7 CVT - Push Belt Configuration

### 5.7.1 Introduction

General properties of CVT chains and belts have been discussed in chapter 5.4 on the pages 275 and following. We shall refer here to push belts, which have been invented by van Doorne in the Netherlands long time ago, but then pushed forward in the seventies, and which are now developed and manufactured at VDT-Bosch, Tilburg, The Netherlands. Nearly 70 different car types are equipped with such van-Doorne-belts, which requires a production of more than three million push belts per year, with reference to the year 2007. The torques transmitted by push belts reach in the meantime values of more than 500 Nm.

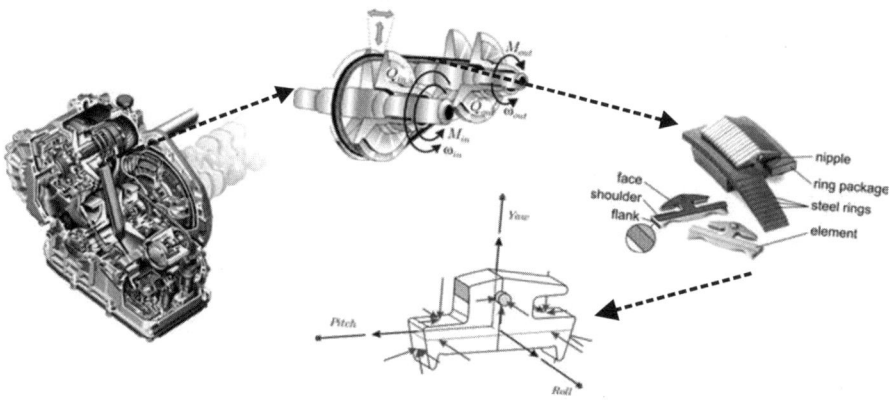

Fig. 5.103: CVT Power Transmission by Push Belts (courtesy Mercedes Benz and VDT-Bosch)

Figure 5.103 depicts the main features of a CVT power transmission with push belts, which are applied also in the A-class of Mercedes. The specialty of the push belt in comparison to the rocker pin chain consists in its special structure. As all other CVT's the belt moves between two pulleys, an input and an output pulley with the corresponding torques and rotational speeds. But the internal belt structure is completely different from chains. Some hundred small elements with a typical thickness of 1-2 mm and a typical width of 30 mm are arranged along the ring packages. These elements come into contact with the two sheaves of each pulley, altogether they are kept together by the two ring packages each with 9-12 sheet metal rings. Going from the driving to the driven wheel and watching the free strand behaviour we recognize, that on the one side the elements are pressed together forming a kind of a pressure bar, and that on the other side the elements are more or less loose on the rings, but the rings act as a tensile bar. During these processes within and outside the pulleys the elements are exposed to maximal 17 forces, which in the plane case reduce to 10 forces due to the symmetry.

To give an impression of the size of such systems we recall some typical data. The element length is about 1.8 mm, its width around 30 mm. The belt has typically 350-420 elements and thus 1050-1260 DOF in the plane and 2100-2520 DOF in the spatial modeling case. In addition the belt possesses 9-11 steel rings and about 3600 unilateral contacts, where the element-ring-contacts include the problem of a continuum with distributed unilateral contacts. At a speed of 1000 rpm with a transmission ratio of 1.0 about 3000 elements enter and leave the pulleys in every second thus generating a polygonal frequency of 3 kHz. The lowest "eigenfrequency" is around 120 Hz for the overall system.

In the following we shall consider only the plane case, the spatial case is still a matter of ongoing research. For establishing a realistic model we need multibody system theory including rigid and elastic components together with bilateral and unilateral constraints. The contacts of push belts are numerous and more complicated than chain contacts, see Figure 5.104. We have a uni-

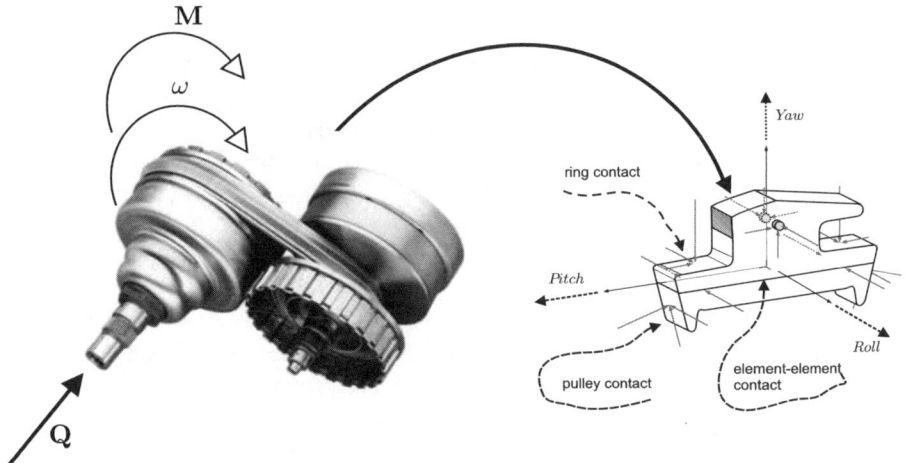

Fig. 5.104: Contact Configuration of Push Belt Elements

lateral spatial contact between the element and the pulley, a plane unilateral contact between the element and the ring and five contacts between the elements themselves. The last ones are covered by an empirical nonlinear force law. All other contacts are unilateral contacts described by complementarities, then converted and solved by prox-functions [82], [81].

## 5.7.2 Models

The push belt CVT consists of a force transmitting push belt and two pulleys (*prim*ary and *sec*ondary pulley). The pulleys are pairs of sheaves, where one sheave is fixed onto the shaft and the second sheave can be shifted hydraulically in axial direction. Figure 5.104 shows the functionality of the CVT for two extreme transmission ratios, where $\omega$ denotes the angular velocity, $M$ the torque and $Q$ the hydraulic force. Power is induced into the CVT from the combustion engine over a torque converter. Within the CVT, power is transmitted from the primary to the secondary pulley in the following way: at the primary pulley, power is conveyed by friction in the contact pulley - push belt. Next a transformation by tension and push forces within the push belt itself takes place. At the secondary pulley, power is transfered from the push belt onto the sheaves, finally leaving the CVT. By applying hydraulic pressures $Q_1$ and $Q_2$ onto the loose sheaves of each pulley, clamping forces acting onto the elements can be varied. As an outcome, the belt is running on variable radii within each pulley changing the ratio of transmission continuously.

In order to allow for the calculation of the dynamics of the push belt CVT, a two-dimensional model of the system is established according to the procedure described in sections to come, see [80], [81], [282], [191]. The build-up of the simulation model is performed in two steps: First the system is split up into three subsystems: pulleys, elements and ring package. In a first step these subsystems will be modeled without considering the type of interactions. In a second step the interactions between these subsystems will be taken into account in form of constraints. Finally, the overall model of the CVT will be established. The equations are shown only for the dynamics between impacts ($t \neq t_i$), the corresponding impact equations can be derived analog to section 3.5 on page 158.

The **elastic pulley model** follows more or less the same ideas as those used for the rocker pin chain. Pulley deformations and the contacts with the belt elements are similar. The pulley sheaves are modeled by rigid cones. For both cones we approximate the deformation by quasistatic force laws given by the Maxwell numbers. External excitations coming from the CVT environment and acting on the pulleys are taken into account. At the primary pulley a kinematic excitation is given by an angular velocity $\omega_{prim}$ (see Fig. 5.104). Accordingly, this pulley has no degree of freedom. At the secondary pulley a kinetic excitation is applied in form of an external torque $M_{sec}$. Accordingly, this pulley has one degree of freedom $\boldsymbol{q}_p = (\alpha_{sec})^T$, which is an angle of rotation. The pulley equations of motion write

$$\boldsymbol{M}_p \dot{\boldsymbol{u}}_p = \boldsymbol{h}_p + \boldsymbol{W}_p \boldsymbol{\lambda} \qquad (5.148)$$

with the positive definite, constant and diagonal mass matrix $\boldsymbol{M}_p$. The vector $\boldsymbol{h}_p$ is only depending on the time $t$: $\boldsymbol{h}_p = \boldsymbol{h}_p(t)$. Thus the matrices $\frac{\partial \boldsymbol{h}_p}{\partial \boldsymbol{q}_p}$ and $\frac{\partial \boldsymbol{h}_p}{\partial \boldsymbol{u}_p}$ used for the numerical integration are zero matrices [81].

The **elements** are modeled by rigid bodies, describing each element $m$ by three degrees of freedom $\boldsymbol{q}_m = (y_m, z_m, \alpha_m)^T$. The model of one single element is depicted in Fig. 5.105, where the center $S$ of gravity is determined by the translational positions $y$ and $z$. The modeling of the whole CVT is performed in the plane containing the axes $AA$ of all elements. The $M$ elements are

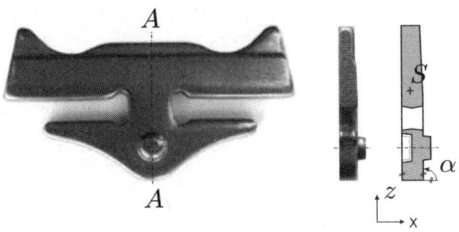

Fig. 5.105: Model of an Element

described by the generalized coordinates $\boldsymbol{q}_e = \left(\boldsymbol{q}_1^T, \ldots, \boldsymbol{q}_M^T\right)^T$, resulting in the equations of motion

$$\boldsymbol{M}_e \dot{\boldsymbol{u}}_e = \boldsymbol{h}_e + \boldsymbol{W}_e \boldsymbol{\lambda} \tag{5.149}$$

with the positive definite, constant and diagonal mass matrix $\boldsymbol{M}_e$. The vector $\boldsymbol{h}_e$ is constant and summarizes forces due to gravity. Thus the matrices $\frac{\partial \boldsymbol{h}_e}{\partial \boldsymbol{q}_e}$ and $\frac{\partial \boldsymbol{h}_e}{\partial \boldsymbol{u}_e}$ are zero matrices.

For the plane model the **two ring packages** of the push belt are considered as one virtual ring package having double width. The 9 to 12 layers of each ring package are homogenized using $EA$, $EI$ and $A\rho$ as parameters representing the longitudinal stiffness, the bending stiffness and the mass per length, respectively. The magnitudes $E$ and $\rho$ give the modulus of elasticity and the density, $A$ and $I$ denote the cross sectional area and the moment of inertia of the ring, which is treated as a one-dimensional continuum. Curling up for an unstressed configuration of the beam model is described by the curl-radius of the relaxed structure. This property is implemented within the elastic components of the equations of motion.

Due to changes in the transmission ratio, no reference path of the push belt can be given. Therefore, the model of the ring package must be able to describe free motion, including large translations and deflections. The model aims for a good approximation of the global dynamics, local errors in the description of local stresses are accepted if the model stiffness can be reduced significantly.

Many common approaches for flexible multi-body systems describe deformations in one global *moving frame of reference* (MFR). The degrees of freedom are chosen usually close to the described physics like rigid body movement and bending or longitudinal deformation. This leads to equations of motion in a compact and mainly decoupled form. Generally, the coordinate set used for this attempt cannot be used for coupling different elastic bodies to one discretized structure.

On the other hand, *finite element* (FE) approaches offer coordinate sets designed for coupling adjacent elements. These different advantages of both, MFR and FE, can be maintained, if two coordinate sets are used. The following formalism is described in detail in [282], see also [81].

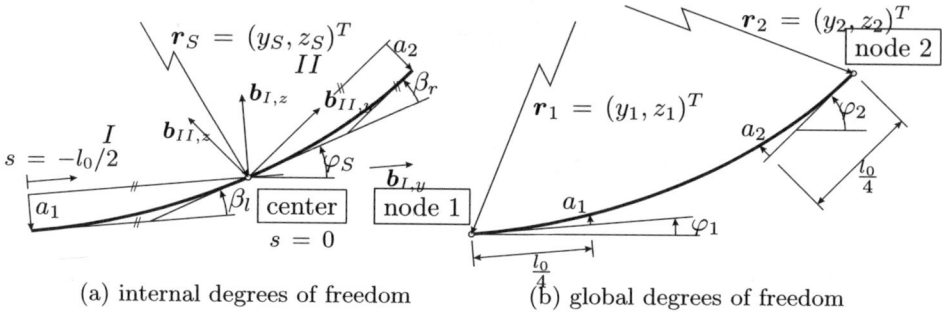

Fig. 5.106: Coordinate sets of one finite element [282].

The *internal* coordinate set $\mathbf{q}_i = (y_S, z_S, \varphi_S, \tilde{\varepsilon}, a_l, \beta_l, a_r, \beta_r)^T \in \mathbb{R}^8$ (see Figure 5.106a) is used to evaluate the equations of motion for one single finite element. This set is inspired by MFR-ideas: $y_S, z_S, \varphi_S$ describe the rigid body movements of the finite element, $\tilde{\varepsilon}$ gives an approximation for the longitudinal strain and $a_l, \beta_l, a_r, \beta_r$ describe the bending deflections.

The *global* coordinate set $\mathbf{q}_g = (y_1, z_1, \varphi_1, a_1, a_2, y_2, z_2, \varphi_2)^T$ is used as second coordinate set (see Figure 5.106b) for coupling different finite elements to a discretized description of one structure. This FE-inspired set is used for time integration of the entire dynamical system.

The correlation between both coordinate sets is shortly explained below. A constraint on position level is developed and derived to gain dependencies on velocity and acceleration level. These equations are used to transform the equations of motion of one single finite element into a form in terms of the global coordinates.

The coordinate sets $\mathbf{q}_i$ and $\mathbf{q}_g$ of one single finite element are subjected to an explicit equality constraint:

$$\mathbf{q}_i = \mathbf{Q}(\mathbf{q}_g) \qquad \in \mathbb{R}^8 \tag{5.150}$$

The derivatives of this equation with respect to the time $t$ for the transformation of the equations of motion are given by:

$$\frac{d\mathbf{q}_i}{dt} = \mathbf{u}_i = \frac{\partial \mathbf{Q}}{\partial \mathbf{q}_g}\mathbf{u}_g = \mathbf{J}_{ig}\mathbf{u}_g \qquad \text{with} \quad \mathbf{J}_{ig} = \mathbf{J}_{ig}(\mathbf{q}_g) = \frac{\partial \mathbf{q}_i}{\partial \mathbf{q}_g} = \frac{\partial \mathbf{u}_i}{\partial \mathbf{u}_g},$$

$$\dot{\mathbf{u}}_i = \frac{d\mathbf{J}_{ig}}{dt}\mathbf{u}_g + \mathbf{J}_{ig}\dot{\mathbf{u}}_g = \dot{\mathbf{J}}_{ig}\mathbf{u}_g + \mathbf{J}_{ig}\dot{\mathbf{u}}_g$$

$$\dot{\mathbf{J}}_{ig} = \dot{\mathbf{J}}_{ig}(\mathbf{q}_g, \mathbf{u}_g) = \frac{d\mathbf{J}_{ig}}{dt}. \tag{5.151}$$

The above relations describe the linear dependencies between the global and internal velocities $\frac{d\mathbf{q}_g}{dt}$ and $\frac{d\mathbf{q}_i}{dt}$ given by the JACOBIAN-matrix $\mathbf{J}_{ig}$. The development of the equations of motion for a single finite element follows common approaches for flexible multi-body systems and is described in detail in [282]. These equations are given in terms of the internal coordinates $\mathbf{q}_i$ and are extended by the decomposition $\mathbf{r} = \mathbf{W}_i(\mathbf{q}_i)\lambda$ of the constraint forces:

$$\mathbf{M}_i(\mathbf{q}_i)\dot{\mathbf{u}}_i - \mathbf{h}_i(\mathbf{q}_i, \mathbf{u}_i, t) - \mathbf{W}_i(\mathbf{q}_i)\lambda = \mathbf{0} \tag{5.152}$$

Using the last three relations we can derive the equations of motion in terms of the global coordinates $\mathbf{q}_g$:

$$\underbrace{\mathbf{J}_{ig}^T\mathbf{M}_i(\mathbf{q}_i)\mathbf{J}_{ig}}_{\mathbf{M}_g} \cdot \dot{\mathbf{u}}_g = \underbrace{\mathbf{J}_{ig}^T\left[\mathbf{h}_i(\mathbf{q}_i, \mathbf{u}_i, t) - \mathbf{M}_i(\mathbf{q}_i)\dot{\mathbf{J}}_{ig}\mathbf{u}_g\right]}_{\mathbf{h}_g} + \underbrace{\mathbf{J}_{ig}^T\mathbf{W}_i(\mathbf{q}_i)}_{\mathbf{W}_g}\lambda \tag{5.153}$$

Maintaining this formalism derived for single finite elements, the global coordinates can be extended to the state of the entire structure. The degrees of freedom $\mathbf{q}_g$ of each single finite element are part of all degrees of freedom $\mathbf{q}_r$ of the ring package, whereas every node is node of two adjacent finite elements. The equations of motion for the entire structure are calculated as the sum of all equations of motion of the $k$ finite elements transformed in the space of all degrees of freedom $\mathbf{q}_r$:

$$\underbrace{\sum_{j=1}^{k}\left[\tilde{\mathbf{J}}_{ig}^T\mathbf{M}_i(\mathbf{q}_i)\tilde{\mathbf{J}}_{ig}\right]_j}_{\mathbf{M}_r} \cdot \dot{\mathbf{u}}_r = \underbrace{\sum_{j=1}^{k}\left\{\tilde{\mathbf{J}}_{ig}^T\left[\mathbf{h}_i(\mathbf{q}_i, \mathbf{u}_i, t) - \mathbf{M}_i(\mathbf{q}_i)\dot{\tilde{\mathbf{J}}}_{ig}\mathbf{u}_g\right]\right\}_j}_{\mathbf{h}_r} +$$

$$+ \underbrace{\sum_{j=1}^{k}\left[\tilde{\mathbf{J}}_{ig}^T\mathbf{W}_i(\mathbf{q}_i)\right]_j}_{\mathbf{W}_r} \lambda \tag{5.154}$$

Here, the extended Jacobian-matrix $\tilde{\mathbf{J}}_{ig}^T = \mathbf{J}_{ir} = \frac{\partial \mathbf{q}_i}{\partial \mathbf{q}_r}$ mainly holds zero entries separate from entries $\mathbf{J}_{ig}$ connected to global degrees of freedom $(\mathbf{q}_g)_j$ which are part of the $j$-th finite element. This transformation is performed during runtime for every finite element in every timestep. The resulting mass-matrix $\mathbf{M}_r$ as well as the derivatives $\frac{\partial \mathbf{h}_r}{\partial \mathbf{q}_r}$ and $\frac{\partial \mathbf{h}_r}{\partial \mathbf{u}_r}$ of the generalized forces $\mathbf{h}_r$ have a cyclic-blockdiagonal structure. Using this formalism, different bodies can be coupled without introducing equality constraints.

The equations given so far hold for classical beams which have a start- and an endpoint for the description as a continuum. A closed ring structure can be seen in the same manner, but the start- and the endpoint need to coincide in an arbitrarily defined origin of the ring structure. For the discretized structure with $k$ finite elements the start node (1) and the end node $(k+1)$ need to coincide. Since a complete relative revolution between these two nodes takes place, the angular difference between their tangent directions is $2\pi$. The constraints can be written explicitly as

$$(y_{k+1}, z_{k+1}, \varphi_{k+1})^T = (y_1, z_1, \varphi_1 + 2\pi)^T \qquad (5.155)$$

and are implemented within the transformation 5.151 for the $k$-th finite element. Finally, the equations of motion 5.154 of the ring package can be written as

$$\mathbf{M}_r \dot{\mathbf{u}}_r = \mathbf{h}_r + \mathbf{W}_r \lambda \qquad \in \mathbb{R}^{5k}, \qquad (5.156)$$

where $k$ is the number of finite elements used for the discretization of the ring package. The constraints $\mathbf{W}_r \lambda$ between the elements and the ring package are explained in the sections dealing with the contacts.

Starting with the **contacts between the elements and the ring package** we have to consider, that the constraints on elastic structures can be formulated using the schemes derived for rigid body constraints in classical multi-body systems [282]. Figure 5.107 displays the contact points on the interacting bodies: the contact point is fixed on the element but sliding on the ring package, whereas the position is defined by the material coordinate $s_c$. It should be mentioned that uni- as well as bilateral constraints can be treated following the same geometrical ideas. The constraint is specified by the corresponding proximal function [65],[135]. In case of the push belt, the contact between the ring package and the elements is assumed to be free of gaps in normal direction of the ring package, so that it can be modeled by

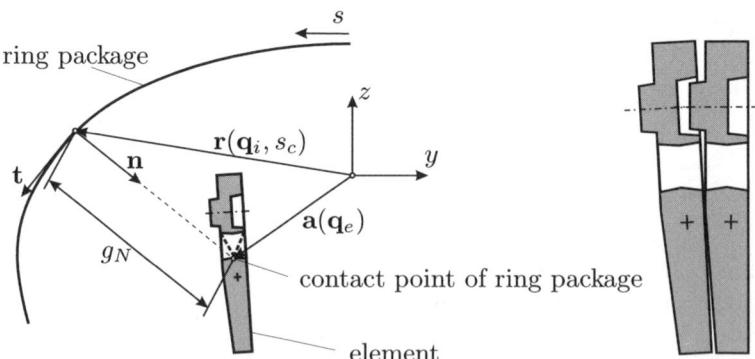

Fig. 5.107: Contact between ring structure and element.

Fig. 5.108: Contact between two elements.

a bilateral constraint. In tangential direction Coulomb friction is considered, which allows unilaterally defined stick-slip processes. Altogether we get the following equations

$$\lambda_{B,er} = \text{prox}_{C_B}(\lambda_{B,er} - r\mathbf{g}_{B,er}),$$
$$\lambda_{T,er} = \text{prox}_{C_T(\lambda_{B,er})}(\lambda_{T,er} - r\dot{\mathbf{g}}_{T,er}). \qquad (5.157)$$

where the indices "B", "T" and "er" stand for bilateral, tangential and element/ring package, respectively.

The **contacts between the elements and the pulley** sheaves are unilateral in normal and in tangential directions. In spite of the fact that within the belt wrap no detachment of an element in normal direction takes place, this kind of events is very important for the belt/pulley zones, where the belt enters or leaves the pulley. We may have contact/detachment in normal and stick-slip by Coulomb friction in tangential directions. Therefore the contact laws write in these cases

$$\lambda_{U,pe} = \text{prox}_{C_U}(\lambda_{U,pe} - r\mathbf{g}_{U,pe}),$$
$$\lambda_{T,pe} = \text{prox}_{C_T(\lambda_{U,pe})}(\lambda_{T,pe} - r\dot{\mathbf{g}}_{T,pe}). \qquad (5.158)$$

**Contacts between the elements** are illustrated in Figure 5.108. Only contacts in normal direction are considered, because tangential relative motion is circumvented by the guidance of the elements by the ring package and the nipple-hole connection between adjacent elements. The contact between adjacent elements is modeled by a rigid body contact

$$\lambda_{U,ee} = \text{prox}_{C_U}(\lambda_{U,ee} - r\mathbf{g}_{U,ee}) \qquad (5.159)$$

with the gap distances $\mathbf{g}_{U,ee}$ between the considered contact points of the interacting elements. For further details see [80] and [81].

Before presenting the overall equations of the **CVT push belt system** as a whole it makes sense to consider the evolution of the system dynamics for the unilateral cases. The dynamics of the rocker pin chains has been represented by complementarities and solved by applying the Lemke-algorithm [163], which requires the interpolation of the beginning and end of the contact events, but is then nearly exact with respect to these events. The newer approach following the ideas of [2] and applying the prox-concept is faster and safer, at least in most large applications, but not so "exact" with respect to the contact events. Nevertheless, the results do not differ for the two methods, but computing time is much smaller for the Augmented Lagrange with the prox-method. From the point of view of chains and belts we experienced, that without the newer approach a simulation of push belts with reasonable computing times would not be feasible.

To show the difference we shall first repeat here the dynamics of the **rocker pin chain**. The equations of motion write principally

$$M(q,t)\ddot{q}(t) + h(q,\dot{q}t) - [(W_N + W_R) \quad W_T]\begin{pmatrix}\lambda_N(t)\\ \lambda_T(t)\end{pmatrix} = 0 \quad \in \mathbb{R}^f,$$

$$\ddot{g}_N = W_N^T \ddot{q} + \bar{w}_N \quad \in \mathbb{R}^{n_N},$$

$$\ddot{g}_T = W_T^T \ddot{q} + \bar{w}_T \quad \in \mathbb{R}^{2n_T},$$

$$\ddot{g}_N \geq 0, \quad \lambda_N \geq 0, \quad \ddot{g}_N^T \lambda_N = 0,$$

$$\lambda_{T0} \geq 0, \quad \ddot{g}_T \geq 0, \quad \lambda_{T0}^T \ddot{g}_T = 0. \tag{5.160}$$

The abbreviations are clear, $q \in \mathbb{R}^f$ are the generalized coordinates, $M \in \mathbb{R}^{f,f}$ the symmetric and positive definite mass matrix, $W$ the constraint matrices, $\lambda$ the constraint forces, $\ddot{g}$ the relative accelerations in the contacts, $\bar{w}$ some external (excitation) terms and $\lambda_{T0}$ the friction reserves with respect to the boundary of the friction cones. The above equations for the dynamics of a rocker pin chain are solved numerically by evaluating the beginning and the end of each contact event, interpolating these points and solving the complementarity problems by Lemke's algorithm, rearranging the constraint matrices and treating the smooth equations between contact events by a Runge-Kutta 4/5 -scheme.

The modeling of the **push belt CVT** by single components and their interconnections leads to the equations of motion for the plane overall system, consisting of the differential equations

$$\begin{pmatrix}M_p & 0 & 0\\ 0 & M_e & 0\\ 0 & 0 & M_r\end{pmatrix}\begin{pmatrix}\dot{u}_p\\ \dot{u}_e\\ \dot{u}_r\end{pmatrix} = \begin{pmatrix}h_p\\ h_e\\ h_r\end{pmatrix} + W_{B,er}\lambda_{B,er} + W_{T,er}\lambda_{T,er}$$
$$+ W_{U,pe}\lambda_{U,pe} + W_{T,pe}\lambda_{T,pe}$$
$$+ W_{U,ee}\lambda_{U,ee} \tag{5.161}$$

and the constraints:

$$\lambda_{B,er} = \operatorname{prox}_{C_B}(\lambda_{B,er} - r g_{B,er}) \quad ; \quad \lambda_{T,er} = \operatorname{prox}_{C_T(\lambda_{N,er})}(\lambda_{T,er} - r\dot{g}_{T,er})$$
$$\lambda_{U,pe} = \operatorname{prox}_{C_U}(\lambda_{U,pe} - r g_{U,pe}) \quad ; \quad \lambda_{T,pe} = \operatorname{prox}_{C_T(\lambda_{N,pe})}(\lambda_{T,pe} - r\dot{g}_{T,pe})$$
$$\lambda_{U,ee} = \operatorname{prox}_{C_U}(\lambda_{U,ee} - r g_{U,ee}) \tag{5.162}$$

The indices (p, e, r) stand for pulley, element and ring, respectively. The indices (U, B, T) indicate unilateral, bilateral and tangential, respectively. The $M$ are the mass matrices, $W$ the constraint matrices, $u = \dot{q}$ are velocities, and $\lambda$ are constraint forces. The magnitudes $g, \dot{g}$ indicate relative displacements and velocities in the contacts, r is an auxiliary variable which has been optimized by [65] with the goal to reduce computing time.

The equations (5.161) are solved numerically by a time stepping scheme including some prox-algorithms [81]. The modular configuration of the model comprising subsystems and constraints enables a refinement or even a substitution of models for single components and interactions in an easy manner. By this, both bodies and contacts can be modeled rigidly or flexibly in a hybrid way. The resulting differential equations have stiff character.

### 5.7.3 Some Results

Rocker pin chains are manufactured by various companies with a certain focus at LUK, Germany. Push belt systems are mainly manufactured by VDT-Bosch in the Netherlands, which also operate production facilities in Japan. Measurements are up to now rare. Very good experiments have been performed by Honda in Japan (see [74], [75] and [123]), which are available by the cited papers, and equally good measurements have been performed by VDT-Bosch, which in the meantime are also available, see for example [234]. Therefore we shall give some general results of simulations and a comparison of theory with the Honda measurements.

Geier [80] evaluated many cases by simulation and performed a comprehensive study of comparisons with measurements. We shall give only two examples. The first one depicts some simulations for the contacts between the elements themselves and between pulley and element. Figure 5.109 illustrates the contact forces. Zone 1 corresponds to the free trum, zone 2 to the primary pulley, zone 3 to the pushing trum and zone 4 to the secondary pulley.

The force $\lambda_{N,ee}$ reflects the normal force in the element/element contact. This force is zero in the free trum and remains zero within the beginning of the primary pulley due to the fact, that the elements remain separated when entering the primary pulley. But then the contact force is built up and leaves the primary pulley with a value, which is maintained during the complete push trum phase making this trum to a kind of pressure bar. Entering the secondary pulley (zone 4) results in a maximum of the force $\lambda_{N,ee}$ firstly due to the influence of the pressure transmitted to the belt by the deformed pulley and secondly due to a prestressing effect by the ring package. This is all very plausible frome physical arguments and also confirmed by the simulations.

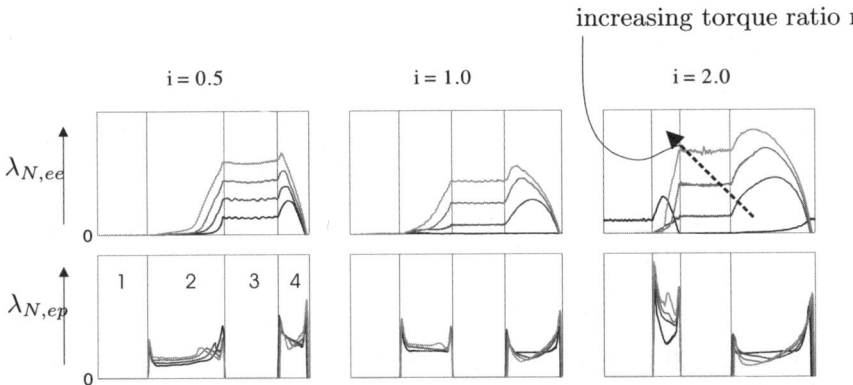

Fig. 5.109: Simulation Results for two Contact Forces $\lambda_{N,ee}$, $\lambda_{N,ep}$ and four Torque Ratios r (N=normal, ee=contact element/element, ep=contact element/pulley, 1 ≡ free trum, 2 ≡ primary pulley, 3 ≡ pushing trum, 4 ≡ secondary pulley)

The contact force $\lambda_{N,ep}$ between pulley and element is of course zero for the two trums in the zones 1 and 3 and non-zero within the pulleys in the zones 2 and 4. The results illustrate the typical characteristics known also from rocker pin chains. In entering and leaving a pulley we always get a sharp rise of the contact forces between pulley and the elements, which are then a bit reduced within the pulley wrap arc. These forces do not depend very much on the torque ratio r, whereas the element/element contact force $\lambda_{N,ee}$ increases considerably with the torque ratio.

The second example taken from [80] concerns a comparison between simulations and the Honda measurements published in [74] and [123]. Figure 5.110 gives a comparison, which confirm the models. It depicts the contact forces at the element shoulder for the plane case. For all torque ratios r the agreement is very good, which has to be seen before the background of a large transmission ratio of i=2. The belt is therefore operating on one side with a large and on the other side with a small radius. More comparisons may be seen from [80].

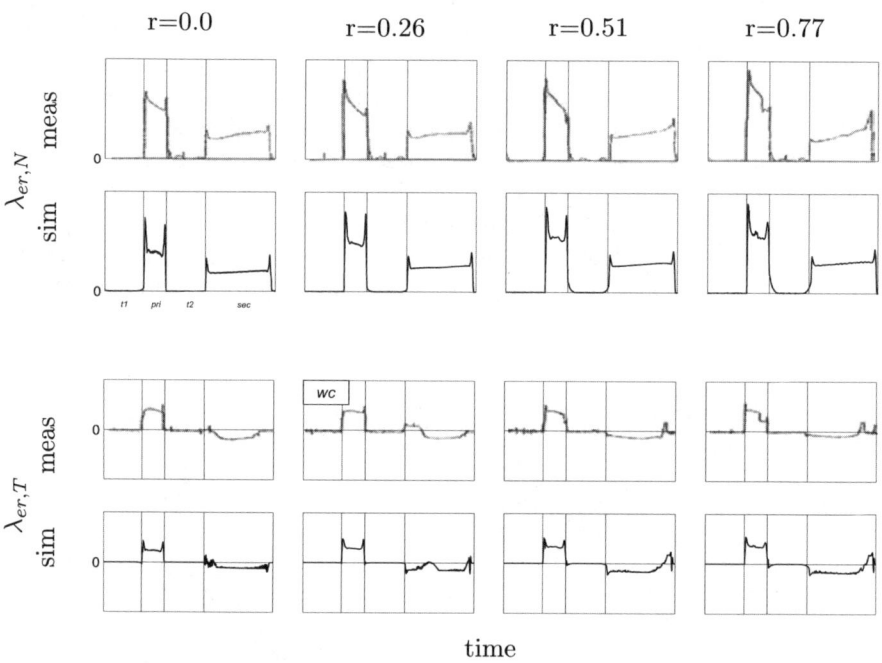

Fig. 5.110: Verification of the Element/Ring Contact Forces $\lambda_{er,N}$ and $\lambda_{er,T}$ for four Torque Ratios r (N=normal, T=tangential, er=contact element/ring, sim=simulation, meas=measurement)

# 6

# Timing Equipment

> *Wir schreiben "P" für Problem, "VL" für vorläufige Lösung, "FE" für Fehlerelimination; dann können wir den Grundablauf der Ereignisse bei der Evolution folgendermaßen beschreiben: $P \to VL \to FE \to P$. (Karl Popper, Objektive Erkenntnis, 1984)*
>
> *Using 'P' for problem, 'TS' for tentative solution, 'EE' for error elimination, we can describe the fundamental evolutionary sequence of events as follows: $P \to TS \to EE \to P$.*
>
> *(Karl Popper, Objective Knowledge, 1972)*

## 6.1 Timing Gear of a Large Diesel Engine

Machines and mechanisms are characterized by rigid or elastic bodies interconnected in such a way that certain functions of the machines can be realized. Couplings in machines are never ideal but may have backlashes or some properties which lead to stick-slip phenomena. Under certain circumstances backlashes generate a dynamical load problem if the corresponding couplings are exposed to loads with a time-variant character. A typical example can be found in gear systems of diesel engines, which usually must be designed with large backlashes due to the operating temperature range of such engines, and which are highly loaded with the oscillating torques of the injection pump shafts and of the camshafts. Therefore, the power transmission from the crankshaft to the camshaft and the injection pump shaft takes place discontinuously by an impulsive hammering process in all transmission elements [217], [71]. We shall start this chapter with such an example, which was also presented in [200].

Figure 6.1 indicates how the process works. A typical gear unit contains several meshes with backlashes, in the case shown two meshes with backlashes between crankshaft and injection pump shaft and three meshes with backlashes between crankshaft and camshafts. Due to periodical excitations mainly from the injection pumps and subordinately from the crankshaft and the camshaft, the tooth flanks separate, generating a free-flight period within the backlash which is interrupted by impacts with subsequent penetration. The driven flank (working flank) usually receives more impacts than the non-

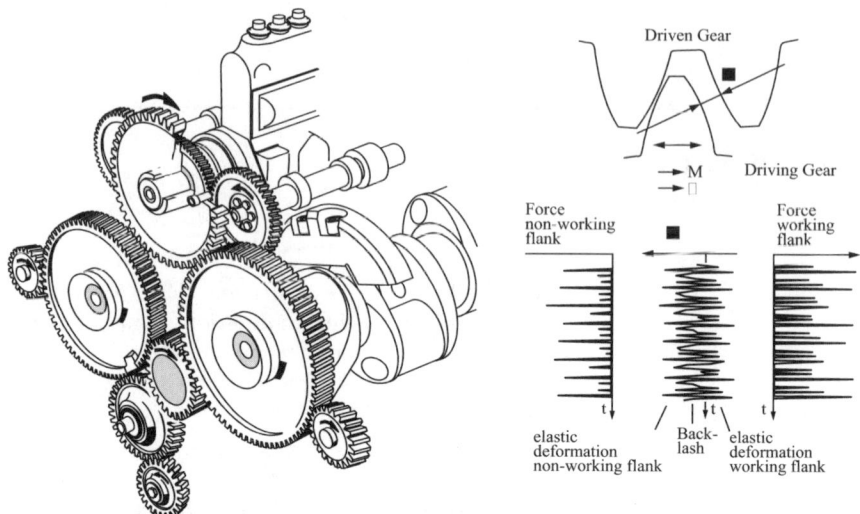

Fig. 6.1: Typical Diesel Engine Driveline System and Forces in a Mesh Meshes of Gears

working flank (Figure 6.1). Additionally, in all other backlashes of the gear unit similar processes take place, where the state and the impacts in one mesh with backlash influence considerably the state in all other meshes. This behavior must be accounted for by the mathematical model. As a definition we use the word "hammering" for separation processes within backlashes where high loads cause large impact forces with deformation. Motion within backlashes without loads is called "rattling." This represents a noise problem without load problems. It will be not considered here (see [200]).

As a rule, such vibrations may be periodic, quasiperiodic or chaotic with a tendency to chaos for large systems. Considering the driveline gear unit as a multibody system with $f$ degrees of freedom and $n_p$ backlashes in the gear meshes, we model the backlash properties by a nonlinear force characteristic with small forces within the backlash and a linear force law in the case of contact of the flanks. The event of a contact is determined by an evaluation of the relative distance in each backlash, which serves as an indicator function. The indicator function for leaving the contact, i.e., flank separation, is given with the normal force in the point of contact, which changes sign in the case of flank separation. These unsteady points (switching points) must be evaluated very carefully to achieve reproducible results. The time series of impact forces will be reduced to load distributions in a last step. They might serve as a basis for lifetime estimates.

The first activities on impulsive processes at the author's former Institute started in 1982 and led to a series of contributions on rattling and hammering processes in gearboxes and drivelines. The fundamental starting point was a general theoretical approach to mechanical systems with unsteady transitions in 1984 [182], which was very quickly extended to rattling applications. The dissertations [124] and [133] deepened the rattling theory and compared one-stage rattling with laboratory tests. From the very beginning all theoretical research focused on general mechanical systems with an arbitrary number of degrees of freedom and with an arbitrary number of backlashes. Application fields are drivelines of large diesel engines, which, due to a large temperature operating range, are usually designed with large backlashes [217].

### 6.1.1 Modeling

#### 6.1.1.1 Body Models

Rigid bodies are characterized by six degrees of freedom: three translational and three rotational ones. We combine these magnitudes in an $\mathbb{R}^6$-vector (Figure 6.2 and [217])

$$p = \begin{pmatrix} \boldsymbol{\Phi} \\ \boldsymbol{r}_H \end{pmatrix} \in \mathbb{R}^6 \tag{6.1}$$

with

$$\boldsymbol{\Phi} = (\varphi_x, \varphi_y, \varphi_z)^T , \qquad \boldsymbol{r}_H = (\Delta x_H, \Delta y_H, \Delta z_H)^T \in \mathbb{R}^3 . \tag{6.2}$$

Accordingly, the velocities are

$$v = \begin{pmatrix} \dot{\boldsymbol{\Phi}} \\ \dot{\boldsymbol{v}}_H \end{pmatrix} = \begin{pmatrix} \boldsymbol{\omega} \\ \boldsymbol{v}_H \end{pmatrix} \in \mathbb{R}^6 . \tag{6.3}$$

Elastic bodies in gear or driveline units usually are shafts with torsional

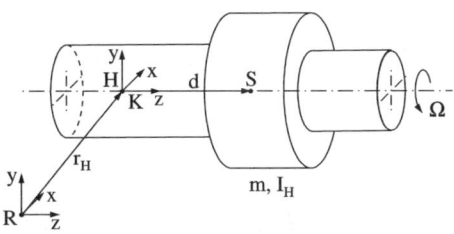

Fig. 6.2: Rigid Body Model [217]

and/or flexural elasticity. In the following we consider only torsion by applying a Ritz approach (see section 3.3.4 on page 124) to the torsional deflection $\varphi$ (Figure 6.3):

$$\varphi(z,t) = \boldsymbol{w}(z)^T \boldsymbol{q}_{el}(t) \quad \text{with} \quad \boldsymbol{w}, \boldsymbol{q}_{el} \in \mathbb{R}^{n_{el}} \tag{6.4}$$

where the subscript $el$ stands for "elastic."

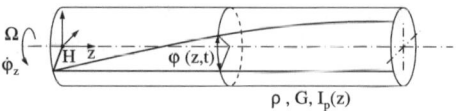

Fig. 6.3: Shaft with Torsional Elasticity

### 6.1.1.2 Coupling Components

An ideal joint couples two bodies. It can be described kinematically. According to the notation used in multibody theory the free directions $f_i$ of motion of a joint $i$ are given by a matrix $\boldsymbol{\Phi}_i \in \mathbb{R}^{6,f_i}$. The spatial possible motion of the joint can be described by $(\boldsymbol{\Phi}_i \boldsymbol{q}_{J_i})$ with relative displacements $\boldsymbol{q}_{J_i}$ in the nonconstrained directions of the joint $i$. A complementary matrix $\boldsymbol{\Phi}_i^c \in \mathbb{R}^{6,6-f_i}$ exists for the constrained directions of the joint. Always $\boldsymbol{\Phi}_i^T \boldsymbol{\Phi}_i^c = 0$, and the constraint forces are written as

$$\boldsymbol{f}_{J_i} = \boldsymbol{\Phi}_i^c \boldsymbol{\lambda}_i, \quad \boldsymbol{\lambda}_i \in \mathbb{R}^{6-f_i}. \tag{6.5}$$

The Lagrange multipliers $\boldsymbol{\lambda}$ follows from d'Alembert's principle and a Lagrangian treatment of the equations of motion.

Elastic couplings in drivelines are characterized by some force law in a given direction between two bodies (Figure 6.4). The relative displacement and displacement velocity may be expressed by

$$\gamma_k = \boldsymbol{\psi}_k^T \left( -\boldsymbol{C}_{ki} \boldsymbol{p}_i + \boldsymbol{C}_{kj} \boldsymbol{p}_j \right), \quad \dot{\gamma}_k = \boldsymbol{\psi}_k^T \left( -\boldsymbol{C}_{ki} \boldsymbol{v}_i + \boldsymbol{C}_{kj} \boldsymbol{v}_j \right) \tag{6.6}$$

$$\boldsymbol{C}_{ki} = \begin{pmatrix} \boldsymbol{E} & \boldsymbol{0} \\ \tilde{\boldsymbol{c}}_{ki}^T & \boldsymbol{E} \end{pmatrix} \in \mathbb{R}^{6,6}.$$

The vector $\boldsymbol{c}_{ki}$ follows from Figure 6.4, and $\tilde{\boldsymbol{c}}_{ki}$ is the relevant skew-symmetric tensor ($\tilde{\boldsymbol{a}} \boldsymbol{b} \equiv \boldsymbol{a} \times \boldsymbol{b}$). The vector $\boldsymbol{\psi}_k \in \mathbb{R}^6$ represents a unit vector in the direction of relative displacement. In that direction we have a scalar force magnitude $\zeta_k$ according to the given force law. It can be expressed in the body coordinate frames $H_i, H_j$ (Figure 6.4) by the generalized forces (see section 3.3.5 on page 128)

$$\boldsymbol{f}_i = \boldsymbol{C}_{ki}^T \boldsymbol{\psi}_k \zeta_k, \quad \boldsymbol{f}_j = -\boldsymbol{C}_{kj}^T \boldsymbol{\psi}_k \zeta_k. \tag{6.7}$$

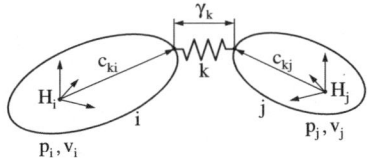

Fig. 6.4: Elastic Coupling

As a simple example a linear force law would be written as

$$\zeta_k = c\gamma_k + d\dot\gamma_k \tag{6.8}$$

Of course, any nonlinear relationship might be applied as well, such as the force law with backlash according to Figure 6.5. Gear meshes with backlash are modeled with the characteristic of Figure 6.5. In this case care has to be taken with respect to the two possibilities of flank contact on both sides of each tooth. Oil reduces the impact forces of the hammering process. Some

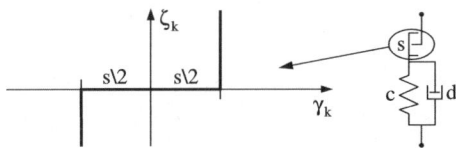

Fig. 6.5: Backlash Force Law

application tests have been performed with the nonlinear oil model of Holland [108]. On this basis a simplified model has been derived to approximate the oil influence (Figure 6.6). The model assumes an exponential damping behavior within the backlash $s$.

Bearings are very important coupling elements in drivelines. Roller bearings are approximated by the force laws in the $(x,y)$-directions:

$$\zeta_{kx} = c_x \gamma_x + d_x \dot\gamma_x$$
$$\zeta_{ky} = c_y \gamma_y + d_y \dot\gamma_y \,. \tag{6.9}$$

For journal bearings with a stationary load we apply the well-known law

$$\begin{pmatrix}\zeta_{kx}\\ \zeta_{ky}\end{pmatrix} = \begin{pmatrix}c_{11} & c_{12}\\ c_{21} & c_{22}\end{pmatrix}\begin{pmatrix}\gamma_{kx}\\ \gamma_{ky}\end{pmatrix} + \begin{pmatrix}d_{11} & d_{12}\\ d_{21} & d_{22}\end{pmatrix}\begin{pmatrix}\dot\gamma_{kx}\\ \dot\gamma_{ky}\end{pmatrix}. \tag{6.10}$$

For journal bearings with nonstationary loads the Waterstraat solution [272] of the Reynolds equations is applied.

334    6 Timing Equipment

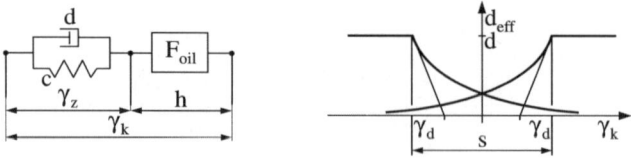

Fig. 6.6: Oil Model

### 6.1.1.3 Excitation System

Various excitation sources must be considered in simulating diesel engine driveline vibrations. First of all the crankshaft excites the system with some harmonics related to the motor speed and depending on the motor design. As a realistic approximation to driveline systems we may assume that the crankshaft motion itself is not influenced by driveline dynamics. Secondly, the valve mechanisms generate a parametric excitation which may be expressed approximately by a time-variant moment of inertia of the camshaft. As a matter of fact, in most applications this influence is fairly small, only a few percent. A

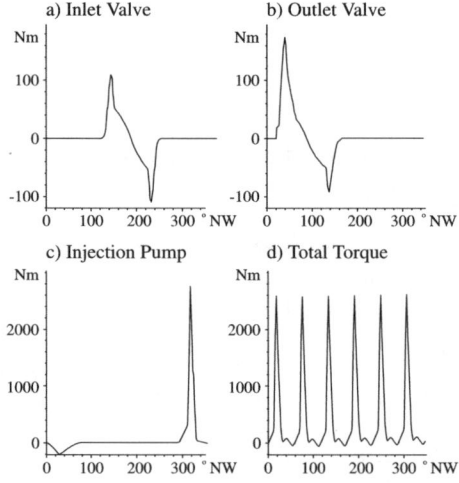

Fig. 6.7: Torques at the Camshaft

dominant influence comes from the third effect, namely the torques generated by the injection pumps. These torques directly counterbalance the driving torque of the crankshaft, thus inducing the hammering process within the gear meshes. A typical example is given by Figure 6.7, which represents the situation in a 12-cylinder diesel engine where valves and injection pumps are controlled by one shaft for each of the two cylinder banks.

## 6.1.2 Mathematical Models

### 6.1.2.1 Component Models

The theory of rigid and elastic multibody systems is applied rigorously. It starts with d'Alembert's principle, which states that passive forces produce no work or, according to Jourdain, generate no power, see chapter 3.3. This statement can be used to eliminate passive forces (constraint forces) and to generate a set of differential equations for the coupled machine system under consideration.

We start with the equations of motion for a single rigid body. Combining the momentum and moment of momentum equations and considering the fact that the mass center $S$ has a distance $d$ from the body-fixed coordinate frame in $H$ (Figure 6.2) we obtain

$$\boldsymbol{I}\dot{\boldsymbol{v}} = -\boldsymbol{f}_K + \boldsymbol{f}_B + \boldsymbol{f}_E = \boldsymbol{f}_S(\boldsymbol{p}, \boldsymbol{v}, t) \tag{6.11}$$

where

$$\boldsymbol{I} = \begin{pmatrix} \boldsymbol{I}_H & m\tilde{\boldsymbol{d}} \\ -m\tilde{\boldsymbol{d}} & m\boldsymbol{E}_3 \end{pmatrix} \in \mathbb{R}^{6,6}$$

$$\boldsymbol{d} = (0,\ 0,\ d)^T \in \mathbb{R}^3$$

$$\boldsymbol{I}_H = \mathrm{diag}\,(A,\ A,\ C) \in \mathbb{R}^{3,3}$$

$$-\boldsymbol{f}_K = \left[(\tilde{\boldsymbol{\omega}}\boldsymbol{I}_H \Omega \boldsymbol{e}_3)^T,\ \boldsymbol{0}\right]^T \in \mathbb{R}^6$$

$$\boldsymbol{f}_B = \left[\left(-\boldsymbol{I}_H \dot{\Omega} \boldsymbol{e}_3\right)^T, \boldsymbol{0}\right]^T \in \mathbb{R}^6$$

$$\boldsymbol{f}_E = \left[\left(\sum \boldsymbol{M}_{Hk}\right)^T, \left(\sum \boldsymbol{F}_k\right)^T\right]^T \in \mathbb{R}^6. \tag{6.12}$$

The magnitudes $A, C$ are moments of inertia; $\boldsymbol{f}_K$, $\boldsymbol{f}_B$ and $\boldsymbol{f}_E$ are gyroscopic, acceleration and applied forces, respectively, and $\Omega$, $\dot{\Omega}$ are prescribed values of angular velocity and acceleration, respectively. The unit vector $\boldsymbol{e}_3$ is body-fixed in $H$ (Figure 6.2).

By adding components with torsional elasticity we can influence the rigid body motion only with respect to the third equation of (6.11). Therefore torsional degrees of freedom can be included in a simple way. The equation of motion for a shaft with torsional flexibility is

$$\varrho I_p(z)\frac{\partial^2 \overline{\varphi}}{\partial t^2} - \frac{\partial}{\partial z}\left[GI_p(z)\frac{\partial \overline{\varphi}}{\partial z}\right] - M_k \delta(z - z_k) = 0. \tag{6.13}$$

($\varrho$ density, $I_p$ area moment of inertia, $G$ shear modulus, $M_k$ torque at location $z_k$). The total angle $\overline{\varphi}$ has three parts:

$$\overline{\varphi}(z,t) = \int \Omega(t)dt + \varphi_z(t) + \varphi(z,t) , \tag{6.14}$$

where $\Omega(t)$ is the angular velocity program, $\varphi_z(t)$ is the $z$-component of $\Phi$ (eq. 6.2) and $\varphi(z,t)$ is the torsional deflection (Figure 6.3). Approximating $\varphi(z,t)$ by eq. (6.4) and applying a Galerkin approach to eq. (6.13) result in the set

$$\boldsymbol{h}(\dot{\Omega} + \ddot{\varphi}_z) + \boldsymbol{M}_{el}\ddot{\boldsymbol{q}}_{el} + \boldsymbol{K}_{el}\boldsymbol{q}_{el} = \sum M_k \boldsymbol{w}(z_k) , \tag{6.15}$$

where

$$\begin{aligned}
\boldsymbol{M}_{el} &= \int_0^l \varrho I_p(z)\boldsymbol{w}(z)\boldsymbol{w}(z)^T dz &&\in \mathbb{R}^{n_{el},n_{el}} \quad \text{mass matrix} \\
\boldsymbol{K}_{el} &= \int_0^l GI_p(z)\boldsymbol{w}'(z)\boldsymbol{w}'(z)^T dz &&\in \mathbb{R}^{n_{el},n_{el}} \quad \text{stiffness matrix} \\
\boldsymbol{h} &= \int_0^l \varrho I_p(z)\boldsymbol{w}(z)dz &&\in \mathbb{R}^{n_{el}} \quad \text{coupling vector.}
\end{aligned} \tag{6.16}$$

In agreement with physical arguments eqs. (6.15) are coupled with the rigid body motion only by the term $\boldsymbol{h}\left(\dot{\Omega} + \ddot{\varphi}_z\right)$. To include torsional coupling in the third equation of (6.11), we only have to complete the angular momentum of the rigid part $C\left(\Omega + \dot{\varphi}_z\right)$ by an elastic part $\int_0^l \varrho I_p(z)\left[\partial\varphi(z,t)/\partial t\right]dz$ and evaluate its time derivative. This results in an additional coupling term $\boldsymbol{h}^T \ddot{\boldsymbol{q}}_{el}$. Equations (6.11) with elastic expansion and eqs. (6.15) can then be combined into

$$\boldsymbol{I}\dot{\boldsymbol{v}} + \boldsymbol{H}^T\ddot{\boldsymbol{q}}_{el} = \boldsymbol{f}_S, \qquad \boldsymbol{H}\dot{\boldsymbol{v}} + \boldsymbol{M}_{el}\ddot{\boldsymbol{q}}_{el} = \boldsymbol{f}_{S_{el}} \tag{6.17}$$

where all terms not containing accelerations are collected in $\boldsymbol{f}_S$ and $\boldsymbol{f}_{S_{el}}$. For formal convenience we define $\boldsymbol{H} = (0,0,\boldsymbol{h},0,0,0)$.

Proceeding now from a single-body to a multibody system, we consider the constraint (passive) forces $\boldsymbol{f}_{J_i}$ (eq. 6.5). In a form corresponding to eqs. (6.17) we get

$$\boldsymbol{I}_i\dot{\boldsymbol{v}}_i + \boldsymbol{H}_i^T\ddot{\boldsymbol{q}}_{el_i} = \boldsymbol{f}_{S_i} + \boldsymbol{f}_{J_i} - \sum_{k\in S(i)} \boldsymbol{C}_k^T \boldsymbol{f}_{J_k} \tag{6.18}$$

$$\boldsymbol{H}_i\dot{\boldsymbol{v}}_i + \boldsymbol{M}_{el_i}\ddot{\boldsymbol{q}}_{el_i} = \boldsymbol{f}_{S_{el_i}} - \sum_{k\in S(i)} \boldsymbol{D}_k^T \boldsymbol{f}_{J_k} . \tag{6.19}$$

These equations have to be supplemented by a kinematical relation in the form

$$\dot{\boldsymbol{v}}_i = \boldsymbol{C}_i \dot{\boldsymbol{v}}_{p(i)} + \boldsymbol{D}_i \ddot{\boldsymbol{q}}_{el_{p(i)}} + \boldsymbol{\Phi}_i \ddot{\boldsymbol{q}}_i . \tag{6.20}$$

The first term on the right-hand side represents an acceleration resulting from the absolute acceleration of the predecessor body (index $p(i)$), the second term represents an acceleration resulting from a torsional deformation of the predecessor body (if any) with a matrix $D_i$ containing the torsional shape functions, and the third term is an acceleration resulting from the relative motion of body $i$ and the predecessor body. $S(i)$ is the set of all bodies following body $i$.

### 6.1.2.2 Order(n) Considerations

For computational reasons eqs. (6.19) and (6.20) must be solved recursively by an order-$n$ algorithm (computing time $\sim$ degrees of freedom, see chapter 3.3.3 on page 119). This algorithm will be repeated here. It works as follows: Organize the multibody system under consideration as a treelike structure with a base body connected to the inertial environment and a series of final bodies possessing no successor body. This will be possible for any case, because for closed kinematical loops these may be cut with an additional closing condition as constraint. Start with the last type of bodies and consider their equations of motion (from eqs. 6.20):

$$I_i \dot{v}_i + H_i^T \ddot{q}_{el_i} = f_{S_i} + f_{J_i}, \qquad H_i \dot{v}_i + M_{el_i} \ddot{q}_{el_i} = f_{S_{el_i}}. \tag{6.21}$$

Eliminate from these equations the elastic coordinates to get

$$\hat{I}_i \dot{v}_i = \hat{f}_{S_i} + f_{J_i} \tag{6.22}$$

with

$$\hat{I}_i = I_i - H_i^T M_{el_i}^{-1} H_i, \qquad \hat{f}_{S_i} = f_{S_i} - H_i^T M_{el_i}^{-1} f_{S_{el_i}}. \tag{6.23}$$

Combine eq. (6.22) with eqs. (6.20) and (6.5) and apply d'Alembert's principle to eliminate the constraint forces (premultiplication with $\Phi_i^T$). We get the generalized relative accelerations

$$\ddot{q}_i = M_i^{-1} \Phi_i^T \left( \hat{f}_{S_i} - \hat{I}_i C_i \dot{v}_{p(i)} - \hat{I}_i D_i \ddot{q}_{el_{p(i)}} \right) \tag{6.24}$$

with $M_i = \Phi_i^T \hat{I}_i \Phi_i$, which can be used together with eqs. (6.22), (6.20) to obtain the elastic deformation accelerations

$$\ddot{q}_{el_i} = -M_{el_i}^{-1} H_i \left( C_i \dot{v}_{p(i)} + D_i \ddot{q}_{el_{p(i)}} + \Phi_i \ddot{q}_i \right) + M_{el_i}^{-1} f_{S_{el_i}}. \tag{6.25}$$

The accelerations $\ddot{q}_i$ and $\ddot{q}_{el_i}$ still depend on those of the predecessor body. Passing to that body requires determination of the joint forces, which are evaluated from eqs. (6.20, 6.22, 6.24):

$$f_{J_i} = N_i C_i \dot{v}_{p(i)} - L_i f_{S_i} + N_i D_i \ddot{q}_{el_{p(i)}} \tag{6.26}$$

with

$$L_i = E - \hat{I}_i \Phi_i M_i^{-1} \Phi_i^T, \qquad N_i = L_i \hat{I}_i . \tag{6.27}$$

With the joint force of eq. (6.26) we enter the equations of motion of the predecessor body and establish a set which corresponds formally to the equations of a final body:

$$(I_{p(i)} + C_i^T N_i C_i)\dot{v}_{p(i)} + (H_{p(i)}^T + C_i^T N_i D_i)\ddot{q}_{el_{p(i)}} =$$
$$= f_{J_{p(i)}} + (f_{S_{p(i)}} + C_i^T L_i f_{S_i}) - \sum_{k \in S(p(i)) - \{i\}} C_k^T f_{J_k} \tag{6.28}$$

$$(H_{p(i)} + D_i^T N_i C_i)\dot{v}_{p(i)} + (M_{el_i} + D_i^T N_i D_i)\ddot{q}_{el_{p(i)}} =$$
$$= (f_{S_{el_{p(i)}}} + D_i^T L_i f_{S_i}) - \sum_{k \in S(p(i)) - \{i\}} D_k^T f_{J_k} \tag{6.29}$$

with the magnitudes

$$\begin{aligned}
I_{p(i)} &:= I_{p(i)} + C_i^T N_i C_i \\
H_{p(i)} &:= H_{p(i)} + D_i^T N_i C_i \\
f_{S_{p(i)}} &:= f_{S_{p(i)}} + C_i^T L_i f_{S_i} \\
M_{el_{p(i)}} &:= M_{el_{p(i)}} + D_i^T N_i D_i \\
f_{S_{el_{p(i)}}} &:= f_{S_{el_{p(i)}}} + D_i^T L_i f_{S_i} .
\end{aligned} \tag{6.30}$$

These equations are analoguous to eqs. (6.21). With this procedure we recursively can proceed to the last base body for which eqs. (6.24) are evaluated in an elementary way. Knowing the accelerations of this base body, we again can go forward step by step to the final bodies and determine all accelerations of each body. If necessary we may add recursions for the evaluation of constraint forces.

Equations (6.19) describe a multibody system with $n$ bodies and a maximum number of ($f = 6n$) rigid degrees of freedom and, according to the Ritz Ansatz (eq. 6.4) a certain number of elastic degrees of freedom (number of shape functions per body times number of elastic bodies). In practical applications, however, the number of degrees of freedom might be reduced drastically. For example, driveline units with straight-tooth bevels may be sufficiently modeled by rotational degrees of freedom only.

### 6.1.2.3 Backlash Management

To include backlashes we have to implement an algorithm which controls the contact events (Fig. 6.8), by considering contact kinematics and contact forces.

## 6.1 Timing Gear of a Large Diesel Engine

A contact at a tooth flank occurs if the relative distance in contact $k$ becomes zero (eq. 6.7):

$$\gamma_k \left( \boldsymbol{p}_i, \boldsymbol{p}_j, \boldsymbol{\psi}_k \right) = 0. \qquad (6.31)$$

The subsequent deflection of both teeth follows the force laws of the Figs. 6.5 and 6.6, but the end of the contact is not reached, when we again get $\gamma_k = 0$. The correct condition consists of the requirement that the normal force ($\boldsymbol{\zeta}_k \boldsymbol{e}_{ki}$, $i = 1, 2$) (Fig. 6.8) vanish.

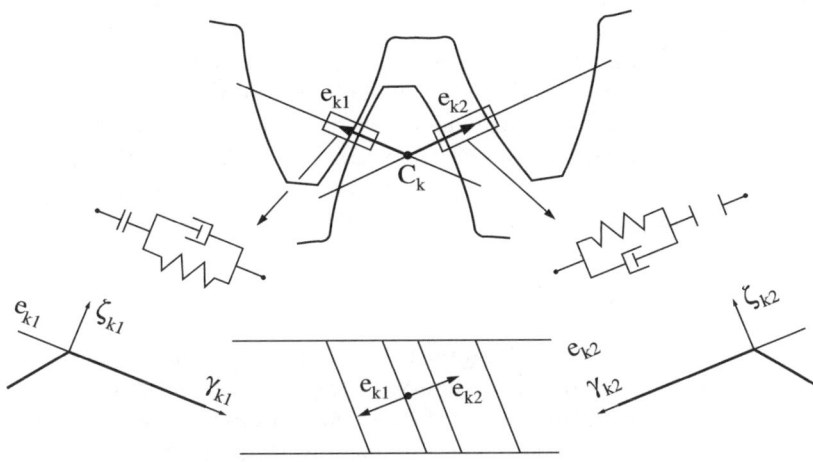

Fig. 6.8: Mesh of Gears with Backlash and Relevant Force Laws

As we have a unilateral contact problem, separation takes place when the normal force changes sign, which is not the case if $\gamma_k = 0$. Due to the dynamics of the contacting bodies and to the damping influence of the contact oil model (Fig. 6.6), the normal force changes sign before $\gamma_k = 0$, which means the tooth separation takes place when the teeth are still deflected. For separation we therefore must interpolate the force condition (Fig. 6.8)

$$\zeta_{kn} = -\boldsymbol{e}_{ki}\boldsymbol{\zeta}_k = -\boldsymbol{e}_{ki}\left(c_k \gamma_k + \zeta_{kD}\left(\gamma_k, \dot{\gamma}_k\right)\right) = 0 \quad (i = 1, 2), \qquad (6.32)$$

where $\zeta_{kn}$ is the normal force vector and $\zeta_{kD}$ the damping force law due to oil and structural damping.

Considering several backlashes we need an additional algorithm to determine the shortest time step to the next contact or separation event. This means formally that we have to evaluate in all existing backlashes the following equations:

$$\Delta t_{FC} = \min_{k \in n_p} \{\Delta t_k |\, \gamma_k(\boldsymbol{p}_i, \boldsymbol{p}_j, \psi_k) = 0 \wedge \zeta_{kn}(\gamma_k, \dot{\gamma}_k, \boldsymbol{e}_n) < 0\},$$

$$\Delta t_{CF} = \min_{k \in n_p} \{\Delta t_k |\, \zeta_{kn}(\gamma_k, \dot{\gamma}_k, \boldsymbol{e}_n) \geq 0\}, \qquad (6.33)$$

where $FC$ means transition from free flight to contact and $CF$ transition from contact to free flight. Equations (6.33) express the situation that during the recursive solution process as presented above at each integration step one must prove the possibility of a contact event or a separation in any of the $n_p$ backlashes. If this proof turns out to be true, the conditions of eqs. (6.33) together with the state of the complete system must be interpolated. The numerical integration process is then started anew at such a switching point. Note that each of the backlash zones of the multibody system might be in a free flight or in a contact state where the transitions are controlled by eqs. (6.33). Note further that the time steps $\Delta t_{FC}$ or $\Delta t_{CF}$ are the macrosteps between two events usually taking place in different backlashes; the microsteps for numerical integration are of course considerably smaller.

### 6.1.2.4 Numerical Models

The main problem in dealing with unsteady dynamical systems consists of the numerical management of unsteadiness. In the dissertation [217] three possibilities were considered and tested. Firstly, numerical integration routines of first order, such as the Euler method, include no problems with unsteadiness but show stability and convergence difficulties. Secondly, the numerical method of Shampine-Gordon has been realized with good results for certain cases. But the determination of switching points for an unsteady event may break down abruptly, thus generating numerical instability. Thirdly, a direct search and interpolation of switching points has been implemented which includes the fewest problems and works quite reliably. Details of this method may be found in [71], [217].

It might be of interest to indicate some computing time aspects (see Table). From this the best method turned out to be a direct search of switching points together with a Runge-Kutta integration method of order 2/3. Even before the background of more modern algorithms for unilateral systems, this result is still of interest and possesses some value for special cases. During the nineties and partly up into our times the interpolation of the beginning and the end of contact events was the rule, and for these procedures it still makes sense to apply between the contact events, for the smooth parts, Runge-Kutta 2/3. This statement also holds for models, which instead of unilateral theories apply nonlinear force laws with kinks. In contrast to these event-driven solutions we use today for equations of motion with complementarities the prox-algorithms in connection with time stepping on the basis of a one-step time dicretization. Time dicretized solutions are much more robust with

| Method | Relative Computing Time |
|---|---|
| • Direct search of switching points | |
| – Runge-Kutta 2./3. order | 1.0 |
| – Runge-Kutta 5./6. order | 1.9 |
| – Gear method | 1.4 |
| • Not direct search | |
| – Adams-Pece integration | 1.4 |

**Table:** Some Comparisons of Computing Time

better stability properties than event-driven solutions, which certainly will play a future role only for special applications.

### 6.1.3 Evaluation of the Simulations

The problem in evaluating the vibrations of unsteady systems consists of a realistic examination of the unsteady events, especially in the further evaluation of the gear forces, which, as a consequence of the hammering process, appear as impulsive forces. In any case we may apply an FFT procedure, and from this we get approximate information about frequencies and amplitudes which represents in steady problems a very powerful tool. In unsteadiness it is different because we do not know with sufficient reliability what force amplitudes will influence lifetime and strength of the gears. If we consider a typical hammering process in the gear mesh (Fig. 6.9) we may conclude that a statistical consideration would be the most appropriate one. By doing so, some principal ideas of Buxbaum [34] will be applied. A hammering impact can be characterized by the magnitude of the tooth forces' amplitude and by the time behavior (Fig. 6.10). Analyzing the forces in the gear meshes, we see that the time behavior follows a Gamma distribution quite well (see Fig. 6.11). Of more interest in practical problems is the load distribution. Again

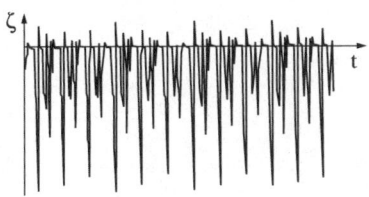

Fig. 6.9: Typical Force Sequence of a Hammering Process

## 342   6 Timing Equipment

it turns out that the loads in the gear meshes can be well approximated by classical rules valid for different stochastic processes:

$$H(x) = H_0 \exp\left[-ax^n\right], \tag{6.34}$$

where $a$ and $n$ might be determined from the simulation results by using a least square fit. The magnitude $H_0$ is the number of events of passing the nominal load. For hammering processes the nominal load is zero. Figure 6.12 gives an excellent comparison of formula (6.34) with simulated results.

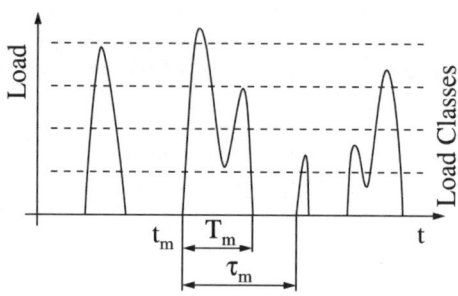

Fig. 6.10: Hammering Force Structure

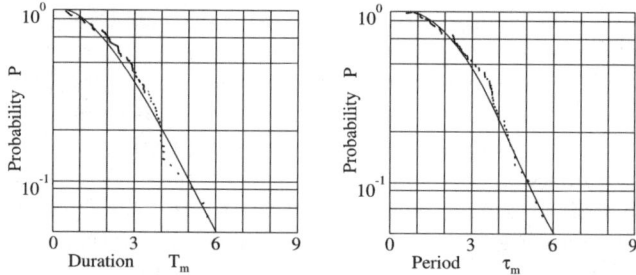

Fig. 6.11: Gamma Distribution for the Duration $T_m$ and the Period $\tau_m$ of a Hammering Process

### 6.1.4 Results

As an application of the theory presented, a four-stroke diesel engine with 12 cylinders and a power of about 3000 kW has been considered. The nonsymmetrical gear system driving the combined camshaft/injection pump shaft is shown in Fig. 6.13 on the left-hand side. On the right-hand side we see the corresponding mechanical model. All gears are spur gears; they will be modeled

## 6.1 Timing Gear of a Large Diesel Engine

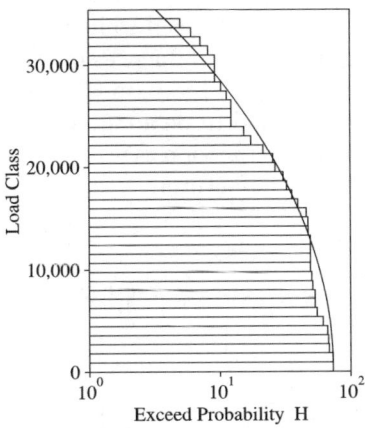

Fig. 6.12: Load Distribution for a Typical Hammering Process

as rigid bodies according to the section on body models. The camshaft will be described as an elastic body considering torsional elasticity only (see the coupling components section). As an additional option we regard a camshaft damper. The simulations focus on side A of the gear system due to the more complicated dynamical properties. Side A possesses one more stage than side B.

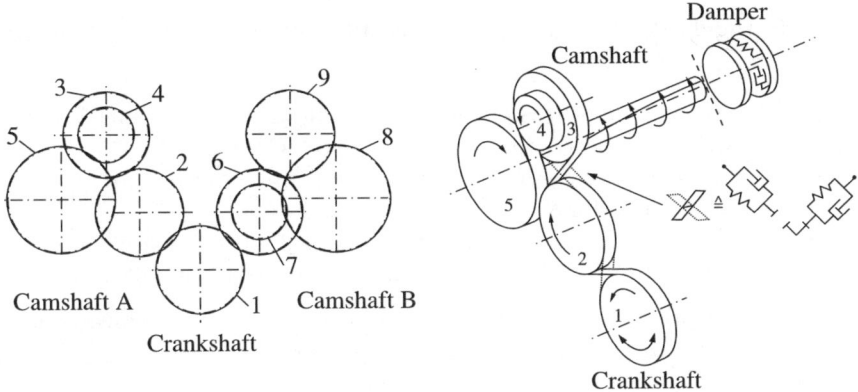

Fig. 6.13: Gear System and Equivalent Mechanical Model for a 4-Stroke 12-Cylinder Diesel Engine

Different models have been established starting from a model with only 5 degrees of freedom and ending with a model with 13 degrees of freedom. In all cases the torsional elasticity of the camshaft was described by two elastic degrees of freedom only, which turned out to be sufficient. As usual the determination of all stiffnesses proved to be rather difficult. Uncertainties mainly come from unknown flexural influences of the motor housing and from the more or less unknown stiffness reductions in one flanged joint and in a press fit. Therefore, stiffnesses had to be adapted to measurements. After this the models compared well with vibration measurements which were performed by a German diesel-engine manufacturer. A comparison is given in Fig. 6.14.

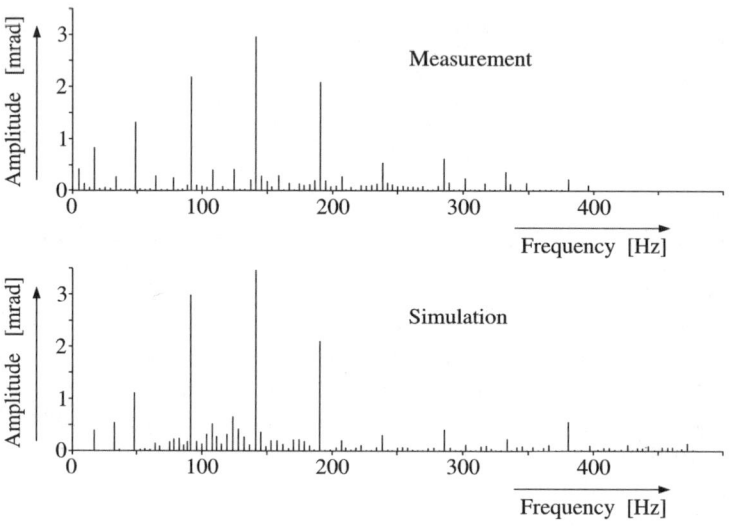

Fig. 6.14: Angular Vibrations of the Camshaft, Measurements and Simulation

After the verification of the theory a series of parameter simulations were performed. Although the results obtained relate to the special diesel engine under consideration, they might nevertheless indicate some general parameter tendencies in such drivelines.

First we consider the loads in the gear meshes and compare these loads with statical and quasi-statical cases, which we define in the following way:

- The static load is that one which would be generated by a transmission of the averaged camshaft torque. This mean value would be $\overline{T}_{CS} = 430$ Nm (Fig. 6.7).
- The quasi-static extremum load is that one which would be generated by a static transmission of the maximum torque values for the camshaft (peak values in Fig. 6.7). This maximum torque is $\overline{T}_{\max} = 2520$ Nm.
- The dynamical maximum load is evaluated from the load distribution (see the section on evaluation of the simulations) under the assumption that these loads will be realized with a probability of 99%.

The results for the three gear meshes of Fig. 6.13 are given in the Table. A systematic investigation of parameter tendencies allows the conclusions [217]:

## 6.1 Timing Gear of a Large Diesel Engine

| Gear Mesh (Fig. 11.13) | Gear Loads [kN] | | | |
|---|---|---|---|---|
| | Static Mean Load | Quasi-static Maximum Load | Dynamical Loads | |
| | | | Backlash = 0 | Nom. Backlash |
| 1/2 | 1.96 | 11.4 | 34.5 | 42.4 |
| 2/3 | 1.96 | 11.4 | 31.3 | 34.8 |
| 4/5 | 3.17 | 18.5 | 37.4 | 38.8 |

**Table:** Gear Loads in Gear Meshes

- Excitation Sources
  - Crankshaft excitation is small.
  - Camshaft excitation dominates, especially due to the injection pump loads.
- Gear System
  - Mass parameters show no much influence on the hammering process.
  - Increasing backlashes produce slightly increasing force amplitudes, but significantly increasing camshaft angular vibrations.
  - Increasing gear stiffness and damping gives decreasing force amplitudes.
- Camshaft
  - Increasing stiffness leads to largely decreasing force amplitudes.
  - Damping is of minor influence.
- Bearings
  - Slightly decreasing force amplitudes with increasing damping (all bearings).
  - No large difference exists between journal and roller bearings with respect to vibrations and force amplitudes.
- Camshaft Vibration Damper
  - Damper considerably reduces force amplitudes in the gear meshes.
  - Damper suppresses the hammering between two flanks and supports force transmission at the working flank.

## 6.2 Timing Gear of a 5-Cylinder Diesel Engine

### 6.2.1 Introduction

Modern Diesel engines of automotive industry possess direct injection, which needs less fuel and produces less emissions with respect to pollutant and noise. Additionally such motors exhibit a better torque performance in comparison to naturally aspirated engines. For achieving such advantages the designers have to deal with improved carburation systems requiring higher and higher pressures, which finally results in higher loads for all components participating primarily in the engine's operation, the crankshafts, the camshafts and the timing control system. We shall focus here on the last component.

For injection pressures larger than 1000 bar the timing chains and belts come to an end of their load carrying abilities, because very large pressures generate large torques at the camshaft thus influencing directly the loads of the timing components. Therefore timing gear wheel systems become an alternative to chains and belts, they enlarge the pressure potentials considerably, but on the other hand they are more complicated and more expensive than belts or chains. During the design and development of some new Diesel engines with five and ten cylinders Volkswagen decided in the nineties to apply for timing control wheel sets, though with the additional reqirement to keep the axial dimensions of such a timing gear set as small as possible. The ignition system consists of a pump-nozzle-system for each cylinder, which is operated by the cams and which are able to generate very large injection pressures. In the following we shall present the two configurations of the VW-5-cylinder engine R5 and the VW-10-cylinder engine V10 with timing gear sets and pump-nozzle-systems for injection.

Timing gears have to control the camshaft or the camshafts, and via the cams they must assure the correct ignition points for each of the cylinders as accurately as possible, For this purpose designers apply today additional automatic valve adjustment systems, which we shall not discuss here. Controling ignition is the main task of timing systems, but they also drive and control various ancillary systems like a generator, an oil pump, a power steering pump, a water pump and an airconditioning compressor. With increasing requirements with respect to comfort these ancillary systems grow in number and performance, and as a matter of fact they influence the dynamics of the timing system considerably.

### 6.2.2 Structure and Model of the 5-Cylinder Timing Gear

The general structure of the timing gear set of the Volkswagen R5 Diesel engine is illustrated by Figure 6.16. The whole set is located at the engine rear between engine and clutch. It includes 12 helical gears forming one direct branch to the camshaft and altogether three smaller branches to ancillary components.

6.2 Timing Gear of a 5-Cylinder Diesel Engine     347

The camshaft is driven by the main branch with wheel 1 of the crankshaft, with the three indermediate wheels 2, 3a3b, 4 and with the wheel 5 of the camshaft. The transmission ratio camshaft/crankshaft $= 1/2$ is realized by the double wheel 3a3b. The valves and the pump-nozzle-elements for the five cylinders are operated by the cams on the camshaft. Figure 6.15 gives an impression of the real world timing gear, and Figure 6.16 depicts the corresponding details, the camshaft, the rocker arms,the pn-elements (pn=pump-nozzle) and the valves. The oil pump is driven by wheel 10, which is the only

Fig. 6.15: Timing Gear of the VW R5 TDI (courtesy VW)

wheel of the side branch 1. The power steering pump and the airconditioning compressor are driven by wheel 23 and the generator by wheel 22, which together form the side branch 2. Finally the side branch 3 consists only of wheel 30 for the water pump. The crankshaft and all ancillary equipment are not shown in Figure 6.16.

The helical gears with an helix angle of 15° are pivoted on case-fixed pins, with the exception of wheel 31. Wheel 31 is shrinked on the water pump shaft, and this shaft is pivoted on a roller bearing in the water pump housing. Therefore all wheels, with the exception of wheel 31, are taper bore mounted, which has to be examined for possible vibration enlargement of the overall system. We place the inertial coordinate system in the origin of the crankshaft

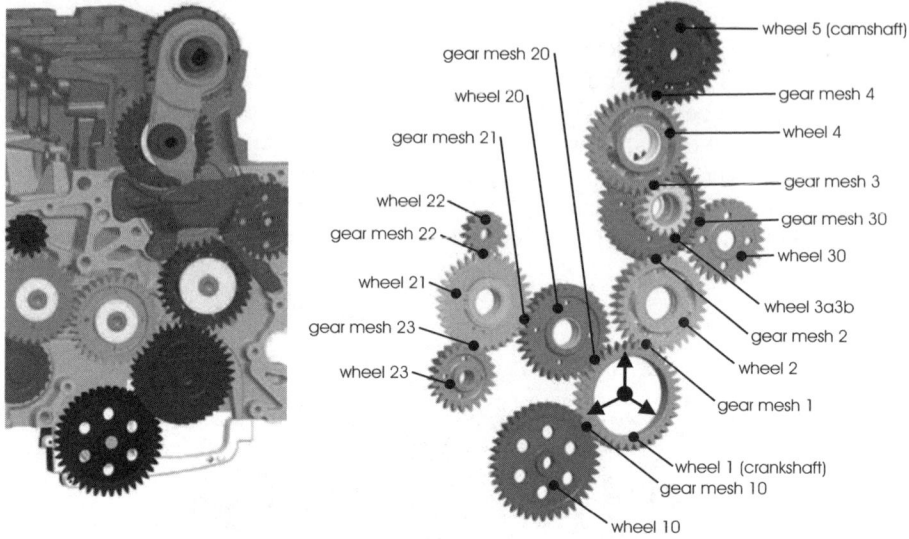

Fig. 6.16: Structure of the Timing Gear R5 (courtesy VW)

wheel, wheel 1, with the z-axis in direction of the crankshaft axis (see the Figures 6.16 and 6.17).

Figure 6.17 depicts only the main features of the mechanical model, which comprises 56 bodies with altogether 97 degrees of freedom being interconnected by 137 force law elements. With respect to the model details we refer to the methods presented in chapter (6.1) on page 329, which we also shall apply in the following without presenting the formal details. We shall give only a description.

The **gear wheels** are modeled with three degrees of freedom, a rotational one for the wheel rotation around the z-axis and two translational ones for translational motion in the bearings. Tilting is neglected, because the backlashes of the bearings are too small for producing tilt and thus for influencing the motion. The elasticity of the wheels is also not considered, because the lowest elastic eigenfrequency of any wheel is far beyond the operation frequencies. The following table 6.1 illustrates the geometry and the mass geometry of the wheel elements, which are already optimized with respect to these parameters.

The gear wheels of the timing gear set are *involute gears* with a transverse pressure angle of 20° and a helix angle of 15°. The module for all wheels is 2.85 mm. The maximum value of the gear backlash is 0.20 mm for a temperature of 130°. The jerky ignition pressure course as well as the irregularities of the crankshaft excite the system to produce vibrations, which are characterized by impulsive contacts between the meshing gears at the front and at the rear flanks thus generating a hammering effect. Hammering consists

## 6.2 Timing Gear of a 5-Cylinder Diesel Engine

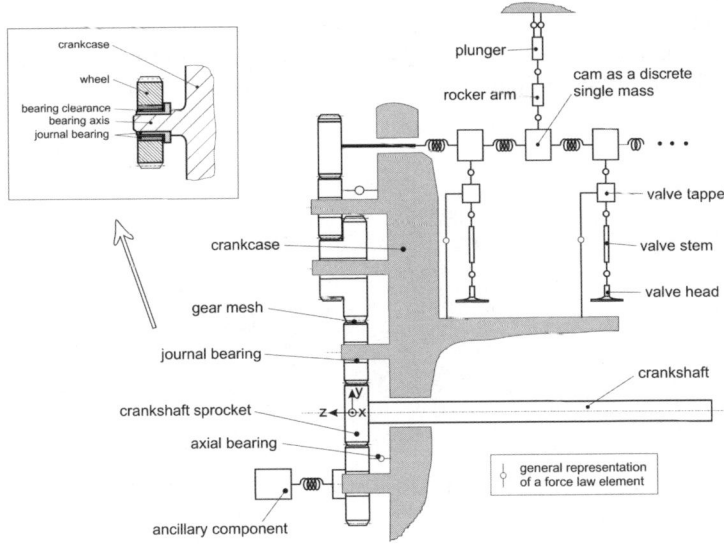

Fig. 6.17: Mechanical Model of the R5 Timing Gear Train

in contact and detachment under load and results in very large load peaks.

Table 6.1: Geometry and Mass Geometry of the R5 Gear Wheels (mass mi = mass moment of inertia)

| wheel number | teeth number | teeth width [mm] | mass [kg] | mass mi [$10^{-4}$ kg·m$^2$] |
|---|---|---|---|---|
| 1 | 37 | 20,0 | 1,49 | 22,78 |
| 2 | 33 | 19,8 | 0,58 | 9,54 |
| 3a3b | 37/17 | 20,0/24,0 | 0,90 | 15,48 |
| 4 | 35 | 23,8 | 0,70 | 12,45 |
| 5 | 34 | 20,0 | 1,02 | 13,77 |
| 10 | 41 | 8,0 | 0,44 | 9,66 |
| 20 | 34 | 10,0 | 0,40 | 5,60 |
| 21 | 30 | 10,0 | 0,32 | 3,44 |
| 22 | 17 | 10,0 | 0,12 | 0,49 |
| 23 | 32 | 10,0 | 0,33 | 4,39 |
| 30 | 28 | 8,0 | 0,19 | 1,92 |

Consequently, this load is also influenced by the backlash itself, it increases with increasing backlashes due to larger kinetic energy in large plays. Therefore we must choose a force law based on two-sided force elements with play, see Figure 6.18 and see also Figure 6.8. The gear construction of this figure is well-known (see for example [170]), and according to that the tooth forces act in the directions of the lines of contact, which are the connecting tangential lines to the two base circles. Assuming a linear spring damper force law we may write for the contact forces $\mathbf{F}_{mesh} = (c \cdot g + d \cdot \dot{g})\mathbf{e}_{1,2}$, where $(g, \dot{g})$

are the deformation and deformation velocity in the contact. The spring and damper coefficients must be evaluated by a FEM-model or by the standard model of Ziegler [285], which due to its excellent approximations is still in use in industry.

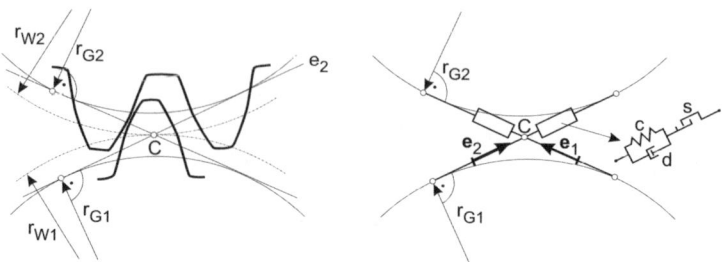

Fig. 6.18: R5 - Gear Mesh with Backlash and Model

For gear meshes these forces depend on the contact ratio, which is a measure of the, not necessarily integral, number of meshing gear teeth. This has to be calculated by a superposition of the spring-damper-coefficients according to the contact ratio. Figure 6.19 depicts a typical example with its maxima around 0-0.75 and its minima around 0.75-1.0, in the first case meshing taking place with three, in the second case with two wheels. Damping is mainly caused by friction in the contacts and not by material damping, which is very small for gear wheels. In practice damping is estimated by the formula

$$d(\varepsilon) = 2 \cdot d_{Lehr} \sqrt{\frac{c(\varepsilon) \cdot I_1 \cdot I_2}{I_1 \cdot r_2^2 + I_2 \cdot r_1^2}}, \qquad (6.35)$$

where c is the tooth stiffness, $I_{1,2}$ the mass moments of inertia and $r_{1,2}$ the base circle radii. According to our experience with systems of that kind the damping measure $d_{Lehr}$ has values around 0.15-0.25 [217]. For the oil model we apply the same model as in section 6.1, see Figure 6.6 on page 334.

Most of the **bearings** are journal bearings. We could take the bearing coefficients from tables, which are well known but usually evaluated for stationary loads, for example gravity loads for large power plant shafts. This does not fir to our case of a timing gear wheel set. Therefore a detailed bearing model has to be evaluated, if necessary, at each integration time step. Such a detailed model has been established by [229] and is applied for problems like the present one.

It works as follows: We describe the fluid motion within the oil layer by the Reynolds equations, a specialized form of the Navier Stokes equations, which we solve in an approximate way by first establishing some functional relations of the oil gap and by an approximation of the oil pressure in the gap by Tschebyscheff polynomials. The oil forces on the shaft are then determined

## 6.2 Timing Gear of a 5-Cylinder Diesel Engine

conditioning compressor. We shall give some indications on modeling and on simulations with respect to loads and noise.

Figure 6.21 illustrates the scheme of these side branches and the corresponding mechanical model. It differs slightly with respect to Figure 6.16, where the two split wheels are missing and replaced by just one wheel. The introduction of the two-wheeled gear rig was an experiment to reduce noise. One parameter influencing noise considerably is backlash, and one possible measure to overcome backlash is such a set of two split wheels. Therefore we find between (w20) and (w22, w23) the two gear wheels (w21a, w21b), Figure 6.21. Following the Figure 6.21 we have to model three ancillary branches of

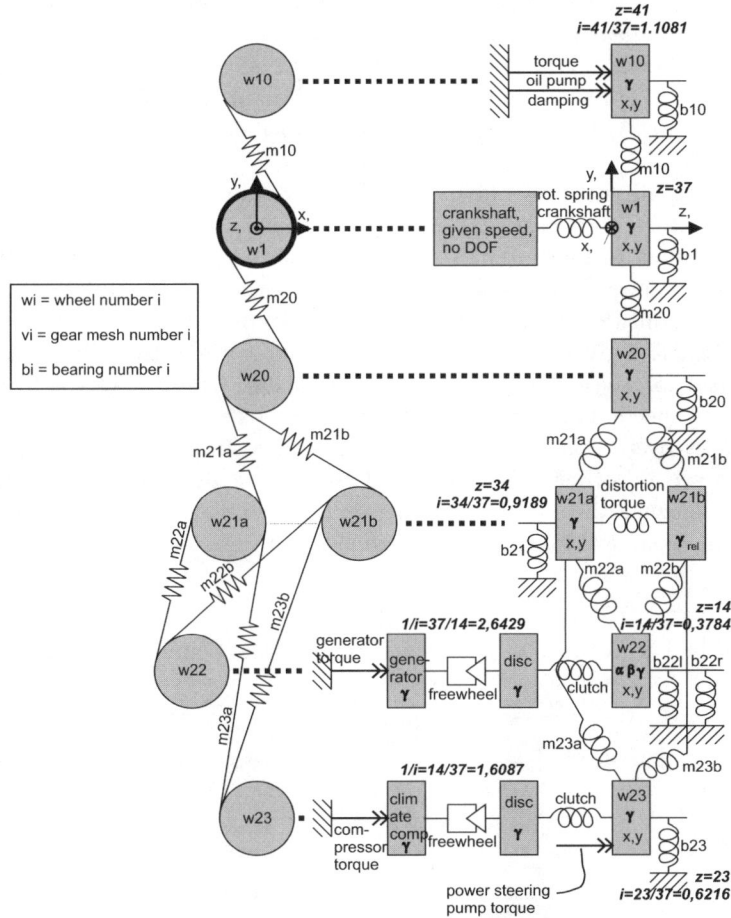

Fig. 6.21: R5 - Gear System for the Ancillary Equipment

the timing gear wheel set. The simplest one is the ***oil pump drive*** including only wheel (10). Therefore the corresponding model starts with the given motion of wheel (1) of the crankshaft, going via the mesh 10 to the wheel 10 and additionally regarding the bearing properties b10. Oil pump torques depending on loads and speed and oil pump damping are given.

The gear sets to the generator, to the climate compressor and to the power steering pump require more complicated models. The **generator drive** starts again at wheel (w1) and proceeds via wheel (20), the split wheels (21a, 21b) to the gear (w22), which drives the generator. All force elements and all degrees of freedom along this path are indicated in Figure 6.21. But additionally we need to model the generator itself with its clutch, disc, freewheel and the direct generator components, which must be given from the manufacturer. The connecting force elements are usually highly nonlinear, which requires corresponding measurements.

From the crankshaft wheel (w1) we go to the **climate branch** following (w20, w21a, w21b, w23) and from there to climate compressor and the power steering pump. Again, the climate compressor includes a clutch, a freewheel, a disc and itself.

Modeling all these ancillary components requires a large amount of data and parameters, to a large extent depending on loads, speeds and temperature. They must be measured or are part of the component's delivery. It would exceed the scope of such a book to include all these data. But to present an example we consider some typical data of the climate compressor and the generator. Table 6.3 gives the torques for the generator and the climate compressor, typical for the car size under consideration. Clutches characteristically depend on speed, load and temperature, but exhibit also strong nonlinear damping by hysteresis. Additionally the clutch stiffness is also nonlinear, very near to a second order curve, but nonlinearity usually does not become very efficient, because the operating range is approximately well near the linear behaviour around the origin, with some exceptions of course. It should be

Table 6.3: Torques of the Generator and the Climate Compressor

| crankshaft speed [rpm] | generator torque [Nm] | climate compressor torque [Nm] |
|---|---|---|
| 0750 | 10.0 | 22.55 |
| 1000 | 11.5 | 19.74 |
| 1500 | 9.3 | 15.92 |
| 2000 | 7.6 | 14.39 |
| 2500 | 6.5 | 14.39 |
| 3000 | 5.8 | 14.39 |
| 3500 | 5.2 | 14.39 |
| 4000 | 4.7 | 14.39 |
| 4250 | 4.5 | 14.39 |

noted, however, that clutch stiffness depends on the frequency applied to the clutch during its operation. This well-known effect has to be measured for different motor speeds and for different engine orders. With respect to the last mentioned parameter the 2.5 order is of the most important relevancy with respect to a 5-cylinder engine. Table 6.4 includes the engine speed for stationary simulations and the resulting frequencies with an excitation of the dominant 2.5 engine order. The stiffnesses are the values corresponding to these frequencies. They are evaluated according to the empirical formula (see for example [36]).

$$c = c'_{Ref} \log_{10}(f) + c_{Ref} = 621.4 \log_{10}(f) + 780 [Nm/rad] \tag{6.36}$$

Very extensive measurements of such Gates-clutches have been performed at

Table 6.4: Influence of Frequency on Clutch Stiffnes at 20°C

| engine speed [rpm] | 2.5. order frequency [Hz] | clutch stiffness $c_{Ref}$ [Nm/rad] |
|---|---|---|
| 750 | 31 | 1709 |
| 1000 | 42 | 1787 |
| 1500 | 63 | 1896 |
| 2000 | 83 | 1974 |
| 2500 | 104 | 2034 |
| 3000 | 125 | 2083 |
| 3500 | 146 | 2125 |
| 4000 | 167 | 2161 |
| 4250 | 177 | 2177 |

the Technical University of Aachen [8]. These results have been used for our case.

### 6.2.4 Simulation Results

#### 6.2.4.1 Complete Timing Gear Train

During the development of the models and before performing extensive parameter simulations a couple of ***comparisons model/experiments*** were carried through applying the following data: temperature of the transmission housing 90°, averaged backlash at this temperature 0.15 [mm], engine speeds of 1000, 2000, 3000, 4000 rpm. Selecting two examples from a large amount of such measurements we show the torques of the camshaft for the driving and braking cases. Figure 6.22 confirms the models with about an accuracy of 10%, which is excellent before the background of a rather complicated system dynamics with clearly recognizable hammering effects. In addition we have to consider the elasticity of the bearings, which influence the results in a clear way. So the irregularities of the rotational camshaft speed becomes twice as large for elastic bearings in comparison to rigid bearings. Therefore the final design of the timing gear train as shown in Figure 6.16 includes in addition a plate arranged before the housing wall with the goal to support the gear wheels not only one-sided at the housing but two-sided in the housing and the plate.

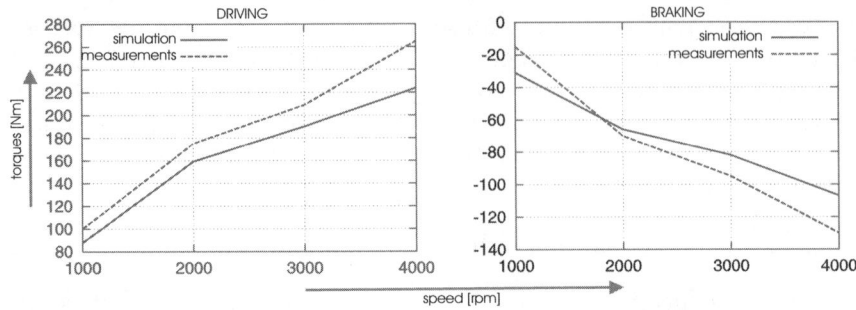

Fig. 6.22: R5 - Comparison Simulation/Measurements

After these verifications the model is used to perform many parameter variations for design improvement. In a first run the torques at the camshaft and the forces of the gear meshes and the bearings turned out to be very large. For example, the maximum torques of the camshaft were about 240 Nm for the driving and -130 Nm for the braking case., the maximum forces in the meshes of the gears about 10 kN and the maximum bearing forces about 16 kN. The reasons are clear: the injection pressure is extremely large (Figure 6.20) leading to a separation of the tooth flancs. Depending on the backlash size large impacts occur on both sides of the flancs generating large forces in the tooth mesh. This effect could be detected also in the example of chapter 6.1, where we got an overload of more than a factor of 10.

#### 6.2.4.2 Ancillary Components

One of the most important problems of the dynamics of the ancillary components consists in the dynamic and in the parameter behaviour of the clutches with some rubber elements, which depend very much on temperature influences. All simulations are started with a reference configuration according to Figure 6.21 and to the appropriate data set, which refers to a speed of 1000 rpm and can be adapted to any other speed. We investigate stationary dynamics for speeds of 750 rpm to 4250 rpm in relatively small steps, and we consider as reference stiffness especially for the clutches a the values at a temperature of 20°C.

By practical experience from the VW test bed it was known, that the case without climate compressor load and with nearly no generator load was the worst case with some damage of the Gates- clutch. It turned out furtheron, that the spanning of the gears (w21a,b) has no influence on this situation, which is determined mainly by the dynamics of the whole ancillary branch system and not so much by the backlashes. Even with only one gear wheel (21) according to Figure 6.16 we got similar results. Figure 6.23 illustrates a typical simulation result in form of a time series of the gear meshes. Not all gears of Figure 6.21 are shown. The results illustrate the characteristic hammering of such timing gears under load, for our case with a the 2.5 th engine order (f=50 Hz, 2.5f= 125 Hz, T=0.008 s).

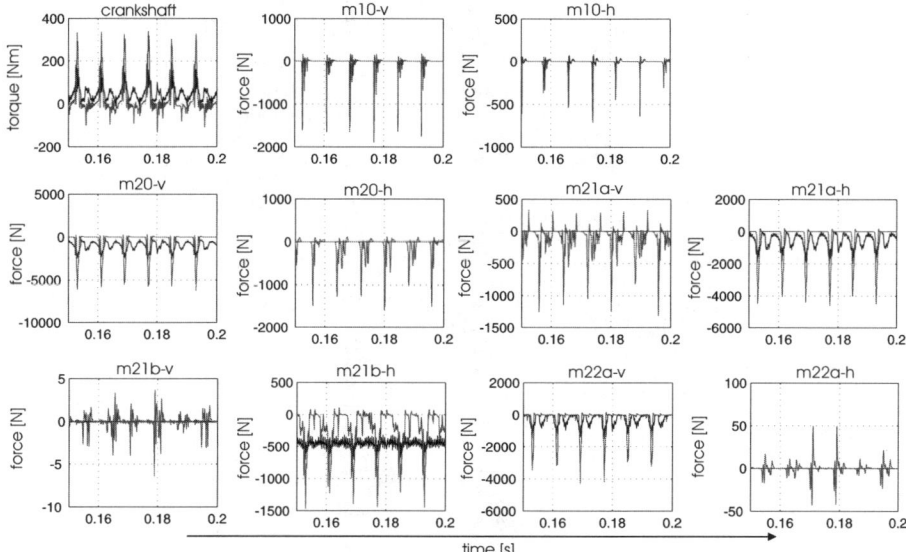

Fig. 6.23: R5 - Simulations of the Ancillary Branch (v=vertical, h=horizontal, abbreviations of the gear meshes (mi) according to Figure 6.21), 3000 rpm, without loads on generator and climate compressor

For evaluating the parameter influences with respect to the clutch load a large amount of simulations have been performed, which came out with the following results.

## Table 6.5: Results of the Parameter Variations

| parameter variation | changes with respect to reference | effect with respect to clutch torque |
|---|---|---|
| clutch stiffness, with compressor, temperature | significant  90°C to −30°C | significant |
| climate compressor | without compressor | significant worst case |
| stiffness clutch without compressor | very significant | significant 4250 rpm |
| generator load | 0 to 12 Nm, constant | significant 4250 rpm |
| total load of the ancillary branch | reduced and applied to generator | very significant 4250 rpm |
| climate compressor generator load | without without | significant 3000 rpm worst case |
| split wheel 21b | without | no effect |
| spanning torque | 0 to 30 Nm | no effect |
| climate compressor generator load stiffness clutch | without 0 to 12 Nm -50% bis +600% | significant |
| climate compressor generator load stiffness clutch | without no load +600% | significant for the whole high speed range |
| moment of inertia of the generator disc | -50% to +100% | significant |
| moments of inertia of gear wheels | +30% bis +80% | moderate |

## 6.3 Timing Gear of a 10-Cylinder Diesel Engine

### 6.3.1 Introduction

The VW 10-cylinder Diesel engine is based on the 5-cylinder concept, an engine, which has proved very worthwile in the course of the years, but it is not the sum of two 5-cylinder engine. It is an individual configuration. Nevertheless many of the aspects, which we have discussed in chapter 6.2 apply also here making modeling a bit easier. So we are able to take over most of the modeling details for the gears and the gear meshes, for the bearings and for the injection pump system. The crankshaft produces a kinematical excitation, and the camshafts behave similar to the 5-cylinder case.

Again, mechanical and mathematical modeling follows the paths presented in chapter 6.1, where detailed mathematical models have been established. Mainly due to the backlashes within the meshes of the gears, but also due to some components like clutches or ancillary equipment, the system dynamics is highly nonlinear and as a rule cannot be linearized around some operating point. We tried to do that for the 5-cylinder engine timing train, but without success.

### 6.3.2 Structure and Model of the 10-Cylinder Timing Gear

The *structure* of the VW V10 Diesel engine is illustrated by Figure 6.24. The timing gear with gear wheels only is located at the backside of the engine, between engine housing and the drive train clutch. The VR-angle between the two 5-cylinder motor blocks is 90°. The left camshaft drives the valves and the injection pump pistons of the cylinders (1 to 5) and the right camshaft those of the cylinders (6 to 10). The crankshaft, here the wheel (w1), drives the two camshafts, on the left side via the gears (2, 3, 4a/4b, 5, 6) and on the right side by the gears (10, 11a/11b, 12, 13).

The necessary transmission ratio for the camshafts is realized by the double gear wheels (4a/4b) on the left and (11a/11b) on the right side. All gear wheels are helical gears with a helix angle of 15°. With the exception of gear (w31), which is shrinked on the water pump shaft, all gears are supported by journal bearings. Ancillary components are driven in the following way: water pump by gear (w31), oil pump, power steering pump and climate compressor by the wheels (w20) and (w22), the generator with freewheel and elastic clutch by wheel (w40).

The origin of the inertial coordinate system is the axis of gear wheel (w1), Figure 6.24. with the z-axis along the crankshaft axis and with the x- and y-axes in the plane of the timing gear system.

The *mechanical model* includes 62 rigid bodies with altogether 137 degrees of freedom, which are interconnected by 207 force elements. Figure 6.25 depicts the main features. The wheels are modeled as rigid bodies with four degrees of freedom, one for rotation and three for translation, which allows

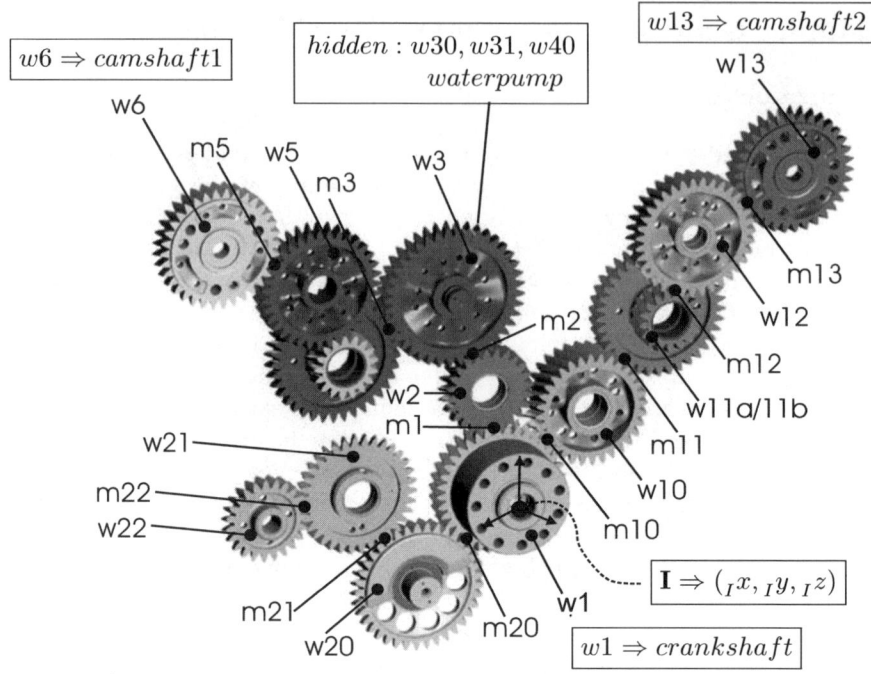

Fig. 6.24: Structure of the V10 Timing Gear Train (wi = gear wheel number i, mi = gear mesh number i, $(_I x, _I y, _I z)$ = inertial coordinates), (courtesy VW)

to take into account the influence of elastic bearings. As in the 5-cylinder case we neglect tilting motion of the gear wheels, because the bearing tolerances make such tilting nearly impossible. Table 6.6 depicts the masses and the mass moments of inertia of the wheels. The gears possess a helix angle of 15° and a pressure angle of 20°. The module is 2.85 with the exception of the gears (w30, w31), where it is 2.00. The maximum backlash of 0.20 mm appears at a temperature of 130°. Due to the large injection pressures and from there due to the large loads hammering will be generated comparable to the R5-engine.

Gear meshes, bearings, camshafts, injection pumps and valve trains data are taken in an appropriate way from the R5-timing gear (chapter 6.2). In addition and between the camshaft gear wheels and the camshaft itself momentum absorbers are included, which are more or less rather stiff torsional springs. They drive the camshaft, and they reduce the camshaft vibrations after each injection impulse.

The crankshaft output excites the system by speed irregularities, which must be measured. Figure 6.26 illustrates some typical measurements. The graph indicates the usual behaviour, namely that for small speeds irregularities become large, an effect requiring certain damping measures for this speed range.

## 6.3 Timing Gear of a 10-Cylinder Diesel Engine

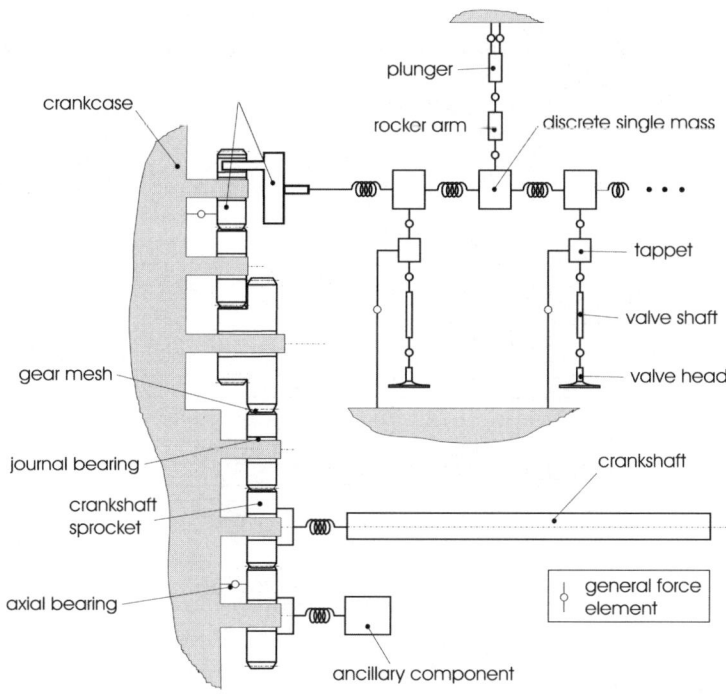

Fig. 6.25: Mechanical Model of the V10 Timing Gear

The ancillary equipment has similar torque requirements as in the 5-cylinder case, for example the oil pump about 2 Nm, the generator from 5 to 15 Nm, the power steering pump about 22 Nm, the climate compressor about 15 Nm and the water pump 2 Nm.

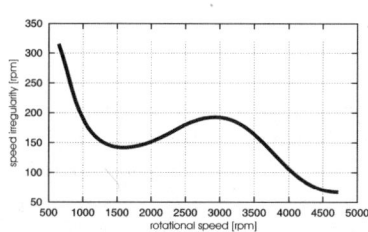

Fig. 6.26: Typical Rotational Speed Irregularities (V10)

Table 6.6: Geometry and Mass Geometry of the V10 Gear Wheels (mass mi = mass moment of inertia)

| gear number | number of teeth | facewidth [mm] | mass [kg] | mass mi $[10^{-4}\ kg \cdot m^2]$ |
|---|---|---|---|---|
| 1 | 37 | 20,0 | 1,49 | 22,3 |
| 2 | 24 | 20,0 | 0,48 | 4,0 |
| 3/30 | 40/26 | 20,0/10,0 | 0,75 | 16,0 |
| 4a4b | 37/17 | 20,0 | 0,9 | 15,5 |
| 5 | 35 | 24,0 | 0,66 | 11,1 |
| 6 | 34 | 20,0 | 0,60 | 11,3 |
| 10 | 32 | 37,0 | 1,12 | 16,7 |
| 11a11b | 37/17 | 20,0 | 0,9 | 15,5 |
| 12 | 35 | 24,0 | 0,66 | 11,1 |
| 13 | 34 | 20,0 | 0,60 | 11,3 |
| 20 | 37 | 10,0 | 0,47 | 9,1 |
| 21 | 31 | 10,0 | 0,39 | 4,5 |
| 22 | 30 | 10,0 | 0,33 | 4,6 |
| 31 | 17 | 10,0 | 0,2 | 0,2 |
| 40 | 14 | 10,0 | 0,3 | 0,4 |

### 6.3.3 Simulation Results

The simulations are performed applying a computer code based on the mathematical models of chapter 6.1 on page 329. The results presented correspond to an early stage of the iterative process for improving and stabilizing the design. Nevertheless they include some important indications with respect to the maximum forces and torques as a basis for a first sizing up of the design ideas.

The maximum torques of the two camshafts are about 260 Nm, approximately the same value for both shafts. The maximum tooth forces in the gear meshes depend also on the loads of the ancillary equipment. According to table 6.7 the loads on the left side are a bit larger than those of the right side, which is due to some ancillary components and in addition to one more wheel in that branch (see also Figure 6.24). The maximum loads on the bearings are also illustrated by table 6.7, where in addition the angle with respect to the positive x-axis is given designating the direction of the corresponding maximum force. The gears of the two main timing gear branches are heavily loaded by the large injection pressures and as the consequence by large torques. As in the case R5 this leads also here to hammering situations, which are always accompanied by very large forces on the gear teeth. The journal bearings thus move always between two extreme positions, which are both equally and heavily loaded. Table 6.7 gives also the two corresponding angles for these two force directions.

Comparing the above mentioned magnitudes with those of the 5-cylinder engine we recognize the same order. Therefore many general aspects concerning the loads of the R5 can appropriately tranfered to the V10-design.

It might be of interest with respect to similar cases to investigate the effect of speed irregularities of the engine on the loads of teeth and bearings. The simulation was performed for a speed run-up from 2200 rpm to 4600 rpm

## 6.3 Timing Gear of a 10-Cylinder Diesel Engine

Table 6.7: Maximum Forces in the Gear Meshes and on the Bearings (mi=gear mesh number i, bi=bearing number i)

| geer teeth number | tooth force [kN] | gear trains | bearing | bearing force [kN] | angle of bearing force |
|---|---|---|---|---|---|
| m1 | 15,5 | to water pump and generator | b1 | 15,6 | 0,3 / 2,1 |
| m2 | 14,3 | left bank | b2 | 27,4 | 0,3 / 3,3 |
| m3 | 17,0 | left bank | b3 | 15,9 | 1,1 / 4,3 |
| m4 | 18,9 | left bank | b4 | 17,2 | 1,1 / 3,1 |
| m5 | 14,0 | left bank | b5 | 22,9 | 0,5 / 4,0 |
|  |  |  | b6 | 13,6 | 0,5 / 4,4 |
| m10 | 14,2 | right bank | b10 | 22,4 | 2,3 / 5,5 |
| m11 | 13,8 | right bank | b11 | 22,4 | 2,8 / 6,0 |
| m12 | 17,7 | right bank | b12 | 26,4 | 2,5 / 5,6 |
| m13 | 13,5 | right bank | b13 | 13,1 | 2,3 / 5,0 |
| m20 | 7,5 | to ancillary components | b20 | 4,8 | 5,2 |
| m21 | 6,6 | oil pump, power steering | b21 | 6,4 | 0,8 |
| m22 | 6,1 | pump and climate compressor | b22 | 5,9 | 5,0 |
| m30 | 4,8 | to generator | b30 | 4,7 | 3,1 / 0,5 |
|  |  |  | b31 | 4,6 | roller bearing |
| m40 | 4,5 | generator | b40 | 4,4 | 2,8 |

using Figure 6.26. Table 6.8 depicts some results, which in some cases indicate large deviations with respect to the two cases "with" and "without" rotational irregularities; and this behaviour inspite of the fact, that the excitation due to the very large camshaft torques are still included. Therefore care has to be taken in modelling the inputs at the system cuts.

## 6 Timing Equipment

Table 6.8: Maximum Forces and Torques in a Comparison

|  | maximum value **with** rotational irregularities | maximum value **without** rotational irregularities | difference |
|---|---|---|---|
| camshaft left side | 271,9 Nm | 221,2 Nm | 23% |
| camshaft right side | -254,7 Nm | -252,2 Nm | 1 % |
| toothing 1 | 15,5 kN | 11,4 kN | 36% |
| toothing 2 | 14,3 kN | 8,7 kN | 64% |
| toothing 3 | 17,0 kN | 8,9 kN | 91% |
| toothing 4 | 18,9 kN | 15,0 kN | 26% |
| toothing 5 | 14,0 kN | 11,9 kN | 18% |
| toothing 10 | 14,2 kN | 10,6 kN | 34% |
| toothing 11 | 13,8 kN | 8,3 kN | 66% |
| toothing 12 | 17,7 kN | 12,2 kN | 45% |
| toothing 13 | 13,5 kN | 10,8 kN | 23% |
| toothing 20 | 7,5 kN | 5,3 kN | 43% |
| toothing 21 | 6,6 kN | 4,5 kN | 47% |
| toothing 22 | 6,1 kN | 4,2 kN | 45% |
| toothing 30 | 4,8 kN | 3,2 kN | 50% |
| toothing 40 | 4,5 kN | 3,3 kN | 36% |
| bearing 1 | 15,6 kN | 12,4 kN | 26% |
| bearing 2 | 27,4 kN | 18,4 kN | 49% |
| bearing 3 | 15,9 kN | 9,6 kN | 66% |
| bearing 4 | 17,2 kN | 13,8 kN | 25% |
| bearing 5 | 22,9 kN | 21,0 kN | 9% |
| bearing 6 | 13,6 kN | 11,4 kN | 19% |
| bearing 10 | 22,4 kN | 16,9 kN | 33% |
| bearing 11 | 22,4 kN | 15,4 kN | 45% |
| bearing 12 | 26,4 kN | 20,6 kN | 28% |
| bearing 13 | 13,1 kN | 10,6 kN | 24% |
| bearing 20 | 4,8 kN | 3,7 kN | 30% |
| bearing 21 | 6,4 kN | 4,5 kN | 42% |
| bearing 22 | 5,9 kN | 4,1 kN | 44% |
| bearing 30 | 4,7 kN | 3,1 kN | 52% |
| bearing 31 | 4,6 kN | 3,0 kN | 53% |
| bearing 40 | 4,4 kN | 3,2 kN | 38% |

## 6.4 Bush and Roller Chains

### 6.4.1 Introduction

The classical solution for controling the camshaft consists in timing belts, which in the course of the years have been steadily improved to meet also requirements in connection with high performance combustion engines. Nevertheless and especially in Europe the chains are more and more gaining ground due to valve trains of greater sophistication and due to injection pumps with very high pressure loads, to name two important influences. Modern Diesel engines with their extremely large output torques put also very large loads on the timing system. Belts come in these cases to the limit of their performance. Therefore chains, inspite of their larger mass and thus of their larger inertial forces, get a chance.

Fig. 6.27: Timing Chain Drive of the BMW M62 V8 (1996)

The Figures 6.27 and 6.28 give an impression of two German timing chain systems for high performance cars. Obviously the design and the arrangement of the chains have to follow very restricted space considerations. Widely distributed chains are the bush and roller chains, which we shall consider in this chapter, and with respect to the inverted tooth chains we refer to [111]. The first configurations are applied more in Europe, the second ones in the US.

366    6 Timing Equipment

Fig. 6.28: Timing Chain Drive of the Porsche 911 Carrera

Figure 6.29 depicts a roller chain and a bush chain together with their models we shall use. In the case of the bush chains the teeth of the sprocket contact directly the pins fixed in the tab plates, whereas in the case of roller chains we have an additional rolling element reducing the friction in the tooth-roller-contact by rolling without sliding, at least approximately. The external elements comprising two tab plates and two pins do not have contact with the sprocket teeth. Typical chain pitches are 7 mm, 8 mm and 9.525 mm. Double roller chains are applied for larger loads. They possess an additional tab plate in the middle of the double configuration.

The link model follows the dissertation [69] of Fritz, who was the first one to present a detailed simulation model of roller chain systems. In the following we shall give some more details (see also [70], [111] and [201]).

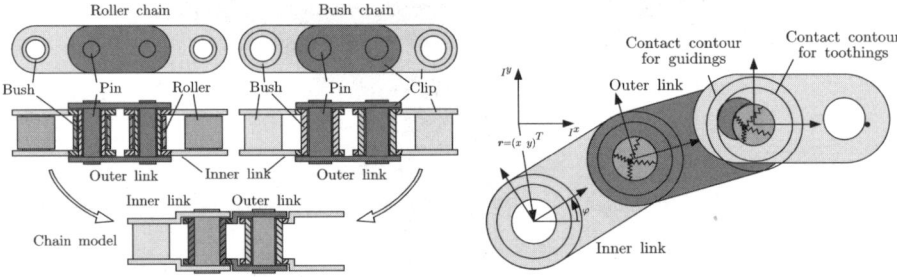

Fig. 6.29: Bush and Roller Chain Elements (left), Corresponding Model (lower element of the left side and right)

### 6.4.2 Mechanical and Mathematical Modeling

#### 6.4.2.1 Sources of Excitation

Chains are loaded by internal and external excitations [69]. The structure comprising plates and pins generates polygonal effects with their frictional

## 6.4 Bush and Roller Chains

impact behaviour in the inlet and outlet of the sprocket wheels. In contrast to belts the chains form a polygonal structure contacting the teeth of the sprocket wheels in one point only (see Figure 6.30). To get an estimate for the velocities and the polygonal frequency we assume that the joint centers

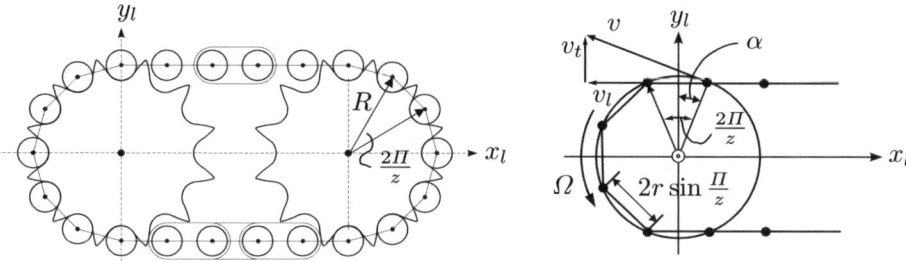

Fig. 6.30: Polygon Effect for Roller Chains

are positioned on the generated pitch circle. If then the sprocket wheel with z teeth and a pitch circle radius R rotates with the constant velocity $\Omega$ we get the following velocity variations with respect to the free strand

transverse $\quad \Delta v_t = R\Omega \sin(\Omega \alpha),$

longitudinal $\quad \Delta v_l = R\Omega [\cos(\Omega \alpha) - (\frac{\sin \frac{\Pi}{z}}{\frac{\Pi}{z}})], \quad (-\frac{\Pi}{z} \leq \alpha < \frac{\Pi}{z}).$ (6.37)

where the second term in the longitudinal equation represents the averaged velocity in longitudinal direction (see section 5.4 on the pages 275 ff., especially equation 5.103). The angle $\alpha$ is always within the pitch angle of the sprocket wheel. Longitudinal oscillations take place along the chain and transverse oscillations perpendicular to it. Considering here a plane model we do not take into account out of plane vibrations. The above velocity oscillations decrease with increasing number of teeth, linearly in transverse and quadratically in longitudinal directions. The polygonal frequency itself is simply the sprocket velocity multiplied with the number of teeth:

$$\Omega_{polygon} = z\Omega \tag{6.38}$$

We always get an impact with friction when a chain element enters or leaves the sprocket wheel. The directions of motion of chain and sprocket are different leading to a vector difference of the velocities, which can only be demounted by a shock between the chain and the sprocket tooth (see Figure 6.31). This vector difference between the element velocity $v_G$ and the sprocket velocity $v_R$ of the contact point writes

$$\Delta v = 2R\Omega \sin(\frac{\Pi}{z}) \cos(\frac{\Pi}{z}), \tag{6.39}$$

which is mainly a vector in transverse direction of the chain. The position of the inlet shock is very important, because it changes significantly the direction **n** and with **n** the direction of the relative velocity $\Delta v$.

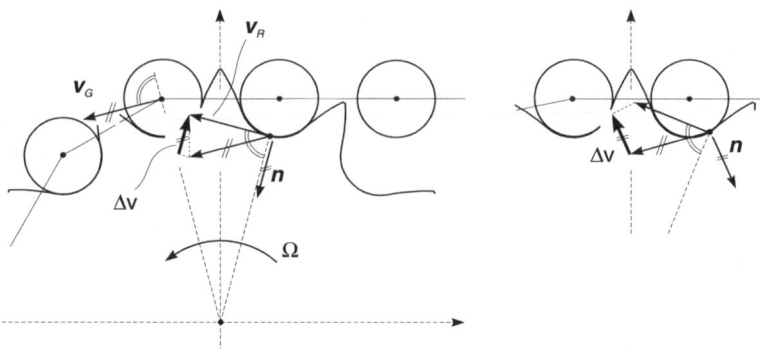

Fig. 6.31: Running-In of a Chain Element into a Sprocket Tooth

The shocks in the running-in area are a main source of noise and wear. Furtheron we know from observations, that especially for large rotational velocities the element already in contact with a sprocket tooth detaches again leading to additional shocks and noise. The running-in shock is always accompanied by an extremely short time distance, within which the chain element has to be accelerated from zero rotational velocity to the sprocket rotational velocity. This effect produces considerable inertial loads.

We get in the running-out zone similar effects, where the element inertia together with some clamping process at the end of the sprocket arc under contact generates a typical running-out arc accompanied by additional longitudinal forces in the order of magnitude of the operational tensile forces.

The internal excitations discussed so far are mainly parameter excited oscillations, which excite the overall system. The external excitations depend on the system boundaries and thus on the system cuts. As a rule we have both, kinematical and kinetic excitations. The output of the crankshaft usually is given by kinematical magnitudes, which are easily to measure. On the other hand, the system's output for example at the driven sprocket wheel is loaded by external sources and thus will be given by measured (or sometimes assumed) load data. In our case of timing chains these chains are loaded by the camshaft or the camshafts, which might lead to a cyclic change of heavily loaded strands and strands with no loads. The same might be true for kinematical excitations with superimposed velocity fluctuations. We shall come back to that.

The system's cuts should be chosen in such a way, that firstly all internal and external excitations come into effect, and that secondly all components within the operational frequency range are taken into consideration, at least approximately. So it makes sense to include a model of the elastic camshafts as a main load of the timing chains. Reasonable cuts are therefore the measured gas pressure in the cylinders of the engine and the kinematical output of the crankshaft. In addition we have inertial connections by the chain tensioners and the guides.

#### 6.4.2.2 Model Properties

*Roller chains*

A roller chain consists of two different kinds of chain links:
- Chain links with two pins.
- Chain links with two bushings.

These two kinds of links differ in their mass, their moment of inertia and their pitch, if wear of the chain is considered. The elasticity of a link is modeled as a force element in the chain joint. Therefore the link can be treated as a rigid body. The dynamics of a chain roller does not influence the vibration of the chain drive in any substantial manner. Hence the roller is not considered as a separate body. The motion of a chain link is given by the translational

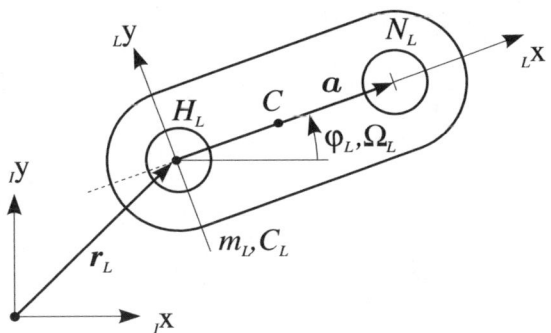

Fig. 6.32: Kinematics of a Link

vector $\boldsymbol{r}_L$ and the rotational vector $\boldsymbol{\varphi}_L$. For the planar motion it is sufficient to consider the displacements in x- and y-direction and the rotation about the z-axis. The following transformation gives a relation between these configuration coordinates $\boldsymbol{q}_L$ and the system coordinates $\boldsymbol{z}_L$ (Figure 6.32):

$$\boldsymbol{r}_L = \begin{pmatrix} x_L \\ y_L \\ 0 \end{pmatrix}, \quad \boldsymbol{\varphi}_L = \begin{pmatrix} 0 \\ 0 \\ \varphi_L \end{pmatrix}, \quad \boldsymbol{z}_L = \begin{pmatrix} \boldsymbol{r}_L \\ \varphi_L \end{pmatrix}, \quad \boldsymbol{q}_L = \begin{pmatrix} x_L \\ y_L \\ \varphi_L \end{pmatrix},$$

## 370  6 Timing Equipment

$$z_L = Q_L q_L = \begin{pmatrix} 1 & 0 & 0 \\ 0 & 1 & 0 \\ 0 & 0 & 0 \\ 0 & 0 & 0 \\ 0 & 0 & 0 \\ 0 & 0 & 1 \end{pmatrix} q_L . \tag{6.40}$$

The coordinate system of a link is fixed in the reference point $H_L$. A second reference point $N_L$ of a link is displaced by the pitch vector $\boldsymbol{a}$.

*Chain Joints*

Due to the low chain tension and the time varying excitations of the camshafts, we regard both the backlash and the oilwhip in a chain joint. Using the theory for journal bearings, the forces acting on the pin and bushing consist of two parts. In this case we only have to consider the forces due to the oil displacements, produced by the translational motion of the pin towards the bushing. In combination with the material damping of the chain, which can be regarded as constant, we get the damping characteristic shown in Figure 6.33 ([217]). The elasticity of a chain link is concentrated in the joint and behaves like a radial spring with the constant spring coefficient $c_L$. Hence, if pin and bushing are in contact, additional spring forces act on the links.

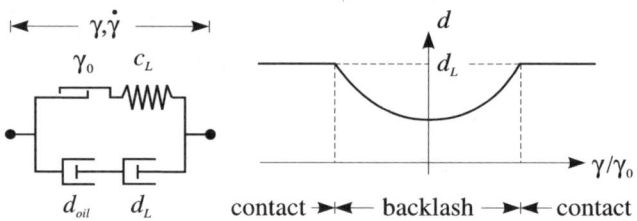

Fig. 6.33: Force Element and Damping of a Chain Joint

*Sprockets*

A sprocket, shown in Figure 6.34, is modeled as a rigid body. With the selection matrix $Q_S$ for the sprocket – like $Q_L$ in equation (6.40) – we can regard optional degrees of freedom

$$z_S = Q_S q_S, \qquad z_S = \begin{pmatrix} r_S \\ \varphi_S \end{pmatrix} . \tag{6.41}$$

For a detailed description of the contact configuration between a link and a sprocket, we have to consider the exact tooth contour. Figure 6.34 illustrates the toothing of a sprocket due to DIN 8196 with the contact areas tooth

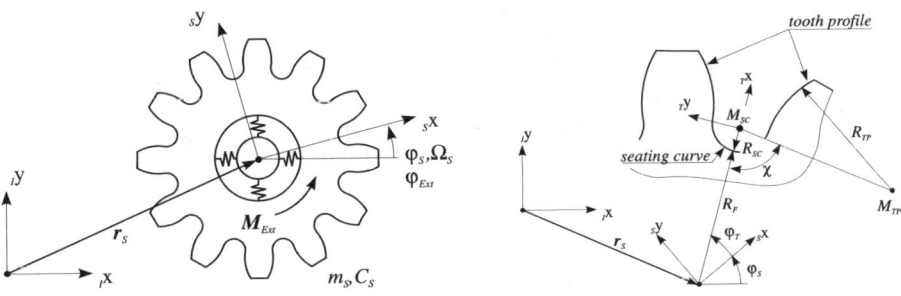

Fig. 6.34: Sprocket Toothing

profile and seating curve. These contact contours are defined by circles. Using a toothing fixed coordinate system the centers of these circles can easily be determined.

Regarding a chain drive in a combustion engine, we have to consider various external excitations from the crankshaft ($\varphi_{Ext}$) and from the camshafts ($M_{Ext}$).

*One Sided and Double Sided Guides*

One sided and double sided guides are applied to reduce the vibrations of the chain strands. Furthermore a tension guide is seated at the slack side, to give the chain drive a definite initial stress. The exact contour of a guide is important. A general formulation of a planar contour is possible by piecewise defined parameter functions, depending on a contour parameter $s$

$$\boldsymbol{r}_K(s) = \begin{pmatrix} x(s) \\ y(s) \end{pmatrix}. \tag{6.42}$$

At the connection points of two functions, it is necessary that the values of the function and those of the first derivation are smooth. The second derivation however can show unsteady behavior, which makes sense of course only, if this effect is not of mathematical nature, but a real consequence of the guide composition by different curve elements. A jump in the curvature of a contour induces a jump in the contact force, which might cause a detachment from the contour. It is a premise to the contact model to make this effects apparent. The direction of the contour parameter $s$ is given by the contour coordinate system $(\boldsymbol{t}, \boldsymbol{n})$ with the condition that the normal vector $\boldsymbol{n}$ directs to the inner part of the guide.

The behavior of the tension device, acting on the tension guide, leads at this point with many simplifications to a description of a force element with a matched damping characteristic. This force element is composed of a spring with the very low spring coefficient $c_{TD}$ and the damping coefficient $d_{TD}$ (Figure 6.35). Depending on the relative velocity in the normal direction $\boldsymbol{n}_{TD}$

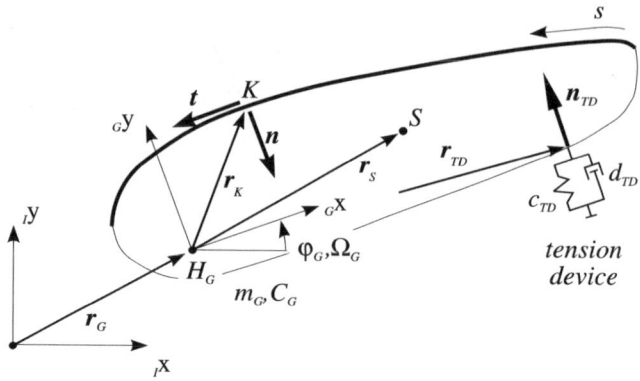

Fig. 6.35: Tension Guide

we assume for negative values a high damping coefficient, for positive values a low damping coefficient. We shall discuss these tension devices in a separate chapter.

*Contact between a Link and a Sprocket*

To describe the contact between a link and a sprocket we have to distinguish the two types of links. For a roller chain the contact is realized by the rollers rotating around the pin of the roller chain (Figure 6.29), for a bush chain the bush comes directly into contact with the sprocket. Keeping in mind that a link is modeled as a rigid body, we have no elasticity between the two contact points of a bushing link. To the next link with contact there are two spring elements of the two joints. In reality however the elasticity of the chain plate acts between each contact. To consider these effects, we suppose that each link has one contact to the sprocket.

Because of the fact, that we do not deal with chain rollers, we define a contour circle with the diameter of a chain roller seated in the reference point $H_L$ of a link (Figure 6.36).

*Contact between a Link and a Guide*

Modeling the contact between a link and a guide the link plate is the contact partner (Figure 6.37). Some varieties of the contact configuration may appear, for example contact in the front or rear side of the link, or along the link plate, which must be considered separately, see [69]. Corresponding to the contact model above, a contour circle with the diameter of the plate width is used.

The high rotation speed of combustion engines induces high relative velocities at the contact points between the links and the guides. Therefore stick-slip processes do not appear. Regarding the contact of a link to a sprocket, there is an additional oilwhip between the roller and the bushing. Neglecting

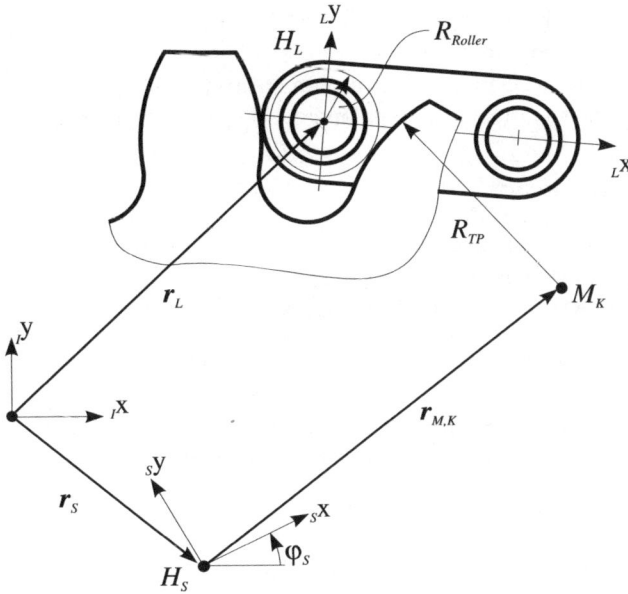

Fig. 6.36: Contact Kinematics for Link and Sprocket

the dynamics of the roller, stick-slip processes are not taken into account. Observations show the fact, that due to the impact of one link, other links

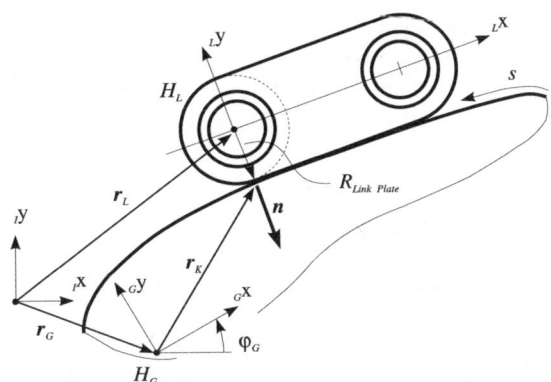

Fig. 6.37: Contact Kinematics for Link and Guide

loose their contact. This effect only can occur, if the contours of the contact partners are modeled without flexibility. Consequently the model has to fulfill the following two requests:

- All contacts are characterized by unilateral constraints, so that a link always can detach from the contour.
- An algorithm, describing the complementarity conditions of the contacts, is used to make the mutual dependence of the contact configuration apparent.

### 6.4.2.3 Mathematical Models

The elements of the motion equations of each individual body are worked out in such a way, that depending on the system's coordinates $z$, the equations of motion can be written in the following form:

$$\mathbf{M}^*\ddot{\mathbf{z}} = \mathbf{h}^* + \sum \mathbf{J}^{*T}\mathbf{F} \quad \in \mathbb{R}^6. \tag{6.43}$$

Using the type of transformation of equation (6.40) we get the equations of motion in configuration space with $f$ degrees of freedom.

$$\mathbf{M}\ddot{\mathbf{q}} = \mathbf{h} + \sum \mathbf{J}^T \mathbf{F} + \mathbf{W}\lambda \quad \in \mathbb{R}^f \tag{6.44}$$

The matrix $\mathbf{M}$ is the symmetric and positive definite mass matrix, the vector $\mathbf{h}$ includes all forces formulated already and directly in the $\mathbf{q}$-space, the sum includes all forces and torques given in the six-dimensional Euclidean space making a transformation with $\mathbf{J}^T$ necessary, and the last term indicates contact forces.

*Chain Elements of the Free Strands*

We start with the chain elements of the free strands, that means with elements without contacts to the sprocket toothing. Figure 6.32 and the equations (6.40) give a first description. From these equations we immediately derive the velocities related to the reference point H of the link:

$$\mathbf{v}_G = \begin{pmatrix} \dot{x}_G \\ \dot{y}_G \\ 0 \end{pmatrix}, \quad \omega_G = \begin{pmatrix} 0 \\ 0 \\ \dot{\varphi}_G \end{pmatrix}, \tag{6.45}$$

with the accompanying Jacobians (see eq.(6.40))

$$\mathbf{J}_T = \frac{\partial \mathbf{v}_G}{\partial \dot{\mathbf{q}}_G} = \begin{pmatrix} 1 & 0 & 0 \\ 0 & 1 & 0 \\ 0 & 0 & 0 \end{pmatrix}, \quad \mathbf{J}_R = \frac{\partial \omega_G}{\partial \dot{\mathbf{q}}_G} = \begin{pmatrix} 0 & 0 & 0 \\ 0 & 0 & 0 \\ 0 & 0 & 1 \end{pmatrix}. \tag{6.46}$$

As we need the pitch vector $\mathbf{a}$ in inertial coordinates we have to evaluate the transformation from the link G to an inertial frame I by

$$\mathbf{A}_{IG} = \begin{pmatrix} \cos\varphi_G & -\sin\varphi_G & 0 \\ \sin\varphi_G & \cos\varphi_G & 0 \\ 0 & 0 & 1 \end{pmatrix}. \tag{6.47}$$

The momentum of a link and its time derivative can easily be determined using the center of mass velocity

$$\mathbf{p} = m(\mathbf{v}_G + \frac{1}{2}\mathbf{a}_{IG}\tilde{\boldsymbol{\omega}}_G \mathbf{a}) = m(\mathbf{J}_T - \frac{1}{2}\mathbf{a}_{IG}\tilde{\mathbf{a}}\mathbf{J}_R)\dot{\mathbf{q}}_G$$
$$\dot{\mathbf{p}} = m(\mathbf{J}_T - \frac{1}{2}\mathbf{a}_{IG}\tilde{\mathbf{a}}\mathbf{J}_R)\ddot{\mathbf{q}}_G + \frac{1}{2}\mathbf{a}_{IG}\tilde{\boldsymbol{\omega}}_G\tilde{\boldsymbol{\omega}}_G\mathbf{a} \tag{6.48}$$

Due to the symmetry of the link the inertia tensor of the link is diagonal. For a change of the reference point we have (Figure 6.32)

$$\mathbf{I}_H = \mathbf{I}_S - \frac{1}{4}m\tilde{\mathbf{a}}\tilde{\mathbf{a}} \tag{6.49}$$

Analogously to the momentum we get for the moment of momentum and its time derivative

$$\mathbf{L} = \mathbf{L}_S + \frac{1}{2}\tilde{\mathbf{a}}\mathbf{p} = \mathbf{I}_S\boldsymbol{\omega}_G + \frac{1}{2}m\tilde{\mathbf{a}}\mathbf{v}_G - \frac{1}{4}m\tilde{\mathbf{a}}\tilde{\mathbf{a}}\boldsymbol{\omega}_G$$
$$\mathbf{L} = \mathbf{I}_H\boldsymbol{\omega}_G + \frac{1}{2}m\tilde{\mathbf{a}}\mathbf{v}_G = \left(\mathbf{I}_H\mathbf{J}_R + \frac{1}{2}m\tilde{\mathbf{a}}\mathbf{J}_T\right)\dot{\mathbf{q}}_G,$$
$$\dot{\mathbf{L}} = \left(\mathbf{I}_H\mathbf{J}_R + \frac{1}{2}m\tilde{\mathbf{a}}\mathbf{J}_T\right)\ddot{\mathbf{q}}_G + \underbrace{\frac{1}{2}m\tilde{\boldsymbol{\omega}}_G\tilde{\mathbf{a}}\,\mathbf{v}_G}_{=0} \tag{6.50}$$

Due to the plane model the representations of the moment of momentum are the same for link and inertial coordinates. The share of the link dynamics with respect to the overall dynamics can now already be summarized. We get for the mass matrix

$$\mathbf{M}_G = m\left(\mathbf{J}_T^T\mathbf{J}_T - \frac{1}{2}\mathbf{J}_T^T\tilde{\mathbf{a}}\mathbf{J}_R + \frac{1}{2}\mathbf{J}_R^T\tilde{\mathbf{a}}\mathbf{J}_T\right) + \mathbf{J}_R^T\mathbf{I}_H\mathbf{J}_R, \tag{6.51}$$

and the centrifugal forces follow from

$$\mathbf{h}_G = -\frac{1}{2}m\mathbf{J}_T^T\tilde{\boldsymbol{\omega}}_G\tilde{\boldsymbol{\omega}}_G\mathbf{a}. \tag{6.52}$$

Furtheron we have gravitational forces $\mathbf{G} = m\mathbf{g}$ and joint forces acting on the link. Including the gravitational force into the equations of motion referenced to the mass center requires a transformation

$$\mathbf{h}_G = \ldots + \mathbf{J}_{TS}^T\mathbf{G} \quad \text{with} \quad \mathbf{J}_{TS} = \frac{\partial\mathbf{v}_S}{\partial\dot{\mathbf{q}}_G} = \mathbf{J}_T - \frac{1}{2}\tilde{\mathbf{a}}\mathbf{J}_R. \tag{6.53}$$

The joints are approximately modeled like a journal bearing (Figure 6.38). For applying the force law given by Figure 6.33 we need the relative displacements and displacement velocities for a link i and a link j. From joint kinematics we get in a first step

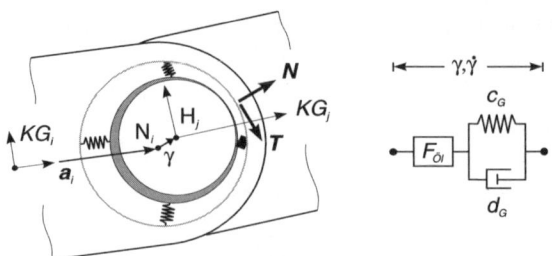

Fig. 6.38: Joint Kinematics between two Links

$$\varepsilon = r_{G,j} - r_{G,i} - a_i, \qquad \dot{\varepsilon} = v_{G,j} - v_{G,i} + \tilde{a}_i \omega_{G,i} \tag{6.54}$$

The joint displacement and displacement velocity follows then in the form

$$\gamma = n^T \varepsilon, \qquad \dot{\gamma} = n^T \dot{\varepsilon}$$
$$n = \frac{\varepsilon}{|\varepsilon|} \quad \text{for} \quad |\varepsilon| \neq 0, \qquad n = \frac{\dot{\varepsilon}}{|\dot{\varepsilon}|} \quad \text{for} \quad |\varepsilon| = 0, \; |\dot{\varepsilon}| \neq 0. \tag{6.55}$$

According to the force law of Figure 6.33 the following formulas are applied

$$\begin{aligned}\lambda_N &= d(\gamma)\dot{\gamma}, & &\text{within backlash,} \\ \lambda_N &= c(\gamma - \gamma_0) + d\dot{\gamma} & &\text{with contact,} \\ N &= n\lambda_N.\end{aligned} \tag{6.56}$$

where $\gamma_0$ is the maximum backlash. The radial joint force acts on the link i with positive sign and in link j with negative sign. The force parts in the equations of motion are therefore augmented by the terms

$$\begin{aligned}h_{G,i} &= \ldots + J_{TNi}^T N & &\text{with} & J_{TNi} &= J_T - \tilde{a}_i J_R \\ h_{G,j} &= \ldots - J_{THj}^T N & &\text{with} & J_{THj} &= J_T.\end{aligned} \tag{6.57}$$

The very small backlashes between pins and bushings allow some simplifications. We assume approximately, that the radial forces cancel out and that only a frictional torque acts in the joints. This torque $M_R$ is evaluated using a smooth and measured friction characteristic. It must also be added to the equations of motion in the form

$$h_{G,i} = \ldots + J_R^T e_z M_R \qquad h_{G,j} = \ldots - J_R^T e_z M_R, \tag{6.58}$$

which has to be evaluated a bit further with respect to the overall structure of the equations of motion [69].

*Elastic Camshafts*

One key component of timing gear systems is the elastic camshaft carrying the sprocket wheel and a couple of additional masses (Figure 6.39). We get correspondingly a number of additive terms in the equations of motion. The elastic displacements are always very small, so that we can linearize with

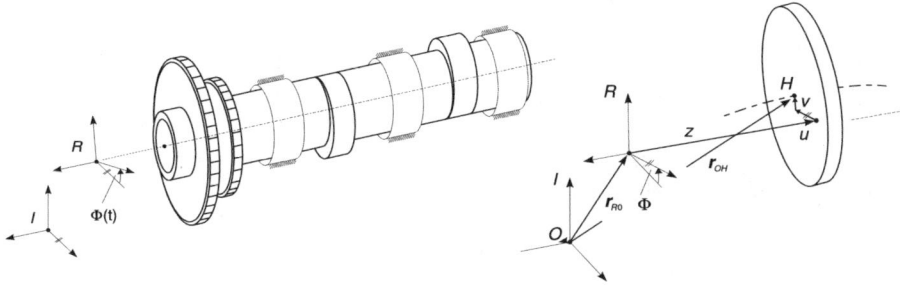

Fig. 6.39: Elastic Camshaft and Sprocket Carrier

respect to elastic motion. This linearized form writes in principle [69]

$$M\ddot{q}_R + \left(D_{el} + D_{Lag} + 2\Omega G\right)\dot{q}_R$$
$$+ \left(K_{el} + K_{Lag} + \Omega N_{Lag} + \dot{\Omega} K_{\dot{\Omega}} + \Omega^2 K_{\Omega^2}\right)q_R + g = h, \qquad (6.59)$$

with the vector and matrix elements to be evaluted in the following:

$M$ : mass matrix,
$G$ : gyroscopic matrix from rotational excitation,
$K_{el}$ : stiffness matrix of elastic deformation,
$D_{el}$ : damping matrix of elastic deformation,
$K_{Lag}$ : stiffness matrix of the bearings,
$N_{Lag}$ : matrix of the non-conservative bearing forces
$D_{Lag}$ : damping matrix of the bearings,
$K_{\dot{\Omega}}$ : matrix due to rotational acceleration,
$K_{\Omega^2}$ : matrix of centrifugal forces (rot. excitation),
$F$ : force vector from rotational excitation,
$h$ : force vector of excitations and external forces.

*Elastic Camshafts - Shaft Element*

As a first component of the camshaft we consider the elastic shaft itself characterized by its kinematics and dynamics. In all cases we refer to the theory presented in the chapters 3.3 on the pages 113 ff. and 3.4 on the pages 131 ff. An elastic shaft can be cut into small discs as indicated in Figure 6.39. The motion of such a disc element includes the reference rotation as given by the engine, the rigid body motion and the elastic deformations, assumed to be small in our case. The kinematics of the reference system is given by the transformation matrix $\mathbf{A}_{RI}$ from the inertial to the reference system and by the rotational velocity and acceleration, which yields

$$\mathbf{A}_{RI} = \begin{pmatrix} \cos\Phi & \sin\Phi & 0 \\ -\sin\Phi & \cos\Phi & 0 \\ 0 & 0 & 1 \end{pmatrix},$$

$$_R\boldsymbol{\omega}_R =\,_I\boldsymbol{\omega}_R = \begin{pmatrix} 0 \\ 0 \\ \Omega \end{pmatrix}, \qquad _R\dot{\boldsymbol{\omega}}_R =\,_I\dot{\boldsymbol{\omega}}_R = \begin{pmatrix} 0 \\ 0 \\ \dot{\Omega} \end{pmatrix}. \qquad (6.60)$$

We choose the following sequence of rotations for the disc element (see chapter 3.3.4 on page 124, [69], [28]):

$$\begin{array}{ll} \text{rotation around the z-axis with} & \varphi(z,t) = \varphi \\ \text{rotation around the x-axis with} & -\frac{\partial v(z,t)}{\partial z} = -v' \\ \text{rotation around the y-axis with} & \frac{\partial u(z,t)}{\partial z} = u' \end{array}$$

All angles are small magnitudes allowing linearization. Therefore the transformation from the reference system R to the body system K comes out with

$$\mathbf{A}_{KR} = \mathbf{E} - \tilde{\boldsymbol{\varphi}}_R = \mathbf{E} - \widetilde{\begin{pmatrix} -v' \\ u' \\ \varphi \end{pmatrix}} = \begin{pmatrix} 1 & \varphi & -u' \\ -\varphi & 1 & -v' \\ u' & v' & 1 \end{pmatrix} \qquad (6.61)$$

From these preliminaries we can evaluate the absolute angular velocity and acceleration formulated in the body-fixed frame by

$$_K\boldsymbol{\omega}_K = \begin{pmatrix} -\dot{v}' - \Omega u' \\ +\dot{u}' - \Omega v' \\ \dot{\varphi} + \Omega \end{pmatrix}$$

$$_K\dot{\boldsymbol{\omega}}_K = \begin{pmatrix} -\ddot{v}' \\ \ddot{u}' \\ \ddot{\varphi} \end{pmatrix} + \Omega \begin{pmatrix} -\dot{u}' \\ -\dot{v}' \\ 0 \end{pmatrix} + \dot{\Omega} \begin{pmatrix} -u' \\ -v' \\ 1 \end{pmatrix} \qquad (6.62)$$

The equations of momentum will be evaluated in the reference system R and therefore we need the position of the center of mass in that coordinates. The radius vector from O to H follows from the vector chain (Figure 6.39)

$$_R\mathbf{r}_{OH} = \mathbf{A}_{RI}\begin{pmatrix} u_{R0} \\ v_{R0} \\ 0 \end{pmatrix} + \begin{pmatrix} 0 \\ 0 \\ z \end{pmatrix} + \begin{pmatrix} u \\ v \\ 0 \end{pmatrix}, \qquad (6.63)$$

from which we get the absolute velocity and acceleration by the expression

$$_R\boldsymbol{v}_H = \begin{pmatrix} \dot{u} \\ \dot{v} \\ 0 \end{pmatrix} + \Omega \begin{pmatrix} -v \\ u \\ 0 \end{pmatrix},$$

$$_R\dot{\boldsymbol{v}}_H = \begin{pmatrix} \ddot{u} \\ \ddot{v} \\ 0 \end{pmatrix} + 2\Omega \begin{pmatrix} -\dot{v} \\ \dot{u} \\ 0 \end{pmatrix} + \Omega^2 \begin{pmatrix} -u \\ -v \\ 0 \end{pmatrix} + \dot{\Omega} \begin{pmatrix} -v \\ -u \\ 0 \end{pmatrix}. \tag{6.64}$$

The elastic deformations are approximated by a RITZ approach, which writes

$$\begin{aligned} u_{el}(z,t) &= \boldsymbol{u}^T(z)\,\boldsymbol{q}_u(t) = \boldsymbol{u}^T\boldsymbol{q}_u : & \text{bending in u-direction} \\ v_{el}(z,t) &= \boldsymbol{v}^T(z)\,\boldsymbol{q}_v(t) = \boldsymbol{v}^T\boldsymbol{q}_v : & \text{bending in v-direction} \\ \varphi_{el}(z,t) &= \boldsymbol{\varphi}^T(z)\,\boldsymbol{q}_\varphi(t) = \boldsymbol{\varphi}^T\boldsymbol{q}_\varphi : & \text{torsion} \end{aligned} \tag{6.65}$$

This ansatz makes a separation of position- and time dependent magnitudes possible. The velocities and accelerations of the equations (6.64) and (6.62) then write

$$\begin{aligned} \boldsymbol{v}_H &= \boldsymbol{J}_T\dot{\boldsymbol{q}}_R + \hat{\boldsymbol{j}}_T, & \dot{\boldsymbol{v}}_H &= \boldsymbol{J}_T\ddot{\boldsymbol{q}}_R + \boldsymbol{j}_T, \\ \boldsymbol{\omega}_K &= \boldsymbol{J}_R\dot{\boldsymbol{q}}_R + \hat{\boldsymbol{j}}_R, & \dot{\boldsymbol{\omega}}_K &= \boldsymbol{J}_R\ddot{\boldsymbol{q}}_R + \boldsymbol{j}_R. \end{aligned} \tag{6.66}$$

The Jacobians contain only position dependent magnitudes thus being functions of the longitudinal coordinate z:

$$\boldsymbol{J}_T = \begin{pmatrix} \boldsymbol{u}^T & 0 & 0 \\ 0 & \boldsymbol{v}^T & 0 \\ 0 & 0 & 0 \end{pmatrix}, \qquad \boldsymbol{J}_R = \begin{pmatrix} 0 & -\boldsymbol{v}'^T & 0 \\ \boldsymbol{u}'^T & 0 & 0 \\ 0 & 0 & \boldsymbol{\varphi}^T \end{pmatrix}. \tag{6.67}$$

The additional vectors depend on the velocities and accelerations due to the reference rotation of the shaft. They write

$$\hat{\boldsymbol{j}}_T = \Omega \begin{pmatrix} -v \\ u \\ 0 \end{pmatrix}, \qquad \boldsymbol{j}_T = 2\Omega \begin{pmatrix} -\dot{v} \\ \dot{u} \\ 0 \end{pmatrix} + \Omega^2 \begin{pmatrix} -u \\ -v \\ 0 \end{pmatrix} + \dot{\Omega} \begin{pmatrix} -v \\ u \\ 0 \end{pmatrix},$$

$$\hat{\boldsymbol{j}}_R = \Omega \begin{pmatrix} -u' \\ -v' \\ 1 \end{pmatrix}, \qquad \boldsymbol{j}_R = \Omega \begin{pmatrix} -\dot{u}' \\ -\dot{v}' \\ 0 \end{pmatrix} + \dot{\Omega} \begin{pmatrix} -u' \\ -v' \\ 1 \end{pmatrix}. \tag{6.68}$$

This has been all kinematics. Going to kinetics we consider again an infinitesimal disc with thickness (dz), the cross section $A(z)$ and the constant density $\rho$. The momentum and moment of momentum derivatives for such a small element is given by

$$\begin{aligned} \dot{\boldsymbol{p}} &= \rho A(z)\dot{\boldsymbol{v}}_H dz = \rho A(z)\Big(\boldsymbol{J}_T\ddot{\boldsymbol{q}}_R + \boldsymbol{j}_T\Big)dz, \\ \dot{\boldsymbol{L}} &= \rho\Big(\hat{\boldsymbol{I}}\dot{\boldsymbol{\omega}}_K + \tilde{\boldsymbol{\omega}}_K\hat{\boldsymbol{I}}\boldsymbol{\omega}_K\Big)dz = \rho\Big(\hat{\boldsymbol{I}}\boldsymbol{J}_R\ddot{\boldsymbol{q}}_R + \hat{\boldsymbol{I}}\boldsymbol{j}_R + \tilde{\boldsymbol{\omega}}_K\hat{\boldsymbol{I}}\boldsymbol{\omega}_K\Big)dz, \end{aligned} \tag{6.69}$$

where the diagonal inertia tensor is simply $\hat{\boldsymbol{I}} = \operatorname{diag}(I_x, I_y, I_p)$ with $I_x = I_y = I_p/2$. The mass matrix and the force vector due to the reference rotation follow from

$$\boldsymbol{M} = \rho \int_0^l \left( A \boldsymbol{J}_T^T \boldsymbol{J}_T + \boldsymbol{J}_R^T \hat{\boldsymbol{I}} \boldsymbol{J}_R \right) dz,$$

$$\boldsymbol{F} = \rho \int_0^l \left( A \boldsymbol{J}_T^T \boldsymbol{j}_T + \boldsymbol{J}_R^T \hat{\boldsymbol{I}} \boldsymbol{j}_R + \boldsymbol{J}_R^T \tilde{\boldsymbol{\omega}}_K \hat{\boldsymbol{I}} \boldsymbol{\omega}_K \right) dz \tag{6.70}$$

We assume that the cross sections of the shaft remain plane (Euler-Bernoulli beam) and determine the deformation potential energy of that beam. The displacement of the mass center with respect to the reference is given by

$$\bar{\boldsymbol{r}} = \begin{pmatrix} u \\ v \\ 0 \end{pmatrix} - \tilde{\boldsymbol{r}}_P \boldsymbol{\varphi}_R \quad \Longrightarrow \quad \begin{pmatrix} \bar{u} \\ \bar{v} \\ \bar{w} \end{pmatrix} = \begin{pmatrix} u - y\varphi \\ v + x\varphi \\ -yv' - xu' \end{pmatrix} \tag{6.71}$$

The potential energy of an elastic system may be writtrn in the form $\frac{1}{2}\mathbf{e}^T \mathbf{H} \mathbf{e}$ with the linearized strain tensor $\mathbf{e}$ given by

$$\mathbf{e} = \begin{bmatrix} \frac{\partial \bar{u}}{\partial x} \\ \frac{\partial \bar{v}}{\partial y} \\ \frac{\partial \bar{w}}{\partial z} \\ \frac{\partial \bar{u}}{\partial y} + \frac{\partial \bar{v}}{\partial x} \\ \frac{\partial \bar{v}}{\partial z} + \frac{\partial \bar{w}}{\partial y} \\ \frac{\partial \bar{u}}{\partial z} + \frac{\partial \bar{w}}{\partial x} \end{bmatrix} = \begin{bmatrix} 0 \\ 0 \\ -yv'' - xu'' \\ 0 \\ x\varphi' \\ -y\varphi' \end{bmatrix} \tag{6.72}$$

See for example the equations (2.141) on page 51 and [28], [69]. Using the magnitudes of the disc element we come to the potential energy in the form

$$\Delta V^* = \left[ E^* \left( y^2 v''^2 + x^2 u''^2 + 2xy u'' v'' \right) + G \left( x^2 + y^2 \right) \varphi'^2 \right] dx\,dy\,dz$$

$$\text{with} \quad E^* = \frac{E(1-\nu)}{(1+\nu)(1-2\nu)} \tag{6.73}$$

All deformation magnitudes depend only on z, and we can integrate the potential energy over the cross sectional area A. As a result we get the potential energy for an infinitesimal disc element with the thickness (dz):

$$\Delta V = \frac{1}{2} \begin{pmatrix} u'' & v'' & \varphi' \end{pmatrix} \begin{pmatrix} E^* I_x & 0 & 0 \\ 0 & E^* I_y & 0 \\ 0 & 0 & G I_p \end{pmatrix} \begin{pmatrix} u'' \\ v'' \\ \varphi' \end{pmatrix} dz, \tag{6.74}$$

and from there directly the unknown stiffness matrix from the equation (6.59) in the form

$$\boldsymbol{K}_{el} = \int_0^l \begin{pmatrix} E^* I_x \boldsymbol{u}'' \boldsymbol{u}''^T & 0 & 0 \\ 0 & E^* I_y \boldsymbol{v}'' \boldsymbol{v}''^T & 0 \\ 0 & 0 & GI_p \boldsymbol{\varphi}' \boldsymbol{\varphi}'^T \end{pmatrix} dz \qquad (6.75)$$

*Elastic Camshafts - Additional Rigid Masses*

The sprocket carriers are usually camshafts with cams, which we consider to be additional rigid masses. It makes sense to model these additional masses individually, because it simplifies the structure of the elastic camshaft parts. They add certain terms to the mass matrix **M** and the force vector **F** of the equations 6.70 due to the fact, that all additional rigid masses are subject to the effects of the reference motion. For a rigid disc at the longitudinal position $z_0$ with mass m and a diagonal inertia tensor **I** we get these shares in the form

$$\begin{aligned}\boldsymbol{M} &= \ldots + m \boldsymbol{J}_{T0}^T \boldsymbol{J}_{T0} + \boldsymbol{J}_{R0}^T \boldsymbol{I} \boldsymbol{J}_{R0} \\ \boldsymbol{F}_g &= \ldots + m \boldsymbol{J}_{T0}^T \boldsymbol{j}_{T0} + \boldsymbol{J}_{R0}^T \boldsymbol{I} \boldsymbol{j}_{R0} + \boldsymbol{J}_{R0}^T \tilde{\boldsymbol{\omega}}_{K0} \boldsymbol{I} \boldsymbol{\omega}_{K0}\end{aligned} \qquad (6.76)$$

*Linearization of the Elastic Parts*

At this point it make sense to linearize the elastic part of the equations of motion due to the camshaft elasticity and according to the concept, that the elastic vibrations are small deformations superimposed to the reference motion. For this purpose we have to regard the expressions (6.68) in the equations (6.66), to put that into the relations (6.69) and evaluate it together with the equations (6.70). After some calculations we come out with the following integrals, only dependent on spatial coordinates [69]

$$\boldsymbol{M} = \rho \int_0^l \begin{pmatrix} A\boldsymbol{uu}^T + I_x \boldsymbol{u}'\boldsymbol{u}'^T & 0 & 0 \\ 0 & A\boldsymbol{vv}^T + I_y \boldsymbol{v}'\boldsymbol{v}'^T & 0 \\ 0 & 0 & I_p \boldsymbol{\varphi}\boldsymbol{\varphi}^T \end{pmatrix} dz,$$

$$\boldsymbol{G} = \rho \int_0^l \begin{pmatrix} 0 & -A\boldsymbol{uv}^T & 0 \\ A\boldsymbol{vu}^T & 0 & 0 \\ 0 & 0 & 0 \end{pmatrix} dz,$$

$$\boldsymbol{K}_{\dot{\Omega}} = \rho \int_0^l \left[ \begin{pmatrix} 0 & -A\boldsymbol{uv}^T - I_x \boldsymbol{u}'\boldsymbol{v}'^T & 0 \\ A\boldsymbol{vu}^T + I_y \boldsymbol{v}'\boldsymbol{u}'^T & 0 & 0 \\ 0 & 0 & 0 \end{pmatrix} + \begin{pmatrix} 0 & 0 & 0 \\ I_p \boldsymbol{v}'\boldsymbol{u}'^T & 0 & 0 \\ 0 & 0 & 0 \end{pmatrix} \right] dz,$$

$$\boldsymbol{K}_{\Omega^2} = \rho \int_0^l \begin{pmatrix} -A\boldsymbol{uu}^T + I_x \boldsymbol{u}'\boldsymbol{u}'^T & 0 & 0 \\ 0 & -A\boldsymbol{vv}^T + I_y \boldsymbol{v}'\boldsymbol{v}'^T & 0 \\ 0 & 0 & 0 \end{pmatrix} dz,$$

$$\boldsymbol{F} = \rho \int_0^l \dot{\Omega} \begin{pmatrix} 0 \\ 0 \\ I_p \boldsymbol{\varphi} \end{pmatrix} dz. \tag{6.77}$$

For an additional rigid mass we have to add similar terms including the inertia tensor $\boldsymbol{I} = \mathrm{diag}(A, B, C)$ and the following simple integrals

$$\int \rho A dz \to m, \quad \int \rho I_x dz \to A, \quad \int \rho I_y dz \to B, \quad \int \rho I_p dz \to C. \tag{6.78}$$

*External Forces*

External forces acting on the sprocket shaft are gravitational forces, forces entering from the cams and the valve mechanism and finally forces generated by the tensioners for the guides. In some cases gravitational forces might be neglected. Otherwise it is

$$_I \boldsymbol{f} = \rho A \boldsymbol{g} dz \tag{6.79}$$

with the gravitational vector $\boldsymbol{g}$ being inertially fixed (see Figure 6.39). The additional force contribution to the equations of motion results from a transformation into the reference system, from there into the configuration space and finally from an integration along the shaft. If we have in addition a rigid mass, we get a second term representing the gravitational effect of this rigid mass. Altogether this yields

$$\boldsymbol{h}_R = \ldots + \rho \int_0^l A \boldsymbol{J}_T^T dz \boldsymbol{A}_{RI} \boldsymbol{g} + \ldots + m \boldsymbol{J}_{T0}^T \boldsymbol{A}_{RI} \boldsymbol{g} \tag{6.80}$$

At the sprocket shaft boundary we have the forces and torques coming from the individual valve trains; they depend for stationary operation on the cam position. We get

$$F_{excitation} = F(\Phi + \boldsymbol{\varphi}_0^T \boldsymbol{q}_\varphi) \qquad M_{excitation} = M(\Phi + \boldsymbol{\varphi}_0^T \boldsymbol{q}_\varphi) \tag{6.81}$$

The forces are conveniently defined in an inertial and the torques in a body-fixed coordinate system. Performing the corresponding transformations and evaluating these parts in configuration space results in

$$\boldsymbol{h}_R = \ldots + \boldsymbol{J}_{T0}^T \boldsymbol{A}_{RI} \boldsymbol{e}_i F_{excitation} + \ldots + \boldsymbol{J}_{R0}^T \boldsymbol{A}_{KI} \boldsymbol{e}_3 M_{excitation} \tag{6.82}$$

For the evaluation of the tensioner's forces we need to know its relative displacements, which can be determined in a similar way as in the case of the joint bearings. The direction of the active force is $\boldsymbol{n}_T$. We get

$$\gamma = \boldsymbol{n}_T^T \boldsymbol{A}_{IR} \boldsymbol{J}_{T0} \boldsymbol{q}_R, \qquad \dot{\gamma} = \boldsymbol{n}_T^T \boldsymbol{A}_{IR} (\boldsymbol{J}_{T0} \dot{\boldsymbol{q}}_R + \Omega \boldsymbol{J}_{T0}^* \boldsymbol{q}_R). \tag{6.83}$$

The contribution to the equations of motion results from that in the form

$$\boldsymbol{h}_R = \ldots + \boldsymbol{J}_{T0}^T \boldsymbol{A}_{RI} \boldsymbol{n}_T \zeta, \tag{6.84}$$

where $\zeta$ must be calculated from the detailed tensioner analysis [69], [111].

In a preceding chapter we have already discussed the model properties of the sprocket. We shall consider here in a bit more detail the kinematics of the toothing contour following Figure 6.40 (see also Figure 6.34). Starting with the reference point H of the shaft at the sprocket position we calculate the position of a contour point K by considering the eccentricity $\rho$, if any, the vector $\boldsymbol{r}_{VZ}$ to the origin of the toothing system and the vector $\boldsymbol{r}_K(s)$ to the contour point. This writes in disc-fixed coordinates

$$\boldsymbol{r}_{HK} = \boldsymbol{\varrho} + \boldsymbol{r}_{VZ} + \boldsymbol{r}_K(s) \tag{6.85}$$

The absolute position and the kinematics follow from these equations in the form (O = center of inertial coordinate system)

$$\begin{aligned}
{}_I\boldsymbol{r}_{OK} &= \boldsymbol{A}_{IR}[\boldsymbol{r}_{OH} + \boldsymbol{A}_{RK}\boldsymbol{r}_{HK}], \\
\boldsymbol{v}_K &= \boldsymbol{A}_{IR}[\boldsymbol{J}_{TK}\dot{\boldsymbol{q}}_R + \hat{\boldsymbol{j}}_{TK}], \qquad \dot{\boldsymbol{v}}_{KG} = \boldsymbol{A}_{IR}[\boldsymbol{J}_{TK}\ddot{\boldsymbol{q}}_R + \boldsymbol{j}_{TK}],
\end{aligned}$$

with
$$\begin{aligned}
\boldsymbol{J}_{TK} &= \boldsymbol{J}_T - \boldsymbol{A}_{RK}\tilde{\boldsymbol{r}}_{HK}\boldsymbol{J}_R \\
\hat{\boldsymbol{j}}_{TK} &= \hat{\boldsymbol{j}}_T - \boldsymbol{A}_{RK}\tilde{\boldsymbol{r}}_{HK}\hat{\boldsymbol{j}}_R \\
\boldsymbol{j}_{TK} &= \boldsymbol{j}_T - \boldsymbol{A}_{RK}\tilde{\boldsymbol{r}}_{HK}\boldsymbol{j}_R + \boldsymbol{A}_{RK}\tilde{\boldsymbol{\omega}}_{K0}\tilde{\boldsymbol{\omega}}_{K0}\boldsymbol{r}_{HK}
\end{aligned} \tag{6.86}$$

These equations hold for every point of the sprocket shaft, they appear in the same form for the chain guides. Linearizing we get

$${}_I\boldsymbol{r}_{OK} = \boldsymbol{r}_{OH}^0 + \boldsymbol{A}_{IR}\boldsymbol{J}_{TK}^0 \boldsymbol{q}_R \tag{6.87}$$

After some calculations we come to a direct form of the contour point position

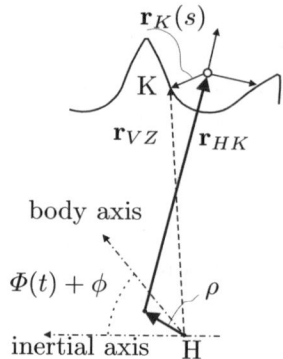

Fig. 6.40: Contour Geometry

$$\begin{aligned}
{}_I r_{OK} &= r_{R0} + \begin{pmatrix} 0 \\ 0 \\ z \end{pmatrix} + A_{IR}\begin{pmatrix} u \\ v \\ 0 \end{pmatrix} + A_{IR}(E - \tilde{\varphi}_R) r_{HK} \\
&= \underbrace{r_{R0} + z e_3 + A_{IR} r_{HK}}_{r^0_{OH}} + A_{IR} \underbrace{(J_T - E\tilde{r}_{HK} J_R)}_{J^0_{TK}} q_R
\end{aligned} \tag{6.88}$$

#### 6.4.2.4 Contact Processes

The chain timing systems include a large variety of contact processes, which we have to describe. For calculating the contact forces and for determining a valid contact configuration we shall proceed with the steps discussed in chapter 3.4 on 131 ff., namely checking the contacts to be active or passive, where at the beginning of a contact event we always have kinematical and at the end kinetic indicators. The example we present here was still treated by Lemke's algorithm.

*Contact of a Link with a Guide*

For the contact of a link with a guide we start with the description of the guide contour by equation (6.42) in the form $r_K(s) = (x_K(s) \;\; y_K(s))^T$, from which we can derive a contact coordinate system $(t, n, b)$ from the following relationships, using $\kappa$ as the curvature of the contour (see Figure 6.41 and section 2.2.6 on the pages 31 ff.):

$$\begin{aligned}
&t = \frac{\partial r_K}{\partial s}, & &n = \frac{1}{\kappa} \frac{\partial^2 r_K}{\partial s^2}, & &b = t \times n, \\
&t = t(q_G, s), & &n = n(q_G, s), & &b = const.
\end{aligned} \tag{6.89}$$

## 6.4 Bush and Roller Chains

In a guide fixed system the vectors $t$ and $n$ depend only on the parameter $s$. Applying the formulas of Frenet

$$\frac{\partial t}{\partial s} = \kappa n, \qquad \frac{\partial n}{\partial s} = -\kappa t, \tag{6.90}$$

and the notation

$$\tilde{\Omega}_G n = -tb^T \Omega_G, \qquad \tilde{\Omega}_G t = nb^T \Omega_G \tag{6.91}$$

we obtain the time derivatives of $t$ and $n$:

$$\begin{aligned}\dot{t} &= \tilde{\Omega}_G t + \frac{\partial t}{\partial s}\dot{s} = nb^T \Omega_G + \kappa \dot{s} n, \\ \dot{n} &= \tilde{\Omega}_G n + \frac{\partial n}{\partial s}\dot{s} = -tb^T \Omega_G - \kappa \dot{s} t.\end{aligned} \tag{6.92}$$

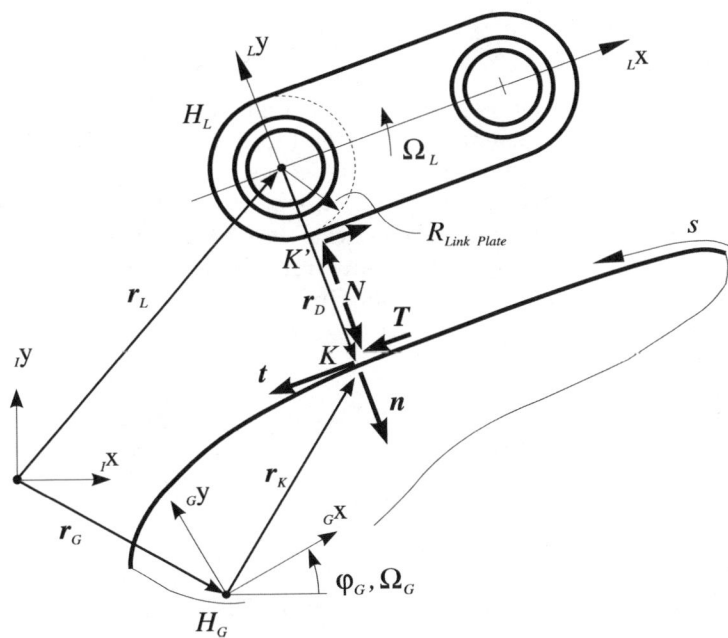

Fig. 6.41: Contact between a Link and a Guide

In the following we look at the vector $r_D$ from the reference point $H_L$ to the body fixed contact point $K$:

$$\begin{aligned}r_D &= r_G + r_K - r_L, \\ \dot{r}_D &= \dot{r}_G + \tilde{\Omega}_G r_K - \dot{r}_L, \\ \ddot{r}_D &= \ddot{r}_G + \tilde{\Omega}_G \tilde{\Omega}_G r_K + \dot{\tilde{\Omega}}_G r_K - \ddot{r}_L.\end{aligned} \tag{6.93}$$

In a further step we have to compute the contour parameter $s$ in such a manner, that the vector $r_D$ is perpendicular to the tangential vector $t$, which means

$$g_t = t^T r_D = 0. \tag{6.94}$$

In general this equation is nonlinear in $s$. The calculation of the normal acceleration needs the time derivative of $s$. Keeping in mind that $r_D$ also depends on $s$, we arrive at

$$\begin{aligned}\frac{dg_t}{dt} &= t^T(\dot{r}_D + \frac{\partial r_D}{\partial s}\dot{s}) + r_D^T \dot{t} \\ &= t^T \dot{r}_D + r_D^T n b^T \Omega_G + \dot{s}(t^T t + \kappa r_D^T n) \\ &= t^T \dot{r}_D + n^T r_D b^T \Omega_G + \dot{s}(1 + \kappa n^T r_D) = 0,\end{aligned}$$

$$\dot{s} = -\frac{t^T \dot{r}_D + n^T r_D b^T \Omega_G}{1 + \kappa n^T r_D}. \tag{6.95}$$

During the integration we use the value $\dot{s}$ to find the time-dependent contact positions. Therefore the nonlinear equation (6.94) must be solved only once, when we deal with the contact for the first time.

To determine the direction of the frictional forces we have to regard the tangential relative velocities of the contact points for all active contacts ($K = K'$). The contact vector of the link results from the vector $r_L$ to the reference point $H_L$ and the radius of the contour circle.

$$r_{L,K'} = r_L + R_{linkplate} n \tag{6.96}$$

With that relation we get the velocity of the body fixed contact points $K'$ and $K$ and finally the relative velocity, projected into the tangential direction.

$$\begin{aligned}v_{L,K'} &= \dot{r}_L + R_{linkplate} \tilde{\Omega}_L n, & v_{G,K} &= \dot{r}_G + \tilde{\Omega}_G r_K, \\ v_{rel} &= t^T(v_{G,K} - v_{L,K'}).\end{aligned} \tag{6.97}$$

A positive relative velocity induces frictional forces in the negative tangential direction $t$.

To calculate the normal distance between the link and the guide, we use the relation

$$g_n = n^T r_D \tag{6.98}$$

For determining the indicator with respect to contact/detachment we have to subtract the radius of the link contour. The derivative with respect to time leads to the normal velocity of the contour points. Applying equation (6.93) and the condition (6.94) we get

$$\begin{aligned}
\dot{g}_n &= \boldsymbol{n}^T(\dot{\boldsymbol{r}}_D + \frac{\partial \boldsymbol{r}_D}{\partial s}\dot{s}) + \boldsymbol{r}_D^T \dot{\boldsymbol{n}}, \\
&= \boldsymbol{n}^T \dot{\boldsymbol{r}}_D - \boldsymbol{r}_D^T \boldsymbol{t} \boldsymbol{b}^T \boldsymbol{\Omega}_G + \dot{s}(\boldsymbol{n}^T \boldsymbol{t} - \kappa \boldsymbol{r}_D^T \boldsymbol{t}), \\
&= \boldsymbol{n}^T \dot{\boldsymbol{r}}_D,
\end{aligned} \qquad (6.99)$$

and from there by an additional time derivation the normal acceleration

$$\begin{aligned}
\ddot{g}_n &= \boldsymbol{n}^T(\ddot{\boldsymbol{r}}_D + \frac{\partial \dot{\boldsymbol{r}}_D}{\partial s}\dot{s}) + \dot{\boldsymbol{r}}_D^T \dot{\boldsymbol{n}} \\
&= \boldsymbol{n}^T \ddot{\boldsymbol{r}}_D + \boldsymbol{n}^T(\tilde{\boldsymbol{\Omega}}_G \frac{\partial \boldsymbol{r}_K}{\partial s})\dot{s} - \dot{\boldsymbol{r}}_D^T \boldsymbol{t} \boldsymbol{b}^T \boldsymbol{\Omega}_G - \kappa \dot{s} \dot{\boldsymbol{r}}_D^T \boldsymbol{t} \\
&= \boldsymbol{n}^T \ddot{\boldsymbol{r}}_D - \boldsymbol{t}^T \dot{\boldsymbol{r}}_D \boldsymbol{b}^T \boldsymbol{\Omega}_G + \dot{s}(\boldsymbol{b}^T \boldsymbol{\Omega}_G - \kappa \boldsymbol{t}^T \dot{\boldsymbol{r}}_D).
\end{aligned} \qquad (6.100)$$

According to Figure 6.41 the contact forces $\boldsymbol{N}$ and $\boldsymbol{T}$, acting on the guide and with negative sign on the link, can be written as

$$\boldsymbol{N} = \boldsymbol{n}\lambda, \qquad \boldsymbol{T} = -\frac{v_{rel}}{|v_{rel}|}\mu \boldsymbol{t} \lambda \qquad (6.101)$$

With the known directions $\boldsymbol{n}$ and $\boldsymbol{t}$ we use the vector $\boldsymbol{w} = \boldsymbol{J}^T \boldsymbol{n}$ to describe the influence of these forces on the motion of the guide and the link.

$$\begin{aligned}
\boldsymbol{J}_G^* &= \left(\boldsymbol{E}^{3\times 3}, -\tilde{\boldsymbol{r}}_K\right), \qquad \boldsymbol{J}_L^* = \left(\boldsymbol{E}^{3\times 3}, -R_{rollerplate}\tilde{\boldsymbol{n}}\right) \\
\boldsymbol{w}_G \lambda &= \boldsymbol{Q}_G^T \boldsymbol{J}_G^{*T}\left(\boldsymbol{n} - \frac{v_{rel}}{|v_{rel}|}\mu \boldsymbol{t}\right)\lambda, \\
\boldsymbol{w}_L \lambda &= -\boldsymbol{Q}_L^T \boldsymbol{J}_L^{*T}\left(\boldsymbol{n} - \frac{v_{rel}}{|v_{rel}|}\mu \boldsymbol{t}\right)\lambda.
\end{aligned} \qquad (6.102)$$

*Contact between a Sprocket and a Link*

Regarding the contact between a sprocket and a link we achieve very simplified relations for the normal velocity and acceleration. Due to the fact that the toothing contour is composed of circles, it is sufficient to consider the center of these contour circles. For this purpose we use the vector $\boldsymbol{d}$ from the reference point $H_L$ to the contour center $M_K$ (Figure 6.42). Applying the radius $R_K$ of the contour in the contact areas, according to Figure 6.34 with $R_{K,toothprofile} = R_{TP}$ and $R_{K,seatingcurve} = -R_{SC}$, the distance vector $\boldsymbol{r}_D$ results in

$$\begin{aligned}
\boldsymbol{r}_D &= \boldsymbol{d} - \boldsymbol{n} R_K, \\
\dot{\boldsymbol{r}}_D &= \dot{\boldsymbol{d}} + R_K \boldsymbol{t} \boldsymbol{b}^T \boldsymbol{\Omega}_S, \\
\ddot{\boldsymbol{r}}_D &= \ddot{\boldsymbol{d}} + R_K \boldsymbol{n}(\boldsymbol{b}^T \boldsymbol{\Omega}_S)^2 + R_K \boldsymbol{t} \boldsymbol{b}^T \dot{\boldsymbol{\Omega}}_S
\end{aligned} \qquad (6.103)$$

## 6 Timing Equipment

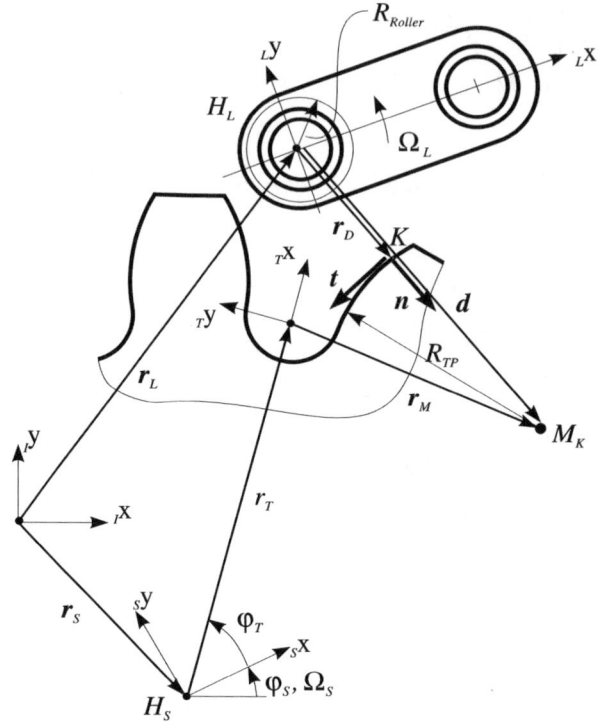

Fig. 6.42: Contact between Link and Sprocket

After some transformations the equations (6.98), (6.99) and (6.100) can be written as

$$g_n = \boldsymbol{n}^T \boldsymbol{d} - R_K, \qquad \dot{g}_n = \boldsymbol{n}^T \dot{\boldsymbol{d}}, \qquad \ddot{g}_n = \boldsymbol{n}^T \ddot{\boldsymbol{d}} + \frac{\left(\boldsymbol{t}^T \dot{\boldsymbol{d}}\right)^2}{\boldsymbol{n}^T \boldsymbol{d}}. \tag{6.104}$$

*Contact Configuration*

For establishing the complementarities we have to evaluate forces and accelerations, which characterize the relevant contact configuration. In order to combine the kinematic constraint (6.100) with the equations of motion we have to write the normal acceleration in the following form:

$$\ddot{g}_n = \boldsymbol{w}_G^T \ddot{\boldsymbol{q}}_G + \boldsymbol{w}_L^T \ddot{\boldsymbol{q}}_L + \bar{w}. \tag{6.105}$$

Transforming the acceleration of the vector $\boldsymbol{r}_D$ (6.93) with the items:

$$\boldsymbol{J}^*_{G,n} = \frac{\partial \ddot{\boldsymbol{r}}_D}{\partial \ddot{\boldsymbol{z}}_G} = \left(\boldsymbol{E}^{3\times 3}, -\tilde{\boldsymbol{r}}_K\right), \qquad \boldsymbol{J}^*_{L,n} = \frac{\partial \ddot{\boldsymbol{r}}_D}{\partial \ddot{\boldsymbol{z}}_L} = \left(-\boldsymbol{E}^{3\times 3}, \boldsymbol{O}^{3\times 3}\right), \quad (6.106)$$

we can write

$$\ddot{\boldsymbol{r}}_D = \boldsymbol{J}^*_{G,n} \boldsymbol{Q}_G \ddot{\boldsymbol{q}}_G + \boldsymbol{J}^*_{L,n} \boldsymbol{Q}_L \ddot{\boldsymbol{q}}_L + \bar{\boldsymbol{j}}, \quad \text{with} \quad \bar{\boldsymbol{j}} = \tilde{\boldsymbol{\Omega}}_G \tilde{\boldsymbol{\Omega}}_G \boldsymbol{r}_K. \qquad (6.107)$$

Comparing this relation with equation (6.100), we finally get the vectors $\boldsymbol{w}_{G,n}, \boldsymbol{w}_{L,n}$ and the value $\bar{w}$ in the form

$$\boldsymbol{w}^T_{G,n} = \boldsymbol{n}^T \boldsymbol{J}^*_{G,n} \boldsymbol{Q}_G,$$
$$\boldsymbol{w}^T_{L,n} = \boldsymbol{n}^T \boldsymbol{J}^*_{L,n} \boldsymbol{Q}_L,$$
$$\bar{w} = \boldsymbol{n}^T \bar{\boldsymbol{j}} - \boldsymbol{t}^T \dot{\boldsymbol{r}}_D \boldsymbol{b}^T \boldsymbol{\Omega}_G + \dot{s}(\boldsymbol{b}^T \boldsymbol{\Omega}_G - \kappa \boldsymbol{t}^T \dot{\boldsymbol{r}}_D). \qquad (6.108)$$

With the normal acceleration of the two bodies in the notation of equation (6.105) we are able to compute the unknown contact forces. Therefore we have to regard the complete equations of motion of a guide and a link.

$$\boldsymbol{M}_G \ddot{\boldsymbol{q}}_G = \boldsymbol{h}_G + \boldsymbol{W}_G \boldsymbol{\lambda}, \qquad \boldsymbol{M}_L \ddot{\boldsymbol{q}}_L = \boldsymbol{h}_L + \boldsymbol{w}_L \boldsymbol{\lambda}. \qquad (6.109)$$

The vector $\boldsymbol{h}$ contains all forces computed by using the state $\boldsymbol{q}, \dot{\boldsymbol{q}}$ of the system. These forces are gravitational forces, joint forces, forces from the tension device, and so on. All additional forces, depending on the generalized accelerations $\ddot{\boldsymbol{q}}$, are collected in the term $\boldsymbol{W}_G \boldsymbol{\lambda}$ or the term $\boldsymbol{w}_L \boldsymbol{\lambda}$, if there is only one additional constraint. The matrix $\boldsymbol{W}_G$ is composed of the vectors $\boldsymbol{w}_G$ of equation (6.102), regarding each contact. The vector $\boldsymbol{\lambda}$ consists of all unknown contact forces $\lambda$ to this guide.

To compute these forces, we have to consider all links, which have an active contact to this guide. To simplify the notation, we collect all $n_K$ equations of motion of these links into one equation, using the sequence of $\boldsymbol{\lambda}$ in the equations (6.109). We then come out with

$$\boldsymbol{M}_L \ddot{\boldsymbol{q}}_L = \boldsymbol{h}_L + \boldsymbol{W}_L \boldsymbol{\lambda}$$

with
$$\ddot{\boldsymbol{q}}^T_L = (\ddot{\boldsymbol{q}}^T_{L,1}, \ldots, \ddot{\boldsymbol{q}}^T_{L,n_K}),$$
$$\boldsymbol{M}_L = \operatorname{diag}(\boldsymbol{M}_{L,1}, \ldots, \boldsymbol{M}_{L,n_K}),$$
$$\boldsymbol{h}^T_G = (\boldsymbol{h}^T_{L,1}, \ldots, \boldsymbol{h}^T_{L,n_K}),$$
$$\boldsymbol{W}_G = (\boldsymbol{w}_{L,1}, \ldots, \boldsymbol{w}_{L,n_K}). \qquad (6.110)$$

In the same manner with the same sequence, we collect the secondary conditions on the acceleration level.

$$\ddot{\boldsymbol{g}}_n = \boldsymbol{W}^T_{G,n} \ddot{\boldsymbol{q}}_G + \boldsymbol{W}^T_{L,n} \ddot{\boldsymbol{q}}_L + \bar{\boldsymbol{w}} \qquad (6.111)$$

From there it is easy to formulate a relation containing the complementarities of all our unilateral constraints. From the equations of motion we get by rearranging a bit these equations

$$\ddot{q}_G = M_G^{-1} h_G + M_G^{-1} W_G \lambda, \qquad \ddot{q}_L = M_L^{-1} h_L + M_L^{-1} W_L \lambda, \tag{6.112}$$

which we can put into the relation (6.111) to achieve the form

$$\ddot{g}_n = A\lambda + b$$

with
$$A = W_{G,n}^T M_G^{-1} W_G + W_{L,n}^T M_L^{-1} W_L,$$
$$b = W_{G,n}^T M_G^{-1} h_G + W_{L,n}^T M_L^{-1} h_L + \bar{w}. \tag{6.113}$$

The links are either in contact with the guides, then $g_{n,i} = 0$ and also $\ddot{g}_{n,i} = 0$, but $\lambda_{n,i} \neq 0$, or vice versa. This establishes a complementarity

$$\ddot{g}_n \geq 0, \qquad \lambda \geq 0, \qquad \ddot{g}_n^T \cdot \lambda = 0, \tag{6.114}$$

which must be evaluated by components. Solving the system of equations (6.113) with these inequality conditions by applying the algorithm of Lemke, we determine in a first step the contact forces thus being able to calculate the generalized accelerations, and in a second step a valid contact configuration. The numerical procedure behind it is not always stable, mainly due to the application of the accelerations instead of the positions. For the example under consideration we applied some projection algorithm for numerical stabilization [69].

*Impacts*

When a chain link enters the sprocket or when a link comes into contact with a guide an impact is generated. The system configuration requires only impacts in normal direction, which makes an analysis quite simple. Stick-slip processes do not occur, for all contacts we have only sliding or in the case of the pin/sprocket or bushing/sprocket contact we get after some sliding a kind of form closure. According to chapter 3.5 on page 158 and the impact relations (3.212) on page 170 we need to take into consideration only the normal components of these equations. Applied to the combination guide/link we get

$$M_G(\dot{q}_G^+ - \dot{q}_G^-) = W_G \Lambda, \qquad M_L(\dot{q}_L^+ - \dot{q}_L^-) = w_L. \tag{6.115}$$

The vectors $\dot{q}_G^-$ and $\dot{q}_G^+$ denote the generalized velocities shortly before and after the impact. On the right side values on the impulse level appear. The relative normal velocity for the contact is given with the equations (6.99) and (6.104), and the normal velocity between link and guide can be written in the same form as equation (6.111), only on a velocity level. It is

$$\dot{g}_n^+ = w_{G,n}^T \dot{q}_G^+ + w_{L,n}^T \dot{q}_L^+ + \tilde{w} \tag{6.116}$$

In a further step we can derive similar complementarity conditions as above on the velocity level and obtain thus finally the equations containing all active contacts.

$$\dot{g}_n^+ = A_{impact}\Lambda + b_{impact},$$

$$\dot{g}_{n,i}^+ \geq 0, \qquad \Lambda_i \geq 0, \qquad \dot{g}_{n,i}^+ \Lambda_i = 0. \tag{6.117}$$

The solution of the above equations results in the necessary impact magnitudes, but also in an information of the passive or active contact state.

Exchanging (G=guide and L=link) by the couple (S=sprocket and P=pin or B=bushing) we can apply the above equations for these combinations.

### 6.4.3 Results

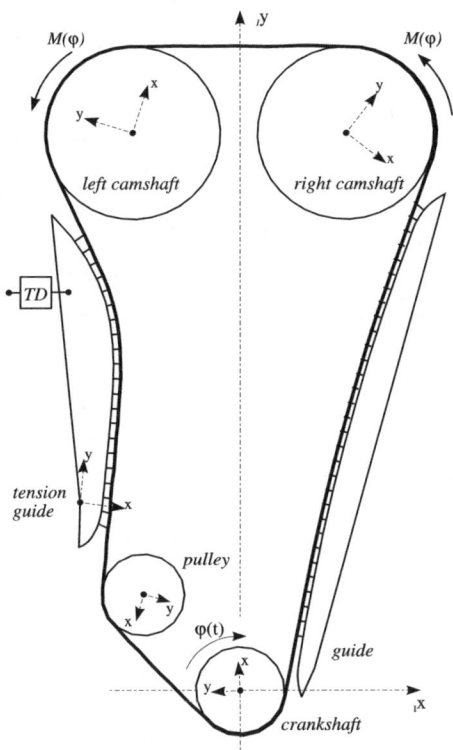

Fig. 6.43: Configuration of the Chain Drive [69]

Figure 6.43 illustrates the chain drive for the numerical simulation.

- The chain consists of 120 links with altogether 360 degrees of freedom.
- The crankshaft is excited by a time dependent rotation speed, resulting from measurements. Hence this sprocket has no degrees of freedom.

- On the camshafts act angle dependent torques, so we have to apply at least one rotational degree of freedom.
- One guide is inertially fixed. The contact contour of this body consists of three arcs.
- The tension guide with the additional tension device act by rotation. This contour is composed of three parameter functions, an arc of an ellipse, a straight line and a polynomial function.

To verify our chain model Figure 6.44 depicts the motion of the tension guide in a comparison of simulation and experiment. The experiments have been performed by [127]. For low rotation speeds at 1980 rpm we achieve good agreement in the amplitudes and frequencies of the guide motion. The main amplitude of the second order with a frequency about 80 Hz results from the torque excitation coming from the camshaft. The polygonal frequency is in that case a bit larger than 600 Hz. At higher rotation speeds with 3000 rpm

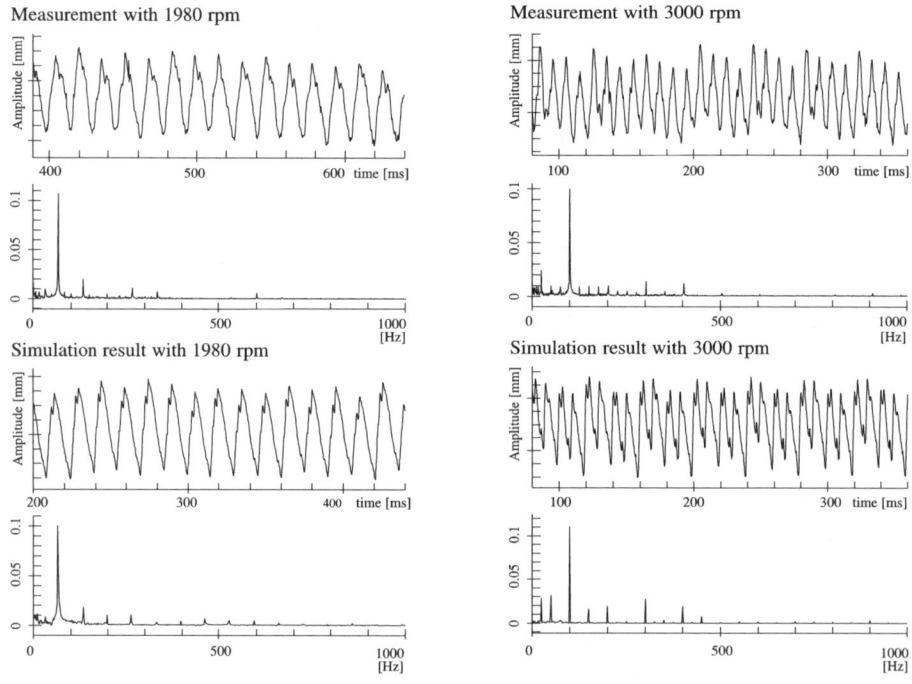

Fig. 6.44: Comparisons between Measurements and Simulations [69]

the simulation still agrees well with measurements. All important frequencies are also evaluated by the simulation. It should be noted however, that the influence of the tensioner model decides to a very high degree the quality of the results. Simulations of timing chain systems without a sophisticated

## 6.4 Bush and Roller Chains

model of the tensioners are always very problematic. We come back to these problems in a separate chapter.

In Figure 6.45 one cycle of a link around the chain drive with the contact forces, the impact values and the chain stress is presented. The time representation of the contact force shows the peaks due to the impacts. In spite of the tangential inlet of a link on a guide, though accompanied by impact-like processes, we reach forces with very high values. Considering the nonlinear behavior and the unsteady structure of the chain model, the forces and the impacts of an incoming link also differ considerably when regarding various cycles.

All beginning contacts are accompanied by large oscillations (sometimes a few mm) due to the impact behaviour of a link entering a sprocket or a link entering a guide, which is always accompanied by the necessity that the incoming link has to carry a significant part of the strand force. These oscillations are then damped by friction, they rise again when the link leaves the sprocket carrying a large amount of the following free strand.

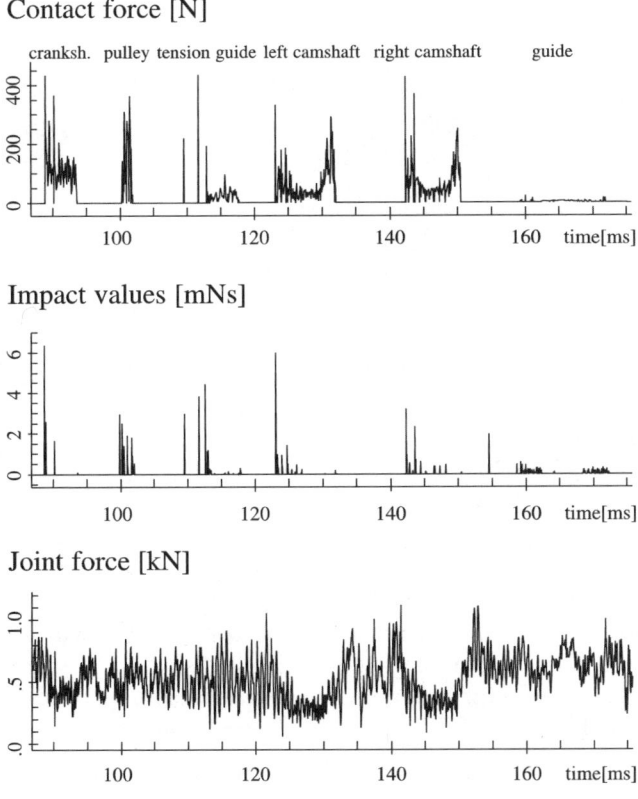

Fig. 6.45: One Cycle of a Link [69]

The graph of the impact values in Figure 6.45 also illustrates the changing contact configuration. Particularly at the beginning and at the end of the guide we see many impacts. The contact configuration between the link and the sprocket changes very often. The small chain stress combined with the small mass and high stiffness of a link leads to joint forces with high frequencies. In the frequency spectrum of this graph the polygonial excitation appears as the dominant amplitude. The vibrations of the chain and especially the vibrations of the joint forces transport directly noise.

Looking at the contact forces of a link in more detail according to Figure 6.46 we recognize several zero points of the graphs indicating a complete release of the contact load and thus nearly a detachment of the contact partner chain/sprocket. It might happen for larger rotational speeds. The time delay between these points corresponds to the polygonial frequency. The different shape of the two curves results from different torque excitations of the sprockets as given by the camshafts.

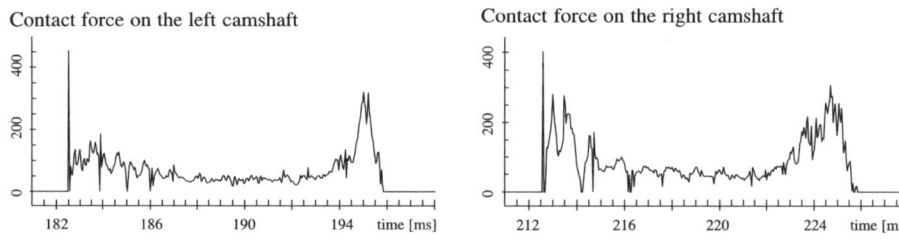

Fig. 6.46: Contact Forces on the Camshaft [69]

The discusion of some results indicate already, that the numerical simulations of timing chains are able to provide substantial informations of

- the dynamic behavior of the camshafts, which is very important for combustion engines,
- the coupled transversal and longitudinal oscillations of the chain at arbitrary points,
- the contact forces, including frictional forces, between the links and the sprockets or guides,
- the relative velocities of the contact points,
- the impacts as the main excitation source of the chain vibrations and the influence on the contacts of neighbouring links
- and finally the summarized impacts as an approximate overall measure for wear and noise.

These factors decide amongst others a good design of a timing system, which has been shown in many examples from industry, for example those of the Figures 6.27 and 6.28, where significant design improvements could be achieved.

## 6.5 Hydraulic Tensioner Dynamics

### 6.5.1 Introduction

As indicated already in the preceeding chapter, all timing chains of modern combustion engines are tensioned by certain hydraulic tensioners, which are applied at the chain guides. By providing these guides with some rotational joint the tensioner can act on the guide, and via the guide it can tighten the chain. Figure 6.47 illustrates a typical application, and it depicts also the curvature of the guides. These curvatures are useful for a space-adapted design, but moreover for a good contact of all chain links along the guide and thus for a good efficacy of the tensioners with regard to the chain vibrations. The larger influence on wear can be counterbalanced by suitable materials of the guides.

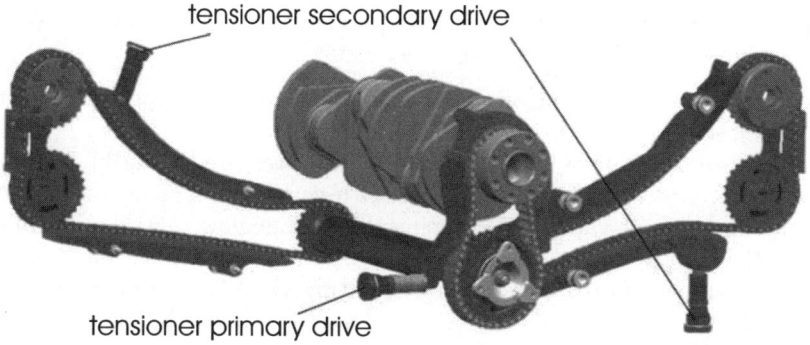

Fig. 6.47: Tensioners of the Timing Chain of the Porsche 996 Engine

The action principle of a hydraulic tensioner comes from the combination of a spring-loaded piston in combination with an oil-hydraulic system. The tensioner force can be controlled hydraulically, and a system of leakage arrangements produces damping. Figure 6.48 depicts a simple tensioner with leakage and thus damping generated by the gap between piston and housing. It shows the principle, which is typical for a large variety of tensioners. As basic elements we always have a housing, a piston, a check valve, piston spring and a leakage, in this case avery simple one.

The piston spring is good for a static force on the guide necessary for some chain tension without engine operation. Going into operation the check valve will be open until a balance of pressure and spring forces is reached. The tensioner is under load from some small engine orders, but sometimes also by resonances, which leads to a closure of the check valve and a generation of a large pressure in the high pressure chamber. This causes highly nonlinear oil flow through the leakage gap producing damping, but at the same time also a reduction of the oil pressure in the high pressure chamber. Again, the check valve can open, and the whole cycle starts anew. Oil must be modelled as a mixture of fluid and gas (air), because the central supply pump cannot avoid a small amount of air being included in the oil.

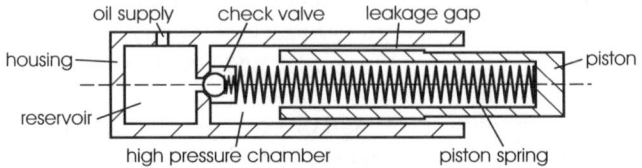

Fig. 6.48: Components of a Simple Tensioner

A large variety of tensioner types have been developed in the last years, mainly with the goal to achieve special performance characteristics. We give a few examples of tensioner's design elements:

- leakage gaps, the lenght of which depend on the piston position, for controlling some progressive performance,
- excess-pressure valves for limitation of the pressure in the high pressure chamber, mostly for timing systems with very large loads,
- more channels or gaps for increasing the leakage flow, partly with a reflux into the supply system,
- additional pressure chambers for generating variable damping effects,
- abandonment of the check valve to reduce the number of components and compensation by a very "soft" tensioner design.

With respect to tensioner theory we follow the dissertation of Hösl [111] and with respect to tensioner experiments we shall present some results from Engelhardt [55]. All fundaments we need from hydraulics are presented already in chapter 4 on page 187 ff., which is mainly based on the work of Borchsenius [23]. In the following we shall consider all tensioner components necessary to compose a tensioner of nearly any type.

### 6.5.2 Piston/Cylinder Component

We consider a piston element, which can be connected to any other hydraulic or mechanical force element being represented as an autonomous node. For the case of two piston elements we have to define the relative kinematics in such a way, that the translational degrees of freedom point in the same direction. The equation of motion for one piston comes out rather simple, but we have to pay attention not to forget some forces. According to Figure 6.49 we get the following relation

$$m_K \ddot{x} = F_{p_1} + F_{p_2} - F_{p_3} - F_{p_4} + F_c - F_d + F_{A_1} - F_{A_2} + F_{R_1} + F_{R_2} - F_g, \quad (6.118)$$

where $F_p$ are pressure forces from node connections, $F_d$ are damping forces, $F_A$ forces coming from mechanical stops, $F_R$ friction forces and $F_g$ gravitational forces, which might become important for inclined tensioners. We come back to the corresponding models lateron.

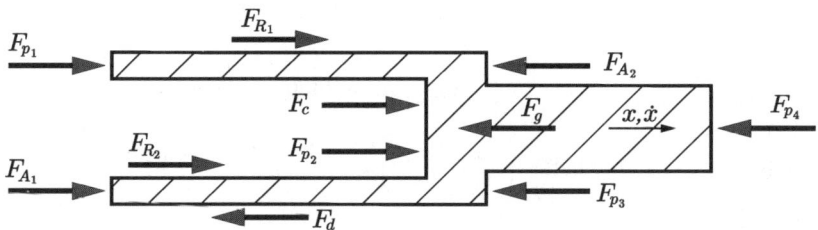

Fig. 6.49: Piston Forces

### 6.5.3 Tube Models

The tubes in connection with hydraulic tensioners are exposed to relatively small pressures in the order of magnitude of maximum 5-6 bars, which allows rather simple tube models (see also section 4 on the pages 187 ff. and [23], [111], [55]). This model takes into consideration oil inertia, friction and pressure losses. Figure 6.50 depicts the simple model of a tube for one-dimensional and, as we assume, incompressible tube flow. Applying the derivation of the

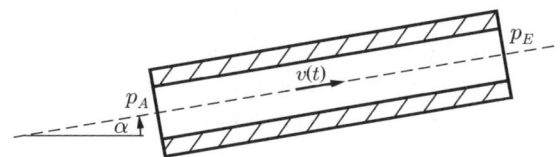

Fig. 6.50: Simple Tube Model

equations (4.22) to (4.26) on page 198 we get for our simplified case

$$m_l \dot{v} = A(p_A - p_E) - f_g + f_r, \tag{6.119}$$

where $m_l = \rho A l$ is the fluid mass in a tube of length l, $v$ the averaged fluid velocity, $f_g = mg \sin(\alpha)$ the gravitational force and $f_r$ the pressure losses. They might be produced by various effects. For straight tubes we have to consider the friction at the tube wall,

$$\Delta p = -\lambda \frac{\rho l}{2D} |v| v. \tag{6.120}$$

The loss coefficient $\lambda$ depends on the Reynolds number and can be determined from standard tables [115]. Alternatively, for the calculation of the $\lambda$-values we apply the well known formulas for laminar and turbulent stationary flows (see [14], [13], [35])

$$\lambda = \frac{64}{Re} \qquad \text{for} \quad Re < 2300 \quad \text{(laminar)}$$

$$\lambda = 8 \left( \left(\frac{8}{Re}\right)^{12} + \frac{1}{(A+B)^{\frac{3}{2}}} \right)^{\frac{1}{12}} \qquad \text{for} \quad Re > 2300 \quad \text{(turbulent)}$$

$$\text{with} \quad A = \left( 2,457 \ln \frac{1}{\left(\frac{7}{Re}\right)^{0,9} + 0,27 \frac{k_s}{d}} \right)^{16}, \quad B = \left(\frac{37530}{Re}\right)^{16} \qquad (6.121)$$

The magnitude $k_s$ is the averaged height of the wall roughness, and d is the tube diameter.

In addition to these losses we get pressure reduction by tube bends or by changes of the sectional areas of piping. These losses are usually described by a similar formula as equation (6.120), namely by

$$\Delta p = -\zeta_l \frac{\rho}{2} |v| v. \qquad (6.122)$$

Again, the empirical $\zeta_l$-values are given by tables [35]. The above formulas can be used to model very different configurations including also labyrinth forms of piping [55].

### 6.5.4 Leakage Models

Leakage gaps of all tensioners are those design elements, which generate damping. Therefore the fantasy for devoloping leakage configurations is extremely large. From all these variants we consider here four basic elements, plane and circular gaps, gaps with variable length and gaps due to eccentricity and due to tilting effects. They represent the most frequently applied elements for hydraulic tensioners.

#### 6.5.4.1 Plane Leakage Gaps

For all tensioners the radius difference of cylinder and piston is so small in comparison with the radii themselves, that the gap resulting from this radius difference very often can be approximated by a plane gap. In addition we have laminar flow, which simplifies the evaluation. Figure 6.51 illustrates the situation. Due to the above remarks we use for the Reynolds number the one for laminar flow in a ring gap, which according to [61] writes

$$Re = v \frac{2h}{\nu}, \qquad (6.123)$$

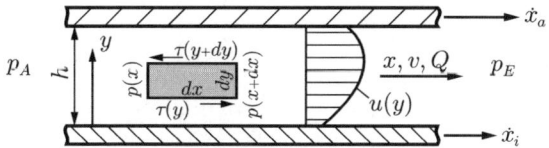

Fig. 6.51: Flow Properties in a Plane Gap

with the gap-averaged velocity v and the kinematical viscosity $\nu$. For gaps of this type we may assume a Reynolds number $Re < 1100$ [61], furtheron incompressibility due to the very small gap volume and a constant pressure perpendicular to the flow direction with $(dp/dy = dp/dz = 0)$. According to Figure 6.51 and Newton's law for the shear stress $\tau = -\eta \frac{du}{dy}$ we get for the fluid equations

$$\frac{dp}{dx} + \frac{d\tau}{dy} = 0 \quad \Rightarrow \quad \frac{dp}{dx} - \eta \frac{d^2 u}{dy^2} \approx \frac{\Delta p}{l} - \eta \frac{d^2 u}{dy^2} = 0, \tag{6.124}$$

which gives us the velocity profile $u(y)$ dependent on the linearly decreasing gap pressure $p(x)$. Integrating twice and including the boundary conditions $u(0) = \dot{x}_i$ and $u(h) = \dot{x}_a$ results in the well known parabolic profile

$$u(y) = \dot{x}_i + \frac{\dot{x}_a - \dot{x}_i}{h} y + \frac{\Delta p}{2\eta l}\left(y^2 - hy\right), \tag{6.125}$$

and from there in the volume flow evaluated by averaging over the cross sectional area $A = 2\pi r_m h$

$$Q = Av = A(\frac{\dot{x}_a + \dot{x}_i}{2} - \frac{h^2 \Delta p}{12 \eta l}) = Q_{\dot{x}} + Q_{\Delta p} \quad \text{with} \quad Q_{\Delta p} = \alpha_{\Delta p} \Delta p \tag{6.126}$$

The magnitude $Q_{\dot{x}}$ denotes the volume flow due to the wall motion and $Q_{\Delta p}$ the part due the pressure difference. The value of $\alpha_{\Delta p}$ depends on the gap geometry alone. The momentum equation 6.119 of the preceding section writes with $f_r = A \Delta p_r$

$$m_G \dot{v} = A(p_A - p_E) - f_g + A \Delta p_r \tag{6.127}$$

and can be combined with equation 6.126 to give

$$\Delta p_r = \frac{Av - Q_{\dot{x}}}{\alpha_{\Delta p}}. \tag{6.128}$$

Regarding the small gap mass $m_G$ we can neglect in most cases the inertia terms and then evaluate the volume flow directly from equation 6.126, which reduces computing time seen before the background, that tensioner dynamics and the gaps involved are located at a low level structure of a large system dynamics.

### 6.5.4.2 Ring-Shaped Leakage Gaps

For plane gaps we made the assumption of being able to neglect the real curvature of tensioner gaps and replace it approximately by a plane configuration. Though for many problems this assumption is quite realistic, it may fail in cases with small cylinder radii. Therefore we have to consider the geometry and the resulting equations in another though related way. Figure 6.52 depicts

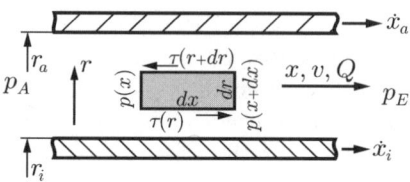

Fig. 6.52: Flow Properties in an Annular Gap

the situation for an annular gap. A statical force balance applied to an infinitesimal volume element comes out after some manipulations with a relation similar to equation (6.124) of the plane case. It is

$$\frac{dp}{dx} + \frac{d(r\tau)}{r\,dr} = 0 \quad \Rightarrow \quad \frac{dp}{dx} - \frac{\eta}{r}\frac{d}{dr}\left(r\frac{du}{dr}\right) \approx \frac{\Delta p}{l} - \frac{\eta}{r}\frac{d}{dr}\left(r\frac{du}{dr}\right) = 0. \quad (6.129)$$

Applying the same steps as before, integration, boundary conditions, averaging, we get the following set of equations

$$u(r) = \dot{x}_a + (\dot{x}_i - \dot{x}_a)\frac{\ln\frac{r}{r_a}}{\ln\frac{r_i}{r_a}} + \frac{\Delta p}{4\eta l}\left[r^2 - r_a^2 + (r_a^2 - r_i^2)\frac{\ln\frac{r}{r_a}}{\ln\frac{r_i}{r_a}}\right],$$

$$Q = \pi\left[\dot{x}_a r_a^2 - \dot{x}_i r_i^2 + (\dot{x}_a - \dot{x}_i)\frac{r_a^2 - r_i^2}{2\ln\frac{r_i}{r_a}}\right] -$$

$$- \frac{\pi \Delta p}{8\eta l}\left[(r_i^2 - r_a^2)^2 + (r_a^2 - r_i^2)\left(2r_i^2 + \frac{r_a^2 - r_i^2}{\ln\frac{r_i}{r_a}}\right)\right]$$

$$= Q_{\dot{x}} + Q_{\Delta p} \quad \text{with} \quad Q_{\Delta p} = \alpha_{\Delta p}\Delta p. \quad (6.130)$$

The flow velocity $u(r)$ follows again a parabolic profile, and the volume $Q$ includes as before two parts, the volume flow $Q_{\dot{x}}$ by drag effects of the moving wall and the volume flow $Q_{\Delta p}$ due to the pressure difference.

### 6.5.4.3 Gaps of Variable Length

For the realization of progressive tensioner characteristics we may design gaps of variable length or gaps with a variable cross section over the length. The second solution is expensive, therefore the first one is quite often applied. Figure 6.53 illustrates one possible solution, which is realized by an assignment of different radii to different lengths, a simple but effective solution. With respect to the body-fixed coordinates the gap limits are given by $x_{min}$ and $x_{max}$. The two coordinate systems themselves possess at the beginning a rel-

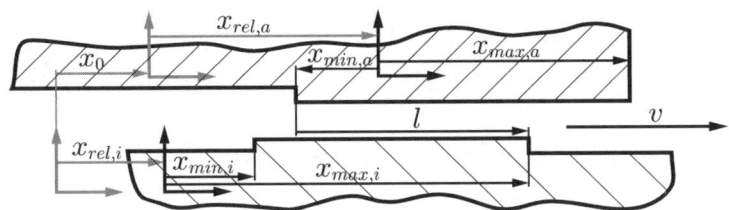

Fig. 6.53: Gap with Variable Length

ative distance $x_0$, during motion they shift by $x_{rel}(t)$. From this the length at time t is given by the relation

$$l(t) = \min[(x_{max,i}+x_{rel,i}), \ (x_0+x_{max,a}+x_{rel,a})] - \\ - \max[(x_{min,i}+x_{rel,i}), \ (x_0+x_{min,a}+x_{rel,a})], \quad (6.131)$$

which for example allows the determination of the fluid mass within the gap, though in most cases it is a good approximation to neglect the gap masses in tensioners. The remaining calculations have to consider the equations (6.126) and (6.127) or (6.130), depending on the model we have chosen, plane or annular.

### 6.5.4.4 Gaps by Eccentricity

Gaps generated by eccentricity are gaps, which we do not want to have, but which we nevertheless must consider, at least if experiments indicate an influence of such phenomena. As tensioners are usually cheap components with sometimes spacious manufacturing tolerances, and as additionally the piston of a tensioner is only supported by the cylinder, we get linear and angular displacements influencing the tensioner characteristic. The asymmetric position and orientation of the piston in the cylinder is mainly caused by the motion of the guide, where the piston is fixed by a rotary joint thus following this motion [111]. This induces eccentricities (see the examples of Figure 6.54).

In gaps of tensioners the volume flow by pressure difference dominates the motion. Experience shows that for these cases the influence of gap eccentricity can be approximated by simply modifying the geometrical coefficient $\alpha_{\Delta p}$ according to [61] in the following form

$$\alpha_{\Delta p} = -A \frac{h^2}{12\eta l} \left(1 + 1{,}5\epsilon^3\right) \quad \text{with} \quad \epsilon = \frac{e}{h}. \quad (6.132)$$

$\epsilon$ is the relative eccentricity with $0 \leq \epsilon \leq 1$. As the influence of the gap height h is with $h^3$ very large, we get for a displaced piston with contact on one side an increase of the volume flow in the gap by a factor of 2.5.

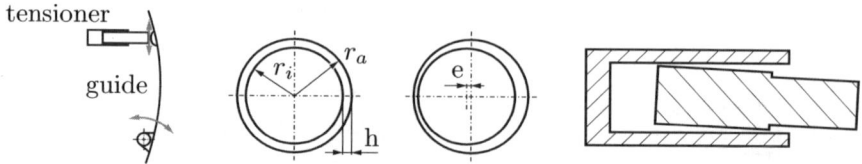

Fig. 6.54: Gaps due to Eccentricity

### 6.5.5 Check Valves

Many tensioners apply check valves to avoid a reflow of the oil into the supply system. Check valves typically have a spherical closure configuration in addition with a spring, which is sometimes omitted depending on the tensioner position. Figure 6.55 illustrates a typical design.

Tensioner dynamics must be described very carefully, because it influences the chain dynamics in a dominant way. From this we have to take into consideration all motion elements, solid and fluid, the motion of the sphere, the fluid motion through the annular areas produced by the sphere motion and all fluid deviations. From many experiments with large and small check valve models we take the following route: We model the most important loss represented by the annulus flow between the sphere and the housing in detail and regard additional losses by a contraction coefficient $\alpha_V$ related to the orifice behaviour of the valve. This theoretical-empirical combination reduces data processing but at the same time makes experiments necessary to measure the coefficients $\alpha_V$.

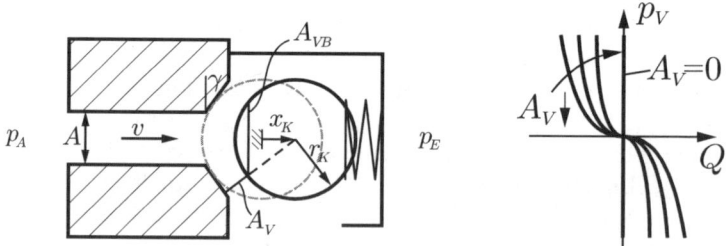

Fig. 6.55: Check Valve Principle and Chracteristic

Going back to Figure 6.55 we write the momentum equation in the form

$$m_V \dot{v} = A(p_A - p_E) + A p_V, \quad (6.133)$$

with the fluid mass $m_V$ of the valve volume under consideration. The nonlinear pressure loss $p_V$ depends on the flow velocity v and on the annular area given by the sphere position. We apply for that the orifice equation

$$p_V = -\frac{\rho}{2}\frac{A^2}{\alpha_V^2 A_V^2}|v|v = -\zeta_V \frac{\rho}{2}|v|v, \quad (6.134)$$

where A is the input cross section, and the area $A_V(x_K)$ can be calculated from Figure 6.55

$$A_V(x_K) = 2\pi x_K r_K \frac{\sin(\gamma)}{\cos(\gamma)} + \pi x_K^2 \frac{\sin(\gamma)}{\cos^2(\gamma)} \quad (6.135)$$

With decreasing flow areas $A_V$ the characteristics become steeper and reach in the limiting case the form of a bilateral constraint. On the other hand, these characteristics can be approximated quite realistically by a complementarity. See for both possibilities the chapters 4.2.2.1 and 4.2.2.2 on the pages 193 and following.

Practical observations of tensioners indicate a pressure rise at the beginning of the piston's motion with a closing check valve. The sphere moves in an oil environment with more or less large viscosity values retarding the closing process. For considering this effect we have to leave the statical force balance, and we have to introduce an individual degree of freedom for the sphere, which moves under the influence of a pressure force on the base area $A_{VB}$ of the sphere, of a pre-tensioned spring force $F_{VF_0}$ with spring stiffness $c_{VF}$ and finally under the influence of damping forces for the closure and opening motion ($d_{zu}$ with ($\dot{x}_K < 0$) for closure and $d_{auf}$ with ($\dot{x}_K \geq 0$) for opening). Without inertia forces we get

$$\dot{x}_K = \frac{1}{d_{auf,zu}}(-c_{VF}x_K - F_{VF_0} + A_{VB}(p_A - p_E)) \quad (6.136)$$

With inertia forces we get

$$m_K \ddot{x}_K = -d_{auf,zu}\dot{x}_K - c_{VF}x_K - F_{VF_0} + A_{VB}(p_A - p_E). \quad (6.137)$$

If the sphere is going to hit a stop, we treat it as a plastic impact. The corresponding point is detected by indicator point search. If the sphere is located already at a stop, we allow only velocities or accelerations leading to a separation. This avoids contact modeling for the price of some additional decisions.

### 6.5.6 Tensioner System

A tensioner represents a combined mechanical-hydraulic system including mechanical and hydraulic components and their individual dynamics, which have to be integrated to come out with the complete tensioner system dynamics. It should be kept in mind, that on a higher level the tensioner dynamics as a whole has also to be included into the dynamics of the chain system with its components chain, sprockets, guides and tensioners, which was presented in the preceding chapters. The methods applied to combine hydraulic and mechanical components to systems are discussed in the original work of Borchsenius [23], then taken and extended by Hösl [111] and Engelhardt [55] and finally included in chapter 4 on page 187 and the following. We refer to these contributions.

The composition of components follow the rule, that all state variables, that means all position/orientation states and all velocity states, are collected in the two vectors **x** and **v**. The equations of the vector $x$ with $n_x$ variables are described by a set of ordinary nonlinear differential equations of first order

$$\dot{x} = f(t, x, v), \tag{6.138}$$

which do not require any constraints. These relations comprise positions of pistons or spheres, the velocities of elastically coupled components like piping elements and pressures or expansion volumes of compressible nodes.

Velocity coordinates including unilateral or bilateral constraints are collected in the vector **v**. The corresponding momentum equations write

$$M\dot{v} = W_P \lambda_P + W_B \lambda_B + W_V \lambda_V + W_T \lambda_T + W_A \lambda_A + h(t, v, x). \tag{6.139}$$

The force vector $h(t, v, x)$ includes all given forces, bearing forces and the like as well as forces originating from contacts with sliding friction only. The $\lambda$-vectors denote pressures and forces due to active set-valued force laws, which are projected into the corresponding coordinate space by the matrices **W**. These magnitudes comprise the following force elements and constraints:

- $n_P$ incompressible hydraulic nodes with the unilateral constraints of cavitation,
- $n_B$ incompressible hydraulic nodes represented by bilateral constraints,
- $n_V$ potential bilateral constraints of closed valves, $n_{Va}$ if active,
- $n_T$ frictional contacts with stick/slip,
- $n_A$ mechanical stops and components with unilateral properties.

Before combining all this to make a system we shall provide some formalisms to deal with the constraints. Let us first consider bilateral constraints. They appear for the $n_B$ incompressible hydraulic nodes and the $n_{Va}$ closed valves, for example the check valves, and they can formally be written

$$W_B^T v + w_B(t) = 0, \qquad W_B \in \mathbb{R}^{n_B, n_v}, \quad v \in \mathbb{R}^{n_v}, \tag{6.140}$$

where the volume flows contained in the rows of the first term must be supplemented by corresponding magnitudes in the second term, if we have in a node also volume flow sources. The number $n_{Va} = n_{Va}(t)$ of closed valves depends of course on time, the appropriate constraints are

$$\boldsymbol{W}_V^T \boldsymbol{v} = \boldsymbol{0}, \qquad \boldsymbol{W}_V \in \mathbb{R}^{n_{Va}, n_v}. \tag{6.141}$$

The matrix $\boldsymbol{W}_V$ contains, according to the constraints, the pressure-exposed valve areas. Combining the above two relations yields

$$\boldsymbol{W}_G^T \boldsymbol{v} + \boldsymbol{w}_G(t) = \boldsymbol{0}, \quad \text{with} \quad \boldsymbol{W}_G = (\boldsymbol{W}_B \; \boldsymbol{W}_V); \; \boldsymbol{w}_G(t) = \begin{pmatrix} \boldsymbol{w}(t) \\ \boldsymbol{0} \end{pmatrix}. \tag{6.142}$$

The number of the independent constraint equations corresponds to the rank r of the matrix $\boldsymbol{W}_G \in \mathbb{R}^{(n_B + n_{Va}), n_v}$. Accordingly the number of independent and minimum velocity degrees of freedom reduces to $n_{min} = n_v - r$. In a further step we express the velocity $\boldsymbol{v}$ by the minimum velocities $\boldsymbol{v}_m$

$$\boldsymbol{v} = \boldsymbol{v}(\boldsymbol{v}_m, t) = \boldsymbol{J} \boldsymbol{v}_m + \boldsymbol{b}(t). \tag{6.143}$$

The matrix $\boldsymbol{J}$ can be evaluated by singular value decomposition, see for example [23] and [283]. Combining the relations (6.139) and (6.143) and taking into account $\boldsymbol{J}^T \boldsymbol{W}_B = \boldsymbol{0}$ and $\boldsymbol{J}^T \boldsymbol{W}_{V_a} = \boldsymbol{0}$, which eliminates the bilateral constraints, we finally get the form

$$\boldsymbol{J}^T \boldsymbol{M} \left( \boldsymbol{J} \dot{\boldsymbol{v}}_m + \dot{\boldsymbol{b}} \right) = \boldsymbol{J}^T \boldsymbol{W}_P \boldsymbol{\lambda}_P + \boldsymbol{J}^T \boldsymbol{W}_T \boldsymbol{\lambda}_T + \boldsymbol{J}^T \boldsymbol{W}_A \boldsymbol{\lambda}_A + \boldsymbol{J}^T \boldsymbol{h}(t, \boldsymbol{v}, \boldsymbol{x}). \tag{6.144}$$

After we have eliminated the bilateral constraints we come to the unilateral constraints. For this purpose we start with a reduced set of equations of motion, which include only parts concerning unilateral constraints. They can be written in the two forms

$$\begin{aligned} \boldsymbol{M} \dot{\boldsymbol{v}} &= \boldsymbol{W}_P \boldsymbol{\lambda}_P + \boldsymbol{W}_T \boldsymbol{\lambda}_T + \boldsymbol{W}_A \boldsymbol{\lambda}_A + \boldsymbol{h}(t, \boldsymbol{v}, \boldsymbol{x}), \\ \dot{\boldsymbol{v}} &= \boldsymbol{M}^{-1} (\boldsymbol{W}_P \boldsymbol{\lambda}_P + \boldsymbol{W}_T \boldsymbol{\lambda}_T + \boldsymbol{W}_A \boldsymbol{\lambda}_A + \boldsymbol{h}) \end{aligned} \tag{6.145}$$

In addition to these equations of motion we have some conditions for the relative acceleration

$$\begin{aligned} \ddot{\boldsymbol{g}}_P &= -\boldsymbol{W}_P \dot{\boldsymbol{v}} && \text{from unilateral hydraulic nodes} \\ \ddot{\boldsymbol{g}}_T &= \boldsymbol{W}_T \dot{\boldsymbol{v}} && \text{from potential sticking contacts} \\ \ddot{\boldsymbol{g}}_A &= \boldsymbol{W}_A \dot{\boldsymbol{v}} && \text{from unilateral force laws} \end{aligned} \tag{6.146}$$

With respect to unilateral constraints we refer to the chapters 3.1.2 on page 89 and 3.4 on page 131. Especially the tangential constraints make a decomposition necessary, which splits the double corner law into four unilateral primitives [87] or according to Figure 3.5 on page 95 into two unilateral primitives [226].

The friction reserve defines within the friction cone the distance of the friction forces to the cone surface, which represents the limiting static friction force. For our case we have the relation for the friction reserve

$$\boldsymbol{\lambda}_{T0,i} = \begin{pmatrix} \lambda_{T01,i} \\ \lambda_{T02,i} \end{pmatrix} = \begin{pmatrix} F_{0,i} \\ F_{0,i} \end{pmatrix} - \begin{pmatrix} 1 \\ -1 \end{pmatrix} \lambda_{T,i} \geq \mathbf{0}, \quad \text{with} \quad F_{0,i} = F_{R0,i} + A_{p,i}|\Delta p_i|, \tag{6.147}$$

which is the admissible regime for sticking. We also split the relative tangential acceleration by $\boldsymbol{\kappa}_i = (\kappa_{i_1} \;\; \kappa_{i_2})^T$. Together with the friction reserve we then come out with the complementarity

$$-\ddot{g}_{T,i} = \kappa_{i_1} - \kappa_{i_2} \quad \text{with} \quad \boldsymbol{\lambda}_{T0,i} \geq \mathbf{0}; \;\; \boldsymbol{\kappa}_i \geq \mathbf{0}; \;\; \boldsymbol{\kappa}_i^T \boldsymbol{\lambda}_{T0,i} = 0, \tag{6.148}$$

confirming that either the friction reserve is not zero for sticking, and then the relative tangential acceleration is zero, or vice versa. The friction $\lambda_{T,i}$, needed in the equations of motion, follows from the first row of eq. (6.147)

$$\lambda_{T,i} = F_{0,i} - \lambda_{T01,i}, \tag{6.149}$$

and together with the second row we get

$$\lambda_{T02,i} = 2F_{0,i} - \lambda_{T01,i}. \tag{6.150}$$

For all tangentially active contacts we then have the corresponding relations in the form

$$\boldsymbol{\lambda}_T = \boldsymbol{G}_0 - \boldsymbol{\lambda}_{T01}, \quad \boldsymbol{\lambda}_{T02} = 2\boldsymbol{G}_0 - \boldsymbol{\lambda}_{T01}, \quad \boldsymbol{\kappa}_1 = -\ddot{\boldsymbol{g}}_T + \boldsymbol{\kappa}_2.$$

$$\boldsymbol{\lambda}_T = \begin{pmatrix} \lambda_{T,1} \\ \vdots \\ \lambda_{T,n_T} \end{pmatrix}, \quad \boldsymbol{G}_0 = \begin{pmatrix} F_{0,1} \\ \vdots \\ F_{0,n_T} \end{pmatrix}, \quad \boldsymbol{\lambda}_{T01} = \begin{pmatrix} \lambda_{T01,1} \\ \vdots \\ \lambda_{T01,n_T} \end{pmatrix},$$

$$\boldsymbol{\lambda}_{T02} = \begin{pmatrix} \lambda_{T02,1} \\ \vdots \\ \lambda_{T02,n_T} \end{pmatrix}, \quad \ddot{\boldsymbol{g}}_T = \begin{pmatrix} \ddot{g}_{T,1} \\ \vdots \\ \ddot{g}_{T,n_T} \end{pmatrix}, \quad \boldsymbol{\kappa}_i = \begin{pmatrix} \kappa_{i,1} \\ \vdots \\ \kappa_{i,n_T} \end{pmatrix}. \tag{6.151}$$

Combining now the equations (6.145), (6.146) and (6.151) we generate a form more suitable for a formulation of the final complementarity conditions [111]

$$\begin{aligned}\boldsymbol{\kappa}_1 = &-\ddot{\boldsymbol{g}}_T + \boldsymbol{\kappa}_2 = -\boldsymbol{W}_T^T \dot{\boldsymbol{v}} + \boldsymbol{\kappa}_2 \\ = &-\boldsymbol{W}_T^T \boldsymbol{M}^{-1} \boldsymbol{W}_P \boldsymbol{\lambda}_P + \boldsymbol{W}_T^T \boldsymbol{M}^{-1} \boldsymbol{W}_T \boldsymbol{\lambda}_{T01} - \boldsymbol{W}_T^T \boldsymbol{M}^{-1} \boldsymbol{W}_A \boldsymbol{\lambda}_A - \\ & - \boldsymbol{W}_T^T \boldsymbol{M}^{-1} \boldsymbol{G}_0 - \boldsymbol{W}_T^T \boldsymbol{M}^{-1} \boldsymbol{h} + \boldsymbol{\kappa}_2 \end{aligned} \tag{6.152}$$

The unilateral hydraulic nodes and the mechanical devices following unilateral laws have the same structural equations comparable to the complementarities in normal direction of a contact. We get by applying the same procedure as above the following sets:

$$\ddot{\boldsymbol{g}}_P \geq \mathbf{0}; \;\; \boldsymbol{\lambda}_P \geq \mathbf{0}; \;\; \ddot{\boldsymbol{g}}_P^T \boldsymbol{\lambda}_P = 0; \quad \text{with}$$

$$\begin{aligned}\ddot{\boldsymbol{g}}_P = &- \boldsymbol{W}_P^T \dot{\boldsymbol{v}} \\ = &- \boldsymbol{W}_P^T \boldsymbol{M}^{-1} \boldsymbol{W}_P \boldsymbol{\lambda}_P + \boldsymbol{W}_P^T \boldsymbol{M}^{-1} \boldsymbol{W}_T \boldsymbol{\lambda}_{T01} - \\ & - \boldsymbol{W}_P^T \boldsymbol{M}^{-1} \boldsymbol{W}_A \boldsymbol{\lambda}_A - \boldsymbol{W}_P^T \boldsymbol{M}^{-1} \boldsymbol{W}_T \boldsymbol{G}_0 - \boldsymbol{W}_P^T \boldsymbol{M}^{-1} \boldsymbol{h}\end{aligned} \tag{6.153}$$

for incompressible nodes with cavitation, for example, and we get furtheron

$$\ddot{g}_A \geq 0; \quad \lambda_A \geq 0; \quad \ddot{g}_A^T \lambda_A = 0; \qquad \text{with}$$

$$\begin{aligned}
\ddot{g}_A &= W_A^T \dot{v} \\
&= + W_A^T M^{-1} W_P \lambda_P - W_A^T M^{-1} W_T \lambda_{T01} + \\
&\quad + W_A^T M^{-1} W_A \lambda_A + W_A^T M^{-1} W_T G_0 + W_A^T M^{-1} h
\end{aligned} \quad (6.154)$$

for mechanical stops or unilateral models of hydraulic components like check valves.

Combining the non-smooth laws (6.148), (6.153), (6.154) with (6.151) and (6.152) we finally arrive at the system equations represented by an overall complementarity condition

$$\ddot{g} = A\lambda + b \qquad \ddot{g} \geq 0; \quad \lambda \geq 0; \quad \ddot{g}^T \lambda = 0; \qquad \text{with the details}$$

$$\underbrace{\begin{pmatrix} \ddot{g}_P \\ \kappa_1 \\ \lambda_{T02} \\ \ddot{g}_A \end{pmatrix}}_{\ddot{g}} = \underbrace{\begin{pmatrix} -W_P^T M^{-1} W_P & W_P^T M^{-1} W_T & 0 & -W_P^T M^{-1} W_A \\ -W_T^T M^{-1} W_P & W_T^T M^{-1} W_T & E & -W_T^T M^{-1} W_A \\ 2G_P^T & -E & 0 & 0 \\ W_A^T M^{-1} W_P & -W_A^T M^{-1} W_T & 0 & W_A^T M^{-1} W_A \end{pmatrix}}_{A} \underbrace{\begin{pmatrix} \lambda_P \\ \lambda_{T01} \\ \kappa_2 \\ \lambda_A \end{pmatrix}}_{\lambda} +$$

$$+ \underbrace{\begin{pmatrix} -W_P^T M^{-1} W_T G_0 - W_P^T M^{-1} h \\ -W_T^T M^{-1} G_0 - W_T^T M^{-1} h \\ 2G_0 \\ W_A^T M^{-1} W_T G_0 + W_A^T M^{-1} h \end{pmatrix}}_{b}$$

$$\begin{pmatrix} \ddot{g}_P \\ \kappa_1 \\ \lambda_{T02} \\ \ddot{g}_A \end{pmatrix} \geq 0; \quad \begin{pmatrix} \lambda_P \\ \lambda_{T01} \\ \kappa_2 \\ \lambda_A \end{pmatrix} \geq 0; \quad \begin{pmatrix} \ddot{g}_P \\ \kappa_1 \\ \lambda_{T02} \\ \ddot{g}_A \end{pmatrix}^T \begin{pmatrix} \lambda_P \\ \lambda_{T01} \\ \kappa_2 \\ \lambda_A \end{pmatrix} = 0 \qquad (6.155)$$

These equations may be solved either by applying Lemke's algorithm or more conveniently by using the Augmented Lagrange Method including the prox-functional approach.

### 6.5.7 Experiments and Verification

Some systematic tests have been performed by Engelhardt [55]. The basic idea is simple, see Figure 6.56. The tensioner is excited by a cam disc with a sinusoidal boundary, and to avoid detachment of the roller a spring presses the piston rod against the cam. Details of this test set-up can be found in [55]. The comparison of measurements and simulations was very succesful for all cases considered. We give a few examples.

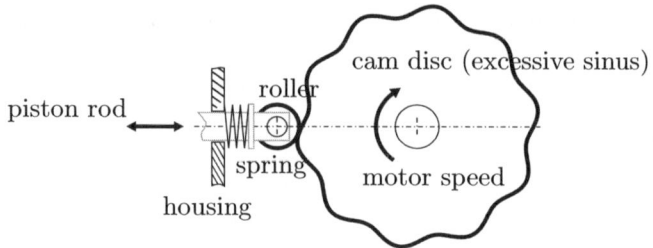

Fig. 6.56: Basic Idea of the Tensioner Test Set-Up [55]

For one series of tests the piston forces of the tensioner were determined for excitation frequencies from 20 Hz to 100 Hz and room temperature and for 70° C. The excitation was 0.2 mm. Figure 6.57 illustrates exemplarly these results. All other comparisons look equally well [55]. For another test series

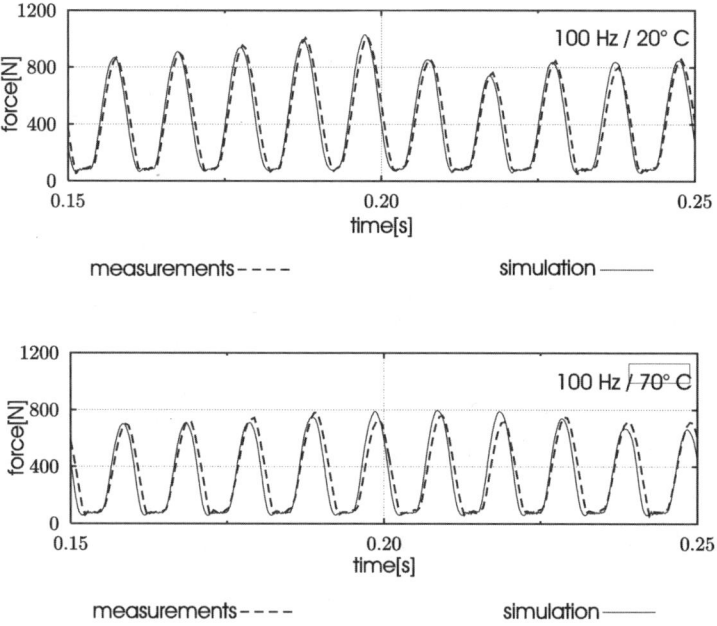

Fig. 6.57: Comparison of Measurements and Simulation for an Industrial Tensioner

also the pressures at the check valve were measured. For a frequency of 100 Hz and room temperature we recognize that the check valve pressure decreases only at two points for a very short time under the supply pressure thus opening the sphere. This dynamics is represented correctly by the theoretical model

(Figure 6.58). Again, various additional comparisons confirm the tensioner model as a further basis of an overall roller chain dynamics analysis.

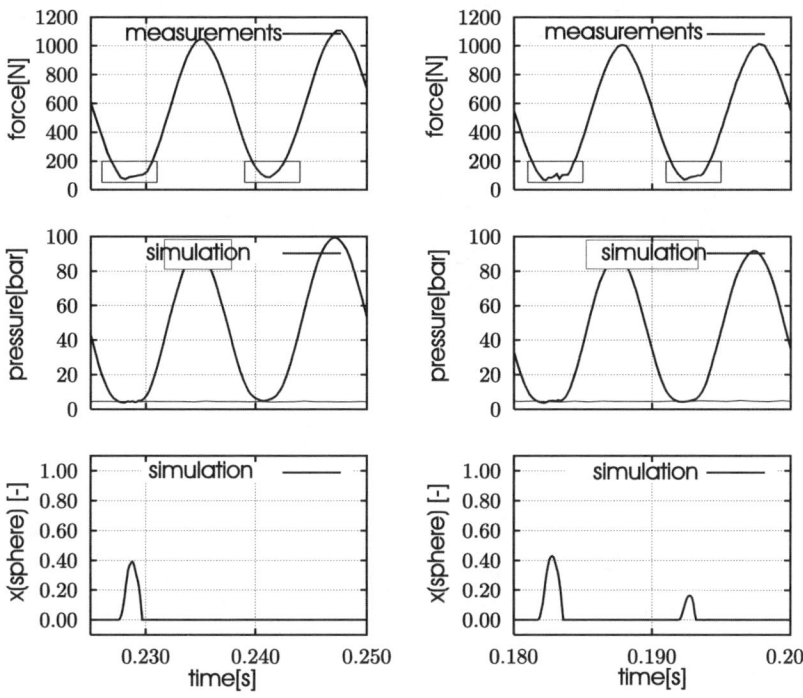

Fig. 6.58: Check Valve Behaviour

# 7

# Robotics

> *Wissenschaft, Philosophie, rationales Denken müssen alle beim Alltagsverstand anfangen. (Karl Popper, Objektive Erkenntnis, 1984)*
>
> *Science, philosophy, rational thought, must all start from common sense.*
>
> *(Karl Popper, Objective Knowledge, 1972)*

## 7.1 Introduction

In the last three decades robotics and walking made dramatic progresses, and especially robots are in the meantime applied not only in industry and surgery, but also in many service areas under water, on earth and in space, with increasing significance. Robots and walking machines are typical products not being realizable before the computer age with all its electronic, sensory and drive train possibilities, necessary to realize such machines. Figure 7.1 gives a characteristic example of robotics, namely robots from the KUKA-company, in action and not in action. KUKA is one of the great producers of industry robots. Figure 7.2 depicts a famous Japanese walking machine with an advanced performance.

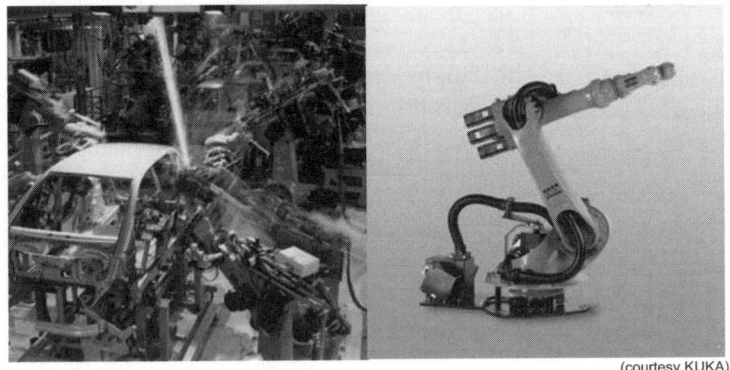

(courtesy KUKA)

Fig. 7.1: KUKA-Robot for Industrial Applications

Robots and walking machines are excellent examples for modeling multibody systems with unilateral and bilateral constraints. As a matter of fact a huge body of literature exists in that field, which we cannot regard in all details. A yearly overview is established during the IEEE Conferences on Robotics and Automation (ICRA) and on Intelligent Robots and Systems (IROS), where most of the leading persons and Institutions of the world can be met. Some classical books on Robotics are for example [7], [37], [3], [275] and [208]. It should be noted, that most of the fundaments with respect to dynamics and control of robots have been developed during the 80ties and the 90ties. Some contributions with respect to technical walking are [221], [262], [162], and with respect to biological walking we have for example [164] and [274], where especially the collection of walking human beings represented by a large foto-series is a very famous one from the 19th century [164]. The CISM course 467 [213] covers biological and technological aspects.

Fig. 7.2: HRP2 Walking Machine from AIST, Japan

From the standpoint of mechanics robots and walking machines are multibody systems, in many cases with a tree-like configuration, but sometimes also with pantograph mechanisms or other more complicated joint structures, which make modeling a bit more complex. Links are rigid or elastic, the consideration of elasticities depends, as always, on the operational frequency range in comparison with the individual link eigenfrequencies. Joint drives, existing in many different realizations, usually have to be modeled with individual degrees of freedom and for many cases also elastically. Contacts with the environment generate unilateral constraints, which need to be regarded. Also here various theories are possible, from unilateral rigid to discretized elastic contacts. We shall give examples. As robots and walking machines are not able to operate without control, the questions of control design and stability have to be answered. Stability problems are a dominant issue for walking, especially for biped walking, and inspite of some intelligent solutions far away from being satifactorily solved. In the framework of this book we shall consider in the following some typical path planning problems of robots, some assembly processes with challenging contact problems of robot and environment, and for walking we shall focus on the dynamics and control of a biped machine.

## 7.2 Trajectory Planning

Trajectory planning for robots and walking machines belongs to the elementary issues, because the problem of going with the end effector or with a foot from one location to another one is a very basic necessity. Manipulation and walking will not be possible without a change of positions and orientations within a given environment, be it that of a manufacturing process or be it a walking environment of a machine. A lot of efforts have been put into obstacle ovoidance [128] by appropriate path planning for classical robots. For walking online processes including vision systems and haptic interferences become more and more important, and the (partly) artificial intelligence connected with these new developments will certainly spill over also into robotics. We shall give here some elementary introduction into the problem and present practical examples.

In the eighties some remarkable methods came up, which all applied some ideas from classical nonlinear dynamics to the path planning problem of robots. Within the framework of dynamics these approaches are still a very valueable example for the solution of a complicated nonlinear dynamics problem, that they are included here. The basic idea consists in the assumption, that the manipulator end effector follows perfectly and in an ideal way a prescribed trajectory, the path coordinate of which is used as the only degree of freedom whatever the robot configuration might be. To realize that we must project all kinematics and all kinetics onto this one degree of freedom resulting in a structure of the equations of motion, which then allows an analytic-topological solution for the time-optimum problem. These ideas were persued nearly at the same time in the US and in Europe, see [21], [48], [109], [243], [204], [186].

### 7.2.1 A Few Fundaments

#### 7.2.1.1 Kinematics

We consider any robot configuration and establish in a first step the kinematics of such a robot. Figure 7.3 gives an example [219], [152]. We refer to chapter 2.2 on page 12, and there especially to the sections 2.2.2 to 2.2.5. It should be noted that for the description of robots also Denavit-Hartenberg coordinates are frequently applied. We did not discuss them in chapter 2.2, but there is a lot of literature dealing with this type of coordinates [42], [3], [208]. We describe the position and orientation of a robot link with respect to its predecessor link (see for example Figures 2.15 on page 26 and 2.16 on page 27). In a similar way we describe also the motor position and orientation with respect to some predecessor motor. For the relevant coordinates we introduce the notation

$$\bar{q} = \begin{pmatrix} \bar{q}_M \\ \bar{q}_L \end{pmatrix} \in \mathbb{R}^{n_f}, \quad \text{with} \quad n_f = n_M + n_L, \tag{7.1}$$

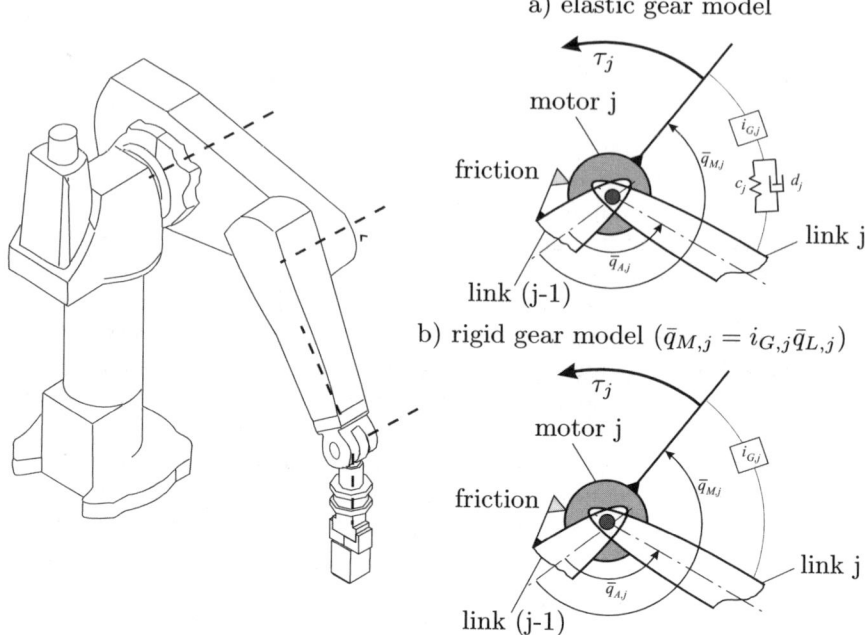

Fig. 7.3: Robot Example and Drive Models

where L stands for link and M for motor. The number of robot degrees of freedom is $n_f$ composed by the link degrees of freedom $n_L$ and the motor degrees of freedom $n_M$. Expressing the position and orientation of a link i by the center of mass vector $\mathbf{r}_i$ regarding in addition the rotation matrix from the inertial system to link i, we come out with (see Figure 7.4)

$$\mathbf{r}_i = \mathbf{r}_{i-1} + \mathbf{r}_{i-1,i}, \qquad \mathbf{r}_{i-1,i} = \mathbf{r}_{SE,i-1} + \mathbf{r}_{ES,i}, \qquad \mathbf{A}_{i,0} = \mathbf{A}_{i,i-1}\mathbf{A}_{i-1,0}, \quad (7.2)$$

where the vector $\mathbf{r}_{SE,i-1}$ connects the mass center $S_{i-1}$ with the joint between the bodies $B_{i-1}$ and $B_i$, and the vector $\mathbf{r}_{ES,i}$ connects the joint with the mass center $S_i$ of link i. From the above relations we get the velocities and the accelerations

$$\omega_i = \omega_{i-1} + \omega_{i-1,i},$$
$$\dot{\mathbf{r}}_i = \dot{\mathbf{r}}_{i-1} + \tilde{\omega}_{i-1}\mathbf{r}_{SE,i-1} + \tilde{\omega}_i \mathbf{r}_{ES,i} + \dot{\mathbf{r}}_{ES,i},$$
$$\dot{\omega}_i = \dot{\omega}_{i-1} + \tilde{\omega}_{i-1}\omega_{i-1,i} + \dot{\omega}_{i-1,i},$$
$$\ddot{\mathbf{r}}_i = \ddot{\mathbf{r}}_{i-1} + \dot{\tilde{\omega}}_{i-1}\mathbf{r}_{SE,i-1} + \tilde{\omega}_{i-1}\tilde{\omega}_{i-1}\mathbf{r}_{SE,i-1} +$$
$$\qquad + \dot{\tilde{\omega}}_i \mathbf{r}_{ES,i} + \tilde{\omega}_i\tilde{\omega}_i\mathbf{r}_{ES,i} + 2\tilde{\omega}_i\dot{\mathbf{r}}_{ES,i} + \ddot{\mathbf{r}}_{ES,i}. \qquad (7.3)$$

For tree-like structures we are able to evaluate the kinematics of a link by that of the predecessor link. For the end-effector with index G of a robot we get from that

$$\mathbf{r}_G = \sum_{i=1}^{n_L}\Big[\prod_{k=1}^{i}\mathbf{A}_{k,k-1}\Big](\mathbf{r}_{ES,i}+\mathbf{r}_{SE,i}), \qquad \mathbf{A}_{G0} = \prod_{i=1}^{n_L}\mathbf{A}_{i,i-1},$$

$$\dot{\mathbf{r}}_G = \sum_{i=1}^{n_L}[\tilde{\omega}_i(\mathbf{r}_{ES,i}+\mathbf{r}_{SE,i})+\dot{\mathbf{r}}_{ES,i}], \qquad \omega_G = \prod_{i=1}^{n_L}\omega_{i,i-1}. \qquad (7.4)$$

These equations describe the so-called forward kinematics representing the end-effector in terms of link and joint coordinates. Going the other way round, namely searching for a given end effector position and orientation the joint and link coordinates, is called the kinematical inverse problem, which for quite a lot of robot configurations can be solved analytically [3], [37]. A very substantial presentation of inverse dynamics is given in [3].

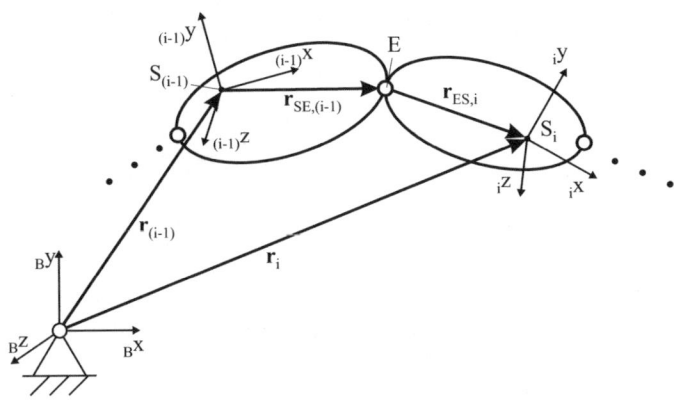

Fig. 7.4: Robot Relative Kinematics

The description of a manipulator in Cartesian space depends of course on the generalized coordinates. A transformation in both directions is necessary due to the fact, that a robot works within a real world environment being described in Cartesian coordinates, but possesses his own dynamics more conveniently described in joint space coordinates. Therefore we need the transformations

$$\dot{\mathbf{r}} = \frac{\partial \dot{\mathbf{r}}}{\partial \dot{\tilde{\mathbf{q}}}}\dot{\tilde{\mathbf{q}}} = \mathbf{J}_T\dot{\tilde{\mathbf{q}}}, \qquad \omega = \frac{\partial \omega}{\partial \dot{\tilde{\mathbf{q}}}}\dot{\tilde{\mathbf{q}}} = \mathbf{J}_R\dot{\tilde{\mathbf{q}}}, \qquad (7.5)$$

where $\mathbf{J}_T \in \mathbb{R}^{3,n_f}$ and $\mathbf{J}_R \in \mathbb{R}^{3,n_f}$ are the Jacobians of translation and rotation, respectively. They project the motion into the not constrained directions, and they will be determined in body-fixed coordinates. They can be calculated recursively, see chapter 2.2.4 on page 25. These considerations yield the following recursions

- with respect to relative translational displacements:

$$\mathbf{J}_{T,i} = \mathbf{A}_{i,i-1}\mathbf{J}_{T,i-1} - [(\mathbf{A}_{i,i-1})_{(i-1)}\tilde{\mathbf{r}}_{SE,i-1} + {}_{(i)}\tilde{\mathbf{r}}_{SE,i}]\mathbf{J}_{R,i} + \frac{\partial_{(i)}\tilde{\mathbf{r}}_{SE,i}}{\partial \dot{\mathbf{q}}},$$
$$\mathbf{J}_{R,i} = \mathbf{A}_{i,i-1}\mathbf{J}_{R,i-1}, \qquad (7.6)$$

- with respect to relative rotational displacements:

$$\mathbf{J}_{T,i} = \mathbf{A}_{i,i-1}(\mathbf{J}_{T,i-1} - {}_{(i-1)}\tilde{\mathbf{r}}_{SE,i-1}\mathbf{J}_{R,i-1}) + \frac{\partial}{\partial \dot{\mathbf{q}}}({}_{(i)}\tilde{\omega}_{i}{}_{(i)}\tilde{\mathbf{r}}_{SE,i}),$$
$$\mathbf{J}_{R,i} = \mathbf{A}_{i,i-1}\mathbf{J}_{R,i-1} + \frac{\partial_{(i)}\omega_{i-1,i}}{\partial \dot{\mathbf{q}}}. \qquad (7.7)$$

The Jacobians for the end effector are of special importance, because all manipulating processes like assembly processes or contact sequences may conveniently be written in end effector coordinates. From the equations above we get

$$\mathbf{J}_{TG} = \frac{\partial_{(G)}\dot{\mathbf{r}}_G}{\partial \dot{\mathbf{q}}} = -\sum_{i=1}^{n_L}[\prod_{k=1}^{n_L}\mathbf{A}_{k,k-1}]({}_{(i)}\tilde{\mathbf{r}}_{ES,i} + {}_{(i)}\tilde{\mathbf{r}}_{SE,i})\frac{\partial_{(i)}\omega_i}{\partial \dot{\mathbf{q}}} \quad \in \mathbb{R}^{3\times n_f},$$
$$\mathbf{J}_{RG} = \frac{\partial_{(G)}\omega_G}{\partial \dot{\mathbf{q}}} = +\sum_{i=1}^{n_L}[\prod_{k=1}^{n_L}\mathbf{A}_{k,k-1}]\frac{\partial_{(i)}\omega_{i-1,i}}{\partial \dot{\mathbf{q}}} \quad \in \mathbb{R}^{3\times n_f}. \qquad (7.8)$$

The Jacobians $\mathbf{J}_{TG}$ and $\mathbf{J}_{TG}$ project the motion in Cartesian space into the space of the generalized coordinates, equation (7.1), which is not the joint space. To go from the generalized coordinate space into the joint space, we need additionally the transformations

$$\mathbf{J}_L = \left[(\frac{\partial_{(G)}\dot{\mathbf{r}}_G}{\partial \dot{\mathbf{q}}_L})^T, \quad (\frac{\partial_{(G)}\omega_G}{\partial \dot{\mathbf{q}}_L})^T\right]^T \quad \in \mathbb{R}^{6\times n_L}. \qquad (7.9)$$

For robots with six joint degrees of freedom the matrix $\mathbf{J}_L$ is invertible, in all cases with $n_L \neq 6$ we have to solve a classical optimization problem [208], which in the case $n_L < 6$ adapts the final effector position and orientation to the given position and orientation as perfect as possible, and which in the case $n_L > 6$ searches the final position and orientation with an additional optimization criterion. In both cases this can be achieved by classical numerical algorithms.

### 7.2.1.2 Dynamics

We have many possibilities to derive the equations of motion, for example by applying the principles of dynamics or by using Lagrange's equations. As

## 7.2 Trajectory Planning

Lagrange's equations are very popular in robotics, we shall focus on the projection method using the principle of d'Alembert/Lagrange (equation (3.38) on page 102). Applying that to our robot case and also looking at the equations (3.99) on page 116 and the relations (3.106) on page 118 we derive for our serial robot the following sets:

$$\sum_{i=1}^{p}\left\{\begin{pmatrix}B_i(\frac{\partial \dot{\mathbf{r}}_i}{\partial \dot{\bar{\mathbf{q}}}})\\B_i(\frac{\partial \omega_i}{\partial \dot{\bar{\mathbf{q}}}})\end{pmatrix}^T\begin{pmatrix}B_i(m_i\ddot{\mathbf{r}}_i + m_i\tilde{\omega}_i\dot{\mathbf{r}}_i - \mathbf{f}_{ai})\\B_i(\mathbf{I}_{Si}\dot{\omega}_i + \tilde{\omega}_i\mathbf{I}_{Si}\omega_i - \tau_{ai})\end{pmatrix}\right\} = \mathbf{0}. \qquad (7.10)$$

The Jacobians are defined by equation (7.5), and p is the number of links (bodies), $m_i$ the mass of the i-th body, $\mathbf{I}_{Si}$ the inertia tensor with respect to the center of mass, and $\mathbf{f}_{ai}, \tau_{ai}$ are active forces and torques. All evaluations take place in body-fixed coordinates $B_i$, which gives a constant inertia tensor. Following the steps given with the derivation of the equations (3.55) from the equations (3.38) we arrive at the equations of motion expressed by generalized coordinates

$$\mathbf{M}(\bar{\mathbf{q}})\ddot{\bar{\mathbf{q}}} + \mathbf{f}(\bar{\mathbf{q}}, \dot{\bar{\mathbf{q}}}) = \bar{\tau} + \mathbf{B}\tau + \bar{\lambda} \qquad (7.11)$$

with the following magnitudes: $\mathbf{M}(\bar{\mathbf{q}}) \in \mathbb{R}^{n_f,n_f}$ the symmetric mass matrix, $\mathbf{f}(\bar{\mathbf{q}}, \dot{\bar{\mathbf{q}}})$ the vector of all gyroscopic and gravitational forces, $\bar{\tau}$ the forces coming from the motor and gear models, $\mathbf{B}\tau$ the control forces and torques with the control input matrix $\mathbf{B}$, $\lambda$ the contact forces with the environment.

To provide the equations of motion (7.11) with the forces and torques on the right hand side we consider in the following the motor and gear forces, the control forces and the contact forces (and/or torques). For motors and gears we shall regard rigid and elastic models. In the first case the drive has no additional degree of freedom, but is connected to the link by a simple transmission ratio acting as an additional condition between drive and link. In the second case elesticity requires an additional degree of freedom (see Figure 7.3) connected to the link by a spring/damper force element. The corresponding torque writes

$$\tau_{L,j} = c_j\left(\frac{\bar{q}_{M,j}}{i_{G,j}} - \bar{q}_{L,j}\right) + d_j\left(\frac{\dot{\bar{q}}_{M,j}}{i_{G,j}} - \dot{\bar{q}}_{L,j}\right), \qquad (7.12)$$

Where $\bar{q}_{M,j}$ is the additional degree of freedom of the j-th joint drive, $i_{G,j}$ is the gear transmission ratio, $\bar{q}_{L,j}$ is the j-th link degree of freedom, and $c_j, d_j$ are the spring and damper coefficients. The torque acting on the motor shaft is the drive torque $\tau_j$ reduced by the spring/damper torque $\frac{\tau_{L,j}}{i_{G,j}}$ and a friction torque $\tau_{R,j}$. We get

$$\tau_{M,j} = \tau_j - \frac{\tau_{L,j}}{i_{G,j}} - \tau_{R,j} \qquad (7.13)$$

If we consider a rigid motor/gear model, the motion of the motor shaft is then connected to the link degree of freedom by the transmission ratio. Therefore the torques at the motor and the shaft moment of inertia are also transmitted to the link by the transmission ratio. We get

$$I_{ML,j} = i_{G,j}^2 I_{M,j}, \qquad \tau_{L,j} = i_{G,j}(\tau_j - \tau_{R,j}), \qquad (j = n_M+1, \cdots, n_L). \quad (7.14)$$

Inspite of the fact, that these formulas are well known, a remark may be allowed with respect to the nonlinear dynamic behaviour of robots. The transmission ratios of the joints are sometimes so large, that the moments of inertia of the motor shafts and not those of the links dominate the dynamics leading to a more or less linear system behaviour, which allows a simple linear and decentralized control.

The friction in the joints depends on the type of the drive and has to be evaluated for each robot configuration individually. We can only dicuss a typical example, which has been treated theoretically and experimentally in [107]. Figure 7.5 depicts such a typical friction characteristic, which can be

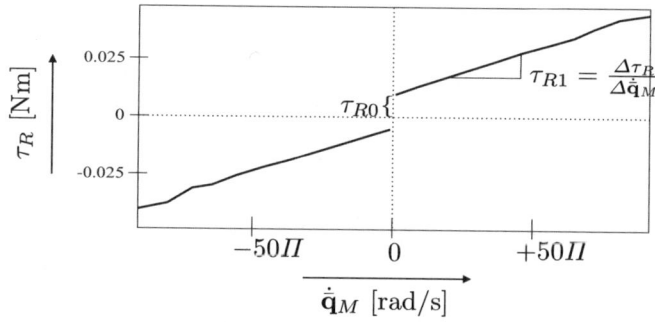

Fig. 7.5: Measured Friction Characteristic of a Robot [107]

easily approximated by the relation

$$\tau_{R,j} = -\tau_{R0,j}\mathrm{sgn}(\dot{\bar{q}}_{M,j}) - \tau_{R1,j}\dot{\bar{q}}_{M,j}, \qquad (7.15)$$

where $\tau_{R0,j}$ is the amount of the constant dry fiction and $\tau_{R1,j}$ the slope of the friction characteristic for the motor shaft. The knowledge of the friction behaviour is also necessary for a compensation by control.

From the equations (7.12) to (7.15) we can then evaluate the torque vector $\bar{\tau}$ and the control input matrix **B**. In a first step we get

$$\bar{\tau} = \Big[(-\frac{\tau_{L,1}}{i_{G,1}} - \tau_{R,1}), \cdots, (-\frac{\tau_{L,n_M}}{i_{G,n_M}} - \tau_{R,n_M}), \vdots \\ \tau_{L,1}, \cdots, \tau_{L,n_M}, -\tau_{R,n_M+1}, \cdots, -\tau_{R,n_L}\Big]^T \quad (7.16)$$

For an elastic drive model the drive torque $\tau_j$ influences directly the drive coordinate $\bar{q}_{M,j}$, whereas for a rigid drive model the torque works on the link coordinates by a transmission matrix $\mathbf{I}_G$. It consists of the matrix $\mathbf{I}_{G,M} \in \mathbb{R}^{n_M,n_M}$ containing the gear ratios and of the matrix $\mathbf{I}_{G,S} \in \mathbb{R}^{n_L-n_M,n_L-n_M}$ containing the rigid joint influence. Together it writes

$$\mathbf{I}_G = \begin{pmatrix} \mathbf{I}_{G,M} & \mathbf{0} \\ \mathbf{0} & \mathbf{I}_{G,S} \end{pmatrix} \in \mathbb{R}^{n_L,n_L} \tag{7.17}$$

From this we get the control input matrix $\mathbf{B}$ in the form

$$\mathbf{B} = \begin{pmatrix} \mathbf{E}_{n_M} & \mathbf{0} \\ \mathbf{0} & \mathbf{0} \\ \mathbf{0} & \mathbf{I}_{G,S} \end{pmatrix} \in \mathbb{R}^{n_f,n_L} \tag{7.18}$$

Many commercial robot control systems apply single joint control subsystems, which work in most cases very well due to the above presented arguments. These controllers usually are simple PD-controllers of the form

$$\tau = -\mathbf{K}_p(\mathbf{B}^T\bar{\mathbf{q}} - \mathbf{I}_G\bar{\mathbf{q}}_S) - \mathbf{K}_d(\mathbf{B}^T\dot{\bar{\mathbf{q}}} - \mathbf{I}_G\dot{\bar{\mathbf{q}}}_S), \tag{7.19}$$

the diagonal matrices $\mathbf{K}_p$ and $\mathbf{K}_d$ contain as PD-parts the control stiffness and damping parameters as an input, and the $\bar{\mathbf{q}}_S$ and $\dot{\bar{\mathbf{q}}}_S$ are the link coordinates and velocities as nominal or desired magnitudes. Usually the matrices $\mathbf{K}_p$ and $\mathbf{K}_d$ can be chosen externally, but nothing else. We come back to that with respect to assembly processes.

Before the decision of how to model contacts at the end effector we have to take into consideration given forces and torques at the gripper, which in most cases of practical relevancy are prescribed by the manipulation process, whatsoever. At this point we must merge the robot and the process dynamics. Anyway, some given forces and torques at the gripper $\mathbf{f}_G \in \mathbb{R}^6$ has to be transformed into the space of the generalized coordinates resulting in the $\bar{\lambda}$-value of equation (7.11). It is

$$\bar{\lambda} = \left[ (\frac{\partial \dot{\mathbf{r}}_G}{\partial \dot{\bar{\mathbf{q}}}})^T (\frac{\partial \omega_G}{\partial \dot{\bar{\mathbf{q}}}})^T \right] \mathbf{f}_G = [\mathbf{J}_{TG}^T \mathbf{J}_{RG}^T] \mathbf{f}_G. \tag{7.20}$$

With these relations all equations are available for describing robot dynamics. We shall use the above forms for the dynamics of assembly processes [219] and in a slightly modified form for path planning considerations [117], [223].

### 7.2.1.3 Solution Variants

We start again with the equations of motion (7.10) and write them in a bit more compact form

$$\mathbf{Q}^T \left[ \begin{pmatrix} \mathbf{M}^* & \mathbf{0} \\ \mathbf{0} & \mathbf{I} \end{pmatrix} \cdot \begin{pmatrix} \ddot{\mathbf{r}} \\ \dot{\omega} \end{pmatrix} + \begin{pmatrix} \mathbf{M}^*\tilde{\omega}\dot{\mathbf{r}} \\ \tilde{\omega}\mathbf{I}\omega \end{pmatrix} - \begin{pmatrix} \mathbf{f}_a \\ \tau_a \end{pmatrix} \right] = \mathbf{0}, \tag{7.21}$$

where the abbreviations are obvious by comparing the equations (7.21) and (7.10). We have generally several possibilities to use these equations with respect to certain tasks of robot performance. Firstly, we can solve the equations of motion directly, which means, given the forces and the torques $\mathbf{f}_a, \tau_a$, solve the equations for getting the motion of the robot. Secondly, we might consider the problem inverse to the first one, namely given the motion, that is the motion kinematics, determine the forces and torques necessary to generate this given motion. And thirdly, but not finally, we want to solve the problem of optimizing a trajectory from one point to another one with regard to certain criteria like time, energy or joint torques under side conditions like the robot configuration, its joint motors, its torque capabilities, collision avoidance and others. We shall consider the last problem in a special section with special assumptions.

The first task for given torques of the joint reduces to an integration of the relation (7.21), which due to the properties of the magnitudes involved is a simple initial value problem, always solvable unambiguously. We remember that (see equation 7.5)

$$\dot{\mathbf{q}} = \begin{pmatrix} \dot{\mathbf{r}} \\ \omega \end{pmatrix} \in \mathbb{R}^{n_f},$$

$$\mathbf{M} = \mathbf{Q}^T \begin{pmatrix} \mathbf{M}^* & \mathbf{0} \\ \mathbf{0} & \mathbf{I} \end{pmatrix} \mathbf{Q} \in \mathbb{R}^{n_f, n_f} \quad \wedge \quad (\bar{\mathbf{q}}^T \mathbf{M} \bar{\mathbf{q}} > 0), \tag{7.22}$$

which confirms the above statement. Many standard algorithms exist to solve this problem [283].

The inverse problem for given kinematics of the robot but unknown forces or torques in the joints is a much more complicated problem, at least for general cases. But it is a very important case, because all manufacturing processes require a certain trajectory behaviour for the realization of a process, and this behaviour has to be performed by the robot, which means solving an inverse kinematical and dynamical problem. In these cases we have a certain amount of unknown forces and torques $\mathbf{f}_u, \tau_u$ and a certain amount of known forces and torques $\mathbf{f}_k, \tau_k$, so that

$$\begin{pmatrix} \mathbf{f}_a \\ \tau_a \end{pmatrix} = \begin{pmatrix} \mathbf{f}_u \\ \tau_u \end{pmatrix} + \begin{pmatrix} \mathbf{f}_k \\ \tau_k \end{pmatrix}, \tag{7.23}$$

and equation (7.21) can be split up accordingly

$$\mathbf{Q}^T \begin{pmatrix} \mathbf{f}_u \\ \tau_u \end{pmatrix} = \mathbf{Q}^T \left[ \begin{pmatrix} \mathbf{M}^* & \mathbf{0} \\ \mathbf{0} & \mathbf{I} \end{pmatrix} \cdot \begin{pmatrix} \ddot{\mathbf{r}} \\ \dot{\omega} \end{pmatrix} + \begin{pmatrix} \mathbf{M}^* \tilde{\omega} \dot{\mathbf{r}} \\ \tilde{\omega} \mathbf{I} \omega \end{pmatrix} - \begin{pmatrix} \mathbf{f}_k \\ \tau_k \end{pmatrix} \right]. \tag{7.24}$$

In many cases and especially for robots with a serial structure each joint corresponds to a degree of freedom, and threrefore the number of joints corresponds exactly to the number of unknown forces and torques. Then the solution of equation (7.24) is straightforward and for given robot motion no problem.

But if these two numbers do not agree, for underactuated robots for example or for the case that equation (7.23) does not apply, which is possible, then the inverse problem should be solved by adding an optimization to the whole equation structure. Having less unkown forces and torques than equations, we may require that the solution is as near as possible to the desired motion resulting in a classical optimization with equality and inequality side conditions, or having more unknown forces and torques as covered by the equations of motion opens the opportunity to apply additional criteria which might be of interest for the robot motion under consideration. Also this gives a classical optimization, in that case with several possible solutions. We shall not go deeper into these very interesting problems (see for example [3], [208]), but concentrate on a specific solution of the path planning problem [117], [223].

### 7.2.2 Parametric Path Planning

#### 7.2.2.1 Models

With respect to a parametric representation of the equations of motion we refer to chapter 2.2.6 on the pages 31, and for robots we refer to [3] and [208]. With respect to the equations of motion we may start with the relations (7.21), or as an alternative we may use Lagrange's equations (3.76) on page 109. We shall go the last way utilizing the kinetic energy

$$T = \frac{1}{2}\dot{\mathbf{q}}^T \left[ \mathbf{Q}^T \begin{pmatrix} \mathbf{M}^* & \mathbf{0} \\ \mathbf{0} & \mathbf{I} \end{pmatrix} \cdot \mathbf{Q} \right] \dot{\mathbf{q}} = \frac{1}{2}\dot{\mathbf{q}}^T \mathbf{M}\dot{\mathbf{q}}, \quad (7.25)$$

and come out with another form of the equations of motion, typical also for robots (see [93], [117], [223], [208], [205], [186]):

$$\sum_i M_{ij}\ddot{q}_j + \sum_i \sum_k \begin{bmatrix} j & k \\ i & \end{bmatrix} \dot{q}_i \dot{q}_k + C_i = T_i, \quad (i,j,k = 1,2,\cdots,f), \quad (7.26)$$

with the Christoffel symbols [283]

$$\begin{bmatrix} j & k \\ i & \end{bmatrix} = \frac{1}{2}\left(\frac{\partial M_{ik}}{\partial q_j} + \frac{\partial M_{ij}}{\partial q_k} - \frac{\partial M_{jk}}{\partial q_i}\right), \quad (7.27)$$

which represent a tensor-like structure without having the transformation properties of a tensor. The masses $M_{ij}$ are the components of the symmetric mass matrix $\mathbf{M} \in \mathbb{R}^{f,f}$, the $q_j$ are components of the vector $\mathbf{q} \in \mathbb{R}^f$, the terms $C_i$ include all external forces or torques and the $T_i$ all joint forces or torques.

The equations of motion (7.26) include the generalized acceleration in linear form and the generalized velocities in a quadratic form. Considering the classical methods of nonlinear, smooth dynamics (see for example [148], [27], [125], [187]) we may express these kinematical magnitudes by a parameter s in the following way:

$$\dot{q}_j = q'_j \dot{s}, \qquad \ddot{q}_j = q'_j \ddot{s} + q''_j \dot{s}^2, \quad \text{with} \quad \dot{q}_j = \frac{dq_j}{dt}, \; q'_j = \frac{dq_j}{ds}, \; \dot{s} = \frac{ds}{dt}. \qquad (7.28)$$

This form of parameterization can easily be interpreted as the path coordinate of a prescribed trajectory, which, together with the, certainly idealized, assumption, that the end effector of the robot should track the given path in an ideal and perfect manner, results in a system with one degree of freedom only, namly this path coordinate, independent of the number of joints and the structure of the robot [117], [186]. Figure 7.6 illustrates the situation. A given

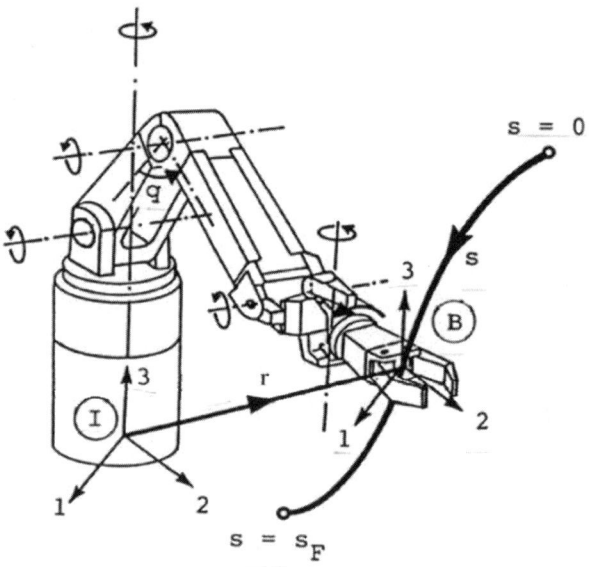

Fig. 7.6: Prescribed Trajectory for a Robot

path starts at $s = 0$ and runs with the path coordinate $s$ to the end point $s = s_F$. The joint coordinates $q_j$ of the robot have to be adapted to the path in a way, so that the end effector coordinate system is exactly located on the trajectory. Applying now the formulas (7.28) to the equations of motion we come out with

$$T_i = A_i(s)(\dot{s}^2)' + B_i(s)(\dot{s}^2) + C_i(s), \qquad (i = 1, 2, \cdots, f)$$

$$A_i(s) = \frac{1}{2} \sum_j M_{ij} q'_j, \qquad B_i(s) = \sum_j M_{ij} q''_j + \sum_j \sum_k \begin{bmatrix} j & k \\ & i \end{bmatrix} q_j q_k,$$

$$C_i(s) = \frac{\partial V}{\partial q_i}, \qquad (7.29)$$

where we have introduced a gravitational potential V, which might of course be extended to additional potential forces. Equation (7.29) represents a linear first-order differential equation, linear with respect to ($\dot{s}^2$), but highly nonlinear considering the coefficients $A_i, B_i, C_i$, which depend on the path coordinate s. It should be noted, that beyond the assumption of an ideally tracked path no further neglections have been made. Also, the structure of the equation (7.29) would be preserved if we add any addional forces or torques depending only on s.

A formal integration of equation 7.29 yields

$$\dot{s}^2(s) = \left[\dot{s}^2(s_0) + \int_{s_0}^{s} \exp\left(\int_{s_0}^{v}(\frac{B_i}{A_i})du\right) \cdot (\frac{T_i - C_i}{A_i})dv\right] \exp\left(\int_{s_0}^{s}(\frac{B_i}{A_i})du\right). \tag{7.30}$$

For achieving optimal solutions for the path planning problem we have more elegant ways than using the relation (7.30), though it might be helpful for special cases. We represent the given path by a vector **r** originating from an inertial system and ending in some path point (Figure 7.6). Such a trajectory point depends on the joint coordinates $\mathbf{q}^T = (q_1, q_2, \cdots, q_f)$ or $\mathbf{r} = \mathbf{r}(\mathbf{q})$, so that the derivatives with respect to s result in

$$\mathbf{q}' = \left\{\left(\frac{\partial \mathbf{r}}{\partial \mathbf{q}}\right)^T \left[\left(\frac{\partial \mathbf{r}}{\partial \mathbf{q}}\right)\left(\frac{\partial \mathbf{r}}{\partial \mathbf{q}}\right)^T\right]^{-1}\right\} \cdot \mathbf{r}'$$

$$\mathbf{q}'' = \left\{\left(\frac{\partial \mathbf{r}}{\partial \mathbf{q}}\right)^T \left[\left(\frac{\partial \mathbf{r}}{\partial \mathbf{q}}\right)\left(\frac{\partial \mathbf{r}}{\partial \mathbf{q}}\right)^T\right]^{-1}\right\} \cdot \left[\mathbf{r}'' - \left(\frac{\partial \mathbf{r}}{\partial \mathbf{q}}\right)\mathbf{q}'\right]. \tag{7.31}$$

The above equations anticipate the fact that in the following we consider motion along the trajectory only and do not take into account the attitude behaviour of the hand, for example. The latter case would demand inclusion of directional terms into equation(7.31). Considering trajectory motion alone, the first equation of (7.31) already includes an adjustment in a least-squares sense for the case of more joints than necessary for a three-dimensional path, resulting always in several solutions which might be selected according to some criteria.

### 7.2.2.2 Time-Minimum Trajectories

The general optimization problem for a manipulator following a prescribed path is the following: evaluate $\dot{s}(t)$ with $s(0) = 0, s(t_F) = s_F, \dot{s}(0) = 0, \dot{s}(t_F) = 0$ in such a way that the performance criterion

$$G = \int_0^{t_F} g[s, \dot{s}^2, (\dot{s}^2)']dt \quad \Rightarrow \quad \text{extr.!} \tag{7.32}$$

has an extremum. At the end of the path the time t achieves its final value $t_F$. If we minimize only the time, we have g=1 and the criterion

$$G = \int_0^{t_F} dt = \int_0^{s_F} \frac{ds}{|\dot{s}|} \quad \Rightarrow \quad \text{min!} \qquad (7.33)$$

The solution of this problem is of course constrained. The most important constraints are the following:

- The joint torques or forces are limited, which means

$$T_{i,min} \leq [A_i(s)(\dot{s}^2)' + B_i(s)(\dot{s}^2) + C_i(s)] \leq T_{i,max}. \qquad (7.34)$$

  For many applications we shall assume $T_{i,min} = -T_{i,max}$.

- The joint angular or translational velocities may be limited due to some maximum speeds of the drive train components,

$$\dot{q}_{i,min} \leq [q'_i \dot{s}] \leq \dot{q}_{i,max}. \qquad (7.35)$$

  Again, we shall assume $\dot{q}_{i,min} = -\dot{q}_{i,max}$.

- The path velocity itself might become constrained by some manufacturing process

$$-v_{max} \leq |\mathbf{r}'|\dot{s} \leq +v_{max}, \qquad (7.36)$$

  with the vector $\mathbf{r}$ from Figure 7.6.

The relations (7.35) and (7.36) define together a maximum velocity $\dot{s}_G$ along the path which must not be exceeded:

$$0 \leq (\dot{s}^2) \leq (\dot{s}^2)_G, \quad \text{with} \quad (\dot{s}^2)_G = \min\left[(\frac{\dot{q}_{i,max}}{q'_i})^2, \ (\frac{v_{max}}{|\mathbf{r}'|})^2\right]. \qquad (7.37)$$

The solution of the time-optimal problem given with the relations (7.33) to (7.37) may be constructed in the following manner: We look at the equations (7.34) for the limiting cases $(T_{i,min}, T_{i,max})$ stating, that minimum time can only be achieved by applying in a maximum number of joints the limiting torques or forces. In operating at the power limits we get the two equations

$$(\dot{s}^2)'_{max} = \frac{-B_i(s)(\dot{s}^2) - C_i(s) + T_{i,max}}{A_i(s)},$$
$$(\dot{s}^2)'_{min} = \frac{-B_i(s)(\dot{s}^2) - C_i(s) - T_{i,min}}{A_i(s)}, \quad (i = 1, 2, \cdots, f), \qquad (7.38)$$

which define two straight lines in the $[(\dot{s}^2)', (\dot{s}^2)]$-plane. The altogether (2f) straight lines form a polygon confined at the left side by the axis $(\dot{s}^2) = 0$

and at the right side by the $(T_{i,min}, T_{i,max})$-straight-lines or by the constraint $(\dot{s}^2)_G$ given by equation (7.37). It should be noted that $A_i(s)$ might become zero generating then a vertical line in the $[(\dot{s}^2)', (\dot{s}^2)]$-plane.

Without violating the constraints, which may by extended without much problems, motion can take place only within or on the polygons as shown in Figure 7.7, where the situation for one path point s is illustrated. These

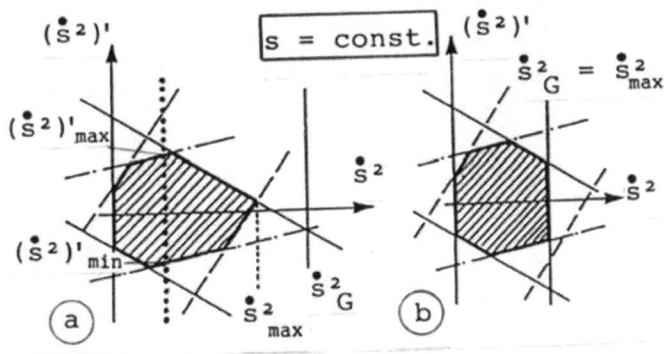

Fig. 7.7: Polygons of Allowed Motion:  $s = const.$,  a)$\dot{s}^2_{max} \leq \dot{s}^2_G$,  b)$\dot{s}^2_{max} = \dot{s}^2_G$

graphs contain the following informations:

Firstly, we immediately obtain the maximum possible velocity $\dot{s}^2_{max}$ for this path point. This maximum velocity may either result from an intersection of the torque-straight-lines as in the left side of Figure 7.7, or it might be a constraint velocity $\dot{s}^2_G$ in the form of (also) a vertical line, as shown by the right part of Figure 7.7.

Secondly, for every path velocity $\dot{s}^2$ smaller than the maximum velocity $\dot{s}^2_{max}$ we can choose two values of the derivative $(\dot{s}^2)'$ from the boundaries of the polygon, which gives us the maximum and minimum possible acceleration, possibly also deceleration, at the path point s under consideration. By means of these extremum values $((\dot{s}^2)'_{min}, (\dot{s}^2)'_{max})$ we are able to construct, in the phase plane $(\dot{s}, s)$, the curves of extreme acceleration or deceleration. We call these curves extremals.

Figure 7.8 illustrates the properties of these extremals for a simple example with constraints for the joint torques only but not with constraints for the velocities. Combining all polygons for all path points s we obtain a constrained phase space bounded by ruled surfaces due to the straight line characteristics of the polygons. These polygons appear as plain cuts perpendicular to the s-axis. The phase plane $\dot{s}(s)$ on the other hand is a vertical plane projection of the constraint phase space on a plane parallel to the $(\dot{s}^2 - s)$-plane with

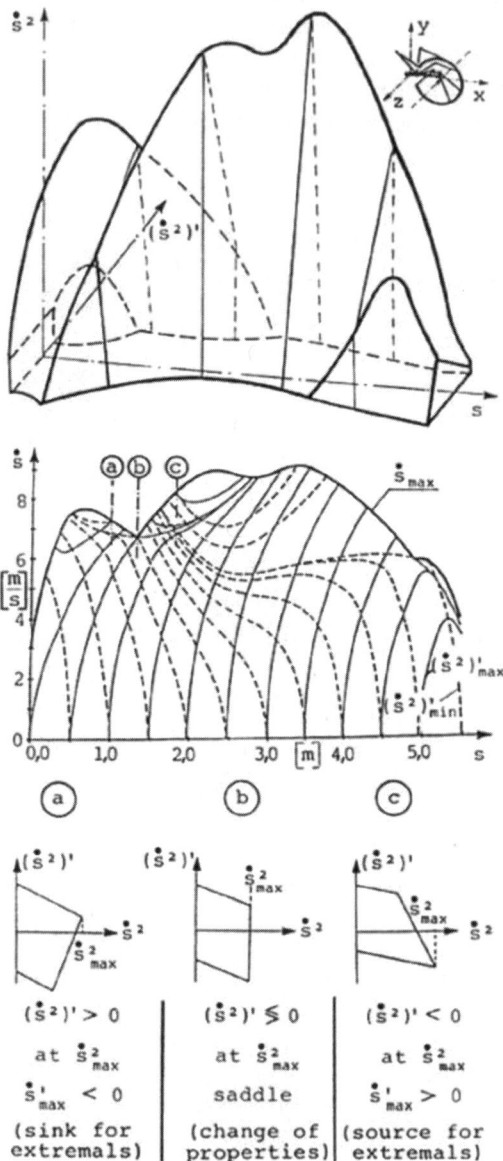

Fig. 7.8: Example Two-Link Robot on a Circular Path, (top) Constrained Phase Space, (middle) Phase Plane with Extremal Field and Boundary of Maximum Velocity

stretched or compressed coordinates due to $\dot{s} = \sqrt{\dot{s}^2}$. At every path point the minimum time solution requires maximum possible acceleration or deceleration. From this we should move with our solution along the boundary of the constrained phase space.

If we move along the constraint space boundary, we must have sometimes the possibility to go from one side of the "hill" to the other one; for example, tracing our solution on the side with $(\dot{s}^2)'_{max}$ we have to look for points, where a transition to the side with $(\dot{s}^2)'_{min}$ will be possible. This maneuver is only possible at saddle points, for example the point ⓑ in Figure 7.8. A saddle exhibits interesting properties. It is generated by equation (7.38) at the points with $A_j(s_S) = 0$, where we consequently get a vetical line bounding the polygon and defining the value of $\dot{s}^2 = \dot{s}^2_{max}$ (point ⓑ of Figure 7.8). The projection of a saddle onto the $(\dot{s}^2 - s)$-plane and from there onto the $(\dot{s} - s)$-plane produces a "critical point", the properties of which follow from the saddle properties. As the minimum-time solution has to pass the saddle, it also has to pass the assigned critical point as well. This necessary and basic requirement agrees with the intuitive idea, that the minimum-time solution must be as near as possible to the boundary curve of the $(\dot{s} - s)$-plane.

A special form of the saddle may appear for $A_j(s) = 0$ in combination with the result that additionally $(\dot{s}^2)'_{max} - (\dot{s}^2)'_{min} = 0$. For this case the vertical boundary line of Figure 7.8b at the point $\dot{s}^2 = \dot{s}^2_{max}$ degenerates to a point, which then marks the change from ⓐ to ⓒ in Figure 7.8. We do not have a jump of the derivative of $\dot{s}^2_{max}$ at that point but a smooth change from negative to positive slopes. In the constraint phase space picture, the saddle degenerates to a deepening of the ridge, something like a kind of a notch. Another special point might appear during the constraint space construction, where all slopes of the velocity curves and the extremal curves coincide. But this type of points does not belong to the time-optimal solution.

The geometrical interpretation of the parameterized equations of motion (7.26) gives us some further insights into the dynamics structure of such systems. The $(\dot{s}^2)'_{max}$ extremals at the positive $(\dot{s}^2)'$ side of the "hill" approach their appropriate ridge from the left and disappear. We shall call a ridge with this property a trajectory-sink. Ridges of this type possess usually a descending character. On the other hand, the $(\dot{s}^2)'_{min}$ extremals at the negative $(\dot{s}^2)'$ side of the "hill" originate from those parts of the ridge with a more or less ascending character. We call these arcs trajectory-sources. More generally, each part of the boundary curve with a sink character may be a sink for $(\dot{s}^2)'_{max}$ extremals as well as for $(\dot{s}^2)'_{min}$ extremals. The same is true for the source parts of the $\dot{s}_{max}$ curve: They may be the origin of both types of extremals. This classification can easily be verified by considering the derivative $(\dot{s}^2)'_{B,max}$ of the boundary curve itself and the derivatives $(\dot{s}^2)'_E$ of the extremals directly at this boundary curve. The $\dot{s}_{max}(s)$-curve possesses sink character if its own derivative $(\dot{s}^2)'_{B,max}$ is smaller than the minimum slopes $(\dot{s}^2)'_{E,min}$ of the extremals disappearing in it. It is a source if its own derivative $(\dot{s}^2)'_{B,max}$ is

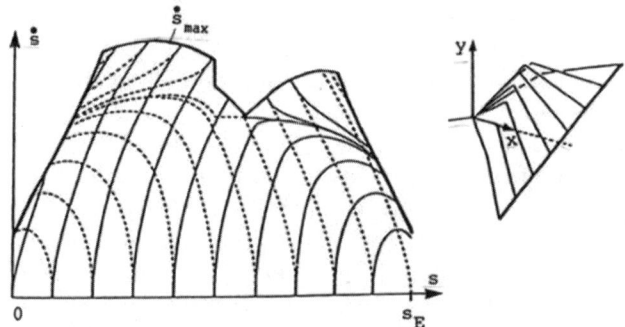

Fig. 7.9: Properties of the Boundary Curve: – – – – – Trajectory Source (eqn.()), – · – · – · – Trajectory Sink (eqn.()), ——— Allowed for Trajectories (eqn.())

larger than the maximum slopes $(\dot{s}^2)'_{E,max}$ of the extremals originating from it. From this we find the following properties of the boundary curve $\dot{s} = \dot{s}_{max}$:

- trajectory sink parts with the property

$$(\dot{s}^2)'_{max} < [(\dot{s}^2)'_{E,min}]_{(\dot{s}^2=\dot{s}^2_{max})}, \tag{7.39}$$

- trajectory source parts with the property

$$(\dot{s}^2)'_{max} > [(\dot{s}^2)'_{E,max}]_{(\dot{s}^2=\dot{s}^2_{max})}, \tag{7.40}$$

- arcs being allowed for trajectories due to special constraints (equations (7.37) and Figure 7.8b ); here we have

$$[(\dot{s}^2)'_{E,min}]_{(\dot{s}^2=\dot{s}^2_{max})} < (\dot{s}^2)'_{B,max} < [(\dot{s}^2)'_{E,max}]_{(\dot{s}^2=\dot{s}^2_{max})} \tag{7.41}$$

Each boundary curve can be partitioned into these three regimes. Figure 7.9 depicts an example of a three-link robot tracking a straight line.

With these preparations we are able to construct the minimum-time solution applying the procedure below (for illustration see also Figure 7.10):

- Start at $s = 0$ and $\dot{s} = 0$, evaluate an extremal with maximum acceleration, follow it to the end of the trajectory or to the point where it disappears at a trajectory sink (Figure 7.10a).
- Start at $s = s_F$ and $\dot{s} = 0$ and track an extremal with maximum deceleration until it meets the acceleration extremal starting from $s = 0$ or until it intersects its originating point at the $\dot{s}_{max}$-curve. If both extremals intersect, the minimum-time solution is obtained; otherwise continue (Figure 7.10a).

- If the acceleration and deceleration extremals both meet the boundary curve $\dot{s} = \dot{s}_{max}$, there must be at least one critical point (saddle point) between these two boundary intersections, because one of them is necessarily a sink, the other one a source, see the arguments above. Search for the critical point starting from the first intersection at a trajectory sink.

- From the first critical point $s = s_{c(1)}$ trace backward an extremal with minimum slope until it intersects with the acceleration extremal. This intersection must exist, because there is no further critical point between $s = 0$ and the first one at $s = s_{c(1)}$; or more generally, between $s = s_{c(j)}$ and $s = s_{c(j-1)}$.

- Trace forward an extremal with maximum slope until it meets the deceleration extremal starting at $s = s_F$ or until it vanishes at a boundary point with sink character. In the first case the calculation is finished, in the second case at least one more critical point must exist, requiring a new start at the third step.

Figure 7.10(b) gives an example of the last two steps, Figure 7.10(c) shows the minimum-time solution, and 7.10(d) depicts the corresponding joint torques.

### 7.2.2.3 Dynamic Programming Approach

The geometrical procedure discussed so far can only be applied to time-minimum trajectories, but not to other criteria like energy, joint torqes or the like. If we consider for example time and torques, we would need a performance criterion

$$g = w_1 + w_2 \sum_i c_i [A_i(s)(\dot{s}^2)' + B_i(s)(\dot{s}^2) + C_i(s)]^2, \qquad (7.42)$$

which would be one possible form. The solution of such problems requires a completely different approach, which may be taken from classical optimization theory, for example [179], [16]. We shall shortly discuss the application of Bellman's dynamic programming theory [17], which is based on the well known principle of optimality. The practical problem with respect to the application of this method consists, still even nowadays, in large computing time and large computer storage, at least for problems of higher dimensions. Due to our projection of the robot motion onto the path coordinate resulting in one degree of freedom only, the equations of motion allow to establish a dynamic programming approach with dimension one only, which does not involve the computer problems as mentioned above. As the Bellman method presupposes a formulation of the optimization problem as a stage process, we first partition the trajectory into (n+1) discrete points, where the optimal solution will be evaluated:

$$\dot{s}_i^2 = \dot{s}^2(s_i), \qquad (\dot{s}_i^2)' = \frac{\dot{s}_i^2 - \dot{s}_{i-1}^2}{d}, \qquad s_i = id = \frac{i}{n} s_F \qquad d = \frac{s_F}{n} \qquad (7.43)$$

430     7 Robotics

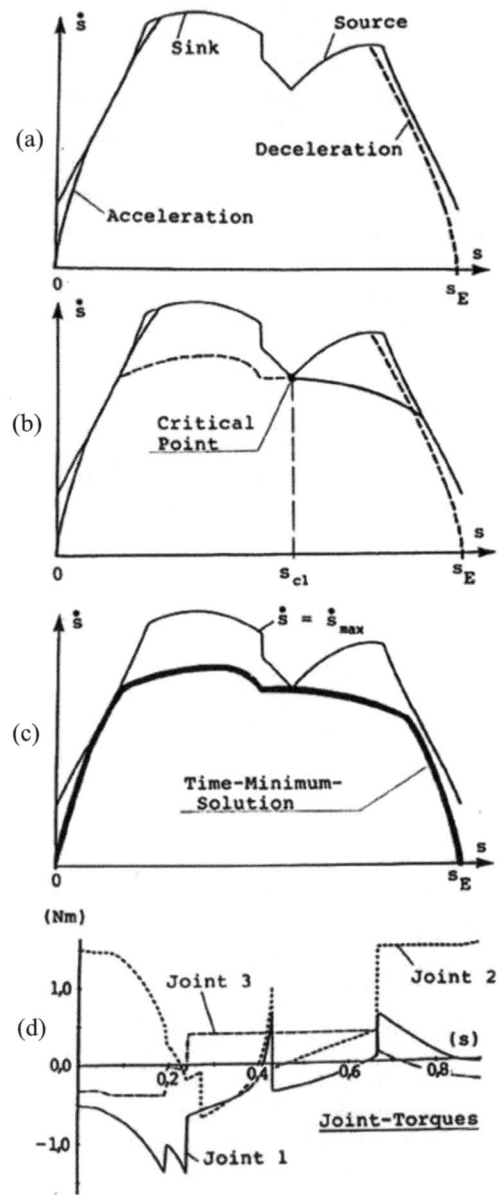

Fig. 7.10: Construction of the Time-Optimal Solution for the Case of Figure 7.9

## 7.2 Trajectory Planning

The performance criterion equation (7.32) can easily be subdivided into stages by forming the sum

$$G = G_n = \sum_{j=1}^{n} \int_{t_{j-1}}^{t_j} g[s, \dot{s}^2, (\dot{s}^2)']dt, \qquad (7.44)$$

which gives us with $dt = ds/|\dot{s}|$ the performance criterion at stage i

$$G_i = G_{i+1} + \Delta G_i$$

$$G_{i+1} = \sum_{j=i+1}^{n} \int_{s_{j-1}}^{s_j} g[s, \dot{s}^2, (\dot{s}^2)']\frac{ds}{|\dot{s}|}, \qquad \Delta G_i = \int_{s_i}^{s_{i+1}} g[s, \dot{s}^2, (\dot{s}^2)']\frac{ds}{|\dot{s}|}. \qquad (7.45)$$

Figure 7.11 illustrates such a stage process [17]. The famous Bellman principle of optimality says [16]: "An optimal policy has the property, that whatever the initial state and initial decision are, the remaining decisions must constitute

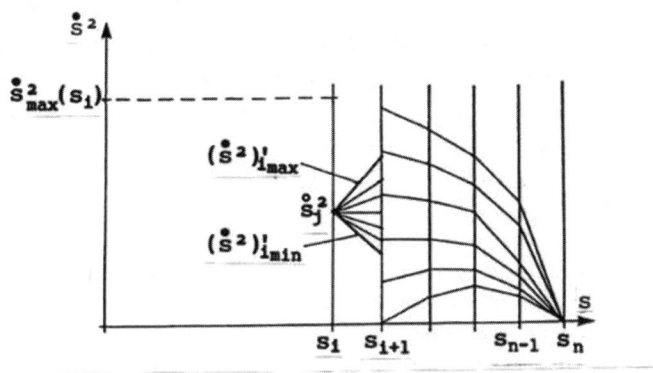

Fig. 7.11: Dynamic Programming Stage Process [17]

an optimal policy with regard to the state resulting from the first decision." Applying this principle to our problem allows us to construct an optimal solution in the following way (Figure 7.11).

We start at the end of the path at $s = s_F = s_n$ and proceed backwards to $s = s_0 = 0$ step by step. As a first step we discretize the velocity at every path point $s_i$ into (m+1) values $\dot{s}_j^2(s_i)$ with $j \in (1, m+1)$ and $0 \leq \dot{s}_j^2(s_i) \leq \dot{s}_{max}^2(s_i)$. This method guarantees that the motion remains in the constrained regime defined by the polygons of Figure 7.7. According to the velocity discretization, we define $G_j \to G_j(s_i) = G_j[s_i, \dot{s}_j^2, (\dot{s}_j^2)']$. We start at

$s = s_{n-1}$ and establish a first optimal table, which contains the magnitudes $s_{n-1}$, $\dot{s}_j^2(s_{n-1})$, $(\dot{s}_j^2)'(s_{n-1})$, $G_j(s_{n-1})$ taking into account that $G_j(s_n) = 0$ and $\dot{s}_j^2(s_n) = 0$. Additional care has to be taken, that the derivatives do not violate the boundary conditions (7.41) of the polygons, see also the Figures 7.7 and 7.11.

Suppose we know already the optimal table at point $s = s_{j+1}$ (Figure 7.11). Within the allowed regimes we then establish the table at the point $s_j$ for all possible $\dot{s}_i^2, (\dot{s}_i^2)'$. For each point $\dot{s}_j^2(s_i)$ with $0 < \dot{s}_j^2(s_i) < \dot{s}_{max}^2(s_i)$ we retain only one value $(\dot{s}_{opt}^2)'$, which gives us an extremum value for $G_j(s_i)$ according to equation (7.45). Additionally we have to store the value of the forward velocity point $\dot{s}_k^2(s_{i+1})$, which produces the optimal $(\dot{s}_{opt}^2)'$ and thus the $G_j(s_i)$. We complete the table by performing the above process for all allowed $\dot{s}_j^2(s_i)$. The final table then contains the following magnitudes: $s_i$, $\dot{s}_j^2(s_i)$, $(\dot{s}_j^2)'(s_i)$, $G_j(s_i)$, $k_j(s_{i+1})$. The last magnitude defines the optimal connection between $\dot{s}_j^2(s_i)$ and $\dot{s}_k^2(s_{i+1})$. The index $j \in (1, J)$ corresponds to the number of dicrete points in the regime $0 < \dot{s}_j^2(s_i) < \dot{s}_{max}^2(s_i)$.

Having reached the second point $s = s_1$ we come to $s = s_0$ by adding to the performance criterion $G_0$ the last $\Delta G$ in such a way, that at the final point of this process or the first point of the trajectory the condition $\dot{s}^2(s_0) = 0$ is assured. It should be noted, that the described process can be evaluated the same manner if we prescibe at the first and/or the final point some velocity $\dot{s}^2 \neq 0$. The last table at $s = s_0$ contains the values of the complete performance criterion summed up over all stages. We choose the extremum value of G and follow then the way back from $s = s_0$ to $s = s_F = s_n$ through all optimal tables with the help of the optimal index list $k_j(s_{i+1})$. The result is the optimal trajectory.

The Bellman method represents a very powerful tool insofar as it offers the possibility to include any boundary conditions and any constraints in a very easy way. A further important advantage of the concept consists in the fact that the differential equations of motion enters into the process only by using the boundaries of the polygons (Figure 7.7), thus assuring that torque and velocity constraints are at no point violated. After the evaluation of the optimal tables the equations of motion are used to calculate forces and torques.

### 7.2.2.4 Results

The above methods have been applied to many examples of practical relevancy. The results confirm the theory and the uniqueness of the solutions, even for complicated trajectory configurations. For some cases laboratory experiments have been performed to verify the simulations. We give two examples.

Figure 7.12 shows a double-parabola trajectory performed by a robot with revolute joints. The parabola lies in a plane parallel to the y-z-plane, the manipulator possesses one vertical and two horizontal axes. The extremal

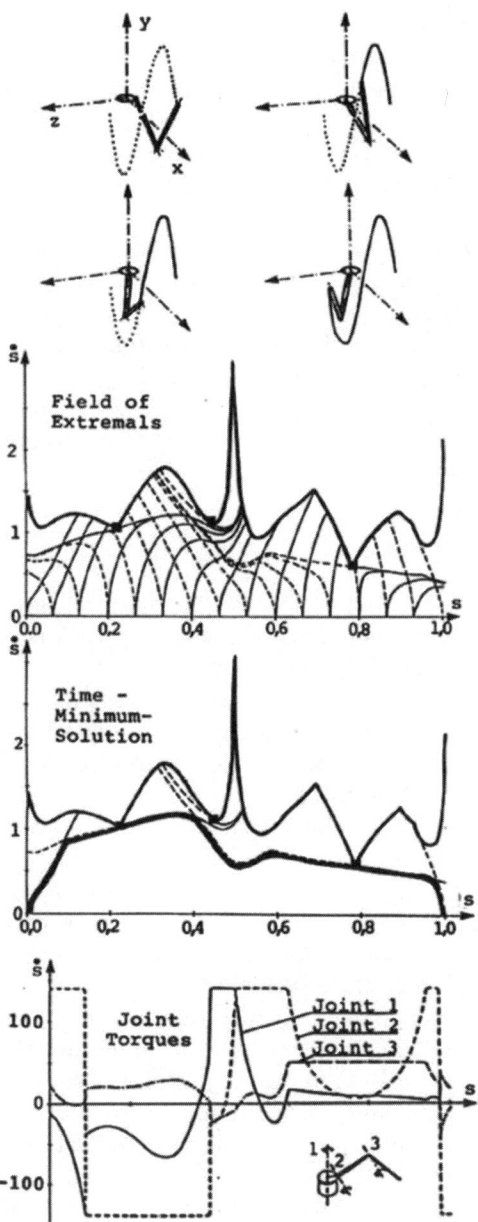

Fig. 7.12: Simulation of a Double Parabola Trajectory for Three-Link Robot, (from top to bottom: Trajectory, Extremal Field, Time-Minimum Solution, Joint Torques)

field is rather complicated with two critical points indicating two saddles of the phase space configuration. A third special point can be seen at $s \approx 0.45$, which is a point where the derivatives of the $\dot{s}_{max}$-curve and the extremals coincide. In most cases such points are not part of a time-optimal solution, in our special case they are. The example considered includes only torque constraints but no velocity constraints. The solution tries to utilize the maximum available torques as frequently as possible. Jumps indicate a change of the "hill"-sides (see arguments above).

Figure 7.13 shows an example, where the Bellman solution is compared with laboratory experiments. A robot with three revolute joints tracks a circular path. Two cases are considered, namely criteria with time only and with time and torques in addition. For an optimization applying a time and torque criterion the torque jumps disappear, and the curves become considerably smoother, which we could have expected of course. The high frequency oscillations come from the measurement system. Simulations and measurements coincide quite nicely [117].

### 7.2.3 Forces at the Gripper

So far we have considered free trajectories without any contact of the manipulator with the environment. For many applications we have to consider trajectories with contact to the environment and then usually with forces and torques at the end effector. Adding such forces and torques to the equations of motion (7.29) of chapter 7.2.2 means, that we have to include in the force term $C_i(s)$ the forces and the torques at the gripper. It means furtheron and with respect to the mountain-like phase space, Figure 7.8 for example, that the ruled surfaces of these phase space mountains are shifted but without changing their slopes. As a result we obtain a somewhat displaced configuration, but still with the same principal properties as before. Therefore the time-optimal solution can be evaluated by the same procedures as used for the path planning problem without contact and without forces.

The idea behind this path problem is the following. Assuming again an ideal motion along the given trajectory with one degree of freedom, the manipulator should perform the given path in minimum time and at the same time generate the prescribed forces at the end effector. In reality, however, a robot with rigid links and joints cannot realize forces at the gripper, and, in addition, modelling uncertainties would influence the force performance to a much larger extent than the path tracking alone. So, what we are doing in this chapter represents a first step to control trajectories and forces with the help of an elastic robot. Assuming small elastic deformations of the links, possibly also of the joints, motivates an approach, which takes the motion of a rigid manipulator as a reference and superimposes small elastic deviations, wherever they may occur. From this it makes sense to consider in this first step the time-optimal solution along a given path with given forces at the gripper for a rigid robot.

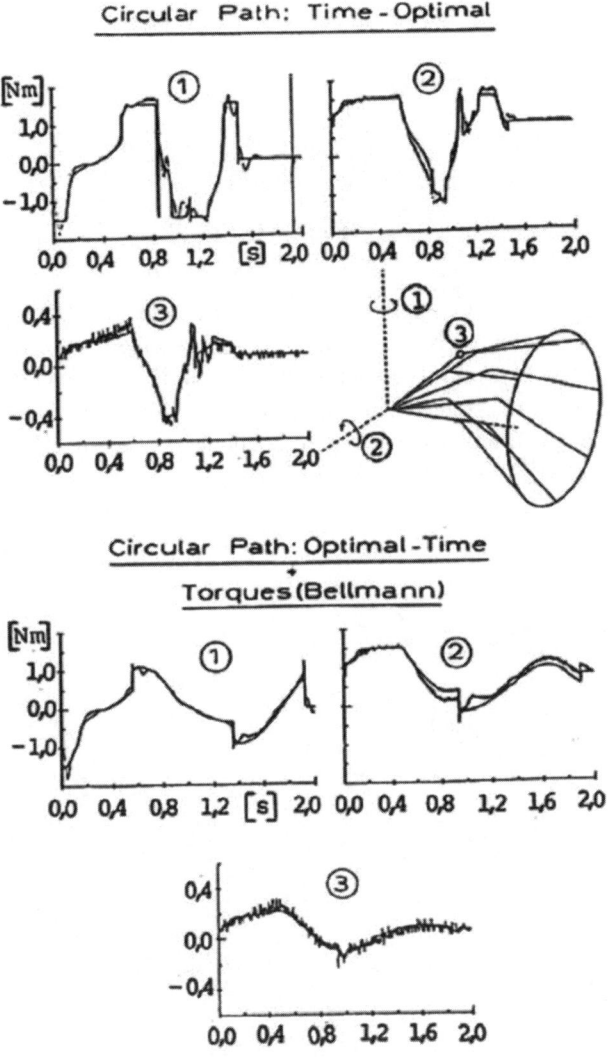

Fig. 7.13: Circular Arc by a Three-link Robot, Comparison Theory (-----) and Measurements (———)

A manipulater with rigid links tracking a prescibed path with prescribed forces at the gripper is a one-degree-of-freedom system described by the path coordinate s and the modified equations of motion (see eqs. (7.29)):

$$T_i(s) = A_i(s)(\dot{s}^2)' + B_i(s)(\dot{s}^2) + C_i(s), \qquad (i = 1, 2, \cdots, f)$$

$$A_i(s) = \frac{1}{2}\mathbf{M}_i\mathbf{q}', \qquad B_i(s) = \mathbf{M}_i\mathbf{q}'' + \mathbf{q}'^T \left[\frac{\partial \mathbf{M}_i^T}{\partial \mathbf{q}} - \frac{1}{2}\frac{\partial \mathbf{M}}{\partial \mathbf{q}_i}\right],$$

$$C_i(s) = \frac{\partial V}{\partial q_i} - \mathbf{J}_{Ti}^T(\mathbf{F}_{G0} + \mathbf{F}_{C0}), \qquad \mathbf{F}_{C0} = (\mathbf{n}^T\mathbf{F}_{G0})\left(-\frac{\mathbf{v}_{rel}}{|\mathbf{v}_{rel}|}\right). \qquad (7.46)$$

The magnitudes are: s path coordinate, $\dot{s}$ path velocity, $\mathbf{M} \in \mathbb{R}^{f,f}$ mass matrix, $\mathbf{M}_i$ the i-th row of $\mathbf{M}$, $\mathbf{q} = \mathbf{q}_r \in \mathbb{R}^f$ generalized coordinates of the "rigid" manipulator, $\frac{\partial V}{\partial q_i}$ mainly gravitaional forces, $\mathbf{F}_{G0}$ forces at the gripper, $\mathbf{J}_{Ti}$ Jacobian from workspace into configuration (joint) space, $\mathbf{F}_{C0}$ friction force along the path, $\mathbf{n}$ surface normal, $\mathbf{v}_{rel}$ relative tangential velocity along the path ($|\mathbf{v}_{rel}| = \dot{s}$), $T_i(s)$ torque in joint (i).

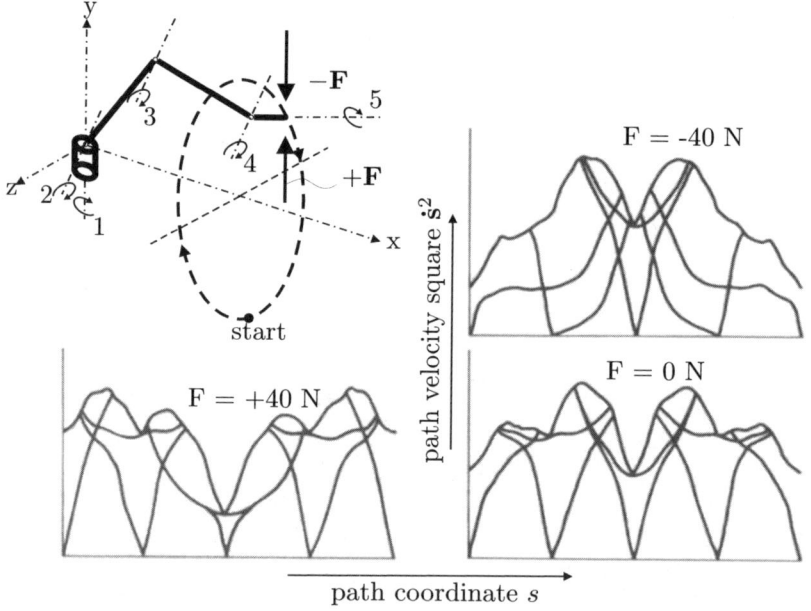

Fig. 7.14: Time-Optimal Topology with Forces at the Gripper for a Circular Path [185]

The relations (7.46) represent a basis for time-optimization according to the preceding sections, but now for given trajectories with prescribed forces at the gripper. Such forces posses a considerable influence on the time-optimal results. Figure 7.14 gives an example for a circular path parallel to the y-z-plane, being tracked by a manipulator with five revolute joints. The numerical results come from [117] and [223]. The zero force solution (F=0) includes three critical points (saddle points). The maximum velocity as well the time-optimal curves are symmetrical with respect to the s=0.5 point. Applying a negative force, which means pushing down the gripper, results in a very slow starting movement of the manipulator due to the negative retarding force. Only one critical point remains, and the deceleration is performed as slowly as the acceleration. On the other hand, lifting the gripper with a positive force (+F) produces a large acceleration and a large deceleration because the positive force helps in generating a steep slope at the beginning and the end of the trajectory by partly compensating the gravity force. In addition such a positive force diminishes considerably the velocity at the symmetry point s=0.5.

### 7.2.4 Influence of Elasticity

#### 7.2.4.1 Elastic Manipulator with Tree-Like Structure

Following the theory presented in the chapter 3.3 on the pages 113 and the following ones we model an elastic manipulator as a multibody system including rigid and elastic components. Furtheron we assume that all elastic deformations are small. Therefore motion can be characterized by a nonlinear gross movement by the rigid robot with elastic joints superimposed by small elastic deformations due to the elasticity of the links. The complete motion of the elastic manipulator, namely gross motion and elastic motion, might be described for each body using three coordinate systems, an inertial one like (I) in Figure 7.15, a conveniently chosen body-fixed frame (R) and an element-fixed coordinate system (E), to give an example. Note that (R) is already rotated and shifted by the joint angles and by all elastic deformation of the predecessor bodies with $(j < i)$. A mass element of the link i is elastically shifted by a vector $\mathbf{r}_{ei}(\mathbf{x}, t)$ and rotated by a vector $\varphi_{ei}(\mathbf{x}, t)$. It has already been pointed out, that the choice of the various and neccessary coordinate systems is to a certain extent arbitrary, but nevertheless it should follow some practical rules of econmy and convenience with respect to all evaluations following from that choice.

We shall proceed here a bit different from the presentation at the beginning (section 7.2.1), where we built up a relative kinematics by considering a body i and its predecessor (i-1). Nevertheless the results are of course comparable. An arbitrary point of an elastic link (Figure 7.15), defined originally in the element-fixed coordinates (E) by $_E\mathbf{r}$ can be written in the body-fixed system (R) by the relation

$$_R\mathbf{r}_{total} = {_R}(\mathbf{r}_0 + \mathbf{x} + \mathbf{r}_e) + \mathbf{A}_{RE} \cdot {_E}\mathbf{r} \tag{7.47}$$

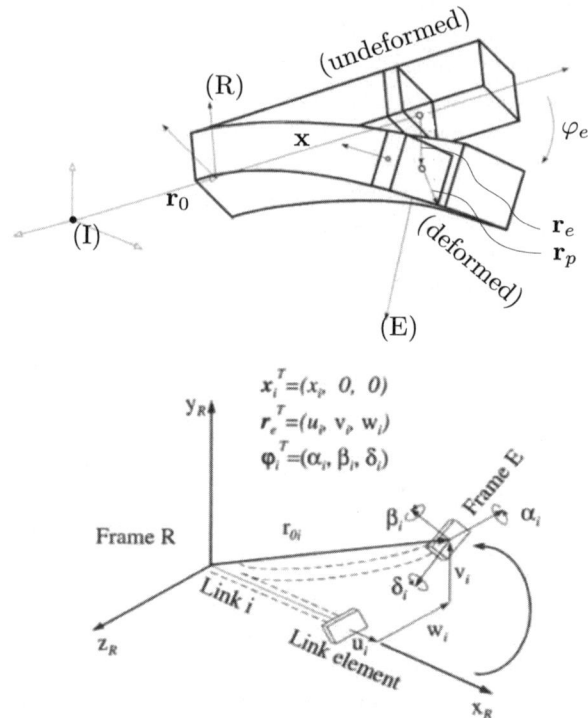

Fig. 7.15: Elastic Link - Coordinates and Deformation Details [196]

with the transformation matrix $\mathbf{A}_{RE}$ from (E) to (R). Considering only small elastic deformations the transformation $\mathbf{A}_{RE}$ can be approximated by (see section 2.2.8 on page 47)

$$\mathbf{A}_{RE} \approx (\mathbf{E} + \tilde{\varphi}_e) + O(2) \quad \in \mathbb{R}^{3,3},$$

$$\mathbf{A}_{RE} \approx \begin{pmatrix} 1 & -\delta & \beta \\ \delta & 1 & -\alpha \\ -\beta & \alpha & 1 \end{pmatrix} \approx \begin{pmatrix} 1 & -v' & -w' \\ v' & 1 & -\alpha \\ w' & \alpha & 1 \end{pmatrix} + O(2) \tag{7.48}$$

with the following magnitudes: $\mathbf{E}$ the identity matrix and $(\cdot)'$ derivation with respect to x, for example $v' = \frac{dv}{dx}$. Considering in a first step manipulators with tree-like structure and altogether j joints (Figure 7.16), the position vector $\mathbf{r}_p$ from the inertial frame to the contact point can be written in the form (see equations (7.4))

$$\mathbf{r}_p = \sum_{i=1}^{j} \left\{ \prod_{k=1}^{i} [\mathbf{E} + \tilde{\varphi}_e]_{k-1} \cdot \mathbf{A}_{E_{k-1} R_k} \right\}_{L_{k-1}} \cdot (\mathbf{x}_i + \mathbf{r}_{ei})_{L_i}, \tag{7.49}$$

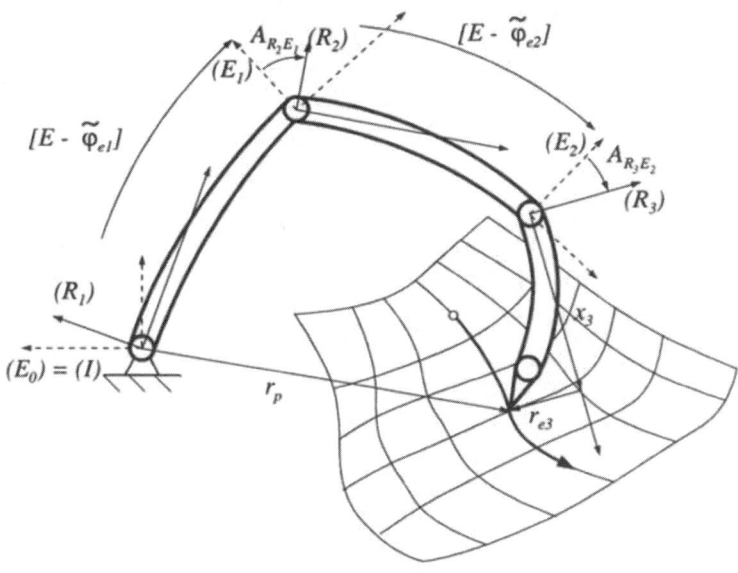

Fig. 7.16: Elastic Manipulator in Contact with a Surface

where $L_i$ refers to the total lenght of the link i and $\mathbf{A}_{E_{k-1}R_k}$ is the transformation from $R_k$ to $E_{k-1}$, which for revolute joints is the usual elementary rotation matrix. With the help of the vectors defined in Figure 7.15 the absolute translational and rotational velocities of a mass element come out with

$$\left.\begin{array}{l}\mathbf{v}_{Ei} = \mathbf{v}_{0i} + \dot{\mathbf{r}}_{ei} + \tilde{\omega}_{0i}(\mathbf{x}_i + \mathbf{r}_{ei}), \\ \omega_{Ei} = \omega_{0i} + \dot{\varphi}_{ei},\end{array}\right\} \quad R_i\text{ - frame} \qquad (7.50)$$

$$\left.\begin{array}{l}\mathbf{v}_i = (\mathbf{E} + \tilde{\varphi}_{ei})^T \cdot \mathbf{v}_{Ei}, \\ \omega_i = (\mathbf{E} + \tilde{\varphi}_{ei})^T \cdot \omega_{Ei}.\end{array}\right\} \quad E_i\text{ - frame} \qquad (7.51)$$

All the magnitudes of these equations depend on the rigid, $\mathbf{q}_r \in \mathbb{R}^{f_r}$, and elastic generalized coordinates, $\mathbf{q}_e \in \mathbb{R}^{f_e}$, where $f_r$ and $f_e$ denote the number of rigid and elastic degrees of freedom. The elastic coordinates enter the equations of motion via a Ritz-approach (equation (3.121) on page 125), which for $\mathbf{r}_{Ei}$ can be written

$$\mathbf{r}_{Ei}(\mathbf{x}_i, t) = \begin{pmatrix} u_i(\mathbf{x}_i, t) \\ v_i(\mathbf{x}_i, t) \\ w_i(\mathbf{x}_i, t) \end{pmatrix} = \begin{pmatrix} \bar{u}_i(\mathbf{x}_i)^T \mathbf{q}_{e,ui}(t) \\ \bar{v}_i(\mathbf{x}_i)^T \mathbf{q}_{e,vi}(t) \\ \bar{w}_i(\mathbf{x}_i)^T \mathbf{q}_{e,wi}(t) \end{pmatrix}, \qquad (7.52)$$

Where $\bar{u}_i, \bar{v}_i, \bar{w}_i$ denote the vector of the shape functions for link i. Applying this approach, collecting all elastic coordinates in a vector $\mathbf{q}_e$, taking into account the dependencies $\mathbf{v}_i(\mathbf{q}_r, \mathbf{q}_e, \dot{\mathbf{q}}_r, \dot{\mathbf{q}}_e)$ and $\omega_i(\mathbf{q}_r, \mathbf{q}_e, \dot{\mathbf{q}}_r, \dot{\mathbf{q}}_e)$, applying the momentum equation to the mass element of Figure 7.15, arranging the forces with respect to active and passive properties, applying Jourdain's principle and regarding the equations (7.50) to (7.52) we arrive at the equations of motion of a linearly elastic robot (see also the relations (3.127) on page 127):

- *rigid body coordinates:*

$$\sum_{i=1}^{n} \int_{B_i} \left\{ \left[ \left(\frac{\partial \mathbf{v}_{0i}}{\partial \dot{\mathbf{q}}_r}\right) + (\tilde{\mathbf{x}}_i + \tilde{\mathbf{r}}_{ei})^T \left(\frac{\partial \omega_{0i}}{\partial \dot{\mathbf{q}}_r}\right) \right]^T \cdot [(\dot{\mathbf{v}}_{Ei} + \tilde{\omega}_{0i}\mathbf{v}_{Ei})dm_i - d\mathbf{F}_{ai}] + \right.$$
$$\left. + \left(\frac{\partial \omega_{0i}}{\partial \dot{\mathbf{q}}_r}\right)^T (\mathbf{E} + \tilde{\varphi}_{ei})[d\mathbf{I}_i\dot{\omega}_i + \tilde{\omega}_i d\mathbf{I}_i \omega_i - d\mathbf{T}_{ai}] \right\} = 0 \quad (7.53)$$

- *deformation coordinates:*

$$\sum_{i=1}^{n} \int_{B_i} \left\{ \left[ \left(\frac{\partial \mathbf{v}_{0i}}{\partial \dot{\mathbf{q}}_e}\right) + (\tilde{\mathbf{x}}_i + \tilde{\mathbf{r}}_{ei})^T \left(\frac{\partial \omega_{0i}}{\partial \dot{\mathbf{q}}_e}\right) + \left(\frac{\partial \dot{\mathbf{r}}_{ei}}{\partial \dot{\mathbf{q}}_e}\right) \right]^T \cdot \right.$$
$$\cdot [(\dot{\mathbf{v}}_{Ei} + \tilde{\omega}_{0i}\mathbf{v}_{Ei})dm_i - d\mathbf{F}_{ai}] +$$
$$\left. + \left[ \left(\frac{\partial \omega_{0i}}{\partial \dot{\mathbf{q}}_e}\right) + \left(\frac{\partial \dot{\varphi}_{ei}}{\partial \dot{\mathbf{q}}_e}\right) \right]^T (\mathbf{E} + \tilde{\varphi}_{ei})[d\mathbf{I}_i\dot{\omega}_i + \tilde{\omega}_i d\mathbf{I}_i \omega_i - d\mathbf{T}_{ai}] \right\} = 0$$
$$(7.54)$$

- *dimensions and abbreviations:*

$$\mathbf{q}_r \in \mathbb{R}^{f_r}, \quad \mathbf{q}_e \in \mathbb{R}^{f_e}, \quad (\mathbf{v}_{0i}, \omega_{0i}, \mathbf{x}_i, \mathbf{r}_{ei}, \varphi_{ei}, \mathbf{F}_{ai}, \mathbf{T}_{ai}) \in \mathbb{R}^3,$$
$$\left(\frac{\partial \mathbf{v}_{0i}}{\partial \dot{\mathbf{q}}_r}\right) \in \mathbb{R}^{3, f_r}, \quad \left(\frac{\partial \dot{\mathbf{r}}_{ei}}{\partial \dot{\mathbf{q}}_e}\right) \in \mathbb{R}^{3, f_e}, \quad \text{etc.}$$
$$dm_i = \left[\int\int \rho dy dz\right]_i dx_i \in \mathbb{R},$$
$$d\mathbf{I}_i = \left[\int\int \rho \tilde{\mathbf{r}} \tilde{\mathbf{r}}^T dy dz\right]_i dx_i \in \mathbb{R}^{3,3}. \quad (7.55)$$

The evaluation of these equations is performed numerically, where in a first step it is advisable to collect all integrals containing only terms with spatial coordinates in a certain set of integral matrices being calculated before the main numerical procedure (see for example [242], [29], [223], [248]). In the Ritz-approach of the equation (7.52) cubic splines are used. With respect to the elastic terms we refer to section 2.2.8 on page 47 and on section 3.3.4 on page 124.

## 7.2 Trajectory Planning

The equations (7.53) and (7.54) represent a set of $f = f_r + f_e$ relations for the f unknowns $\mathbf{q}_r, \mathbf{q}_e$. They can always be reduced to the form

$$\mathbf{M}(\mathbf{q},t)\ddot{\mathbf{q}} + \mathbf{g}(\dot{\mathbf{q}},\mathbf{q},t) = \mathbf{BT}(t) + \mathbf{J}_T^T \mathbf{F}_G + \mathbf{J}_R^T \mathbf{T}_G,$$

$$\mathbf{q} = (\mathbf{q}_r^T, \mathbf{q}_e^T)^T \in \mathbb{R}^f, \quad \mathbf{M} \in \mathbb{R}^{f,f}, \quad \mathbf{g} \in \mathbb{R}^f, \quad \mathbf{B} \in \mathbb{R}^{f,j}, \quad (\mathbf{F}_G, \mathbf{T}_G) \in \mathbb{R}^3,$$

$$\mathbf{J}_T = \left(\frac{\partial \mathbf{v}_G}{\partial \dot{\mathbf{q}}}\right) \in \mathbb{R}^{3,f}, \quad \mathbf{J}_R = \left(\frac{\partial \omega_G}{\partial \dot{\mathbf{q}}}\right) \in \mathbb{R}^{3,f}, \tag{7.56}$$

where j is the number of the manipulator joints, $\mathbf{M}$ its mass matrix, $\mathbf{g}$ a vector with all gyroscopical and gravitational forces, $\mathbf{B}$ the actuator matrix, $\mathbf{T}$ includes the actuators forces and torques, $\mathbf{F}_G$ and $\mathbf{T}_G$ are forces and torques at the gripper and the matrices $\mathbf{J}_T, \mathbf{J}_R$ the Jacobians of translation and rotation at the gripper.

From this we conclude, that without elasticities in the joints we get $f_r = j$ and with joint elasticities $f_r = 2j$ considering only one degree of freedom for each joint. It might be more, of course. In the last case the additional joint equations of motion write

$$I_M \cdot \ddot{q}_{MI} = T_0 - \frac{K_{gear}}{i_{gear}}\left(\frac{q_{MI}}{i_{gear}} - q_{ME}\right). \tag{7.57}$$

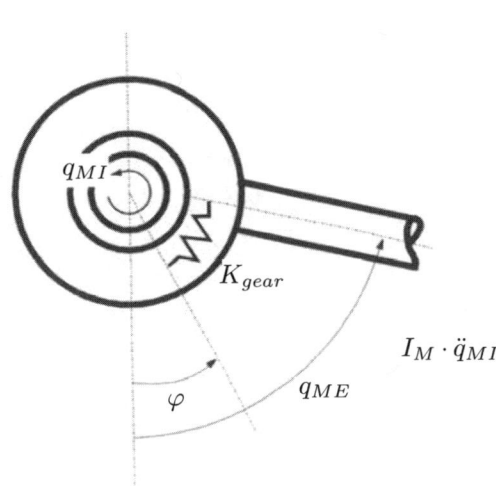

Fig. 7.17: Elastic Joint Model

Figure 7.17 illustrates the corresponding model, see also Figure 7.3. The magnitudes $K_{gear}$ and $i_{gear}$ denote the joint stiffness and the gear transmission ratio, respectively, $q_{MI}$ and $q_{ME}$ are the internal and external joint coordinates, $I_M$ is the motor moment of inertia and $T_0$ the actuating motor torque.

### 7.2.4.2 Closed Loop Conditions

A manipulator coming into contact with its environment forms a closed kinematical loop with special properties (see Figure 7.16). Considering for example manufacturing processes usually requires also the inclusion of some given forces or torques at the end-effector. Let us first consider the holonomic constraint accompanying such a configuration in the form $\boldsymbol{\Phi}(\mathbf{q}, t) = \mathbf{0} \in \mathbb{R}^m$. For later convenience we differentiate this constraint twice with respect to time and come out with

$$\ddot{\boldsymbol{\Phi}} = \left(\frac{\partial \boldsymbol{\Phi}}{\partial \mathbf{q}}\right) \ddot{\mathbf{q}} + \frac{d}{dt}\left(\frac{\partial \boldsymbol{\Phi}}{\partial \mathbf{q}}\right) \dot{\mathbf{q}} + \frac{d}{dt}\left(\frac{\partial \boldsymbol{\Phi}}{\partial t}\right) = \mathbf{0}. \tag{7.58}$$

Substituting $\left(\frac{\partial \boldsymbol{\Phi}}{\partial \mathbf{q}}\right) = (\mathbf{w}_1, \mathbf{w}_2, \cdots \mathbf{w}_m)^T$ we arrive at the general constraint equation on the velocity level

$$\mathbf{w}_k^T(\mathbf{q})\ddot{\mathbf{q}} + \zeta_k(\mathbf{q}, \dot{\mathbf{q}}) = 0, \qquad \mathbf{w}_k \in \mathbb{R}^f, \quad (k \in (1, 2, \cdots m)), \tag{7.59}$$

which applies for m active contacts and the corresponding constraints. This can be included into the equations of motion by (chapter 3.4 on the pages 131)

$$\mathbf{M}\ddot{\mathbf{q}} - \mathbf{W}\lambda = -\mathbf{g} + \mathbf{BT} + \mathbf{J}_T^T \mathbf{F}_G + \mathbf{J}_R^T \mathbf{T}_G = \mathbf{H},$$
$$\mathbf{w}_k^T \ddot{\mathbf{q}} + \zeta_k = 0,$$

$$\mathbf{W} = (\mathbf{w}_1, \mathbf{w}_2, \cdots \mathbf{w}_m) \in \mathbb{R}^{f,m}, \quad \lambda \in \mathbb{R}^m, \quad \zeta \in \mathbb{R}^m. \tag{7.60}$$

The unknowns $\ddot{\mathbf{q}}$ and $\lambda$ are determined from the above equations [200]

$$\ddot{\mathbf{q}} = \mathbf{M}^{-1}(\mathbf{H} + \mathbf{W}\lambda), \qquad \lambda = (\mathbf{W}^T \mathbf{M}^{-1} \mathbf{W})^{-1}(\zeta - \mathbf{W}^T \mathbf{M}^{-1} \mathbf{H}) \tag{7.61}$$

with the effective mass $(\mathbf{W}^T \mathbf{M}^{-1} \mathbf{W})^{-1}$ acting at the contacts. For our special case of a manipulator in contact with its environment we have only one contact and one vector constraint for the closed kinematical loop (i.e. m=1), which can be represented by (equations (7.4) and (7.49))

$$\boldsymbol{\Phi}(\mathbf{q}) = \mathbf{r}_p - \sum_{i=1}^{j} \left\{ \prod_{k=1}^{i} [\mathbf{E} + \tilde{\varphi}_e]_{k-1} \cdot \mathbf{A}_{E_{k-1}R_k} \right\}_{L_{k-1}} \cdot (\mathbf{x}_i + \mathbf{r}_{ei})_{L_i} = \mathbf{0} \in \mathbb{R}^3. \tag{7.62}$$

The constraint allows a motion only in the tangential plane of the constraining surface and constrains the motion in the direction perpendicular to that. This corresponds to the classical robot statement, that for a manipulator with contact to the environment we have to apply force control in normal and path control in tangential direction of the constraining surface. Projecting the above constraint into normal direction results in

$$\Phi_n = \mathbf{n}^T \boldsymbol{\Phi}(\mathbf{q}) = 0 \in \mathbb{R}. \tag{7.63}$$

The normal vector $\mathbf{n}^*$ in the space of generalized coordinates is the given by

$$\mathbf{n}^* = \left(\frac{\partial \Phi_n}{\partial \mathbf{q}}\right)^T = \left[\left(\frac{\partial \Phi_n}{\partial \mathbf{r}_p}\right)\left(\frac{\partial \mathbf{r}_p}{\partial \mathbf{q}}\right)\right]^T = \mathbf{J}_{T_p}^T \mathbf{n}. \tag{7.64}$$

We shall use it to transform also the elastically generated deviations $\mathbf{y}$ from the "rigid" reference into the configuration space, for example $(\mathbf{n}^*)^T \mathbf{y} = 0$.

### 7.2.4.3 Linearization around a Reference Trajectory

Tracking a given path by a manipulator is mainly a control problem, where many concepts may apply. Out of the various possibilities with respect to nonlinear control ([174], [246], [68]), the nonlinear decoupling schemes became and still are very popular for robot control. The basic idea is simple. In the case of prescribed trajectories with prescribed forces at the gripper we establish a feedforward control concept, which forces, as good as possible, the end effector along the given path with the given forces. All deviations from that reference are usually small and can be controlled by an additional linear controller. From the dynamical standpoint of view we then have to evaluate in a first step the prescibed path and forces from the process under consideration, for example a manufacturing process, and in a second step to linearize the equations of motion around that reference.

Linearizing the equations of motion (7.53) and (7.54) requires to develop by a Taylor expansion all terms of these equations up to first order, which means with respect to the Jacobians that the velocities $\mathbf{v}_{0i}$ and $\omega_{0i}$, for example, need to be developed up to second order terms. We obtain two sets of equations, one for the reference motion without elasticities and one for the elastic deviations:

- *reference path:*

$$\mathbf{q}_R \equiv \mathbf{q}_{ME,R} \in \mathbb{R}^j \tag{7.65}$$

- *reference motion:*

$$\mathbf{M}_R(\mathbf{q_R})\ddot{\mathbf{q}}_R + \mathbf{g}_R(\dot{\mathbf{q}}_R, \mathbf{q}_R) = \mathbf{B}_R\mathbf{T}_0 + \mathbf{J}_{TR}^T\mathbf{F}_{G0} + \mathbf{J}_{RR}^T\mathbf{T}_{G0} \tag{7.66}$$

- *linearization:*

$$\mathbf{q} = \mathbf{q}_0 + \mathbf{y}, \quad \mathbf{T} = \mathbf{T}_0 + \mathbf{\Delta T}, \quad \mathbf{F}_G = \mathbf{F}_{G0} + \mathbf{f}_G, \quad \mathbf{T}_G = \mathbf{T}_{G0} + \mathbf{t}_G, \tag{7.67}$$

- *deviation dynamics:*

$$\mathbf{M}_0(\mathbf{q}_0)\ddot{\mathbf{y}} + \mathbf{P}_0(\dot{\mathbf{q}}_0, \mathbf{q}_0)\dot{\mathbf{y}} + \mathbf{Q}_0(\ddot{\mathbf{q}}_0, \dot{\mathbf{q}}_0, \mathbf{q}_0)\mathbf{y} = \mathbf{h}_0(\ddot{\mathbf{q}}_0, \dot{\mathbf{q}}_0, \mathbf{q}_0) + \mathbf{B}_0\mathbf{\Delta T} \tag{7.68}$$

- *dimensions and abbreviations:*

$$\mathbf{J}_{ref}^T = \left(\frac{\partial \mathbf{q}}{\partial \mathbf{q}_R}\right)^T = [\operatorname*{diag}_{k=1,\cdots j}\{i_{gear}, k\}, \quad \mathbf{E}_j, \quad \mathbf{0}] \in \mathbb{R}^{j,f},$$

$$\mathbf{M}_R = \mathbf{J}_{ref}^T\mathbf{M}\mathbf{J}_{ref} \in \mathbb{R}^{j,j}, \quad \mathbf{g}_R = \mathbf{J}_{ref}^T\mathbf{g}(\dot{\mathbf{q}}_R, \mathbf{q}_R) \in \mathbb{R}^j,$$

$$\mathbf{B}_R = \mathbf{J}_{ref}^T\mathbf{B}_0 \in \mathbb{R}^{j,j},$$

$$\mathbf{J}_{TR}^T = \mathbf{J}_{ref}^T\mathbf{J}_{T0}^T = \left(\frac{\partial \mathbf{v}_G}{\partial \dot{\mathbf{q}}_R}\right)^T, \quad \mathbf{J}_{RR}^T = \mathbf{J}_{ref}^T\mathbf{J}_{R0}^T = \left(\frac{\partial \omega_G}{\partial \dot{\mathbf{q}}_R}\right)^T \in \mathbb{R}^{j,3},$$

$$\mathbf{y} = \left(\mathbf{y}_{MI}^T, \mathbf{y}_{ME}^T, \mathbf{y}_e\right)^T \in \mathbb{R}^f, \quad \mathbf{q}_0 = \mathbf{J}_{ref}\mathbf{q}_R \in \mathbb{R}^f, \tag{7.69}$$

where ($f = 2j + f_e$). $\mathbf{E}_j \in \mathbb{R}^{j,j}$ is a identity matrix. The above equations apply for the reference motion of a manipulator following a given trajectory with given forces and for the disturbances around this reference motion. Note that the reference motion is realized in the joint coordinate space $\mathbf{q}_R \in \mathbb{R}^j$, whereas the linearized deviation motion moves in an augmented space with $\mathbf{q}_0 \in \mathbb{R}^f$ and $\mathbf{y} \in \mathbb{R}^f$.

If the origin of the gripper coordinate system and the contact point do not coincide but have a distance $\mathbf{r}_{GP}$, then we get $\mathbf{F}_{G0} = \mathbf{F}_{P0}, \mathbf{T}_{G0} = \tilde{\mathbf{r}}_{GP}\mathbf{F}_{P0}$. The right hand side of the equation (7.66) changes accordingly to yield

$$\mathbf{M}_R(\mathbf{q_R})\ddot{\mathbf{q}}_R + \mathbf{g}_R(\dot{\mathbf{q}}_R, \mathbf{q}_R) = \mathbf{B}_R \mathbf{T}_0 + \left[\frac{\partial(\mathbf{v}_G + \tilde{\mathbf{r}}_{GP}\omega_G)}{\partial \dot{\mathbf{q}}_R}\right]^T \mathbf{F}_{P0} \qquad (7.70)$$

The contact force $\mathbf{F}_{P0}$ is assigned to the closing loop condition of equation (7.63) in the following way. Within a first order expansion

$$\Phi_n(\mathbf{q}) \approx \Phi_n(\mathbf{q}_0) + \left(\frac{\partial \Phi_n}{\partial \mathbf{q}}\right)_0 \mathbf{y} = 0. \qquad (7.71)$$

The first term of this equation is zero, $\Phi_n(\mathbf{q}_0) = 0$, because for the reference motion the idealized constraint is perfectly fulfilled. Some deviations due to elasticity must meet the requirement of the second term, which means, that equation (7.68) has to be supplemented by a term $\left(\frac{\partial \Phi_n}{\partial \mathbf{q}}\right)_0^T \lambda_1$. With regard to equation (7.60) the term $\left(\frac{\partial \Phi}{\partial \mathbf{q}}\right)^T = \mathbf{W} \in \mathbb{R}^{f,m}$ reduces to (m=1) $\left(\frac{\partial \Phi_n}{\partial \mathbf{y}}\right)^T = \mathbf{w}_1 \equiv \mathbf{n}^* \in \mathbb{R}^f$, whereas $\lambda \in \mathbb{R}^m$ is left with one component $\lambda_1$ only. For this special case we get the deviation dynamics including the contact effects in the form (see eq. (7.68))

$$\mathbf{M}_0\ddot{\mathbf{y}} + \mathbf{P}_0\dot{\mathbf{y}} + \mathbf{Q}_0\mathbf{y} = \mathbf{h}_0 + \mathbf{B}_0\Delta\mathbf{T} + \mathbf{J}_P^T\Delta\mathbf{F}_P, \qquad \text{with} \quad \Delta\mathbf{F}_P = \mathbf{n}\lambda_1,$$

$$\lambda_1 = m_1\left\{\mathbf{w}_1^T\mathbf{M}_0^{-1}[\mathbf{P}_0\dot{\mathbf{y}} + \mathbf{Q}_0\mathbf{y} - \mathbf{h}_0 - \mathbf{B}_0\Delta\mathbf{T}] - \ddot{\mathbf{w}}_1^T\mathbf{y} - 2\dot{\mathbf{w}}_1^T\dot{\mathbf{y}}\right\}$$

$$\mathbf{w}_1 = \mathbf{J}_P^T\mathbf{n} = \mathbf{n}^* \in \mathbb{R}^f, \qquad m_1 = \left(\mathbf{w}_1^T\mathbf{M}_0^{-1}\mathbf{w}_1\right)^{-1} \in \mathbb{R}. \qquad (7.72)$$

Note that firstly $\mathbf{w}_1(\mathbf{q}) = \mathbf{J}_{T_p}^T(\mathbf{q})\mathbf{n}$ and that secondly the surface normal $\mathbf{n}$ changes along the path. Due to these two properties a time derivative exists.

### 7.2.4.4 Control of Elastic Manipulators

To go into the problem of controlling the elasticity of manipulators in all details would be much beyond the scope of this book. But we shall give some principal ideas and considerations and refer for details to [223], [224], [209], [196]. An elastic manipulator tracking a path with prescribed forces and torques requires the solution of a complicated problem, namely to control the gripper position and orientation with the help of the drives alone acting on the elastic links, which on their side must put the gripper into the wanted position. This means that control is executed only indirectly not directly. In addition we have spill-over effects due to the elasticities. By measuring the drive kinematics alone the robot is not observable, additional measurements of the elastic deformations are mandatory. Some detailed investigations came out with the theoretical and practical experience, that measuring the curvatures of the elastic links gives an excellent basis for reducing the elastic vibrations [131], [223]. This can be performed by strain gauges.

Combined path and force control of an elastic manipulator implies in our case a closed kinematical loop in the presence of elastic links and joints. The control concept follows a classical concept with feedforward and feedback control (see Figure 7.18), where new aspects enter into the method how control activities are distributed in both loops. The feedforward control loop includes

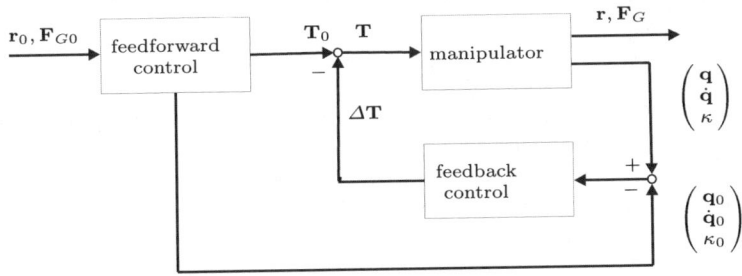

Fig. 7.18: Basic Control Concept

- an optimal trajectory planning procedure for a manipulator with rigid links tracking a prescribed trajectory with given forces at the end effector, for example the methods introduced in the above sections 7.2.2,
- a correction procedure with respect to the stationary elastic deviations as generated along this reference trajectory,
- a calculation scheme for the reference joint kinematics and kinetics as well as for the reference elastic deviations and the curvatures.

The underlying idea consists in putting as much control efforts as possible into the feedforward loop being evaluated off-line and in leaving only minor activities in the feedback loop being evaluated on-line. Two design steps realize this idea. Firstly, the stationary elastic deviations cover already most of the overall path deviations, because the elastic vibrations are small and usually quite nicely damped. Secondly, including the prescribed forces at the gripper into the gross reference motion means, that tracking the correct path generates automatically the correct force. Therefore, putting back the gripper to its reference path yields also the correct given force.

From these arguments we conclude, that it is sufficient to control within the feedback loop only

- the elastic vibrations taking into account controllability in the presence of elasticity and maximum damping on the basis of a realistic closed loop model of the elastic manipulator.

We start with the feedforward loop by establishing the reference using the relations (7.65) and (7.66). This provides us with the reference trajectory including the forces at the gripper. In a second step we consider the elastic deformations, which contain two parts: one part coming from the nominal forces along the given path and another part emerging from the vibration capability of the manipulator system. Therefore we subdivide the complete elastic deformations in $\mathbf{y} = \mathbf{y}_s + \mathbf{y}_v$, where $\mathbf{y}_s$ represents the stationary first part and $\mathbf{y}_v$ the vibratory second part. The first part we call quasi-static corrections and evaluate it from equation (7.68)

$$\mathbf{Q}_0 \mathbf{y}_s = \mathbf{h}_0 + \mathbf{B}_0 \mathbf{\Delta T}_s, \tag{7.73}$$

where the inertia effects due to the elastic vibrations have been neglected, which makes sense because they are usually very small. The evaluation of equation (7.73) goes straightforward and gives the nominal elastic deviations for a robot along a given path with given forces.

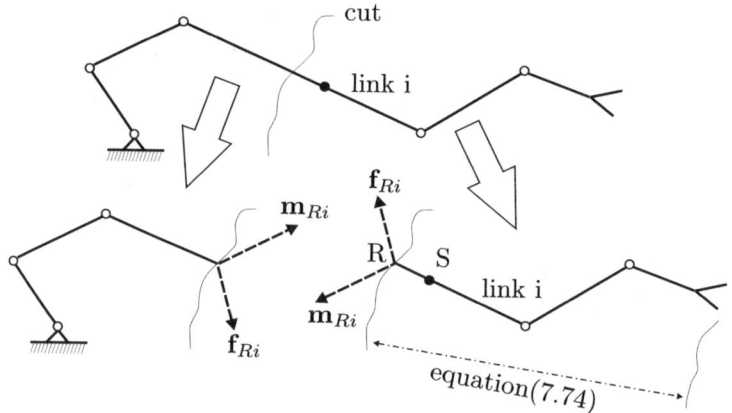

Fig. 7.19: Cutting the Link i of a Manipulator

It should be noted, that for the curvature control to follow we also need the nominal curvature values, which are calculated by cutting the corresponding link i at the location of the strain gauges and applying the multibody formulas for the evaluation of the cut forces and torques for the nominal magnitudes of the rigid robot (see [223] and Figure 7.19). The internal forces and torques at the cut point R are then determined by applying the projection equations (7.53) for a rigid robot and without any elastic components to that part of the manipulator, which have been cut

$$\left(\frac{\partial \mathbf{v}_{Ri}}{\partial \dot{\mathbf{q}}_r}\right)^T \left\{ m_i \left[ (\dot{\mathbf{v}}_{Ri} + \tilde{\omega}_{Ri} \mathbf{r}_{RSi}) + \tilde{\omega}_{Ri}(\mathbf{v}_{Ri} + \tilde{\omega}_{Ri} \mathbf{r}_{RSi}) \right] - \mathbf{f}_{Ri} \right\} +$$
$$+ \left(\frac{\partial \omega_{Ri}}{\partial \dot{\mathbf{q}}_r}\right)^T \left\{ \left(\mathbf{I}_{Si} + m_i \tilde{\mathbf{r}}_{RSi} \tilde{\mathbf{r}}_{RSi}^T\right) \dot{\omega}_{Ri} + \right.$$
$$\left. + \tilde{\omega}_{Ri} \left(\mathbf{I}_{Si} + m_i \tilde{\mathbf{r}}_{RSi} \tilde{\mathbf{r}}_{RSi}^T\right) \omega_{Ri} + m_i \tilde{\mathbf{r}}_{RSi}(\dot{\mathbf{v}}_{Ri} + \tilde{\omega}_{Ri} \mathbf{r}_{RSi}) - \mathbf{m}_{Ri} \right\} = \mathbf{0},$$
(7.74)

for all links starting with the link i up to the end effector. The magnitudes are: $\mathbf{v}_{Ri}$ the absolute velocity of point R of link i, $\omega_{Ri}$ the absolute angular velocity, $\mathbf{r}_{RSi}$ the distance vector from the cutting point R to the center of mass S, $\mathbf{I}_{Si}$ the inertia tensor with respect to point $S_i$, $m_i$ the mass of link i and $\mathbf{f}_{Ri}, \mathbf{m}_{Ri}$ the internal forces and torques at the point R.

The above procedure has a couple of advantages. Firstly, we can do that evaluation for the completely rigid robot without generating large errors, secondly with the knowledge of the internal cut forces and torques $\mathbf{f}_{Ri}, \mathbf{m}_{Ri}$ we immediately can calculate from linear elasticity theory the elastic curvatures and the warp coefficient, which are both important for controlling elasticity providing us with the nominal values of the elastic deformations, for example

$$\kappa_{0i} = \mathbf{K}_i^{-1} \cdot \mathbf{m}_{Ri} \tag{7.75}$$

with the stiffness matrix $\mathbf{K}_i$ of link i [223].

With the knowledge of all nominal magnitudes we evaluate in a further step the vibrations by the relation (7.68) and by using $\mathbf{y} = \mathbf{y}_s + \mathbf{y}_v$ and $\mathbf{h}_0 = \mathbf{J}_T^T \mathbf{f}_G + \mathbf{J}_R^T \mathbf{t}_G$. We get

$$\mathbf{M}_0 \ddot{\mathbf{y}}_v + \mathbf{P}_0 \dot{\mathbf{y}}_v + \mathbf{Q}_0 \mathbf{y}_v = \mathbf{h}_v + \mathbf{B}_0 \mathbf{\Delta T}_v \tag{7.76}$$

Forces are controlled in a plane perpendicular to the plane of contact, and the trajectory is controlled in the contact plane. In both cases we use the curvatures as a control input. They can easily be measured and serve as a basis for the evaluation of path and force deviations due to elasticity. Figure illustrates such a hybrid path and force control system, which has been applied for the control of a very elastic manipulator ([223], [209]). From this Figure we can derive the following control law

$$\mathbf{\Delta T}_v = -\mathbf{K}_P(\mathbf{q} - \mathbf{q}_0) - \mathbf{K}_D(\dot{\mathbf{q}} - \dot{\mathbf{q}}_0) - \mathbf{K}_K(\kappa - \kappa_0), \tag{7.77}$$

where $\mathbf{q}$ is measured by angular encoders, $\dot{\mathbf{q}}$ is determined by differentiation of the q-measurements, and $\kappa$ is measured by strain gauges. An attempt to evaluate the derivative $\dot{\kappa}$ by differentiating the $\kappa$-measurements failed, because the $\kappa$-measurements were too noisy. The dependence of the curvatures $\kappa$ on the generalized elastic coordinates ($\mathbf{q}_e = \mathbf{y}_e$) are approximated by the linear relationship

$$\kappa = \mathbf{C} \mathbf{q}_e. \tag{7.78}$$

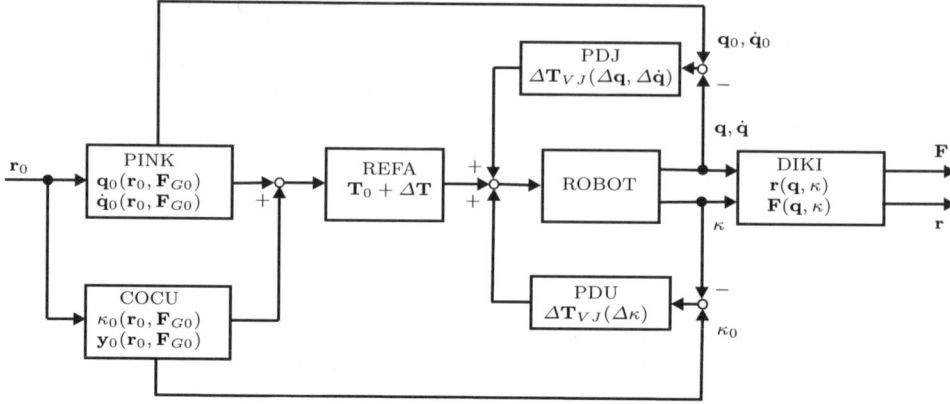

Fig. 7.20: Hybrid Control of an Elastic Robot

The matrix $\mathbf{C} \in \mathbb{R}^{f_e, f_e}$ contains the second derivatives of the shape functions (equation (7.52)) in the case of bending and the first derivatives in the case of torsion, and this at the location of the strain gauges. So altogether the scheme of Figure 7.20 includes the following control activities.

- PINK:
  - trajectory planning according to chapters 7.2.2, 7.2.3 and 7.2.4.3 with the equations (7.29), (7.46), (7.65) to (7.69) together with the appropriate methods and algorithms,
  - inverse kinematics, standard recursive procedures,
  - transformations, Jacobians,
- COCU:
  - evaluation of the elastic manipulator via elastic multibody theory, chapters 7.2.4.1 and 7.2.4.3 with the equations (7.53) to (7.55) and (7.65) to (7.69),
  - stationary elastic corrections, equations (7.73),
  - reference curvatures and warp coefficients, see equation (7.74),
- REFA:
  - combination of the reference magnitudes with the reference elastic values, as in equation (7.67) for the complete deviations,
- ROBOT:
  - hardware with two very elastic links and three revolute joints,
  - measurements by angular encoders and strain gauges, two types of bending and torsion,
  - power supply, interface and control electronics, A/D- and D/A-channels, sampling rate $\approx 1000 Hz$,
- PDJ, PDU:

- joint and elasticity control, equation (7.77),
- on-line controller with a controller design following criteria of controllability and stability,
• DIKI:
- direct kinematics from joint space to work space, transformations, Jacobians.

### 7.2.4.5 Some Results

To get some confidence in the theory various experiments with a laboratory robot were carried through ([223], [196]). They all confirm modelling and the control concept. Elastic manipulators might be used for polishing tasks, for detecting forms by scanning the shape with the help of sensitive contact forces or for measuring contact forces directly with a robot. From these examples we consider the polishing problem, where the manipulator had to perform a typical polishing motion.

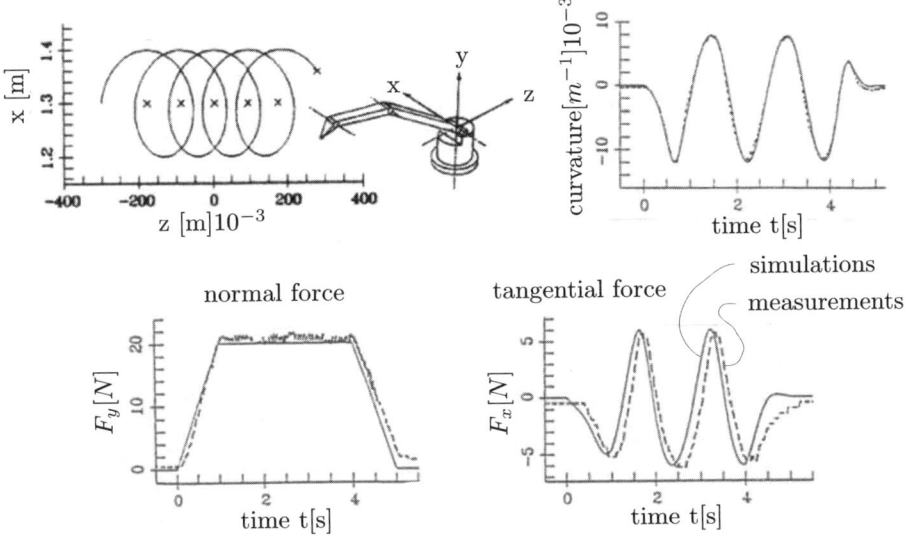

Fig. 7.21: Comparison of Simulation and Measurements

Figure 7.21 illustrates a typical example [196]. A three DOF robot with elastic links performs a polishing motion with a prescribed path and with a prescribed normal force (upper left of Figure 7.21). The upper right graph shows a comparison of the precalculated nominal curvature for bending of the lower link with coressponding measurements by strain gauges. The agreement is nearly perfect. The lower part of Figure 7.21 depicts the contact forces in normal and tangential directions, where the normal force is prescibed. Again the agreement is very good, where ths slight differences for the tangential force come from some errors in the sliding friction force at the gripper, which is not perfectly adapted via the corresponding friction coefficient.

## 7.3 Dynamics and Control of Assembly Processes with Robots

### 7.3.1 Introduction

Dealing with assembly processes means combining the dynamics and control of one or more manipulators with the dynamics and control of the assembly process under consideration. Typical for such processes is the contact of the robot with its environment through the parts to be assembled. The arising contact forces influence the motion of the robot, and, as a matter of fact they also influence the assemply procedure itself. In the following we shall consider dynamical models of a manipulator together with models for some mounting processes. Different methods are applied depending on the mechanical properties of the parts. With compliant mating parts, the forces and torques are only dependent on the relative position and the velocity of the robot's gripper with respect to the environment. If the mating parts are very stiff, the dynamics of the robot is characterized by closed loops, which requires special mathematical treatment.

Small tolerances between the mating parts are often characteristic for such mounting tasks. During the automatic assembly with a robot, the parts will contact each other due to uncertainties in the manipulators position and orientation and due to tolerances of the parts' geometry. Undesirable high strains on the workpieces or even the unfeasibility of the task, for example due to jamming, may result from that. A well known example for parts mating is the peg-in-hole problem. Many assembly processes can be reduced to this example, so we shall study this pecific problem in some detail.

There are three different approaches to handle the above mentioned problems. One possible solution is the development of special passive compliant mechanisms, based on the Remote Center of Compliance (RCC) measure. Applying this measure gives more safety and tolerances for typical mating processes. Mechanisms of that kind were first developed for the two-dimensional peg-in-hole problem. An extension to three dimensions may be found in [254], [256].

A second solution is the additional use of sensor information. Nearly all concepts follow the same principle: First an initial contact state has to be identified. Then the peg has to be moved towards the hole. This phase can be called peg-on-hole. Afterwards the peg is aligned and inserted into the hole. Thereby, especially in the case of two point contacts, jamming has to be avoided. For this phase an optimized controller is presented in [271]. If the operation fails due to sensing, model or control errors, the method of error detection and recovery is applied in order to complete the given task [251].

A third approach is a theoretical investigation. The assembly task is described by geometrical and mechanical models, where uncertainties from the robot's position, the robot's trajectory and the parts' geometry can also be taken into account. If the problem is solved by geometrical considerations and simple assumptions for the contact forces one talks of fine motion planning [141]. The dynamics of the complete parts mating process including a complete dynamic model of the manipulator is presented in [241]. Every contact point is considered closing a kinematical loop between the manipulator and its environment. For an efficient numerical implementation, the contact laws and the transitions between the different contact states (contact, no contact, sticking or sliding) are formulated as a Linear Complementarity Problem (LCP). The same method is applied to the three–dimensional peg-in-hole problem in [152].

Different questions arise when regarding flexible workpieces. Through the compliance of the parts, effects like jamming and wedging are not likely to occur. In this case the forces and moments arising during the mating process are more interesting in order to evaluate the stresses and strains on the mating parts. We shall come back to this problem in connection with snap joints and pistons with elastic rings mated to a cylinder.

In modelling the dynamics and control of manipulators and of the assembly process together, so-to-say as one integrated model with the appropriate theory, we automatically arrive at questions like parameter and control optimization for achieving a best solution not only with respect to robot dynamics and control, but also with respect to process dynamics and control, where we must consider the interaction between robot and environment as an essential part of the problem. In order to deal, for example, with impacts, oscillations and constrained motion, a model-based optimization approach will be necessary, which relies on a detailed dynamic model of the manipulator incorporating finite gear stiffnesses and damping. These models are used to define an optimization problem, which is then solved using numerical programming methods.

The intention of such an optimizing consideration consists in providing a tool for the planning stage of an assembly task with respect to increased productivity and reliability. This will be illustrated by a typical assembly task, namely inserting a rigid peg into a hole. Application of the proposed approach to such a rigid peg-in-hole insertion under practical constraints reduces the measure for impact sensitivity by 17 %, that for mating tolerances by 78 % and the damping of end-effector oscillations and motor torques by up to 79 %. These improvements are shown to be reproducible experimentally. The result is a planning tool, which allows any industrial robot to be optimized for the specific needs of any manipulation task.

In the following we shall investigate mating of compliant and of rigid parts using a PUMA 560 robot. The findings presented are mainly based on the publications [271], [241], [152], [219], [154], [220].

## 7.3.2 Mating with a Manipulator

### 7.3.2.1 Manipulator Model

We start with the specific model for the PUMA 560 robot. It possesses six axes and is modeled as a tree-like multibody system with rigid bodies and ideal links. As generalized coordinates we take the relative angles between the bodies:

$$\boldsymbol{\gamma}_A = (\gamma_{A1}, \gamma_{A2}, \gamma_{A3}, \gamma_{A4}, \gamma_{A5}, \gamma_{A6})^T \quad \in \mathbb{R}^6, \tag{7.79}$$

as shown in Fig.1a. Since the natural frequency of oscillations due to the stiffness in the first three joints are in the range of interest, an elastic joint model is introduced. A link–joint unit consists of two bodies, the drive and

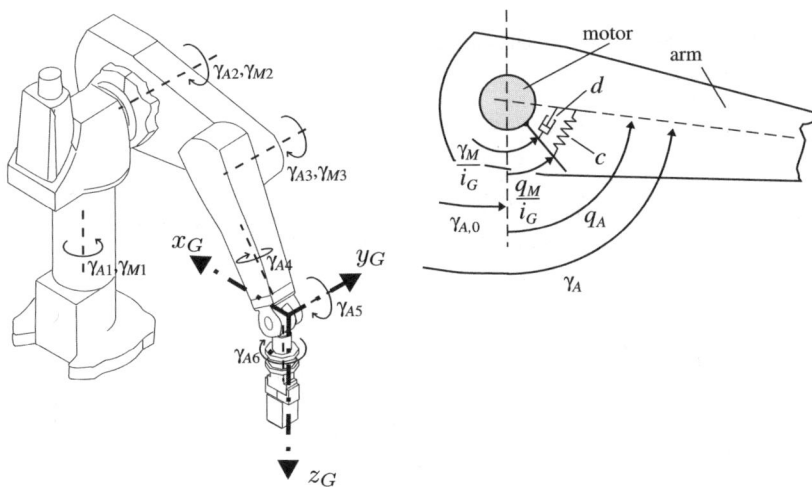

Fig. 7.22: The Industrial Robot PUMA 560 and its Joint Model

the arm segment (Figure 7.22). They are coupled by a gear model which is composed of the physical elements stiffness $c$ and damping $d$ (see equation (7.12) on page 417). Thus three additional degrees of freedom are introduced between motor shafts and arm segments:

$$\boldsymbol{\gamma}_M = (\gamma_{M1}, \gamma_{M2}, \gamma_{M3})^T. \quad \in \mathbb{R}^3. \tag{7.80}$$

In the remaining links no joint model is necessary, because their stiffnesses are high compared to the acting forces, and the elasticity of these joints have no effect on the system dynamics under consideration. Thus we come to altogether nine degrees of freedom $\boldsymbol{\gamma}$:

$$\boldsymbol{\gamma} = (\boldsymbol{\gamma}_M, \boldsymbol{\gamma}_A)^T \quad \in \mathbb{R}^9. \tag{7.81}$$

Specializing the multibody equations (7.10) and (7.11) on page 417 for our manipulator and regarding the above simplifications yields

$$\overline{M}(\gamma)\ddot{\gamma} + \overline{h}(\gamma,\dot{\gamma}) = B\overline{u} \quad \in \mathbb{R}^9, \tag{7.82}$$

with the inertia matrix $\overline{M} \in \mathbb{R}^{9,9}$, with the centrifugal, Coriolois and gravitational forces summarized in the vector $\overline{h} \in \mathbb{R}^9$ and with the control input $B\overline{u}$.

Because most mating tasks are limited to a small area of workpiece interaction the robot motion will be slow and centrifugal and Coriolis forces in equation (7.82) may be neglected compared to gravitational and inertia forces. Hence, the robot dynamics can be linearized around the working point $\gamma_0$, where $\dot{\gamma}_0 \equiv 0$. The vector

$$q = \gamma - \gamma_0 = (q_{M1}, q_{M2}, q_{M3}, q_1, q_2, q_3, q_4, q_5, q_6)^T \quad \in \mathbb{R}^9, \tag{7.83}$$

denotes the deviation from this working point. From that we obtain the following equations of motion:

$$M\ddot{q} + P\dot{q} + Qq = h + Bu \quad \in \mathbb{R}^9, \tag{7.84}$$

with the inertia matrix $M$, the damping matrix $P$ and the stiffness matrix $Q$. The vector $h$ contains the remaining gravitational forces, and $Bu$ regards the influence of the controller. For a PD position control we have the typical form of the vector $u$ [174]:

$$u = -K_P \left( B^T q - q_D \right) - K_D \left( B^T \dot{q} - \dot{q}_D \right) \quad \in \mathbb{R}^6. \tag{7.85}$$

The matrices $K_P$ and $K_D$ contain the positional and velocity feedback gains. Through changing the desired positions and velocities $q_D$ and $\dot{q}_D$ the motion of the manipulator along a trajectory is realized. By integration the equations of motion (7.82) or (7.84) we obtain for every time step the position and orientation of the gripper (**G**–frame) with respect to the inertial fixed system (**I**–frame), as well as its translational and rotational velocity

$$\underbrace{r_{IG}, \quad A_{IG},}_{\text{position and orientation}} \quad \underbrace{v_G, \quad \Omega_G,}_{\text{translational and angular velocity}} \tag{7.86}$$

which form the interface for the assembly process models.

### 7.3.2.2 Mating Models

Characteristic for assembly tasks performed by a manipulator is the contact with the environment through the mating parts. Additional loads act on the gripper and thus influence the motion of the manipulator. For numerical simulation, the dynamic model of the robot and the assembly process models have to be coupled.

## 7.3 Dynamics and Control of Assembly Processes with Robots

Depending on the parts properties, there are two basic approaches to describe these loads, assuming a force law in the contact or assuming a stiff contact. We consider as an example a peg touching a flat surface in one point. If we deal with compliant workpieces, typically made of rubber or plastics, the coupling between robot and environment is given by a smooth force law, and all these laws for contact forces and torques are regularized:

$$f = f(q, \dot{q}), \qquad m = m(q, \dot{q}). \tag{7.87}$$

These contact loads must be regarded on the right-hand side of the equations of motion to yield

$$M\ddot{q} + P\dot{q} + Qq = Bu + \left(J_T^T | J_R^T\right) \begin{pmatrix} f \\ m \end{pmatrix} \tag{7.88}$$

If we model the process by assuming stiff mating parts, like steel, aluminium or the like, we come out with a complementarity problem due to the unilateral properties of the contact manipulation, which means (see chapter 3.4 on page 131)

$$M\ddot{q} + P\dot{q} + Qq = Bu + W\lambda$$
$$g(q) \geq 0, \qquad \dot{g}(q, \dot{q}) \geq 0, \qquad \ddot{g}(q, \dot{q}, \ddot{q}) \geq 0, \tag{7.89}$$

where the vector $\lambda$ contains the constraint forces. The matrix $W$ is the constraint matrix which projects the constraint forces into the degrees of freedom of the robot. The geometrical constraints $g$ are needed on acceleration level in order to combine them with the equations of motion, also describing accelerations. The contact kinematical fundaments for evaluating the magnitudes $g, \dot{g}, \ddot{g}$ are already presented in the chapters 2.2.6 and 2.2.7 on the pages starting with page 31 (see also [154]).

### 7.3.2.3 Compliant Mating Parts

In this section we will consider two examples. The first one is an O–ring mounted on a piston to be inserted into a corresponding cylinder hole. The second one is a snap fastener, which plays an important role in automated assembly. The task is to find the correct relationship between the displacement and orientation $r_{IG}$ and $A_{IG}$ of the gripper with respect to the environment and the forces and moments $f$ and $m$ acting due to the deformation of the parts. For this purpose we have to introduce an additional local frame $\mathbf{L}$, necessary to describe the deformation of the workpiece.

*O-Ring*

Figure 7.23 depicts an elastic ring mounted in a groove of a piston. It is inserted into a hole with a rounded chamfer at the beginning and possibly also at the end. This is a typical application of O–rings in hydraulic cylinders

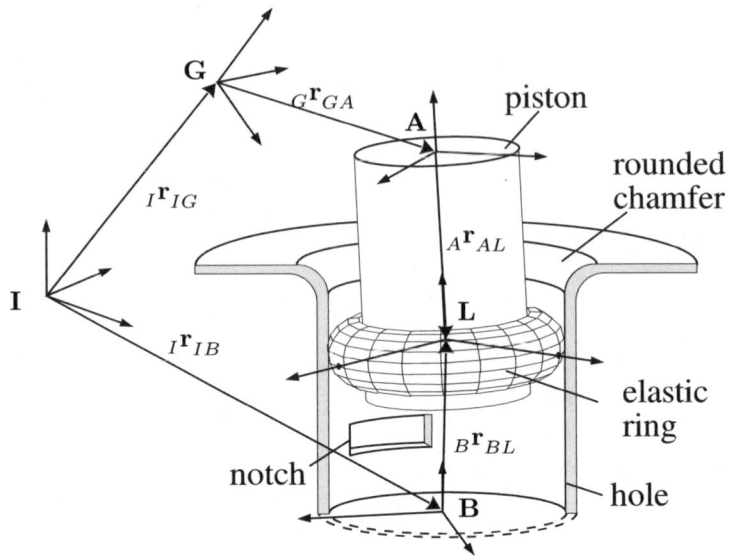

Fig. 7.23: O–Ring in a Groove on a Piston

or pneumatic valves. In the hole there might be notches serving as entrance or outlet.

An analytical solution for the stresses and strains in the elastic ring with approaches from continuum mechanics is not possible because the displacements and also frequently the material laws are nonlinear. Therefore we use a simplified approach, assuming the outer contour of the ring to be circular and rigid, so that there is a line contact between ring and hole. The only compliance taken into account is the radial stiffness of the O–ring, which is not discrete, but continuously distributed over the circumference. Its characteristic is supposed to be nonlinear quadratic. The line load representing the elastic deformation of the ring is $p_i$. A further assumption is that the ring does not tilt in the hole, because errors with respect to the orientation of industrial robots are generally small. According to the special geometry of the elastic ring in the groove, we use two contact line loads: one between the ring and the hole $p_a$ and one between the ring and the groove $p_o$. Frictional loads $p_{a,R}$ and $p_{o,R}$ act perpendicular to this normal loads and opposite the direction of the relative motion along the contact line.

All further considerations are made in the **L**–frame, which is placed at the center of the ring. It has a radial eccentricity $e$ and a rotation $\varphi_{BL}$ with respect to the **B**–frame. Its orientation is chosen such that its $z$ axis is parallel to the **B**–frame: $z_L \parallel z_B$ (because tilting is not regarded), and the $x_L$ axis coincides with the shortest stretch between piston and hole. The eccentricity $e$, the angle $\varphi_{BL}$ and the transformation matrix $\boldsymbol{A}_{BL}$ are defined as:

$$e = \sqrt{{}_B r_{BL,x}^2 + {}_B r_{BL,y}^2}, \qquad \varphi_{BL} = \arctan\frac{{}_B r_{BL,y}}{{}_B r_{BL,x}},$$

$$\mathbf{A}_{BL} = \begin{pmatrix} \cos\varphi_{BL} & \sin\varphi_{BL} & 0 \\ -\sin\varphi_{BL} & \cos\varphi_{BL} & 0 \\ 0 & 0 & 1 \end{pmatrix}. \tag{7.90}$$

Through this special choice all further considerations can be made in two modeling planes, the $x_L$–$y_L$–plane shown in Figure 7.24 on the left side and the $x_L$–$z_L$–plane shown on the right. In the left part of the figure the line load

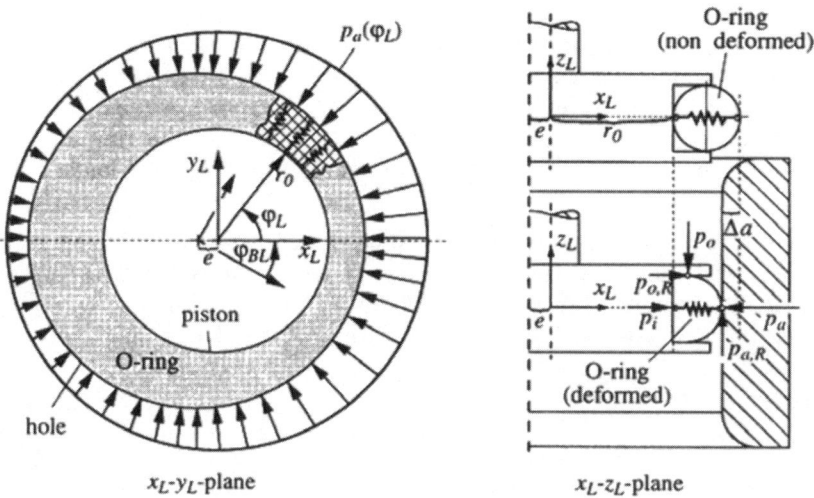

Fig. 7.24: The two Modeling Planes for the Description of the Mounting Task with the O–Ring

$p_a$ between ring and hole is drawn. The non uniform distribution of the load can be recognized.

The detection of the contact area is performed in the $x_L$–$z_L$–plane, where we then have to calculate the contact point between two circles, if the ring is entering or leaving the hole and touching one of the chamfers, or between a circle and a straight line, if the ring is inside the hole. For clarity we have drawn the line loads acting between the bodies in the right modeling plane in Figure 7.24.

If $\Delta a$ is the deformation of the O–ring at the $x_L$–axis, then the deformation around the circumference $a(\varphi_L)$ is:

$$a(\varphi_L) = \Delta a - e + e\cos\varphi_L. \tag{7.91}$$

The line load representing the radial elasticity of the ring has the form:

$$p_i(\varphi_L) = c_1 a(\varphi_L) + c_2 a^2(\varphi_L), \tag{7.92}$$

where the coefficients $c_1$ and $c_2$ are the spring constants. They can be determined by a FEM calculation, where the cross section of the elastic ring is radially deformed. From the resulting force–displacement–diagram the two coefficients are found by curve fitting. Inserting Eq.(7.91) in Eq.(7.92) yields:

$$p_i(\varphi_L) = k_0 + k_1 \cos\varphi_L + k_2 \cos^2\varphi_L,$$

$$k_0 = c_1\left(\Delta a - \frac{e}{2}\right) + c_2\left(\Delta a - \frac{e}{2}\right)^2,$$

$$k_1 = c_1\frac{e}{2} + c_2 e\left(\Delta a - \frac{e}{2}\right), \qquad k_2 = c_2\left(\frac{e}{2}\right)^2. \tag{7.93}$$

With this line load $p_i$ and the other loads $p_a, p_{a,R}, p_o, p_{o,R}$ the force equilibrium at an infinitesimal ring segment is formulated. Solving the equilibrium for the outer load $p_a$ yields:

$$p_a(\varphi_L) = \frac{k_0 + k_1 \cos\varphi_L + k_2 \cos^2\varphi_L}{1 - \mu^2}, \tag{7.94}$$

if the ring is in the hole. $\mu$ is the coefficient of friction between rubber and steel. For the contact with the chamfers the equations are slightly different, because the changing direction of the loads has to be taken into account. This load is split into the three cartesian directions of the **L**–frame:

$$p_x(\varphi_L) = -p_a(\varphi_L)\cos\varphi_L, \quad p_y(\varphi_L) = -p_a(\varphi_L)\sin\varphi_L, \quad p_z(\varphi_L) = \mu p_a(\varphi_L). \tag{7.95}$$

The resulting forces and moments acting on the origin of the **L**–frame are found through integration of $p_x$, $p_y$ and $p_z$ over the circumference of the ring:

$$_L\boldsymbol{f} = \begin{pmatrix} -2\int_0^\pi p_a(\varphi_L)\cos\varphi_L r_0 d\varphi_L \\ -2\int_0^\pi p_a(\varphi_L)\sin\varphi_L r_0 d\varphi_L \\ 2\mu\int_0^\pi p_a(\varphi_L) r_0 d\varphi_L \end{pmatrix}, \quad _L\boldsymbol{m} = \begin{pmatrix} 0 \\ 2\mu\int_0^\pi p_a(\varphi_L) r_0^2 \cos\varphi_L d\varphi_L \\ 2\mu\int_0^\pi p_a(\varphi_L) r_0^2 d\varphi_L \end{pmatrix}$$

$$\tag{7.96}$$

where $r_0$ is the radius of the bolt inside the groove. These two loads are then transformed into the gripper system (**G**–frame), in order to combine the assembly process with the robot model:

$$\begin{pmatrix} _G\boldsymbol{f} \\ _G\boldsymbol{m} \end{pmatrix} = \boldsymbol{A}_{GL}\begin{pmatrix} _L\boldsymbol{f} \\ _L\boldsymbol{m} +_L \boldsymbol{r}_{GL} \times_L \boldsymbol{f} \end{pmatrix}. \tag{7.97}$$

To verify our model we present a comparison between measurement and calculation. In the experiment a piston with a rubber ring $20 \times 3.15$ (ø $20mm$ of the ring, ⌀ $3.15mm$ of the crossection) was inserted into a hole with diameter of $26mm$. For the specific material the spring constants are $c_1 = 16.0[N/mm^2]$ and $c_2 = 43.0[N/mm^3]$. This special experiment was conducted on a force measurement machine, which is very stiff, in order to avoid any disturbances from the manipulator. Thus we can assume ideal conditions. The force–

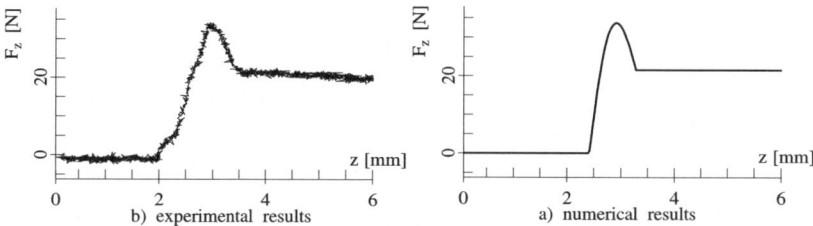

Fig. 7.25: Force–Distance Graph for O–Ring Insertion

distance graphs are shown in Figure 7.25. The left diagram contains the measurement, the right the calculations. The correspondence between both curves is very good. The maximum of the mating force arises, when the ring is entering the hole. Inside the cylinder the load is constant.

*Snap Fastener*

Snap fasteners are widespread fixtures in automated assembly. They consist of three different characteristic parts: the snap hook, the elastic support for the hook and the counterpart or chamfer (see Figure 7.26). The support consists of a beam like in the figure, a plate or an even more complicated structure. We make the assumption that the snap hook and the chamfer are rigid and only the compliant support is flexible.

We have to introduce a local **L**–frame, fixed to the snap hook, to describe the elasticity in the system. The deformations should be linear elastic, so that the vector ${}_A\boldsymbol{r}_{AL} = (w_x, w_y, w_z)^T$ contains the displacement and the vector ${}_A\boldsymbol{\varphi}_{AL} = (\varphi_x, \varphi_y, \varphi_z)^T$ represents the orientation between the **A**– and **L**–system, expressed in the **A**–frame (indicated by the left lower index). With this description the compliance in the support can be reduced to a stiffness matrix $\boldsymbol{K}$ between the **A**– and the **L**–frame. This is symbolized by the spring in Figure 7.26. Generally, the stiffness matrix $\boldsymbol{K}$ has the dimension $\boldsymbol{K} \in \mathbb{R}^{6,6}$. The relationship between the linear deformations and the linear elastic reaction forces has then the following form:

$$\begin{pmatrix} {}_A\boldsymbol{f} \\ {}_A\boldsymbol{m} \end{pmatrix} = \boldsymbol{K} \begin{pmatrix} {}_A\boldsymbol{r}_{AL} \\ {}_A\boldsymbol{\varphi}_{AL} \end{pmatrix} \quad \in \mathbb{R}^6, \tag{7.98}$$

where ${}_A\boldsymbol{f}$ is the vector of the forces and ${}_A\boldsymbol{m}$ is the vector of the torques acting at the origin of the L frame when the deformations ${}_A\boldsymbol{r}_{AL}$ and ${}_A\boldsymbol{\varphi}_{AL}$ are imposed.

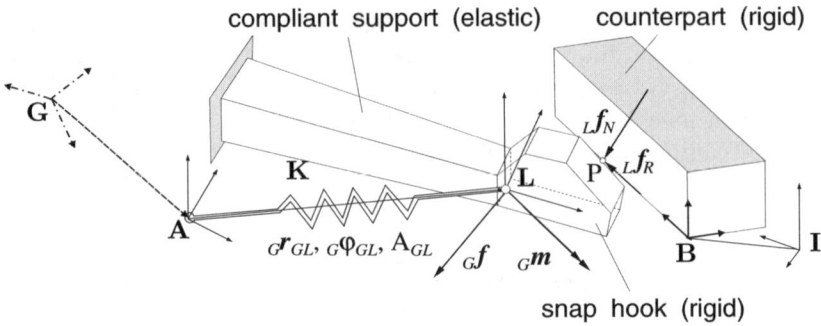

Fig. 7.26: Structure of Snap Fastener

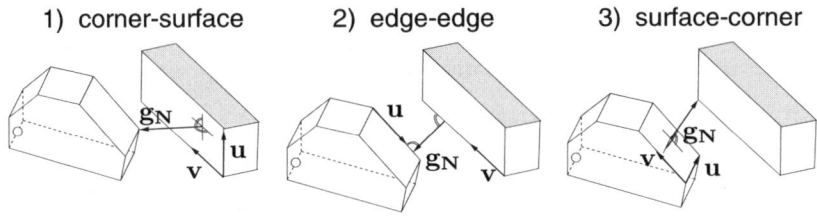

Fig. 7.27: Three Different Types of Contacts Between the Snap Hook and the Chamfer

The description of the geometry is easy for the snap fasteners under consideration. The counterpart is a simple cuboid and the hook a polygonal part with six corners. Thus there are three basic possibilities of contact points between the snap hook and the counterpart as indicated in Figure 7.27: corner–surface (type 1), edge–edge (type 2), and surface–corner (type 3). A contact between the flexible part and the counterpart is not regarded. The location of a contact point is always indicated by two parameters $u$ and $v$, which will be needed later in the equations of the force equilibrium.

When the snap hook and the counterpart get in contact the parts will slide on each other and the hook will be displaced and twisted. In order to determine this movement and the accompanying forces we have to calculate the equilibrium position between these two parts. There is a force equilibrium between the elastic forces $_A\boldsymbol{f}$ and $_A\boldsymbol{m}$ on the one side and the contact forces $_L\boldsymbol{f}_N$ and $_L\boldsymbol{f}_R$ on the other side. The normal contact force $_L\boldsymbol{f}_N$ acts on the touching point. Its direction depends on the type of the contact. For type 1 and 3 (corner–surface) $_L\boldsymbol{f}_N$ is normal to the plane, for type 2 (edge–edge) $_L\boldsymbol{f}_N$ is parallel to the cross product of the two lines. The friction force $_L\boldsymbol{f}_R$ acts perpendicular to $_L\boldsymbol{f}_N$ and opposite to the direction of motion. For the formulation of the equilibrium, the forces have to be transformed into the same coordinate system, here the **L** frame:

7.3 Dynamics and Control of Assembly Processes with Robots  461

$$\begin{pmatrix} A_{LA} & 0 \\ 0 & A_{LA} \end{pmatrix} \begin{pmatrix} {}_A f \\ {}_A m \end{pmatrix} = \begin{pmatrix} {}_L f_N + {}_L f_R \\ {}_L r_{LP} \times ({}_L f_N + {}_L f_R) \end{pmatrix}, \tag{7.99}$$

where ${}_L r_{LP}$ is the vector from the origin of the L frame to the actual contact point P. $A_{LA}$ is the transformation matrix between the **A**– and the **L** frame:

$$A_{LA} = \begin{pmatrix} 1 & \varphi_z & -\varphi_y \\ -\varphi_z & 1 & \varphi_x \\ \varphi_y & -\varphi_x & 1 \end{pmatrix}. \tag{7.100}$$

Inserting eq.(7.99) into eq.(7.98) results in a system of six nonlinear equations:

$$\begin{pmatrix} A_{LA} & 0 \\ 0 & A_{LA} \end{pmatrix} K \begin{pmatrix} {}_A r_{AL} \\ {}_A \varphi_{AL} \end{pmatrix} = \begin{pmatrix} {}_L f_N + {}_L f_R \\ {}_L r_{LP} \times ({}_L f_N + {}_L f_R) \end{pmatrix}. \tag{7.101}$$

These equations are rather complicated due to the multiplication with the transformation matrix. Therefore, for the evaluation the symbolic manipulation program MAPLE V has been applied. From the contact condition we obtain three additional equations, so that we have altogether a set of nine nonlinear algebraic equations. The unknowns are the six parameters for the position and the orientation of the hook ($w_x$, $w_y$, $w_z$, $\varphi_x$, $\varphi_y$, $\varphi_z$), two parameters for the contact between the parts ($u$, $v$) and the magnitude of the normal contact force $|_L f_N|$. Determining the mating forces has been described for one single contact. The problem can be solved for up to three contact points. For every additional contact we obtain three more equations from geometry and three additional unknowns ($u_i$, $v_i$, $|_L f_N|_i$). The system of nonlinear equations has then the dimension $\mathbb{R}^{6+3n}$, where $n$ is the number of contacts $n = 1, 2$ or 3. The transformation into the gripper system is rather simple:

$$\begin{pmatrix} {}_G f \\ {}_G m \end{pmatrix} = A_{GL} \begin{pmatrix} {}_L f_N + {}_L f_R \\ {}_L r_{GP} \times ({}_L f_N + {}_L f_R) \end{pmatrix}. \tag{7.102}$$

The determination of the stiffness matrix $K$ is still missing. It represents the compliances of the parts. The elastic support might consist of a beam (Figures 7.26 and 7.28) or a plate (Figure 7.30). We apply beam and plate theory, respectively. In a first step we regard the snap fastener from Figure 7.28. According to the picture the displacement in the $x_A$ ($w_x$) and $y_A$ ($w_y$) direction and the twist around the $x_A$ ($\varphi_x$) and $z_A$ ($\varphi_z$) axes are constrained. The stiffness in these directions would be very high compared to the other elements of $K$, therefore generating approximately zeros in the first, second, fourth and sixth rows and columns. From beam theory (BERNOULLI–beam) the deflection curve is derived from the following differential equation:

$$EI_y(x) w_z'''(x) = -F_z, \qquad I_y(x) = \frac{1}{12}(a + \frac{x}{l}(b-a))^3 c. \tag{7.103}$$

The parameters $a$, $b$, $c$ and $l$ can be seen from Figure 7.28, and $E$ is the modulus of elasticity. Integrating equation (7.103) three times and using the

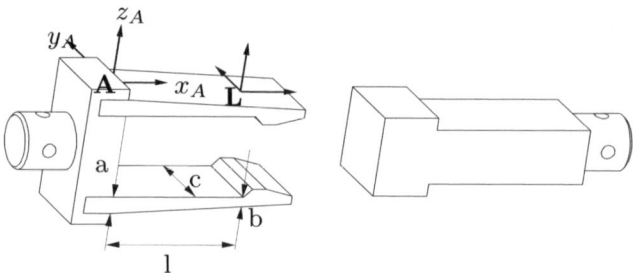

Fig. 7.28: Snap Fastener with a Beam as Elastic Support

boundary conditions $w_z(0) = 0$, $w'_z(0) = 0$ and $EI_y(l)w''_z(l) = -M_y$, yields a relationship between the displacement $w_z$ and the twisting angle $\varphi_y$ ($\varphi_y = -w'_z(l)$) of the beam and the force $F_z$ and the moment $M_y$. With $a = 5mm$, $b = 2.7mm$, $c = 20mm$, $l = 40mm$ and $E = 2700N/mm^2$ we obtain the following stiffness relationship

$$\begin{pmatrix} F_x \\ F_y \\ F_z \\ M_x \\ M_y \\ M_z \end{pmatrix} = \begin{pmatrix} 0 & 0 & 0 & 0 & 0 & 0 \\ 0 & 0 & 0 & 0 & 0 & 0 \\ 0 & 0 & 45.5 & 0 & 638.8 & 0 \\ 0 & 0 & 0 & 0 & 0 & 0 \\ 0 & 0 & 638.8 & 0 & 14285.4 & 0 \\ 0 & 0 & 0 & 0 & 0 & 0 \end{pmatrix} \begin{pmatrix} w_x \\ w_y \\ w_z \\ \varphi_x \\ \varphi_y \\ \varphi_z \end{pmatrix}. \quad (7.104)$$

Our model is verified by a comparison of measurements and calculations. Measurements were made using again the single axis force measurement machine. Figure 7.29 shows the force vs. distance for the insertion of the upper half of the snap fastener from Figure 7.28. $F_x$ is the force in the direction of insertion, and $F_z$ acts perpendicularly. When mating the complete fitting with both parts, $F_x$ becomes twice as large, and $F_z$ disappears because of symmetry.

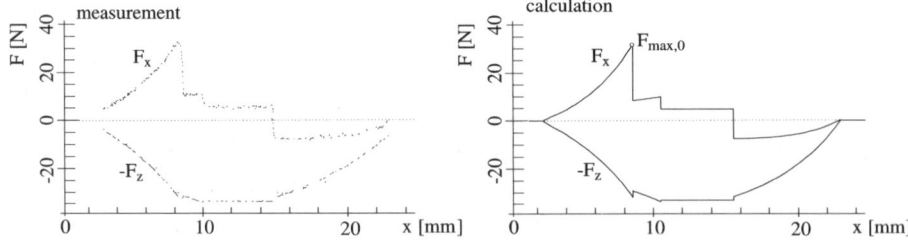

Fig. 7.29: Snap Fastener Insertion with Beam - Force vs. Distance

## 7.3 Dynamics and Control of Assembly Processes with Robots

Our second example is a snap fastener with a plate as elastic support from Figure 7.30. Here the displacement in the $x_A$ ($w_x$) and $y_A$ ($w_y$) direction and the twist around the $z_A$ ($\varphi_z$) axes are constrained. Therefore $\boldsymbol{K}$ contains zeros in the first, second and sixth rows and columns. We assume a KIRCHHOFF plate. The bending is approximated by a Ritz–series $w_z(x,y) = \boldsymbol{q}^T \boldsymbol{w}(x,y)$, with the coordinates $\boldsymbol{q}$ and the shape functions $\boldsymbol{w}$. As shape functions we use piecewise defined cubic splines which satisfy the boundary conditions. The coordinates $\boldsymbol{q}$ of the shape functions are found by minimizing the potential $\Pi = W_i - W_a$. We therefore have to solve the variational problem $\left(\dfrac{\partial \Pi}{\partial \boldsymbol{q}}\right)^T = \boldsymbol{0}$ with $W_i$, the elastic energy in the plate:

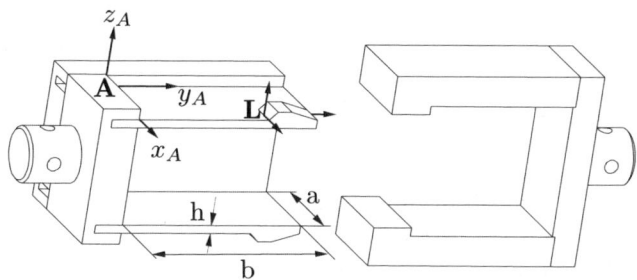

Fig. 7.30: Snap Fastener with a Plate as Elastic Support

$$W_i = \frac{1}{2} D\, \boldsymbol{q}^T \int_0^a \int_0^b \boldsymbol{W}^T \begin{pmatrix} 1 & \nu & 0 \\ \nu & 1 & 0 \\ 0 & 0 & 2(1-\nu) \end{pmatrix} \boldsymbol{W}\, dy\, dx\, \boldsymbol{q},$$

$$D = \frac{Eh^3}{12(1-\nu^2)}, \qquad \boldsymbol{W} = [\boldsymbol{w}_{,xx} | \boldsymbol{w}_{,yy} | \boldsymbol{w}_{,xy}]^T. \tag{7.105}$$

In these equations $a$, $b$ and $h$ describe the geometry of the plate according to Figure 7.30, $E$ is the modulus of elasticity and $\nu$ is Poisson's ratio. $W_a$ is the work done by the loads $F_z$, $M_x$ and $M_y$. Let $x_L$ and $y_L$ be the coordinates of the origin of the **L**–frame, then

$$W_a = [F_z | M_x | M_y] \begin{pmatrix} +\boldsymbol{w}^T(x_L, y_L) \\ +\boldsymbol{w}_{,y}^T(x_L, y_L) \\ -\boldsymbol{w}_{,x}^T(x_L, y_L) \end{pmatrix} \boldsymbol{q}. \tag{7.106}$$

After differentiating the potential with respect to $\boldsymbol{q}$, we get a system of linear equations. Let $\hat{\boldsymbol{q}}$ be the solution of the equations. The dimension of the system depends on the number of shape functions we use. We then calculate the deformation of the point $x_L$, $y_L$ using

$$\begin{pmatrix} w_z \\ \varphi_x \\ \varphi_y \end{pmatrix} = \begin{pmatrix} +\boldsymbol{w}^T(x_L, y_L) \\ +\boldsymbol{w}_{,y}^T(x_L, y_L) \\ -\boldsymbol{w}_{,x}^T(x_L, y_L) \end{pmatrix} \widehat{\boldsymbol{q}}. \tag{7.107}$$

For $a = 32mm$, $b = 66mm$, $h = 3mm$, $E = 2700N/mm^2$, $\nu = 0.3$ and $x_L = 27mm$, $y_L = 48mm$, $\boldsymbol{K}$ becomes

$$\begin{pmatrix} F_x \\ F_y \\ F_z \\ M_x \\ M_y \\ M_z \end{pmatrix} = \begin{pmatrix} 0 & 0 & 0 & 0 & 0 & 0 \\ 0 & 0 & 0 & 0 & 0 & 0 \\ 0 & 0 & 168.5 & -455.8 & 2105.1 & 0 \\ 0 & 0 & -455.8 & 26794.7 & -1443.3 & 0 \\ 0 & 0 & 2105.1 & -1443.3 & 37700.4 & 0 \\ 0 & 0 & 0 & 0 & 0 & 0 \end{pmatrix} \begin{pmatrix} u \\ v \\ w \\ \alpha \\ \beta \\ \gamma \end{pmatrix}. \tag{7.108}$$

In Figure 7.31 the results from the experiments and the calculations are shown, for the insertion of the snap fastener from Figure 7.30.

Altogether we recognize a good correspondence between theory and experiment for both cases. The force vs. distance graphs show an unsteady shape because of the nonsmooth contour of the snap hooks. It is also observed that the jumps in the mating force are sharper in the calculation than in the measurement. This results from local deformations of the snap hook especially when the contact forces become very high, for example at about $25mm$ in Figure 7.31.

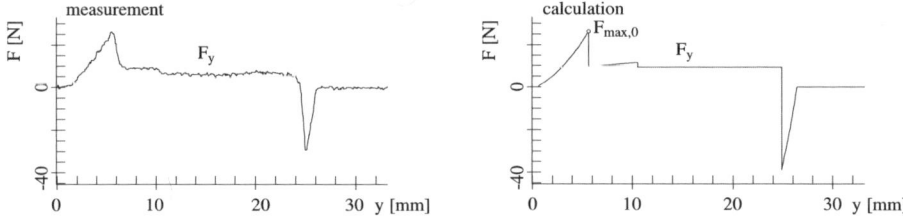

Fig. 7.31: Snap Fastener Insertion with Plate - Force vs. Distance

### 7.3.2.4 Rigid Mating Parts

*Some Fundaments*

For this topic we refer to the contributions of [152] and [154]. The basic theory of relative contact kinematics is presented by the section 2.2.7 on page 36 and the following pages. For our specific case under consideration we shall repeat some of the most important formulas.

We regard rigid workpieces, where the deformation during assembly is very small. Then, every contact point represents a constraint with respect to

## 7.3 Dynamics and Control of Assembly Processes with Robots

the robot dynamics. A constraint in normal direction is the relative distance between two points going to get into contact. This relative distance $g_N$ and its time derivative $\dot{g}_N$ write

$$g_N = \boldsymbol{n}_1^T \boldsymbol{r}_D, \quad \Longrightarrow \quad \dot{g}_N = \dot{\boldsymbol{n}}_1^T \boldsymbol{r}_D + \boldsymbol{n}_1^T \dot{\boldsymbol{r}}_D. \tag{7.109}$$

Considering section 2.2.7 on page 36 and especially the equations (2.127) on page 45 with the appropriate definitions we come out with

$$\dot{g}_N = ((\boldsymbol{\Omega}_G \times \boldsymbol{n}_1) + (\alpha \boldsymbol{u} + \beta \boldsymbol{v})\dot{u}_1 + (\alpha' \boldsymbol{u} + \beta' \boldsymbol{v})\dot{v}_1)^T \boldsymbol{r}_D + \\ \boldsymbol{n}_1^T(\underbrace{\dot{\boldsymbol{r}}_{IG}}_{\boldsymbol{v}_G} + \boldsymbol{\Omega}_G \times \underbrace{(\boldsymbol{r}_{GA} + \boldsymbol{r}_{\Sigma 1})}_{\boldsymbol{r}_{GC1}} + u_1\dot{u}_1 + v_1\dot{v}_1 - u_2\dot{u}_2 - v_2\dot{v}_2). \tag{7.110}$$

This equation can also be written in the form (see equation(2.122) on page 44)

$$\dot{g}_N = \boldsymbol{n}_1^T \boldsymbol{v}_{C1}, \quad \text{where: } \boldsymbol{v}_{C1} = \boldsymbol{v}_G + \boldsymbol{\Omega}_G \times \boldsymbol{r}_{GC1}. \tag{7.111}$$

The normal constraint is active, if the distance is zero, $g_N = 0$. $\boldsymbol{v}_{C1}$ is the velocity of the potential contact point on the upper body, which is connected to the gripper (see Figure 2.21 on Page 43). In the two tangential directions a constraint is active, if sticking occurs, which means that the relative sliding velocities $\dot{g}_U$ and $\dot{g}_V$ at a contact point vanish $\dot{g}_U = 0$, $\dot{g}_V = 0$. $\dot{g}_U$ and $\dot{g}_V$ are defined as the projection of $\boldsymbol{v}_{C1}$ on the two tangents $\boldsymbol{u}_1$ and $\boldsymbol{v}_1$

$$\dot{g}_U = \boldsymbol{u}_1^T \boldsymbol{v}_{C1}, \qquad \dot{g}_V = \boldsymbol{v}_1^T \boldsymbol{v}_{C1}. \tag{7.112}$$

In order to combine these constraint equations with the equations of motion of the manipulator we put them on an acceleration level. Differentiating the equations (7.109) and (7.111) once with respect to time yields

$$\begin{aligned} \ddot{g}_N &= \boldsymbol{n}_1^T \dot{\boldsymbol{v}}_{C1} + \dot{\boldsymbol{n}}_1^T \boldsymbol{v}_{C1}, \\ \ddot{g}_U &= \boldsymbol{u}_1^T \dot{\boldsymbol{v}}_{C1} + \dot{\boldsymbol{u}}_1^T \boldsymbol{v}_{C1}, \\ \ddot{g}_V &= \boldsymbol{v}_1^T \dot{\boldsymbol{v}}_{C1} + \dot{\boldsymbol{v}}_1^T \boldsymbol{v}_{C1}. \end{aligned} \tag{7.113}$$

The time derivatives of the surface normal $\dot{\boldsymbol{n}}_1$ and the surface tangents $\dot{\boldsymbol{u}}_1$, $\dot{\boldsymbol{v}}_1$ are defined as

$$\begin{aligned} \dot{\boldsymbol{n}}_1 &= \boldsymbol{\Omega}_G \times \boldsymbol{n}_1 + \frac{\partial \boldsymbol{n}_1}{\partial u_1}\dot{u}_1 + \frac{\partial \boldsymbol{n}_1}{\partial v_1}\dot{v}_1, \\ \dot{\boldsymbol{u}}_1 &= \boldsymbol{\Omega}_G \times \boldsymbol{u}_1 + \frac{\partial \boldsymbol{u}_1}{\partial u_1}\dot{u}_1 + \frac{\partial \boldsymbol{u}_1}{\partial v_1}\dot{v}_1, \\ \dot{\boldsymbol{v}}_1 &= \boldsymbol{\Omega}_G \times \boldsymbol{v}_1 + \frac{\partial \boldsymbol{v}_1}{\partial u_1}\dot{u}_1 + \frac{\partial \boldsymbol{v}_1}{\partial v_1}\dot{v}_1, \end{aligned} \tag{7.114}$$

where the partial derivatives $\frac{\partial \boldsymbol{n}_1}{\partial u_1}, \frac{\partial \boldsymbol{n}_1}{\partial v_1}, \frac{\partial \boldsymbol{u}_1}{\partial u_1}, \frac{\partial \boldsymbol{u}_1}{\partial v_1}, \frac{\partial \boldsymbol{v}_1}{\partial u_1}, \frac{\partial \boldsymbol{v}_1}{\partial v_1}$ are known from the equations (2.84) on page 35. The velocity $\boldsymbol{v}_{C1}$ and thus its time derivative

$\dot{v}_{C1}$ can be expressed by the generalized coordinates of the manipulator $q$, $\dot{q}$, because the body with point $C1$ is connected to the robot's gripper

$$v_{C1} = v_G + \Omega_G \times r_{GC1} = J_T \dot{q} - \tilde{r}_{GC1} J_R \dot{q} = J_{C1} \dot{q}, \tag{7.115}$$

with $J_{C1}$ being the translational Jacobian with respect to the contact point $C1$. The matrix $\tilde{r}_{GC1}$ substitutes the crossproduct $\tilde{r}_{GC1} \Omega_G = r_{GC1} \times \Omega_G$. Differentiating this equation yields

$$\dot{v}_{C1} = \dot{v}_G + \dot{\Omega}_G \times r_{GC1} + \Omega_G \times v_{C1} + \Omega_G \times (u_1 \dot{u}_1 + v_1 \dot{v}_1)$$

$$= \underbrace{(J_T - \tilde{r}_{GC1} J_R)}_{J_{C1}} \ddot{q} +$$

$$+ \underbrace{\left(\dot{J}_T - \tilde{r}_{GC1} \dot{J}_R\right)}_{\dot{j}_{C1}} \dot{q} + \Omega_G \times v_{C1} + \Omega_G \times (u_1 \dot{u}_1 + v_1 \dot{v}_1). \tag{7.116}$$

With the help of the equations (7.114) and (7.83) we can rewrite equation (7.113) in the following form

$$\ddot{g}_N = n_1^T \left(J_{C1} \ddot{q} + j_{C1} + \Omega_G \times v_{C1} + \Omega_G \times (u_1 \dot{u}_1 + v_1 \dot{v}_1)\right) +$$
$$+ v_{C1}^T (\Omega_G \times n_1) + v_{C1}^T [(\alpha_1 u_1 + \beta_1 v_1) \dot{u}_1 + (\alpha_1' u_1 + \beta_1' v_1) \dot{v}_1]$$

$$\ddot{g}_U = u_1^T [J_{C1} \ddot{q} + j_{C1} + \Omega_G \times v_{C1} + \Omega_G \times (u_1 \dot{u}_1 + v_1 \dot{v}_1)] +$$
$$+ v_{C1}^T (\Omega_G \times u_1) + v_{C1}^T [(\Gamma_{11,1}^1 u_1 + \Gamma_{11,1}^2 v_1 + L_1 n_1) \dot{u}_1 +$$
$$+ (\Gamma_{12,1}^1 u_1 + \Gamma_{12,1}^2 v_1 + M_1 n_1) \dot{v}_1]$$

$$\ddot{g}_V = v_1^T \left(J_{C1} \ddot{q} + j_{C1} + \Omega_G \times v_{C1} + \Omega_G \times (u_1 \dot{u}_1 + v_1 \dot{v}_1)\right) +$$
$$+ v_{C1}^T (\Omega_G \times v_1) + v_{C1}^T [(\Gamma_{12,1}^1 u_1 + \Gamma_{12,1}^2 v_1 + M_1 n_1) \dot{u}_1 +$$
$$+ (\Gamma_{22,1}^1 u_1 + \Gamma_{22,1}^2 v_1 + N_1 n_1) \dot{v}_1]. \tag{7.117}$$

A simplification is possible if we substitute the scalar products $v_{C1}^T n_1$, $v_{C1}^T u_1$ and $v_{C1}^T v_1$ through the constraints on velocity level $\dot{g}_N$, $\dot{g}_U$ and $\dot{g}_V$ from the equations (7.109) and (7.111). We also know that $\dot{g}_N$ disappears, when the normal constraint is active. The relative sliding velocities $\dot{g}_U$ and $\dot{g}_V$ vanish, if the tangential constraints are active (stiction). Regarding all this, the constraint equations have the final form:

## 7.3 Dynamics and Control of Assembly Processes with Robots

$$\ddot{g}_N = \underbrace{\boldsymbol{n}_1^T \boldsymbol{J}_{C1}}_{\boldsymbol{w}_N^T} \ddot{\boldsymbol{q}} +$$

$$+ \underbrace{\boldsymbol{n}_1^T \boldsymbol{j}_{C1} + \boldsymbol{n}_1^T \left( \boldsymbol{\Omega}_G \times (\boldsymbol{u}_1 \dot{u}_1 + \boldsymbol{v}_1 \dot{v}_1) \right) + \dot{\boldsymbol{g}}_U (\alpha_1 \dot{u}_1 + \alpha_1' \dot{v}_1) + \dot{\boldsymbol{g}}_V (\beta_1 \dot{u}_1 + \beta_1' \dot{v}_1)}_{\tilde{w}_N},$$

$$\ddot{g}_U = \underbrace{\boldsymbol{u}_1^T \boldsymbol{J}_{C1}}_{\boldsymbol{w}_U^T} \ddot{\boldsymbol{q}} + \underbrace{\boldsymbol{u}_1^T \boldsymbol{j}_{C1} + \boldsymbol{u}_1^T \left( \boldsymbol{\Omega}_G \times \boldsymbol{v}_1 \right) \dot{v}_1}_{\tilde{w}_U},$$

$$\ddot{g}_V = \underbrace{\boldsymbol{v}_1^T \boldsymbol{J}_{C1}}_{\boldsymbol{w}_V^T} \ddot{\boldsymbol{q}} + \underbrace{\boldsymbol{v}_1^T \boldsymbol{j}_{C1} + \boldsymbol{v}_1^T \left( \boldsymbol{\Omega}_G \times \boldsymbol{u}_1 \right) \dot{u}_1}_{\tilde{w}_V}.$$

(7.118)

The terms linearly dependent on the generalized accelerations $\ddot{\boldsymbol{q}}$ are summarized in the constraint vectors $\boldsymbol{w}_N$ in normal and $\boldsymbol{w}_U$, $\boldsymbol{w}_V$ in tangential direction. The remaining parts are abbreviated with the scalar values $\tilde{w}_N$, $\tilde{w}_U$, $\tilde{w}_V$.

The only unknowns in equation (7.118) are the time derivatives of the parameters $\dot{u}_1, \dot{v}_1, \dot{u}_2, \dot{v}_2$. They describe the motion of the contact point on the two surfaces during the simulation. To evaluate these derivatives we require, that the equations specifing the contact point always have to be fulfilled. Their time derivatives have to disappear:

$$\frac{d}{dt}\left(\boldsymbol{n}_1^T \boldsymbol{u}_2\right) = 0, \quad \frac{d}{dt}\left(\boldsymbol{n}_1^T \boldsymbol{v}_2\right) = 0, \quad \frac{d}{dt}\left(\boldsymbol{r}_D^T \boldsymbol{u}_2\right) = 0, \quad \frac{d}{dt}\left(\boldsymbol{r}_D^T \boldsymbol{v}_2\right) = 0.$$

(7.119)

Figuring out equation (7.119) we obtain a system of equations, which are linear in the derivatives of the contour parameters:

$$\begin{pmatrix} \boldsymbol{u}_2^T (\alpha_1 \boldsymbol{u}_1 + \beta_1 \boldsymbol{v}_1) & \boldsymbol{u}_2^T (\alpha_1' \boldsymbol{u}_1 + \beta_1' \boldsymbol{v}_1) & L_2 & M_2 \\ \boldsymbol{v}_2^T (\alpha_1 \boldsymbol{u}_1 + \beta_1 \boldsymbol{v}_1) & \boldsymbol{v}_2^T (\alpha_1' \boldsymbol{u}_1 + \beta_1' \boldsymbol{v}_1) & M_2 & N_2 \\ -\boldsymbol{u}_1^T \boldsymbol{u}_2 & -\boldsymbol{u}_1^T \boldsymbol{u}_2 & E_2 & F_2 \\ -\boldsymbol{u}_1^T \boldsymbol{v}_2 & -\boldsymbol{u}_1^T \boldsymbol{v}_2 & F_2 & G_2 \end{pmatrix} \begin{pmatrix} \dot{u}_1 \\ \dot{v}_1 \\ \dot{u}_2 \\ \dot{v}_2 \end{pmatrix} =$$

$$= \begin{pmatrix} -\boldsymbol{\Omega}_1^T (\boldsymbol{u}_2 \times \boldsymbol{n}_1) \\ -\boldsymbol{\Omega}_1^T (\boldsymbol{v}_2 \times \boldsymbol{n}_1) \\ \boldsymbol{u}_2^T \boldsymbol{v}_{C1} \\ \boldsymbol{v}_2^T \boldsymbol{v}_{C1} \end{pmatrix}.$$

(7.120)

This linear problem has to be solved at every time step of the numerical integration.

Between mating parts more than one sliding or sticking contact point may exist, so that a variable number of constraints is active during the simulation. Let $n_n$ be the number of contact points and $n_T$ the number of sticking contact points. Then the constraint equations in vector form are:

$$\ddot{g}_N = W_N^T \ddot{q} + \tilde{w}_N, \quad W_N = (w_{N,1}, \ldots, w_{N,n_N}) \in \mathbb{R}^{f,n_N},$$
$$\tilde{w}_N = (\tilde{w}_{N,1}, \ldots, \tilde{w}_{N,n_N})^T \in \mathbb{R}^{n_N},$$
$$\ddot{g}_U = W_U^T \ddot{q} + \tilde{w}_U, \quad W_U = (w_{U,1}, \ldots, w_{U,n_T}) \in \mathbb{R}^{f,n_T},$$
$$\tilde{w}_U = (\tilde{w}_{U,1}, \ldots, \tilde{w}_{U,n_T})^T \subset \mathbb{R}^{n_T},$$
$$\ddot{g}_V = W_V^T \ddot{q} + \tilde{w}_V, \quad W_V = (w_{V,1}, \ldots, w_{V,n_T}) \in \mathbb{R}^{f,n_T},$$
$$\tilde{w}_V = (\tilde{w}_{V,1}, \ldots, \tilde{w}_{V,n_T})^T \in \mathbb{R}^{n_T}. \tag{7.121}$$

They are combined with the equations of motion and thus form a system of differential algebraic equations:

$$M\ddot{q} = \hat{h} + (W_N + W_F)\lambda_N + W_U \lambda_U + W_V \lambda_V,$$
$$\text{with: } \hat{h} = h + Bu - P\dot{q} - Qq, \tag{7.122}$$

$$\underbrace{\begin{pmatrix} \ddot{g}_N \\ \ddot{g}_N \\ \ddot{g}_N \end{pmatrix}}_{\ddot{g}} = \underbrace{\begin{pmatrix} W_N^T \\ W_U^T \\ W_V^T \end{pmatrix}}_{W^T} \ddot{q} + \underbrace{\begin{pmatrix} \tilde{w}_N \\ \tilde{w}_U \\ \tilde{w}_V \end{pmatrix}}_{\tilde{w}} ., \tag{7.123}$$

The components of the vectors $\lambda_N = (\lambda_{N,1}, \ldots, \lambda_{N,n_N})^T$, $\lambda_U = (\lambda_{U,1}, \ldots, \lambda_{U,n_T})^T$ and $\lambda_V = (\lambda_{V,1}, \ldots, \lambda_{V,n_T})^T$ correspond to the unknown constraint forces normal and tangential to the respective tangent plane. The term $W_F \lambda_N$ considers frictional forces in all contact points where sliding occurs. The sliding direction is given by the relative tangential velocity of the contact point under consideration, $v_{C1}$, the magnitude by the normal contact force $\lambda_N$, applying Coulomb's friction law. The vector $v_{C1}$ can be split into two tangential directions, which are assumed to be perpendicular

$$v_{C1} = \frac{u_1^T v_{C1}}{u_1^T u_1} u_1 + \frac{v_1^T v_{C1}}{v_1^T v_1} v_1 \quad \Longrightarrow$$

$$v_{C1} = \frac{\dot{g}_U}{E_1} u_1 + \frac{\dot{g}_V}{G_1} v_1, \quad |v_C| = \sqrt{\frac{\dot{g}_U^2}{E_1} + \frac{\dot{g}_V^2}{G_1}}. \tag{7.124}$$

The vector of the friction force $f_R$ at a sliding contact point is then defined as

$$f_R = -\frac{v_{C1}}{|v_{C1}|} \mu \lambda_N = u_1 \frac{-\mu \dot{g}_U \lambda_N}{E_1 \sqrt{\frac{\dot{g}_U^2}{E_1} + \frac{\dot{g}_V^2}{G_1}}} + v_1 \frac{-\mu \dot{g}_V \lambda_N}{G_1 \sqrt{\frac{\dot{g}_U^2}{E_1} + \frac{\dot{g}_V^2}{G_1}}} =$$

$$= \left(u_1 \frac{\dot{g}_U}{E_1} + v_1 \frac{\dot{g}_V}{G_1}\right) \frac{-\mu \lambda_N}{\sqrt{\frac{\dot{g}_U^2}{E_1} + \frac{\dot{g}_V^2}{G_1}}}. \tag{7.125}$$

The projection of $f_R$ into the space of the generalized coordinates is realized by multiplication with the Jacobian with respect to the contact point $J_{C1}$

$$\boldsymbol{w}_F = \boldsymbol{J}_{C1}^T \left( \boldsymbol{u}_1 \frac{\dot{g}_U}{E_1} + \boldsymbol{v}_1 \frac{\dot{g}_V}{G_1} \right) \frac{-\mu \lambda_N}{\sqrt{\frac{\dot{g}_U^2}{E_1} + \frac{\dot{g}_V^2}{G_1}}} =$$
$$= \left( \boldsymbol{w}_U \frac{\dot{g}_U}{E_1} + \boldsymbol{w}_V \frac{\dot{g}_V}{G_1} \right) \frac{-\mu \lambda_N}{\sqrt{\frac{\dot{g}_U^2}{E_1} + \frac{\dot{g}_V^2}{G_1}}}. \qquad (7.126)$$

The matrix $\boldsymbol{W}_F$ is then composed by the sliding friction constraint vectors $\boldsymbol{w}_F$, as defined above. At the contact points with stiction the elements $\boldsymbol{w}_F$ are zero. We have

$$\boldsymbol{W}_F = (\boldsymbol{w}_{R,1}, \ldots, \boldsymbol{w}_{R,n_N}) \in \mathbb{R}^{f,n_N}. \qquad (7.127)$$

The system of differential algebraic equations (7.122) and (7.123) can easily be evaluated, if all active constraints are known and do not change, $\ddot{\boldsymbol{g}} = 0$. We solve equation (7.122) for the accelerations $\ddot{\boldsymbol{q}}$ and insert these accelerations into the algebraic equation (7.123), which yields

$$\underbrace{\boldsymbol{W}^T \boldsymbol{M}^{-1} ((\boldsymbol{W}_N + \boldsymbol{W}_F) | \boldsymbol{W}_U | \boldsymbol{W}_V)}_{A} \begin{pmatrix} \lambda_N \\ \lambda_U \\ \lambda_V \end{pmatrix} + \underbrace{\boldsymbol{W}^T \boldsymbol{M}^{-1} \widehat{\boldsymbol{h}} + \widetilde{\boldsymbol{w}}}_{b} = 0. \qquad (7.128)$$

This system of linear equations has to be solved at every time step of numerical integration, to evaluate the constraint forces. They are then inserted into the constrained equations of motion (7.123) to simulate the robot in contact with the environment.

Special treatment is necessary, if the constraints are changing during the insertion. For this purpose we have defined special indicators, that notify a transition in the state of a the contact points. The resulting sliding velocity in the tangential plane at the contact point is $\dot{g}_T = |\boldsymbol{v}_{C1}| = \sqrt{\frac{\dot{g}_U^2}{E} + \frac{\dot{g}_V^2}{G}}$. A summary of all indicators and possible transitions is shown in the following table (see also the equations (3.154) and (3.150) on page 141)

| constraint | change | indicator | typ of indicator |
|---|---|---|---|
| getting active | no contact → contact | $g_N = 0$ | kinematic |
| | sliding → sticking | $\dot{g}_T = 0$ | |
| getting passive | contact → no contact | $\lambda_N = 0 \wedge \ddot{g}_N > 0$ | kinetic |
| | sticking → sliding | $\lambda_{T0} = 0 \wedge \ddot{g}_T > 0$ | |

*Round Peg in Hole*

The experimental setup with the PUMA, the environment and the sensors is shown in Fig.16. The force–torque sensor between the last joint of the manipulator and the gripper is used to measure the mating forces. Six laser sensors are utilized to observe the position and orientation of the gripper. In this specific setup the peg was cylindrical with a diameter of ø $39.9mm$ and a round chamfer with radius $r_1 = 4mm$, shown in Fig.17. The hole had a diameter of ø $40mm$ and also a round chamfer with radius $r_2 = 6mm$. Thus the clearance between the peg and the hole is only $0.1mm$. Mathematically the peg and the hole are cylinders, each chamfer is modeled as a torus. The parameterization of the peg is for example:

$$\text{cylinder: } \boldsymbol{r}_{\Sigma 1} = \begin{pmatrix} R_1 \cos u_1 \\ R_1 \sin u_1 \\ v_1 \end{pmatrix},$$

$$\text{torus: } \boldsymbol{r}_{\Sigma 1} = \begin{pmatrix} (R_1 + r_1(\cos v_1 - 1)) \cos u_1 \\ (R_1 + r_1(\cos v_1 - 1)) \sin u_1 \\ \sin v_1 \end{pmatrix}, \quad (7.129)$$

where $R_1 = 19.95mm$ and $r_1 = 4.0mm$. The description of the hole is the same, we only have to replace the index $()_1$ by the index $()_2$, where $R_2 = 20.0mm$ and $r_2 = 6.0mm$. Between the to mating partners three potential contact points exist:

| number | peg | | hole |
|---|---|---|---|
| 1 | torus 1 | $\longleftrightarrow$ | torus 2 |
| 2 | cylinder 1 | $\longleftrightarrow$ | torus 2 |
| 3 | torus 1 | $\longleftrightarrow$ | cylinder 2 |

The position of the robot for the insertion task was $\boldsymbol{\gamma}_0 = (-8.4°, -152.8°, 17.9°, 0.0°, -44.9°, -8.4°)^T$. The initial displacement of the robot with respect to the axis of the hole was $2.1mm$ in the $x_G$–direction and $1.3mm$ in the $y_G$–direction (see **G**–frame in Figure 7.32 for detailed explanation). The mating trajectory was $80mm$ along the $z_G$–axis.

In Figure 7.33 we see the first results from the insertion. The peg and the hole are displayed from two sides. On the two parts we recognize the trace of the contact points. On the left side we see the point of the first contact between the two chamfers, due to the initial displacement. The workpieces are then sliding along the chamfers, until there is a transition of the contact point to the cylinder of the peg. In this situation, the peg touches the chamfer along a straight line, as long as only one point is in contact. After about $5cm$ of insertion, a second contact point arises at the other side, displayed in Figure 7.33 on the right. Through this additional constraint the peg is moving in such a way, that both contact points are moving to the middle of the peg with respect to the displayed viewpoint.

In the next Figure (Fig. 7.34) we have plotted the constraint forces of the three possible contact point combinations, which prevent the parts from penetrating each other, and the mating force in the direction of insertion. In the left diagram we recognize, that the force between the two chamfers is comparatively small. As the two chamfers lose contact, the force is transferred

## 7.3 Dynamics and Control of Assembly Processes with Robots

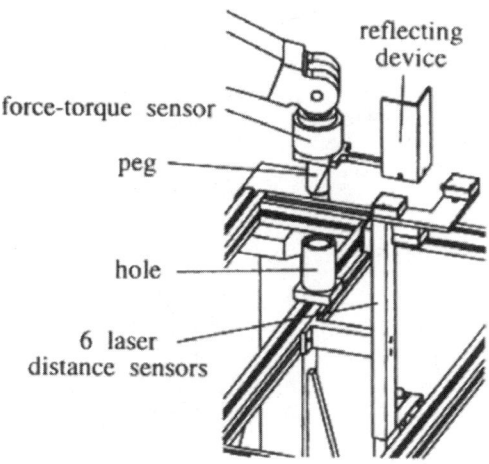

Fig. 7.32: Mating Experiments

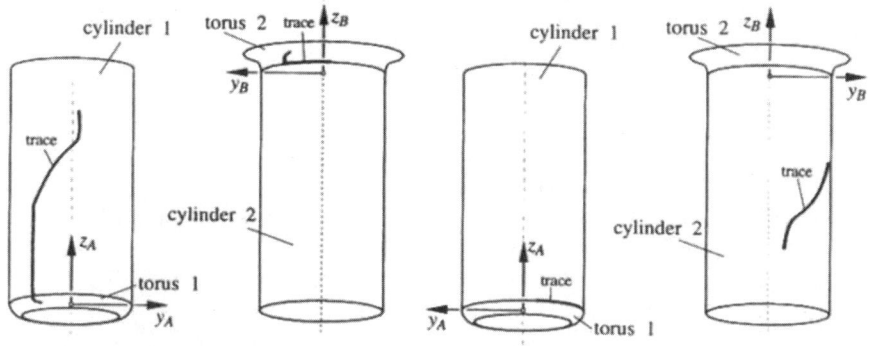

Fig. 7.33: Contact Point Traces for Round Peg-in-Hole Configuration

by the next constraint between cylinder$_1$ and torus$_2$. When the second contact point arises, at about 1.3s, the load starts rising. As the peg moves deeper into the hole two opposing forces are acting at different sides of the peg, achieving values of more than 100$N$.

Here we can show both, measurement and calculation. Even though, the contact forces reach very high values, the mating forces stay on a lower level. At the beginning we see clearly the peak, when the two chamfers get in contact. As only one constraint is active, the load is small afterwards. The highest values appear again in the two point contact situation. But they do not rise to such a high level as the contact forces themselves, because the latter act in

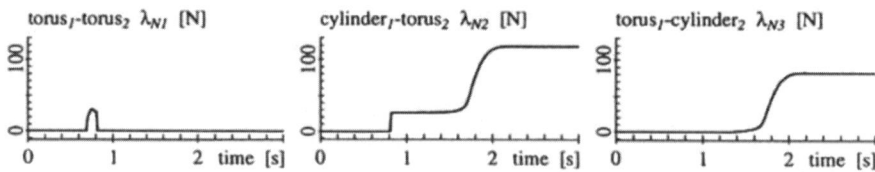
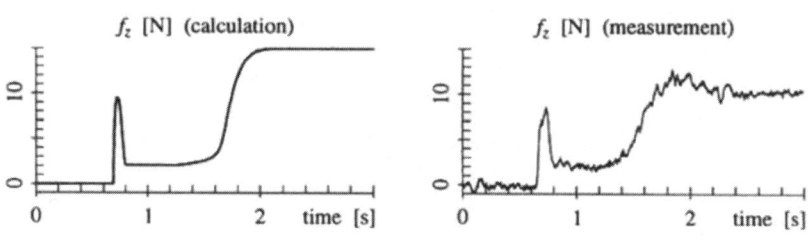

Fig. 7.34: Constraint Forces in the Contacts and Insertion Force

different directions. This part of the insertion is mainly governed by friction between the parts.

*Rectangular Peg in Hole*

Finally we consider a rectangular peg with a chamfer inserted into a rectangular hole, where the geometry is shown in Fig.20 on the left. If we introduce four modeling planes, we can investigate this spatial example in a planar manner. In every side of the peg one such plane is introduced, which is displayed in Figure 7.35 on the right. Two sides are situated within the $xy$–plane, and two in the $xz$–plane. There are two different types of contact points: point–plane and edge–edge. Let the letters $a, b, c, d$ in Figure 7.35 denote points and the numbers $1', 2', 3', 4', 5'$ denote planes. Then there are four possible contact points of the type point–plane: $a - 2'$, $b - 1'$, $c - 5'$, $d - 4'$. For the other type edge–edge the numbers $1, 2, 3, 4, 5$ denote edges on the peg and the letters $a', b'$ denote edges of the hole. There exist six potential contacts of this type: $1 - a'$, $2 - a'$, $3 - a'$, $3 - b'$, $4 - b'$, $5 - b'$. Thus we have altogether 40 potential constraints between the peg and the hole with the sketched geometry in the spatial case.

Measurements were again conducted with the PUMA 560 manipulator inserting the rectangular peg with the chamfer into the rectangular hole. The starting position of the manipulator was $\gamma_0 = (4.6°, -157.2°, 27.5°, 0.0°, -50.3°, 4.6°)^T$. The equations of motion of the robot were linearized around this working point. The mating parts can be seen in Figure 7.35, where the peg had the measures $a = 45.2mm, b = 45.4mm$ with a chamfer $45° \times 4mm$ and the hole had the dimensions $a' = 46.0mm, b' = 45.8mm$. The robot's path during the mating task was $80mm$ in positive $x_A$ direction. We show

## 7.3 Dynamics and Control of Assembly Processes with Robots

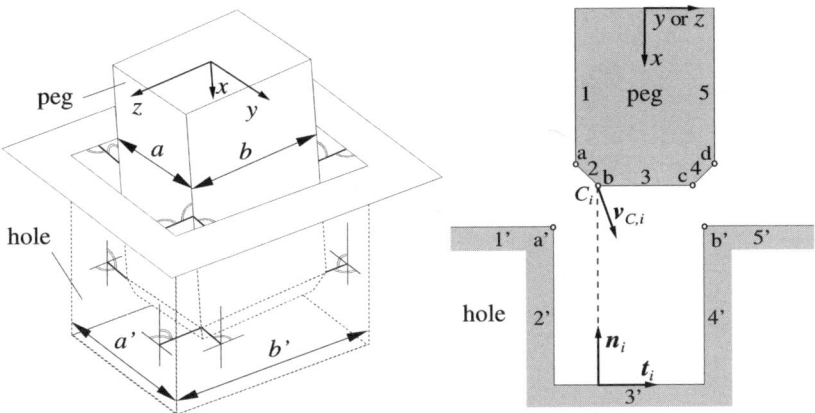

Fig. 7.35: Rectangular Peg in Hole Configuration

here the results of four experiments compared to numerical simulations. The initial lateral displacement between the peg and the hole was set to $\pm 4mm$ in the two cartesian directions $y_A$ and $z_A$.

Let us first regard the experiments, where the displacement was $\Delta y_A = \pm 4mm$. In Figure 7.36 the gripper forces during insertion $F_x$ and $F_y$ are shown. The upper plots are measurements, the lower plots are the calculated results for the same starting configuration. In both cases there is a peak of $F_x$ versus the manipulator motion, when the chamfer of the peg comes in contact with the upper edge of the hole (see Figure 7.35 on the right, contact points of type $4 - b'$ in case of positive or $2 - a'$ in case of negative displacement). After having passed the edge, it is sliding downwards, having contact with one side of the hole (see Figure 7.35, contact points of type $5 - b'$ in case of positive or $1 - a'$ in case of negative displacement). The force $F_y$ due to this contact acts towards the center of the hole.

More interesting are the experiments, where the displacement was varied in the $z_A$ direction: a) $\Delta z_A = +4mm$, b) $\Delta z_A = -4mm$. Here the behavior of the manipulator is different for both cases, see Figure 7.37 (top: measurement, bottom: calculation). If there is a displacement $\Delta z_A = +4mm$, there is again a force peak in $F_x$ at the first contact (contact points of type $4 - b'$), when the chamfer slides at the upper edge of the hole. The peg is then sliding into the hole, having contact with the upper edge $(5 - b')$, similar to the first two experiments. A completely different behavior can be observed, when the lateral displacement is $\Delta z_A = -4mm$. Here only the beginning of the insertion is similar to the above cases $(2 - a'$ and $1 - a')$. But as the peg goes deeper into the hole, there are additional contact points (of type $d - 4'$) inside the hole, after about $2.7s$. The contact forces and thus the mating forces $F_x$ and $F_z$ become very large, due to jamming. The insertion finally succeeds, because the drive torques are increased by the controller.

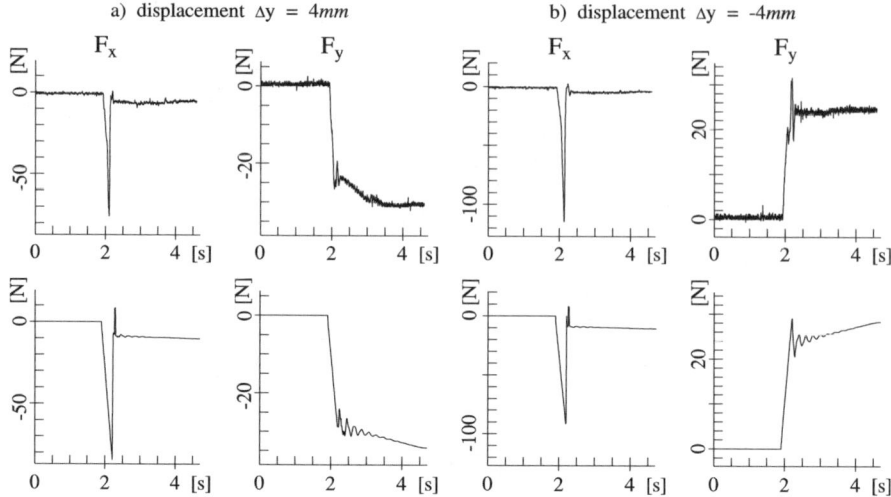

Fig. 7.36: Mating Forces for a Displacement in $y_A$-Direction (top measurements, bottom simulations)

The reason for the unsymmetric behavior of the robot in cases a) and b) can be found in the robots starting configuration. If a force in negative $x_A$ direction is applied, the manipulator is not only displaced in the same direction $(-x_A)$, but also in negative $z_A$ direction because of couplings in the stiffness matrix. This means for the example with initial displacement $\Delta z_A = +4mm$ (Figure 7.37), that the gripper is moved towards the center of the hole, when the peg is in contact with the hole. Therefore the contact forces are reduced. The opposite happens, if the lateral displacement is $\Delta z_A = -4mm$. As mating forces act on the gripper, the gripper moves away from the hole, whereas the mating forces additionally increase.

All measurements confirm very well the models and the theory.

7.3 Dynamics and Control of Assembly Processes with Robots 475

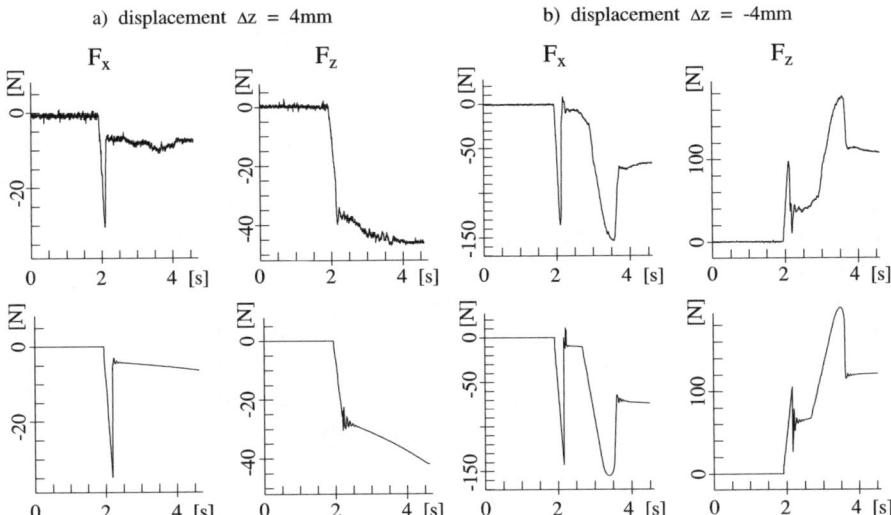

Fig. 7.37: Mating Forces for a Displacement in $z_A$-Direction (top measurements, bottom simulations)

## 7.3.3 Combined Robot and Process Optimization

### 7.3.3.1 Introduction

Machines perform processes, and also robots perform processes. Consequently the consideration of one of these two parts alone is not enough, one must deal with both. We are interested in dynamics and control of assembly processes in manufacturing executed by robots. For achieving a realistic presentation of the real world we have to model the assembly process in combination with the robot's dynamics in its entirety, meaning a consideration of dynamics and control for the manipulator and the assembly process together [219], [220], [271]. Mating parts by a manipulator will be always accompanied by local impacts, by oscillations and by tolerance problems, which influence the assembly process or, in the worst case, make mating impossible. A suitable parameter optimization of robot, process and control can avoid this and assure more productivity and reliability. In the following we shall consider such an optimization.

### 7.3.3.2 System Models

A typical feature in robotic manipulation tasks is a frequent change of contact configurations between the corresponding workpieces. The resulting forces and moments acting on the end-effector influence the motion of the manipulator during the task. When modeling such processes, we can distinguish between the dynamic model of the robot and that of the process dynamics, but we have to consider it in a realistic combination.

Industrial robots suitable for complex assembly tasks have to provide at least six degrees of freedom and, to ensure flexible operation, a large workspace. We will therefore focus on manipulators with 6 rigid links and 6 revolute joints, which are very popular in industry. Such a robot can be modelled as a tree-structured multibody system (Figure 7.38).

The joints of the first three axes are considered elastic in order to take the finite gear stiffnesses and damping into account, which play an important role in precision assembly. For this purpose a linear force law consisting of a spring-damper combination $c_j, d_j$, combined with the gear ratio $i_{G,j}$, $j = 1, \ldots, 3$, is assumed. The gears of the hand axes are considered stiff and the motion of one arm and its corresponding motor is kinematically coupled. Nevertheless we should keep in mind, that in reality the gears of the hand axes are elastic as well. However, the masses of the wrist bodies are comparatively small and thus the associated natural frequencies are out of the range of interest for our purposes. Backlash does not play an important role in modern robots [107], therefore we shall neglect such effects.

The robot possesses therefore 9 degrees of freedom, 6 arm angles and 3 free motor angles. The vector $q$ of generalized coordinates writes accordingly

## 7.3 Dynamics and Control of Assembly Processes with Robots

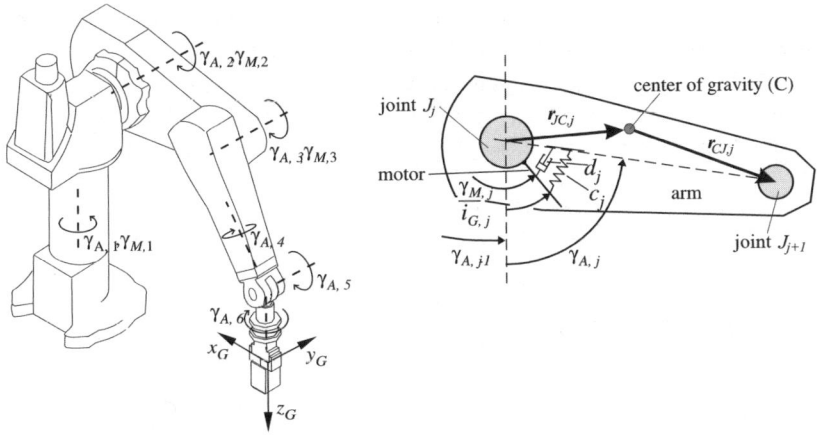

Fig. 7.38: Model of the Manipulator

$$\bar{q} = \begin{pmatrix} q_M \\ q_A \end{pmatrix} \in \mathbb{R}^9,$$

$$q_M = \begin{bmatrix} \gamma_{M,1} & \gamma_{M,2} & \gamma_{M,3} \end{bmatrix}^T \in \mathbb{R}^3, \qquad q_A = \begin{bmatrix} \gamma_{A,1} & \cdots & \gamma_{A,6} \end{bmatrix}^T \in \mathbb{R}^6, \qquad (7.130)$$

where $\gamma_{M,j}, \gamma_{A,j}$ denote the angle of the $j$-th motor and arm, respectively, relative to the previous body, see Figure 7.38. Translational and rotational velocities $\dot{r}_j$ and $\omega_j$ of the center of gravitiy of each link can thus be calculated recursively starting at the robot's base (see also equation (7.3)),

$$\begin{aligned} \omega_j &= \omega_{(j-1)} + \Delta\omega_{(j-1),j}, \qquad \Delta\omega_{(j-1),j} = \dot{\gamma}_{A,j} \\ \dot{r}_j &= \dot{r}_{(j-1)} + \left(\omega_{(j-1)} \times r_{CJ,(j-1)}\right) + \left(\omega_j \times r_{JC,j}\right), \end{aligned} \qquad (7.131)$$

where $r_{CJ,(j-1)}$ and $r_{JC,j}$ denote the distance vectors between center of gravity (C) and joint position (J), as depicted in Figure 7.38. Particularly, translational and angular velocity of the gripper end point are

$$\omega_G = \sum_{j=1}^{6} \omega_j,$$

$$\dot{r}_G = \sum_{j=1}^{6} \left[ (\omega_{j-1} \times r_{GJ,j-1}) + (\omega_j \times r_{JG,j}) \right] + (\omega_6 \times r_{GJ,6}). \qquad (7.132)$$

The gripper Jacobians, which relate the Cartesian motion of the gripper to the generalized coordinates, write

$$\begin{aligned} J_{TG} &= \frac{\partial(A_{60}\dot{r}_G)}{\partial \dot{\bar{q}}} \in \mathbb{R}^{3\times 9}, \\ J_{RG} &= \frac{\partial(A_{60}\omega_G)}{\partial \dot{\bar{q}}} \in \mathbb{R}^{3\times 9}. \end{aligned} \qquad (7.133)$$

In this operation, $\dot{r}_G$ and $\omega_G$ are transformed by $A_{60}$ into a gripper-fixed coordinate frame, so that $J_{TG}$ and $J_{RG}$ are related to the gripper system $G$ (Figure 7.38), which is convenient in manipulation tasks.

In the three basic axes the arm angles and the motor angles are connected by a linear force law representing the finite gear stiffnesses:

$$\tau_{A,j} = c_j \left( \frac{\gamma_{M,j}}{i_{G,j}} - \gamma_{A,j} \right) + d_j \left( \frac{\dot\gamma_{M,j}}{i_{G,j}} - \dot\gamma_{A,j} \right), \qquad (j=1,2,3) \qquad (7.134)$$

$c_j$ and $d_j$ are stiffness and damping factors of the $j$-th gear and $i_{G,j}$ is the gear ratio. The equations of motion of the robot with forces acting on the gripper can be written as

$$g := M(\overline{q})\ddot{\overline{q}} + f(\overline{q},\dot{\overline{q}}) - B\tau_C - W(\overline{q})\lambda = 0, \qquad (7.135)$$

with $M \in \mathbb{R}^{9\times 9}$ being the inertia matrix, $B \in \mathbb{R}^{9\times 6}$ and $W = [J_{TG}^T \ J_{RG}^T] \in \mathbb{R}^{9\times 6}$ are the input matrices for the 6 motor torques ($\tau_C \in \mathbb{R}^6$) and gripper forces ($\lambda \in \mathbb{R}^6$), respectively. $\lambda$ contains all contact forces and torques acting on the gripper, reduced to the gripper reference point. $f(\overline{q},\dot{\overline{q}})$ is a vector containing the gravitational and centrifugal forces as well as the moments in the gear elements, described in equation (7.134). Let us assume the robot to be controlled by six PD joint controllers, one for each joint, which are represented by

$$\tau_C = -K_p \left( \overline{q}_M - \overline{q}_{Md} \right) - K_d \left( \dot{\overline{q}}_M - \dot{\overline{q}}_{Md} \right), \qquad (7.136)$$

where

$$K_p = \text{diag}\,[K_{p,1}, \ldots, K_{p,6}] \in \mathbb{R}^{6\times 6},$$
$$K_d = \text{diag}\,[K_{d,1}, \ldots, K_{d,6}] \in \mathbb{R}^{6\times 6},$$
$$\overline{q}_{Md} = [\gamma_{M1d}, \ldots, \gamma_{M6d}]^T \in \mathbb{R}^6,$$
$$\overline{q}_M = B^T \overline{q} \in \mathbb{R}^6,$$

with $K_{p,j}, K_{d,j}$ being the stiffness and damping control coefficients of the j-th axis referring to motor angles as inputs and motor torques as outputs. $\gamma_{Mjd}$ is the motor angle of the $j$-th motor desired for a given position.

Taking into account that the length of a trajectory for mating two parts together is small compared to the robot's characteristic measures, equation (7.135) can be linearized around a working point $\overline{q} = q_0 + q, \dot{q}_0 = \ddot{q}_0 = 0$, with $q_0$ being

$$q_0 = [i_{G,1}\gamma_{A,1,0}\ i_{G,3}\gamma_{A,3,0}\ |\ \gamma_{A,1,0}\ \cdots\ \gamma_{A,6,0}]^T, \qquad (7.137)$$

which gives the position of the robot within its workspace $[\gamma_{A,1,0} \cdots \gamma_{A,6,0}]^T$ and the respective motor drive positions $[i_{G,1}\gamma_{A,1,0} \cdots i_{G,3}\gamma_{A,3,0}]^T$ of the three base axes. The linearized equations of motion are then derived from equation (7.135)

## 7.3 Dynamics and Control of Assembly Processes with Robots

$$M(q_0)\ddot{q} + \tilde{P}(q_0, K_d)\dot{q} + \tilde{Q}(q_0, K_p)q =$$
$$= h(q_0) + BK_p q_{Md} + BK_d \dot{q}_{Md} + W(q_0)\lambda(q, \dot{q}),$$

$$\tilde{P}(q_0, K_d) = P(q_0) + BK_d B^T, \qquad P(q_0) = \left.\frac{\partial g}{\partial \dot{\bar{q}}}\right|_{\bar{q}=q_0; \dot{\bar{q}}=0; \ddot{\bar{q}}=0},$$

$$\tilde{Q}(q_0, K_p) = Q(q_0) + BK_p B^T, \qquad Q(q_0) = \left.\frac{\partial g}{\partial \bar{q}}\right|_{\bar{q}=q_0; \dot{\bar{q}}=0; \ddot{\bar{q}}=0}, \qquad (7.138)$$

with the damping matrix $P(q_0) \in \mathbb{R}^{9 \times 9}$ and the stiffness matrix $Q(q_0) \in \mathbb{R}^{9 \times 9}$. The magnitude $h(q_0) = f|_{\bar{q}=q_0; \dot{\bar{q}}=0}$ contains only the gravitational forces needed to balance the system statically.

According to equation (7.138) the robot dynamics is linear and time-invariant as long as the robot moves within a small domain around the linearization point. This is the case for most manipulation tasks, since in most cases they can be divided into a phase of free motion (large motion, nonlinear) and a manipulation phase with small (linearizable) movements.

Table 7.1: Optimization concept

| domain | robot response behavior |
|---|---|
| **criteria** | • impact sensitivity<br>• maximal robot force<br>• mating tolerance<br>• vibrational behavior |
| **constraints** | **robot design:**<br>• joint angle limitations<br>• joint torque limitations<br><br>**robot control:**<br>• controller stability<br>• singularities<br><br>**practical demands in cell:**<br>• workspace restrictions |

The goal is now to find $q_0$, $K_p$ and $K_d$ such that the system described by the equations (7.138) behaves optimally with respect to the criteria relevant for a specific process, where $q_0$, $K_p$ and $K_d$ can be optimized already with only a rough knowledge of the process to be carried out. As can be seen from table 7.1, optimization of robotic manipulation processes, particularly assembly, is essentially a trade-off between different, sometimes contradictory, aims. Optimizing for reduced sensitivity against gripper impacts, for example, may deteriorate the behavior with respect to the maximal applicable gripper

480    7 Robotics

force and vice versa. Thus, a set of criteria is established, with which a vector optimization problem can be stated. The specific needs of different processes are taken into account by the correct choice of criteria and weighting factors. For this purpose, the functional-efficient set of solutions is calculated, from which an optimal trade-off between the criteria can be chosen.

It should be noted, that the above approach can be extended to robots with any kinematics. The tree structure of many industrial robots is not necessarily needed for successful application. To apply the method the following quantities describing the robot's kinematics and kinetics must be known:

- forward kinematics $\boldsymbol{r}_G(\overline{\boldsymbol{q}})$, $\boldsymbol{A}_{60}(\overline{\boldsymbol{q}})$,
- gripper Jacobians of translation and rotation $\boldsymbol{J}_{TG}(\overline{\boldsymbol{q}})$, $\boldsymbol{J}_{RG}(\overline{\boldsymbol{q}})$,
- equations of motion (7.135): $\boldsymbol{M}(\overline{\boldsymbol{q}})$, $\boldsymbol{f}((\overline{\boldsymbol{q}},(\dot{\overline{\boldsymbol{q}}}))$, $\boldsymbol{B}(\overline{\boldsymbol{q}})$.

#### 7.3.3.3 Criteria and Constraints

The effects, which influence the robot's behaviour during an assembly task have been worked out by a quasistatic and dynamic analysis of the robot dynamics in conjunction with a detailed modeling of different mating processes, see [153], [211], [241]. Scalar optimization criteria are derived from them, the minimization of which yields an improvement of the system's performance with regard to the respective effect.

*Optimization Criterion - Impact Sensitivity*

When the mating parts are getting in contact with each other, impacts are unavoidable. However, their intensity is proportional to the effective mass $m_{red} = \left[\boldsymbol{w}^T \boldsymbol{M}^{-1} \boldsymbol{w}\right]^{-1}$, reduced to the end effector, where $\boldsymbol{w}$ is the projection of the impact direction into the generalized coordinates and depends on the robot's position as well as on the Cartesian impact direction (see also the relations (3.160) and (3.161) on the pages 144). Figure 7.39 shows an ellipsoid

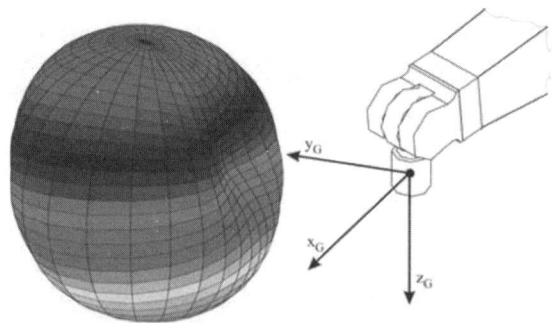

Fig. 7.39: Impact Sensitivity in the Cartesian Directions [271]

## 7.3 Dynamics and Control of Assembly Processes with Robots

at the robot's gripper, from which the reduced end-point inertia for each Cartesian direction can be seen. In order to reduce the impact sensitivity, the volume of that ellipsoid must be minimized. Thus, we define the reduced endpoint inertia matrix $M_{red}$ as

$$M_{red} = \left( \begin{bmatrix} J_{TG} \\ J_{RG} \end{bmatrix} M^{-1} \begin{bmatrix} J_{TG} \\ J_{RG} \end{bmatrix}^T \right)^{-1} \in \mathbb{R}^{6 \times 6} . \tag{7.139}$$

Depending on the specific needs of the mating process, a $6 \times 6$ diagonal positive semidefinite matrix $g_M$ of weighting factors is introduced for the trade-off between the cartesian directions. In $g_M$ the directions, in which impacts will occur during manipulation, can be emphasized. Thus, geometrically, the ellipsoid will be squeezed or rotated during an optimization from directions, in which the mating process considered is sensitive against impacts into directions, in which impacts are not likely to occur. Therefore, as a first optimization criterion for the minimization of impact intensities in the sensitive directions

$$G_1 = \| g_M^T M_{red} g_M \| \tag{7.140}$$

is stated, with $\|A\| = \sqrt{\text{trace}\left(A^T A\right)}$ being the Frobenius-norm of $A$. The matrix $g_M$ contains the proper physical units, so that $G_1$ is dimensionless.

*Optimization Criterion - Maximum Applicable Mating Force*

The maximum applicable mating force is $\lambda_{max}$ in the direction of insertion. The upper bound for the applicable mating force $\lambda$ is defined by the maximum torque of each motor multiplied by the resulting lever arms. $\lambda_{max}$ can thus be evaluated from

$$\lambda_{max} = \min_i \left\{ \max \left\{ \begin{array}{l} \max \left( \dfrac{h_i - \tau_{i,max}}{J_{TG}^T n} \, ; \, 0 \right) \\ \max \left( \dfrac{h_i - \tau_{i,min}}{J_{TG}^T n} \, ; \, 0 \right) \end{array} \right\} \right\}, \quad (i = 1, \ldots, n_M), \tag{7.141}$$

where $n_M$ is the number of driven axes and $n$ is a unit vector denoting the cartesian insertion direction. $h_i(q_0)$ is the torque necessary at joint $i$ to balance the gravitational forces, it is equal to the $i$-th component of $h(q_0)$. For a maximization of $\lambda_{max}$, its inverse is taken as the second criterion

$$G_2 = \dfrac{1}{\lambda_{max}} . \tag{7.142}$$

*Optimization Criterion - Mating Tolerance*

The deviation $\Delta x_G$ from the desired path for a given static force depends on the endpoint stiffness, reduced to the Cartesian gripper coordinates, see Figure 7.40. A quasistatical force equilibrium at the gripper yields

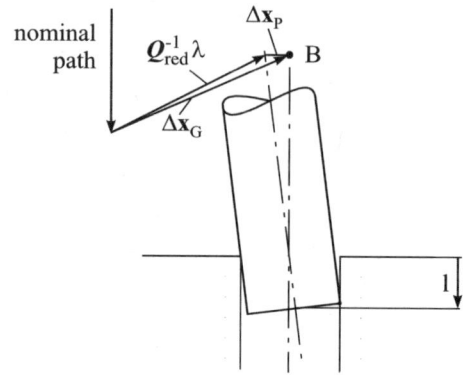

Fig. 7.40: Force Equilibrium for Mating Tolerance

$$\Delta x_G = Q_{red}^{-1}\lambda + \Delta x_P, \qquad Q_{red} = \left( \begin{bmatrix} J_{TG} \\ J_{RG} \end{bmatrix} \tilde{Q}^{-1} \begin{bmatrix} J_{TG} \\ J_{RG} \end{bmatrix}^T \right)^{-1}. \qquad (7.143)$$

$\Delta x_P$ is the deviation resulting from the clearance between the two parts and from the stiffness of the parts themselves. Therefore it depends only on the mating process itself and needs not to be considered here. For a maximization of $\Delta x_G$ the reduced stiffnesses $Q_{red}$ in the lateral directions must be minimized. Together with a weighting factor $g_Q$, which contains the cartesian directions, in which the tolerances are critical, this forms the criterion for the maximization of the mating tolerance. It is the third criterion

$$G_3 = \|g_Q^T Q_{red} g_Q\|. \qquad (7.144)$$

$G_3$ is also made dimensionless by proper physical units in $g_Q$. It should be noted, that for translational deviations, mainly the directions perpendicular to the insertion direction should be emphasized by $g_Q$, and all rotational directions can possibly be contained, whereas the cartesian stiffness in the insertion direction does not contribute to the mating tolerance and should be high for a reduced path deviation.

*Optimization Criterion - Disturbance and Tracking Properties*

When being excited by disturbances (e.g. by impacts), the gripper performs oscillations, the amplitude and damping of which depend strongly on the robot's position and on the joint controller. On the other hand, a desired force or

## 7.3 Dynamics and Control of Assembly Processes with Robots

motion must be transmitted to the end effector as directly as possible. All this has to be performed with as little expenditure of energy as possible. To meet these requirements, the equations of motion (7.138) are evaluated by time simulations for certain test inputs and three integral criteria are formulated:

- *Damping of force induced gripper oscillations*

$$G_4 = \int_0^\infty \boldsymbol{x}^T(t)\boldsymbol{g}_S\boldsymbol{x}(t)\,\mathrm{dt}\,, \qquad (7.145)$$

where $\boldsymbol{x}(t) = \begin{bmatrix} \boldsymbol{J}_{TG} \\ \boldsymbol{J}_{RG} \end{bmatrix} \boldsymbol{q}(t)$ and $\boldsymbol{\lambda}(t) = [0,\,0,\,\delta(t),\,0,\,0,\,0]^T$ represents a unit impulse input to the equations (7.138) in the direction of insertion. $\boldsymbol{g}_S$ in equation (7.145) gives the trade-off between end-effector oscillations in the different cartesian directions.

- *Transmission of a desired force to the end-effector:*

$$G_5 = \int_0^\infty (\boldsymbol{\lambda} - \boldsymbol{\lambda}_d)^T \boldsymbol{g}_F (\boldsymbol{\lambda} - \boldsymbol{\lambda}_d)\,\mathrm{dt}\,, \qquad (7.146)$$

where $\boldsymbol{\tau}_{C,d}(t) = -\boldsymbol{I}_G^{-1}\boldsymbol{W}(\boldsymbol{q}_0)^T\boldsymbol{\lambda}_d(t)$ contains the motor torques needed to exert the desired end effector forces. $\boldsymbol{W}$ is the projection from working space into configuration space and $\boldsymbol{I}_G$ denotes the matrix of gear ratios. As a test signal $\boldsymbol{\lambda}_d(t) = [0,0,\sigma(t),0,0,0]^T$ is used, where $\sigma(t)$ is the unit step function.

- *Joint torques:* A perfect damping of gripper oscillations and perfect tracking properties would require infinite joint torques. Thus, as soon as control coefficients are being optimized, the necessary torques must be considered. The performance criterion to be minimized is

$$G_6 = \int_0^\infty \boldsymbol{\tau}_C^T \boldsymbol{g}_\tau \boldsymbol{\tau}_C\,\mathrm{dt} \qquad (7.147)$$

with the same disturbance as in equation (7.145) and $\boldsymbol{\tau}_C$ being the joint torques from equation (7.136).

The above list of optimization criteria is of course not a complete list of possible objectives for robotic optimization. However, for a large class of manipulation tasks, a combination can be found suitable for the specific properties of the process. For example the assembly of snap fasteners is characterized by comparatively high mating forces and jumpy force trajectories, which might excite gripper oscillations. The latter must be damped by the robot controller. On the other hand impacts occur when the parts are getting in contact to each other. For these reasons the criteria for impact sensitivity $G_1$, for mating force $G_2$ and a mixed criterion for gripper oscillations and motor torques $G_4 + G_6$ would make sense in this case. Mating tolerance, i. e. gripper compliance, is not requested because the compliance of the snap itself is sufficient to compensate for possible position errors.

For any process the cartesian weighting factors $\boldsymbol{g}_M, \boldsymbol{g}_Q, \boldsymbol{g}_S, \boldsymbol{g}_F$ and $\boldsymbol{g}_\tau$ can be chosen from physical evidence: e. g. the impact intensities during insertion of a rigid peg into a hole will be worst in the cartesian $z_G$-direction. Thus, a large weight must be imposed on it in $\boldsymbol{g}_M$. The cost functions are normalized using the values $G_{i,0}$ referring to a reference configuration, see Figure 7.41.

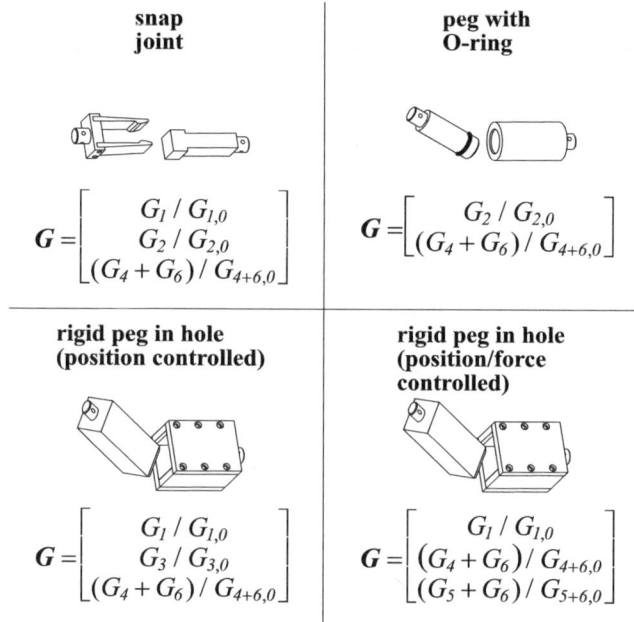

Fig. 7.41: Cost Functions for Different Processes

*Constraints*

In order to obtain sensible results, which can be utilized in practice, certain constraints have to be imposed on the optimization problem. We shall also see from the sensitivity analysis that the highly nonlinear programming problem defined by the equations (7.140) to (7.147) shows good convergence only if it is "properly" constrained, i. e. that the parameters are restricted to a domain, where a minimum of the criteria can be reliably found. We consider the following constraints:

7.3 Dynamics and Control of Assembly Processes with Robots     485

- *The linearized equations of motion* (7.138),
- *Joint angle limitations:*

$$\boldsymbol{q}_{min} \leq \boldsymbol{q}_0 \leq \boldsymbol{q}_{max}, \tag{7.148}$$

- *Joint torque limitations:*

$$\boldsymbol{\tau}_{C,min} \leq \boldsymbol{\tau}_C \leq \boldsymbol{\tau}_{C,max}. \tag{7.149}$$

Joint angle and joint torque limitations for our example are chosen according to those of the PUMA 562 robot. Joint speed limitations due to motor current limitations need not to be considered in the optimization of robot position and controller coefficients. But it is an issue in trajectory planning tasks as described in [219].

- *Stability of the controller used:* As soon as control coefficients are being optimized, stability of the resulting system must be assured by suitable constraints. Since the robot dynamics is linearized around a working point and therefore has linear time-invariant characteristics (see equation (7.138), the eigenvalues of the dynamic system matrix derived from equation (7.138) are calculated and their real parts are restricted to be negative.

$$\operatorname{Re}\left\{\operatorname{eig}\begin{bmatrix} \mathbf{0} & \mathbf{I} \\ -\boldsymbol{M}^{-1}\tilde{\boldsymbol{Q}} & -\boldsymbol{M}^{-1}\tilde{\boldsymbol{P}} \end{bmatrix}\right\} < 0. \tag{7.150}$$

- *The proximity to singularities* must be avoided. In such positions the robot would not be able to move in the desired manner and the obtained results would be without any practical relevance. Furthermore, some of the optimization criteria tend to infinity at singular positions. Thus, punching out finite regions around them would improve the condition of the optimization problem. As a measure the condition number $\kappa$ of the end effector's Jacobian is used, which is defined by $\kappa(\boldsymbol{A}) = \|\boldsymbol{A}\|\|\boldsymbol{A}^{-1}\|$ and tends to infinity as the Jacobian becomes singular.

$$\kappa\left(\begin{bmatrix} \boldsymbol{J}_{TG}(\boldsymbol{q}_0) \\ \boldsymbol{J}_{RG}(\boldsymbol{q}_0) \end{bmatrix}\right) < \varepsilon. \tag{7.151}$$

In the example $\varepsilon$ is chosen $\varepsilon = 20$.

- *Position and orientation of the gripper* are restricted by external constraints, such as obstacles within the working space, or the requirement that the parts should be assembled on a workbench with a given height. Position and orientation are calculated using the robot's forward kinematics, so that geometrical constraints can be stated in Cartesian space. For simplicity in our example, we restrict the robot's position to a cube, the edges of which are parallel to the base coordinate frame $B$ of the robot, see Figure 7.42:

$$_B\boldsymbol{r}_{min} \leq_B \boldsymbol{r}_G(\boldsymbol{q}_0) \leq_B \boldsymbol{r}_{max}. \tag{7.152}$$

Orientation restrictions are expressed using the rotational gripper transform $\boldsymbol{A}_{GB}$. In the example, we choose the orientation to be restricted such that the $z_G$ direction should have a negative component in the $x_B$- and the $z_B$-directions, which means that the mating direction points downwards away from the robot's base.

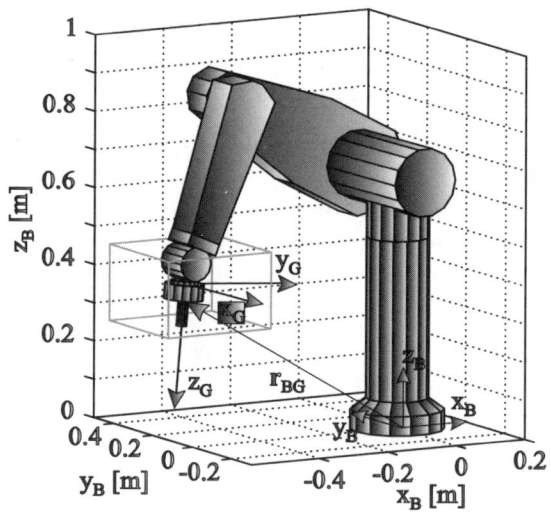

Fig. 7.42: Work Space Restrictions

*Example: Rectangular Peg-in-Hole Insertion*

We illustrate the method with the position controlled insertion of a rigid rectangular peg into a hole using a PUMA 562 manipulator, starting from the reference configuration $\boldsymbol{q}_{0,ref}$, $\boldsymbol{K}_{p,ref}$, $\boldsymbol{K}_{d,ref}$ defined in equation (7.153). This configuration is characterized by short effective lever arms that disturbances can work on, and small control coefficients, which increase gripper compliance for improved mating tolerance:

$$\boldsymbol{q}_{0,ref} = \begin{pmatrix} 2 & -152 & -4 & 0 & -19 & 179 \end{pmatrix} [°]$$

$$\boldsymbol{K}_{p,ref} = \begin{pmatrix} 1.604 & 1.304 & 2.608 & 0.395 & 0.556 & 0.390 \end{pmatrix} \left[\frac{Nm}{rad}\right]$$

$$\boldsymbol{K}_{d,ref} = \begin{pmatrix} 0.055 & 0.013 & 0.019 & 0.00263 & 0.00280 & 0.00195 \end{pmatrix} \left[\frac{Nms}{rad}\right] \tag{7.153}$$

## 7.3 Dynamics and Control of Assembly Processes with Robots

Rigid peg-in-hole insertion is mainly characterized by rigid body contacts, the occurrence of which can not be predicted because of the limited positioning accuracy of the gripper. Thus, peg and hole will show lateral and angular offsets between each other. This causes impacts between the peg and the chamfer, which result in gripper oscillations. On the other hand, compliance in the lateral directions is required in order to compensate for positioning errors. Therefore, for a rectangular peg-in-hole process, the criteria for impact sensitivity, mating tolerance and damping of gripper oscillations are the most relevant ones. For our case study we choose the vector of objective functions

$$\boldsymbol{G} = \begin{bmatrix} \left(\frac{G_1}{G_{1,0}}\right) \\ \left(\frac{G_3}{G_{3,0}}\right) \\ \left(\frac{G_4+G_6}{G_{4,0}+G_{6,0}}\right) \end{bmatrix} \in \mathbb{R}^{n_G} \qquad (7.154)$$

normalized to the objective function values $G_{i,0}, i = 1, 3, 4, 6$ of the reference configuration $\boldsymbol{q}_{0,ref}$, $\boldsymbol{K}_{p,ref}$ and $\boldsymbol{K}_{d,ref}$.

Let us for an appropriate choice of Cartesian weighting factors assume the lateral clearance in $x$-direction between the two parts to be smaller than the robot's positioning accuracy. In $y$-direction the clearance is assumed to be large enough to avoid contact with the chamfers. Thus, impacts will occur mainly in the $x$- and $z$-directions, and optimizing for impact intensities means to find a position, where the effective end-effector masses in $x$ and $z$ are minimized. The mating tolerance for this process is determined by $x$-translational and $\varphi_y$-rotational Cartesian stiffnesses. Also the vibration behavior is most critical in $x$- and $z$-direction and the weights for the motor torques are chosen according to the maximum motor torques of the PUMA 562 robot. Thus, the cartesian weights for our example problem write

$$\begin{aligned}
\boldsymbol{g}_M &= \text{diag}\left[1\tfrac{1}{\sqrt{kg}} \ 0 \ 1\tfrac{1}{\sqrt{kg}} \ 0 \ 0 \ 0\right], \\
\boldsymbol{g}_Q &= \text{diag}\left[1\sqrt{\tfrac{m}{N}} \ 0 \ 0 \ 0 \ 1\sqrt{\tfrac{rad}{Nm}} \ 0\right], \\
\boldsymbol{g}_S &= \text{diag}\left[1\tfrac{1}{m^2} \ 0.1\tfrac{1}{m^2} \ 1\tfrac{1}{m^2} \ 0 \ 0.1\tfrac{1}{rad^2} \ 0.1\tfrac{1}{rad^2}\right], \\
\boldsymbol{g}_\tau &= \text{diag}\left[0.2 \ 0.2 \ 0.2 \ 6.2 \ 6.2 \ 6.2\right]\tfrac{1}{Nm^2}.
\end{aligned} \qquad (7.155)$$

### 7.3.3.4 Sensitivity Analysis

*Objectives*

Before optimization, a sensitivity analysis of the objective functions with respect to $\boldsymbol{q}_0$, $\boldsymbol{K}_p$ and $\boldsymbol{K}_d$ is performed, and an impression of the shape of the objectives is given. Such an analysis has to be very specific about the robot used and the cartesian weighting factors, because the characteristics of the objectives depend strongly on the manipulator kinematics and the task to be carried out. We regard a PUMA 562 robot performing the position controlled peg-in-hole task described before.

488   7 Robotics

The optimization problem possesses 18 degrees of freedom, 6 joint angles and 12 coefficients $K_{p,i}$, $K_{d,i}$. Furthermore, the objectives are nonlinear and nonconvex functions of the parameters. Thus, a sensitivity analysis can only give limited insight into the shapes of the objectives, because only projections into some parameter directions can be plotted. However, some characteristic properties of the functions can be worked out. For this purpose the functions are discussed starting from the reference configuration from equation (7.153) by variation of one or two parameters each.

As can be seen from Figure 7.43 the position of the manipulator has a strong influence on the cost functions and therefore on the dynamic behavior. Thus, performing a process in an unfavorable position may deteriorate performance significantly or even be prohibitive for a successful completion. The kinematics of the PUMA 562 is such that a variation of the base axis

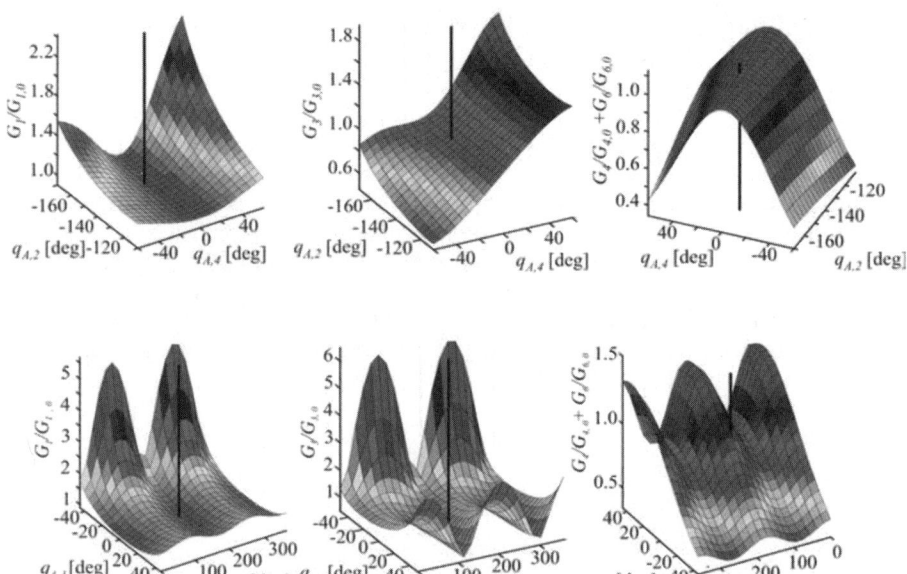

Fig. 7.43: *Graphs Top:* Cost Functions over Shoulder Angle $\gamma_2$ and Wrist Angle $\gamma_4$, *Graphs Bottom:* Cost Functions over Elbow Angle $\gamma_3$ and Gripper Angle $\gamma_6$, $(\gamma_i = q_{A,i})$, [219]

has no influence on the dynamic behavior, because there is no change in the configuration connected with it. Nevertheless, $\gamma_1$ must be considered during optimization, because it influences the working space constraints (7.151). It can be seen from Figure 7.43 that in the reference position ($\gamma_4 = 0$), where the mating direction is in a common plane with axis 1, variation of $\gamma_2$ has no influence on any of the cost functions. Kinematically, $\gamma_2$ affects only the effective lever arm between gripper and axis 1. Thus, if its length is already zero, no further influence is possible. If, on the other hand, $\gamma_4$ is twisted by turning the tool axis out of the vertical plane, variation of $\gamma_2$ has significant influence on either of the objectives.

In contrast to this, the influence of the elbow angle $\gamma_3$ and that of the tip angle $\gamma_6$ on the cost functions are largely decoupled. Significant in Figure 7.43 is the periodic nature of the objectives with respect to $\gamma_6$. The reason is that $\gamma_6$ gives the relative position of the directions emphasized in the Cartesian weighting factors relative to the robot arm. Figure 7.43 shows also that the influence of $\gamma_6$ on $G_1$ and $G_3$ is stronger for smaller values of $\gamma_3$, hence for the mating position being close to the robot's base. The contrary applies for $G_4 + G_6$. From this we can conclude that volume and shape of the impact ellipsoid described by the definitions (7.140) and of the stiffness norm (7.144) are mainly determined by $\gamma_2$ to $\gamma_5$, whereas they are rotated into directions favoured by the Cartesian weights through variation of $\gamma_6$. The value of $\gamma_5$ was shown to have significant influence only on $G_3$ and $G_4 + G_6$, apart from a singularity at $\gamma_5 = 0$, which causes a pole in the functions $G_1$ and $G_3$.

In general, Figure 7.43 shows the objective for gripper oscillations and motor torques $G_4 + G_6$ to be highly competitive with impact sensitivity $G_1$ and mating tolerance $G_3$. On the other hand $G_1$ and $G_3$ show a similar behavior over the entire workspace, which means that they can well be optimized simultaneously. The reason for this is that oscillation amplitudes can be minimized by choosing high end-effector stiffness and damping, whereas the mating tolerance requires a low gripper stiffness. This trade-off is well known in robotics research from the development of compliance elements.

As shown in Figure 7.44, the objective for mating tolerance is a monotonically increasing function in all proportional gains $K_{p,i}$, which have a direct influence on the end-effector stiffness. However, the controller stiffnesses act on the gripper via the forward kinematics, which is strongly dependent on the robot position. As a consequence, the position determines largely how strong the control gains influence performance. Figure 7.44 shows that in the reference position $G_3$ is mainly influenced by $K_{p,2}$, $K_{p,3}$ and $K_{p,5}$, hence the joint controllers of the axes, which allow motion in the desired compliant Cartesian direction. With the 6th axis twisted by $90°$, compliance in the desired direction is mainly governed by $K_{p,4}$, $K_{p,5}$ and $K_{p,6}$. It is this dependency between controller gains and robot position, which makes it sensible to optimize $\boldsymbol{q}_0$, $\boldsymbol{K}_p$ and $\boldsymbol{K}_d$ simultaneously.

From the above considerations we can conclude that the considered objectives are in accordance with physical evidence. Hence they reflect the physical behavior of the system in the sense that they become a minimum at the locations, where performance is best. On the other hand, they are highly nonlinear functions of the optimization parameters. This forms a nonlinear, nonconvex optimization problem. Moreover, some of the cost functions tend to infinity at singularities showing very high curvatures. Thus, for an efficient optimization, analytical derivatives of the objective functions, for which a calculation

490    7 Robotics

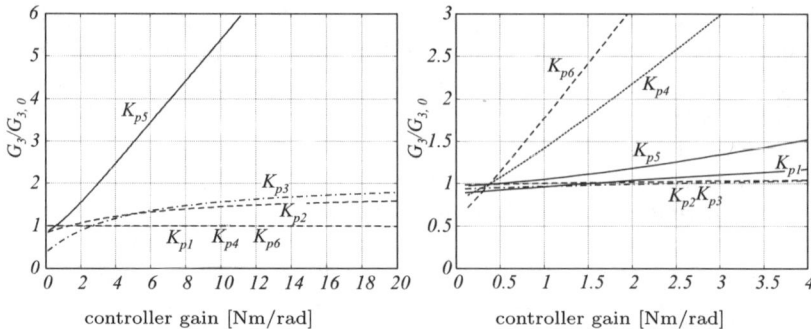

Fig. 7.44: Cost Function $G_3$ over Proportional Controller Gains $\mathbf{K}_p$ (left: reference position; right: 6th axis twisted by 90°)

is possible at a reasonable cost may significantly improve convergence. This is done for $G_1$, $G_2$ and $G_3$ and for the singularity (7.151) and working space constraints (7.152) using analytical calculation software. The objective functions possess local minima, in which the optimization routine may converge, so that an optimization of the position makes sense only, if the problem is constrained to a certain region within the working space. However, in most cases practical considerations in a real environment yield working space restrictions anyway.

*Analytic Cost Function Derivatives*

In order to improve numerical convergence of an optimization procedure, analytic function derivatives of these functions are calculated, for which this can be done at a reasonable computational expense, namely $G_1$, $G_2$ and $G_3$ from the equations (7.140), (7.142) and (7.144). In constraint space analytic gradients are supplied for singularity (7.151) and working space constraints in position, (7.152), and orientation, $\mathbf{A}_{GB}$. All other derivatives are calculated by finite differences during optimization.

Let $\frac{d\mathbf{M}}{dq_{0i}}$, $\frac{d\mathbf{P}}{dq_{0i}}$, $\frac{d\mathbf{Q}}{dq_{0i}}$ and $\frac{d\mathbf{h}}{dq_{0i}}$ be the elementwise derivatives of the system matrices in the relation (7.138) with respect to the $i$-th component of $\mathbf{q}_0$. $\frac{d\mathbf{P}}{d\mathbf{K}_{di}}$ and $\frac{d\mathbf{Q}}{d\mathbf{K}_{pi}}$ are analogously the elementwise derivatives with respect to the control coefficients. $\frac{d\mathbf{J}_G}{dq_{0i}}$ and $\frac{d\mathbf{J}_G^{-1}}{dq_{0i}}$ denote the elementwise derivatives of $\mathbf{J}_G = (\frac{\partial \mathbf{A}_{60}\dot{\mathbf{r}}_G}{\partial \dot{\mathbf{q}}_A} \quad \frac{\partial \mathbf{A}_{60}\omega_G}{\partial \dot{\mathbf{q}}_A}) \in \mathbb{R}^{6\times 6}$ and $\mathbf{J}_G^{-1}$. All of these expressions can be calculated from the relation (7.138) using analytic calculation software. We achieve the following results.

- *Impact Sensitivity:*

With $\mathbf{M}_{red}$ being established in section 7.3.3.3 and under the assumption that $\mathbf{J}_G$ is regular, equation (7.139) can be rewritten as

## 7.3 Dynamics and Control of Assembly Processes with Robots

$$M_{red} = \left(J_G^{-1}(q_0)\right)^T M_A(q_0) J_G(q_0), \tag{7.156}$$

where $M_A \in \mathbb{R}^{n_A \times n_A}$ is the mass-matrix related to the arm coordinates $q_A$, which can be calculated using equation (7.130)

$$M_A(q_0) = \left(\frac{\partial \dot{q}}{\partial \dot{q}_A}\right)^T M(q_0) \left(\frac{\partial \dot{q}}{\partial \dot{q}_A}\right), \quad \text{where} \quad \left(\frac{\partial \dot{q}}{\partial \dot{q}_A}\right) = \begin{pmatrix} 0 \\ E_{n_A} \end{pmatrix},$$

$$\frac{dM_A(q_0)}{dq_{0i}} = \left(\frac{\partial \dot{q}}{\partial \dot{q}_A}\right)^T \frac{dM(q_0)}{dq_{0i}} \left(\frac{\partial \dot{q}}{\partial \dot{q}_A}\right). \tag{7.157}$$

Regularity of $J_G$ can be assumed for any 6 DOF robot, which is in a nonsingular position. Thus, the derivative of $M_{red}$ with respect to the joint angles $q_0$ is

$$\frac{dM_{red}}{dq_{0i}} = \left(\left(\frac{dJ_G^{-1}}{dq_{0i}}\right)^T M_A + \left(J_G^{-1}\right)^T \frac{dM_A}{dq_{0i}}\right) J_G + \left(J_G^{-1}\right)^T M_A \frac{dJ_G}{dq_{0i}}. \tag{7.158}$$

From equation (7.140) the derivative of $G_1$ with respect to $q_0$ is then

$$\frac{dG_1}{dq_{0i}} = \frac{\text{trace}\left(g_M \left(\frac{dM_{red}}{dq_{0i}}\right)^T g_M g_M M_{red} g_M\right)}{\|g_M^T M_{red} g_M\|}. \tag{7.159}$$

Note that, obviously, the reduced end point mass does not depend on the control coefficients.

- *Maximal Mating Force:*

The derivative of $G_2$ with respect to $q_0$ can be calculated from equation (7.141).

$$\frac{d\lambda_{max}}{dq_{0i}} = \frac{\left(\frac{dh}{dq_{0i}}\right)_j}{\left(\left(\frac{dJ_{TG}}{dq_{0i}}\right)^T n\right)_j}, \tag{7.160}$$

where $j$ is the number of the axis, which gives the minimum in the sense that it fulfills equation (7.141). Equation (7.160) yields together with equation (7.142)

$$\frac{dG_2}{dq_{0i}} = \left(\frac{d\lambda_{max}}{dq_{0i}}\right)^{-1}. \tag{7.161}$$

From equation (7.141) follows that $\frac{dG_2}{dq_{0i}}$ has discontinuities at the points where the 'active' joint as a joint, which gives the minimal possible force in equation (7.141), changes. This can cause convergence problems.

- *Mating Tolerance:*

The derivative of $G_3$ can be calculated analogously to that of $G_1$, see the equations (7.143) and (7.144). Thus, the derivative with respect to $q_{0i}$ is

$$\frac{d\boldsymbol{Q}_{red}}{dq_{0i}} = \left(\left(\frac{d\boldsymbol{J}_G^{-1}}{dq_{0i}}\right)^T \boldsymbol{Q}_A + \left(\boldsymbol{J}_G^{-1}\right)^T \frac{d\boldsymbol{Q}_A}{dq_{0i}}\right) \boldsymbol{J}_G + \left(\boldsymbol{J}_G^{-1}\right)^T \boldsymbol{Q}_A \frac{d\boldsymbol{J}_G}{dq_{0i}}, \quad (7.162)$$

with $\boldsymbol{Q}_A \in \mathbb{R}^{n_A \times n_A}$ and $\frac{d\boldsymbol{Q}_A}{dq_{0i}} \in \mathbb{R}^{n_A \times n_A}$ being

$$\boldsymbol{Q}_A(\boldsymbol{q}_0, \boldsymbol{K}_p) = \left(\frac{\partial \dot{\boldsymbol{q}}}{\partial \dot{\boldsymbol{q}}_A}\right)^T \boldsymbol{Q}(\boldsymbol{q}_0, \boldsymbol{K}_p) \left(\frac{\partial \dot{\boldsymbol{q}}}{\partial \dot{\boldsymbol{q}}_A}\right),$$

$$\frac{d\boldsymbol{Q}_A(\boldsymbol{q}_0, \boldsymbol{K}_p)}{dq_{0i}} = \left(\frac{\partial \dot{\boldsymbol{q}}}{\partial \dot{\boldsymbol{q}}_A}\right)^T \frac{d\boldsymbol{Q}(\boldsymbol{q}_0, \boldsymbol{K}_p)}{dq_{0i}} \left(\frac{\partial \dot{\boldsymbol{q}}}{\partial \dot{\boldsymbol{q}}_A}\right). \quad (7.163)$$

The derivative with respect to $K_{pi}$ is

$$\frac{d\boldsymbol{Q}_{red}}{dK_{pi}} = \left(\boldsymbol{J}_G^{-1}\right)^T \frac{d\boldsymbol{Q}_A}{dK_{pi}} \boldsymbol{J}_G^{-1},$$

$$\frac{d\boldsymbol{Q}_A(\boldsymbol{q}_0, \boldsymbol{K}_p)}{dK_{pi}} = \left(\frac{\partial \dot{\boldsymbol{q}}}{\partial \dot{\boldsymbol{q}}_A}\right)^T \frac{d\boldsymbol{Q}(\boldsymbol{q}_0, \boldsymbol{K}_p)}{dK_{pi}} \left(\frac{\partial \dot{\boldsymbol{q}}}{\partial \dot{\boldsymbol{q}}_A}\right). \quad (7.164)$$

With respect to the derivatives of $G_3$ it follows from the relations (7.162) and (7.164)

$$\frac{dG_3}{dq_{0i}} = \frac{\operatorname{trace}\left(\boldsymbol{g}_Q^T \left(\frac{d\boldsymbol{Q}_{red}}{dq_{0i}}\right)^T \boldsymbol{g}_Q \boldsymbol{g}_Q^T \boldsymbol{Q}_{red} \boldsymbol{g}_Q\right)}{\|\boldsymbol{g}_Q^T \boldsymbol{Q}_{red} \boldsymbol{g}_Q\|};$$

$$\frac{dG_3}{dK_{pi}} = \frac{\operatorname{trace}\left(\boldsymbol{g}_Q^T \left(\frac{d\boldsymbol{Q}_{red}}{dK_{pi}}\right)^T \boldsymbol{g}_Q \boldsymbol{g}_Q^T \boldsymbol{Q}_{red} \boldsymbol{g}_Q^T\right)}{\|\boldsymbol{g}_Q^T \boldsymbol{Q}_{red} \boldsymbol{g}_Q\|}. \quad (7.165)$$

- *Singularity Constraint:*

7.3 Dynamics and Control of Assembly Processes with Robots    493

With equation (7.151) and the definition of the condition number of a matrix, the gradients of $\|\boldsymbol{J}_G\|$ and $\|\boldsymbol{J}_G^{-1}\|$ are

$$\frac{d\|\boldsymbol{J}_G\|}{dq_{0i}} = \frac{\operatorname{trace}\left(\left(\frac{d\boldsymbol{J}_G}{dq_{0i}}\right)^T \boldsymbol{J}_G\right)}{\|\boldsymbol{J}_G\|}, \qquad \frac{d\|\boldsymbol{J}_G^{-1}\|}{dq_{0i}} = \frac{\operatorname{trace}\left(\left(\frac{d\boldsymbol{J}_G^{-1}}{dq_{0i}}\right)^T \boldsymbol{J}_G^{-1}\right)}{\|\boldsymbol{J}_G^{-1}\|}.$$
(7.166)

The derivative of the product of the two norms is then

$$\frac{d\kappa}{dq_{0i}} = \frac{d\|\boldsymbol{J}_G\|}{dq_{0i}}\|\boldsymbol{J}_G^{-1}\| + \|\boldsymbol{J}_G\|\frac{d\|\boldsymbol{J}_G^{-1}\|}{dq_{0i}}.$$
(7.167)

- *Workspace Restrictions:*

The workspace restrictions are evaluated from the robot's gripper transform $\boldsymbol{r}_G(\boldsymbol{q}_0)$, $\boldsymbol{A}_{GB}(\boldsymbol{q}_0)$. Therefore, the gradients of the respective constraints can be directly calculated from differentiating the forward kinematics with respect to $\boldsymbol{q}_0$.

### 7.3.3.5 Vector Optimization Problem

The above mentioned criteria and constraints form a nonlinear vector problem for the position/controller optimization. Thereby the manipulation task to be carried out is characterized by a specific combination of cost functions and weighting factors, which can be chosen by physical evidence, as shown before. Thus, the complete vector problem for our example writes

$$\min_{\boldsymbol{q}_0,\boldsymbol{K}_p,\boldsymbol{K}_d} \{\boldsymbol{G} : \boldsymbol{f}_1 = \boldsymbol{0}; \boldsymbol{f}_2 \leq \boldsymbol{0}\}$$
(7.168)

with $\boldsymbol{f}_1(\boldsymbol{q}_0, \boldsymbol{K}_p, \boldsymbol{K}_d)$ coming from equation (7.138) and $\boldsymbol{f}_2(\boldsymbol{q}_0, \boldsymbol{K}_p, \boldsymbol{K}_d)$ being

$$\boldsymbol{f}_2 = \begin{bmatrix} \boldsymbol{q}_{min} - \boldsymbol{q}_0 \\ \boldsymbol{q}_0 - \boldsymbol{q}_{max} \\ \boldsymbol{\tau}_{C,min} - \boldsymbol{\tau}_C \\ \boldsymbol{\tau}_C - \boldsymbol{\tau}_{C,max} \\ \operatorname{Re}\left\{\operatorname{eig}\begin{bmatrix} \boldsymbol{0} & \boldsymbol{I} \\ -\boldsymbol{M}^{-1}\tilde{\boldsymbol{Q}} & -\boldsymbol{M}^{-1}\tilde{\boldsymbol{P}} \end{bmatrix}\right\} \\ \kappa\left(\begin{bmatrix} \boldsymbol{J}_{TG}(\boldsymbol{q}_0) \\ \boldsymbol{J}_{RG}(\boldsymbol{q}_0) \end{bmatrix}\right) - \varepsilon \\ {}_B\boldsymbol{r}_{min} - {}_B\boldsymbol{r}(\boldsymbol{q}_0) \\ {}_B\boldsymbol{r}(\boldsymbol{q}_0) - {}_B\boldsymbol{r}_{max} \\ \boldsymbol{A}_{GB;1,3} \\ \boldsymbol{A}_{GB;3,3} \end{bmatrix}$$
(7.169)

It is known from the theory of vector optimization that the equations (7.168) can not be uniquely solved, if any of the components of $\boldsymbol{G}$ are competing [39]. Rather, the solution of the equation (7.168) is a subspace of the parameter space of dimension $\mathbb{R}^{n_G-1}$ and denotes the Pareto-optimal set of all possible solutions of $\boldsymbol{f}_1 = 0$, $\boldsymbol{f}_2 \leq 0$. Pareto-optimality is reached if none of the objective functions can be improved without deteriorating at least one of the other criteria. Using the method of objective weighting [57], a substitute problem is stated with a scalar preference function $P$ to be minimized. For this, a vector of weighting factors $\boldsymbol{w} = [w_1 w_3 w_{46}] \in \mathbb{R}^{n_G}$ is introduced such that

$$0 \leq w_i \leq 1; \quad \sum_{i=1,3,46} w_i = 1;$$
$$P(\boldsymbol{G}(\boldsymbol{q}_0, \boldsymbol{K}_p, \boldsymbol{K}_d), \boldsymbol{w}) = \boldsymbol{w}\boldsymbol{G}(\boldsymbol{q}_0, \boldsymbol{K}_p, \boldsymbol{K}_d).$$
(7.170)

The Pareto-optimal set of solutions is then obtained by solving the scalar substitute problem

$$\min_{\boldsymbol{q}_0, \boldsymbol{K}_p, \boldsymbol{K}_d} \{P : \boldsymbol{f}_1 = \boldsymbol{0}; \boldsymbol{f}_2 \leq \boldsymbol{0}\} \tag{7.171}$$

for each vector $\boldsymbol{w}$, which fulfills equation (7.170). Equation (7.171) is solved for a systematic variation of $\boldsymbol{w}$ using a Sequential Quadratic Programming (SQP) algorithm with the Hessian matrix of the Lagrangian function being updated at each iteration by a quasi-Newton approximation.

#### 7.3.3.6 Results

*Numerics and Convergence*

The treated optimization problem possesses 18 parameters ($\boldsymbol{q}_0 \in \mathbb{R}^6$, $\boldsymbol{K}_p \in \mathbb{R}^6$, $\boldsymbol{K}_d \in \mathbb{R}^6$) and – for the position controlled peg-in-hole insertion – 3 objectives ($G_1$, $G_3$, $G_4 + G_6$). The problem itself is nonlinear and nonconvex, as we have seen in the discussion above. Therefore, if the robot's workspace is not restricted during optimization, suboptimal solutions are likely to occur. However, practical demands in the work-cell require workspace restrictions anyway, as explained.

To ensure a good condition of the optimization problem both optimization parameters and objectives are normalized by appropriate scaling factors so that their different orders of magnitude are balanced. If the problem is well posed in that sense and if analytic function derivatives are used, the SQP-algorithm converges versus the solution within approximately 30 function and gradient evaluations, see Figure 7.45. This number is approximately the same for all criterion combinations.

The problem of producing suboptimal solutions by numerical optimization can in the vector problem be adressed by regarding the cost function values at

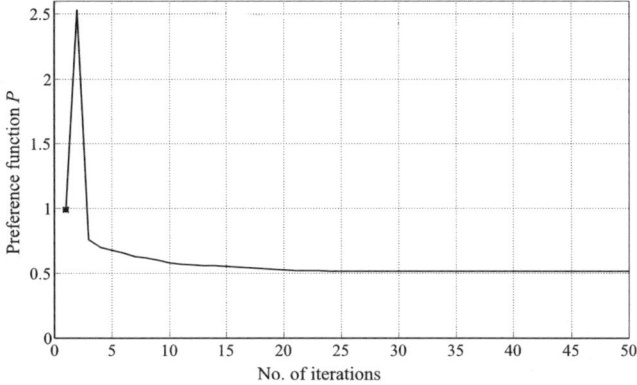

Fig. 7.45: Convergence Rate of Optimization

the Pareto-optimal set of solutions: All cost functions must be monotonically increasing when their weight in the preference function (7.170) is reduced. If not, this is a strong indication that the solver jumped from one local minimum to another one. This problem can be overcome by deriving appropriate starting guesses from the previously found solutions.

*Optimization Results*

In the following considerations, the reference configuration from the given data, equation (7.153) being considered suitable for the regarded process, is used as starting point for the optimization. It is then compared to an 'optimal trade-off' configuration, which is chosen from the set of Pareto-optimal solutions. As shown in Figure 7.46, the single cost-functions $G_i$ can be considerably diminished with respect to the reference configuration if they are emphasized in the preference function $P$.

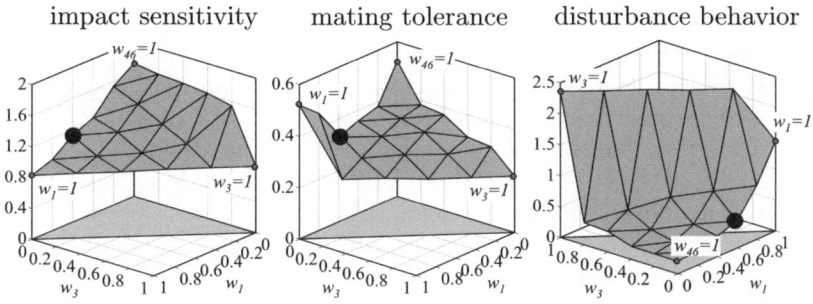

Fig. 7.46: Pareto-Optimal Set of Solutions (• point of optimal trade-off)

The criterion for impact intensities can be reduced by at most 17 %, that for mating tolerances by 78 % and the damping of end-effector oscillations and motor torques by up to 79 % with respect to the reference point. However, it is evident that such large improvements cause deteriorations in other criteria. For example, optimizing for damping of oscillations only deteriorates the impact sensitivity by 72 % and optimization for mating tolerance only increases the oscillation criterion by 136 %. But there are also regions within the Pareto-optimal area, where all criteria are improved with respect to the reference configuration. Figure 7.46 shows that over a wide range of possible weighting factors the criteria for impact sensitivity and for mating tolerance are not contradictory to each other: A simultaneous improvement of both cost functions can be observed for a large number of possible weights. Only if $G_1$ is strongly weighted in $P$, $G_3$ becomes worse. On the other hand, $G_{46}$ is found to impose completely different demands on the position and on the controller. $G_1$ and $G_3$ have their largest value at the point, where $G_{46}$ is minimized.

From these considerations it is clear that an 'optimal trade-off' must be found, which gives a satisfactory improvement in each of the criteria. The process of finding this optimal trade-off can hardly be formalized in a mathematical sense, because the trade-off between the cost-function weights is generally governed by criteria, which require human expertise. Thus, the Pareto-optimal region in Figure 7.46 has to be judged in order to find an optimal solution. However, finding the best configuration out of this set of possible solutions is not trivial. Each configuration has thus to be judged by an appropriate analysis, which figures out the physical properties, from which the optimization criteria are derived.

To judge the improvement in **impact sensitivity** the impact sensitivity ellipsoid, introduced in section 7.3.3.3 is investigated, which indicates the effective mass at the gripper during impacts in different cartesian directions.

**Mating tolerance** is analyzed by calculating the tolerance area for lateral and angular path deviations using the methods presented in [271]. The tolerance area indicates the maximal translatory and angular deviation, for which a given static mating force ($\lambda_{max} = 10N$) is not exceeded, see the tolerance equation (7.143).

The **disturbance behavior** of a linear time-invariant system can be comprehended by frequency response functions, which relate the oscillation amplitude of the gripper to the gripper forces, as shown in Figure 7.49. Natural frequencys and resonance peaks give physical evidence about the vibrational behavior.

These analysis tools provide an engineering feeling for the physical quality of the solutions, so that the final trade-off can be found. Finally, using the methods described in [153] and [189] a detailed simulation of the process including all effects as unilateral contacts, impacts, sliding friction and stiction as well as all interactions with the robot dynamics can be performed. Physical evidence can thus be given on different abstraction levels to help the engineer finding the best solution.

In our peg-in-hole example $\boldsymbol{w} = \begin{bmatrix} 0.6 & 0 & 0.4 \end{bmatrix}$ is chosen, which yields the following configuration:

$$\boldsymbol{q}_{0,opt} = \begin{bmatrix} 3.3 & -165.9 & 1.1 & 92.3 & -39.0 & 72.5 \end{bmatrix}^\circ$$

$$\boldsymbol{K}_{p,opt} = \begin{bmatrix} 1.226 & 1.219 & 1.000 & 0.132 & 0.139 & 0.133 \end{bmatrix} \frac{Nm}{rad}$$

$$\boldsymbol{K}_{d,opt} = \begin{bmatrix} 0.0715 & 0.0331 & 0.0145 & 0.0016 & 0.0024 & 0.0074 \end{bmatrix} \frac{Nms}{rad}$$

$$\boldsymbol{G}_{opt} = \begin{bmatrix} 1.24 & 0.36 & 0.33 \end{bmatrix}^T \tag{7.172}$$

It can be seen from $\boldsymbol{G}_{opt}$ from the data (7.172) that significant improvements

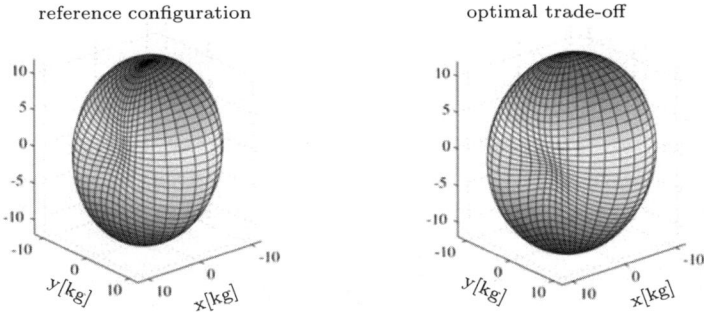

Fig. 7.47: Impact Sensitivity Ellipsoid for 'optimal trade-off' Compared to Reference Configuration

in $G_3$ and $G_{46}$ can be expected, which must be paid for by a slight deterioration in $G_1$. This can be fully comprehended in the Figures 7.47 to 7.49. The volume of the impact ellipsoid has slightly increased and the main axis is rotated with respect to the $y$-axis by a small angle. The tolerance area for a given maximal mating force of $\lambda_{max} = 10N$, calculated from a quasistatic force equilibrium is significantly enlarged, Figure 7.48. For the judgement of the disturbance behavior, the amplitude frequency response function for $z_G$ gripper displacement related to $z_G$ gripper force is depicted in Figure 7.49. The resonance peak at the first natural frequency vanishes completely, which indicates gripper oscillations to be well damped. However, the starting amplitude is increased. This is due to the trade-off with the mating tolerance criterion, which reduces the cartesian end-effector stiffness.

*Experimental Verification*

The optimized configuration is verified experimentally using the test-setup depicted in Figure 7.50. A rigid rectangular peg is assembled into a rigid hole with a clearance of 0.3 mm in $x_G$-direction and an $x_G$ offset of 2 mm

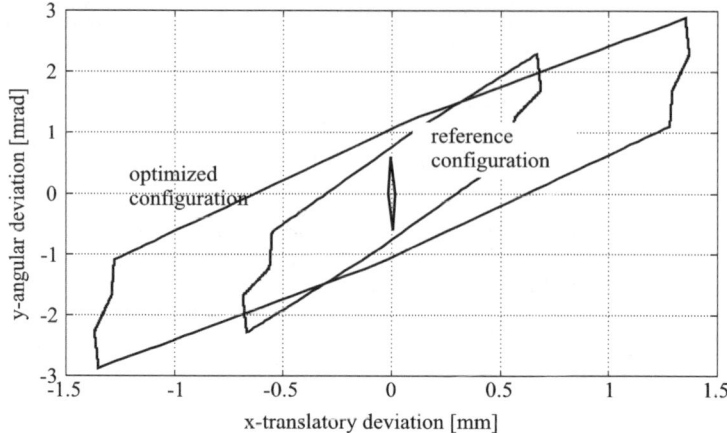

Fig. 7.48: Tolerance Area for $\lambda_{max} = 10N$ for 'optimal trade-off' Compared to Reference Configuration

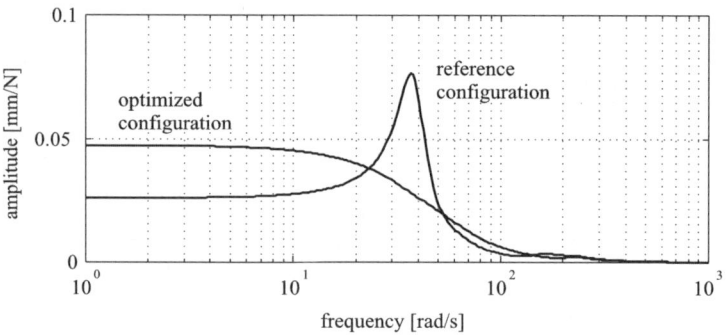

Fig. 7.49: Transfer Function $f = \frac{Gx_z}{G\lambda_z}$ for 'optimal trade-off' Compared to Reference Configuration

using a PUMA 562 manipulator. The desired mating path is a straight line in $z_G$-direction with a length of 60 mm and an assembly time of 0.4 s. Forces are measured using a Schunk FTS 330/30 force-torque sensor installed at the robot's end-effector. The gripper position is reconstructed by measuring the joint encoder angles using the robot's forward kinematics.

For the judgement of impacts, the time histories of the $z_G$ gripper force is considered, Figure 7.51. Significantly, a force peak occurs at the time where the two parts are getting in contact for the first time, the height of which gives a measure for the impact intensity. Since the relative velocity, with which the parts meet, is almost equal in both cases (about 150 mm/s), the peak height gives a direct measure of the effective mass acting on the impacting bodies. Figure 7.51 shows that in the regarded direction similar impact intensities can be expected and thus, the 'optimal trade-off' yields no improvement with

## 7.3 Dynamics and Control of Assembly Processes with Robots

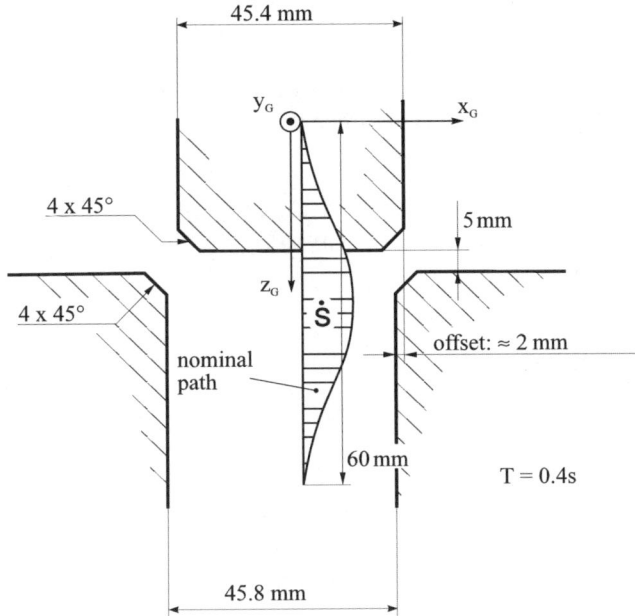

Fig. 7.50: Experimental Test Set-Up for Verification

respect to the impact behavior, which expresses itself also in $\boldsymbol{G}_{opt}$ and in Figure 7.47.

In contrast to this, according to the values in $\boldsymbol{G}_{opt}$ and the tolerance area of Figure 7.48, the optimal trade-off must show significant advantages with regard to the mating tolerance. This is verified with time histories of the $x_G$ lateral gripper force during insertion, Figure 7.52. Although the lateral offset is approximately equal in both cases, the reduced lateral end-effector stiffness of the optimized system allows compliant motion and thus reduces the strains on the manipulator and on the mating parts significantly. Most of the improvement in this criterion is due to the change in control coefficients, since the end-effector stiffness is essentially determined by $\boldsymbol{K}_p$ and $\boldsymbol{K}_d$. The manipulator position $\boldsymbol{q}_0$ defines the gripper's Jacobian and thus the lever arms the compliant controllers can work on.

In Figure 7.53 the $z_G$ path deviation due to external forces is depicted for both the reference and the optimized configurations. In fact, two different sources exist, which excite the manipulator dynamics, external contact forces and the desired movement. In order to separate those two effects in the experiment, the trajectory is measured twice for each configuration. First, the desired trajectory is performed without any external forces, particularly with no contacts. The resulting path deviation is then subtracted from the path deviation measured during manipulation, Figure 7.53. This ensures that only the path deviation resulting from contact events show up in Figure 7.53. It can be seen that the first amplitude peak resulting from the initial impact is reduced by a factor of 3. Furthermore, the transient behavior is much better damped in the optimized configuration and shows no oscillations. Thus, also the vibrational performance is significantly improved by the optimization.

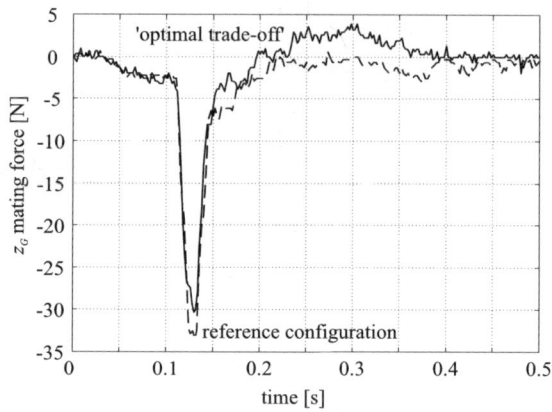

Fig. 7.51: Gripper Force in Mating Direction during Insertion

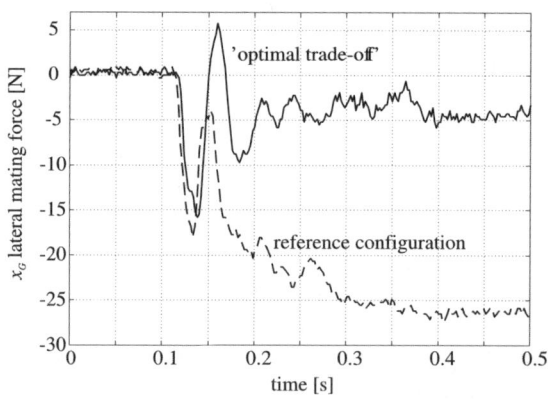

Fig. 7.52: Lateral Gripper Force during Insertion

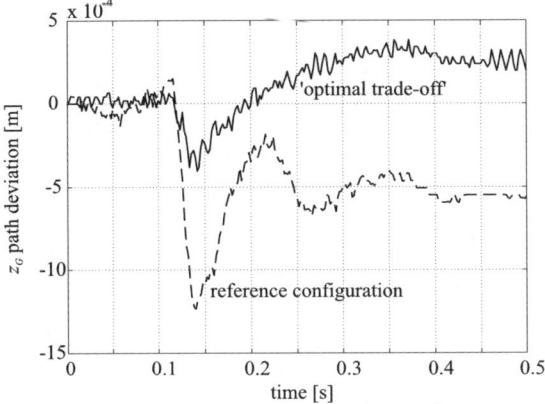

Fig. 7.53: $z_G$ Path Deviation in Mating Direction due to Mating Forces during Insertion

# 8
# Walking

> *Das Spiel Wissenschaft hat grundsätzlich kein Ende; wer eines Tages beschließt, die wissenschaftlichen Sätze nicht weiter zu überprüfen, sondern sie etwa als endgültig verifiziert zu betrachten, der tritt aus dem Spiel aus. (Karl Popper, Logik der Forschung, 1935)*
>
> *The game of science is, in principle, without end. He who decides one day that scientific statements do not call for any further tests, and that they can be regarded as finally verified, retires from the game.*
>
> *(Karl Popper, The Logic of Scientific Discovery, 1959)*

## 8.1 Motivation, Technology, Biology

### 8.1.1 Motivation

Walking is a fascinating invention of nature[1]. It is versatile, flexible and perfectly adapted to a natural environment. Walking in its various realizations enables the biological systems to have access to all the natural structures of the earth. Walking performance means actuating the typical walking components like legs, muscles, sensors, signal processing lines, nerves and brains. This generates motion of the complete biological system. Walking realizes motion, and motion with motion planning is the basis for intelligence, as modern biologists state. If intelligence is defined as the ability to deal with unknown and new situations, such as the possibility to find solutions for new problems, biological movement, both mental and physical, can be considered as a manifestation of intelligence. Therefore, motion and intelligence might be regarded as the prerequisites for animals and men to conquer the earth. All this makes walking research so extremely interesting for biologists and engineers.

Walking has been detected by engineers some 20-30 years ago, although before that time numerous trials had been made to realize some mechanisms with walking capabilities. Nowadays the computer age and a large variety of sophisticated technologies give walking machine realizations a high probability of success. Up to now the technical world of artefacts has come out with a large variety of machines and transportation systems mainly based on the invention

---

[1] Section 8.1 is based on the text of [203], see also [202]

of the wheel, a true human artefact and as a matter of fact a really basic one. But be it cars, trains, ships or airplanes, they all need highly organized areas, at least for starting and landing, they need roads, tracks, harbours, airports, which might be seen as a price for high speed and comfort. Access to non-organized areas of the earth is still difficult, even today. Walking machines will help to change this situation.

Most walking machines take some biological systems as templates, with two, four or with six legs. This makes sense, because natural evolution has created a large variety of excellent solutions, of course under the given evolutionary constraints, historically and environmentally. For several reasons it makes no sense to copy biology one by one. Evolution had to find its solutions within the framework of its possibilities: no wheels, muscles instead of rotating motors, nerves with synapses instead of wires and various types of sensors on a biological analogue basis. However, the number of sensors applied is sometimes extremely large.

But on the other hand technology has to offer also some positive constructive solutions: computer and computer science, a sophisticated design methodology, excellent motor-gear combinations with a high power to weight ratio, a highly developed sensor technology and very advanced light-weight solutions. What should be done is to combine the design principles of biological evolution with the best of technological possibilities. This results in two requirements. Firstly, biological research should be able to depict as many design principles invented by biological evolution as possible. Secondly, technology should make feasible the application of the most advanced software and hardware available with respect to cognition, mechanics and control with its sensors and actors. Realizing walking machines is a challenge and a technology-pushing issue by itself.

From the technological standpoint of view we have a large variety of aspects to develop walking robots. A decade ago arguments started with applications in hazardous environment, in areas where human beings have no access to. In the meantime walking robot technology made enormous progress including very perfect mechanical systems, sensor and control concepts and astonishing advanced technologies, which at least are rather near to that what might be one day artificial intelligence (see Fujita, Sony, [76]). This development establishes a confidence level allowing us to say, that we are able to realize a walking biped robot being able to interact with humans without boring them. Biped machines, because the infrastructure of our societies is designed for humans. Everything around us possesses human measures. Therefore in designing humanoids with human measures spares additional investments for special walking topologies. It is a strong motivation to persue this concept (see Hirukawa, [106]).

## 8.1.2 Technologies

In recent times the first one, possibly worldwide the first one, to start with scientifically oriented research on walking robots, especially on bipeds, was Professor Ichiro Kato (1925-1994) in Japan, who built in 1967 his artificial biped walker WL-1 (see Lim, [139]). Since then many activities all over the world pushed forward walking machine technology, though from the very beginning there was a clear focus in Japan due to really significant support of the Japanese Government and the Japanese industry. This fact is underlined by Honda and Sony, where Honda started already in the year 1986 with its first walking robot E0 (see Hirose and Ogawa, [104]). The E0 was followed by a whole series of 11 bipeds, the last one being the third version of the worldwide acknowledged ASIMO, the astonishing capabilities of which are well known. Honda, as a car manufacturer, persues three goals with respect to its walking bipeds, namely to create new mobility, to co-exist and cooperate with humans and to make general purpose robots.

Sony started its walking machine activities in the second half of the nineties with a clear focus on entertainment robotics and as a consequence on pet-type robot configurations like AIBO or QRIO, by the way with significant commercial success. Pet-type robots need autonomy, which is achieved by modern control and cognition mechanisms. For example AIBO includes a behavior-based architecture with an action-selection mechanism, a stochastic state-machine realizing context-sensitive responses and to a certain extent some kind of instinct-emotion-generator. It also includes learning abilities by reinforcement learning into the architecture. Sony (see Fujita, [76]) calls that Intelligence Dynamics with the goal of realizing an ever-developing Open-Ended-System.

The industrial activities were and are accompanied by a broad variety of University research. We give a few characteristic examples. Kato's Institute at Waseda University has developed a series of bipeds, the last one is WABIAN-RIV. The WABIAN-family has been very successful, its last member can walk forward, backward and sideways, it can dance and it can carry heavy goods (see Lim and Takanishi, [139]). The National Institute of Advanced Industrial Science and Technology (AIST) in Tsukuba, Japan, is involved in significant improvements of the HRP-robot, originally developed by Honda for maintenance tasks of industrial plants and for guard tasks of homes and offices. AIST developed new control concepts enabling HRP-2 to drive industrial vehicles, to perform more sophisticated maintenance tasks, to take care of patients in bed and to cooperate with human workers. HRP-2P may fall down, then it can get up from the floor autonomously (see Hirukawa, [106]).

One of the very advanced University robots is the H7, developed by the group of Professor Inoue at the University of Tokyo (Nishiwaki et al., [172]). The H7 was developed over several years as an experimental platform for walking, autonomy and human interaction research. With respect to the body mechanism for biped walk, lie down and stand up, support body by hand, pick itself up a couple of features become important like the arrangement and the number of degrees of freedom, the rotation range and the maximum torqes of the joints, the self-containedness, the ease of maintenance and smooth surfaces for attaching tactile sensors. Sensor availability of the H7-robot includes as a standard all joint sensors, force sensors, some foot sensors, but also tactile sensors for contacting the world and a 3D-vision system for seeing the world. This is accompanied by real time computational and software systems in order to process from low-level software such as servo-loop and sensor-processing to high-level software such as motion planning and behaviour control. Therefore high computational performance and most sophisticated software design are required.

The much more modest activities in Europe and especially in Germany may be characterized by the walking machines, which have been developed at the Technical University in Munich and in Karlsruhe. The group at the Institute B of Mechanics (see Pfeiffer, [193]), Technical University of Munich, started in 1989 to design and to realize a six-legged walking machine MAX, which followed very closely the, at that time new, neurobiological findings by Professor Cruse in Bielefeld and Professor Bässler in Kaiserslautern with respect to the control of walking stick insects. Although this machine is already 12 years old, its control concept is still very modern and in the area of neurobiology a matter of ongoing research. Without any central surveillance the control of MAX is completely autonomous, also in uneven terrain.

The six-legged machine MAX was followed by the eight-legged machine MORITZ, which was able to walk in tubes of any position and orientation, and the control of which possesses on a lower level that of MAX supplemented by a high-level structure being able to manage the contact forces at the foot-contacts at the inner tube wall. After these research results a large priority project of the German Research Foundation (DFG) enabled the Institute to develop a biped machine with a certain degree of autonomy, which could be achieved by the combination of the two-legged machine JOHNNIE with a 3-D-vision system developed by the group of Professsor Günther Schmidt at the Technical University of Munich. JOHNNIE is a light-weight design with 17 joints, a height of 1.80 m and a weight of about 40 kg.

The same idea of a close cooperation of biologists and engineers has been pursued at the University of Karlsruhe (see Dillmann et al., [45]) by Professor Dillmann and his group in developing the four-legged machine BISAM, the morphology and behavior based control of which follows as near as possible biological findings, especially those of Fischer in Jena (see Fischer, [62]). Considering also some ideas from researchers in the US and in Japan the behaviour based architecture forms a behaviour coordination network by connecting the inputs and the outputs of the behaviours. These connections transport control and sensor informations as well as loop-back informations. BISAM is driven by pneumatical joints, which on the one side show a good performance quite similar to that of biological muscles, but which on the other side generate control problems due to their compressibility and temperature sensitivity. BISAM is walking, but improvements are still necessary.

A very successful four-legged robot was developed during the last years by Kimura and his co-workers (see Kimura, [129]). The walking performance of his TEKKEN-series is really impressive, due to some good ideas concerning mechanical design, but mainly due to a very advanced design of the control system, which is based on biological findings. Kimura applies a limit-cycle based control for legged locomotion. It includes the adjustment of joint torques within a single step cycle, the adjustment of initial conditions of the legs at the transition from swing to stance and an adjustment of the stance or swing phases. His "coupled-dynamics-based motion generation" can interact with the environment emergingly and adaptively. The gait pattern generator deviates from that described by Cruse for the stick insect (see MAX), but it is equally adaptive and autonomous.

An important contribution concerning the actuators is given by Arikawa (see Arikawa, Hirose, [4]). Actuators represent the heart of walking machines, their power-to-weight ratio is the key for being able to establish many degrees of freedom (Inoue) and to provide the machine with sufficiently large torques for the walking process. This problem can be neutralized at least partly by the introduction of two ideas, the concepts of Gravitationally Decoupled Actuation (GDA) and of Coupled Drives. The GDA decouples the driving system against the gravitational field to suppress the generation of negative power and to improve the energy efficiency. Negative power is defined as a result of braking. The Coupled Drive couples the driving system with the goal to distribute the load as equally as possible among all actuators, thus maximizing the utilization of the installed actuator power. Some hardware examples confirm these two ideas.

A very nice review of the historical developments of that field are given in [245].

## 8.1.3 Biology

Every technological development and progress starts with examples, which in most cases are available by corresponding technical products, cars, cameras or machines. The technology of walking is a very young one without examples coming from the technical world itself. Therefore it makes sense to look for templates in Biology, where the evolution of millions of years has generated astonishing solutions. Nevertheless by looking for such solutions one must keep in mind, that Nature works with completely different materials and concepts, which only in some cases might be transfered to technology. The basic ideas applied in biological systems are indeed very attractive, as far as they are known, but a one-by-one transposition makes sense only for a very few cases. In the following we shall give some examples from various fields of Biology, where walking performance has been investigated, including also a very "biological" design of a walking machine.

Fischer and Witte [62] reinforce some of these ideas stating that the evolution of legs comes late in phylogeny and that "biologically inspired" technologies might be a better and more moderate approach than bionics or biomimetics. They continue that on the one side technological structures are always designed anew, but that on the other side biological structures are always the result of a past, permanent and ongoing process, which means "derived from ancestors". This antithesis is as a matter of fact only partly correct, because not only biology but also technology are carrying their special evolutionary burden, in classical technologies more than in walking technology. This results in the logical consequence that an adaptation of evolutionary ideas makes only sense, if the past and the recent functional requirements are identical, which under the logically necessary prerequisite of a constant environment will even be more an exception than the rule. Therefore applying biological concepts to technology requires a thorough investigation of goals and environments. Two-legged machines are a good solution, because it fits perfectly into the environments generated by man through thousands of years (Hirukawa [106], Hirose [104]).

The basic matter of concern of the contribution on intelligence by mechanics (Blickhan et al., [20]) consists in facilitating control by self-stability, which depends on the global system properties as well as on all major building blocks of the body. Technological design should realize as much self-stability as possible. The existing actuator systems are certainly a problem with respect to such a transposition, but the idea is convincing inspite of the fact that only simple and small mechanical configurations have been considered up to now. During biological walking potential energy is stored in the muscles, which is released during certain phases of the walking process. Together with the mass and stiffness properties of the legs of humans or animals the cyclic behaviour of potential and kinetic energies generate a very simple but at the same time a very robust self-stability. Experimental models composed from masses, springs and dampers are able to demonstrate quite well "natural walking", but under the guidance of a rotating lever arm.

The control mechanisms of human walking are not very well understood, they are a matter of ongoing intensive and worldwide research. Humans belong to the mammals, and mammals possess much more complex control behaviour than insects, which are easier to investigate and where as a consequence quite a lot about walking and walking control is known. Cruse [38], Büschges [79] and their research groups are working since years on the problems of walking and walking control of stick insects. The control concept of these insects is widely decentralized including more or less three layers, the lowest layers for stabilization purposes, the middle layer in a similar form as a finite state machine for controling the cycles and regarding all unexpected obstacles and the upper layer for organizing the gaits by influencing neighboring legs shifting their phase in the right direction, inhibiting some actions or set actions going.

An important property of these insect legs is the possibility to parametrize the segments of the legs. All segments move according to the coxa-femur control following some given (and environmentally adapted) laws. Cruse has developed a computer code on the basis of neural nets which gives a good mapping of the insect's behaviour. Büschges investigates the influence of the stepping velocity on the insect's walking control. Speed is simply generated by increasing the cycles applied to the joint connecting the first segments (coxa, femur) to the body. This is a bit in contradiction to the idea that biology applies different control structures for different speeds. Obviously accelerating the cycles might be the simpler solution. The control structure of the stick insect has been realized in the six-legged machine MAX [193] nearly in the sense of biomimetics.

Another example of a biomimetic robot has been developed by Ayers and Witting with their eight-legged machine NU/DARPA/ONR [9], the concept of which is based on the American Lobster. The robot is designed to achieve the performance advantages of the animal model by adopting the biomechanical features and neurobiological control principles. Three types of controllers are considered. The first is a state machine based on the connectivity and dynamics of the lobster CPG (central pattern generator). It controls myomorphic actuators realized with shape memory alloys, and it responds to environmental perturbation through sensors that employ a labeled-line code. The controller supports a library of action patterns and exteroceptive reflexes to mediate tactile navigation, obstacle negotiation and adaptation to surge. The second type of control is based on synaptic networks of electronic neurons and has been adapted to control the shape memory alloy actuated leg. A rudimentary brain is being developed as a third higher order controller using layered reflexes based on discrete time map-based neurons.

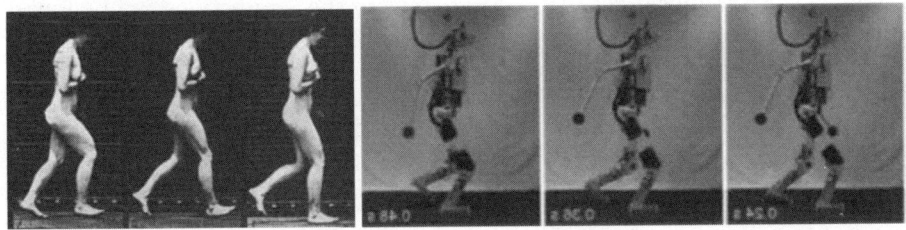

Fig. 8.1: Walking - Biology [164] and Technology [137]

## 8.2 Walking Dynamics

### 8.2.1 Preliminary Comments

As can be evaluated from biology, walking is extremely difficult (Figure 8.1). Seen from the technological point of view, it includes mechanics, sensor and actor technologies, computer soft- and hardware, vision systems and image processing, decision algorithms and advanced control structures. The list is not complete, but it indicates the highly interdisciplinary character of all walking machine developments. They require interdisciplinary cooperation of many technological fields. Within the framework of this book we shall focus on the dynamics (and control) of walking machines, and we shall select as a benchmark problem the dynamics of a biped machine, which represents the most sophicticated case of walking dynamics.

Criteria like stability, energy, loads to the joint drives or a more or less equal distribution of loads on all drives may be chosen for controling walking processes. We shall come to this later. Biological investigations indicate, that at least in the case of disturbances stability possesses highest priority and not energy (see [142] and [197]). This makes sense for any walking system, because without stability as a primary requirement any energy minimization will be in vain.

We have several possibilities for assuring stability. The oldest criterion is the famous zero-moment-point criterion (ZMP) of Vucobrativic, see [268], [267], [120]. It reduces all forces and torques acting between ground and feet to that one point, where forces alone are transmitted and no torques. This point has to be positioned within the supporting polygon of the walking machine. If it leaves that polygon, the machine will tilt, see [268], [137]. The static ZMP-criterion is very popular and, as long as the walking speed is small, an effective criterion. In recent years more dynamics arises, which makes extended stability criteria necessary. As the complete set of equations of motion of a walking machine is highly nonlinear, we may choose various possibilities to investigate stability, for example Ljapunov stability or a search for the fixed points of the dynamical evolution. First steps in those directions can be found in the literature [44].

In the following we shall give a summary of mechanical and mathematical modeling as well as for optimization and control of the walking biped JOHNNIE, which has been developed at the authors former Institute, see [137], [85], [213], [197], [202].

### 8.2.2 Modeling

The biped JOHNNIE has been designed and realized by an iterative process of simulation and construction, which allows a rather fast and safe layout of such a machine. We shall come to this point later. The first step of the design phase is the choice of the joint structure. It has to be ensured that the kinematics allows to realize the planned motion. Figure 8.2 shows the chosen structure of JOHNNIE. Each leg is equipped with six driven joints. With these degrees of freedom, the six degrees of freedom of the upper body can be controlled arbitrarily within the workspace of the joints. Furthermore, the upper body can rotate about its vertical axis and each shoulder is equipped with a pitch and roll joint. The upper body joint is redundant with the two hip yaw joints but allows for a pelvis rotation to increase the step length. With the shoulder joints, the overall moment of momentum about the body vertical axis can be compensated.

JOHNNIE is equipped with 17 joints. Each leg is driven with 6 joints, three in the hip, one in the knee and two (pitch and roll) in the ankle. The upper body has one degree of freedom (DOF) about the vertical axis of the pelvis. To compensate for the overall moment of momentum, each shoulder incorporates 2 DOF. The 6 DOF of each leg allow for an arbitrary control of the upper bodie's posture within the workrange of the leg. Such, the major characteristics of human gait can be realized. The robot's geometry corresponds to that of a male human of a body height of 1.8 m. The total weight is about 40 kg. The biped is autonomous to a far extent, solely power supply and currently a part of the computational power is supplied by cables.

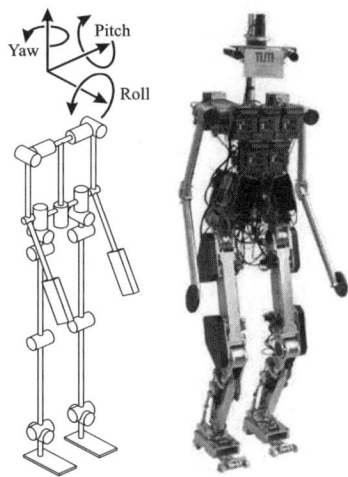

Fig. 8.2: JOHNNIE - Joint Model and Real Machine

### 8.2.2.1 Joint Structure

We start with the joint structure of the biped, which represents the basis of all further models. According to the table 8.1 and Figure 8.3 we define generalized coordinates **q** for all joints, where special care has to be taken for the ankle joint (see [137]). The set of the generalized coordinates writes

$$q = (q_O^T\ q_B\ q_{B0}^T\ q_{B1}^T\ q_{A0}^T\ q_{A1}^T)^T \tag{8.1}$$

with the following magnitudes: the vector $q_O = (\psi\ \vartheta\ \varphi\ x_O\ y_O\ z_O)^T \in \mathbb{R}^6$ includes the degrees of rotation and translation of the trunk, $q_B$ is the degree

Table 8.1: Generalized Coordinates of the Joint structure

| | |
|---|---|
| $\psi$ | rotation trunk around z-axis, inertial system |
| $\vartheta$ | rotation trunk around x-axis after first rotation |
| $\varphi$ | rotation trunk around z-axis after second rotation |
| $x_O$ | translation trunk x |
| $y_O$ | translation trunk y |
| $z_O$ | translation trunk z |
| $q_B$ | rotation pelvis |
| $q_{B00}/q_{B10}$ | rotation pelvis around vertical axis, right/left |
| $q_{B01}/q_{B11}$ | adduction/abduction pelvis, right/left |
| $q_{B02}/q_{B12}$ | flexion pelvis, right/left |
| $q_{B03}/q_{B13}$ | flexion knee, right/left |
| $q_{B04}/q_{B14}$ | adduction/abduction ankle joint, right/left |
| $q_{B05}/q_{B15}$ | flexion ankle joint, right/left |
| $q_{A00}/q_{A10}$ | flexion arm, right/left |
| $q_{A10}/q_{A11}$ | adduction/abduction arm, right/left |

of freedom of the pelvis, $q_{B0} \in \mathbb{R}^6$ and $q_{B1} \in \mathbb{R}^6$ are those of the right and the left leg. $q_{A0} \in \mathbb{R}^2$ and $q_{A1} \in \mathbb{R}^2$ represent the arms. The rest is explained in Table 8.1.

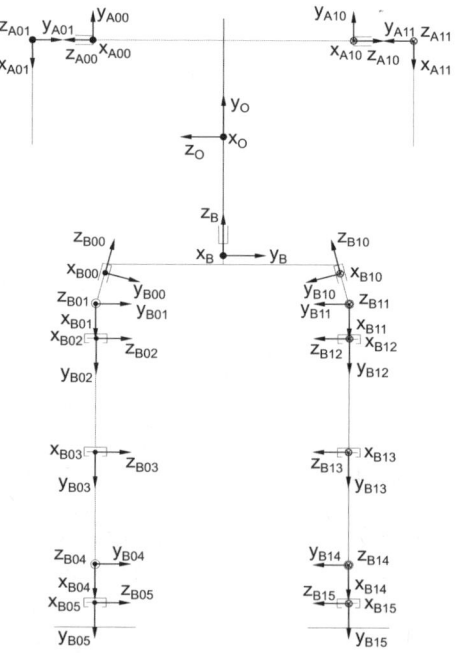

Fig. 8.3: Joint Structure and Coordinates

## 8.2 Walking Dynamics

We express the coordinates $\mathbf{r} = \mathbf{r}(\mathbf{q})$ and the velocities $\omega = \omega(\mathbf{q}, \dot{\mathbf{q}})$ in workspace by the generalized coordinates $\mathbf{q}$ and get thus the derivations

$$\frac{\partial \dot{\mathbf{r}}(\mathbf{q}, \dot{\mathbf{q}})}{\partial \dot{\mathbf{q}}} =: \mathbf{J}_T \quad \Rightarrow \quad \dot{\mathbf{r}} = \mathbf{J}_T \dot{\mathbf{q}}$$

$$\frac{\partial \omega(\mathbf{q}, \dot{\mathbf{q}})}{\partial \dot{\mathbf{q}}} =: \mathbf{J}_R \quad \Rightarrow \quad \omega = \mathbf{J}_R \dot{\mathbf{q}} \tag{8.2}$$

$$\ddot{\mathbf{r}} = \frac{\partial \dot{\mathbf{r}}}{\partial \mathbf{q}} \frac{\partial \mathbf{q}}{\partial t} + \frac{\partial \dot{\mathbf{r}}}{\partial \dot{\mathbf{q}}} \frac{\partial \dot{\mathbf{q}}}{\partial t} = \hat{\mathbf{J}}_T \dot{\mathbf{q}} + \mathbf{J}_T \ddot{\mathbf{q}} \quad \text{with} \quad \frac{\partial \dot{\mathbf{r}}(\mathbf{q}, \dot{\mathbf{q}})}{\partial \mathbf{q}} =: \hat{\mathbf{J}}_T$$

$$\dot{\omega} = \frac{\partial \omega}{\partial \mathbf{q}} \frac{\partial \mathbf{q}}{\partial t} + \frac{\partial \omega}{\partial \dot{\mathbf{q}}} \frac{\partial \dot{\mathbf{q}}}{\partial t} = \hat{\mathbf{J}}_R \dot{\mathbf{q}} + \mathbf{J}_R \ddot{\mathbf{q}} \quad \text{with} \quad \frac{\partial \omega(\mathbf{q}, \dot{\mathbf{q}})}{\partial \mathbf{q}} =: \hat{\mathbf{J}}_R \tag{8.3}$$

The Jacobians $\hat{\mathbf{J}}_T, \hat{\mathbf{J}}_R$ are more conveniently to evaluate than the time derivations of the Jacobians, which we would have to calculate if we derive directly the first two equations with respect to time.

We define the beginning of our kinematical chain by the trunk coordinates

$$\mathbf{q}_O = (\psi \; \vartheta \; \varphi \; x_O \; y_O \; z_O)^T. \tag{8.4}$$

From this we get the transformation matrix from the inertial system into the trunk coordinates by the expression

$$_O\mathbf{A}_{OI} = \begin{pmatrix} \cos\psi \, \cos\varphi - \sin\psi \, \cos\vartheta \, \sin\varphi & \sin\psi \, \cos\varphi + \cos\psi \, \cos\vartheta \, \cos\varphi & \sin\vartheta \, \sin\varphi \\ -\cos\psi \, \sin\varphi - \sin\psi \, \cos\vartheta \, \cos\varphi & -\sin\psi \, \sin\varphi + \cos\psi \, \cos\vartheta \, \cos\varphi & \sin\vartheta \, \cos\varphi \\ \sin\psi \, \sin\vartheta & -\cos\psi \, \sin\vartheta & \cos\vartheta \end{pmatrix} \tag{8.5}$$

The translational and rotational velocities of the trunk given in trunk coordinates then write:

$$_O\mathbf{v}_O =\, _O\mathbf{A}_{OI} \, _I\mathbf{v}_O, \qquad _O\omega_O =\, _O\mathbf{A}_{\vartheta\varphi}(\dot{\psi} \; \dot{\vartheta} \; \dot{\varphi})^T$$

$$\mathbf{J}_{O,R} = (_O\mathbf{A}_{\vartheta\varphi} | \mathbf{0}_{3\times(n-3)}), \qquad \mathbf{J}_{O,T} = (\mathbf{0}_{3\times 3} |_O\mathbf{A}_{OI}| \mathbf{0}_{3\times(n-6)})$$

$$\hat{\mathbf{J}}_{O,R} = \frac{\partial\, _O\omega_O}{\partial \mathbf{q}}, \qquad \hat{\mathbf{J}}_{O,T} = \frac{\partial\, _O\mathbf{v}_O}{\partial \mathbf{q}} \tag{8.6}$$

with the transformation matrix

$$_O\mathbf{A}_{\vartheta\varphi} = \begin{pmatrix} \sin\vartheta \, \sin\varphi & \cos\varphi & 0 \\ \sin\vartheta \, \cos\varphi & -\sin\varphi & 0 \\ \cos\vartheta & 0 & 1 \end{pmatrix} \tag{8.7}$$

According to section 2.2.4 on page 25 we evaluate the walking machine kinematics recursively.

The ankle joint needs special modeling due to the fact that actuation is realized by a ball screw spindle, see Figure 8.4. Due to this mechanism we get a kinematical closed loop not consistent with the tree-like structure of the machine kinematics. Therefore we choose for the generalized coordinates

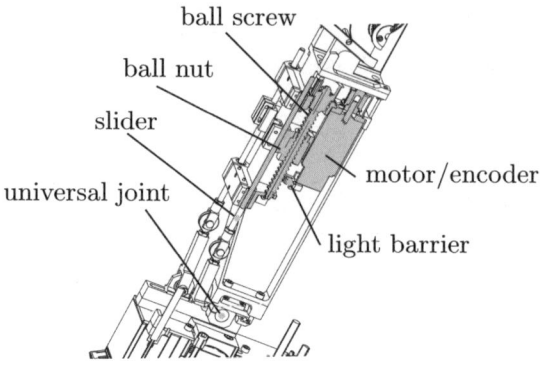

Fig. 8.4: Ankle Joint Actuation

the two angles of the universal joint, from which we are able to calculate the position of the ball screw spindles. Figure 8.5 depicts the principle of the ankle joint kinematics. The feet are symmetrical, therefore we can apply the same form of calculations for both feet. Hence, the generalized feet coordinates $q_{B04}, q_{B05}$ and $q_{B14}, q_{B15}$ can be replaced by $q_4, q_5$. The ball screw spindles are connected to the reference points $P_0, P_1$, they are parallel to the y-axis of the lower leg. The rotation of the spindles shift therefore the points S relative to the points P (Figure 8.5) by the amount

$$_U\boldsymbol{r}_{P_0S_0} = (0\ s_0\ 0)^T, \qquad _U\boldsymbol{r}_{P_1S_1} = (0\ s_1\ 0)^T. \tag{8.8}$$

From this we define the degrees of freedom of the spindles by $\mathbf{s} = (s_0, s_1)^T$. Two rigid rods connect the spindle end points $S_0, S_1$ with the points $A_0, A_1$ at the foot. From this the degrees of freedom of the ankle joint $\mathbf{q}_K = (q_4, q_5)^T$ are given unambiguously by the coordinates $\mathbf{s}$ of the spindles. For the corresponding vector chain we come out with (Figure 8.5)

$$\begin{aligned}_U\boldsymbol{r}_{S_0A_0} &= {_U\boldsymbol{r}_{UK}} + \boldsymbol{A}_{UK}\ _K\boldsymbol{r}_{KF} + \boldsymbol{A}_{UK}\boldsymbol{A}_{KF}\ _F\boldsymbol{r}_{FA_0} - {_U\boldsymbol{r}_{UP_0}} - {_U\boldsymbol{r}_{P_0S_0}}, \\ _U\boldsymbol{r}_{S_1A_1} &= {_U\boldsymbol{r}_{UK}} + \boldsymbol{A}_{UK}\ _K\boldsymbol{r}_{KF} + \boldsymbol{A}_{UK}\boldsymbol{A}_{KF}\ _F\boldsymbol{r}_{FA_1} - {_U\boldsymbol{r}_{UP_1}} - {_U\boldsymbol{r}_{P_1S_1}},\end{aligned} \tag{8.9}$$

with the transformation matrices

$$\boldsymbol{A}_{UK} = \begin{pmatrix} 0 & 0 & 1 \\ \cos q_4 & -\sin q_4 & 0 \\ \sin q_4 & \cos q_4 & 0 \end{pmatrix}, \qquad \boldsymbol{A}_{KF} = \begin{pmatrix} \sin q_5 & \cos q_5 & 0 \\ 0 & 0 & 1 \\ \cos q_5 & -\sin q_5 & 0 \end{pmatrix}. \tag{8.10}$$

With the given lenghts of the connecting rods $l_{V0}, l_{V1}$ we can establish two constraining equations $\boldsymbol{\Phi}$ for the evaluation of the ankle joint angles in the form

$$\boldsymbol{\phi} = \begin{pmatrix} _U\boldsymbol{r}_{V0}^T\ _U\boldsymbol{r}_{V0} - l_{V0}^2 \\ _U\boldsymbol{r}_{V1}^T\ _U\boldsymbol{r}_{V1} - l_{V0}^2 \end{pmatrix} = \boldsymbol{0}. \tag{8.11}$$

8.2 Walking Dynamics    515

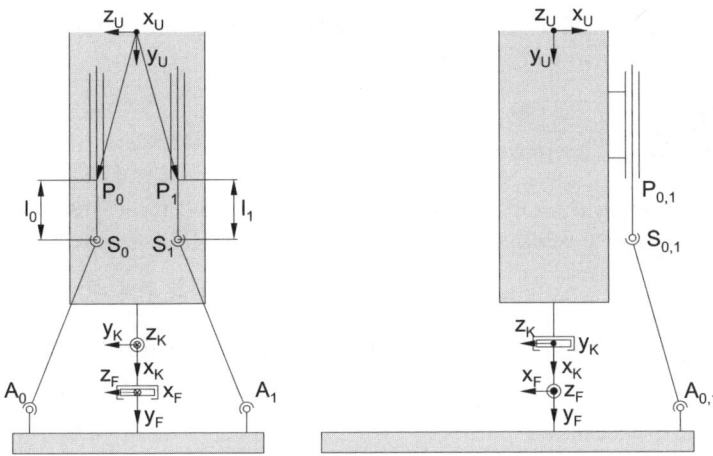

Fig. 8.5: Ankle Joint Kinematics

Given the spindle positions $_U r_{P0S0}, _U r_{P1S1}$, we have now to evaluate the generalized angles $q_K = (q_4 \; q_5)^T$ of the ankle joint. The coresponding equations can only be solved iteratively, for example by the Newton-algorithm:

$$q_{K,i+1} = q_{K,i} - \left(\frac{\partial \phi_i}{\partial q_K}\right)^{-1} \phi_i \tag{8.12}$$

In addition to that we need for the simulation the derivations of the generalized ankle coordinates with respect to the spindle coordinates. This writes

$$\frac{\partial q_K}{\partial s} = \left(\frac{\partial \phi}{\partial q_K}\right)^{-1} \left(\frac{\partial \phi}{\partial s}\right). \tag{8.13}$$

The time derivatives follow from that to

$$\dot{q}_K = \frac{\partial q_K}{\partial s} \dot{s}$$
$$\ddot{q}_K = \left(\frac{\partial \phi}{\partial q_K}\right)^{-1} \left(\frac{\partial \phi}{\partial s}\ddot{s} + (\frac{\partial \dot{\phi}}{\partial s})\dot{s} + (\frac{\partial \dot{\phi}}{\partial q_K})\dot{q}_K\right) \tag{8.14}$$

The above formulas are necessary for the determination of the transmission ratios. According to Figure 8.4 the spindles are driven by an electrical motor, which are coupled to the spindles by a timing belt stage. The rotation of the spindles in combination with the ball nuts transform rotation into translation. The timing belt stage has a transmission ration of $i_Z = \frac{15}{11}$, which gives together with the thread pitch of 5 mm a transmission ratio of

$$i_{SP} = i_Z \; i_S = \frac{15}{11} \frac{1}{5 \text{ mm}} \tag{8.15}$$

for the special case of JOHNNIE. For the transmission from the motors to the ankle joints we have to consider the relation

$$\dot{\mathbf{q}}_K = \frac{\partial \mathbf{q}_K}{\partial \mathbf{s}} \frac{\partial \mathbf{s}}{\partial \mathbf{q}_M} \dot{\mathbf{q}}_M = \frac{\partial \mathbf{q}_K}{\partial \mathbf{s}} \begin{pmatrix} \frac{1}{i_{SP}} & 0 \\ 0 & \frac{1}{i_{SP}} \end{pmatrix} \dot{\mathbf{q}}_M = \mathbf{J}_{SP} \dot{\mathbf{q}}_M \qquad (8.16)$$

with the Jacobian $\mathbf{J}_{SP}$ of the spindle drive. Table 8.2 indicates some typical values of the above relations for the case of JOHNNIE (see also [137]).

Table 8.2: Ankle Joint Transmission Ratios

|  | $q_4$[degree] | $q_5$[degree] | $\frac{\partial \dot{q}_{M0}}{\partial \dot{q}_4}$ | $\frac{\partial \dot{q}_{M0}}{\partial \dot{q}_5}$ | $\frac{\partial \dot{q}_{M1}}{\partial \dot{q}_4}$ | $\frac{\partial \dot{q}_{M1}}{\partial \dot{q}_5}$ |
|---|---|---|---|---|---|---|
| reference | 0.0 | 0.0 | 62.0 | -91.7 | -56.2 | -91.7 |
| $q_4 = q_{4,min}$ | -22.6 | -1.7 | 67.6 | -86.7 | -46.6 | -88.4 |
| $q_4 = q_{4,max}$ | 22.8 | -0.2 | 51.7 | -86.9 | -60.5 | -84.9 |
| $q_5 = q_{5,min}$ | 0.0 | -50.8 | 59.2 | -73.2 | -56.9 | -73.2 |
| $q_5 = q_{5,max}$ | 0.0 | 21.7 | 62.7 | -82.0 | -55.4 | -82.0 |

### 8.2.2.2 Motors and Gears

The drive systems decide more or less completely the walking machine configuration. Inspite of very significant progresses with respect to electrical motors and also with respect to gears, the requirements of large joint torques in combination with rather small joint velocities reduce the search for effective drives to a few technical solutions, for example the combination of DC-motors with or without brushes together with harmonic drive gears or sometimes also together with planetary gear sets. These solutions for power and power transmission can be found all over the world. With respect to our biped machine JOHNNIE we applied the same solution, which we shall discuss in the following. The Table 8.2.2.2 illustrates some typical data for the hip joint.

The motors are pulse-width controled with a frequency of 20 kHz. This means that during one control cycle the motor voltage is approximately constant and proportional to the pulse width. The voltage $U_M$ itself has the three components $U_I, U_R, U_G$ denoting the voltage induced in the coils, the voltage resulting from Ohm's resistance in the windings and finally the mutual induced voltage in the windings proportional to the motor shaft speed, respectively. From that we have

$$U_M = U_I + U_R + U_G, \qquad U_I = L\dot{I}, \quad U_R = RI, \quad U_G = k_M \omega, \qquad (8.17)$$

with L the inductance of the motor windings. The torque $T_M$ of the motor is proportional to the current I applied to the armature windings

Table 8.3: Technical Data of the Hip Joint

|  | Yaw | Roll | Pitch |
|---|---|---|---|
| Motor | Maxon RE40 | Maxon RE40 | 2 x Maxon RE40 |
| Gear | HFUC25-160-UL | HFUC26-160-UL | HFUC26-80 modified |
| Transmission ratio | 160 | 160 | 80 |
| Max. static joint torque | 178Nm | 178Nm | 220Nm |
| Average static joint torque | 22,3Nm | 22,3Nm | 22,3Nm |
| Max. joint velocity | 4,7rad/s | 4,7rad/s | 9,4rad/s |

$$T_M = k_M I, \tag{8.18}$$

with $k_M$ depending on the motor configuration. Combining the above equations results in a well known differential equation for the motor current

$$L\dot{I} = U_M - RI - k_M \omega. \tag{8.19}$$

As the winding inductances are very small we can neglect this kind of dynamics and approximate the motor torque by

$$T_M = \frac{k_M}{R}(U_M - k_M \omega). \tag{8.20}$$

The resistance R depends strongly on the temperature, which implies some problems not being solved alone by the simple relations above (see [137]).

The mostly prefered gears in walking machine applications are Harmonic Drive gears, which are well known [95]. A very concise model of these gears has been developed by Roßmann [226] on the basis of unilateral multibody systems. Inspite of many advantages of Harmonic Drive gears one drawback consists in the large friction losses, which have to be modeled. Roßmann subdivides these friction losses into three parts [137]

$$T_R = T_{R,0} + T_{R,T} + T_\omega, \tag{8.21}$$

where $T_{R,0}$ represents a constant friction share generated between wave generator and flexspline and often called "no-load backdriving torque". It includes also deformation effects of the flexspline and the friction between internal and external gear meshing. The second part $T_{R,T}$ depends on the transmitted torque and is approximately proportional to the load torque $T_{R,T} \approx \mu T_N$. The two parts $T_{R,0} + T_{R,T}$ correspond together to the effect of Coulomb's friction law. The third part $T_\omega$ of the friction is joint damping. It improves joint control, and it can be approximated, due to experimental experiences, by the following law:

$$T_\omega = d\omega + b\omega^3. \tag{8.22}$$

The first term $d\omega$ usually dominates the behaviour. All data necessary for the evaluation of the above equations can be taken from the Harmonic Drive catalogue [95]. Table 8.4 gives some typical data for the Harmonic Drives used in JOHNNIE.

Table 8.4: Typical Data of Harmonic Drives

| type | $i$ | $T_0[Nm]$ | $\eta_{500}$ | $\eta_{2000}$ | $\eta_{3500}$ | $d[Nms]$ | $b[Nms^3]$ | $\mu$ |
|---|---|---|---|---|---|---|---|---|
| HFUC 25 | 160 | 0.055 | 0.775 | 0.650 | 0.600 | $7.547 \; 10^{-4}$ | $-1.620 \; 10^{-9}$ | 0.0651 |
| HFUC 25 | 80 | 0.075 | 0.838 | 0.744 | 0.681 | $7.943 \; 10^{-4}$ | $-6.897 \; 10^{-10}$ | 0.0461 |

For modeling Harmonic Drive gears the friction has to be described by a set-valued force law, because we have two speed directions and a speed reversal after passing some null-speed point. The model in [226] distinguishes in a first step the speed directions and introduces a degree $q_1$ of freedom for the driving and a DOF $q_2$ for the driven direction:

$$J_1\ddot{q}_1 = T_{an} - \frac{1}{i}\lambda_G - (\frac{1}{i}\text{sign}(\dot{q}_1)\mu_{G1}\lambda_G + \frac{1}{i^2}d_1\dot{q}_1 + \frac{1}{i^4}b_1\dot{q}_1^3),$$
$$J_2\ddot{q}_2 = -T_{ab} + \lambda_G - (\text{sign}(\dot{q}_2)\mu_{G2}\lambda_G + d_2\dot{q}_2 + b_2\dot{q}_2^3). \tag{8.23}$$

The magnitude $\lambda_G$ is the transmitted torque, the moments of inertia of the input and output sides are denoted by $J_1, J_2$, and the friction data of the equations 8.21 and 8.22 are split up for the two directions

$$b = b_1 + b_2, \qquad d = d_1 + d_2, \qquad \mu = \mu_{G1} + \mu_{G2}. \tag{8.24}$$

In a further step we consider the Harmonic Drive gear as a unilateral system and apply the reduced mechanical model of Figure 8.6. The pinion with the degree of freedom $q_1$ is driven by the torque $T_{an}$ and meshes with the gear wheel with a degree of freedom $q_2$ and with an output torque $T_{ab}$. The configuration is here represented by one tooth pairing only though the contact ratio of the Harmonic Drives is very large. We have normal forces $F_n$ and friction forces $F_r$, the last ones arising by the relative motion in the tooth contact.

Applying this model to Harmonic Drives requires some approximations:

- Due to the large contact ratios we do not consider every individual tooth pairing, but we consider an average of them.
- All teeth not positioned in the pitch point experience a relative tangential motion and produce tangential friction forces.
- For these frictional forces we apply Coulomb's friction law.

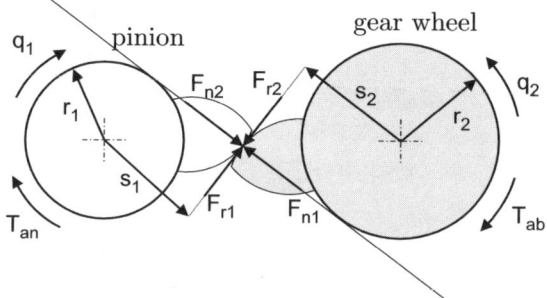

Fig. 8.6: Friction in a Harmonic Drive Stage [137]

- Considering Harmonic Drives as unilateral systems comes from the idea, that either the working flanks have contact and the non-working flanks not, or vice versa, backlashes neglected. This results in a complementarity for the two flank types.

All normal forces $\mathbf{F}_{ni}$ averaged over all teeth contacts are denoted by $\lambda_N$ and are composed of two parts, namely $\lambda_G$ resulting from the transmitted torque and $\lambda_{G0}$ resulting from pretensioning the gear. Equivalently, the magnitude $\lambda_T$ denotes the averaged tangential friction forces. We then may write the gear equetions of motion (see also the relations 8.23)

$$J_1 \ddot{q}_1 = T_{an} - r_1 \lambda_G - s_1 \lambda_T$$
$$J_2 \ddot{q}_2 = -T_{ab} + r_2 \lambda_G - s_2 \lambda_T \qquad (8.25)$$

The friction forces are set-valued forces. For sliding and stiction we get accordingly ([137], [226])

$$\lambda_T = sign(\dot{q}_1)\mu_G(|\lambda_G| + \lambda_{G,0}) \qquad \text{for} \quad \dot{q}_2 = \frac{1}{i}\dot{q}_1 \neq 0,$$
$$|\lambda_T| \leq \mu_G(|\lambda_G| + \lambda_{G,0}) \qquad \text{for} \quad \dot{q}_2 = \frac{1}{i}\dot{q}_1 = 0. \qquad (8.26)$$

The equations of motion will take into consideration this situation by applying the structures of section 3.4 on page 131.

### 8.2.2.3 Feet Contacts

Walking stability depends mainly on the contacts between feet and ground. For JOHNNIE a foot consists of three separate bodies (see Figure 8.7). The two lower foot plates are connected by a rotational joint about the foot longitudinal axis ensuring that the ground contact situation is not overconstrained. The ground contact elements are rounded such that a smooth rolling motion of the foot can be realized during touch down and lift off. The upper foot

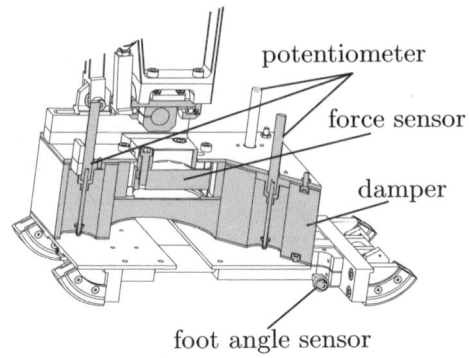

Fig. 8.7: Foot Design [85]

plate is connected to the lower plates by a damping element which absorbes shocks and bridges the time gap between impact and the controller response.

With respect to the foot contacts we have two possibilities, namely modeling the ground forces in a unilateral way, which means rigid body contacts and set-valued forces, or modeling the ground forces by smooth contact laws in the form of spring-damper characteristics. We shall consider both possibilities.

*Unilateral model*

For this case we take the general contact kinematics from section 2.2.6 on page 31, applied to the specific case of the JOHNNIE-feet [137]. The geometry of the foot plates is depicted in Figure 8.8. The foot possesses four cylindrical contact elements in the four corners of the foot plate, which is subdivided into two parts connected by a joint to assure static definiteness. The vector from the inertial coordinate system I to the potential contact point $K_1$ is composed by the three vectors

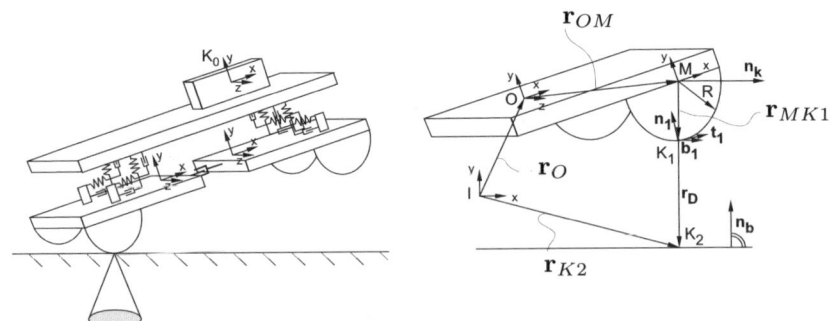

Fig. 8.8: Foot Model and Kinematics [137]

8.2 Walking Dynamics   521

$$r_{OK1} = r_O + r_{OM} + r_{MK1}, \tag{8.27}$$

which can be expressed in any coordinate frame. The direction of $r_{MK1}$, though, has to be oriented in such a way as to fulfill the contact conditions 2.102 and 2.103 on page 39. This results in

$$r_{MK1} = \frac{r^*_{MK1}}{|r^*_{MK1}|} R \quad \text{with} \quad r^*_{MK1} = (n_b \times n_k) \times n_k \tag{8.28}$$

The radius R is explained by Figure 8.8, and the potential contact point results from a projection of the point $K_1$ onto the ground plane with the direction $n_b$. For simulation purposes the ground plane is oriented parallel to the x-y-plane of the inertial coordinate system resulting in a simple form of the contact equations. The normal distance between the points $K_1$ and $K_2$ writes

$$g_N = |\mathbf{r}_D| = n_b^T (r_{K1} - r_{K2}). \tag{8.29}$$

Together with the vector $r_{OK1}$ from equation (8.27) we then get for the time derivatives

$$\dot{r}_{K1} = \dot{r}_O + \tilde{\omega}_1 r_{O,K1} = J_{O,1} \dot{q} + (\tilde{J}_{R,1} r_{O,K1}) \dot{q} = W_{K1} \dot{q} \tag{8.30}$$
$$\dot{r}_{K2} = 0 \tag{8.31}$$

The above equations are based on the assumption, that the contact velocities depend only on the generalized coordinates, and that they include no additional applied velocity terms, which is reasonable. The velocity of the ground-fixed contact point is of course zero. Using the equations (2.110) to (2.113) from the pages 41 we get for the relative contact velocities

$$\dot{g}_N = n_b^T (\dot{r}_{K1} - \dot{r}_{K2}) = n_b^T (J_{O,1} \dot{q} + (\tilde{J}_{R,1} r_{O,K1}) \dot{q}) = W_N \dot{q},$$
$$\dot{g}_T = n_T^T (\dot{r}_{K1} - \dot{r}_{K2}) = n_T^T (J_{O,1} \dot{q} + (\tilde{J}_{R,1} r_{O,K1}) \dot{q}) = W_T \dot{q}, \tag{8.32}$$

and from there also the relative contact accelerations in the form [137]

$$\ddot{g}_N = n_b^T ((J_{O,1} + \tilde{J}_{R,1} r_{O,K1}) \ddot{q} + (\dot{J}_{O,1} + \dot{\tilde{J}}_{R,1} r_{O,K1} + \tilde{J}_{R,1} \dot{r}_{O,K1}) \dot{q})$$
$$= W_N \ddot{q} + \dot{W}_N \dot{q},$$
$$\ddot{g}_T = n_T^T ((J_{O,1} + \tilde{J}_{R,1} r_{O,K1}) \ddot{q} + (\dot{J}_{O,1} + \dot{\tilde{J}}_{R,1} r_{O,K1} + \tilde{J}_{R,1} \dot{r}_{O,K1}) \dot{q})$$
$$= W_T \ddot{q} + \dot{W}_T \dot{q}, \tag{8.33}$$

which are sufficient to build up the system equations of motion including the contacts at the feet of the walking machine, where we have established here the models for rigid body contacts. The unilateral contact laws can also be taken from the kinematics chapter in a suitable form for our specific application

$$\ddot{g}_N \geq 0, \quad \lambda_N \geq 0, \quad \ddot{g}_N \lambda_N = 0,$$

$$|\dot{\mathbf{g}}_T| = \mathbf{0} \quad \Rightarrow \quad |\boldsymbol{\lambda}_T| \leq \mu \lambda_N,$$
$$|\dot{\mathbf{g}}_T| \neq \mathbf{0} \quad \Rightarrow \quad \dot{\mathbf{g}}_T = -\kappa \boldsymbol{\lambda}_T; \quad \kappa \geq 0; \quad |\boldsymbol{\lambda}_T| = \mu \lambda_N. \tag{8.34}$$

*Smooth Force Laws*

Another possibility to model contacts of the feet consists in assuming a single-valued and smooth force law between feet and ground. In many cases of walking machines this corresponds quite well to the real situation, because the feet very often are equipped with some soft rubber material. Figure 8.9 illustrates such a contact with springs and dampers in normal and tangential directions. The determination of the necessary spring and damper values may sometimes be cumbersome, they can be measured or evaluated using FEM, for example. Considering the relative kinematics $(g_N, \dot{g}_N)$ in normal contact direction we

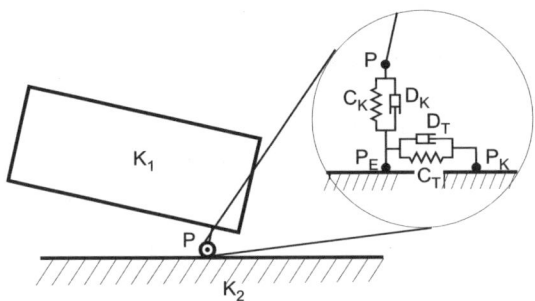

Fig. 8.9: Contact with Smooth Force Laws

get for the normal force

$$F_N = -C_K g_N - D_K \dot{g}_N. \tag{8.35}$$

In tangential direction we need the point $P_K$ where the tangential force applies. We choose for that point the body-fixed position at the beginning of the contact. According to Figure 8.9 we come out with the tangential force

$$\boldsymbol{F}_T = \boldsymbol{C}_T(\boldsymbol{r}_{PK} - \boldsymbol{r}_{PE}) + \boldsymbol{D}_T(\dot{\boldsymbol{r}}_{PK} - \dot{\boldsymbol{r}}_{PE}) \tag{8.36}$$

If the force amount $|\boldsymbol{F}_T|$ reaches the friction reserve $\mu_0 F_N$, then we have sliding with a tangential friction force following Coulomb's law, namely $\boldsymbol{F}_T = \mu F_N$. The point $P_K$ slides also and with a velocity

$$\dot{\boldsymbol{r}}_{PK} = \boldsymbol{D}_T^{-1}(\boldsymbol{F}_T^* - \boldsymbol{C}_T(\boldsymbol{r}_{PK} - \boldsymbol{r}_{PE}) + \dot{\boldsymbol{r}}_{PE}), \tag{8.37}$$

where the sliding friction force will be always directed along the spring-damper force element

$$\boldsymbol{F}_T^* = \mu F_N \frac{\boldsymbol{r}_{PE} - \boldsymbol{r}_{PK}}{|\boldsymbol{r}_{PE} - \boldsymbol{r}_{PK}|}. \tag{8.38}$$

Both states, sliding and sticking, must continuously be controled to discover possible changes from sliding to sticking or vice vera. Such changes can be detected by considering the forces and relative velocities according to the equations (8.34), where every beginning of a contact event is indicated by kinematical and every end by kinetic magnitudes (sections 2.2.6 and 3.4).

The model with spring-damper elements can easily be applied also to such cases like surface to surface contacts or line to surface contacts, including also the case, that surfaces may roll on each other with or without sliding. Some corresponding relations are given in [137].

### 8.2.3 Equations of Motion

#### 8.2.3.1 System Equations

For a general derivation of the equations of motion for multibody systems with bilateral and unilateral constraints we refer to the chapters 3.3 and 3.4 on the pages 113 and 131, respectively. Regarding all bodies of the walking machine we have the following equations of motion (see [137] and [280])

$$M(q,t)\ddot{q} - h(q,\dot{q},t) = Q_g + Q_{ng}, \qquad q \in \mathbb{R}^f, \tag{8.39}$$

with the mass matrix $M(q,t)$ and the gyroscopical force vector $h(q,\dot{q},t)$ in the following form

$$M = \sum_i \left\{ \begin{pmatrix} J_{To} \\ J_R \end{pmatrix}^T \begin{pmatrix} mE & m\tilde{r}_s^T \\ m\tilde{r}_s & I_o \end{pmatrix} \begin{pmatrix} J_{To} \\ J_R \end{pmatrix} \right\}_i$$

$$h = \sum_i \left\{ \begin{pmatrix} J_{To} \\ J_R \end{pmatrix}^T \begin{pmatrix} mE & m\tilde{r}_s^T \\ m\tilde{r}_s & mI_o \end{pmatrix} \begin{pmatrix} J_{Toq} \\ J_{Rq} \end{pmatrix} \dot{q} \right\}_i +$$

$$+ \sum_i \left\{ \begin{pmatrix} J_{To} \\ J_R \end{pmatrix}^T \left[ \begin{pmatrix} mE & m\tilde{r}_s^T \\ m\tilde{r}_s & I_o \end{pmatrix} \begin{pmatrix} \dot{\omega}\dot{r} \\ 0 \end{pmatrix} + \begin{pmatrix} m\tilde{\omega}\tilde{\omega}r_s \\ \tilde{\omega}I_o\omega \end{pmatrix} \right] \right\}_i \tag{8.40}$$

with the following abbreviations

$$J_R = \frac{\partial_K \omega_K}{\partial \dot{q}}, \qquad J_{To} = \frac{\partial_K v_0}{\partial \dot{q}},$$

$$J_{Rq} = \frac{\partial_K \omega_K}{\partial q}, \qquad J_{T0q} = \frac{\partial_K v_0}{\partial q}, \tag{8.41}$$

and with the following magnitudes for the body i: m maas, $I_o$ inertia tensor, $J_{To}, J_R$ Jacobians of translation and rotation, $\omega$ angular velocity vector, $r_s$ radius vector to the center of mass. The forces $Q_g$ are gravity forces, and $Q_{ng}$ contains general external forces but also the forces at the feet acting on the robot. They can be projected into the free directions by

$$Q_g = \sum_i \left\{ \begin{pmatrix} J_{To} \\ J_R \end{pmatrix}^T \begin{pmatrix} m\mathbf{g} \\ m\tilde{\omega}\mathbf{g} \end{pmatrix} \right\}_i ,$$

$$Q_{ng} = \sum_i \left\{ \begin{pmatrix} J_{TF} \\ J_{RM} \end{pmatrix}^T \begin{pmatrix} F \\ M \end{pmatrix} \right\}_i \qquad (8.42)$$

The Jacobians $J_{TF}, J_{RM}$ are those for the force and torque application points.

### 8.2.3.2 Gear Equations

Modeling Harmonic Drive Gears always includes many questions of modeling bilateral constraints with friction, which represents a difficult problem. In the following we shall consider two approaches, a more or less complete one and a simplified one. The calculation of such gears require large computing times, hence an adequate simplification makes sense.

We start with the complete model and take note of the fact, that the gear equations of motion have to be integrated into the system equations. According to Figure 8.6 each of the Harmonic Drive Gears possesses at least two degrees of freedom, an input DOF $q_1$ and an output DOF $q_2$, which corresponds to the joint degree of freedom. Thus we have to add for every gear an additional degree of freedom. Having that in mind and considering the gear equations of motion (8.25) these equations may be written for the sliding case as

$$M\ddot{q} = h + W_{GN}\lambda_G + H_G(\lambda_G + \lambda_{G,0}). \qquad (8.43)$$

The Jacobian matrix $W_{GN}$ projects the averaged normal forces $\lambda_G$ onto the degrees of freedom of the gear $q = (q_1 \; q_2)^T$. The matrix $H_G$ includes the friction coefficients for the determination of the friction forces and torques from the averaged normal forces. For the stiction case we get a similar relation

$$M\ddot{q} = h + W_{GN}\lambda_G + W_{GT}\lambda_T, \qquad (8.44)$$

with the same properties of the matrix $W_{GT}$. For a moving walking machine the gears change very quickly their direction of motion, which results in frequent sliding/sticking changes and vice versa. Therefore it would be necessary to switch rather often between the two equations of motion (8.43) and (8.44). To avoid it, we arrange the matrices of these equations in such a form that they can be combined into one equation:

$$M\ddot{q} = h + W_{GN}\lambda_G + H_G(\lambda_G + \lambda_{G,0}) + W_{GT}\lambda_T \qquad (8.45)$$

For the Harmonic Drives the wheels and pinions are coupled very tightly and without backlash generating a bilateral constraint with friction. To deal with these features Rossmann [226] subdivides the bilateral constraint into two unilateral ones by splitting the corresponding contact forces into two parts [137]

$$\boldsymbol{\lambda}_G = \boldsymbol{\lambda}_G^+ - \boldsymbol{\lambda}_G^- \quad \text{with} \quad \boldsymbol{\lambda}_G^+ \geq \mathbf{0}, \quad \boldsymbol{\lambda}_G^- \geq \mathbf{0}, \quad \boldsymbol{\lambda}_G^{+T}\boldsymbol{\lambda}_G^- = 0. \tag{8.46}$$

These relations possess a simple interpretation. They state, that either the front face of the gear tooth has contact exhibiting the appropriate forces and torques, then the back face has contact without forces and torques, or vice versa. The product of the corresponding forces and torques has to be always zero, which defines an exclusiveness requirement in the form of a complementarity. In our case the magnitudes $\boldsymbol{\lambda}_G^+, \boldsymbol{\lambda}_G^-$ are the torques in one of the tooth sides, front or back flanks. These two constraints are accompanied by friction torques $\boldsymbol{\lambda}_T^+, \boldsymbol{\lambda}_T^-$, which write

$$\boldsymbol{\lambda}_T^+ = \mu_{d,H}(\boldsymbol{\lambda}_G^+ + \boldsymbol{\lambda}_{G,0}) - \boldsymbol{\lambda}_{T01}^+, \qquad \boldsymbol{\lambda}_T^- = \mu_{d,H}(\boldsymbol{\lambda}_G^- + \boldsymbol{\lambda}_{G,0}) - \boldsymbol{\lambda}_{T01}^-. \tag{8.47}$$

The magnitude $\mu_{d,H}$ is a friction coefficient, and $\boldsymbol{\lambda}_{T01}$ is the friction reserve (see Figure 3.4 on page 94). The matrices $\boldsymbol{W}_{GN}, \boldsymbol{H}_G$ and $\boldsymbol{W}_{GT}$ have to be split also into these two unilateral constraints.

The idea of introducing an exclusiveness condition for the tooth pairings in Harmonic Drives allows to construct a complementarity inequality in standard form. The details of the evaluation of this LCP may be found in [226], they will not be presented here. The result writes

$$\begin{pmatrix} \boldsymbol{\lambda}_G^- \\ \kappa_{T1}^+ \\ \boldsymbol{\lambda}_{T02} \end{pmatrix} =$$

$$= \begin{pmatrix} \boldsymbol{E} - 2(\boldsymbol{W}_{GT}^{+'}\bar{\boldsymbol{\mu}}_{d,H} + \boldsymbol{H}_G^{+'}) & \boldsymbol{W}_{GT}^{+'} & \boldsymbol{0} \\ -2\boldsymbol{W}_{GN}^{+T}\boldsymbol{M}^{-1}(\boldsymbol{W}_{GT}^{+''}\bar{\boldsymbol{\mu}}_{d,H} + \boldsymbol{H}_G^{+''}) & \boldsymbol{W}_{GN}^{+T}\boldsymbol{M}^{-1}\boldsymbol{W}_{GT}^{+''} & \boldsymbol{E} \\ 4\bar{\boldsymbol{\mu}}_{d,H}(\boldsymbol{E} - \boldsymbol{W}_{GT}^{+'}\bar{\boldsymbol{\mu}}_{d,H} - \boldsymbol{H}_G^{+'}) & 2\bar{\boldsymbol{\mu}}_{d,H}\boldsymbol{W}_{GT}^{+'} - \boldsymbol{E} & \boldsymbol{0} \end{pmatrix} \begin{pmatrix} \boldsymbol{\lambda}_G^+ \\ \boldsymbol{\lambda}_{T01} \\ \kappa_{T2}^+ \end{pmatrix} +$$

$$+ \begin{pmatrix} -\boldsymbol{h}' - 2(\boldsymbol{W}_{GT}^{+'}\bar{\boldsymbol{\mu}}_{d,H} + \boldsymbol{H}_G^{+'})\boldsymbol{\lambda}_{G,0} \\ \boldsymbol{W}_{GN}^{+T}\boldsymbol{M}^{-1}(\boldsymbol{h}'' + 2(\boldsymbol{W}_{GT}^{+''}\bar{\boldsymbol{\mu}}_{d,H}) + \boldsymbol{H}_G^{+''})\boldsymbol{\lambda}_{G,0} \\ 2\bar{\boldsymbol{\mu}}_{d,H}(-\boldsymbol{h}' + 2(\boldsymbol{E} - \boldsymbol{W}_{GT}^{+'}\bar{\boldsymbol{\mu}}_{d,H}) - \boldsymbol{H}_{GN}^{+'})\boldsymbol{\lambda}_{G,0} \end{pmatrix}$$

$$\begin{pmatrix} \boldsymbol{\lambda}_G^- \\ \kappa_{T1}^+ \\ \boldsymbol{\lambda}_{T02} \end{pmatrix} \geq 0 \quad \begin{pmatrix} \boldsymbol{\lambda}_G^+ \\ \boldsymbol{\lambda}_{T01} \\ \kappa_{T2}^+ \end{pmatrix} \geq 0 \quad \begin{pmatrix} \boldsymbol{\lambda}_G^- \\ \kappa_{T1}^+ \\ \boldsymbol{\lambda}_{T02} \end{pmatrix} \begin{pmatrix} \boldsymbol{\lambda}_G^+ \\ \boldsymbol{\lambda}_{T01} \\ \kappa_{T2}^+ \end{pmatrix} = 0 \tag{8.48}$$

$\kappa_{T1}^+, \kappa_{T2}^+$ are the tangential accelerations of the tooth pairings, which correspond to the averaged sliding acceleration of the gears. The (')-magnitudes can be calculated by (as an example for $\boldsymbol{h}'$)

$$\boldsymbol{h}' = \boldsymbol{M}^*\boldsymbol{h},$$
$$\boldsymbol{M}^* = -[\boldsymbol{W}_{GN}^{+T}\boldsymbol{M}^{-1}(\boldsymbol{W}_{GN}^+ - \boldsymbol{W}_{GT}\bar{\boldsymbol{\mu}}_{d,H} + \boldsymbol{H}_G^+)]^{-1}\boldsymbol{W}_{GN}^{+T}\boldsymbol{M}^{-1}, \tag{8.49}$$

and all the rest the same way. The (")-magnitudes follow from

$$\boldsymbol{h}'' = \boldsymbol{h} + (\boldsymbol{W}_{GN}^+ - \boldsymbol{W}_{GT}^+\bar{\boldsymbol{\mu}}_{d,H} + \boldsymbol{H}_G^+)\boldsymbol{h}'. \tag{8.50}$$

The complementary quantities of equation (8.48) are the transmitted torques $\lambda_G^+, \lambda_G^-$ in positive and negative directions of the tooth pairings, the tangential accelerations $\kappa_{T1}^+$ and the friction reserves $\lambda_{T01}$ and the friction reserves $\lambda_{T02}$ together with the tangential accelerations $\kappa_{T2}^+$. The solution of the Linear Complementarity Problem (LCP), equations (8.48), gives us a possible set of the unknown magnitudes for the gears. It regards transitions from sliding to stiction and vice versa. The relevant force laws are set-valued.

As for all other modeling problems it makes sense to investigate possibilities for simplifications. Simulations with the complete model as presented above demonstrate, that the nonlinearities due to transitions from stick to slip and vice versa do not influence much the overall system behaviour. The reasons for that are threefold: Firstly, for Harmonic Drives the torques to initiate such transitions are very small in comparison to the transmitted torques. Secondly, moving the walking machine along some reference trajectories requires the motion of all joints and therefore of all gears. During such a motion we do not have very frequent changes of the angular speed directions in the gears, and moreover, if we have such changes, they do not appear at the same time instant in several joints. Thirdly, joint control works with rather low time constants and is designed very stiff; as a result some disturbances coming from speed changes in the joints can be damped away by the control system.

One of the main sources of large computing times are the complementarity conditions (8.48), which should be avoided. The above arguments allow a simple approximation by smooth force laws in the contacts, because these stick-slip events do not influence dynamics so much, and thus a smooth approximation is still an excellent approach. According to the equations (8.20) to (8.22) we can express the friction torque by

$$T_R = \mu(T_{G,0} + T_G) + d\omega + b\omega^3. \tag{8.51}$$

Generally, the term $\mu(T_{G,0} + T_G)$ represents that part of the friction, which depends on the speed direction, and which, in the case of stiction, becomes set-valued. To avoid that, we choose the following set of friction coefficients

$$\mu = \mu_0(\frac{\omega}{\omega_0}) \quad \text{for} \quad -\omega_0 \leq \omega \leq \omega_0,$$
$$\mu = +\mu_0 \quad \text{for} \quad \omega \geq +\omega_0, \qquad \mu = -\mu_0 \quad \text{for} \quad \omega \leq -\omega_0. \tag{8.52}$$

The friction coefficient increases linearly up to a limiting speed $\omega_0$ and then remains constant. This means a replacement of the vertical line of Coulomb's law by an inclined line. For this type of regularization we get some small errors in the range $|\omega| \leq \omega_0$, and we cannot describe stick-slip phenomena. Practical experiments with the walking machine JOHNNIE, though, have shown, that this approximation does not influence the walking performance.

As a next step we model each of the motors and the gears with three degrees of freedom, one for a motor and two for a gear. The walking machine JOHNNIE has 17 joints and 6 DOF of the trunk. Altogether this sums up to

## 8.2 Walking Dynamics

$f_{total} = 17 \times 3 + 6 = 57$ degrees of freedom. But we are able to reduce that. The motors and gears are coupled with given transmission ratios, which allow a reduction of the degrees of freedom. For example, the coordinates $q_{j,mot}$ of a motor, $q_{j,get}$ of a gear and $q_j$ of a joint with respect to the degree of freedom $j$ are coupled by the transmission ratio i in the following form:

$$\boldsymbol{q}_j = (q_j \ q_{j,get} \ q_{j,mot})^T = (1 \ i \ i)^T q_j = \boldsymbol{J}_j q_j. \tag{8.53}$$

Summarizing all joint Jacobians to a combined Jacobian $\boldsymbol{J} = (\boldsymbol{J}_1^T \ \boldsymbol{J}_2^T \ ... \ \boldsymbol{J}_n^T)^T$ we are able to project the equations of motion onto the new minimal coordinates by

$$\boldsymbol{M}\ddot{\boldsymbol{q}} = \boldsymbol{h} + \boldsymbol{Q}_e, \qquad \boldsymbol{J}^T \boldsymbol{M} \boldsymbol{J} \ddot{\boldsymbol{q}}_r = \boldsymbol{J}^T \boldsymbol{h} + \boldsymbol{J}^T \boldsymbol{Q}_e, \tag{8.54}$$

from which we derive the equations of motion utilizing these new minimal coordinates:

$$\boldsymbol{M}^* \ddot{\boldsymbol{q}}_r = \boldsymbol{h}^* + \boldsymbol{Q}_e^*,$$
$$\text{with} \quad \boldsymbol{M}^* = \boldsymbol{J}^T \boldsymbol{M} \boldsymbol{J}, \quad \boldsymbol{h}^* = \boldsymbol{J}^T \boldsymbol{h}, \quad \boldsymbol{Q}_e^* = \boldsymbol{J}^T \boldsymbol{Q}_e. \tag{8.55}$$

These relations diminish the original number $f_{total} = 57$ degrees of freedom to the reduced number $f_{total} = 23$, which has a significant effect on the computing time.

## 8.3 Walking Trajectories

### 8.3.1 The Problem

Every walking process requires trajectory planning including all details, the system's structure, the joints, the body, all interconnections and many criteria. Obviously we could look at the biological systems and try to copy their trajectories. This has been done with much success for the six-legged machine MAX (see [202], [273], [54]) by a biologically oriented development. But it cannot be done equally for biped systems, in spite of many investigations of human walking.

Human beings and humanoid robots possess for reasons discussed before (section 8.1.2) many structural similarities, but their specific body and extremity data are completely different. Human beings concentrate their mass in the trunk, and their arms and legs possess relatively small masses. For walking machines it is the other way around. They have heavy arms and legs but a lightweight trunk. The reason is clear: muscles as drive systems are much more effective than motor-gear-combinations. Moreover, the human trunk contains all internal components, whereas the walking machine trunk contains only the computers [137].

The difference in design, in spite of all structural similarities, make a 1:1 tranfer of measured human walking trajectories impossible. The machine would not be stable, it would not even walk. Therefore we have the task to develop our own trajectories, as perfect as possible adapted to the specific walking machine under consideration and with the requirement of many criteria to be fulfilled. This leads to costly optimizations, which we shall consider in the following.

For the optimization of walking machines we shall apply criteria like stability, as the most important criterion, and additionally aspects like energy and velocities together with a large bunch of constraints. They have to refer to the limits of the motors and the gears, the kinematical limits of the machine, the limits of computer speeds and control performance, to name a few.

### 8.3.2 Trajectory Generation

#### 8.3.2.1 Trajectory Coordinates

We have the choice of Cartesian world coordinates and the walking machine's joint coordinates, of work-space and of joint-space. And we need both. Additionaly we have to watch the requirement to evaluate during walking certain transformations in real-time, from work-space into joint-space and vice versa. For control purposes and with respcet to real-time requirements we choose for JOHNNIE the following treatment.

The trajectories will be prescribed in world coordinates $\mathbf{x}$ and transformed by real-time computing into joint coordinates $\mathbf{q}$, where the two coordinate sets

## 8.3 Walking Trajectories

$\mathbf{x}, \mathbf{q}$ have to be chosen in such a way, that the transformations are unique in both directions. Thus the vector $\mathbf{x}$ includes the six trunk coordinates $\mathbf{x}_O$, one pelvis degree of freedom $q_B$, the arm degrees of freedom $\mathbf{x}_A$ and the foot degrees of freedom $\mathbf{x}_F$. This makes altogether

$$\mathbf{x} = (\mathbf{x}_O^T \; q_B \; \mathbf{x}_F^T \; \mathbf{x}_A^T)^T \quad \in \mathbb{R}^{17}. \tag{8.56}$$

We first consider the six trunk coordinates, where the translations are elementary, but the rotations need special treatment. The usual representation of rotations by Euler and Cardan angles or by quaternions is not very convenient due to certain control requirements. The control concept of JOHNNIE anticipates the possibility, that not all degrees of freedom can be controlled. Such a case appears, if the maximum torques being transmitted by the feet to the machine, exceed their limits. Then controllability is lost, and we have to define, which of the coordinates may deviate from the reference trajectory. For example, this might be the inclination of the trunk. In addition to this situation we cannot assign to the angle coordinates of the classical representations inertial coordinate axes.

For a description of the trunk of JOHNNIE we therefore develop an individual angular coordinate system. Figure 8.10 illustrates these new definitions [137]. The orientation of the coordinate system K (with primes) is defined by

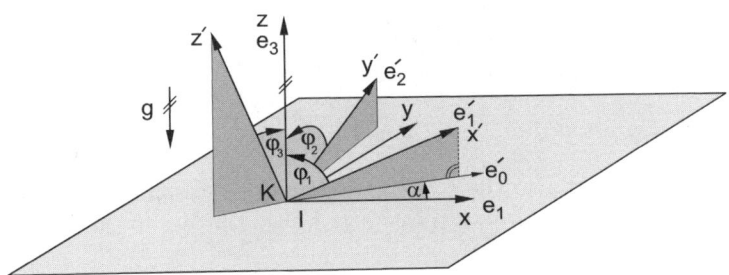

Fig. 8.10: Definition of the Trunk Angular Coordinates [137]

the angles $\alpha$, $\varphi_1$ and $\varphi_2$. The angle $\alpha$ describes a rotation around the z-axis of the inertial system. It is the angle between the $e_1$-axis and the $e_1'$-axis projected onto the inertial x-y-plane. By the angles $\varphi_1$ and $\varphi_2$ spanned between the $e_1'$-, the $e_2'$-axis and the z-axis we are able to define independently the inclinations of the trunk in the forward and sidewards directions. This is a very important issue. The angle $\varphi_3$ is redundant. An additional advantage of the above definitions consists in a measurement convenience, all three angles $\varphi_1$, $\varphi_2$ and $\varphi_3$ can be directly measured by the inclination sensor.

Of course we have to pay a price, namely in the form of a bit more costly transformations. For a mapping from a body-fixed frame into an inertial system we may write generally

$$_I\boldsymbol{x} = \begin{pmatrix} a_{00} & a_{10} & a_{20} \\ a_{01} & a_{11} & a_{21} \\ a_{02} & a_{12} & a_{22} \end{pmatrix} {}_K\boldsymbol{x} = \begin{pmatrix} \boldsymbol{a}_0^T \\ \boldsymbol{a}_1^T \\ \boldsymbol{a}_2^T \end{pmatrix}^T {}_K\boldsymbol{x} \tag{8.57}$$

With the angles $\alpha$, $\varphi_1$ and $\varphi_2$ we can directly calculate the values of

$a_{00} = \sin\varphi_1 \cos\alpha \qquad a_{01} = \sin\varphi_1 \sin\alpha \qquad a_{02} = \cos\varphi_1$
$a_{12} = \cos\varphi_2 \qquad\qquad (a_{22} = \cos\varphi_3)$

and for the rest of the unknown coefficients we have the conditions

$$\boldsymbol{a}_0^T \boldsymbol{a}_1 = \boldsymbol{0} \quad \wedge \quad |\boldsymbol{a}_1| = 1 \quad \wedge \quad \boldsymbol{a}_2 = \boldsymbol{a}_0 \times \boldsymbol{a}_1$$

These equations can be solved in an unambiguous way [137]. The Jacobian of this rotation can be determined with the help of the velocities $\dot\alpha$, $\dot\varphi_1$ and $\dot\varphi_2$ and with the Cartesian angular velocity $\boldsymbol{\omega}$. We come out with

$$\dot\alpha = (\boldsymbol{\omega} \times {}_I\boldsymbol{e}'_0)^T \frac{\boldsymbol{n}_\alpha}{|\boldsymbol{n}_\alpha|} \qquad \text{with} \qquad \boldsymbol{n}_\alpha = {}_I\boldsymbol{e}'_0 \times ({}_I\boldsymbol{e}'_0 \times {}_I\boldsymbol{e}_1)$$
$$\dot\varphi_1 = (\boldsymbol{\omega} \times {}_I\boldsymbol{e}'_1)^T \frac{\boldsymbol{n}_{\varphi_1}}{|\boldsymbol{n}_{\varphi_1}|} \qquad \text{with} \qquad \boldsymbol{n}_{\varphi_1} = {}_I\boldsymbol{e}'_1 \times ({}_I\boldsymbol{e}'_1 \times {}_I\boldsymbol{e}_3)$$
$$\dot\varphi_2 = (\boldsymbol{\omega} \times {}_I\boldsymbol{e}'_2)^T \frac{\boldsymbol{n}_{\varphi_2}}{|\boldsymbol{n}_{\varphi_2}|} \qquad \text{with} \qquad \boldsymbol{n}_{\varphi_2} = {}_I\boldsymbol{e}'_2 \times ({}_I\boldsymbol{e}'_2 \times {}_I\boldsymbol{e}_3) \tag{8.58}$$

The Jacobian for this special coordinate definition and its time derivation results directly from the above relations. We get

$$\dot{\boldsymbol{q}}_R = (\dot\alpha \; \dot\varphi_1 \; \dot\varphi_2)^T = \boldsymbol{J}_R^{-1} \boldsymbol{\omega}, \qquad \dot{\boldsymbol{J}}_R = -\boldsymbol{J}_R \, \dot{\boldsymbol{J}}_R^{-1} \, \boldsymbol{J}_R \tag{8.59}$$

The stability and also the controllability of the walking machine depends to a large extent on the torques transmitted from the feet to the machine. They

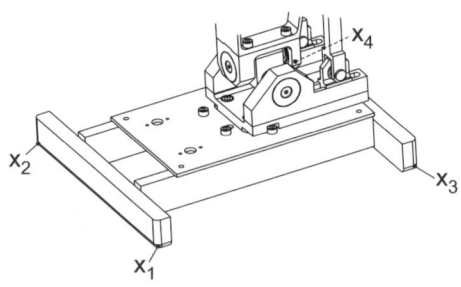

Fig. 8.11: Foot Design

are directly proportional to the lenght and width of a foot characterized by its four corner points. The reference trajectory will be described by the inertial position of these four corner points, see Figure 8.11. The vector with respect to inertial coordinates to the four corner points is defined as $\boldsymbol{x}_{K11}...\boldsymbol{x}_{K14}$ for the right and $\boldsymbol{x}_{K21}...\boldsymbol{x}_{K24}$ for the left foot. The position of the complete foot is given with the middle point of the four corners

$$\boldsymbol{x}_{F,i} = \frac{1}{4}(\boldsymbol{x}_{Ki1} + \boldsymbol{x}_{Ki2} + \boldsymbol{x}_{Ki3} + \boldsymbol{x}_{Ki4}), \quad i \in 1,2. \tag{8.60}$$

The orientation of the feet can be calculated by the relative position of the corners to each other. We get

$$\kappa_{i,1} = \frac{z_{Ki1} + z_{Ki2} - z_{Ki3} - z_{Ki4}}{x_{Ki1} + x_{Ki2} - x_{Ki3} - x_{Ki4}},$$

$$\kappa_{i,2} = \frac{z_{Ki1} - z_{Ki2} + z_{Ki3} - z_{Ki4}}{x_{Ki1} - x_{Ki2} + x_{Ki3} - x_{Ki4}},$$

$$\kappa_{i,3} = \frac{y_{Ki1} + y_{Ki2} - y_{Ki3} - y_{Ki4}}{x_{Ki1} + x_{Ki2} - x_{Ki3} - x_{Ki4}}, \tag{8.61}$$

where $\boldsymbol{x}_{Kij} = (x_{Kij}, y_{Kij}, z_{Kij})^T$ are the corner coordinates. The magnitude $\kappa_{i,1}$ describes the foot inclination forward/backwards (pitch), $\kappa_{i,2}$ is the inclination to the side (roll) and $\kappa_{i,3}$ the orientation around the vertical axis (yaw).

### 8.3.2.2 Motion of the Center of Mass

For robots and walking machines we always have to regard two aspects with respect to modeling. For simulation purposes usually applied for design and verification we need a detailed model considering all influences like contact events or temperature dependencies of the drives. Computing time is then not so important. But for model-based control we need real-time models being able to give at least the most important informations of the system's dynamics within very short time instants. In the last sections we have presented detailed descriptions and a few simplification aspects. We shall continue this section with simplified models, though complete models are of course available.

The approximate motion of the mass center will be described by a combination of the zero-moment-point model (ZMP) [267] and the inverted pendulum model. Both models are very well known since long time, and they are applied in nearly all modern walking machines considering the arguments just given above. The ZMP model is nothing else but the application of the equivalence principle of classical statics generating from force-couples a force alone. The inverted pendulum model is a very popular example for advanced course exercises together with laboratory models. The relation with respect to walking is nicely presented by some Japanese papers, for example [156], [78], [121].

For the zero moment point we get the simple relation (see Figure 8.12)

$$\boldsymbol{F}_{ZMP} = \boldsymbol{F}_F, \quad \boldsymbol{T}_{ZMP} = \boldsymbol{T}_F + \boldsymbol{F}_F \times \boldsymbol{r}_{F,ZMP} = \boldsymbol{0}. \tag{8.62}$$

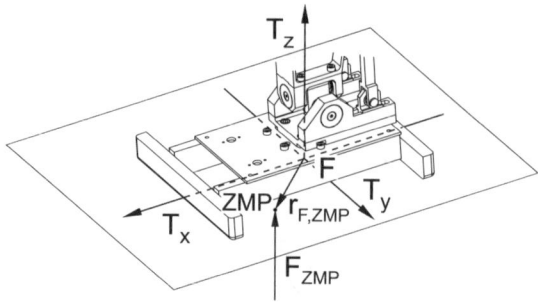

Fig. 8.12: ZMP Forces and Torques

The index F denotes the point F, where we know the forces and torques at the feet. The last equation can be solved for the unknown vector $r_{F,ZMP}$ to the zero moment point. The condition, that this point has to be positioned within the foot supporting area, provides us only with a static stability information, which is sometimes helpful. In the following we shall discuss that a bit.

Human walking is characterized by two phases, not considering here the phase without ground contact taking place only for running. The two phases are the single- and the double-support phases. For the single-support phase we have stance for one foot and swing for the other one. For the double-support phase we have stance for both feet. During walking we have a periodic change of single- and double-support phases.

The inverted pendulum model provides a simple model for a also simple control design, which for small speeds might be a first approach. Figure 8.13 illustrates the model. The momentum equation for the total system writes

$$M\ddot{x}_S = -G + F_{ext}, \qquad (8.63)$$

which includes the gravity forces $G$ and some external forces $F_{ext}$. The vector $x_S$ represents the coordinates of the machine's center of mass. As long as the machine mass is constant, we get from equation (8.63) the correct motion of the center of mass without simplifications. With respect to the moment of momentum equations we are not able to reduce the machine's overall moment of inertia to a constant value. Therefore we take as a rough approximation

$$J\dot{\omega}_S = T_{ext}, \qquad (8.64)$$

which is a simplification, because on the one side the moments of inertia of the machine are not constant, and on the other side the moment of momentum of the machine might change even for a not moving trunk but with moving legs or arms. Astonishingly enough, the rough approximation as introduced above is quite nicely confirmed by experiments, and it is therefore used by many research teams around the world. For JOHNNIE we apply that for real-time computations, but not for simulations.

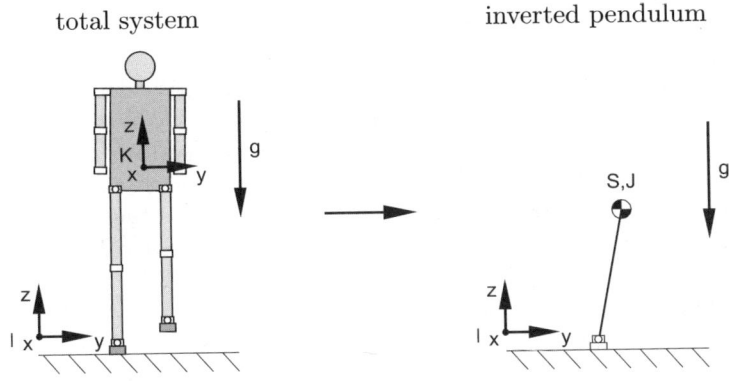

Fig. 8.13: Inverted Pendulum Model

For the following trajectory consideration we assume, again approximately, that we have an upright gait for $\boldsymbol{T}_{ext} = \boldsymbol{0}$. According to Figure 8.13 only forces are acting in the zero-moment-point and no torques. From this we get also only forces along the connecting line from the ZMP to the system center of mass. With a prescribed vertical acceleration $\ddot{z}_{S,ref}$ of the center of mass the horizontal accelerations can be calculated by

$$\begin{pmatrix} \ddot{x}_S \\ \ddot{y}_S \\ \ddot{z}_S \end{pmatrix} = \begin{pmatrix} (g + \ddot{z}_{S,ref})\frac{x_S - x_{ZMP}}{z_S - z_{ZMP}} \\ (g + \ddot{z}_{S,ref})\frac{y_S - y_{ZMP}}{z_S - z_{ZMP}} \\ \ddot{z}_{S,ref} \end{pmatrix} \tag{8.65}$$

These equations can be solved for some prescribed vertical motion, for exam-

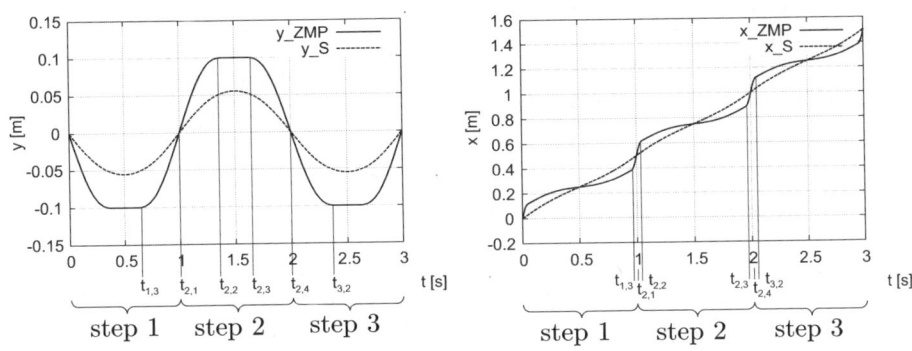

Fig. 8.14: Motion of the Center of Mass and of the ZMP (see Figure 8.13)

ple one of the magnitudes $z_{S,ref}, \dot{z}_{S,ref}, \ddot{z}_{S,ref}$. Figure 8.14 depicts the lateral motion of the center of mass and the ZMP for a constant height of the center of mass [137].

The single- and double-support phases are uniquely determined by the trunk and the arm motion and by the movements of the feet. For the mathematical description of these movements we define certain reference points of the trajectories, where the kinematics is prescibed, and we interpolate between these points by polynomials of fifth order to avoid jerk effects. Figure 8.15 depicts a typical foot trajectory. The foot detaches vertically from the ground, and at the same time the heel is lifted a bit. Then we have a swing phase, where the heel is rotated to meet the necessary stance position. After that we have the stance phase. The details are adapted by the control system during walking.

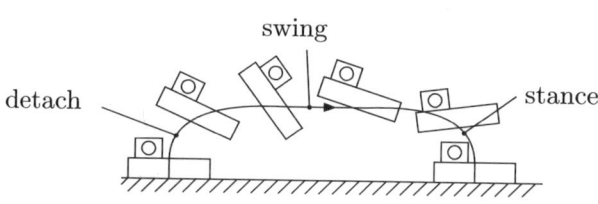

Fig. 8.15: Foot Trajectory

It should be noted, however, that this simplified approach, though quite successful for many machines, will eventually be replaced by more sophisticated ones without these simplifications. The ideal solution would consist in an automatic generation of trajectories coming out from an analysis of the nonlinear equations of motion together with a stability investigation. Probably the gaits would be the fixed-point solutions of the complete set of the nonlinear biped equations, but at the time being this is still a matter of ongoing research [44]. The problem with regard to such considerations is, that we do not know very much about the criteria according to which walking is governed in biological systems. But this situation might change in the years to come.

### 8.3.2.3 Trajectory Optimization

Optimizations are a necessity for both, robotics and walking. With respect to trajectories we not only have a big bunch of constraints, which are needed to watch, but also a couple of criteria, which make sense. The matter has already been discussed in the robot chapter, especially in section 7.3.3 on page 476, where we consider combined trajectory and process optimization. The optimization theory applied for this topic are not far away from those for walking trajectories, as a matter of fact they have influcnced very much walking optimization, see [271], [219], [220], [211].

As usual we start with the equations of motion, which give us the motion of all joints, for JOHNNIE the motion trajectories of 17 joints. The necessary optimization criteria are not very clear in the case of walking, but at least we know from neurobiological investigations that stability has to be the number one criterion [165]. In addition to that we may take into consideration criteria like total energy consumption, equally distributed torques over all joints, an averaged value of all torques, the knee angle or the amount of power consumed. For a vector optimization problem they must be combined with a smaller valuation than stability.

A crucial problem of any optimization consists always in computing time, because one simulation of the system corresponds then only to one point in the parameter space [179]. The problem can be reduced considerably by evaluating the gradients of the criteria with respect to the parameters, if it is really possible. In many cases it is not, but for walking machines it is, though a little bit costly [137], [33].

The problems of optimizing walking trajectories are still a matter of ongoing research. As a first step the authors of [137], [33] performed a trajectory optimization not including the control system and the underactuated degrees of freedom, but regarding the complete system's equations. The result is also of practical relevancy, because it contains informations of optimal trajectories without leaving the allowed range of parameters, especially forces and torques. These optimized trajectories have been implemented in JOHNNIE.

But there remain some open questions. For a next step such an optimization should be performed for the complete system including control. First trials indicate, that at the time being the computing time problems might be too large. This is true also for an optimization including control and underactuated degrees of freedom, for example feet degrees of freedom due to visco-elastic material. But very likely all possibilities to achieve optimal trajectories will be realized, especially for fast gaits.

## 8.4 The Concept of JOHNNIE

### 8.4.1 Requirements

From a large project at the author's former Institute concerning normal and hemiparetic human walking it was easy to take for a first layout the quite well-known data of human walking [2]. These data as a matter of fact refer more to mechanical properties like kinematics, masses, moments of inertia, torques and forces and not so much on human walking control, sensors and actuators. The requirements for the sensors come from the technical control concept, which is a combined position-force-control system. The state of all joints must be known, the force/torque situation at each foot must be measured, and an inertial reference must be given. Therefore, encoders and tachometers, six-component-force-torque-sensors and an inertial platform are needed. Quantities like friction, which are not measured, can be estimated by observers. Altogether the most important requirements are as follows:

| | |
|---|---|
| size | 1.80 m |
| weight | < 50 kg |
| max. speed | 0.5 – 1.0 m/s |
| configuration | humanlike |
| degrees of freedom | |
| leg | 6 DOF |
| foot (internal) | 4 – 8 DOF |
| sensors | encoders |
| | force-torque-sensors |
| | inertial platform |
| actuators | Neodym-Bor-DC-Motors |
| | Harmonic Drives Gears |
| | Ball Screws |

### 8.4.2 Mechanical Models

In the following we shall consider some more details concerning the hardware selection. Figure 8.16 shows a sectional view of the final version of the hip joint. The actuation for the yaw and roll axis are arranged coaxially with the joint axis and are integrated in the aluminum structure. The yaw joint is inclined 15 degrees with respect to the pelvis. This leads to a better power distribution among the four hip motors. The pitch joint is actuated with two motors via a timing belt. The employed gear has a modified Circular Spline, which is T-shaped in order to reduce weight. Further, an aluminum Wave Generator with optimized shape is included. Its moment of inertia is 50% lower than that of the standard series. The shank includes the PWM-amplifiers for the knee joint actuation (PWM - pulse width modulation). The table on page 517 shows technical data of the joints.

---

[2] Section 8.1 is based on the text of [202].

8.4 The Concept of JOHNNIE    537

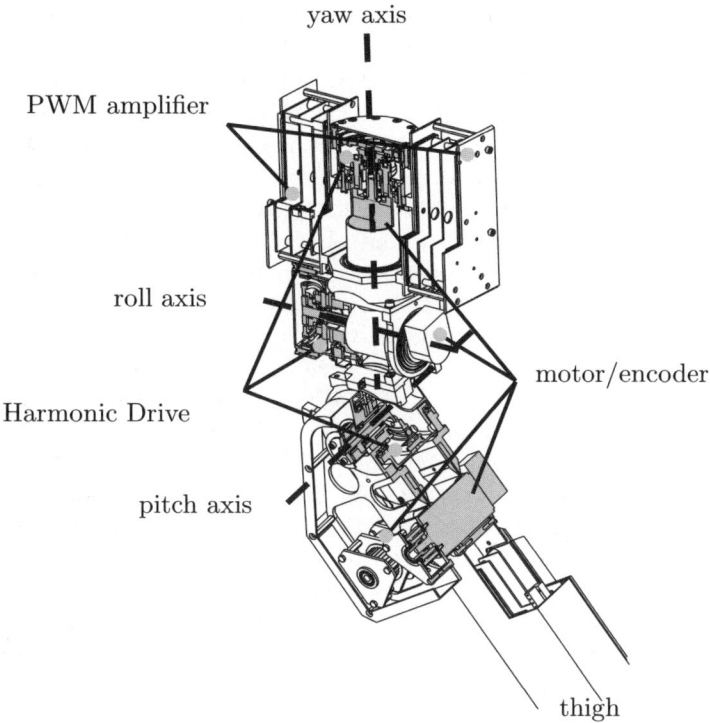

Fig. 8.16: Hip Joint Design [85]

The design of the knee joint corresponds to that of the hip pitch joint. The actuation of the ankle joint is realized with two linear drives based on ball screws (see Figure 8.4 on page 514). Two motors drive the ballscrews via timing belt. A motion of the sliders in the same direction leads to a pitch motion of the foot, the roll motion is realized by moving the sliders in reverse direction.

The foot consists of three separate bodies (see Figure 8.7 on page 520). The two lower foot plates are connected by a rotational joint about the foot longitudinal axis ensuring that the ground contact situation is not overconstrained. The ground contact elements are rounded such that a smooth rolling motion of the foot can be realized during touch down and lift off. The upper foot plate is connected to the lower plates by a damping element which absorbes shocks and bridges the time gap between impact and the controller response.

### 8.4.3 Sensors

The joint angles and joint angular velocities are measured by incremental encoders (HP 5550 HDSL) that are attached to the motor shafts. They have

500 lines such that an accuracy of 1/2000 of a revolution can be achieved with the microcontroller hardware. In addition, a reference line is evaluated. In order to obtain a reference position, light barriers are positioned in the work range. Before operation, the robot has to perform an initialization where all joints pass the light barriers. This position is used as the basis to find the next reference line which is the reference position. As the Harmonic Drive gears are very stiff, the error due to elastic gear deformation is small. The high resolution allows for an exact control of the joint position at a short settling time. The joint velocity is identified by numerical differentiation of the joint position. To avoid damage of the robot, each joint is equipped with switches that confine the minimum and maximum joint angle. When the workspace is exceeded, the PWM signal for the corresponding joint is turned off.

For controlling the ground contact, especially tilting or slipping, a six-component-force-torque-sensor has been developed [85]. Its design is based on requirements resulting from simulations of the controlled jogging motion. As commercial sensors meeting these requirements were not available, mainly with respect to weight and size, an especially adapted sensor was realized (Figure 8.17). Its performance with regard to the measurement range and to measurement errors is excellent. The final version is based on a classical sensor design with three deformation beams holding strain gauges. Thin membranes make sure that defined stresses occur at the strain gauge positions. These membranes decouple the force directions to a certain extent.

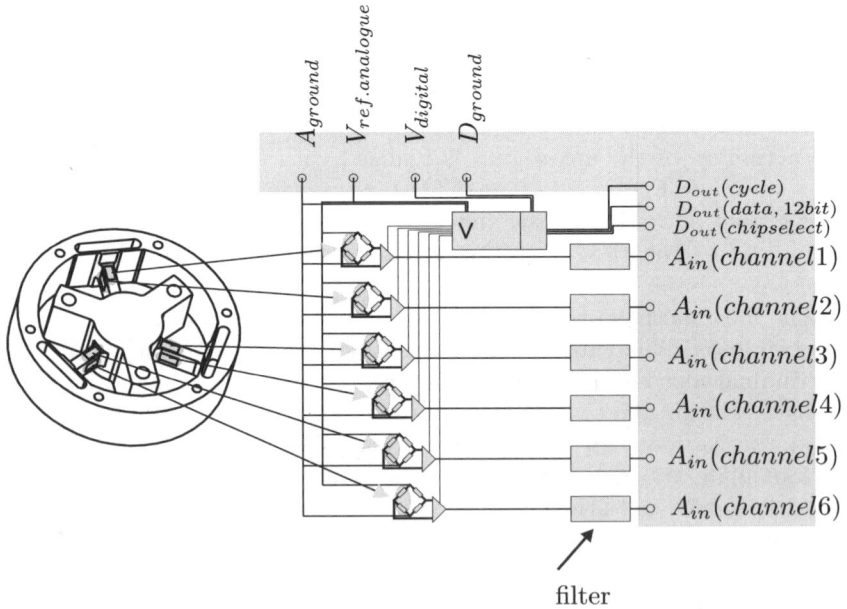

Fig. 8.17: Force Sensor [85]

The detailed layout of the sensor has been performed employing the method of Finite Elements. Based on the simulated force-torque information, calculations have been made in order to ensure that the maximum van Mises stresses are below the durability stress of the sensor material. Strain gauges are applied as half bridges on the deformation bars. The amplifier is included in the sensor housing.

The control of the robot requires a precise information about the orientation of the upper body in space. Since it cannot be determined from the joint angles with sufficient accuracy, an inertial orientation sensor system is included. The upper body motion is characterized by high linear accelerations in vertical direction and high oscillations (2 Hz at a jogging speed of 5 km/h). Therefore, the application of an inclination sensor leads to poor results due to their poor dynamic properties. A set of three gyroscopes is used to compensate their dynamic behaviour. As the integrated angular velocity information of the gyros cannot be computed without drift, a sensor fusion method is used to combine both sensor data from acceleration sensors and gyroscopes to obtain the best performance. The sensor fusion methods often employed for such systems are the drift estimation using a Kalman Filter or fusing the information with a complementary filter.

### 8.4.4 Control Concept

Various control concepts for the complete machine and for local controllers have been considered and tested for JOHNNIE [137]. We shall present here only one of these configurations, the feedback linearization, inspite of the fact that for fast walking this concept is too slow. The control scheme is a three-layer-concept as shown in Figure 8.18. If the biped is in addition vision-controlled we get a fourth layer deciding on the walking tasks and requirements. In Figure 8.18 the lowest layer includes a feedback linearization scheme mainly applied to stabilize the machine and to assure a safe basis for the higher layers. As the realization of a feedback linearization is rather sensitve with respect to parameter uncertainties we need some observers to estimate friction, gravity, position and orientation. It is one of the drawbacks of feedback linearization resulting in relatively large computing times. Figure 8.19 illustrates a bit more in detail the concept. Altogether 39 measurement signals enter the control block, 27 signals for the joint status and 12 signals for forces and torques at the feet. After being processed in the control loop the power signals for the joint actuators leave the control block. The complete processing from measurements to the power signals takes about 4 ms, which is too long for fast walking or jogging.

The second layer in Figure 8.18 concerns the process of trajectory generation for normal and fast walking and for jogging, where the control and trajectory parameters are evaluated, the reference values are determined, and where finally the feedback linearization will be activated. The computation of

## 8 Walking

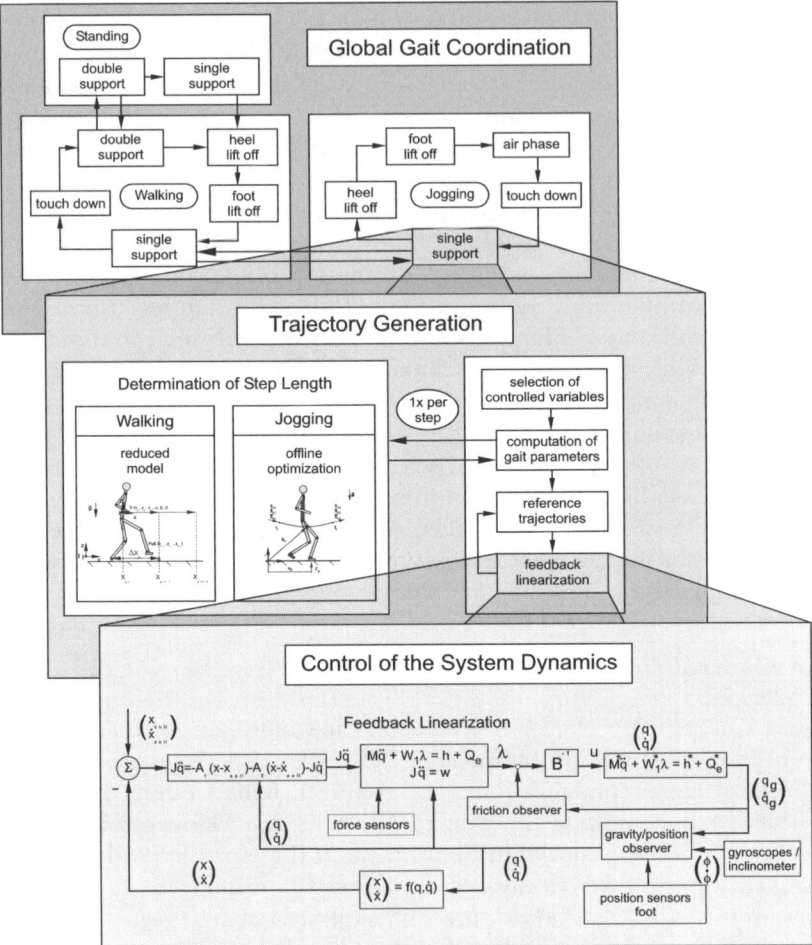

Fig. 8.18: Control Concept of JOHNNIE [137]

the reference trajectories is crucial for a stable motion of the robot. In particular all existing constraints have to be satisfied throughout the entire gait cycle. Nevertheless the trajectories are not uniquely defined by these constraints. An infinite number of trajectories is possible for a given walking speed, such that the most suitable trajectory has to be determined by an optimization.

While an optimized trajectory leads to a very good system performance when tracked exactly, it is not necessarily the best solution for a real walking machine. Highly optimized trajectories are computed as spline curves in terms of the joint angles. It is very difficult to adapt these trajectories in case of disturbances and to change the gait pattern in an unknown environment. A modification of the trajectories would require a huge data base or an online optimization of the trajectories. Presently both solutions work only in simulations since they require extensive computational power and cannot be used for a system operating in real time. Biological systems do not track a given set of trajectories extremely exact, but adapt their motion to upcoming disturbances and can compensate for a great part of sensor errors, inaccurate tracking and disturbances. We therefore use a reduced model for the computation of dynamically stable reference trajectories. The solution is not completely exact, but it can be computed in real time and allows for an adaptation of the trajectories during walking. This way it becomes possible to compensate model inaccuracies as well as external disturbances.

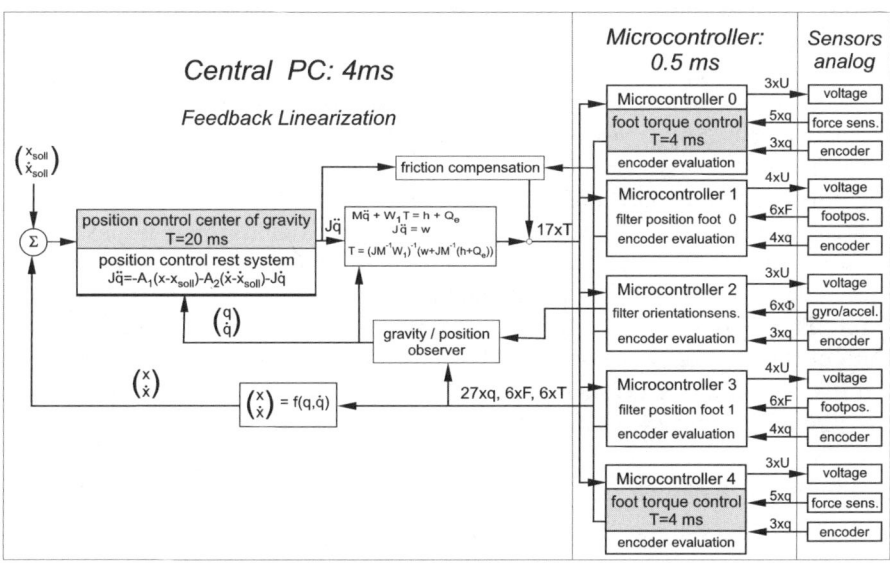

Fig. 8.19: The Feedback Linearization Concept [137]

The highest layer in Figure 8.18 deals with the global walking coordination including features like walking, jogging and standing, the last one requiring certain control measures. The various phases of these features have to be coordinated correctly and transfered to the next layer of trajectory generation. A supervisory layer has been realized for vision control (Prof. G. Schmidt, TU-Munich (see [140], [236], [235])).

Two additional aspects should be finally mentioned, the problem of constraints and the important properties of foot dynamics. One of the main difficulties in the control of dynamically walking robots result from constraints that limit the applicability of conventional control concepts. Two groups of constraints need to be considered. Firstly the workspaces of the joints, the maximum rotor velocities and the joint torques are limited. These are typical constraints for industrial robots and can be satisfied by an adequate design and an appropriate choice of the trajectories. However critical control problems result from the second group of constraints that describe the unilateral contact between the feet and the ground. Depending on the normal force that is transmitted from foot to the ground, the maximum transmissible torques as well as the tangential forces are limited by the corresponding friction cone. While practical experiments show that the robot usually does not start slipping, the limits of the torques in the lateral and frontal direction lead to a small margin of stability.

From human walking we know that foot dynamics is a crucial point for any walking or running process [171]. Forces and torques at the feet contribute significantly to the stability of the system. For example when the orientation of the upper body deviates slightly from its reference, the foot torques are increased to bring the orientation back to its reference value. Depending on the time constants that were chosen for tracking of the orientation, the foot torques can easily exceed their maximum limits. The feet would tilt even though the robot is very close to its reference trajectory. Therefore a direct measurement and control of the foot torques is inevitable when the motion of the robot is based on an orientation sensor. For our robot it is particularly easy to control the foot torques with a high bandwidth.The torques of the feet depend only on the forces of the ball screw drives that actuate the ankle joint. These are controlled by the same microcontroller that also reads in the data of the six axes force sensor. The controller operates at a sampling rate of 0.4 ms. Steady state errors due to gear friction are compensated by a friction observer. The control scheme has been verified in experiments.

### 8.4.5 Some Results

#### 8.4.5.1 Simulations

JOHNNIE walks, but does not reach jogging velocities. We use multibody simulations to test the performance of the controller and to optimize the mechanical design of the robot. Particular emphasis is put on the simulation of the contact between foot and ground. The foot contacts consist of four cylindrical elements with which the foot can perform a rolling motion at touch down and lift off. The contact between these elements and the ground is modeled as a rigid body contact leading to a complementarity problem, which can be solved by well-known standard algorithms. Another important issue is the simulation of the friction of the Harmonic Drive gears. The friction is modeled with a nonlinear characteristic, while the stick-slip transitions are also implemented with LCPs (LCP - Linear Complementarity Problem). The simulations show that the controller is suitable to generate a stable gait pattern. External disturbances can be handled effectively with the described strategy. The simulation results are used to optimize the design. Figure 8.20 shows typical simulation results that were otained in an optimization of the geometric arrangement of the hip joint.

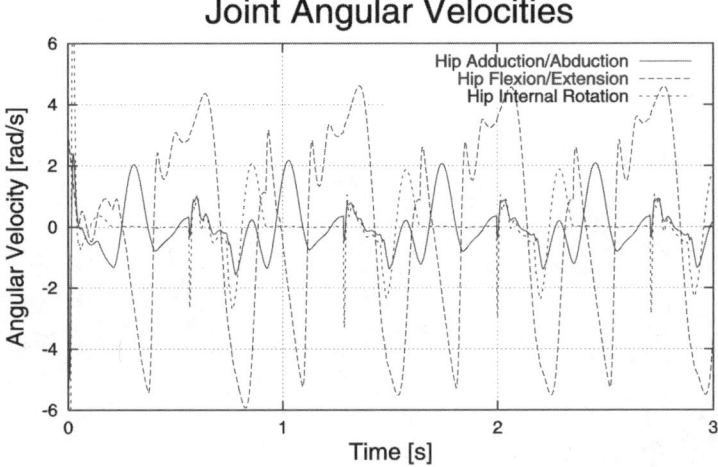

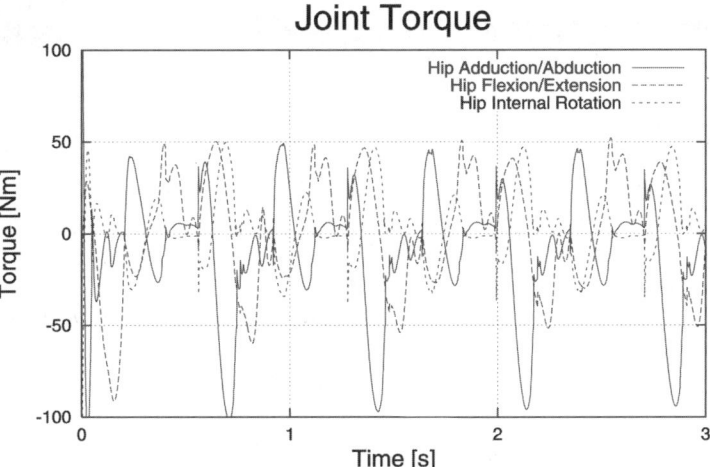

Fig. 8.20: Hip Joint Velocities and Torques for Jogging

**8.4.5.2 Walking Experiments**

Two types of experiments have been carried through, a large number of walking tests on a conveyor belt and some tests, where a certain amount of autonomous walking was realized. The robot's speed on the conveyor belt can be adjusted manually by the operator. In addition the walking direction is controled in such a way, that the machine remains centered on the belt even for long term experiments. It has been found that the belt acceleration influences very little the walking stability allowing therefore a fast transition to the maximum speed of the belt. The walking speed realized by the machine

544    8 Walking

control as presented above is 2.4 km/h with a maximum step length of 60 cm. A new biped being just now developed will exceed these values. Figure 8.21 illustrates a walking process on the conveyor belt.

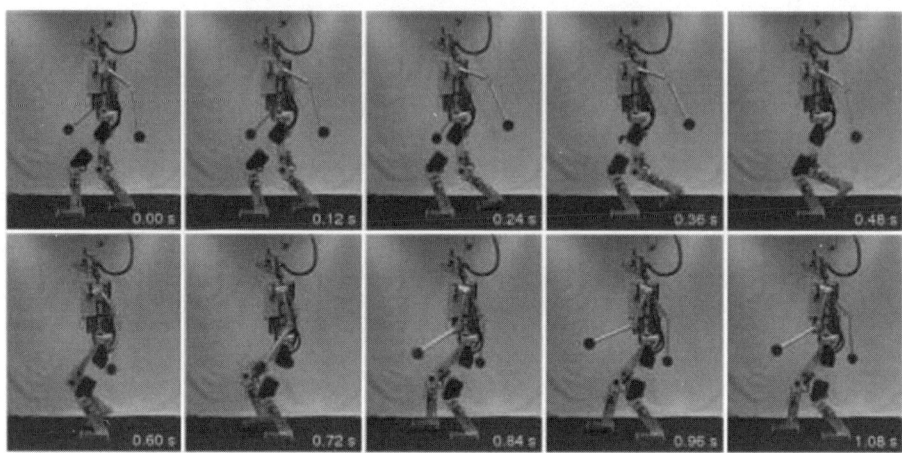

Fig. 8.21: Walking JOHNNIE

The second test including autonomous walking has been prepared for the Hannover Fair 2003 and has been presented there with the following scenario. Within an area of 5 by 7 meters the robot starts in one corner, comes to an obstacle and decides by himself to step over it. He comes then to a second obstacle, which is too large for the robot, therefore he decides to go around. JOHNNIE then walks along the external limits of the area meeting finally some stairs. He decides to go upstairs to the conveyor belt, where he performs some walking with large speed. The decision capabilities were achieved by a vision system developed by Professor Günther Schmidt in Munich. The vision results were combined with the walking possibilities of JOHNNIE to realize the appropriate walking process. All decesions for avoiding obstacles and for climbing the stairs were based on an external world model resulting from the vision process. From this JOHNNIE could see, decide and walk without any operator's support. Figure 8.22 depicts the stair case walking.

8.4 The Concept of JOHNNIE 545

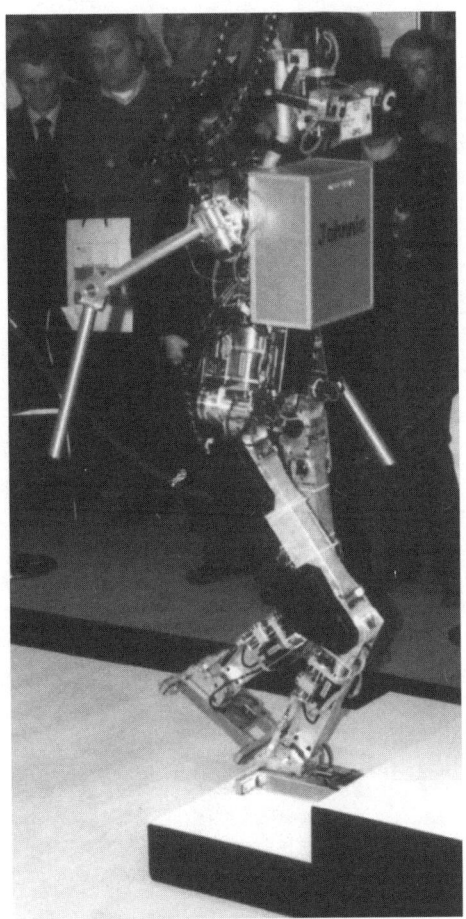

Fig. 8.22: JOHNNIE on Stairs

# References

[1] ABRAHAM, Ralph ; MARSDEN, Jerrold E.: *Foundations of Mechanics*. Westview Press, 1978
[2] ALART, P. ; CURNIER, A.: A mixed formulation for frictional contact problems prone to Newton-like solution methods. In: *Comp. Methods Appl. Mech. Engrg.* 92 (1991), S. 353–375
[3] ANGELES, Jorge: *Fundamentals of Robotic Mechanical Systems*. Springer Verlag New York Inc., 1997
[4] ARIKAWA, Keisuke ; HIROSE, Shigeo: Mechanical Design of Walking Machines. In: *in: Walking Machines, Phil. Trans. R. Soc. A, vol 365, no 1850 (eds. Pfeiffer, Inoue)* (2007), S. 171–183
[5] ARNOLD, M.: *Zur Theorie und zur numerischen Lösung von Anfangswertproblemen für differential-algebraische Systeme von Höherem Index*. Fortschrittberichte VDI, Reihe 20, Nr. 264, VDI-Verlag Düsseldorf, 1998
[6] ARNOLD, Martin ; BREEDVELD, Peter ; BRÜLS, Olivier ; CUADRADO, Javier ; EBERHARD, Peter ; SCHIEHLEN, Werner ; VALASEK, Michael ; ARNOLD, Martin (Hrsg.) ; SCHIEHLEN, Werner (Hrsg.): *Simulation Techniques for Applied Dynamics*. CISM Course 2007, 2007
[7] ASADA, H. ; SLOTINE, J.-J.E.: *Robot Analysis and Control*. Wiley, New York, 1986
[8] ASSMANN, C ; GOLD, P.W.: Kennwertermittlung der Eurogrip Kupplung, Size 28, Ausführung VW. In: *RWTH Aachen, IME* (2002)
[9] AYRES, Joseph ; WITTING, Jan: Biomimetic Approaches to the Control of Underwater Walking Machines. In: *in: Walking Machines, Phil. Trans. R. Soc. A, vol 365, no 1850 (eds. Pfeiffer, Inoue)* (2007), S. 273–295
[10] BANERJEE, A. K. ; DICKENS, J. M.: Dynamics of Flexible Bodies in Large Rotation and Translation with Small Vibration. In: *AIAA Journal of Dynamics & Control* 13 (2) (1990), S. 221–227
[11] BANERJEE, A. K. ; KANE, T. R.: Dynamics of a Plate in Large Overall Motion. In: *Journal of Applied Mechanics* 56 (1989), S. 887–892
[12] BATCHELOR, G. K.: *An Introduction to Fluid Dynamics*. Cambridge University Press, New York, 1973
[13] BECKER, Ernst: *Technische Strömungslehre*. Teubner Studienbücher Mechanik, B.G. Teubner Stuttgart, 1982

[14] BECKER, Ernst ; BÜRGER, Wolfgang: *Kontinuumsmechanik*. Teubner Studienbücher Mechanik, B. G. Teubner Stuttgart, 1975
[15] BEITELSCHMIDT, Michael: *Reibstöße in Mehrkörpersystemen*. Fortschritt-Berichte VDI, Reihe 11, Nr. 275, 1999
[16] BELLMAN, R.-E.: *Dynamic Programming*. Princeton University Press, Princeton, 1957
[17] BELLMAN, R.E. ; DREYFUS, S.E.: *Applied Dynamic Programming*. Princeton University Press, Princeton, New Jersey, 1962
[18] BERNOULLI, Johann: *Opera Omnia*. Bd. TOM. II. 1742
[19] BETTEN, Josef: *Kontinuumsmechanik*. Springer Verlag Berlin, Heidelberg, New York, 2001
[20] BLICKHAN, Reinhard ; SEYFARTH, Andre ; GEYER, Hartmut ; GRIMMER, Sten ; WAGNER, Heiko ; GÜNTHER, Michael: Intelligence by Mechanics. In: *in: Walking Machines, Phil. Trans. R. Soc. A, vol 365, no 1850 (eds. Pfeiffer, Inoue)* (2007), S. 199–220
[21] BOBROW, J.E. ; DUBOWSKY, S. ; GIBSON, J.S.: Time optimal control of robotic manipulators along specified paths. In: *Int. J. Robot. Res.* 4 (1985), S. 3–17
[22] BOLTZMANN, Ludwig: *Vorlesungen über die Prinzipe der Mechanik*. Verlag von Johann Ambrosius Barth, T I, T II, 1922
[23] BORCHSENIUS, Fredrik: *Simulation ölhydraulischer Systeme*. Fortschritt-Berichte VDI, Reihe 8, Nr. 1005, VDI-Verlag Düsseldorf, 2003
[24] BORK, H. ; SRNIK, J. ; PFEIFFER, F. ; NEGELE, E. ; HEDDERICH, R.: Modellbildung, Simulation und Analyse eines leistungsverzweigten Traktorgetriebes. In: *Tagungsband Simulation im Maschinenbau, Institut für Werkzeugmaschinen, TU-Dresden* (2000), S. 329–347
[25] BRANDL, H. ; JOHANNI, R. ; OTTER, M.: A very efficient Algorithm for the Simulation of Robots and similar Multibody Systems without Inversion of the Mass Matrix. In: *IFAC/IFIP/IMACS Int. Syposium on the Theory of Robots, Vienna* (1986)
[26] BRAUN, Jürgen: Einfluß von Preßverbänden auf die Dynamik von Antriebssträngen. In: *Fortschrittberichte VDI, Reihe 11, Nr. 231, VDI-Verlag Düsseldorf* (1996)
[27] BREMER, Hartmut: *Dynamik und Regelung mechanischer Systeme*. B. G. Teubner Stuttgart, 1988. – Teubner Studienbücher Mechanik
[28] BREMER, Hartmut ; PFEIFFER, Friedrich: *Elastische Mehrkörpersysteme*. B. G. Teubner Stuttgart, 1992. – Teubner Studienbücher Mechanik
[29] BREMER, Hartmut ; PFEIFFER, Friedrich: Experiments with flexible manipulators. In: *IFAC Robot Control (SYROCCO'94), Pergamon* (1994), S. 515–522
[30] BRENAN, K. E. ; CAMPBELL, S. L. ; PETZOLD, L. R.: *Numerical Solution of Initial-Value Problems in Differential-Algebraic Equations*. North-Holland, New York, Amsterdam, London, 1989
[31] BULLINGER, M. ; PFEIFFER, F. ; ULBRICH, H.: Elastic Modelling of Bodies and Contacts in Continuous Variable Transmissions. In: *Multibody System Dynamics* 13 (2005), S. 175–194
[32] BULLINGER, Markus: *Dynamik von Umschlingungsgetrieben mit Schubgliederband*. Fortschrittberichte VDI, Reihe 12, Nr. 593, VDI-Verlag Düsseldorf, 2005
[33] BUSCHMANN, Thomas: *Trajektorienoptimierung einer zweibeinigen Laufmaschine*, Diplomarbeit am Lehrstuhl für Angewandte Mechanik, Technische Universität München, Diplomarbeit, 2003

[34] BUXBAUM, O.: *Betriebsfestigkeit.* Verlag Stahleisen mbH., Düsseldorf, 1986
[35] CHURCHILL, S.: Friction-factor equation spans all fluid flow regimes. In: *Chemical Engineering* (1977)
[36] CLARK, A.J. ; ROSSBURCH, F. ; GOLD, P.W. ; ASSMANN, C.: Kennwertermittlung an Elastomer-Manschettenkupplungen. In: *Antriebstechnisches Kolloqium ATK* (1999)
[37] CRAIG, J. J.: *Introduction to Robotics.* Addison-Wessley Publishing Company, 1986
[38] CRUSE, Holk ; DÜRR, Volker: Insect Walking based on decentralized architecture revealing a simple and robust controller. In: *in: Walking Machines, Phil. Trans. R. Soc. A, vol 365, no 1850 (eds. Pfeiffer, Inoue)* (2007), S. 221–250
[39] CUNHA, N. O. D. ; POLAK, E.: Constrained minimization under vector-valued criteria in finite dimensional spaces. In: *J. Math. Anal. Appl.* vol. 19 (1967), S. 103–124
[40] D'ALEMBERT, Jean le R.: *Traité de Dynamique.* Chez David, Libraire, rüe & vis-à-vis la grille des Mathurins, a Paris, 1743
[41] DE LAGRANGE, Joseph Louis C.: *Mécanique Analytique.* Chez la Veuve Desant, Libraire, rue du Foin S. Jacques, a Paris, 1788
[42] DENAVIT, J. ; HARTENBERG, R.S.: *Kinematic Synthesis of Linkages.* McGraw-Hill Book Company, New York, 1964
[43] DÜHRING, E.: *Principien der Mechanik.* Dr. Martin Sändig oHG., Wiesbaden, 1970. – Reprint of the original edition from 1887
[44] DIEHL, Moritz ; MOMBAUR, Katja: *Fast Motions in Biomechanics and Robotics.* Springer-Verlag Berlin Heidelberg, 2006
[45] DILLMANN, Rüdiger ; ALBIEZ, Jan ; GASSMANN, Bernd ; KERSCHER, Thilo ; ZÖLLNER, Marius: Biologically Inspired Walking Machines - Design, Control and Perception. In: *in: Walking Machines, Phil. Trans. R. Soc. A, vol 365, no 1850 (eds. Pfeiffer, Inoue)* (2007), S. 133–151
[46] DITTRICH, O.: Theorie des Umschlingungsgetriebes mit keilförmigen Reibscheibenflanken. In: *Dissertation. Technische Hochschule Karlsruhe* (1953)
[47] DRESIG, Hans: *Schwingungen mechanischer Antriebssysteme.* Springer-Verlag, Berlin, Heidelberg, New York, 2001
[48] DUBOWSKY, S. ; SHILLER, Z.: Optimal dynamic trajectories for robotic manipulators. In: *Proc. 5th Symp. on Theory and Practice of Robots and Manipulators* (1985)
[49] DUDITZA, Florea: *Kardangelenkgetriebe und ihre Anwendungen.* VDI- Verlag GmbH., Düsseldorf, 1973
[50] DUGAS, René: *A History of Mechanics.* Dover Publications, Inc., New York, 1988
[51] DUSCHEK, Adalbert ; HOCHRAINER, August: *Tensorrechnung in analytischer Darstellung.* Bd. Teil I, II, III. Springer Verlag Wien, 1960-1965
[52] EBERHARD, Peter ; SCHIEHLEN, Werner: Computational Dynamics of Multibody Systems: History, Formalisms, and Applications. In: *Transactions of the ASME, Journal of Computational and Nonlinear Dynamics* 1 (2006), S. 1–12
[53] EICH-SOELLNER, Edda ; FÜHRER, Claus: *Numerical Methods in Multibody Dynamics.* B.G.Teubner Stuttgart, 1998
[54] ELTZE, Jürgen: *Biologisch orientierte Entwicklung einer sechsbeinigen Laufmaschine.* Fortschritt-Berichte VDI, Reihe 17, Nr. 110, VDI-Verlag Düsseldorf, 1994

[55] ENGELHARDT, Thomas: *Dynamik von Steuer- und Ventiltrieben*. Dissertation TU-München, Lehrstuhl für Angewandte Mechanik, 2007
[56] ERDMANN, Arthur G.: *Modern Kinematics - Developments in the last Fourty Years*. John Wiley & Sons, INC, New York, 1993
[57] ESCHENAUER, H. H. ; KOSKI, J. ; OSYCZKA, A.: Multicriteria Optimization - Fundamentals and Motivation. In: *In: Multicriteria Design Optimization, Eds. H. Eschenauer, J. Koski, A. Osyczka, Springer, Berlin* (1990), S. 1–32
[58] EULER, Leonhard: Nova methodus motum corporum rigidorum determinandi. In: *Novi Commentarii Academiae Scientiarum Petropolitanae* 20 (1776), S. 208–238
[59] EYTELWEIN, J. A.: Handbuch der Statik fester Körper. In: *Band 1, Berlin, S. 21–23, Reimer* (1808)
[60] FIDLIN, Alexander: *Nonlinear Oscillations in Mechanical Engineering*. Springer-Verlag Berlin Heidelberg New York, 2006
[61] FINDEISEN, D. ; FINDEISEN, D.: *Ölhydraulik - Handbuch für die hydrostatische Leistungsübertragung in der Fluidtechnik*. Springer Verlag Berlin Heidelberg, 1994
[62] FISCHER, Martin ; WITTE, Hartmut: Legs evolved only at the end. In: *in: Walking Machines, Phil. Trans. R. Soc. A, vol 365, no 1850 (eds. Pfeiffer, Inoue)* (2007), S. 185–198
[63] FISCHER, U. ; STEPHAN, W.: *Prinzipien und Methoden der Dynamik*. VEB Fachbuchverlag, Leipzig, 1972
[64] FOERG, Martin: *Mehrkörpersysteme mit mengenwertigen Kraftgesetzen - Theorie und Numerik*. Fortschrittberichte VDI, Reihe 20, Nr. 411, VDI-Verlag Düsseldorf, 2007
[65] FOERG, Martin ; GEIER, Thomas ; NEUMANN, Lutz ; ULBRICH, Heinz: r-factor strategies for the augmented Lagrangian approach in multi-body contact mechanics. In: *III European Conference on Computational Mechanics; Solids, Structures and Coupled Problems in Engineering, Editors C. A. Mota Soares et al. Lisboa, Portugal, June 2006* (2006)
[66] FOURIER, J. B.: Mémoire sur la Statique, contenant la démonstration du principe des vitesses virtuelles et la théorie des moments. In: *Journal de l'Ecole polytechniques* 5 th cahier (1798)
[67] FÖPPL, August: *Technische Mechanik*. B. G. Teubner, Leipzig und Berlin, 1925. – achte Auflage 1925, erste Auflage 1898
[68] FREUND, E.: Fast Nonlinear Control with Arbitrary Pole Placement for Industrial Robots and Manipulators. In: *Int. J. Robot. Research* 1 (1982), S. 65–78
[69] FRITZ, Peter: *Dynamik schnellaufender Kettentriebe*. Fortschritt-Berichte VDI, Reihe 11, Nr. 253, VDI Verlag Düsseldorf, 1998
[70] FRITZ, Peter ; PFEIFFER, Friedrich: Dynamics of high speed roller chain drives. In: *Proceedings 1995 Design Engineering Technical Conference, Boston* Vol3, Part A, De-Vol.84-1 (1995), S. 471–497
[71] FRITZER, Anton: *Nichtlineare Dynamik von Steuertrieben*. Fortschrittberichte VDI, Reihe 11, Nr. 176, VDI Verlag Düsseldorf, 1992
[72] FRÉMOND, Michel: *Non-Smooth Thermomechanics*. Springer Verlag Berlin, Heidelberg, New York, 2002
[73] FÖRSTER, H.-J.: *Automatische Fahrzeuggetriebe*. Springer-Verlag, Berlin Heidelberg, 1991

[74] FUJII, T. ; KURAKAWA, T. ; KANEHARA, S.: A Study of Metal-Pushing V-Belt Type CVT - Part 1: Relation between Transmitted Torque and Pulley Thrust. In: *Int. Congress and Exposition Detroit, SAE Technical Paper Series, Nr. 930666* (1993), S. 1–11

[75] FUJII, T. ; TAKEMASA, K. ; KANEHARA, S.: A Study of Metal-Pushing V-Belt Type CVT - Part 2: Compression Force between Metal Blocks and Ring Tension. In: *Int. Congress and Exposition Detroit, SAE Technical Paper Series, Nr. 930667* (1993), S. 13–22

[76] FUJITA, M.: Autonomous Behavior Control by Design and Emergence. In: *in: Walking Machines, Phil. Trans. R. Soc. A, vol 365, no 1850 (eds. Pfeiffer, Inoue)* (2007), S. 21–47

[77] FUNK, Kilian: *Simulation eindimensionaler Kontinua mit Unstetigkeiten.* Fortschrittberichte VDI, Reihe 18, Nr. 294, VDI-Verlag Düsseldorf, 2004

[78] FURUSHO, J. ; MASUBUCHI, M.: A Theoretically Motivated Reduced Order Model for the Control of Dynamic Biped Locomotion. In: *J. of Dynamic Systems, Measurement, and Control, Trans. od ASME* Vol. 109 (June 1987), S. 155–163

[79] GABRIEL, Jens P. ; BÜSCHGES, Ansgar: Control of stepping velocity in a single insect leg during walking. In: *in: Walking Machines, Phil. Trans. R. Soc. A, vol 365, no 1850 (eds. Pfeiffer, Inoue)* (2007), S. 251–271

[80] GEIER, Thomas: *Dynamics of Push Belt CVTs.* Fortschritt-Berichte VDI, Reihe 12, Nr. 654, VDI-Verlag Düsseldorf, 2007

[81] GEIER, Thomas ; FOERG, Martin ; ZANDER, Roland ; ULBRICH, Heinz ; PFEIFFER, Friedrich ; BRANDSMA, Arjen ; VELDE, Arie van d.: Simulation of a push belt CVT considering uni- and bilateral constraints. In: *ZAMM, Z. Angew. Math. Mech.*, 86, no. 10 (2006), S. 795–806

[82] GEIER, Thomas ; FOERG, Martin ; ZANDER, Roland ; ULBRICH, Heinz ; PFEIFFER, Friedrich ; BRANDSMA, Arjen ; VELDE, Arie van d.: Simulation of Large Multibody Systems with Numerous Contacts. In: *Proc. 8th Int. Conf. on Motion and Vibration Control (MOVIC 2006)* (2006)

[83] GERBERT, B. G.: Force and Slip Behaviour in V-Belt Drives. In: *Mechanical Engineering Series, Acta Polytechnical Paper Series No.930667, Lund Techn. University* (1972)

[84] GERBERT, B. G. ; OLSSON, J.: Deformation of Pulley in a V-Belt Transmission. In: *Report 1989-01-37, Machine and Vehicle Design, Chalmers University of Technology, Göteborg* (1989)

[85] GIENGER, Michael: *Entwurf und Realisierung einer zweibeinigen Laufmaschine.* Fortschritt-Berichte VDI, Reihe 1, Nr. 378, VDI-Verlag Düsseldorf, 2005

[86] GLOCKER, Christoph: *Dynamik von Starrkörpersystemen mit Reibung und Stößen.* Fortschrittberichte VDI, Reihe 18, Nr.182, VDI-Verlag Düsseldorf, 1995

[87] GLOCKER, Christoph: *Set-Valued Force Laws - Dynamics of Non-Smooth Systems.* Springer Berlin Heidelberg New York, 2001

[88] GLOCKER, Christoph: An introduction to impacts. In: *in "Nonsmooth Mechanics of Solids", eds. Haslinger, Stavroulakis, CISM courses and lectures, Springer Wien New York* No. 485 (2006), S. 45–102

[89] GLOCKER, Christoph ; STUDE, Christian: Formulation and Preparation for Numerical Evaluation of Linear Complementarity Systems in Dynamics. In: *Multibody System Dynamics* 13 (2005), S. 447–463

[90] GRASHOF, F.: *Theoretische Maschinenlehre*. Verlag Leopold Voss, Hamburg, Band 2, pp. 304-324, 1883
[91] HAJ FRAJ, Ali: *Dynamik und Regelung von Automatikgetrieben*. Fortschrittberichte VDI, Reihe 12, Nr. 489, VDI-Verlag Düsseldorf, 2002
[92] HAJ FRAJ, Ali ; PFEIFFER, Friedrich: Optimization of Gear Shift Operations in Automatic Transmission. In: *Proc. 6th Int. Workshop on Advanced Motion Control (AMC), Nagoya, Japan* (2000), S. 469–473
[93] HAMEL, Georg: *Theoretische Mechanik*. Springer Verlag, Berlin, Göttingen, Heidelberg, 1949
[94] HAMROCK, B.: *Fundamentals of Fluid Film Rheology*. McGraw Hill Inc., 1994
[95] HARMONIC DRIVE, : : *Precision in Motion*. Catalogue of Harmonic Drive AG., Limburg, 2007/2008
[96] HARTMANN, W.: Beitrag zur Ermittlung der für die Konstruktion von Umschlingungsgetrieben mit keilförmigen Reibscheibenflanken maßgebenden Faktoren. In: *Dissertation, Karl-Marx-Stadt (Chemnitz)* (1964)
[97] HERBERTZ, R.: Untersuchung des dynamischen Verhaltens von Föttinger-Getrieben. In: *Dissertation TU-Hannover, Faculty of Mechanical Engineering* (1973)
[98] HERTZ, Heinrich: *Die Prinzipien der Mechanik, in:Gesammelte Werke, Band III*. Verlag J. A. Barth, Leipzig, 1894
[99] HESTENES, M.: Multiplier and gradient methods. In: *J. Opt. Th. Appl. 4* (1969), S. 303–320
[100] HÖHN, B.-R. ; PFLAUM, H. ; KRASTEV, I. ; LECHNER, C.: Bereichsumschaltung und Verbrennungsmotorstart im optimierten CVT-Hybrid-Antriebsstrang. In: *VDI-Berichte Nr. 1943, Getriebe in Fahrzeugen* (2006), S. 383–405
[101] HÖHN, B.-R. ; PFLAUM, H. ; TOMIC, D.: Energiebilanzierung und Verbrauchsabschätzung für den optimierten CVT-Hybrid-Antriebsstrang. In: *VDI-Berichte Nr. 1943, Getriebe in Fahrzeugen* (2006), S. 363–381
[102] HILLER, M. ; KECSKEMETHY: A Computer-Oriented Approach for the Automatic Generation and Solution of the Equations of Motion for Complex Mechanisms. In: *Proceedings of the 7th IFFTOMM World Congress on the Theory of Machines and Mechanisms, Sevilla* (1987), S. 425–430
[103] HILLER, M. ; WOERNLE, C.: The characteristic pair of joints - an effective approach for the inverse kinematics problem for robots and for complex mechanisms. In: *Proceedings of the IEEE Int. Conf. on Robotics and Automation* (1988)
[104] HIROSE, Masato ; OGAWA, Kenichi: Honda Humanoid Robots Development. In: *in: Walking Machines, Phil. Trans. R. Soc. A, vol 365, no 1850 (eds. Pfeiffer, Inoue)* (2007), S. 11–19
[105] HIRSCHMANN, V.: *Tragfähigkeitsuntersuchungen an stufenlosen Umschlingungsgetrieben*. Dissertation, FZG (Forschungsstelle für Zahnradgetriebe), TU-München, 1997
[106] HIRUKAWA, Hirohisa: Walking Biped Humanoids that Perform Manual Labor. In: *in: Walking Machines, Phil. Trans. R. Soc. A, vol 365, no 1850 (eds. Pfeiffer, Inoue)* (2007), S. 65–77
[107] HÖLZL, Josef: *Modellierung, Identifikation und Simulation der Dynamik von Industrierobotern*. Fortschritt-Berichte VDI, Reihe 8, Nr. 372, VDI-Verlag Düsseldorf, 1994
[108] HOLLAND, J.: Die instationäre Elstohydrodynamik. In: *Konstruktion Vol. 30*, Nr. 9 (1978), S. 363–369

[109] HOLLERBACH, J.M.: Dynamic scaling of manipulator trajectories. In: *ASME J. Dynamic. Systems, Measurement and Control* 106 (1984), S. 102–106
[110] HOOKER, W. W. ; MARGULIS, G.: The Dynamical Attitude Equations for an n-Body Satellite. In: *Journal of Astronautical Sciences* 12 (1965), S. 123–128
[111] HÖSL, Andreas: *Dynamiksimulation von Steuerkettentrieben*. Fortschritt-Berichte VDI, Reihe 12, Nr. 618, VDI-Verlag Düsseldorf, 2006
[112] IDE, T. ; TANAKA, H.: Contact Force Distribution Between Pulley Sheave and Metal Pushing V-Belt. In: *VDI-Bericht 1709 - CVT 2002 Congress Munich* (2002), S. 343–355
[113] IDE, T. ; UCHIYAMA, H. ; KATAOKA, R.: Experimental Investigation on Shift Speed Characteristics of a Metal V-Belt CVT. In: *CVT'96 SAE Technical Papers Series [9636330]* (1996), S. 59–64
[114] IDE, T. ; UDAGAWA, A. ; KATAOKA, R.: Simulation Approach to the Effect of the Ratio Changing Speed of a Metal V-Belt CVT on the Vehicle Response. In: *Vehicle System Dynamics* 24(4-5) (1995), S. 377–388
[115] IDELCHICK, I. E.: *Handbook of Hydraulic Resistance*. Springer Verlag Berlin, 1986
[116] JAMMER, Max: *Der Begriff der Masse in der Physik*. Wissenschaftliche Buchgesellschaft, Darmstadt, 1964. – German Translation of "Concepts of Mass", Harvard-University Press, Cambridge/USA, 1960
[117] JOHANNI, Rainer: *Optimale Bahnplanung bei Robotern*. Fortschritt-Berchte VDI, Reihe 18, Nr. 51, VDI-Verlag Düsseldorf, 1988
[118] JOHNSON, K. L.: *Contact Mechanics*. Cambridge University Press, 1996
[119] JOURDAIN, P. G. B. (Hrsg.): *Abhandlungen der Prinzipien der Mechanik von Lagrange, Rodrigues, Jacobi und Gauss*. Ostwalds Klassiker, Nr, 167, Wilhelm Engelmann Verlag, Leipzig, 1908
[120] JURICIC, D. ; VUCOBRATOVIC, M. ; FRANK, A.A.: On the Stability of Biped Locomotion. In: *IEEE Transactions on Biomedical Engineering, BME-17, No. 1* (1970), S. 25–36
[121] KAJITA, Shuuji ; TANI, Kazuo: Experimental Study of Biped Walking in the Linear Inverted Pendulum Mode. In: *Proc. of IEEE Int Conf. on Robotics and Automation, 0-7803-1965-6/95* (1995), S. 2885–2891
[122] KANE, T. R. ; RYAN, R. R. ; BANNERJEE, A. K.: Dynamics of a Cantilever Beam Attached to a Moving Base. In: *Journal of Guidance, Control and Dynamics* 10 (2) (1987), S. 139–151
[123] KANEHARA, S. ; FUJII, T. ; KITAGAWA, T.: A Study of Metal-Pushing V-Belt Type CVT - Part 3: What Forces act on Metal Blocks. In: *Int. Congress and Exposition Detroit, SAE Technical Paper Series, Nr. 940735* (1994), S. 95–105
[124] KARAGIANNIS, Konstantinos: *Analyse stoßbehafteter Schwingungssysteme mi A nwendung auf Rasselschwingungen in Getrieben*. Fortschrittberichte VDI, Reihe 11, Nr. 125,VDI Verlag Düsseldoerf, 1989
[125] KAUDERER, Hans: *Nichtlineare Mechanik*. Springer-Verlag Berlin, Göttingen, Heidelberg, 1958
[126] KÜCÜKAY, Ferit: *Dynamik der Zahnradgetriebe*. Springer-Verlag, Berlin Heidelberg, 1987
[127] KELL, Thomas J.: *Experimentelle Schwingungsuntersuchungen an Kettentrieben*. Dissertation TU-München, Fakultät für Maschinenwesen, Lehrstuhl B für Mechanik, 1999
[128] KHATIB, O.: Real-time Obstacle Avoidance for Manipulators and Mobile Robots. In: *Int. J. of Robotics Research* Vol. 5, Nr. 1 (1986), S. 90–98

[129] KIMURA, Hiroshi ; FUKUOKA, Yasuhiro ; COHEN, Avis H.: Biologically inspired adaptive walking of a quadruped robot. In: *in: Walking Machines, Phil. Trans. R. Soc. A, vol 365, no 1850 (eds. Pfeiffer, Inoue)* (2007), S. 153–170
[130] KIRCHHOFF, G.: *Vorlesungen über mathematische Physik, Mechanik.* B. G. Teubner Verlag, Leipzig, 1883
[131] KLEEMANN, Ulrich: *Regelung elastischer Roboter.* Fortschrittberichte VDI, Reihe 8, Nr. 191, VDI-Verlag Düsseldorf, 1989
[132] KORN, Granino A. ; KORN, Theresa M.: *Mathematical Handbook for Scientists and Engineers.* Second Edition. McGraw-Hill Book Company, New York, 1968
[133] KUNERT, Andreas: *Dynamik spielbehafteter Maschinenteile.* Fortschrittberichte VDI, Reihe 11, Nr. 175, VDI Verlag Düsseldorf, 1992
[134] LACHENMAYR, Georg: *Schwingungen in Planetengetrieben mit elastischen Hohlrädern.* Fortschrittberichte VDI, Reihe 11, Nr. 108, VDI-Verlag Düsseldorf, 1988
[135] LEINE, I. ; NIJMEIJER, Henk: *Dynamics and Bifurcations of Non-Smooth Mechanical Systems.* Springer Berlin Heidelberg New York, 2004
[136] LEINE, R.I. ; CH.GLOCKER: A set-valued force law for spatial Coulomb-Contensou friction. In: *Eur. J. Mech. A/Solids* 22 (2003), S. 193–216
[137] LÖFFLER, Klaus: *Dynamik und Regelung einer zweibeinigen Laufmaschine.* Fortschrittberichte VDI, Reihe 8, Nr. 1094, VDI-Verlag Düsseldorf, 2006
[138] LICHTENBERG, J. ; LIEBERMAN, A.: *Regular and Stochastic Motion.* Springer Verlag, New York, Heidelberg, Berlin, 1983
[139] LIM, Hun-Ok ; TAKANISHI, Atsuo: Biped Walking Robots Created at Waseda University: WL and Wabian Family. In: *in: Walking Machines, Phil. Trans. R. Soc. A, vol 365, no 1850 (eds. Pfeiffer, Inoue)* (2007), S. 49–64
[140] LORCH, O. ; DENK, J. ; BUSS, M. ; FREYBERGER, F. ; SCHMIDT., G.: Coordination of Perception and Locomotion Planning for Goal-Oriented Walking. In: *Proc. of the Int. Conf. on Climbing and Walking Robots (CLAWAR), Madrid, Spain* (2000), S. 183–192
[141] LOZANO-PÉREZ, T. ; MASON, M. T. ; TAYLOR, R. H.: Automatic Synthesis of Fine–motion Strategies for Robots. In: *Int. J. Robotics Research* 3(1) (1984)
[142] LUDWIG, O. ; MÖHL, B. ; NACHTIGALL, W. ; DILLMANN, U.: Human walking stability and reaction to disturbances. In: *in: Gantchev N., Gantchev G. (eds.) From basic motor control to functional recovery, Academic Publishing House, Sofia* (1999)
[143] LURIE, Anatolii I. ; ENGINEERING MECHANICS, Wittenberg Springer Series on Foundations o. (Hrsg.): *Analytical Mechanics.* Springer Verlag, Berlin, Heidelberg, New York, 2002
[144] LUTZ, O.: Zur Theorie des Keilscheiben-Umschlingungsgetriebes. In: *Konstruktion im Maschinen-, Apparate- und Gerätebau* 12 (1960), S. 265–267
[145] LUTZ, O. ; SCHLUMS, K. D.: Selbsthemmung in kraftschlüssigen Keilscheiben-Umschlingungsgetrieben. In: *Konstruktion* 17, Nr. 9 (1965), S. 365–368
[146] MAGNUS, Kurt: *Kreisel - Theorie und Anwendungen.* Springer Verlag, Berlin, Heidelberg, New York, 1971
[147] MAGNUS, Kurt ; MÜLLER, Hans H.: *Grundlagen der Technischen Mechanik.* Teubner Studienbücher Mechanik, B.G.Teubner Stuttgart, 1984
[148] MAGNUS, Kurt ; POPP, Karl: *Schwingungen.* Teubner Studienbücher Mechanik, B.G. Teubner Stuttgart, 1997
[149] MAISSER, P.: Analytical dynamics of multibody systems. In: *Computer Methods in Applied Mechanics and Engineering* 91 (1991), S. 1391–1396

[150] MAISSER, P.: Differential-Geometric Methods in Multibody Dynamics. In: *Nonlinear Analysis, Theory, Methods & Applications* 30(91) (1997), S. 5127–5133
[151] MANGASARIAN, O. L.: Equivalence of the Complementarity Problem to a System of Nonlinear Equations. In: *SIAM Journal of Applied Mathematics* 31(1) (1976), S. 89–92
[152] MEITINGER, Thomas: *Dynamik automatisierter Montageprozesse.* Fortschrittberichte VDI, Reihe 2, Nr.476, VDI-Verlag Düsseldorf, 1998
[153] MEITINGER, Thomas ; PFEIFFER, Friedrich: Dynamic Simulation of Assembly Processes. In: *Proc. IEEE/RSJ Intl. Conf. on Intelligent Robots and Systems, Pittsburgh, PA* (1995), S. 198–304
[154] MEITINGER, Thomas ; PFEIFFER, Friedrich: Modelling and Simulation of Assembly Processes with Robots. In: *Appl. Math. and Comp. Sci.* Vol. 7, No.2 (1997), S. 343–375
[155] MEYBERG, Kurt ; VACHENAUER, Peter: *Höhere Mathematik.* Bd. Volumes 1 and 2. Springer Lehrbuch, Springer Verlag, Berlin, Heidelberg, New York, 1990/1991
[156] MIYAZAKI, Fumio ; ARIMOTO, Suguru: A Control Theoretic Study on Dynamical Biped Locomotion. In: *J. of Dynamic Systems, Measurement, and Control, Trans. of ASME* Vol. 102 (December 1980), S. 233–239
[157] MÜLLER, P.C. ; SCHIEHLEN, W.O.: *Lineare Schwingungen.* Akademische Verlagsgesellschaft, Wiesbaden, 1976
[158] MOON, C.: *Chaotic and Fractal Dynamics.* John Wiley & Sons, Inc., New York, 1992
[159] MOON, Francis C.: *Applied Dynamics.* John Wiley & Sons, Inc., New York, 1998
[160] MOREAU, Jacques: *Une formulation du contact a frottement sec; application au calcul numerique.* C. R. Acad. Sci. Paris,, 1986 (Series II, Technical Report 13)
[161] MOREAU, Jean J.: *Unilateral Contact and Dry Friction in Finite Freedom Dynamics. Non-Smooth Mechanics and Applications.* CISM Courses and Lectures, Volume 302, Springer Verlag, Wien, 1988. – CISM Courses and Lectures, Volume 302
[162] MORECKI, A. (Hrsg.) ; WALDRON, K.J. (Hrsg.): *Human and Machine Locomotion.* CISM Course No. 375, Springer Wien New York, 1997
[163] MURTY, K. G.: *Linear Complementarity, Linear and Nonlinear Programming.* Sigma Series in Applied Mathematics, Heldermann Verlag, Berlin, 1988
[164] MUYBRIDGE, Eadweard: *The Human Figure in Motion.* Dover Publications Inc., 1955
[165] NACHTIGALL, Werner ; MÖHL, Bernhard: Dynamik und Anpassungsvorgänge bei der Laufkoordination des Menschen. In: *in: Pfeiffer, Cruse (Eds.), Autonomes Laufen,* Springer Verlag, Berlin, Heidelberg (2005), S. 97–106
[166] NAYFEH, Ali ; BALACHANDRAN, Balakumar: *Applied Nonlinear Dynamics.* Wiley Series in Nonlinear Science, John Wiley &Sons, Inc., New York, 1995
[167] NEUMANN, Lutz: *Optimierung von CVT-Ketten.* Dissertation Lehrstuhl für Angewandte Mechanik, TU-München, 2007
[168] NEUMANN, Lutz ; ULBRICH, Heinz ; PFEIFFER, Friedrich: New Model of a CVT Rocker Pin Chain with Exact Joint Kinematics. In: *Journal of Computational and Nonlinear Mechanics, Transactions of the ASME* Vol. 1 (2006), S. 143–149

[169] NEWTON, Isaac ; WOLFERS, J. P. (Hrsg.): *Mathematische Prinzipien der Naturlehre*. Verlag von Robert Oppenheim, 1687. – Deutsche Übersetzung der Principia, herausgegeben von Wolfers 1872
[170] NIEMANN, Gustav ; WINTER, Hans: *Maschinenelemente*. Bd. Band 2: Getriebe. Springer Verlag, Berlin, Heidelberg, New York, 2002
[171] NISHIWAKI, K. ; SUGIHARA, T. ; KAGAMI, S. ; INABA, M. ; INOUE, H.: Online Mixture and Connection of Basic Motion for Humanoid Walking Control by Footprint Specification. In: *Proc. of Conference on Robotics and Automation, Seoul, Korea* (2001)
[172] NISHIWAKI, Koichi ; KAGAMI, Satoshi ; KUFFNER, James ; INABA, Masayuki ; INOUE, Hirochika: The Experimental Humanoid Robot H7: A Research Platform for Autonomous Behavior. In: *in: Walking Machines, Phil. Trans. R. Soc. A, vol 365, no 1850 (eds. Pfeiffer, Inoue)* (2007), S. 79–107
[173] NITESCU, G.: Geometrische Verhältnisse in stufenlos verstellbaren Keilscheiben-Umschlingungsgetrieben. In: *Antriebstechnik* Vol. 23, Nr. 7 (1984), S. 54–56
[174] OGATA, Katsuhiko: *Modern Control Engineering*. Prentice Hall International, Inc., Englewood Cliffs, 1970
[175] OSGOOD, William F.: *Mechanics*. Dover Publications, Inc., New York, 1965
[176] PANAGIOTOPOULOS, P. D.: *Inequality Problems in Mechanics and Applications*. Birkhäuser, Boston, Basel, Stuttgart, 1985
[177] PANAGIOTOPOULOS, P. D.: *Hemivariational Inequalities*. Springer Verlag, Berlin, Heidelberg, New York, 1993
[178] PANG, J. S. ; TRINKLE, J. C.: Complementarity Formulations and Existence of Solutions of Dynamic Multi-Rigid-Body Contact Problems with Coulomb Friction. In: *Mathematical Programming* 30 (1996), S. 199–226
[179] PAPAGEORGIOU, Markos: *Optimierung*. Oldenbourg Verlag München Wien, 1991
[180] PAPASTAVRIDIS, John G.: *Analytical Mechanics*. Oxford University Press, Oxford, New York, 2002
[181] PARS, L. A.: *Analytical Dynamics*. Ox Bow Press, Woodbridge, Connecticut 06525, 1979. – First Published 1965
[182] PFEIFFER, Friedrich: Mechanische Systeme mit unstetigen Übergängen. In: *Ingenieur-Archiv, Springer Verlag* Vol. 54 (1984), S. 232–240
[183] PFEIFFER, Friedrich: Über die Bewegung spielbehafteter Maschinenteile - Modelle und Berechnung. In: *VDI-Berichte* Nr. 596 (1986), S. 225–246
[184] PFEIFFER, Friedrich: On Unsteady Dynamics in Machines with Plays. In: *Proceedings IFTOMM Conference, Sevilla, Spain* (1987)
[185] PFEIFFER, Friedrich: Combined Path and Force Control for Elastic Manipulators. In: *Mechanical Systems and Signal Processing* 6(3) (1990), S. 237–249
[186] PFEIFFER, Friedrich: Optimal Trajectory Planning for Manipulators. In: *Systems and Control Encyclopedia, Pergamon Press, Oxford, New York* Supplementary Volume 1 (1990), S. 445–452
[187] PFEIFFER, Friedrich: *Einführung in die Dynamik*. B. G. Teubner, Stuttgart, 1992. – Teubner Studienbücher Mechanik
[188] PFEIFFER, Friedrich: Assembly processes with robotic systems. In: *Robotics and Autonomous Systems* 19 (1996), S. 151–166
[189] PFEIFFER, Friedrich: Robots with Unilateral Constraints. In: *Proc. of 5th IFAC Symposium on Robot Control (SYROCO '97), Nantes* (1997), S. 539–550

[190] PFEIFFER, Friedrich: On the mechatronics of an automatic gear transmission system. In: *Multibody dynamics 2005-ECCOMAS thematic conference, Madrid, Spain* (2005)
[191] PFEIFFER, Friedrich: CVT - A Large Application of Nonsmooth Mechanics. In: *ECCOMAS, Multibody Dynamics 2007, Milano* (2007)
[192] PFEIFFER, Friedrich: Deregularization of a smooth system - example hydraulics. In: *Nonlinear Dynamics* 47 (2007), S. 219–233
[193] PFEIFFER, Friedrich: The TUM Walking Machines. In: *in: Walking Machines, Phil. Trans. R. Soc. A, vol 365, no 1850 (eds. Pfeiffer, Inoue)* (2007), S. 109–131
[194] PFEIFFER, Friedrich: Dynamics of a Ravigneaux Gear. In: *Journal of Vibration and Control, Sage Publications Ltd.*, 14(1-2) (2008), S. 181–196
[195] PFEIFFER, Friedrich ; BORCHSENIUS, Fredrik: New Hydraulic System Modelling. In: *Journal of Vibration and Control* 10 (2004), S. 1493–1515
[196] PFEIFFER, Friedrich ; BREMER, Hartmut ; FIGUEIREDO, J.: Surface polishing with flexible link manipulators. In: *European Journal of Mechanics, A/Solids* Vol. 15, No.1 (1996), S. 137–153
[197] PFEIFFER, Friedrich (Hrsg.) ; CRUSE, Holk (Hrsg.): *Autonomes Laufen*. Springer Verlag, Heidelberg, New York, 2005
[198] PFEIFFER, Friedrich ; FOERG, Martin ; ULBRICH, Heinz: Numerical Aspects of Non-Smooth Multibody Dynamics. In: *Computer methods in applied mechanics and engineering 195, Elsevier* (2006), S. 6891–6908
[199] PFEIFFER, Friedrich ; FRITZ, Peter ; SRNIK, Jürgen: Nonlinear Vibrations of Chains. In: *Journal of Vibration and Control* 3 (1997), S. 397–410
[200] PFEIFFER, Friedrich ; GLOCKER, Christoph: *Multibody Dynamics with Unilateral Contacts*. John Wiley & Sons, INC., New York, 1996. – Within the Wiley Series of Nonlinear Science (Ed. Ali Nayfeh)
[201] PFEIFFER, Friedrich ; HÖSL, Andreas: Geräuschoptimale Auslegung von Komponenten bei Steuertrieben von Motoren. In: *Abschlussbericht Vorhaben 734, Forschungsvereinigung Verbrennungskraftmaschinen (FVV)* Heft R 519 (2003)
[202] PFEIFFER, Friedrich (Hrsg.) ; INOUE, Hirochika (Hrsg.): *Walking Machines*. Philosophical Transactions A of the Royal Society, London , volume 365, number 1850, 2007
[203] PFEIFFER, Friedrich ; INOUE, Hirochika: Walking: technology and biology. In: *in: Walking Machines, Phil. Trans. R. Soc. A, vol 365, no 1850 (eds. Pfeiffer, Inoue)* (2007), S. 3–9
[204] PFEIFFER, Friedrich ; JOHANNI, Rainer: A Concept for Manipulator Trajectory Planning. In: *Proc. IEEE Int. Conf. Robotics and Automation* (1986), S. 1399–1405
[205] PFEIFFER, Friedrich ; JOHANNI, Rainer: A Concept for Manipulator Trajectory Planning. In: *IEEE Journal of Robotics and Automation* Vol.RA-3, No.2 (1987), S. 115–123
[206] PFEIFFER, Friedrich ; KUNERT, Andreas: Rattling Models from Deterministic to Stochastic Processes. In: *Nonlinear Dynamics* Vol. 1 (1990), S. 63–74
[207] PFEIFFER, Friedrich ; LEBRECHT, Wolfram ; GEIER, Thomas: State-of-the-Art of VCT-Modelling. In: *2004 SAE International, 04CVT-46* (2004)
[208] PFEIFFER, Friedrich ; REITHMEIER, Eduard: *Roboterdynamik*. Teubner Studienbücher Mechanik, B.G. Teubner, Stuttgart, 1987

[209] PFEIFFER, Friedrich ; RICHTER, Klaus: Optimal Path Planning Including Forces at the Gripper. In: *J. of Intelligent and Robotic Systems* 3 (1990), S. 251–258
[210] PFEIFFER, Friedrich ; STIEGELMEYR, Andreas: Damping Towerlike Structures by Dry Friction. In: *Proceedings of DETC'97, 1997 ASME Design Engineering Technical Conferences, Sacramento* DETC97/VIB-4013 (1997)
[211] PFEIFFER, Friedrich ; WAPENHANS, Henner ; SEYFFERTH, Wolfgang: Dynamics and Control of Automated Assembly Processes. In: *IN: Smart Structures, Nonlinear Dynamics and Control, Eds. A. Guran, D. J. Inman, Prentice Hall PTR, New Jersey* (1994), S. 190–225
[212] PFEIFFER, Friedrich ; WOLFSTEINER, Peter: Relative Kinematics of Multibody Contacts. In: *Active/Passive Vibration Control and Nonlinear Dynamics of Structures, Editors: W.W.Clark, W.C.Xie, D.Allaei, Y.F.Hwang, N.S.Namachchivaya, O.M.O'Reilly* DE-Vol.95, AMD-Vol.223, Book No. H01110 (1997), S. pp. 107–114
[213] PFEIFFER, Friedrich (Hrsg.) ; ZIELINSKA, Teresa (Hrsg.): *Walking: Biological and Technological Aspects*. CISM Course No. 467, Springer Wien New York, 2004
[214] POINCARÉ, Henri: *Les Methodes Nouvelles de la Mécanique Celèste*. Bd. vols. 1-3. Gauthier-Villars, Paris, 1899
[215] POST, Joachim: *Objektorientierte Softwareentwicklung zur Simulation von Antriebssträngen*. Fortschrittbericht VDI, Reihe 11, Nr. 317, VDI-Verlag Düsseldorf, 2003
[216] POWELL, M.J.D.: A method for nonlinear constraints in minimization problems. In: *Optimization, Academic Press, London, New York* (1969), S. 283–298
[217] PRESTL, Willibald: *Zahnhämmern in Rädertrieben*. Fortschrittberichte VDI, Reihe 11, Nr. 145, VDI-Verlag GmbH., Düsseldorf, 1991
[218] PRIGOGINE, Ilya: *Is Future Given?* World Scientific, New Jersey, London, Singapore, Hong Kong, 2003
[219] PROKOP, Günther: *Optimale Prozeßdynamik bei Manipulation mit Robotern*. Fortschritt-Berichte VDI, Reihe 8, Nr.713, VDI Verlag Düsseldorf, 1998
[220] PROKOP, Günther ; PFEIFFER, Friedrich: Synthesis of Robot Dynamics Behavior for Environmental Interaction. In: *IEEE Trans. on Robotics and Automation* Vol. 14, No. 5 (1998), S. 718–731
[221] RAIBERT, M. H.: *Legged Robots that Balance*. MIT Press Cambridge, 1986
[222] REITHMEIER, Eduard: *Die numerische Behandlung nichtlinearer Schwingungssysteme mit geschlossener Struktur und Anwendungen*. VDI Fortschrittberichte, Reihe 11, Nr. 62, VDI-Verlag GmbH, Düsseldorf, 1984
[223] RICHTER, Klaus: *Kraftregelung elastischer Roboter*. Fortschritt-Berichte VDI, Reihe 8, Nr. 259, VDI-Verlag Düsseldorf, 1991
[224] RICHTER, Klaus ; PFEIFFER, Friedrich: A Flexible Link Manipulator as a Force Measuring and Controlling Unit. In: *Proc. of the 1991 IEEE Int. Conf. on Robotics and Automation* (1991), S. 1214–1219
[225] ROCKAFELLER, R.T.: *Convex Analysis*. Princeton Landmarks in Mathematics, Princeton University Press, 1970
[226] ROSSMANN, Thomas: *Eine Laufmaschine für Rohre*. Fortschrittberichte VDI, Reihe 8, Nr. 732, VDI-Verlag Düsseldorf, 1998
[227] RUMBAUGH, J. ; BLAHA, M. ; PREMERLANI, W. ; EDDY, F. ; LORENSEN, W.: *Object-Oriented Modeling and Design*. Prentice-Hall International, Inc., 1991

[228] SAADA, Adel S.: *Elasticity - Theory and Applications*. Pergamon Unified Engineering Series, Pergamon Press Inc., New York, 1974
[229] SANTOS, Ferreira: *Aktive Kippsegmentlagerung - Theorie und Experiment*. Fortschrittberichte VDI, Reihe 11, Nr. 189, VDI Verlag Düsseldorf, 1993
[230] SATTLER, H.: Stationäres Betriebsverhalten stufenlos verstellbarer Metallumschlingungsgetriebe. In: *Dissertation, Universität Hannover* (1999)
[231] SAUER, G.: *Grundlagen und Betriebsverhalten eines Zugketten-Umschlingungsgetriebes*. Fortschritt-Berichte VDI, Reihe 12, Nr. 293, VDI-Verlag Düsseldorf, 1996
[232] SAUER, G. ; RENIUS, K. T. ; KARDELKE, J.: Zur Mechanik des Zugkettenwandlers, Teil1: Leistungsübertragung - Messung und Berechnungen. In: *Antriebstechnik* (1996)
[233] SCHIEHLEN, Werner: *Technische Dynamik*. Teubner Studienbücher Mechanik, B. G. Teubner Verlag Stuttgart, 1986
[234] SCHINDLER, Th. ; GEIER, Th. ; ULBRICH, H. ; PFEIFFER, F. ; VELDE, A. van d. ; BRANDSMA, A.: Dynamics of Pushbelt CVTs. In: *VDI-Berichte 1997 - Umschlingungsgetriebe, Berlin, VDI-Verlag Düsseldorf* (2007), S. 389–396
[235] SCHMIDT, G.: Perzeptionsbasiertes humanoides Gehen. In: *In:Pfeiffer, Cruse(Hrsg.), Autonomes Laufen, Springer Berlin Heidelberg New York* (2005), S. 161–171
[236] SCHMIDT, G. ; LORCH, O. ; SEARA, J. ; DENK, J. ; FREYBERGER, F.: Investigations into Goal-Oriented Vision-Based Walking of a Biped Humanoid Robot. In: *Proc. of the 10th int. Workshop on Robotics in Alpe-Adria-Danube Region (RAAD'01), Vienna, Austria* (2001)
[237] SCHWERTASSEK, Richard ; WALLRAPP, Oskar: *Dynamik flexibler Mehrkörpersysteme*. Friedr. Vieweg & Sohn, Verlagsgesellschaft mbH., Braunschweig, Wiesbaden, 1999
[238] SEDLMAYR, M. ; PFEIFFER, F.: Spatial Contact Mechanics of CVT Chain Drives. In: *18th ASME Bien. Conf. on Mech. Vibration and Noise DETC2001/VIB-21511* (2001)
[239] SEDLMAYR, Martin: *Räumliche Dynamik von CVT-Keilkettengetrieben*. Fortschrittberichte VDI, Reihe 12, Nr. 558, VDI-Verlag Düsseldorf, 2003
[240] SEHERR-TOSS, Hans C. ; SCHMELZ, Friedrich ; AUCKTOR, Erich: *Universal Joints and Driveshafts - Analysis, Design, Applications*. second, enlarged edtion. Springer-Verlag Berlin Heidelberg New York, 2006
[241] SEYFFERTH, Wolfgang: *Modellierung unstetiger Montageprozesse mit Robotern*. Fortschrittberichte VDI, Reihe 11, Nr. 199, VDI-Verlag Düsseldorf, 1993
[242] SHABANA, A.: *Dynamics of Multibody Systems*. Third Edition. Cambridge University Press, Cambridge, New York, 2005
[243] SHIN, K.G. ; MCKAY, N.D.: Minimum-time control of robotic manipulators with geometric path constraints. In: *IEEE Trans. Autom.* 30 (1985), S. 531–541
[244] SIGNORINI, A.: Sopra alcune questioni di elastostatica. In: *Atti Soc Ital per il progresso delle Science* (1933)
[245] SILVA, Manuel F. ; MACHADO, J. A. T.: A Historical Perspective of Legged Robots. In: *Journal of Vibration and Control, SAGE Publications, London* 13 (9-10) (2007), S. 1447–1486
[246] SLOTINE, J.-J. ; LI, W.: *Applied Nonlinear Control*. Prentice Hall, Eaglewood Cliffs, New Jersey, 1991
[247] SOMMERFELD, Arnold: *Mechanik der deformierbaren Medien*. Bd. Band II. Verlag Harry Deutsch, Thun, Frankfurt/M., 1978

[248] SORGE, Kai: *Mehrkörpersysteme mit starr-elastischen Subsystemen*. Fortschrittberichte VDI, Reihe 11, Nr.184, VDI-Verlag Düsseldorf, 1993
[249] SRNIK, J. ; PFEIFFER, F.: Dynamics of CVT Chain Drives. In: *Int. Journal of Vehicle Design, Special Edition* Vol. 22, No. 1/2 (1999), S. 54–72
[250] SRNIK, Jürgen: *Dynamik von CVT-Keilkettengetrieben*. Fortschrittberichte VDI, Reihe 12, Nr. 372, VDI-Verlag Düsseldorf, 1999
[251] STEINLE, Joachim: *Entwicklung einer prozeßangepaßten Roboterregelung für Montagevorgänge*. Fortschrittberichte VDI, Reihe 8, Nr. 548, Düsseldorf, 1996
[252] STIEGELMEYR, Andreas: A Time Stepping Algorithm For Mechanical Systems with Unilateral Contacts. In: *ASME Proc. of DETC'99, Las Vegas 1999* (1999)
[253] STIEGELMEYR, Andreas: *Zur numerischen Berechnung strukturvarianter Mehrkörpersysteme*. VDI Fortschrittberichte, Reihe 18, Nr. 271, VDI Verlag Düsseldorf, 2001
[254] STRIP, D.: A Passive Mechanism for Insertion of Convex Pegs. In: *Proc. of IEEE Conf. on Robotics and Automation, Scottsdale, AZ* (1989), S. 242–248
[255] STRONGE, W. J.: *Impact Mechanics*. Cambridge University Press, 2000
[256] STURGES, R. H. ; LAOWATTANA, S.: Passive Assembly of Non-Axisymmetric Rigid Parts. In: *Proc. of IEEE/RSJ Int. Conf. on Int. Robots and Systems, Munich, Germany* (1994), S. 1218–1225
[257] SYNGE, John L. ; FLÜGGE, S. (Hrsg.): *Classical Dynamics, in Encyclopedia of Physics, Volume III/1: Principles of Classical Mechanics and Field Theory*. Springer-Verlag, Berlin, Göttingen, Heidelberg, 1960
[258] SZABÓ, István: *Einführung in die Technische Mechanik*. Springer-Verlag Berlin, Heidelberg, New York, 1975
[259] SZABÓ, István: *Geschichte der mechanischen Prinzipien*. Birkhäuser Verlag, Basel, Boston, Stuttgart, 1979
[260] TENBERGE, P. ; LEESCH, M. ; MÜLLER, J. ; VORNEHM, M.: Hybridantrieb mit Umschlingungs-CVT. In: *VDI-Berichte Nr. 1943, Getriebe in Fahrzeugen* (2006), S. 343–362
[261] THOMPSON, J.M.T. ; STEWART, H.B.: *Nonlinear Dynamics and Chaos*. Second Edition. John Wiley & Sons, Ltd., West Sussex, England, 2002
[262] TODD, D. J.: *Walking Machines: An Introduction to Legged Robotics*. Kogan Page, 1985
[263] TRINKLE, J. C. ; PANG, J. S. ; SUDARSKY, S. ; LO, G.: On Dynamic Multi-Rigid-Body Contact Problems with Coulomb Friction. In: *Zeitschrift für Angewandte Mathematik und Mechanik (ZAMM)* 77(4) (1997), S. 267–279
[264] TRUESDELL, C. ; FLÜGGE, S. (Hrsg.) ; TRUESDELL, C. (Hrsg.): *Encycopedia of Physics, Mechanics of Solids II*. Bd. Volume VIa/2. Springer Verlag, Berlin, Heidelberg, New York, 1972
[265] VDI-GESELLSCHAFT, EKV (Hrsg.): *Welle-Nabe-Verbindungen*. VDI-Verlag Düsseldorf, 2003
[266] VDI-GESELLSCHAFT, EKV: Damping of materials and members (Werkstoff- und Bauteildämpfung). In: *VDI-Richtlinien, VDI 3830, Part 1-5, Verein Deutscher Ingenieure, Düsseldorf (Beuth Verlag, Berlin)* (2004-2005)
[267] VUCOBRATOVIC, M. ; BOROVAC, B. ; SURLA, D. ; STOKIC, D.: *Biped Locomotion: Dynamics, Stability, Control and Applications*. Springer-Verlag, Berlin, Heidelberg, New York, 1990
[268] VUCOBRATOVIC, M. ; STEPANENKO, J.: On the Stability of Anthropomorphic Systems. In: *Mathematical Biosciences* 15 (1972), S. 1–37

[269] WAGNER, Gerhard: Customer Benefit and Potential of Advancements of Automatic Transmission Systems. In: *VDI-Berichte Nr. 1943, VDI-Verlag Düsseldorf* (2006), S. 73–92

[270] WAGNER, U. ; TEUBERT, A. ; ENDLER, T.: Entwicklung von CVT-Ketten für PKW-Anwendungen bis 400 Nm. In: *VDI-Berichte, VDI Verlag Düsseldorf* 1610 (2001), S. 223–242

[271] WAPENHANS, Henner: *Optimierung von Roboterbewegungen bei Manipulationsvorgängen*. Fortschrittberichte VDI, Reihe 2, Nr. 304, VDI-Verlag Düsseldorf, 1994

[272] WATERSTRAAT, A.: *Das instationär belastete Gleitlager endlicher Breite*. Dissertation Universität Hannover, 1981

[273] WEIDEMANN, Hans-Jürgen: *Dynamik und Regelung von sechsbeinigen Robotern und natürlichen Hexapoden*. Fortschritt-Berichte VDI, Reihe 8, Nr. 362, VDI-Verlag Düsseldorf, 1993

[274] WINTER, D. A.: *Biomechanics of Human Movement*. John Wiley, New York, 1979

[275] WIT, Carlos Canudas d. ; SICILIANO, Bruno ; BASTIN, Georges: *Theory of Robot Control*. Springer-Verlag London, 1996

[276] WITTENBURG, Jens: *Dynamics of Systems of Rigid Bodies*. B. G. Teubner, Stuttgart, 1977

[277] WOLFSTEINER, Peter: *Dynamik von Vibrationsförderern*. Fortschritt-Berichte VDI, Reihe 2, Nr. 511, VDI-Verlag Düsseldorf, 1999

[278] WRIGGERS, Peter: *Computational Contact Mechanics*. John Wiley & Sons, Ltd., Chichester, England, 2002

[279] WÖSLE, Markus: *Dynamik von räumlichen strukturvarianten Starrkörpersystemen*. Fortschrittberichte VDI, Reihe 18, Nr. 213, VDI-Verlag Düsseldorf, 1997

[280] ZAGLER, Andreas: *Dynamik und Regelung eines Rohrkrabblers*. Fortschritt-Berichte VDI, Reihe 8, Nr. 1112, VDI-Verlag Düsseldorf, 2006

[281] ZANDER, R. ; SCHINDLER, T. ; FRIEDRICH, M. ; HUBER, R. ; FÖRG, M. ; ULBRICH, H.: Non-smooth dynamics in academia and industry: recent work at TU München. In: *Acta Mechanica, Springer Verlag, Wien* (2007)

[282] ZANDER, R. ; ULBRICH, H.: Reference-free mixed FE-MBS approach for beam structures with constraints. In: *Nonlinear Dynamics, Kluwer Academic Publishers, Dordrecht Netherlands* 46, No. 4 (2006), S. 349–361

[283] ZEIDLER, E. ; SCHWARZ, R. ; HACKBUSCH, W. ; ZEIDLER, E. (Hrsg.): *Teubner-Taschenbuch der Mathematik*. Teubner Verlag, Wiesbaden, 2003

[284] ZF GETRIEBE, Saarbruecken: Funktionsbeschreibung Automatikgetriebe 5HP24. In: *Technical Report, ZF Getriebe GmbH* (1996)

[285] ZIEGLER, H.: *Verzahnungssteifigkeit und Lastverteilung schrägverzahnter Stirnräder*. Dissertation RWTH-Aachen, 1971

# Index

academia and industry 213
acceleration 19
  absolute 24
  angular 24
  applied 24, 39
  centrifugal 24
  centripedal 32
  Coriolis 24
  relative 24, 41
  rotational 28, 36
  tangential 32
  translational 28
acceleration level 93, 138
action-selection 505
active contact state 135
Agco/Fendt 257
agricultural work 257
AIST 505
AIST, Japan 412
ancillary equipment 361
annulus 220
arc of wrap 279, 294
art of neglects 114
ASIMO 505
atmospheric density 222
Augmented Lagrangian Method 98, 148, 325
automatic transmission 208, 213
automotive industry 187
axial drum bearing 269
axle, rear, front 259

backlash 242, 283, 329, 338, 370

base pitch 243
bearing
  journal bearing 333, 350
  roller bearing 333
Bellman 429
belt 275
Bernoulli
  Johann 1
Bernoulli equation 193
Betti/Maxwell 302
biomimetics 508
bionics 508
biped
  ankle joint 513
  ball screw spindle 513
  controllability 530
  drive system 516
  feet contact 519
  feet torques 530
  foot pitch 531
  foot roll 531
  foot yaw 531
  generalized coordinates 511
  inverted pendulum 531
  joint structure 511
  legs 512
  model-based control 531
  pelvis 512
  spindles 515
  stability 530
  trajectory coordinates 528
  trajectory optimization 534
  trunk 511

trunk coordinates   529
ZMP   531
Block-Gauss-Seidel relaxation scheme
   148
body
  deformed   64, 308
  elastic   114, 251, 289, 331
  predecessor   121
  rigid   114, 251, 289, 307, 331
  rotationsl symmetry   35
  undeformed   66
Boltzmann   3
Boltzmann's axiom   53
Borg Warner chain   276
boundary layers   76
brake torque   222
branching problem   123
bulk modulus   191
bush chain   365

cable winch   205
camshaft   329, 377
canonical equations of Hamilton   111
cardan shaft   215, 262
Carnot   175
Cauchy Green tensor   49
Cauchy' stress tensor   53
chain   120, 275
  link   286
  drive   288
  guides   395
  joint   370
  links with bushings   369
  links with pins   369
  tensioners   369
chain of bodies   27
chain performance   314
chaos   330
Christoffel   34, 45, 421
closure   89
clutch
  multiple disc clutch   245
  wet   214
  multiple   226, 229
  one-way   214, 225, 229
coefficient
  of damping   244
  of restitution   71
  of sliding friction   70, 92

  of static friction   70, 92
collisions   68
combinatorial problem   133
combustion engine   216
complementarity   69
complementary pairs   143
complementary partner   147
component
  composition   404
  piston/cylinder   396
composition   16
  multiple   16
  successive   17
compressibility module   188
compression   68, 132, 159
computer application   VIII
computing time   119
configuration
  deformed   49
  reference   49
conservative system   63
constrained phase space   427
constraint   4
constraint equations   14
constraint matrices   137
constraint vectors   137
constraints   3, 8, 404
  acceleration level   322
  active   194
  bilateral   VIII, 8, 85, 114, 405
  concept   85
  element/ring   324
  forces   85
  holonomic   9, 31, 85
  hypersurfaces   86
  kinematical   8
  non-holonomic   9, 31, 85
  normal force   70
  passive   194
  position level   322
  redundant   88
  rheonomic   9, 85
  scleronomic   9, 85, 172
  surface   101
  tangential force   70
  unilateral   VIII, 8, 89, 114, 131, 405
contact
  parameters   40
  points   40

active  3, 70
closed  89
configuration  388
constraint  70
continual  174
element/element  325
element/pulley  325
element/ring  324
event  326
force  134
laws  173
line of  349
link/guide  372, 384
link/sprocket  372
multiple  42, 90, 133
passive  3, 70
ratio  243
regularization  526
smooth approximation  526
spatial  42
sprocket/link  387
torque  303, 310
zone  68
zones  43
contour
　body  37
　convex  37
　curvature  384
　guide  371
　planar  36
　smooth  36
　tooth  370
contour velocity  40
control concept  508
convex analysis  71, 95, 170
convex set  91
coordinate transformation  14
coordinate-free  15
coordinates  9
　curvilinear  12, 31
　generalied  11
　generalized  12, 30, 86
　inertial  12, 54
　minimal  11
　non-inertial  12
　of surfaces  14, 33
　orientation  9
　orthogonal  12, 33
　parameterized  31

parametric form  14, 421
path  32
position  9
selection  185
Coriolis  37
Coriolis equation  24
corner law  73, 89
Coulomb  71, 164
Coulomb friction  70
CPG, central pattern generator  509
crankshaft  216, 329
curvature  33, 301
curve
　spatial  32
cut  6
　position  6
cut principle  4, 6, 185
CVT, Continuous Variable Transmission
　　213, 275
CVT-belt gear  113

d'Alembert  1
damping  76
　by oil  339
　complete  77
　linear  77
　mechanical  76
　nonlinear  81
　penetrating  77
　structural  339
damping matrix  77
Darboux  33
DCT, Dual Clutch Transmission  213
deadband  197
decomposition  93, 171
deformations  1
　by impact  158
　elastic  47, 65, 289
　gradient  49
　local  89
　measure  49
　normal  89
　pulley  295
　small  47
　small elastic  124
　tangential  89
deformed element  125
degrees of freedom  11
design concepts  IX

design methodology  504
Dessauer  VII
detachment  89, 166
detachment-contact  40
Diesel engine  83, 258
   10-cylinder  359
   5-Cylinder  346
   four stroke, 12 cylinders  342
differential gear  215
differential geometry  33
differential-algebraic-equations  14
direction
   normal  71
   tangential  71
discrete state equations  234
discretization  74
disengagement  215
displacements  1
   virtual  11, 86, 101
distance
   relative  44
distance function  97
distance vector  39
Dittrich  277
DME, digital engine electronics  216
downshifting  234
drift  138
drive flange  265
drive train  215, 227
driving resistance  215
dynamic programming  231, 429
dynamics  VII
   contact  3
   geometric  VII
   longitudinal  229
   nonlinear  VIII

eccentricity  80
edge carrying  307
efficiency  316
EGS, gear control electronics  216
eiegenbehaviour  272
eigenform  271, 285
eigenfrequency  78, 271
elastic degrees of freedom  247
elastic model ring gear  246
elastic vibrations  47, 445
elasticity  47
   linear  47, 114

electro-hydraulic converter  208
electronic transmission control system  210
elementary rotations  17
elliptic friction cone  291
encapsulation of components  253
encoder  176
end effector
   forces and torques  434
   Jacobians  416
   orientation  415
   position  415
energy
   conservation  63
   considerations  173
   consumption  76
   conversion  76
   deformation  66
   kinetic  62, 108
   loss  63, 172
   potential  66
   total  63
engagement  215
engine  215
engine model  216
engine speed  216
engineering  VII
equations of motion  117, 184, 250
equilibrium  103
Euler  1, 6, 277
Euler angles  17
Euler formula
   explicit  146
   implicit  146
Euler-Bernoulli beam  380
event-driven schemes  145
excitation
   external  31
   sources  334, 366
expansion  68, 132, 159
experiments  IX, 176, 407, 497
Eytelwein  277

FEM  282, 302, 322
finite elements  VIII
first fundamental form  34
fixed point equation  150
flanged joint  344
flexibility  124

flexible workpiece   452
fluid
   capacitance   188
   compressibility   187
   density   188
   pressure   188
   volume   188
fluid flow   76
   annulus   402
   bearings   80
   gap   79
   Hagen-Poiseuille   200
   laminar   76, 200
   lossless   194
   one-dimensional   198
   oscillatory   200
   rates   192
   slide way   79
   turbulent   76
fluid mechanics   187
force   6
   acceleration   335
   active   7, 53, 100
   applied   7, 56, 335
   conservative   109
   constraint   3, 7
   contact   8, 291
   damping forces   397
   elastic   56
   element   231
   external   6, 7
   field   62
   friction forces   397
   generalized   109
   gravitational   382, 417
   gyroscopic   335, 417
   internal   6, 7
   law, multi-valued   113
   law, single-valued   113
   laws   128
   lost   3
   non-conservative   109
   nonlinear characteristic   330
   passive   7, 100
   pressure forces   397
   set-valued   56, 131, 226
   single-valued   56
   smooth force law   522
   surface   8

tensioner   383
   volume   8
   weight   8
forestry   257
Fourier   3, 125
frame
   body-fixed   24
   coordinates   15
   inertial   23
free direction   14, 117
free fall tower   204
free strand   280, 318
free wheel   218
Frenet   33
friction   76
   coefficients   165
   cone   70, 96, 139
   effects   76
   frequency dependent   198
   in pipe   198
   in the joints   418
   isotropic   141
   local   89
   reserve   70, 93, 143
   rolling   215
   sliding   132
   static   70, 132
   surplus   70
friction cone linearization   139
fuel efficiency   257
fuel injection pump   351
fun ride   204
functionality   2

Galerkin's method   199
Galilei   1
Gamma distribution   341
gap   396
   annular   400
   by eccentricity   401
   plane   398
   pressure   399
   variable cross section   400
   variable length   400
gas pressure   369
Gauss   1, 34, 44
Gauss integral theorem   56
gear
   helical   347

568  Index

inner planets  245
involute  348
meshes  261
outer planets  245
planet carrier  245
ring gear  245
ring gear coupling  247
shift operations  214
stage combinations  244
sun gear  245
train  242
wheel stages  113
gears  76
geometrical stiffness  50
geometry of motion  12
gradient  86
Grashof  277
gravitational potential  423
gripper
   contact forces  478
   contact torques  478
   coordinates  478
   insertion direction  482
   joint torques  483
   orientation constraints  486
   oscillations  483
   positioning errors  487
   prescribed force  483
   reference configuration  487
   reference point  478
guides  369
gyroscopic matrix  77

Hamilton  1
hammering  330, 348, 360
Harmonic Drive  517
Harmonic Drive model  524
Heaviside function  243
helix angle  348
heredity  253
Hertz  183
homokinematical configuration  264
Honda  505
Hooke's law  57, 67, 302
hoses  198
hydraulic
   amplifier  196
   components  190
   control unit  209

degrees of freedom  207
lines  198
networks  201
safety brake  204
systems  187
hydrodynamic converter  217
hydromotor  258, 266
hydropump  258, 266
hydrostat  264
hydrostatic drive  257

IEEE
   ICRA  412
   IROS  412
ignition  346
impact  158
   details  69
   hydraulic  203
   ideal  69
   inelastic  69
   intensity  481
   models  74
   normal direction  163, 390
   phenomena  68
   single  68
   sliding/sliding  174
   sliding/sticking  175
   sticking/sliding  175
   sticking/sticking  174
   tangential direction  163
   with friction  69, 181, 367
impeller  218
impulse level  160
impulsive motion  68
impulsive process  132
index set  90
indicator function  97
injection
   pressure  346, 360
   pump  329
input  6
integral matrices  252
interpolation  145
inversion
   mass matrix  119

Jacobi  1
Jacobian  117, 185
Jenkins element  83

jerk   231, 234
JOHNNIE   506, 510, 536
   ankle joint   537
   constraints   541
   control concept   539
   encoders   536
   experiments   543
   feedback linearization   539
   foot dynamics   542
   ground contact elements   537
   hip joint   536
   knee joint   537
   orientation sensor system   539
   position-force control   536
   simulation   542
   tachometers   536
   trajectory generation   539
junctions   190
   compressible   191
   incompressible   191

Kato   505
kinematics   12, 185
   absolute   122
   contact   36
   forward kinematics   415
   inverse problem   415
   relative   3, 36, 122
kinetics   1
Kirchhoff   1
Kirchhoff's nodal equation   189
KUKA company   411

Lagrange   1, 185
Lagrange multiplier   135, 332
Lagrange's Central equation   110
Lagrange's equations of first kind   105
Lagrange's equations of second kind   107
Lagrangian   111
Laplace transform   77
leakage   395
leakage models   398
least constraints   104
least squares   104
line of action   129, 242
linear system dynamics   77
linearization   184
   elastic components   381

kinematic   50
   reference trajectory   443
LUK/PIV rocker pin chain   276, 282

machine dynamics   83
machines   VII
machines and mechanisms   329
manipulation process   419
manufacturing   476
mass   5
   center   12
   constant   5
   effective   442
   elastic   5
   positive   5
   rigid   5
mass action matrix   150, 163, 203
mass matrix   77, 380
mass-element   47
material damping   76
MAX   506
Maxwell   282
measure
   differential equations   170
   atomic   171
   continuous   170
   discontinuous   171
   Lebesgue   171
   singular   171
measurements   177, 230, 297, 327, 407, 458
mechanic-hydraulic system   260
mechanical interaction   56
mechanics
   classical   1
   non-smooth   VIII
   real   1
mechanisms   VII
mesh of teeth   129
mesh structure   241
misalignment   301
model
   concept   173
   discretized   74
   elastic   47
   mathematical   183
   mechanical   2, 114, 183, 359
   non-smooth   74
   nonlinear oil model   333

rigid   47
rigid body   75, 158
smooth   74
model-based optimal control   233
models   IX
moment of momentum   4, 53
momentum   4, 53
momentum balance   58
Moreau   VIII, 170
MORITZ   506
motion   1
motion planning   503
moving point   21
moving trihedron   35
mulching   264
multibody system   VIII, 90
multibody theory   30

natural frequency
  damped   78
  undamped   78
network   190
Newton   1, 71, 159
Newton's laws   54
Newton-Euler-equations   115, 186
nodes   190
nodes, hydraulic   406
non-smooth mechanics   71
non-smoothness   131
Nonlinear Dynamics   110
nonlinearities
  kinematic   50
normal cone   96
normal spaces   87, 102
numerical aspects   115, 145
numerical implementation   452
numerical solution   149, 340

object oriented method   253
oil hydraulics   269
oil pressure   283
oil pump   347
oilwhip   370
optimal
  control   452
  index list   432
  parameters   452
  robotic manipulation process   479
  solution   423

table   432
time-minimum trajectory   423
optimal control strategy   234
optimality principle   431
optimization   100, 149, 476
  constraints   484
  cost function derivatives   490
  criteria   480
  disturbances   482
  functional efficient set   480
  gripper impact sensitivity   480
  gripper mating tolerance   482
  maximal applicable mating force
    481
  nonlinear, nonconvex   489
  results   494
  sensitivity analysis   487
  vector problem   493
order-(n)-algorithm   119, 337
orientation   13
orientation level   149
orifices   193
orthogonal point   278
orthogonal trihedron   32
out-of-plane effects   301
output   6
output shaft   222
output train   215, 222

Panagiotopoulos   VIII
parameter excitation   243, 278, 334
parameter influences   IX
parameter study   314
parameter-excited vibrations   241, 262
Parseval   125
passive separation state   135
path of motion   296
peg-in-hole
  compliant mating parts   455
  contour parameters   467
  elastic O-ring   455
  elastic ring deformation   457
  feedbeck gains   454
  mating models   454
  normal constraint   465
  parameterization peg   470
  problem   451
  rectangular peg   472, 486
  rigid mating parts   464

robot/workpiece interaction   454
   round peg   469
   tangential constraint   465
pendulum   10
penetration   161
performance function   234, 423
phase shift   245, 249
physics   VII
piston   216, 395
   spring-loaded   395
piston drum   265
pitch   280
pitch circle   367
pitch vector   370
pivoting algorithm   148
planar eigenmodes   254
plane case   36, 278
planetary gear systems   219
planetary set   220
planetary sets   214
plate assemblage   314
ploughing   264
point mapping   149
Poisson   71, 162, 165
Poisson's ratio   67
polar decomposition theorem
   rotational   49
   stretching   49
polygon effect   278
polygon excitation   315
polygon frequency   279, 319, 367
polygonal pyramid   139
polytope   140
position   13
position level   149
potential   63, 109
potential energy   380
power converter   113
power train model   231
power transmission   6, 213
power transmission hydraulics   207
power, mechanical, hydraulic   258
power-to-weight ratio   507
predecessor body   27
predictive control   236
press fit   344
press fits   81
pressure angle   348
pressure drop   193

pressure jump   207
pressure losses   397
pressure wave phenomena   198
principle   100
   d'Alembert   102
   differential   100
   Gauss   104
   Hamilton   110
   Jourdain   104, 123
   minimal   100
   virtual work   102
process
   assembly process   451, 476
   iterative   510
   mating process   451
   mating tolerance   487
   mounting process   451
projections   14, 117
projective method   115
prox-algorithm   326
prox-function   98
proximal point   97
PTO, power take off   258
pulley   275
   clamping force   298
   deformation   302
   driven   282
   driving   282
   elastic   285
   elastic model   320
   misalignment   316
   pair of sheaves   320
   primary   320
   secondary   320
   set   282
   sheave   275, 301
   subsystem   320
   thrust ratio   316
pump wheel   218
pump-nozzle system   346
push belt   276
   configuration   318
   element   320
   power transmission   318
   ring package   320
push belt CVT   326

quantity of matter   6
quaternions   17

R&D-problems 184
radial movements 296
radius of curvature 250
rattling 330
Ravigneaux gear 221
Rayleigh dissipation function 77
RCC, Remote Center of Compliance 451
recurrence relations 25, 122
recursion
  backward 122
  first forward 120
  robotics 416
  second forward 124
recursive algorithm 25, 119
refelections 14
regularization 293, 455
Reimers PIV 282
release mechanism 167
release unit 176
reluctance 5
resonance 78
reversible impulse 167
Reynolds equations 350
Reynolds number 398
Reynolds-Sommerfeld-theory 80
Riemann space 34
rigid body approach 89
Ritz approach 125, 247, 252
RITZ-approach 284, 379
Ritz-approach 439
road inclination 222
robotics 411
  arm segment 453
  contacts 442
  control 444
  control system 419
  elastic joint 441, 453
  elastic manipulator 437
  elastic robot 434
  gear model 417
  gear stiffness and damping 476
  gripper 434
  gripper Jacobians 477
  inverse problem 420
  joint space 416
  manipulation task 476
  motor shafts 453
  path coordinate 422

PD joint controllers 478
process dynamics 419
PUMA 560 453
ruled surfaces 434
spill-over 445
transmission ratio 418
tree-like structure 438
working space constraints 488
rocker pin 276
  angled joint 304
  chain - plane model 282
  chain - spatial model 301
  curvature pin ends 315
  gap function 309
  halves 304
  inner contour 304
  joint kinematics 305
  offset 305
  pairs 303
  pulley contacts 307
  rolling joint 304
roll resistance 222
roller chain 365
rolling disc 10
rotation 12, 64
rubber disc 180

screws 81
second fundamental form 34
self-stability 508
separation velocity 174
servovalve 196
shaft-pulley-system 284
shafts
  elastic 223, 261
  rigid 223, 261
shape functions 247, 282
shape memory alloys 509
shear modulus 67
sheave 276
  fixed 282
  movable 283
shift control 231
shift elements 215
shifting process 231
shock
  running in 368
  running out 368
Signorini 3

Signorini's law   89
simulation   230
simulation/experiment   270, 392, 434
singular value decomposition   202
sledge   10
slide ways   76
sliding   69
sliding angle   313
snap fastener
  assembly   459
  chamfer   459
  hook   459
  stiffness matrix   459
  with beam   461
  with plate   463
solid-fluid interactions   76
spatial case   42, 278
spatial rotations   88
spatial shape functions   199
spectral radius   149
split wheels   354
spring tension   194
sprocket   370
  carrier   377
  shaft   382
  toothing   374
  wheels   367
stage combinations   219
stage process   429
static equilibrium   291
statics   1
steel on steel   180
steep characteristics   187
stick insect   508
stick-slip   40
sticking   69
stiff differential equations   74
stiffness distribution   284
stiffness matrix   77, 380
stochastic process   342
Stokes theorem   63
stored impulses   165
strain tensor   49, 66
strands   368
stress tensor   66
stretch, right, left   49
Stribeck curve   137, 226
stroboscopic exposure   177
structures
  mathematical   2
subdifferential   97
successor body   27
sun wheel   220
surface normal vector   56
surface unit vectors   42
swaying   242
switching elements   208
synchronous point   231, 235
Synge   2
system
  ancillary   346
  biological   503
  body-fixed   22
  coordinates   30
  cut   368
  dynamics   131
  inertial   22
  large   185
  mechanical-hydraulic   403
  oil-hydraulic   395

tangent spaces   87, 102
Taylor expansion   443
tensioner
  dynamics   402
  hydraulic   395
  system   403
tensor components   34
test set-up   177
throttle   195
throwing machine   177
throwing machine control   177
tilt angle   284
time increments   234
time-discretization   146
time-stepping schemes   146, 326
timing
  belts   365
  chains   365
  chains and belts   346
  equipment   329
  gear   359
  gear wheels   346
toothing   242
  elasticity   242
  stiffness   243, 249
  theory   243
torque

incoming 276
outgoing 276
torque converter 215
torsion 33
tractor 257
trajectory 20
  contact force 444
  controllability 446
  critical point 429
  curvature control 446
  elastic curvature 447
  extremals 425
  feedback control 445
  feedforward control 445
  planning 413
  point 21
  prescribed 20
  reference 233
  saddle point 427
  sink 427
  source 427
  spatial 33
  time-minimum solution 429
  warp 447
transformation 59
  chains 25
  linear 16
  matrix 16
  triangle 15
transition points 293
translation 12, 64
transmission
  ratio 275, 347
travel range selector 258
tree 120
tree-like structure 25
tripod joint 265
tube
  incompressible flow 397
  laminar flow 398
  models 397
  turbulent flow 398
turbine wheel 218

UML, Unified Modeling Languange 253
unilateral primitives 73, 93, 405
unit vector 12, 22, 130
unit-matrix 16

upshifting 234

valve train 352
valves
  check valves 194, 395, 402
  complex 193
  elementary 193
  multistage 196
  one-stage 4-way 196
van Doorne 318
variation
  bounded 170
  Gauss 103
  Jourdain 103
variational calculus 111
variational inequality 91
VARIO system 258
VDT push belt 276
VDT-Bosch 318
vector
  binormal 32
  deformation 49
  displacement 49
  normal 32, 40, 86, 102
  surface 44
  tangent 32, 40
vector chain 15
vector-matrix-notation 28
vehicle
  mass 215
velocity 19
  absolute 21, 55
  angular 25
  applied 38
  generalized 30
  profile 399
  relative 22, 39, 41
  rotational 36
  virtual 11, 106
velocity level 149
Verein Deutscher Ingenieure 76
verification 254
verification of impacts 176
vis inertiae 6
VW R5 TDI 347
VW V10 TDI 359

walking 411, 503
  dynamics 509

machine   503
performance   507
walking machine
  control   510
  dynamics   510
  gear equations   524
  Harmonic Drive Gears   517
  pulse width control   516
  supporting polygon   510
  system equations   523
  trajectories   528
  transmission ratio   515
  zero moment point   510
wall roughness   398
wave process   158

Weingarten   34, 44
wild mouse   183
woodpecker
  free flight phase   154
  limit cycle   150
  pole   153
  sequence os events   151
  sleeve   152
  toy   150
work   62
  virtual   101

Young's modulus   57, 67

Zentralgleichung   108

Printing: Krips bv, Meppel, The Netherlands
Binding: Stürtz, Würzburg, Germany